Epistemic Game Theory

Reasoning and Choice

Andrés Perea

CAMBRIDGE
UNIVERSITY PRESS

University Printing House, Cambridge CB2 8BS, United Kingdom

Cambridge University Press is part of the University of Cambridge.

It furthers the University's mission by disseminating knowledge in the pursuit of education, learning and research at the highest international levels of excellence.

www.cambridge.org
Information on this title: www.cambridge.org/9781107401396

First published 2012

A catalogue record for this publication is available from the British Library

Library of Congress Cataloguing in Publication data
Perea, Andrés.
Epistemic game theory : reasoning and choice / Andrés Perea.
pages cm
Includes bibliographical references and index.
ISBN 978-1-107-00891-5 (hardback) – ISBN 978-1-107-40139-6 (paperback)
1. Game theory. 2. Epistemic logic. I. Title.
QA269.P446 2012
519.3–dc23 2012007500

ISBN 978-1-107-00891-5 Hardback
ISBN 978-1-107-40139-6 Paperback

To my children Maria and Lucas

Contents

Figures

Tables

Acknowledgments

The idea for writing this book came to me during my Christmas holiday on Mallorca in 2006. A few weeks later, when I wrote up my first sentences, I was suddenly asked to give a mini-course on epistemic game theory at the Max Planck Institute of Economics in Jena. The lectures I prepared for that course have shaped this book in a very important way, as the structure of this book closely corresponds to the structure of that first mini-course. In fact, the course in Jena marked the beginning of a continuous and fruitful cross-fertilization between my book on the one hand, and my epistemic game theory course on the other hand – I have often used examples and ideas from the course for my book, whereas at other times I have used new ingredients from the book to improve the course. Moreover, the various courses I have given at universities across Europe have served as an extremely useful test case for the book. I would therefore like to thank the following universities and institutes for allowing me to give a course on epistemic game theory: the Max Planck Institute of Economics in Jena (Germany), Maastricht University (The Netherlands), Universidad Carlos III de Madrid (Spain), the University of Amsterdam (The Netherlands), the University of Lausanne (Switzerland) and Aarhus University (Denmark). The feedback I received from the various audiences at these places has helped me to substantially improve parts of this book. I would therefore like to thank all the students and researchers who have attended some of these courses.

Among the many people who have contributed to this book there are two who have played an extraordinary role. First, Geir Asheim, who introduced me to the wonderful world of epistemic game theory some thirteen years ago, and who guided me during my first steps on that planet. Without Geir, I would probably not have written this book. I am also very grateful to my colleague and dear friend Christian Bach, who has carefully read the entire book – and probably knows the book better than I do, who continuously provided me with fruitful comments and suggestions on the book, with whom I had the pleasure to teach the epistemic game theory course in Maastricht, and with whom I have had many inspiring discussions on epistemic game theory. Without Christian, the book would not have been the same.

During the writing process I have received very valuable feedback from the following people who have read parts of the book (in alphabetical order): Luca Aberduci, Geir Asheim, Christian Bach, Pierpaolo Battigalli, Christine Clavien, János Flesch, Amanda Friedenberg, Herbert Gintis, Jens Harbecke, Aviad Heifetz, Willemien Kets, Simon Koesler, Jiwoong Lee, Topi Miettinen, Christian Sachse, Elias Tsakas, Leopoldo Vilcapoma, Alexander Vostroknutov and some anonymous referees for Cambridge University Press. Thank you all for your input! I am particularly indebted to Christian Bach and Aviad Heifetz for their very extensive and detailed remarks on the book.

The cooperation with Cambridge University Press has been a very happy one right from the beginning. A special word of appreciation goes to Chris Harrison for his support during and after the refereeing procedure, and for some valuable advice on the title of this book.

Last but not least I would like to thank the following people for giving me so much positive energy during the writing process: my dear friends Christian Bach, Frédérique Bracoud, Nadine Chlaß and János Flesch, my grandmother Tonnie, my brother Juan, my sister Toñita, my father Andrés, my mother Ans, my children Maria and Lucas, and of course my wife Cati. Thank you all for providing such a warm basis upon which I could build this book!

1 Introduction

One thing I learned from Pop was to try to think
as people around you think.
And on that basis, anything's possible.
Al Pacino alias Michael Corleone
in *The Godfather – Part II*

What is this book about? In our lives we are continuously asked to make choices – small choices about daily issues, but also big choices with important consequences. Often, the final outcome of a choice does not only depend on our own decision, but also on decisions made by other people surrounding us. For instance, if you are about to meet a couple of friends in a pub this evening, then whether you have to wait for your friends or not depends on your own arrival time, but also on the times your friends arrive. And if you are at a flea market negotiating about the price of a rare Beatles record, then the final price will not only depend on your own bargaining strategy, but also on that of the seller.

Such situations, in which the final outcome does not only depend on your own choice but also on the choices of others, are called *games*. The discipline that studies such situations is called *game theory*. The name *game theory* is perhaps a bit misleading as it suggests that its main application is to recreational games – such as chess or poker – but this is not true. In fact, game theory can be applied to *any* situation where you must make a choice, and in which the final outcome also depends on the decisions of others. The people whose choices directly influence the final outcome are called *players* – so you are one of the players in the two real-life situations sketched above – and we usually refer to the *other* players as your *opponents*, even if these are your friends.

In order to evaluate the possible consequences of your own choice, it is important to form some *belief* about the likely choices of your opponents, as these will affect the final result. Moreover, to make a *good* choice it is necessary to form a *reasonable* belief about your opponents. But in general not every belief about your opponents will be reasonable: your opponents will have in general some choices that seem more plausible than others. But to determine which choices of your opponent are plausible and which are not, you must put yourself in the shoes of the opponent, and think about the possible beliefs *he* can have about *his* opponents. That is, you must *reason about your opponents* before you can form a reasonable belief about them. And this reasoning process will precisely be the main topic of this book.

More precisely, this book studies several plausible ways in which you can reason about your opponents before you make your final choice in a game. As different people reason differently, we do not believe in a *unique* way of reasoning. Instead, we offer the reader a *spectrum* of plausible reasoning patterns in this book. We also investigate how your eventual decision will be affected by the type of reasoning you use. The discipline that studies these patterns of reasoning, and how they influence the eventual choices of the players, is called *epistemic game theory*. This explains the title of this book, *Epistemic Game Theory: Reasoning and Choice*.

Why this book? We have just seen that reasoning about your opponents is a crucial step towards making a good choice in a game. In fact, Oskar Morgenstern – one of the early founders of game theory – had stressed the importance of this reasoning process, in particular to form beliefs about the beliefs of others, in his paper Morgenstern (1935). But strangely enough it is exactly this reasoning step that has largely been overlooked by the game theory literature – including the game theory textbooks – during the last sixty years! This immediately raises the question: Why? In our opinion, this phenomenon is largely due to the concept of *Nash equilibrium* and its various refinements, which have dominated the game theory literature for so many decades. Nash equilibrium describes just one possible way – and not even a very plausible one – in which the players in a game may choose, or form a belief about the opponents' choices. Yet many game theory textbooks and articles *assume* that the players' reasoning process will eventually lead them to choose in accordance with Nash equilibrium, without explicitly describing this reasoning process.

We find this approach unsatisfactory for two reasons. First, we believe that the reasoning is an essential part of the decision-making process for a player in a game, and hence deserves to be discussed explicitly. Second, we will see in this book that Nash equilibrium is based on some rather implausible assumptions about the way players reason, which makes Nash equilibrium a rather unnatural concept to use when analyzing the reasoning of players in a game. That is also the reason why Nash equilibrium plays only a minor role in this book – it is only discussed in Chapter 4.

Things started to change with the rise of *epistemic game theory*, some twenty-five years ago. This relatively young discipline attempts to bring game theory back to its basic elements – namely the reasoning by players about their opponents. In recent years it has developed a whole spectrum of concepts that are based on more plausible assumptions than Nash equilibrium. But to date there is no textbook on epistemic game theory, nor is there any other game theory textbook that focuses on the reasoning process of the players. The aim of this book is to fill this gap, by providing a text that concentrates on the way people can reason about their opponents before making a choice in a game. This book will thus be the first textbook on epistemic game theory. We feel there is a need for such a textbook, because reasoning about your opponents is such an important and natural ingredient of the decision-making process in games.

Moreover, for researchers it will also be valuable to have a book that discusses the various concepts from epistemic game theory in a systematic and unified way.

While this book was being written, Pierpaolo Battigalli, Adam Brandenburger, Amanda Friedenberg and Marciano Siniscalchi were also working on a book on *epistemic game theory*, and so were Eric Pacuit and Olivier Roy but their books were not finished at the time this introduction was written. So, shortly there will be two new books on epistemic game theory on the market, which is very good news for the field. The first book mentioned above will be rather complementary to ours, as it will treat topics like games with incomplete information, infinite games and psychological games, which are not covered by our book.

Intended audience. This textbook is primarily written for advanced bachelor students, master students and Ph.D. students taking a course in game theory. This course could either be an introductory course or a more specialized follow-up course. In fact, the book does not presuppose any knowledge about game theory, and should thus be accessible also for students who have not studied game theory before. Moreover, the mathematics that we use in this book is very elementary, and the book can therefore be used within any program that has a game theory course in its curriculum. But the book can also be used for self-study by researchers in the field, or people who want to learn about epistemic game theory.

Structure of the book. The book has been divided into three parts, according to the type of game and type of beliefs we consider.

In Part I we assume that the game is *static* – that is, all players choose only once, and in complete ignorance of the opponents' choices. In this part we assume moreover that the belief of a player about the opponents is a *standard belief*, represented by a *single* probability distribution. This is the type of belief that is most commonly used for static games. Part I includes Chapters 2–4.

In Part II, which contains Chapters 5–7, we still assume that the game is static, but now we model the belief of a player about the opponents by a *lexicographic belief*, consisting of *various* probability distributions instead of only one. Lexicographic beliefs are particularly useful if we want to model *cautious reasoning* – that is, a state of mind in which you do not completely rule out *any* opponent's choice from consideration. This type of belief is not as well known as standard beliefs, but we believe it is a very natural way to describe cautious reasoning about your opponents.

Part III, which contains Chapters 8 and 9, is dedicated to *dynamic games*, where players may choose one after the other, and may fully or partially observe what the opponents have done in the past before making a choice themselves. For such games we model the belief of a player about the opponents by a sequence of *conditional beliefs* – one at every point in time where the player must make a choice. So, instead of holding just one belief once and for all, the player holds a separate belief at every instance when a choice has to be made, and the player's belief about the opponents may *change* as the game proceeds.

The first two chapters of Part I – Chapter 2 and Chapter 3 – form the basis for this book, as they introduce the central idea of *common belief in rationality*. This concept

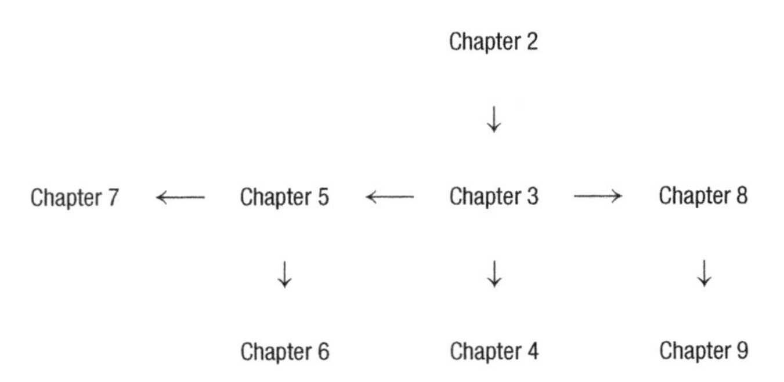

Figure 1.1 Logical connection between the chapters

is at the heart of epistemic game theory, and all other concepts in this book may be viewed as variations on the idea of common belief in rationality. Every other chapter is about exactly one such variation, describing one particular way of reasoning about your opponents. Some of these chapters can be read independently of each other, whereas others cannot. Figure 1.1 gives an overview of the logical connection between the chapters. Here, the arrow from Chapter 3 to Chapter 4 means that Chapter 4 builds upon Chapter 3. As there is no arrow between Chapter 4 and Chapter 5, you should be able to understand Chapter 5 without having read Chapter 4. The other arrows are to be interpreted in a similar fashion.

Players can be male or female, but for simplicity we will often refer to a player as "he" or "him."

Structure of the chapters. Every chapter in this book is on one particular concept, describing one possible way of reasoning about your opponents. The main structure of each chapter is as follows: We begin the chapter with one or more examples, illustrating the way of reasoning we will discuss in that chapter. Subsequently, we show how this particular way of reasoning can be formally described within an *epistemic model* – that is, a model that describes the belief a player holds about the opponents' choices, the belief held about the opponents' beliefs about the other players' choices, and so on. In most of the chapters we then provide a simple *existence proof* for this concept, showing that this particular way of reasoning is always *possible* within any game, so never leads to logical contradictions. For every concept – except Nash equilibrium – we design an *algorithm* that yields for every game precisely those choices you can rationally make if you reason in accordance with that concept. Such an algorithm typically proceeds by iteratively removing choices or strategies from the game. An exception is the algorithm we discuss in Chapter 6. These algorithms are relatively easy to use, and greatly facilitate the task of finding those choices that are selected by the concept. All formal proofs are given in a separate section at the end of the chapter, and these proofs are mainly for Ph.D. students and researchers. In every chapter we also provide seven practical problems and three theoretical problems that the student or reader can work on. Every chapter concludes with a literature section in which we discuss the relevant literature for that chapter, and provide references to relevant articles, books and book chapters. We have decided not to include these references in the main text of the chapter, as we believe this would distract the student's attention too much from the main ideas in that chapter. After all, this is a textbook and not a monograph or survey.

One-person perspective. Throughout this book we take a *one-person perspective* to analyze game-theoretic situations. That is, we always view the game from the perspective of *one single player*, and put restrictions only on the beliefs of this particular player – including beliefs about the opponents' beliefs – without imposing restrictions on the *actual* beliefs of the opponents. We believe this approach to be plausible, as we cannot look inside the minds of our opponents at the time we make a choice. So, you can only base your choice on your *own* beliefs about the opponents, and not on the actual beliefs and choices of your opponents, since these are not known to you. But then, if we want to analyze the reasonable choices a player can make in a game, it is sufficient to concentrate only on the beliefs of this particular player, as they encompass everything that can be used to make a decision. Although we believe the one-person perspective to be very natural, it crucially differs from the usual approach to games in books and articles, which typically proceed by imposing restrictions on the beliefs of *all* players, and not only one player.

Descriptive versus normative approach. In this book we do not tell people what they should do or believe in a certain game. We only explore different intuitive ways of reasoning that you could use in a game, and see how these various ways of reasoning would affect the final choice – or choices – you could rationally make *if* you were to use this particular way of reasoning. But it is eventually up to the reader, or the modeler, to choose the preferred way of reasoning. That is, we take a purely *descriptive* approach in this book, and *not* a *normative* one.

In Chapters 6 and 7 of this book we even introduce two ways of reasoning that in some games lead to completely opposing choices! We do not believe that one of these two concepts is better – or more intuitive – than the other, they are just different. In fact, we believe that both ways of reasoning are very plausible, so we see no problem in presenting both concepts next to each other, even if in some games they lead to completely different choices. The same actually holds for the two concepts we discuss in Chapters 8 and 9.

As a consequence, we do *not* believe in a *unique* concept for game theory. In my opinion we must accept that different people tend to reason differently in the same game-theoretic situation, and to me there is simply no best way of reasoning in a game – only *different* ways of reasoning.

Rational and reasonable choices. The word *rational* has often led to confusion in game theory. What do we mean precisely when we say that a player chooses *rationally*? In this book, we say that a player chooses *rationally* if some belief is formed about the opponents' choices, and then the player makes a choice that is optimal under this belief.

However, the belief held about the opponents may be completely unreasonable, and hence – within our terminology – a rational choice is not necessarily a *reasonable* choice. Of course, the meaning of a *reasonable* choice is very subjective, as it depends on the way of reasoning one has in mind. As we have already argued above, there are intuitive concepts in this book that in some games lead to completely opposed choices. Hence, what is a "reasonable" choice under one concept may be "unreasonable" under another concept. Since we do not believe in a unique concept for game theory, we also do not believe in a unique definition of what is a reasonable choice in a game. To a large extent, it is up to the reader to decide. The book is only there to help the reader make this choice.

Examples and problems. The book is full of examples and practical problems, and in my opinion they constitute the most important ingredient of this book. Each of the examples and practical problems is based on a story – mostly inspired by everyday life situations – in which *you* (the reader) are the main character. The reason we choose scenarios from everyday life is that it makes it easier for the reader to identify with such situations. We could also have chosen scenarios from professional decision making, such as business, politics, economics or managerial decision making, but such situations would probably be more distant for some readers.

All of the stories in the examples and practical problems take place in an imaginary world in which you experience some adventures together with your imaginary friends Barbara and Chris. These stories often have a humorous side, and thereby also serve as points of relaxation in the book. The one-person perspective we discussed above is very strongly present in these examples and practical problems, as *you* always play the main role in these stories, and the situation is always analyzed completely from *your* perspective. The crucial question is always: "Which choice would *you* make in this situation, and why?"

The examples are there to illustrate the main ideas behind the concepts, and to show how the various concepts and algorithms can be applied to concrete situations. As every example is based on a particular story, it will make it easier for the reader to remember the various examples, and to keep these examples in the back of the mind as a benchmark.

The theoretical problems in the book are of a completely different type compared to the practical problems. They do not refer to any story, but rather discuss some general theoretical issues related to the concept of that chapter. Usually, these problems require the reader to formally prove some statement. These theoretical problems are primarily meant for Ph.D. students and researchers who wish to deepen their theoretical insights.

Beliefs diagrams. In Chapters 2, 3 and 4 of this book we use a *beliefs diagram* to graphically represent the belief hierarchy of a player – that is, the belief about the opponents' choices, the belief about the opponents' beliefs about the other players' choices, and so on. We invented this beliefs diagram because a belief hierarchy may seem rather abstract and complicated when stated formally – certainly for readers that are new to epistemic game theory. However, belief hierarchies are crucial for developing the various concepts in this book. By visualizing the various levels of a belief hierarchy – by means of a *beliefs diagram* – we hope the reader will find it easier to work with such belief hierarchies, and to understand what a belief hierarchy really represents. Moreover, the beliefs diagrams also play a crucial role in the examples and practical problems of Chapters 2, 3 and 4.

Choices and beliefs. In this book we always make a very clear distinction between the *choices* and the *beliefs* of a player. The reason we raise this issue here is that in some books and articles this distinction is not very clear. Some books, when they introduce the concept of Nash equilibrium for instance, state that the *mixed strategy* of a player can either be interpreted as the actual choice by this player, or as the belief his opponents hold about the player's choice. But what is then the real interpretation of a mixed strategy? This often remains unclear. But if the meaning of a mixed strategy remains ambiguous, it is likely to cause confusion, which is of course very undesirable and unnecessary. Such confusion can easily be avoided by always being clear about the interpretation of the various objects that are being introduced. In particular, we believe we must always make a very clear distinction between the choices and the beliefs of a player, as these are completely different objects. And that is precisely what we do in this book.

Randomized choices. The concept of a *mixed strategy* – or *randomized choice* – is still used as an object of choice in many books and articles in game theory. Strictly speaking, a *randomized choice* means that a player, before making a choice, uses a randomization device and bases the actual choice on the outcome of the randomization device. For instance, the player rolls a dice, and chooses based on the outcome of the dice roll. We believe, however, that decision makers do not randomize when making serious decisions! The reason is that there is nothing to gain for a player by randomizing. Namely, randomizing between two choices a and b can only be optimal for the player if there is the belief that a and b yield the same maximal utility. But in that case, the player could just as well choose a or b – without randomizing – and save the trouble of having to roll a dice.

In this book we take seriously the fact that people typically do not randomize when making choices. Throughout the book, we assume that players do not use randomized choices, but always go for one particular choice (with probability one). Randomized choices are only used in this book as *artificial auxiliary* objects, used to characterize *rational* and *irrational choices* – see Theorem 2.5.3 in Chapter 2. Within that theorem, randomized choices are not interpreted as real objects of choice, but are rather used as an abstract mathematical tool to verify whether a given choice is rational or not.

Using the book for a course. This book is well suited for a course in game theory at any university. Moreover, it can be used at different levels – for advanced bachelor students, master students and Ph.D. students. Depending on the type, the level and the length of the game theory course, one can use all chapters or selected chapters from this book – of course respecting the arrows in Figure 1.1. As we mentioned above, Chapters 2 and 3 present the central idea in epistemic game theory – *common belief* *in rationality* – and these two chapters should thus be part of any course based on this book. Chapter 2 is a preparatory chapter, meant to make the reader familiar with the main ingredients of a game, and which informally introduces the first reasoning steps that will eventually lead to the concept of *common belief in rationality*. Chapter 3 shows how these reasoning steps can be modeled formally, and how these first few steps can be completed to arrive at the concept of *common belief in rationality*. When designing a course in game theory one could decide to discuss Chapter 2 only briefly, or to merge it with Chapter 3, depending on the type of course. For instance, for a trimester course of seven weeks, one could teach Chapters 2 and 3 in week 1, and dedicate every subsequent week to one of the remaining six chapters.

Irrespective of the level of the course, we think that the examples should play a prominent role in class. From my own experience I know that these examples are a powerful tool for revealing the main ideas behind the concepts. Moreover, the examples are likely to stick in the students' heads as they are all based on some particular story. If time allows, we would also strongly advise dedicating at least one session per week to discussing some of the practical problems at the end of each chapter. By working on these problems, the student will be trained in applying the concepts and algorithms to concrete situations, and we believe this is the best way for a student to master the various concepts. Besides, we really hope – and do believe – that the practical problems are a lot of fun. I, at least, had a lot of fun inventing them.

As an example of how to use the book for a course, we will briefly outline the course in epistemic game theory that Christian Bach and I will give at Maastricht University during the academic year 2011/2012. The course is designed for master students and Ph.D. students, and lasts seven weeks. Every week there are two theory lectures of two hours, and one problem session of two hours. In week 1 we cover Chapters 2 and 3 of the book, whereas every subsequent week covers one of the remaining Chapters 4–9. Every week the students must work on three practical problems from the associated chapter, and the solutions to these problems are discussed during the problem session of that week. But this is just an example – every teacher can design the course depending on the time available and the teaching method used.

Limitations of this book. As with any book, this book has its limitations. First, we only consider *finite* games, which means that for a static game, every player only has finitely many possible choices, whereas in the context of dynamic games we additionally assume that the game will stop after finitely many moves by the players. By doing so, we exclude dynamic games of possible infinite duration such as infinitely repeated games, evolutionary games or stochastic games. But the concepts we develop in this book should be applicable to such infinite games.

We also assume throughout the book that the players' utility functions are completely transparent to all the players involved. That is, players do not have any uncertainty about their opponents' utility functions. We thereby exclude games with *incomplete information*, in which some or all of the players do not completely know their opponents' utility functions. Some of the concepts in this book have been extended to games with incomplete information, but we do not discuss these extensions here.

Recently, game theorists have started to study situations in which some of the players are *unaware* of certain elements in the game. For instance, you may be unaware of some of the opponent's choices. In this book we do not study such situations, but this is certainly a very interesting line of research. In particular, it would be interesting to see whether – and if so how – the various concepts in this book can be extended to such games with unawareness.

Finally, we have taken the concept of *common belief in rationality* as the basis for this book. Every other concept in this book can be seen as a *refinement* of common belief in rationality, since they are obtained by taking the concept of common belief in rationality and imposing additional restrictions. There are also intuitive concepts in the literature that *violate* some of the conditions in common belief in rationality – especially within the bounded rationality literature – but such concepts are not discussed in this book.

Surveys on epistemic game theory. As we mentioned above, this is the first ever textbook on epistemic game theory. In the literature there are, however, survey papers that give an overview of some of the most important ideas in epistemic game theory. The interested reader may consult the surveys by Brandenburger (1992a), Geanakoplos (1992), Dekel and Gul (1997), Battigalli and Bonanno (1999), Board (2002) and Brandenburger (2007).

Part I

Standard beliefs in static games

2 Belief in the opponents' rationality

2.1 Beliefs about the opponent's choice

In everyday life we must often reach decisions while knowing that the outcome will not only depend on our own choice, but also on the choices of others. If you meet with a couple of friends, for instance, whether your friends have to wait or not depends on the time you arrive yourself, but also on the time your friends arrive. If you prepare for an exam, then the grade you will obtain depends on your effort, but also on the specific questions chosen by the teacher. When playing a chess match with your friend, the winner is obviously determined not only by your own strategy, but also by the strategy of your friend.

Each of these situations is called a *game*, and the people involved are called *players*. These terms are perhaps a bit misleading, since "games" and "players" usually refer to recreational games, sports or other competitions with winners and losers. However, the term "game" as we will use it in this book is much broader – it may be used for any situation in which two or more individuals make choices, and the final outcome depends not only on the choice of one individual, but on the choices of all. Also, the outcome of a game may be any situation, not just a win, a draw, or a loss. In fact, in the first two games described above there are no winners or losers, and the situations at hand do not correspond to any form of competition. The chess example is different as it represents a game in the classical sense.

In the first two parts of this book we will concentrate on a special class of games, namely those situations where every player, while making a choice, does not receive any information about any of the opponents' choices. Such games are called *static games*. If, in contrast, a player learns about some of the opponents' choices during the game, then this situation is called a *dynamic game*. These games will be studied in the third part of this book.

The first two examples above are static games – when deciding at what time to leave your home, you do not know whether your friends will arrive early or late, and when preparing for the exam you are unsure about the questions that will be asked in the exam. However, if the teacher were to announce in advance the topics for the questions that are to be asked, then this would be a dynamic game since you may now focus your effort on these topics. Chess is clearly a dynamic game – whenever it is your turn to move, you know exactly which moves your opponent has already made.

The key questions are: Which choices in a static game may be considered reasonable, and why? A naive answer could be: Simply make the choice that would yield the best outcome. However, games are not as simple as that, since the outcome depends on your own choice, but also on the choices of the other players. Hence, the choice you think will yield the best outcome will crucially depend on what you believe the others will do. To illustrate this problem, consider the following example.

Example 2.1 Where to locate my pub?

You and your friend Barbara both want to open a new pub in your street. There are 600 houses in the street, equally spaced. You and Barbara can both choose from seven possible locations for the pub: a, b, c, d, e, f and g. See Figure 2.1 for an illustration. Between every two neighboring locations there are exactly one hundred houses, and the father in every house regularly visits a pub. People in your street are lazy: they will always visit the pub that is closest to them. If Barbara happens to choose the same location as you, then half of the fathers will go to your pub, and half will visit Barbara's pub.

Now, if you wish to maximize the number of customers, which location would you choose? This will depend crucially on the location you think Barbara will choose. Suppose you believe that Barbara will choose location a. Then, by choosing location b you would obtain 550 customers, namely everybody living between b and g, and half of the fathers living between a and b. In order to see this, note that everybody between b and g is closer to b than to a. From the fathers between a and b, exactly 50 live closer to a, and 50 live closer to b. By choosing location c you would only obtain 500 customers, namely all fathers between b and g. Notice that all fathers between a and b are closer to a than to c, and all fathers between b and g are closer to c than to a. By choosing any of the other locations, you would obtain even fewer than 500 customers. Table 2.1 gives an overview of the number of customers you would obtain if you believe that Barbara chooses location a. From this table we can see that, if you believe that Barbara chooses location a, then it is optimal to choose location b, as it maximizes the number of customers you believe you can obtain.

Suppose now you believe that Barbara will choose location f instead of a. Then, the numbers of customers you would obtain by choosing each of the seven possible locations can be found in Table 2.2. Hence, if you believe that Barbara chooses location f, your *optimal* choice would be location e. In a similar way, you may verify that

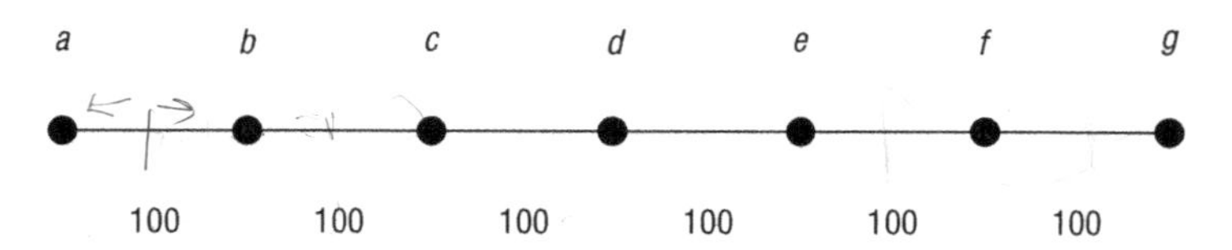

Figure 2.1 Where to locate my pub?

Table 2.1. *Number of customers you would obtain if you believe that Barbara chooses a*

a	b	c	d	e	f	g
300	550	500	450	400	350	300

Table 2.2. *Number of customers you would obtain if you believe that Barbara chooses f*

a	b	c	d	e	f	g
250	300	350	400	450	300	50

- location c is optimal for you if you believe that Barbara chooses location b,
- location d is optimal for you if you believe that Barbara chooses location d as well,
- location f is optimal for you if you believe that Barbara chooses location g.

So we see that each of the locations b, c, d, e and f is optimal for you for some specific belief you can hold about Barbara's choice. We call such choices *rational*. In general, we say that a choice is *rational* if it is optimal for some belief you can hold about the opponent's choice. In this example we thus see that b, c, d, e and f are all rational choices for you.

But what can we say about your choices a and g? Is there a belief about Barbara's choice for which choosing location a, or choosing location g, is optimal? The answer is "no"! To see this, let us compare your choice a with your choice b. It may be verified that, whatever location Barbara chooses, choosing location b will always give you more customers than choosing a. We say that your choice b *strictly dominates* your choice a. So, it is clear that choosing a can never be optimal for you for *any* belief you may hold about Barbara's choice. That is, a is an *irrational* choice. In exactly the same fashion it can be verified that your choice g is strictly dominated by your choice f, and hence g is an irrational choice as well.

Summarizing, we see that

- your choice a is irrational, since it is not optimal for any belief about Barbara's choice,
- your choice b is rational, since it is optimal if you believe that Barbara chooses location a,
- your choice c is rational, since it is optimal if you believe that Barbara chooses location b,
- your choice d is rational, since it is optimal if you believe that Barbara chooses location d as well,
- your choice e is rational, since it is optimal if you believe that Barbara chooses location f,

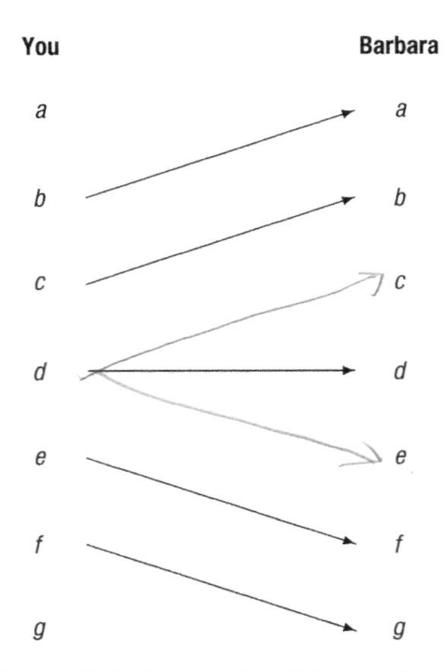

Figure 2.2 A beliefs diagram for "Where to locate my pub?"

- your choice f is rational, since it is optimal if you believe that Barbara chooses location g, and
- your choice g is irrational, since it is not optimal for any belief about Barbara's choice.

Figure 2.2 gives a graphical representation of these facts. We refer to this figure as a *beliefs diagram*. The arrows in this diagram should be read as follows: The arrow from b to a means that your choice b is optimal if you believe that Barbara chooses a, and similarly for the other arrows. So, the arrows represent *beliefs* that you can have about Barbara's choice. In particular, the choices on the left-hand side that have an outgoing arrow to the right-hand side are all *rational choices*, since these are optimal for some belief about Barbara's choice, namely the belief indicated by the arrow.

Note that your choice d is also optimal if you believe that Barbara chooses c or e. So, instead of the arrow from d to d, we could also have chosen an arrow from d to c, or an arrow from d to e, in order to support your choice d. This already indicates that the beliefs diagram we have chosen in Figure 2.2 is not the only beliefs diagram possible – we could also have chosen one with an arrow from d to c or one with an arrow from d to e.

As you can see, the choices a and g do not have an outgoing arrow since these choices are *not* optimal for *any* belief about Barbara's choice. So, the beliefs diagram tells us exactly which choices are rational for you and which are not. Moreover, for each of your rational choices it specifies one particular belief about the opponent's choice for which this choice is optimal, represented by an arrow. □

Table 2.3. *Your utilities in "Going to a party"*

blue	green	red	yellow	same color as Barbara
4	3	2	1	0

2.2 Utility functions

In the example above it was easy to see what your objective is, namely to maximize the number of customers. However, in many other situations it may not be immediately clear what you wish to maximize. Here is an example that illustrates this.

Example 2.2 Going to a party

This evening you are going to a party with your friend Barbara. The big problem is: Which color should you wear tonight? Suppose that in your cupboard you have blue, green, red and yellow suits, and the same holds for Barbara. You prefer *blue* to *green*, *green* to *red* and *red* to *yellow*. However, the situation you dislike most of all is when Barbara wears the same color as you!

We can model these preferences by the numbers in Table 2.3. Here, the numbers represent "utilities," and should be interpreted as follows. If you choose your blue suit, and Barbara wears a different color, then your satisfaction is measured by the number 4. We say that you derive a *utility* of 4 from this situation. If you choose your green suit, and Barbara goes for a different color, then this would give you a utility of 3. Similarly for the other two colors. However, if Barbara happens to choose the same color as you – something you wish to avoid – then your utility is only 0.

So, the utilities 0, 1, 2, 3 and 4 measure your satisfaction for every combination of choices that you and Barbara can possibly make. The higher the utility, the higher your satisfaction. In particular, the utilities indicate that you prefer *blue* to *green*, *green* to *red*, and *red* to *yellow*, as long as Barbara chooses a different color from you, and that you like wearing a different color more than wearing the same color as Barbara.

Which colors are rational for you? Let us start with your most preferred color *blue*. Clearly, *blue* is optimal if you believe that Barbara chooses any other color, for instance *green*. So *blue* is certainly a rational choice for you.

Also your second most preferred color, *green*, is rational. Suppose that you believe that Barbara will choose *blue*. Then, by choosing *blue* yourself, your utility would be 0 since you would be wearing the same color. By choosing *green* you would expect a utility of 3, whereas by choosing *red* or *yellow* you would anticipate a utility of 2 or 1, respectively. So, *green* is optimal for you if you believe that Barbara chooses *blue*.

What about *red*? If you believe that Barbara chooses *blue*, your only optimal choice would be *green*. On the other hand, if you believe that Barbara chooses *green*, or believe she chooses *red*, or believe she chooses *yellow*, then in each of these cases your only optimal choice would be *blue*. But does this mean that *red* cannot be optimal for you? The answer is "no"!

The reason is that up until now, we have been concentrating on rather special beliefs – namely beliefs that assign probability 1 to one particular choice for Barbara. For instance, when we say that "you believe that Barbara chooses *blue*," then what we really mean to say is that you assign probability 1 to the event that Barbara will indeed choose *blue*. So, in a sense it reflects a state of mind in which you are rather certain about what Barbara will do – namely choose a *blue* dress.

But often we are *uncertain* about the choice that our opponent will make, and in these situations such beliefs seem inappropriate. Namely, if you really are uncertain about the color that Barbara will choose, then you could as well hold a *probabilistic* belief that assigns positive probability to *different* choices for Barbara. For instance, you could believe that, with probability 0.6, Barbara chooses *blue*, and that, with probability 0.4, she chooses *green*. So, this would model a state of mind in which you deem both colors *blue* and *green* possible for Barbara, but where you deem *blue* slightly more likely than *green*.

Suppose you hold this probabilistic belief. If you choose *blue* yourself, then with probability 0.6 you expect to wear the same color, yielding a utility of 0, and with probability 0.4 you expect to wear a different color, yielding a utility of 4. So, your *expected utility* from choosing *blue* would be

$$(0.6) \cdot 0 + (0.4) \cdot 4 = 1.6.$$

If you choose *green*, then with probability 0.6 you expect to wear a different color, yielding a utility of 3, and with probability 0.4 you expect to wear the same color, yielding a utility of 0. So, your *expected utility* from choosing *green* would be

$$(0.6) \cdot 3 + (0.4) \cdot 0 = 1.8.$$

Now, if you choose *red* instead then, under your belief above, you would expect – with probability 1 – to wear a different color from Barbara, and hence your expected utility would be 2. Similarly, if you choose *yellow*, then you would also expect – with probability 1 – to wear a different color, but your utility would be less, namely 1.

Hence, we see that among the four colors, *red* gives you the highest expected utility. In other words, choosing *red* is optimal for you if you believe that, with probability 0.6, Barbara chooses *blue*, and that, with probability 0.4, she chooses *green*. So, *red* is also a rational color for you!

Finally, we investigate your least preferred color *yellow*. Is there any belief about Barbara's choice for which choosing *yellow* is optimal? The answer is "no." To see this we distinguish the following two cases, according to the probability that you assign to Barbara choosing *blue*. Suppose first that you assign probability *less than 0.5* to Barbara choosing *blue*. Then, by choosing *blue* yourself, your expected utility will be at least

$$(0.5) \cdot 4 = 2,$$

since you expect that, with probability at least 0.5, you will wear a different color from Barbara. By choosing *yellow*, your expected utility would be at most 1. Suppose next that you assign probability *at least 0.5* to Barbara choosing *blue*. Then, by choosing *green* yourself your expected utility would be at least

$$(0.5) \cdot 3 = 1.5,$$

since you expect that, with probability at least 0.5, you will wear a different color from Barbara. Again, by choosing *yellow* your expected utility would be at most 1. So, for every possible belief you could have about Barbara's choice, either *blue* or *green* will give you a higher expected utility than *yellow*. This means that *yellow* can never be an optimal choice for you.

Summarizing, we see that *blue*, *green* and *red* are rational choices, but *yellow* is not. Moreover,

- *blue* is optimal for you if you believe that, with probability 1, Barbara chooses *green*,
- *green* is optimal for you if you believe that, with probability 1, Barbara chooses *blue*,
- *red* is optimal for you if you believe that, with probability 0.6, Barbara chooses *blue*, and that, with probability 0.4, she chooses *green*.

These finding are shown by the beliefs diagram in Figure 2.3. Note that from your choice *red* there is a *forked* arrow: One end points to Barbara's choice *blue*, with probability 0.6, and the other end points to Barbara's choice *green*, with probability 0.4. This represents the probabilistic belief in which you assign probability 0.6 to Barbara choosing *blue* and probability 0.4 to Barbara choosing *green*. The forked arrow leaving your choice *red* thus indicates that choosing *red* is optimal for you if you believe that, with probability 0.6, Barbara chooses *blue*, and with probability 0.4 she chooses *green*.

Suppose now that your utilities are different, and are given by Table 2.4 instead. Again, these utilities would model a situation in which you prefer *blue* to *green*, *green*

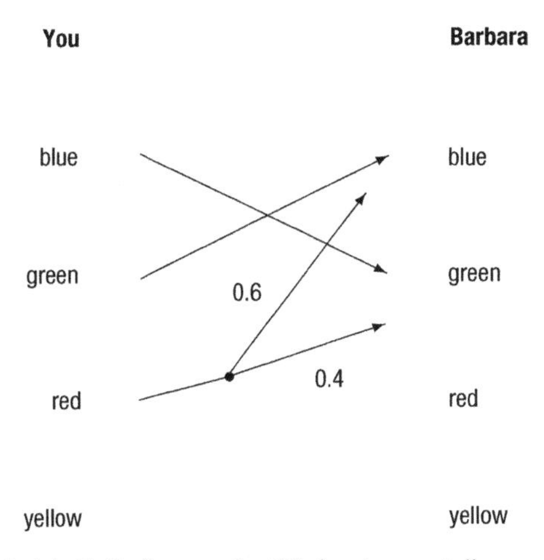

Figure 2.3 A beliefs diagram for "Going to a party"

Table 2.4. *Alternative utilities in "Going to a party"*

blue	green	red	yellow	same color as Barbara
5	4	2	1	0

Table 2.5. *Alternative utilities in "Going to a party" (II)*

blue	green	red	yellow	same color as Barbara
10	9	8	7	0

to *red*, and *red* to *yellow*, as before. At first sight, one may perhaps think that your preferences have not changed, and therefore your rational choices would still be *blue*, *green* and *red*. However, this reasoning is false! In the new situation, *red* is no longer a rational choice for you.

To see this we again distinguish two cases, according to the probability that you assign to Barbara choosing *blue*. Suppose first that you assign probability *at most* 0.55 to the event that Barbara chooses *blue*. Then, your expected utility from choosing *blue* would be at least

$$(0.45) \cdot 5 = 2.25,$$

since you expect, with probability at least 0.45, to wear a different color. By choosing *red* you would get at most 2. If, on the other hand, you assign probability *at least* 0.55 to the event that Barbara chooses *blue*, then by choosing *green* your expected utility would be at least

$$(0.55) \cdot 4 = 2.2,$$

since you expect, with probability at least 0.55, to wear a different color. Again, by choosing *red* you expect at most 2. So, for every possible belief that you can hold about Barbara's choice, either *blue* or *green* gives you a higher expected utility than *red*. Hence, under the new utilities *red* is no longer a rational choice.

The explanation for this phenomenon is that under the new utilities, you still prefer the colors *blue* and *green* to *red*, but you now prefer these colors *more strongly* to *red* than before. In fact, under the new utilities you like *blue* and *green* so much more than *red* that choosing *red* can no longer be optimal for you.

Finally, consider a situation in which your utilities are as in Table 2.5. Again, these utilities represent a situation in which you prefer *blue* to *green*, *green* to *red*, and *red* to *yellow*, as before. Moreover, the "relative ranking" of the colors has not changed compared to the original utilities in Table 2.3 – for any two colors, the difference in utility is the same as in Table 2.3. So, one might be tempted to say that these changes in utility should have no impact on your rational choices. But this argument is again false!

Under the new utilities every color is rational for you, even your least preferred color *yellow*! In order to see this, suppose you believe that, with probability 0.5, Barbara chooses *blue*, with probability 0.3 she chooses *green* and with probability 0.2 she chooses *red*. By choosing *blue* your expected utility would be

$$(0.5) \cdot 10 = 5,$$

by choosing *green* your expected utility would be

$$(0.7) \cdot 9 = 6.3,$$

and by choosing *red* your expected utility would be

$$(0.8) \cdot 8 = 6.4.$$

On the other hand, choosing *yellow* would yield an expected utility of 7, since you believe that, with probability 1, you have chosen a different color from Barbara. So, under the belief above, it would be optimal for you to choose your least preferred color *yellow*. In particular, *yellow* is a rational choice under the new utilities. You may verify that the other three colors are also rational for you.

Summarizing, we see that under the new utilities in Table 2.5 all colors are rational for you, whereas under the original utilities in Table 2.3 *yellow* was not a rational choice. This is despite the fact that the relative ranking of the colors has not changed. How can this be?

The crucial difference between the situations is that in Table 2.5 you are much more afraid of wearing the same color than before. Suppose that you are considering switching from *green* to *blue*. Under the original utilities of Table 2.3, this switch would increase your utility by 1 if Barbara does not choose *blue*, but it would decrease your utility by 3 if she chooses *blue*. Compare this to the situation of Table 2.5: There, this switch would again increase your utility by 1 if Barbara does not choose *blue*, but it would decrease your utility by a much higher amount – 9 – if she chooses *blue*! So, under the new utilities the danger of wearing the same color as your friend has a much larger effect than before.

Consider again your belief above, by which you believe that, with probability 0.5, Barbara chooses *blue*, with probability 0.3 she chooses *green* and with probability 0.2 she chooses *red*. We have seen that under the new utilities your least preferred color *yellow* is optimal for this belief. The reason is that you deem every other color than *yellow* possible for Barbara. Since you are very afraid of wearing the same color, it is optimal for you to avoid the colors *blue*, *green* and *red*, and therefore *yellow* remains your best option. □

2.3 More than two players

In each of the examples discussed so far we only had two players. The reason is that we wanted to keep things as simple as possible at the beginning of this chapter. It is now time to extend the concepts we have learned so far, such as beliefs about the opponent's choices, utility, expected utility, rational choices and beliefs diagrams, to situations with more than two players. As an illustration, let us consider the following example.

Example 2.3 Waiting for a friend

It is Saturday evening 7.15 pm and you want to go to the cinema with two of your friends, Barbara and Chris. The movie starts at 8.30 pm and you have all agreed to meet around 8.00 pm in a pub near the cinema, so that you can have a drink before. Barbara told you a few minutes ago by phone that she still has to take a shower, so she can never make it before 7.40 pm. Chris has to work until 7.30 pm, and he cannot be there before 7.50 pm. If you leave your home now, you could be there at 7.30 pm.

In order to keep things simple, let us assume that you can choose to be at the pub at 7.30, at 7.40, at 7.50, at 8.00, at 8.10, at 8.20 or at 8.30. Of course you do not want to be late for the movie, so arriving later than 8.30 is not an option. Similarly, Barbara can choose any arrival time between 7.40 pm and 8.30 pm with intervals of ten minutes, whereas Chris can choose any arrival time between 7.50 pm and 8.30 pm with intervals of ten minutes.

The question we wish to answer here is: Which arrival times are rational for you? As in the example "Going to a party," the answer depends crucially on how you evaluate the possible outcomes in this situation. For instance, if you could choose between waiting ten minutes for a friend or letting your friends wait for ten minutes, what would you prefer?

Suppose now that you prefer to wait for ten minutes rather than to let your friends wait for ten minutes. However, there is a limit to your patience, and you prefer to let your friends wait for ten minutes rather than to wait for half an hour yourself. More precisely, if you have to wait in the pub for Barbara or Chris or both, then your utility is equal to

$$20 - \text{ time you have to wait,}$$

where the waiting time is expressed in minutes. So, your utility decreases the longer you have to wait. However, if you are the last one to arrive (possibly arriving together with one friend, or with both friends), then your utility would be

$$5 - \text{ time you have let the first entrant wait.}$$

Hence, if you are the last one to enter the pub, then your utility decreases the longer you have let your friends wait. Overall, we see that the best possible scenario for you is to wait for exactly ten minutes until the last one arrives, yielding a utility of 10.

Here are some examples which show how to compute your utility in different situations: If you arrive at 7.50 pm, Barbara arrives at 8 pm and Chris at 8.20 pm, you have to wait for 30 minutes in total, so your utility would be $20 - 30 = -10$. If you arrive at 8 pm, Barbara arrives at 7.50 pm and Chris at 8.10 pm, then you have to wait for ten minutes, so your utility would be $20 - 10 = 10$. If you arrive at 8.10 pm, Barbara arrives at 8.10 pm and Chris arrives at 8.00 pm, then you are the last one to arrive (together with

Barbara), and you let Chris wait for ten minutes. So, your utility would be $5-10=-5$. If you arrive at 8.10 pm, Barbara arrives at 7.50 pm and Chris arrives at 8.00 pm, then you are the last one to arrive. Moreover, the first to arrive was Barbara, and you have let her wait for 20 minutes, so your utility would be $5-20=-15$. Finally, if you all arrive at the same time, then you are the last one to arrive (together with Barbara and Chris) but nobody had to wait, so your utility would be $5-0=5$.

In particular, you see that waiting for ten minutes is more desirable than to be the last one to arrive, because the first outcome gives you a utility of 10, whereas the last outcome gives you a utility of at most 5. On the other hand, waiting for half an hour is less desirable than letting your friends wait for ten minutes, since the first outcome gives you a utility of -10, whereas the last outcome gives you a utility of -5.

Now, if your utilities are as described above, which arrival times are rational for you? Let us simply start by considering the possible arrival times one by one, and see which ones can be supported by some belief about the friends' choices and which ones cannot.

- If you choose to arrive at 7.30 pm, then you must wait for at least 20 minutes because Chris cannot be there before 7.50 pm. But then, you could always shorten your waiting time by choosing 7.40 pm instead, and increase your utility. So, 7.40 pm is always a better choice than 7.30 pm, no matter at what time your friends show up. In particular, 7.30 cannot be a rational choice.
- Choosing 7.40 pm is optimal if you believe, for instance, that both friends will come at 7.50 pm. In this case, you would have to wait for ten minutes and your utility would be 10, which is the highest utility you can ever achieve.
- Similarly, choosing 7.50 pm is optimal if you believe that both friends will come 10 minutes later. In the same way, you can show that choosing 8.00 pm, 8.10 pm and 8.20 pm are optimal choices if you believe that both friends will come exactly 10 minutes later.
- However, 8.30 pm can never be an optimal choice. Suppose that in the best case both friends arrive at 8.30 pm. Even in this case it would have been better for you to have chosen 8.20 pm instead, since this would give you a utility of 10, which is more than the utility of 5 you get by choosing 8.30 pm. If, on the other hand, Barbara arrives before 8.30 pm and Chris arrives at 8.30 pm, then by choosing 8.30 pm you would be the last one to arrive, and you would have let Barbara wait for at least ten minutes, and hence your utility would be at most $5-10=-5$. However, by choosing 8.20 pm instead, you would no longer be the last to arrive, and you would have to wait for only ten minutes, and hence your utility would be $20-10=10$. Finally, suppose that both friends arrive before 8.30 pm. Then, by choosing 8.30 pm you would be the last one to arrive, and you would have let your friends wait for at least ten minutes. If you choose 8.20 pm instead, you would still be the last one to arrive, but you would reduce the waiting time for your friends, and hence increase your utility. So, we see that 8.20 pm is always a better option than 8.30 pm, no matter at what times the others arrive. In particular, 8.30 pm cannot be a rational choice.

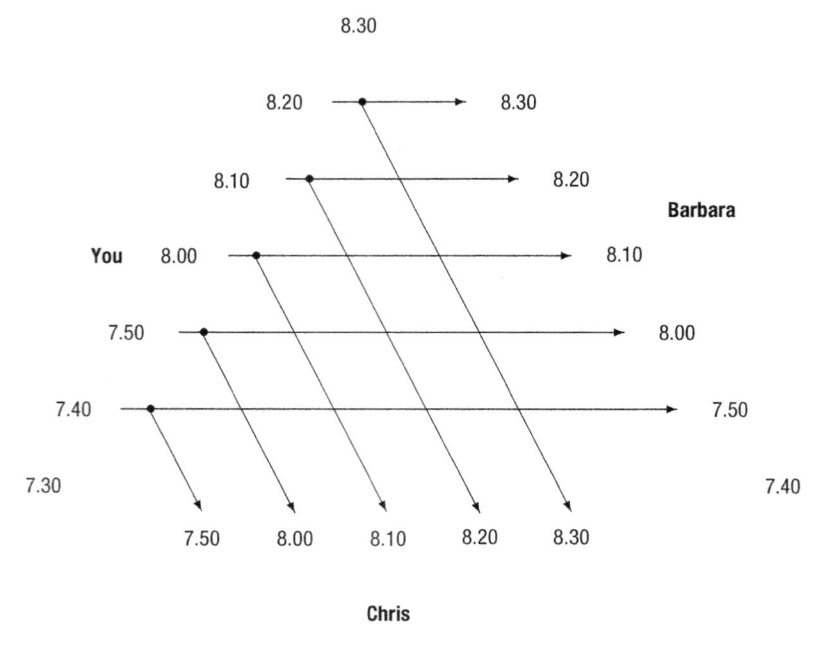

Figure 2.4 Beliefs diagram for "Waiting for a friend"

Summarizing, we see that all arrival times, except 7.30 pm and 8.30 pm, are rational choices for you. We can therefore make the beliefs diagram as depicted in Figure 2.4. The diagram should be read as follows: The forked arrow that leaves your choice "arrive at 7.40 pm" leads to Barbara's choice 7.50 and Chris' choice 7.50, and indicates that arriving at 7.40 pm is optimal for you if you believe that Barbara arrives at 7.50 pm and Chris arrives at 7.50 pm.

The beliefs that are represented in Figure 2.4 are, again, extreme cases. For instance, the forked arrow that leaves your choice to arrive at 7.40 states that you believe, with probability 1, that Barbara and Chris will both arrive at 7.50. It may well be, however, that you are inherently uncertain about the arrival times of Barbara and Chris. For instance, you may believe that, with probability 0.7, Barbara will arrive at 7.50 and Chris at 8.00, and you may believe that, with probability 0.3, Barbara and Chris will both arrive at 8.10. This would be a *probabilistic belief* in which you assign positive probability to various opponents' choice-combinations, indicating that you are truly uncertain about the opponents' choices.

What choice would be optimal for you under this probabilistic belief? In order to answer this question we must compute, for each of your possible arrival times, the expected utility under this particular belief, and then select the choice that gives the highest expected utility. We will show how to compute the expected utility for two of your choices, and leave the computations for the other choices to the reader.

Table 2.6. *Expected utilities for you in "Waiting for a friend"*

arrival time	7.30	7.40	7.50	8.00	8.10	8.20	8.30
expected utility	−13	−3	7	−0.5	−9	−19	−29

- Suppose you arrive at 7.30. You believe that, with probability 0.7, your last friend will arrive at 8.00, in which case you would have to wait for 30 minutes, yielding a utility of −10. With probability 0.3 you believe that both will arrive at 8.10, in which case you would have to wait for 40 minutes, resulting in a utility of −20. So, your expected utility from arriving at 7.30 would be

$$(0.7) \cdot (-10) + (0.3) \cdot (-20) = -13.$$

- Suppose you arrive at 8.00. You believe that, with probability 0.7, Barbara will arrive at 7.50 and Chris will arrive at 8.00, in which case you would be the last one to arrive (together with Chris), letting Barbara wait for 10 minutes, and thus yielding a utility of −5. With probability 0.3 you believe that both will arrive at 8.10, in which case you would have to wait for 10 minutes, giving you a utility of 10. So, your expected utility from arriving at 8.00 would be

$$(0.7) \cdot (-5) + (0.3) \cdot 10 = -0.5.$$

In the same fashion, you can compute the expected utility for each of the other choices. We show all these expected utilities in Table 2.6. So, your optimal arrival time under this particular probabilistic belief is 7.50. In other words, your choice to arrive at 7.50 is also optimal if you believe that, with probability 0.7, Barbara will arrive at 7.50 and Chris at 8.00, and that with probability 0.3, Barbara and Chris will both arrive at 8.10. Within a beliefs diagram, this probabilistic belief is shown in Figure 2.5. □

2.4 Choosing rationally

In the previous sections we have discussed, by means of a few examples, some important notions such as a belief about the opponents' choices, a utility function, the expected utility induced by a choice and choosing rationally. So far we have described these concepts somewhat informally, and now it is time to give formal definitions for these notions.

Let us start with a *belief about the opponents' choices*. The examples "Where to locate my pub?" and "Going to a party" illustrated that, in a game with two players, your belief about the opponent's choice can be of two kinds: It can either assign probability 1 to one particular choice of the opponent, or it can assign positive probability to various choices of the opponent. In the example "Waiting for a friend" we saw that if you have

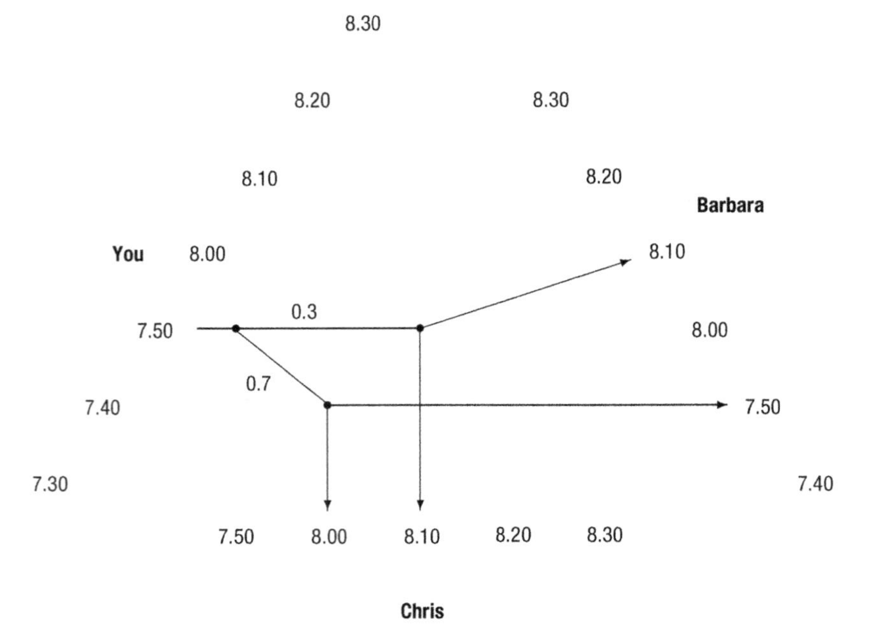

Figure 2.5 A probabilistic belief in "Waiting for a friend"

more than one opponent, then you must hold a belief about the possible opponents' choice-combinations. That is, your belief either assigns probability 1 to one particular choice-combination for the opponents, or it assigns positive probability to various choice-combinations.

In general, let us denote the set of players by $I = \{1, 2, ..., n\}$. That is, we number the players from 1 to n. For every player $i \in I$, let C_i be the set of possible choices for player i. Suppose we focus on a particular player i. The *opponents* for player i are the players $1, ..., i-1, i+1, ..., n$, that is, all the players except player i. A *choice-combination* for i's opponents is a list

$$(c_1, ..., c_{i-1}, c_{i+1}, ..., c_n)$$

of choices for the opponents, where

$$c_1 \in C_1, ..., c_{i-1} \in C_{i-1}, c_{i+1} \in C_{i+1}, ..., c_n \in C_n.$$

So, a choice-combination specifies for every opponent j a choice $c_j \in C_j$. By $C_1 \times ... \times C_{i-1} \times C_{i+1} \times ... \times C_n$ we denote the set of all such opponents' choice-combinations $(c_1, ..., c_{i-1}, c_{i+1}, ..., c_n)$. A *belief* for player i about the opponents' choices assigns a probability to every opponents' choice-combination. More formally, we say that a belief for player i about the opponents' choices is a *probability distribution* b_i over the set $C_1 \times ... \times C_{i-1} \times C_{i+1} \times ... \times C_n$, assigning to every opponents' choice-combination

$$(c_1, ..., c_{i-1}, c_{i+1}, ..., c_n) \in C_1 \times ... \times C_{i-1} \times C_{i+1} \times ... \times C_n$$

some probability

$$b_i(c_1,...,c_{i-1},c_{i+1},...,c_n) \geq 0.$$

Moreover, the sum of all the probabilities $b_i(c_1,...,c_{i-1},c_{i+1},...,c_n)$ should be equal to 1. As a special case, the belief b_i could assign probability 1 to one particular choice-combination, but in general it can assign positive probability to different opponents' choice-combinations. This leads to the following definition.

Definition 2.4.1 *(Belief about the opponents' choices)*
*A **belief for player i about the opponents' choices** is a probability distribution b_i over the set $C_1 \times ... \times C_{i-1} \times C_{i+1} \times ... \times C_n$ of opponents' choice-combinations. For every choice-combination $(c_1,...,c_{i-1},c_{i+1},...,c_n)$, the number $b_i(c_1,...,c_{i-1},c_{i+1},...,c_n)$ denotes the probability that player i assigns to the event that the opponents choose exactly according to the choice-combination $(c_1,...,c_{i-1},c_{i+1},...,c_n)$.*

Let us illustrate this definition by some of the examples we have seen so far. In the example "Going to a party," suppose you are player 1 and Barbara is player 2. Your set of choices is $C_1 = \{blue, green, red, yellow\}$, and the set of choices for Barbara is $C_2 = \{blue, green, red, yellow\}$ as well. Consider the beliefs diagram of Figure 2.3. The arrow from your choice *blue* to Barbara's choice *green* represents the belief where you assign probability 1 to Barbara's choice *green*. More precisely, this arrow represents the probability distribution b_1 on C_2 where $b_1(green) = 1$, and $b_1(c_2) = 0$ for every other choice $c_2 \in C_2$. The forked arrow from your choice *red* to Barbara's choices *blue* and *green*, with probabilities 0.6 and 0.4, represents your probabilistic belief b_1 with $b_1(blue) = 0.6$, $b_1(green) = 0.4$, and $b_1(red) = b_1(yellow) = 0$.

In the example "Waiting for a friend," suppose you are player 1, Barbara is player 2 and Chris is player 3. Then, your set of choices is $C_1 = \{7.30, ..., 8.30\}$, Barbara's set of choices is $C_2 = \{7.40, ..., 8.30\}$, and Chris' set of choices is $C_3 = \{7.50, ..., 8.30\}$. In the beliefs diagram of Figure 2.4, consider the forked arrow that leaves your choice 7.40. This forked arrow represents the belief in which you assign probability 1 to the opponents' choice-combination (7.50, 7.50) in which Barbara and Chris both choose 7.50. Formally, it represents your belief b_1 with $b_1(7.50, 7.50) = 1$ and $b_1(c_2, c_3) = 0$ for every other choice-combination of the opponents $(c_2, c_3) \in C_2 \times C_3$. Consider now the probabilistic belief in the beliefs diagram of Figure 2.5. It represents the probabilistic belief that assigns probability 0.7 to the event that Barbara chooses 7.50 and Chris chooses 8.00, and probability 0.3 to the event that Barbara and Chris both choose 8.10. More formally, it shows the probabilistic belief b_1 with $b_1(7.50, 8.00) = 0.7$, $b_1(8.10, 8.10) = 0.3$, and $b_1(c_2, c_3) = 0$ for every other choice-combination of the opponents $(c_2, c_3) \in C_2 \times C_3$.

The probabilistic belief from Figure 2.5 thus shows that in a game with more than two players, the belief that player 1 has about player 2's choice may very well be *correlated* with his belief about player 3's choice. In Figure 2.5 you believe that, conditional on Barbara arriving at 7.50, Chris will arrive at 8.00, and conditional on Barbara arriving at 8.10, you believe that Chis will arrive at 8.10 as well.

We now turn to the concept of a *utility function*. As the examples have already shown, the *utility* you derive from an outcome measures your degree of satisfaction if this outcome is realized. In the example "Going to a party," for instance, your utility will be 3 if you choose *green* and Barbara chooses *red*. We write $u_1(green, red) = 3$, meaning that the choice-combination $(green, red)$ in which you choose *green* and Barbara chooses *red* gives you a utility of 3. In the example "Waiting for a friend," your utility will be -10 if you arrive at 7.50, Barbara arrives at 8.00 and Chris arrives at 8.20. We formally denote this as $u_1(7.50, 8.00, 8.20) = -10$, indicating that the choice-combination $(7.50, 8.00, 8.20)$, in which you choose 7.50, Barbara chooses 8.00 and Chris chooses 8.20, gives you utility -10. In the example "Where to locate my pub?," your utility is simply equal to the number of customers you obtain. For instance, if you choose location b and Barbara chooses location e, then you will have 250 customers. We write $u_1(b, e) = 250$.

In general, let $C_1 \times ... \times C_n$ denote the set of all choice-combinations $(c_1, ..., c_n)$ for the players, where $c_1 \in C_1, ..., c_n \in C_n$. Then, a utility function assigns to every possible choice-combination a number, representing the utility you receive if this particular choice-combination is realized.

Definition 2.4.2 *(Utility function)*
*A **utility function** for player i is a function u_i that assigns to every choice-combination $(c_1, ..., c_n)$ a number $u_i(c_1, ..., c_n)$. The number $u_i(c_1, ..., c_n)$ denotes the utility that player i derives from the outcome induced by the choice-combination $(c_1, ..., c_n)$.*

If we specify for every player i the set of choices C_i and the utility function u_i, then the resulting object is called a *static game*.

Definition 2.4.3 *(Static game)*
*A **static game** $\Gamma = (C_1, ..., C_n, u_1, ..., u_n)$ specifies for every player i a set of choices C_i and a utility function u_i.*

Suppose player i holds a belief b_i about his opponents' choices, and has utility function u_i. Then, each of the player's choices will generate an *expected utility*. The three examples we have discussed so far illustrate how to calculate the expected utility. Consider, for instance, the example "Waiting for a friend." Suppose you hold the belief b_1 about the opponents' choices where

$$b_1(8.30, 8.00) = 0.5, \; b_1(7.50, 7.50) = 0.3, \; b_1(8.00, 8.10) = 0.2,$$

and $b_1(c_2, c_3) = 0$ for every other choice-combination (c_2, c_3). That is, you believe that, with probability 0.5, Barbara will arrive at 8.30 and Chris at 8.00, you believe that, with probability 0.3, Barbara and Chris will arrive at 7.50, and you believe that, with probability 0.2, Barbara will arrive 8.00 and Chris at 8.10. If you plan to arrive at 8.00, your expected utility would be given by

$$\begin{aligned} u_1(8.00,b_1) &= (0.5)\cdot(20-30)+(0.3)\cdot(5-10)+(0.2)\cdot(20-10) \\ &= (0.5)\cdot u_1(8.00,8.30,8.00)+(0.3)\cdot u_1(8.00,7.50,7.50) \\ &\quad +(0.2)\cdot u_1(8.00,8.00,8.10) \\ &= b_1(8.30,8.00)\cdot u_1(8.00,8.30,8.00) \\ &\quad +b_1(7.50,7.50)\cdot u_1(8.00,7.50,7.50) \\ &\quad +b_1(8.00,8.10)\cdot u_1(8.00,8.00,8.10). \end{aligned}$$

The logic behind the last equation is that, with probability $b_1(8.30,8.00)$, you expect your friends to arrive at 8.30 and 8.00, which would give you a utility of $u_1(8.00,8.30,8.00)$, and similarly for the other two terms. In order to define expected utility formally, we use the abbreviation

$$C_{-i} := C_1 \times ... \times C_{i-1} \times C_{i+1} \times ... \times C_n.$$

So, C_{-i} denotes the set of choice-combinations that are possible for player i's opponents.

Definition 2.4.4 *(Expected utility)*
Suppose player i has utility function u_i and holds the belief b_i about the opponents' choices. Then, the ***expected utility*** *for player i of making a choice c_i is given by* $u_i(c_i,b_i) =$

$$\sum_{(c_1,...,c_{i-1},c_{i+1},...,c_n)\in C_{-i}} b_i(c_1,...,c_{i-1},c_{i+1},...,c_n)\cdot u_i(c_1,...,c_{i-1},c_i,c_{i+1},...,c_n).$$

The expression above should be read as follows: Suppose you hold the belief b_i about the opponents' choices, and make the choice c_i. Then you expect any opponents' choice-combination $(c_1,...,c_{i-1},c_{i+1},...,c_n)$ to occur with probability $b_i(c_1,...,c_{i-1},c_{i+1},...,c_n)$, in which case the final outcome would be $(c_1,...,c_{i-1},c_i,c_{i+1},...,c_n)$ and your utility would accordingly be $u_i(c_1,...,c_{i-1},c_i,c_{i+1},...,c_n)$.

We can now define *optimal choice* and *rational choice*. If you hold a certain belief about your opponents' choices, then a choice c_i is said to be *optimal* for you if it gives you the highest possible expected utility. A choice c_i is called *rational* if it is optimal for *some* belief that you can hold about the opponents' choices.

Definition 2.4.5 *(Optimal and rational choice)*
(a) Suppose player i has utility function u_i and holds the belief b_i about the opponents' choices. Then, the choice c_i^* is ***optimal*** *for player i given his belief b_i if this choice maximizes his expected utility, that is,*

$$u_i(c_i^*,b_i) \geq u_i(c_i,b_i)$$

for all other choices c_i in C_i.
(b) Suppose player i has utility function u_i. Then, a choice c_i^* is ***rational*** *for player i if there is some belief b_i about the opponents' choices for which c_i^* is optimal.*

In the example "Going to a party," for instance, your choice *red* is optimal under the belief b_1 with $b_1(blue) = 0.6$, $b_1(green) = 0.4$ and $b_1(red) = b_1(yellow) = 0$. See the beliefs diagram in Figure 2.3 for an illustration of this fact. In particular, your choice *red* is rational as there is some belief b_1 about Barbara's choice for which choosing *red* is optimal. Your choice *yellow* is not rational, as there is no belief b_1 about Barbara's choice for which choosing *yellow* is optimal.

2.5 Strictly dominated choices

In each of the three examples discussed so far, we have identified the rational choices you can make. We have called a choice *rational* if you can find a belief about the opponents' behavior such that this choice would give you the highest expected utility under this belief. Such a belief about the opponents' choices may be of two kinds – it may assign probability 1 to one particular opponents' choice-combination, or it may assign positive probability to various opponents' choice-combinations. The second type of belief, which we call a *probabilistic belief*, would reveal that you face some serious uncertainty about the opponents' choices.

For every example we gave a beliefs diagram, in which we indicate for each of your rational choices a belief about the opponents' choices that supports it. For every choice that is *not* rational, we have explained why it cannot be optimal for *any* belief about the opponents' choices. In most cases, such an explanation was easy to give.

In the example "Where to locate my pub?," location a is not rational since choosing location b is always better than choosing a, no matter where Barbara wants to locate her pub. We say that choice a is *strictly dominated* by choice b. It can therefore never be optimal to choose a. Similarly, location g is strictly dominated by location f, and hence g cannot be rational.

In the example "Waiting for a friend," the choice to arrive at 7.30 is not rational since it is strictly dominated by arriving at 7.40 pm. Similarly, the choice to arrive at 8.30 is not rational since it is strictly dominated by the choice to arrive at 8.20 pm.

Things are more complicated in the example "Going to a party." Consider the utilities as specified in Table 2.3. We have seen that your least preferred color *yellow* is not a rational choice, since there is no belief about Barbara's choice for which *yellow* is optimal. However, this time there is no easy argument why *yellow* is not rational, since it is not strictly dominated by any of your other choices. For instance, *yellow* is not strictly dominated by your most preferred color *blue* since *yellow* would be better if Barbara were to choose *blue*. Similarly, you can check that *yellow* is also not strictly dominated by *green* or *red*.

Now, imagine a situation in which you toss a coin before choosing your color: If the coin lands "heads" you choose *blue*, and if the coin lands "tails" you choose *green*. We call this a *randomized* choice in which you choose *blue* and *green* with probability 0.5. We show that this randomized choice strictly dominates your choice *yellow*. Namely, if Barbara chooses *blue* then the randomized choice would give you an expected utility of

$$(0.5) \cdot 0 + (0.5) \cdot 3 = 1.5,$$

since with probability 0.5 you choose *blue*, and hence choose the same color as Barbara, and with probability 0.5 you choose *green* and hence choose a different color from Barbara. On the other hand, *yellow* would only give you 1 if Barbara chooses *blue*. If Barbara chooses *green*, the randomized choice would yield the expected utility

$$(0.5) \cdot 4 + (0.5) \cdot 0 = 2,$$

and *yellow* would only yield a utility of 1. Finally, if Barbara chooses *red* or *yellow* then the randomized choice gives expected utility

$$(0.5) \cdot 4 + (0.5) \cdot 3 = 3.5,$$

whereas choosing *yellow* would give at most 1. So, for every possible color that Barbara can choose, the randomized choice is better than *yellow*. Hence, the irrational choice *yellow* is strictly dominated by the randomized choice in which you choose *blue* and *green* with probability 0.5.

This raises the question whether people will actually use such randomized choices in practice. Well, purely theoretically people *could* use randomized choices, but there is nothing to gain compared to making "usual" choices! Suppose, for instance, you believe that, with probability 0.4, Barbara will choose *blue*, and with probability 0.6 she will choose *red*. Then, the expected utility from choosing *blue* yourself would be

$$(0.4) \cdot 0 + (0.6) \cdot 4 = 2.4,$$

your expected utility from choosing *green* would be

$$(0.4) \cdot 3 + (0.6) \cdot 3 = 3,$$

and your expected utility from choosing *red* would be

$$(0.4) \cdot 2 + (0.6) \cdot 0 = 0.8.$$

Hence, with this belief it would be optimal to choose *green*, and this choice would be strictly better than *blue*. Now, assume that you are considering using the randomized choice as described above, in which you choose *blue* and *green* with probability 0.5 each. So, first you toss a coin, and depending on the side that shows up you choose *blue* or *green*. Imagine that the coin shows "heads," in which case the randomized choice tells you to choose *blue*. However, your belief about Barbara's choice has not changed, so after observing that the coin has landed "heads" you still believe that, with probability 0.4, Barbara will choose *blue*, and with probability 0.6 she will choose *red*. So, it would be better for you to choose *green*, despite what the randomized choice tells you to do in this case. Thus, it would be better not to use this randomized choice.

Suppose now, you believe that Barbara, with probability 0.25, will choose *blue* and with probability 0.75 will choose *red*. In this case, your expected utility from choosing *blue* would be

$$(0.25) \cdot 0 + (0.75) \cdot 4 = 3,$$

and your expected utility from choosing *green* would be 3 as well. Hence, both choices *blue* and *green* would be optimal. Imagine you are considering using the randomized choice as described above. If the coin shows "heads," the randomized choice tells you to choose *blue*, and if it lands "tails" you are supposed to choose *green*. In both situations, it would indeed be optimal to do what the randomized choice tells you, so using this randomized choice would be optimal if you believe that, with probability 0.25, Barbara will choose *blue*, and with probability 0.75 she will choose *red*. The expected utility from using this randomized choice would be 3. Namely, choosing *blue* and choosing *green* both give you an expected utility of 3. So, your expected utility would be 3 if the coin lands "heads," and it would be 3 if the coin lands "tails," and hence the expected utility would be 3 on average. However, you could have obtained the same expected utility by simply choosing *blue* with probability 1, or choosing *green* with probability 1, without using a coin! So, even though your randomized choice could be optimal for some beliefs that you could have about Barbara's choice, there is no need to use the randomized choice in this case since you can always get the same expected utility by simply selecting one of your choices with probability 1. Moreover, implementing such a randomized choice is more troublesome than making a "regular" choice, since you first have to toss a coin, look which side is showing, and then remember which choice to make in that case. As such, the randomized choice is merely a "theoretical object," but is not something that people would really use in practice. In real life, people do *not* randomize when reaching important decisions.

This, however, does not mean that randomized choices are useless for our purposes! As we will see, they will help us identify which choices are not rational in a game. Before we come to this, let me briefly summarize what we have learned about the irrational choices in the three examples discussed so far: In the example "Going to a party," we have seen that the choice *yellow* is not rational, and it is strictly dominated by the randomized choice in which you would choose *blue* and *green* both with probability 0.5. In the other two examples, every choice that is not rational is strictly dominated by some other (non-randomized) choice. So, in all the examples above, *every choice that is not rational is either strictly dominated by some other choice, or is strictly dominated by a randomized choice.* In fact, we will see below that this is not only true for the examples discussed above, but it is true in general! So, randomized choices provide a useful technical tool for finding irrational choices in a game, although people do not randomize in practice.

Before we come to this remarkable result, let us first try to formalize these new concepts, such as *being strictly dominated by another choice* and *being strictly dominated by a randomized choice*. Let us start with the general definition of a randomized choice. In the example "Going to a party" we have seen a randomized choice in which, with probability 0.5, you choose *blue*, and with probability 0.5 you choose *green*. Let us assume that you are player 1, and Barbara is player 2. Then, this randomized choice can be written as a probability distribution r_1 over your set of choices $C_1 = \{blue, green, red, yellow\}$ with

$$r_1(blue) = 0.5,\ r_1(green) = 0.5,\ r_1(red) = 0 \text{ and } r_1(yellow) = 0.$$

Another randomized choice would be, for instance, a probability distribution r_1 with

$$r_1(blue) = 0.1,\ r_1(green) = 0.4,\ r_1(red) = 0.2 \text{ and } r_1(yellow) = 0.3,$$

which means that, with probability 0.1, you choose *blue*, with probability 0.4 you choose *green*, with probability 0.2 you choose *red*, and with probability 0.3 you choose *yellow*. This randomized choice can be simulated by taking a deck of ten cards, numbered 1 to 10, from which you randomly draw one card. If the number on the card is 1, you choose *blue*, if the number on the card is 2, 3, 4 or 5 you choose *green*, if the number is 6 or 7 you choose *red*, and if the number is 8, 9 or 10 you choose *yellow*.

In general, a randomized choice for player i in a game is a probability distribution r_i over the set C_i of choices for player i. For every choice c_i, the number $r_i(c_i)$ indicates the probability that you would choose c_i. Such a randomized choice can be implemented, for instance, by tossing a coin, throwing a dice, drawing a card or spinning a roulette wheel, and letting the final choice depend on the outcome of this randomizing device.

Let us go back to the example "Going to a party," and suppose you make the randomized choice r_1 above with

$$r_1(blue) = 0.1,\ r_1(green) = 0.4,\ r_1(red) = 0.2 \text{ and } r_1(yellow) = 0.3.$$

Assume that Barbara chooses *green*. Then, with probability 0.1 you will choose *blue* and earn a utility of 4, with probability 0.4 you will choose *green* and earn a utility of 0, with probability 0.2 you will choose *red* and earn a utility of 2, and with probability 0.3 you choose *yellow* and earn a utility of 1. So, your expected utility by using this randomized choice r_1 if Barbara chooses *green* would be

$$\begin{aligned} u_1(r_1, green) &= (0.1)\cdot 4 + (0.4)\cdot 0 + (0.2)\cdot 2 + (0.3)\cdot 1 \\ &= (0.1)\cdot u_1(blue, green) + (0.4)\cdot u_1(green, green) + \\ &\quad + (0.2)\cdot u_1(red, green) + (0.3)\cdot u_1(yellow, green) \end{aligned}$$

which, in turn, is equal to

$$\begin{aligned} &r_1(blue)\cdot u_1(blue, green) + r_1(green)\cdot u_1(green, green) + \\ &+ r_1(red)\cdot u_1(red, green) + r_1(yellow)\cdot u_1(yellow, green). \end{aligned}$$

The intuition behind the last expression is that with probability $r_1(blue)$ you will choose *blue*, which would give you a utility of $u_1(blue, green)$, and so on.

In general, if player i makes a randomized choice r_i in a game, and the opponents make the choices $c_1, ..., c_{i-1}, c_{i+1}, ..., c_n$, then the expected utility for this randomized choice would be

$$u_i(c_1, ..., c_{i-1}, r_i, c_{i+1}, ..., c_n) = \sum_{c_i \in C_i} r_i(c_i)\cdot u_i(c_1, ..., c_{i-1}, c_i, c_{i+1}, ..., c_n).$$

The logic behind this expression is the same as above: Any choice c_i is chosen with probability $r_i(c_i)$, in which case the outcome would be $(c_1, ..., c_{i-1}, c_i, c_{i+1}, ..., c_n)$ and the utility would thus be $u_i(c_1, ..., c_{i-1}, c_i, c_{i+1}, ..., c_n)$.

Summarizing, the concept of a randomized choice and its associated expected utility can be defined as follows:

Definition 2.5.1 *(Randomized choice and its expected utility)*
(a) A ***randomized choice*** *for player i is a probability distribution r_i over the set C_i of choices. For every choice c_i the number $r_i(c_i)$ denotes the probability that choice c_i is made.*
(b) Suppose that player i makes a randomized choice r_i, and his opponents choose $c_1, ..., c_{i-1}, c_{i+1}, ..., c_n$. Then, the ***expected utility of the randomized choice*** *is*

$$u_i(c_1, ..., c_{i-1}, r_i, c_{i+1}, ..., c_n) = \sum_{c_i \in C_i} r_i(c_i) \cdot u_i(c_1, ..., c_{i-1}, c_i, c_{i+1}, ..., c_n).$$

We now turn to the general definition of a choice being strictly dominated. As we have seen in the examples until now, a choice c_i can either be strictly dominated by another choice c_i', or be strictly dominated by a randomized choice r_i. In both cases, this means that the other choice c_i', or the randomized choice r_i, is always better than the original choice c_i, no matter what the opponents do. In terms of utility, it means that the other choice c_i' always yields a higher utility than c_i against any choice-combination of the opponents. Similarly, if the choice c_i is strictly dominated by the randomized choice r_i this means that, against any choice-combination of the opponents, the expected utility of the randomized choice r_i will always be larger than the utility of choosing c_i.

Definition 2.5.2 *(Strictly dominated choice)*
(a) A choice c_i for player i is ***strictly dominated by another choice*** *c_i' if for every choice-combination $(c_1, ..., c_{i-1}, c_{i+1}, ..., c_n)$ of the opponents*

$$u_i(c_1, ..., c_{i-1}, c_i, c_{i+1}, ..., c_n) < u_i(c_1, ..., c_{i-1}, c_i', c_{i+1}, ..., c_n).$$

(b) A choice c_i for player i is ***strictly dominated by a randomized choice*** *r_i if for every choice-combination $(c_1, ..., c_{i-1}, c_{i+1}, ..., c_n)$ of the opponents*

$$u_i(c_1, ..., c_{i-1}, c_i, c_{i+1}, ..., c_n) < u_i(c_1, ..., c_{i-1}, r_i, c_{i+1}, ..., c_n).$$

(c) A choice c_i is ***strictly dominated*** *if it is either strictly dominated by another choice or strictly dominated by a randomized choice.*

In the examples we have discussed so far, we have seen that every choice that was strictly dominated by another choice, or strictly dominated by a randomized choice, was irrational. In fact, it is not difficult to see that this property holds in general for *every* static game.

Suppose that choice c_i is strictly dominated by choice c_i'. Then, c_i' will always give player i a higher utility than c_i, no matter which choices the opponents make. Hence, for every probabilistic belief that player i can hold about the opponents' choices, c_i' will always yield a higher expected utility than c_i, which means that c_i can never be optimal for any belief.

The more difficult case is when c_i is strictly dominated by a randomized choice r_i. Then, for every possible choice-combination of the opponents, the expected utility generated by r_i will always be higher than the utility induced by c_i. Hence, for every probabilistic belief that player i can hold about the opponents' choices, r_i will always yield a higher expected utility than c_i. This means, however, that for every possible belief b_i about the opponents' choices, there must be a choice c_i' with a positive probability under r_i, which generates a higher expected utility than c_i under the belief b_i. Therefore, c_i can never be optimal for any belief about the opponents' choices. Hence, in general, every choice that is strictly dominated by another choice, or by a randomized choice, is irrational.

In the examples discussed so far we have seen that the converse is also true – namely that every irrational choice is strictly dominated by another choice, or by a randomized choice. It can be shown that this property holds in general, although this is not as easy to see.

So, by combining these two properties we reach the conclusion that in every static game, a choice is irrational, if and only if, it is strictly dominated by another choice or by a randomized choice. Or, equivalently, a choice is rational precisely when it is *not* strictly dominated by another choice *nor* strictly dominated by a randomized choice. We thus obtain the following characterization of rational choices in a static game.

Theorem 2.5.3 *(Characterization of rational choices)*
The rational choices in a static game are exactly those choices that are neither strictly dominated by some other choice nor strictly dominated by a randomized choice.

The proof for this result can be found in the proofs section at the end of this chapter. The theorem above can be very useful if we want to find *all* rational and *all* irrational choices in a game. Suppose that we want to verify whether a given choice c_i is rational or not. By the theorem above, there are only two possibilities: *either* we find a belief about the opponents' choices for which choosing c_i is optimal, in which case we may conclude that c_i is rational, *or* we find another choice or a randomized choice that strictly dominates c_i, in which case we may conclude that c_i is not rational. Now, finding another choice or a randomized choice that strictly dominates c_i is in general a much easier task than directly showing that there is no belief about the opponents' choices for which c_i is optimal. Here is an example which illustrates this.

Example 2.4 The traveler's dilemma

You have been on a trip to China with Barbara, where you have both bought the same Chinese vase on a market. When opening the luggage at the airport, you and Barbara discover that both vases are broken, probably because the employees at the airport have not handled the suitcases with sufficient care. You both decide to negotiate with the airport manager for compensation. Since the manager does not know exactly how much the vases are really worth, he proposes the following procedure: You must both

Table 2.7. *Compensation you receive in "The traveler's dilemma"*

		Barbara			
		100	200	300	400
You	100	100	250	300	350
	200	−75	200	350	400
	300	−25	25	300	450
	400	25	75	125	400

write down a price on a piece of paper. The price must be a multiple of 100 euros, and it should not be higher than 400 euros. You cannot communicate with each other about the price you write down. The person with the lower price will receive compensation equal to the average of both prices, *plus* a bonus of 100 euros for being so honest. The person with the higher price will receive compensation equal to the average of both prices, *minus* a penalty of 225 euros for being dishonest. If you and Barbara choose the same price, you will both receive a compensation equal to that price. Which prices are rational for you, and which prices are not?

In Table 2.7 we give, for each combination of prices that you and Barbara could choose, the compensation that you would receive. Choosing a price of 100 is optimal if you believe that, with probability 1, Barbara will write down a price of 200. As such, choosing a price of 100 is rational. It is optimal to write down a price of 200 if you believe that, with probability 1, Barbara will choose a price of 300, and choosing a price of 300 is optimal if you believe that, with probability 1, Barbara will choose a price of 400. So, we see that 100, 200 and 300 are rational choices.

What about choosing a price of 400? First of all, choosing a price of 400 is never optimal if you believe that, with probability 1, Barbara will choose one particular price. So, there is no "easy" belief for which choosing 400 is optimal. Moreover, you may verify that there is no other choice that strictly dominates choosing 400. So, there is no "easy" way of concluding that choosing a price of 400 is irrational. Can choosing 400 be optimal if you hold a probabilistic belief about Barbara's choice? For instance, can choosing 400 be optimal if you believe that Barbara, with probability p, will choose price 100 and with probability $1-p$ will choose price 400, for some number p? Or, could choosing 400 be optimal if you believe that, with probability p, she will choose price 200, with probability q she will choose price 300, and with probability $1-p-q$ she will choose price 400, for some numbers p and q? You could also try other probabilistic beliefs about Barbara's choice. After trying for some time, you will see that you cannot find any probabilistic belief about Barbara's choice for which choosing 400 is optimal. So, most probably this choice is irrational, but we are not completely sure yet. Now you have two options: You can either try to *show* that there is no probabilistic belief for which choosing 400 is optimal, or you can try to find a randomized choice that

Table 2.8. *Your expected compensation if you choose price 100 with probability 0.45 and price 300 with probability 0.55*

100	200	300	400
31.25	126.25	300	405

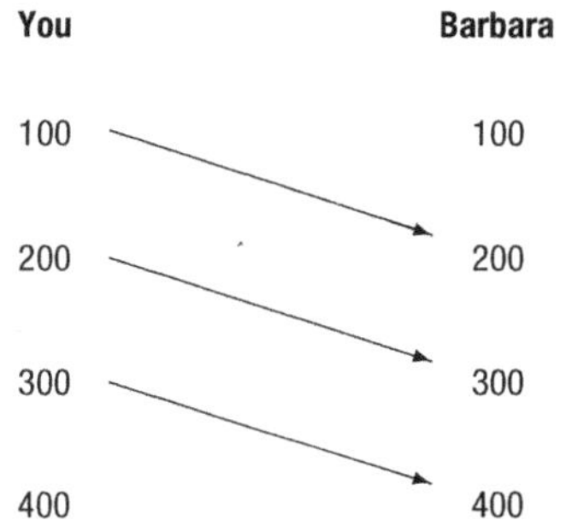

Figure 2.6 A beliefs diagram for "The traveler's dilemma"

strictly dominates choosing 400. In this case, the second option turns out be the easier one. After some trying, you may discover that choosing 400 is strictly dominated by the randomized choice in which you choose price 100 with probability 0.45 and price 300 with probability 0.55. Table 2.8 indicates, for every possible price that Barbara can choose, the expected compensation that this randomized choice would give you. If you compare these numbers with the compensation that you would receive from choosing 400 (see Table 2.7), you will see that choosing 400 is strictly dominated by this randomized choice: For every possible price that Barbara could write down, choosing 400 always gives you less compensation than by using the randomized choice. So, we can conclude that choosing a price of 400 is an irrational choice.

On the other hand, it is rather complicated to show directly that there is no probabilistic belief about Barbara's choice for which choosing 400 is optimal. You may try and you will discover how difficult it is!

Summarizing, we conclude that choosing the prices 100, 200 and 300 are rational options, whereas choosing the price 400 is not. Based on our insights, we can draw the beliefs diagram shown in Figure 2.6. □

2.6 Belief in the opponents' rationality

Remember that we have called a choice *rational* if this choice if optimal for *some* belief about the opponents' behavior. Certainly, any irrational choice must be considered unreasonable as it cannot be justified by *any* belief you can possibly hold about the opponents' choices. The question remains whether every *rational* choice is also *reasonable*. The answer is "no": If a choice can only be justified by *unreasonable beliefs* about the opponents' behavior, then the choice itself is also unreasonable. In order to see what we mean by unreasonable beliefs, let us revisit the example "Where to locate my pub?"

Example 2.5 Where to locate my pub?

Recall the story given in Example 2.1. We have seen that each of your choices b, c, d, e and f is rational, and that the locations a and g are irrational. The beliefs diagram in Figure 2.2 shows for each of these rational choices a belief that justifies this choice. For instance, choosing b is optimal if you believe that Barbara will choose a, choosing c is optimal if you believe that Barbara will choose b, and so on. The question remains whether each of these rational choices is also *reasonable*.

Let us have a look at the rational choice b. The arrow from b to a in the beliefs diagram of Figure 2.2 tells us that choosing b is optimal if you believe that Barbara will choose a. We have already seen that the choice a is irrational for you, since choosing b is always better for you than choosing a, no matter what Barbara does. Since you and Barbara have identical roles in this location problem, the choice a is also *irrational for Barbara* as choosing b is always better for her than choosing a, no matter which location you choose. Therefore, if you believe that Barbara chooses rationally too, you must assign *zero probability* to Barbara choosing a. By a similar argument, you should also assign probability zero to Barbara choosing g, as location g is irrational for her.

In particular, the belief in Figure 2.2 that justifies your choice b is an unreasonable belief, as it assigns probability 1 to Barbara choosing the irrational location a. Does this immediately mean that the choice b is unreasonable? Not yet. Perhaps we can justify the choice b by some other belief that only assigns positive probability to Barbara's rational choices b, c, d, e and f. However, if you believe that Barbara will only choose from b, c, d, e and f, then choosing c is always better for you than choosing b. As such, choosing location b cannot be optimal if you *believe that Barbara chooses rationally*. By a similar reasoning, also location f cannot be optimal if you believe that Barbara chooses rationally.

Hence, if you choose rationally yourself, and believe that Barbara chooses rationally too, you will never choose the locations a, b, f and g. What about the remaining locations c, d and e? Can these be chosen rationally if you believe that Barbara chooses rationally too? The answer is "yes." In order to see why, consider the beliefs diagram in Figure 2.7. In this beliefs diagram, we do not just represent the beliefs that you may hold about Barbara's choice, but we also depict beliefs that Barbara may have about your choice. For instance, the arrow from Barbara's choice b to your choice a, in the right-hand part of the diagram, indicates that the choice b is optimal for Barbara if she believes that you will choose a. We can interpret the other arrows in the right-hand part of the diagram in the same way.

Consider now your choice c. As we can read from the beliefs diagram, choosing c is optimal for you if you believe that Barbara will choose b, and choosing b is optimal

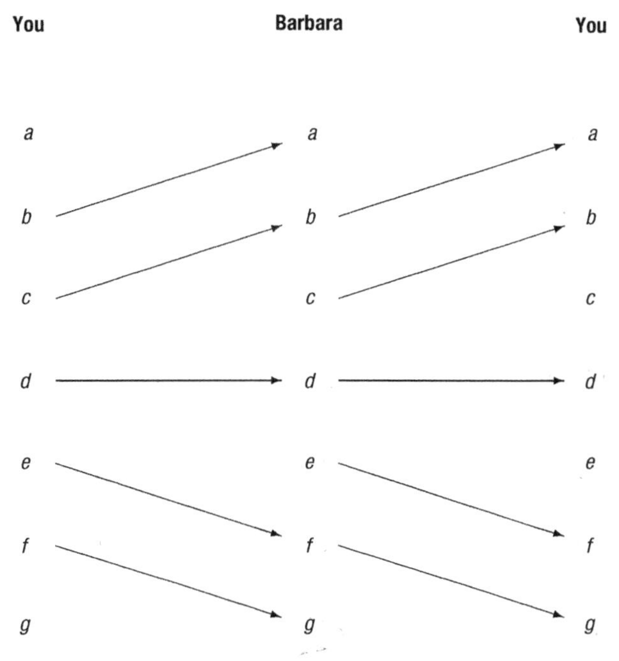

Figure 2.7 A beliefs diagram for "Where to locate my pub?" (II)

for Barbara if she believes that you will choose a. In particular, choosing b is rational for Barbara, and therefore you can rationally choose c while believing that Barbara chooses rationally too. Similarly, e is optimal for you if you believe that Barbara will choose f, and f is optimal for Barbara if she believes that you will choose g. So, you can also rationally choose e while believing that Barbara chooses rationally too. Finally, choosing d is optimal for you if you believe that Barbara will also choose d, and, vice versa, d is optimal for Barbara if she believes that you will choose d too. Therefore, you can also rationally choose d while believing that Barbara chooses rationally too.

Summarizing, we conclude that your choices can be divided into three classes:

- the locations a and g are irrational, since they are never optimal for any belief about Barbara's choice,
- the locations b and f are rational, but can no longer be rationally chosen if you believe that Barbara chooses rationally, and
- the locations c, d and e can be chosen rationally, even if you believe that Barbara chooses rationally as well. □

In the previous example, we have identified the choices you can rationally make if you believe that the opponent chooses rationally as well. In order to find these choices, you must *reason about your opponent*, as you should think of the choices that the opponent could rationally make. More precisely, you must form a belief about your opponent's choice that only assigns positive probability to those choices that are *rational* for your

Table 2.9. *Utilities for you and Barbara in "Going to a party"*

	blue	green	red	yellow	same color as friend
you	4	3	2	1	0
Barbara	2	1	4	3	0

opponent. Let us now turn to the example "Going to a party" and see which colors you can rationally choose if you believe that Barbara acts rationally.

Example 2.6 Going to a party

Recall the story from Example 2.2, and the utilities from Table 2.3. So, according to this table, you prefer *blue* to *green*, *green* to *red* and *red* to *yellow*. We have already seen that the colors *blue*, *green* and *red* are rational for you, as they are all optimal for some belief about Barbara's choice. However, your least preferred color *yellow* is irrational, as it cannot be supported by *any* belief about Barbara's choice. The beliefs diagram in Figure 2.3 summarizes these facts.

Now, will each of the colors *blue*, *green* and *red* also be reasonable choices for you? As we will see, this crucially depends on Barbara's preferences for the colors. Assume that Barbara prefers *red* to *yellow*, *yellow* to *blue* and *blue* to *green*. Suppose, moreover, that Barbara also strongly dislikes wearing the same color as you. Let the utilities for Barbara be given by Table 2.9. For completeness, we have also included your own utilities in that table.

Which colors are rational for Barbara? Note that Barbara's utilities are similar to yours, except for the fact that she has a different ordering of the colors in terms of preferences. So, her choice problem is pretty much the same as yours.

In exactly the same way as before, we can show that her three most preferred colors, *red*, *yellow* and *blue*, are rational, but that her least preferred color, *green*, is irrational. Indeed, *red* is optimal for her if she believes that you will choose *yellow* for sure. Choosing *yellow* is optimal for her if she believes that you will choose *red* for sure. And *blue* is optimal for her if she believes that, with probability 0.6, you will choose *red*, and with probability 0.4 you will choose *yellow*. On the other hand, her least preferred color *green* is irrational, as there is no belief for Barbara about your choice for which choosing *green* is optimal for her. The computations are exactly the same as in Example 2.2, so we leave it to the reader to verify these statements.

If we combine these facts about Barbara with the beliefs diagram of Figure 2.3, we get the beliefs diagram as depicted in Figure 2.8. From this beliefs diagram, we can conclude that you can rationally choose *green* if you believe that Barbara chooses rationally. The beliefs diagram tells us that *green* is optimal for you if you believe that Barbara will choose *blue*, and, in turn, *blue* is optimal for Barbara if she believes that, with probability 0.6, you will choose *red* and with probability 0.4 you will choose *yellow*.

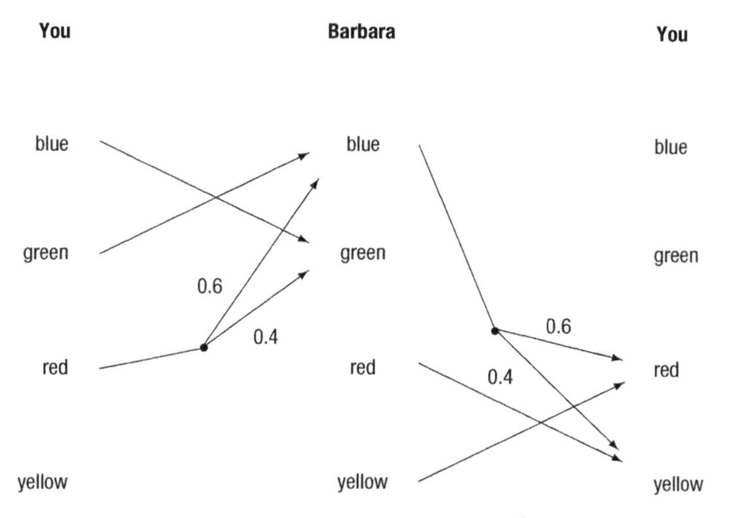

Figure 2.8 A beliefs diagram for "Going to a party" (II)

What about your choice *blue*? According to the beliefs diagram, *blue* is optimal for you if you believe that Barbara will choose *green*, but *green* is an irrational choice for Barbara, since it has no outgoing arrow. Does this mean that you cannot rationally choose *blue* if you believe that Barbara chooses rationally? The answer is "no"! The beliefs diagram in Figure 2.8 only represents *one* possible belief about Barbara's choice for which your choice *blue* is optimal, but there are *many more* beliefs about Barbara's choice for which choosing *blue* is optimal for you! For instance, choosing *blue* is also optimal for you if you believe that Barbara will choose *red*. Moreover, as can be read from the beliefs diagram in Figure 2.8, *red* is optimal for Barbara if she believes that you will choose *yellow*. Hence, you can rationally choose *blue* while believing that Barbara chooses rationally. In order to visualize this we can build an *alternative* beliefs diagram like the one in Figure 2.9. In this new beliefs diagram, the arrow from your choice *blue* to Barbara's choice *green* is replaced by an arrow from *blue* to *red*, indicating that your choice *blue* is optimal if you believe that Barbara will rationally choose *red*. So, from the alternative beliefs diagram in Figure 2.9 we can directly conclude that you can rationally choose *blue* while believing that Barbara chooses rationally!

Let us now turn to your choice *red*. Can you rationally choose *red* while believing that Barbara chooses rationally? The beliefs diagrams in Figures 2.8 and 2.9 tell us that your choice *red* is optimal if you believe that, with probability 0.6, Barbara will choose *blue* and with probability 0.4 she will choose *green*. However, *green* is not a rational choice for Barbara. Does this imply that you cannot rationally choose *red* if you believe that Barbara chooses rationally? Not per se: In principle, there *could* be another belief for you about Barbara's choice, assigning positive probability only to *rational* choices for Barbara, for which choosing *red* would be optimal for you. So, let

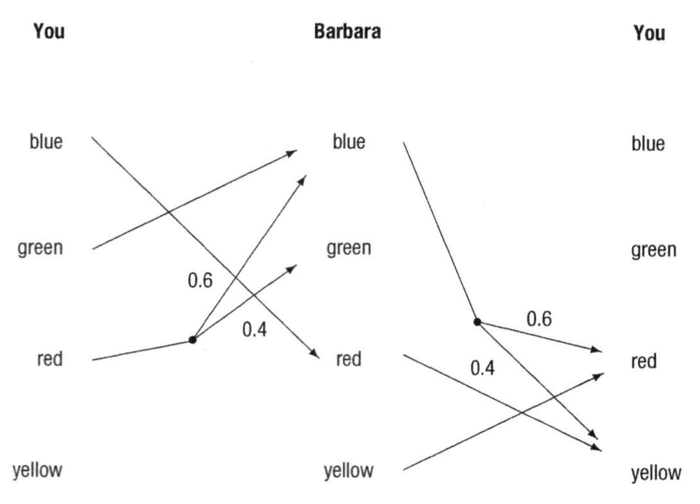

Figure 2.9 An alternative beliefs diagram for "Going to a party"

us see whether we can find such a belief. If you believe that Barbara chooses rationally, then you must assign probability 0 to her irrational choice *green*. However, in that case choosing *green* yourself would give you an expected utility of 3, since you believe that, with probability 1, you would be wearing a different color from Barbara. As choosing *red* gives you at most 2, choosing *red* cannot be optimal for you if you believe that Barbara chooses rationally.

Summarizing, we see that you can rationally choose *blue* and *green* if you believe that Barbara chooses rationally. Moreover, this can be read immediately from the new beliefs diagram in Figure 2.9. On the other hand, your choice *red* can no longer be made rationally if you believe that Barbara chooses rationally. This cannot be concluded graphically on the basis of the beliefs diagrams in Figures 2.8 and 2.9. In order to draw this conclusion one must make an argument as above, stating that if you believe that Barbara chooses rationally, you must believe that Barbara will not choose *green*, and therefore *green* is better than *red* for you. Finally, we see that you cannot rationally choose *yellow* if you believe that Barbara chooses rationally, since *yellow* was not a rational choice for you in the first place. □

So far in this section, we have only investigated games in which you have one opponent. However, if you have several opponents we can still investigate the consequences of believing that *each of your opponents* chooses rationally. Consider, for instance, the example "Waiting for a friend" from before.

Example 2.7 Waiting for a friend

Recall the story from Example 2.3. We have previously seen that arriving between 7.40 and 8.20 are all rational choices for you, whereas arriving at 7.30 or arriving at 8.30 are not. The reasoning, in a nutshell, was as follows: Arriving at 7.40 is always better than arriving at 7.30 since Chris cannot be there before 7.50. Arriving at 8.20 is always better than arriving at 8.30 since by arriving at 8.20 instead you either decrease the time you let the others wait, or you will arrive just ten minutes before the last one, yielding you the highest possible utility. Arriving at any time between 7.40 and 8.20 is optimal if you believe that your friends will both arrive ten minutes later. See the beliefs diagram in Figure 2.4 for an illustration of this.

Now, at which times can you rationally arrive if you believe that both Barbara and Chris choose their arrival times rationally? Let us assume, for simplicity, that Barbara and Chris both have the same utility functions as you. Then, by the same argument as above, we can conclude that arriving at 8.30 can never be rational for Barbara and Chris. On the other hand, Barbara can rationally arrive at any time between 7.40 and 8.20 if she believes that both you and Chris will arrive exactly ten minutes later. Similarly, Chris can rationally arrive at any time between 7.50 and 8.20 if he believes that his two friends will both arrive exactly ten minutes later. We can thus make the beliefs diagram in Figure 2.10. This diagram should be read as follows: The arrows in the upper left triangle represent beliefs that you may have about Barbara's and Chris' arrival times, and that support your rational choices. For instance, the forked arrow from your arrival time 7.50 to Barbara's arrival time 8.00 and Chris' arrival time 8.00 indicates that it is optimal for you to arrive at 7.50 pm if you believe that both Barbara and Chris will arrive at 8.00 pm. The arrows in the upper right triangle represent beliefs that Barbara may have about your arrival time and Chris' arrival time. The arrows in the lower triangle, finally, represent beliefs that Chris may have about your arrival time and Barbara's arrival time.

So, if you believe that Barbara and Chris choose their arrival times rationally, you must believe that Barbara and Chris will both arrive no later than 8.20. But then, it can no longer be optimal to arrive at 8.20, since arriving at 8.10 would always be better in this case. The argument for this fact is similar to the argument that arriving at 8.20 is always better than arriving at 8.30, but now the latest arrival time you expect to occur is not 8.30 but 8.20. On the other hand, any arrival time between 7.40 and 8.10 can still rationally be chosen if you believe that your friends choose rationally, since any arrival time between 7.40 and 8.10 is optimal if you believe that both friends will rationally arrive ten minutes later.

The conclusion is thus that you can rationally choose any arrival time between 7.40 and 8.10 if you believe that your friends choose rationally, but you can no longer rationally choose to arrive at 8.20 in this case. □

To conclude this section we shall try to formalize the concepts we have seen above. In particular, we will formally define what it means to *believe that your opponents choose rationally* and to *choose rationally while believing that your opponents choose rationally too.*

Suppose you are player i in a game, and hold a belief b_i about the opponents' choices. Recall that b_i is a probability distribution over the set $C_1 \times ... \times C_{i-1} \times C_{i+1} \times ... \times C_n$ of opponents' choice-combinations. For every choice-combination $(c_1, ..., c_{i-1}, c_{i+1}, ..., c_n)$ for the opponents, the number $b_i(c_1, ..., c_{i-1}, c_{i+1}, ..., c_n)$ denotes

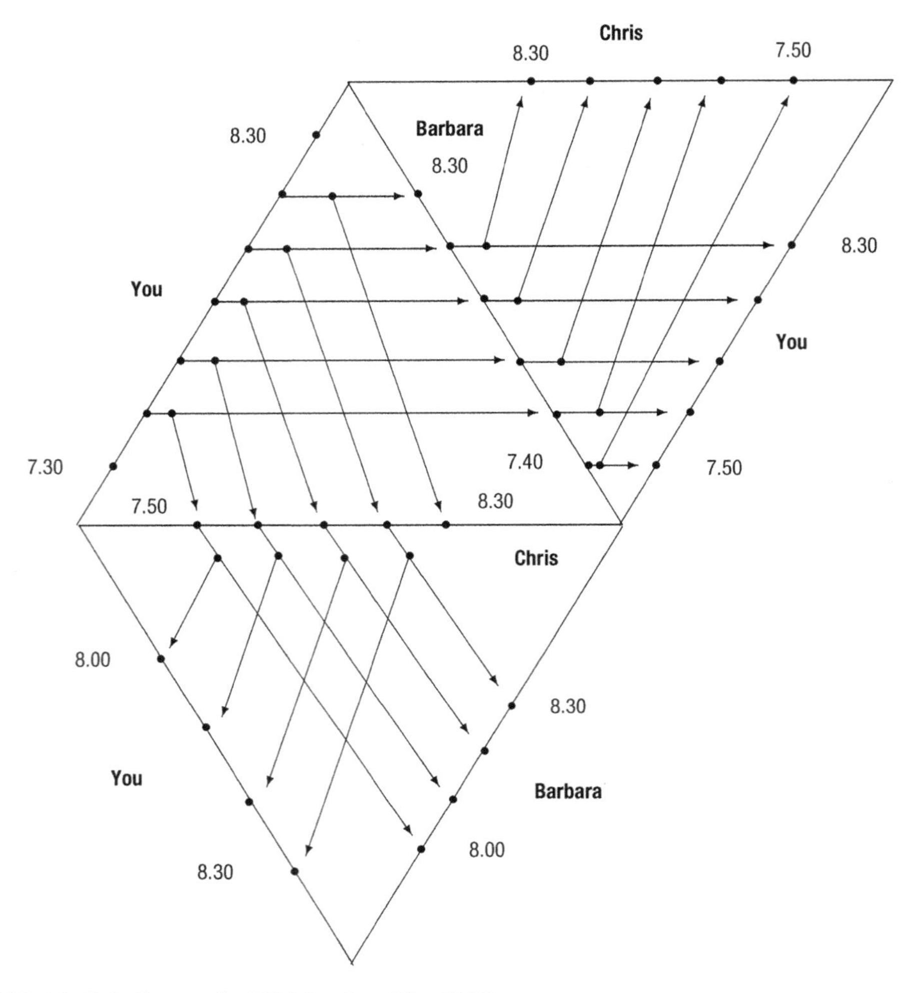

Figure 2.10 A beliefs diagram for "Waiting for a friend" (II)

the probability that you assign to the event that the opponents will choose the choice-combination $(c_1, ..., c_{i-1}, c_{i+1}, ..., c_n)$. If you believe that your opponents choose rationally, your belief b_i should only assign positive probability to the opponents' choices that are rational for them. Recall what it means when a choice c_j is rational for an opponent j: It means that there is a belief b_j for him about the other players' choices for which c_j is optimal, given his utility function u_j. Belief in the opponents' rationality can thus be defined as follows.

Definition 2.6.1 *(Belief in the opponents' rationality)*
Suppose that player i holds a belief b_i about the opponents' choices. Then, player i ***believes in the opponents' rationality*** *if b_i only assigns positive probability to the opponents' choice-combinations $(c_1, ..., c_{i-1}, c_{i+1}, ..., c_n)$ in which every choice is rational*

for the corresponding opponent. In this case, we say that b_i ***expresses belief in the opponents' rationality****.*

With this definition in mind, it is now easy to formalize the sentence that "player i chooses rationally while believing that his opponents choose rationally too." It simply means that player i makes a choice c_i, which is optimal for a belief b_i that expresses belief in the opponents' rationality.

Definition 2.6.2 *(Rational choice under belief in the opponents' rationality)*
Player i can ***rationally*** *make choice c_i while* ***believing in the opponents' rationality,*** *if the choice c_i is optimal for some belief b_i about the opponents' choices that expresses belief in the opponents' rationality.*

In the remainder of this chapter we will focus on two different methods that will help us to determine those choices you can rationally make while believing in the opponents' rationality.

2.7 Graphical method

In the previous section we determined, for three different examples, those choices you can rationally make if you believe that each of your opponents chooses rationally. A possible way to find such choices is by looking at a beliefs diagram. Suppose you have made a beliefs diagram for a game with two players – you and some opponent. Remember that in a beliefs diagram, the rational choices are exactly those choices that have an outgoing arrow. Concentrate on one of your possible choices, say choice a. Assume that in this beliefs diagram there is a unique arrow leaving a, going from a to some choice b of the opponent, and that choice b is rational for your opponent. Then, you can immediately conclude that a can rationally be chosen if you believe that the opponent chooses rationally too. Namely, the arrow from a to b indicates that a is optimal if you believe that the opponent chooses b. Since b is rational for your opponent, you can rationally choose a if you believe that the opponent chooses rationally as well.

It is also possible to have forked arrows in a beliefs diagram. See, for instance, the beliefs diagram in Figure 2.9 in which there is a forked arrow leaving your choice *red*, leading to Barbara's choice *blue* with probability 0.6, and to Barbara's choice *green* with probability 0.4. This forked arrow, together with the probabilities, indicates that your choice *red* is optimal if you believe that with probability 0.6 Barbara will choose *blue* and with probability 0.4 she will choose *green*. However, if you hold this belief then you would not believe in Barbara's rationality, since you assign probability 0.4 to Barbara's choice *green*, which is irrational for her. If both *blue* and *green* were rational choices for Barbara, then you would believe in Barbara's rationality when holding this belief, as you would have assigned positive probability only to rational choices for Barbara.

The same principle can be applied to games with more than two players. Consider, for example, the beliefs diagram for "Waiting for a friend" in Figure 2.10. Note that from your choice 7.50 there is a forked arrow, leading to Barbara's choice 8.00 and to Chris' choice 8.00. Barbara's choice 8.00 and Chris' choice 8.00 are both rational, as they both have outgoing arrows. Hence, you can rationally choose to arrive at 7.50 pm while believing that your friends will both rationally arrive at 8.00 pm. In particular, you can rationally choose 7.50 if you believe that both opponents choose rationally.

The insights above thus lead to the following general conclusion: If there is a beliefs diagram such that your choice c_i has an outgoing arrow leading only to *rational* opponents' choices, then c_i can rationally be chosen if you believe the opponents to choose rationally as well. It is easy to check that the converse of this statement is also true: If you can rationally make choice c_i while believing that the opponents choose rationally as well, then there must be a beliefs diagram such that your choice c_i has an outgoing arrow that leads only to rational choices for the opponents. With these two insights, we arrive at the following *graphical method* to verify whether a given choice c_i can rationally be chosen while believing that your opponents choose rationally too.

Graphical Method 2.7.1 *(Arrow method)*
Consider a choice c_i for player i. If there is a beliefs diagram in which c_i has an outgoing arrow that leads only to rational choices for the opponents, then player i can rationally make choice c_i while believing that his opponents choose rationally as well. If no such beliefs diagram exists, then player i cannot rationally make choice c_i while believing that his opponents choose rationally as well.

One should handle this graphical method with some care though. Suppose you make a beliefs diagram, and find out that within this particular beliefs diagram some arrow leaving choice c_i leads to an irrational choice for an opponent. Then, this does *not necessarily* mean that you cannot rationally choose c_i while believing that your opponents choose rationally as well. Consider, for instance, the beliefs diagram in Figure 2.8 for "Going to a party." Let us concentrate on your choice *blue*. We see that, in this particular beliefs diagram, there is an arrow from your choice *blue* to Barbara's choice *green*, which is an irrational choice for Barbara. This is not an argument, however, to conclude that you cannot rationally choose *blue* if you believe Barbara to choose rationally too. The reason is that there may be *another beliefs diagram* in which the arrow leaving your choice *blue* would only lead to rational choices for Barbara, and this is indeed the case here. Consider, instead, the alternative beliefs diagram in Figure 2.9. There, we have an arrow from your choice *blue* to Barbara's choice *red*, which is a rational choice for Barbara. The conclusion is thus that you *can* rationally choose *blue* if you believe Barbara to choose rationally as well, although there is a beliefs diagram in which the arrow leaving your choice *blue* leads to an irrational choice for Barbara.

2.8 Algorithm

In the previous section we have seen that we can use beliefs diagrams to check whether a choice can rationally be made while believing that the opponents choose rationally as well. If we can make a beliefs diagram in which choice c_i has an outgoing arrow that leads only to rational choices for the opponents, then we can conclude that c_i can rationally be chosen while believing that the opponents choose rationally too. However, if in a certain beliefs diagram the arrow leaving choice c_i leads to an *irrational* choice for an opponent, then we cannot conclude yet that c_i cannot rationally be chosen while believing that the opponents choose rationally. As we have seen above, there may be some *other* beliefs diagram in which the arrow leaving c_i *does* only lead to rational choices for the opponents. So, in order to conclude that a choice c_i *cannot* be chosen rationally if you believe the opponents to choose rationally as well, one must show that there is *no* beliefs diagram in which the arrow leaving c_i only leads to rational choices for the opponents. This would require us to scan through all possible beliefs diagrams – a complicated task in general.

The purpose of this section is to develop a more systematic procedure for finding *all* choices you can rationally make if you believe your opponents to choose rationally too. More precisely, we will present an algorithm that proceeds by eliminating choices from the game, and that eventually yields precisely those choices you can rationally make if you believe your opponents to choose rationally as well.

How does the algorithm work? Our objective is to select, for a given player i, precisely those choices c_i he can rationally make if he believes that his opponents choose rationally. We know from Theorem 2.5.3 that the rational choices are precisely those choices that are not strictly dominated by some other choice, nor by some randomized choice. So, if player i believes that his opponents choose rationally, then he will only assign positive probability to the opponents' choices that are not strictly dominated. Or, stated equivalently, player i will assign probability zero to all choices for the opponents that *are* strictly dominated. This requirement can be mimicked by *eliminating* for every opponent all strictly dominated choices from the game. This will be step 1 of the procedure.

So, after step 1 we obtain a *reduced* game in which we have eliminated, for each of i's opponents, all strictly dominated choices from the game. If player i believes that his opponents choose rationally, then he believes that his opponents will only make choices from the reduced game obtained after step 1. But then, player i can rationally make a choice c_i while believing that his opponents choose rationally, precisely when this choice c_i is optimal for some belief that player i can hold about the opponents' choices in the *reduced* game. That is, when c_i is a rational choice within the reduced game. But we know from Theorem 2.5.3 that choice c_i is rational within the reduced game, if and only if, it is not strictly dominated within this reduced game. So, within the reduced game obtained after step 1, we can eliminate for player i all strictly dominated choices, and the choices that remain for player i will be precisely those that he can rationally choose if he believes that his opponents choose rationally. We have thus developed a two-step procedure that yields precisely those choices that player i can rationally make if he believes his opponents to choose rationally.

But we can still simplify the procedure a bit. Note that in step 1 of the procedure we have only eliminated the strictly dominated choices for i's opponents, but not for player i himself. Suppose now that in step 1 we *also* eliminate the strictly dominated choices for player i. Would this affect the final outcome of the algorithm? The answer is "no"!

To explain why, let us compare the following two procedures: Procedure 1 starts at step 1 by eliminating, for each of i's opponents, those choices that are strictly dominated. This yields a reduced game Γ_1^{red}. Then, at step 2, it eliminates those choices for player i that are strictly dominated in Γ_1^{red}.

Procedure 2, on the other hand, starts at step 1 by eliminating for every player – including player i – those choices that are strictly dominated. This yields a reduced game Γ_2^{red}. Then, at step 2, it eliminates those choices for player i that are strictly dominated in Γ_2^{red}. We will show that Procedure 1 and Procedure 2 eventually yield the same set of choices for player i.

Suppose first that choice c_i survives Procedure 1. Then, c_i is not strictly dominated within the reduced game Γ_1^{red}. Hence, by Theorem 2.5.3, choice c_i is optimal, among all choices for player i in Γ_1^{red}, for some belief b_i about the opponents' choices in Γ_1^{red}. Note that the reduced game Γ_1^{red} still contains *all* choices for player i, and that the set of opponents' choices in Γ_1^{red} is the same as in Γ_2^{red}. But then, choice c_i is also optimal, among the smaller set of choices for player i in Γ_2^{red}, for the belief b_i about the opponents' choices in Γ_2^{red}. So, by Theorem 2.5.3, choice c_i is not strictly dominated within Γ_2^{red}, and hence survives Procedure 2. We have thus shown that every choice c_i that survives Procedure 1 also survives Procedure 2.

We now show that the converse is also true. So, take a choice c_i that survives Procedure 2. Then, c_i cannot be strictly dominated within the reduced game Γ_2^{red}. By Theorem 2.5.3 we then know that choice c_i is optimal, among all choices for player i in Γ_2^{red}, for some belief b_i about the opponents' choices in Γ_2^{red}. As the set of opponents' choices in Γ_2^{red} is the same as in Γ_1^{red}, it follows that the belief b_i is also a belief about the opponents' choices in Γ_1^{red}. We will now show that choice c_i is not only optimal for belief b_i among all choices for player i in Γ_2^{red}, but in fact among all choices for player i in the original game Γ. Suppose that among all choices for player i, the optimal choice for the belief b_i is not c_i, but some choice c_i' that is not in Γ_2^{red}. By the definition of Γ_2^{red}, choice c_i' must then be strictly dominated in the original game Γ, and hence, by Theorem 2.5.3, cannot be optimal for any belief. This, obviously, is a contradiction. Hence, not only is choice c_i optimal for b_i among all choices for player i in Γ_2^{red}, it is even optimal for b_i among all choices for player i in the original game. That is, choice c_i is optimal, among all choices for player i in Γ_1^{red}, for the belief b_i about the opponents' choices in Γ_1^{red}. Hence, by Theorem 2.5.3, choice c_i is not strictly dominated in Γ_1^{red}, and therefore survives Procedure 1. So, we have shown that every choice c_i that survives Procedure 2 also survives Procedure 1.

Summarizing, we see that Procedure 1 and Procedure 2 generate precisely the same set of choices for player i, namely those choices that player i can rationally make if he believes that his opponents choose rationally as well. However, as Procedure 2 is simpler, it is this procedure that we will use. Procedure 2 is formally described below, and we call it *two-fold elimination of strictly dominated choices*.

Algorithm 2.8.1 *(Two-fold elimination of strictly dominated choices)*
Step 1: *Eliminate for every player all choices that are strictly dominated.*
Step 2: *Within the reduced game, eliminate for every player all choices that are strictly dominated.*

The choices for player i that remain after Step 2 are exactly the choices that player i can rationally make if he believes that his opponents choose rationally.

Compared to the graphical method discussed in the previous section, this algorithm has an important disadvantage: The graphical method gives us, for every choice that can rationally be made while believing that the opponents choose rationally, a belief that supports this choice, whereas the algorithm does not. On the other hand, the algorithm immediately gives us *all* choices that can rationally be made while believing that the opponents choose rationally, whereas a beliefs diagram does not necessarily do so. So, both methods have their strong points and their weaknesses.

To illustrate our algorithm, let us first go back to the examples "Where to locate my pub?" and "Going to a party."

Example 2.8 Where to locate my pub?

We already know that for you and Barbara, the location a is strictly dominated by b, and the location g is strictly dominated by f. No other choices are strictly dominated. Hence, in step 1 we eliminate the choices a and g for you and Barbara, and obtain a reduced game in which you and Barbara can only choose from $\{b, c, d, e, f\}$. Within this reduced game, your location b is strictly dominated by c, and f is strictly dominated by e. As no other choices are strictly dominated within this reduced game, the algorithm selects the locations c, d and e for you, which are indeed precisely those locations you can rationally choose if you believe that Barbara chooses rationally too. □

Example 2.9 Going to a party

Consider the utilities for you and Barbara as depicted in Table 2.9. In Section 2.5 we saw that your least preferred color, *yellow*, is strictly dominated by the randomized choice in which you choose your two most preferred colors, *blue* and *green*, with probability 0.5. In exactly the same way, we can show that Barbara's least preferred color, *green*, is strictly dominated for her by the randomized choice in which she chooses her two most preferred colors, *red* and *yellow*, with probability 0.5. So, in step 1 of the algorithm we eliminate your choice *yellow* and Barbara's choice *green*. No other colors are strictly dominated in the original game.

Within the reduced game obtained after step 1, your choice *red* is strictly dominated by your choice *green*: If you choose *green* then, within the reduced game, you obtain utility 3 for sure as you will never choose the same color as Barbara. On the other hand, choosing *red* will always give you a utility strictly less than 3. So, indeed, *green* is always strictly better for you than *red* within the reduced game. On the other hand, *blue* and *green* are not strictly dominated for you within the reduced game. So, after step 2 of the algorithm we are left with your choices *blue* and *green*. The algorithm thus tells us that *blue* and *green* are the only choices you can rationally make if you believe that Barbara chooses rationally too. □

The algorithm can also be applied to games with more than two players, as the following example illustrates.

Example 2.10 Waiting for a friend

We have already seen that for you, arriving at 7.30 is strictly dominated by arriving at 7.40, and arriving at 8.30 is strictly dominated by arriving at 8.20. Moreover, for Barbara and Chris arriving at 8.30 is strictly dominated by arriving at 8.20. Since no other choices are strictly dominated, the reduced game is such that you and Barbara can only choose from $\{7.40,...,8.20\}$ and Chris can only choose from $\{7.50,...,8.20\}$. Within this reduced game, arriving at 8.20 is strictly dominated for you by arriving at 8.10, since you expect Barbara and Chris to arrive no later than 8.20. All other choices for you are not strictly dominated in the reduced game. Hence, the algorithm yields your choices $\{7.40,...,8.10\}$ as a result. Therefore, you can rationally arrive at any time between 7.40 and 8.10 if you believe your friends to act rationally too. □

Let us finally turn to "The traveler's dilemma."

Example 2.11 The traveler's dilemma

Recall the story from Example 2.4. We have already seen that choosing price 400 is strictly dominated by the randomized choice in which you choose price 100 with probability 0.45 and price 300 with probability 0.55. Since Barbara plays exactly the same role as you, we may conclude for the same reason that Barbara's choice 400 is strictly dominated as well. No other choices are strictly dominated in the original game. So, in step 1 we eliminate the choice 400 for you and Barbara, and obtain a reduced game in which you and Barbara only choose from $\{100, 200, 300\}$. The compensation for you in this reduced game is given by Table 2.10. Within this reduced game, we see that your choice 300 is strictly dominated by the randomized choice in which you choose 100 and 200 both with probability 0.5. It can easily be seen that the other two choices, 100 and 200, are not strictly dominated in the reduced game: If Barbara chooses 100, then choosing 100 is better than choosing 200. If Barbara chooses 300, then choosing 200 is better than choosing 100. So, the algorithm yields the choices 100 and 200 for you, and these are exactly the choices you can rationally make if you believe that Barbara chooses rationally too. □

2.9 Proofs

In this section we prove the following theorem, which characterizes the rational choices in a static game.

Theorem 2.5.3 *(Characterization of rational choices)*
The rational choices in a static game are exactly those choices that are neither strictly dominated by some other choice nor strictly dominated by a randomized choice.

Table 2.10. *Reduced game after step 1 in "The traveler's dilemma"*

		Barbara		
		100	200	300
You	100	100	250	300
	200	−75	200	350
	300	−25	25	300

Proof: We must show that the rational choices are precisely those choices that are not strictly dominated. Or, equivalently, we must prove that the *irrational* choices are exactly those choices that *are* strictly dominated. We prove this statement in two parts: In part (a), which is the easier part, we show that every choice that is strictly dominated must be irrational. In part (b), which is more difficult, we prove that every irrational choice must be strictly dominated.

(a) Every strictly dominated choice is irrational

Consider a choice c_i^* that is strictly dominated by another choice or by a randomized choice. We prove that c_i^* is irrational, that is, it is not optimal for any belief b_i about the opponents' choice-combinations.

Suppose first that c_i^* is strictly dominated by another choice c_i. Then, no matter what i's opponents do, choosing c_i is always better for player i than choosing c_i^*. Hence, no matter what belief player i has about the opponents' choice-combinations, choosing c_i will always be better for him than choosing c_i^*, and hence c_i^* is an irrational choice.

Suppose next that c_i^* is strictly dominated by a randomized choice r_i. For every belief b_i about the opponents' choice-combinations, let $u_i(r_i, b_i)$ be the expected utility induced by the randomized choice r_i under the belief b_i. As we assume that c_i^* is strictly dominated by r_i, we must have that

$$u_i(c_i^*, b_i) < u_i(r_i, b_i)$$

for every belief b_i about the opponents' choice-combinations. Since

$$u_i(r_i, b_i) = \sum_{c_i \in C_i} r_i(c_i) \cdot u_i(c_i, b_i)$$

and

$$u_i(c_i^*, b_i) = \sum_{c_i \in C_i} r_i(c_i) \cdot u_i(c_i^*, b_i)$$

it follows that

$$\sum_{c_i \in C_i} r_i(c_i) \cdot u_i(c_i^*, b_i) < \sum_{c_i \in C_i} r_i(c_i) \cdot u_i(c_i, b_i)$$

for every belief b_i about the opponents' choice-combinations. This, in turn, means that for every belief b_i there must be some choice $c_i \neq c_i^*$ with $u_i(c_i^*, b_i) < u_i(c_i, b_i)$. That is, c_i^* is not optimal for any belief b_i about the opponents' choices, and hence c_i^* is irrational.

Summarizing, we have shown that every choice c_i^* that is strictly dominated by another choice, or by a randomized choice, must be irrational. This completes the proof of part (a).

(b) Every irrational choice is strictly dominated

Consider a choice c_i^* that is irrational, meaning that there is no belief b_i about the opponents' choice-combinations for which c_i^* is optimal. We show that c_i^* must be strictly dominated by another choice, or by a randomized choice.

Let C_{-i} denote the set of opponents' choice-combinations, and let $\Delta(C_{-i})$ be the set of all probability distributions on C_{-i}. So, every belief b_i about the opponents' choice-combinations is – formally speaking – a probability distribution in $\Delta(C_{-i})$. For every choice $c_i \in C_i$ and every belief $b_i \in \Delta(C_{-i})$ we define the numbers

$$d(c_i, b_i) := u_i(c_i, b_i) - u_i(c_i^*, b_i)$$

and

$$d^+(c_i, b_i) := \max\{0, d(c_i, b_i)\}.$$

We define a function $f : \Delta(C_{-i}) \to \mathbb{R}$ that assigns to every belief $b_i \in \Delta(C_{-i})$ the number

$$f(b_i) := \sum_{c_i \in C_i} d^+(c_i, b_i)^2.$$

Since the function f is continuous, and the domain $\Delta(C_{-i})$ is compact, it follows from Weierstrass' theorem that the function f has a minimum value at $\Delta(C_{-i})$. Let $b_i^* \in \Delta(C_{-i})$ be a belief for which the function f takes this minimum value, that is,

$$f(b_i^*) \leq f(b_i) \text{ for all } b_i \in \Delta(C_{-i}).$$

We define the randomized choice r_i^* for player i by

$$r_i^*(c_i) := \frac{d^+(c_i, b_i^*)}{\sum_{c_i' \in C_i} d^+(c_i', b_i^*)} \text{ for every } c_i \in C_i.$$

We will prove that r_i^* is indeed a randomized choice and that r_i^* strictly dominates the choice c_i^*.

Let us first verify that r_i^* is indeed a randomized choice. Since the choice c_i^* is irrational, it cannot be optimal under the belief b_i^*. Hence, there must be some choice c_i with $u_i(c_i, b_i^*) > u_i(c_i^*, b_i^*)$, which means that $d^+(c_i, b_i^*) > 0$ for at least some choice c_i. Since, by construction, $d^+(c_i, b_i^*) \geq 0$ for all $c_i \in C_i$, it follows that

$$\sum_{c_i' \in C_i} d^+(c_i', b_i^*) > 0,$$

and hence $r_i^*(c_i)$ is well defined for every c_i. Moreover, as $d^+(c_i, b_i^*) \geq 0$ for every c_i, we have that $r_i^*(c_i) \geq 0$ for every c_i. Finally, it is clear that

$$\sum_{c_i \in C_i} r_i^*(c_i) = 1,$$

which implies that r_i^* is indeed a randomized choice.

We will now show that the randomized choice r_i^* as constructed strictly dominates the choice c_i^*. That is, we must show that

$$u_i(r_i^*, c_{-i}^*) > u_i(c_i^*, c_{-i}^*)$$

for every opponents' choice-combination $c_{-i}^* \in C_{-i}$. Or, equivalently, we must show that

$$u_i(r_i^*, c_{-i}^*) - u_i(c_i^*, c_{-i}^*) > 0$$

for every $c_{-i}^* \in C_{-i}$. Take an arbitrary opponents' choice-combination c_{-i}^*, and let $b_i^{c_{-i}^*} \in \Delta(C_{-i})$ be the belief that assigns probability 1 to c_{-i}^*. Note that

$$\begin{aligned}
& u_i(r_i^*, c_{-i}^*) - u_i(c_i^*, c_{-i}^*) \\
&= u_i(r_i^*, b_i^{c_{-i}^*}) - u_i(c_i^*, b_i^{c_{-i}^*}) \\
&= \sum_{c_i \in C_i} r_i^*(c_i) \cdot u_i(c_i, b_i^{c_{-i}^*}) - \sum_{c_i \in C_i} r_i^*(c_i) \cdot u_i(c_i^*, b_i^{c_{-i}^*}) \\
&= \sum_{c_i \in C_i} r_i^*(c_i) \cdot (u_i(c_i, b_i^{c_{-i}^*}) - u_i(c_i^*, b_i^{c_{-i}^*})) \\
&= \sum_{c_i \in C_i} r_i^*(c_i) \cdot d(c_i, b_i^{c_{-i}^*}).
\end{aligned}$$

Since

$$r_i^*(c_i) = \frac{d^+(c_i, b_i^*)}{\sum_{c_i' \in C_i} d^+(c_i', b_i^*)} \text{ for every } c_i \in C_i,$$

showing that

$$u_i(r_i^*, c_{-i}^*) - u_i(c_i^*, c_{-i}^*) > 0$$

is equivalent to showing that

$$\sum_{c_i \in C_i} d^+(c_i, b_i^*) \cdot d(c_i, b_i^{c_{-i}^*}) > 0. \tag{*}$$

Hence, we will show that

$$\sum_{c_i \in C_i} d^+(c_i, b_i^*) \cdot d(c_i, b_i^{c_{-i}^*}) > 0$$

for every $c^*_{-i} \in C_{-i}$.

Fix an opponents' choice-combination $c^*_{-i} \in C_{-i}$. For every $\lambda \in (0,1]$ we define the belief

$$b_i^\lambda := (1-\lambda) \cdot b_i^* + \lambda \cdot b_i^{c^*_{-i}},$$

that is,

$$b_i^\lambda(c_{-i}) = (1-\lambda) \cdot b_i^*(c_{-i}) + \lambda \cdot b_i^{c^*_{-i}}(c_{-i})$$

for every $c_{-i} \in C_{-i}$.

Now, we choose a number $\epsilon > 0$ small enough such that for all $c_i \in C_i$ whenever $d(c_i, b_i^*) < 0$, then also $d(c_i, b_i^\lambda) < 0$ for all $\lambda \in (0, \epsilon]$. We show for all $c_i \in C_i$ that

$$d^+(c_i, b_i^\lambda)^2 \leq ((1-\lambda) \cdot d^+(c_i, b_i^*) + \lambda \cdot d(c_i, b_i^{c^*_{-i}}))^2 \text{ for all } \lambda \in (0, \epsilon]. \quad (2.1)$$

To prove (2.1), we choose some $c_i \in C_i$, some $\lambda \in (0, \epsilon]$, and we distinguish the following two cases.

Case 1. Suppose that $d(c_i, b_i^\lambda) < 0$. Then $d^+(c_i, b_i^\lambda) = 0$ and (2.1) follows automatically.

Case 2. Suppose that $d(c_i, b_i^\lambda) \geq 0$. Then, by our choice of ϵ, we have that $d(c_i, b_i^*) \geq 0$. So, we conclude that $d^+(c_i, b_i^\lambda) = d(c_i, b_i^\lambda)$ and $d^+(c_i, b_i^*) = d(c_i, b_i^*)$. Therefore,

$$\begin{aligned}
d^+(c_i, b_i^\lambda) = d(c_i, b_i^\lambda) &= d(c_i, (1-\lambda) \cdot b_i^* + \lambda \cdot b_i^{c^*_{-i}}) \\
&= (1-\lambda) \cdot d(c_i, b_i^*) + \lambda \cdot d(c_i, b_i^{c^*_{-i}}) \\
&= (1-\lambda) \cdot d^+(c_i, b_i^*) + \lambda \cdot d(c_i, b_i^{c^*_{-i}}),
\end{aligned}$$

which implies (2.1). In the third equality above we used the fact that $d(c_i, b_i)$ is linear in the argument b_i.

Hence, (2.1) holds for every choice $c_i \in C_i$, and every $\lambda \in (0, \epsilon]$.

Recall that b_i^* is a belief where the function f is minimized. So, we have in particular that

$$f(b_i^*) \leq f(b_i^\lambda) \text{ for every } \lambda \in (0, \epsilon]. \quad (2.2)$$

By definition of the function f, we have for every $\lambda \in (0, \epsilon]$ that

$$\begin{aligned}
f(b_i^\lambda) &= \sum_{c_i \in C_i} d^+(c_i, b_i^\lambda)^2 \\
&\leq \sum_{c_i \in C_i} ((1-\lambda) \cdot d^+(c_i, b_i^*) + \lambda \cdot d(c_i, b_i^{c^*_{-i}}))^2 \\
&= (1-\lambda)^2 \cdot \sum_{c_i \in C_i} d^+(c_i, b_i^*)^2 \\
&\quad + 2\lambda(1-\lambda) \cdot \sum_{c_i \in C_i} d^+(c_i, b_i^*) \cdot d(c_i, b_i^{c^*_{-i}})
\end{aligned}$$

$$+\lambda^2 \cdot \sum_{c_i \in C_i} d(c_i, b_i^{c^*_{-i}})^2. \tag{2.3}$$

Here, the inequality follows from (2.1) above. On the other hand,

$$f(b_i^*) = \sum_{c_i \in C_i} d^+(c_i, b_i^*)^2. \tag{2.4}$$

By combining (2.2), (2.3) and (2.4) we obtain that

$$\begin{aligned}\sum_{c_i \in C_i} d^+(c_i, b_i^*)^2 = f(b_i^*) \le f(b_i^\lambda) \\ \le (1-\lambda)^2 \cdot \sum_{c_i \in C_i} d^+(c_i, b_i^*)^2 \\ + 2\lambda(1-\lambda) \cdot \sum_{c_i \in C_i} d^+(c_i, b_i^*) \cdot d(c_i, b_i^{c^*_{-i}}) \\ + \lambda^2 \cdot \sum_{c_i \in C_i} d(c_i, b_i^{c^*_{-i}})^2\end{aligned}$$

for all $\lambda \in (0, \epsilon]$. Since $(1-\lambda)^2 = 1 - 2\lambda + \lambda^2$, it follows that

$$\begin{aligned}(2\lambda - \lambda^2) \cdot \sum_{c_i \in C_i} d^+(c_i, b_i^*)^2 \le 2\lambda(1-\lambda) \cdot \sum_{c_i \in C_i} d^+(c_i, b_i^*) \cdot d(c_i, b_i^{c^*_{-i}}) \\ + \lambda^2 \cdot \sum_{c_i \in C_i} d(c_i, b_i^{c^*_{-i}})^2\end{aligned}$$

for all $\lambda \in (0, \epsilon]$. But then, by dividing both sides by λ, we have that

$$\begin{aligned}(2-\lambda) \cdot \sum_{c_i \in C_i} d^+(c_i, b_i^*)^2 \le 2(1-\lambda) \cdot \sum_{c_i \in C_i} d^+(c_i, b_i^*) \cdot d(c_i, b_i^{c^*_{-i}}) \\ + \lambda \cdot \sum_{c_i \in C_i} d(c_i, b_i^{c^*_{-i}})^2\end{aligned}$$

for all $\lambda \in (0, \epsilon]$. Now, if we let λ approach 0, we get

$$\sum_{c_i \in C_i} d^+(c_i, b_i^*)^2 \le \sum_{c_i \in C_i} d^+(c_i, b_i^*) \cdot d(c_i, b_i^{c^*_{-i}}).$$

Since we saw above that

$$\sum_{c_i \in C_i} d^+(c_i, b_i^*) > 0,$$

it follows that

$$\sum_{c_i \in C_i} d^+(c_i, b_i^*)^2 > 0,$$

and hence

$$\sum_{c_i \in C_i} d^+(c_i, b_i^*) \cdot d(c_i, b_i^{c^*_{-i}}) > 0.$$

This, as we saw above at (*), is equivalent to

$$u_i(r_i^*, c^*_{-i}) > u_i(c_i^*, c^*_{-i}).$$

As this holds for *every* opponents' choice-combination c^*_{-i}, we may conclude that the randomized choice r_i^* indeed strictly dominates the choice c_i^*.

We thus have shown that for every irrational choice c_i^* we can construct a randomized choice r_i^* that strictly dominates it. This completes the proof. ■

Practical problems

2.1 Where to locate a supermarket?

Suppose you wish to open a supermarket in an area with three little villages: Colmont, Winthagen and Ransdaal. Colmont has 300 inhabitants, Winthagen has 200 inhabitants and Ransdaal has 400 inhabitants. Every inhabitant is a potential customer. There are four possible locations for the supermarket, which we call *a*, *b*, *c* and *d*. Figure 2.11 shows a map, with the scale 1:50 000, which shows where the villages and the possible locations are situated. However, there is a competitor who also wishes to open a supermarket in the same area. Once you and your competitor have chosen a location, every

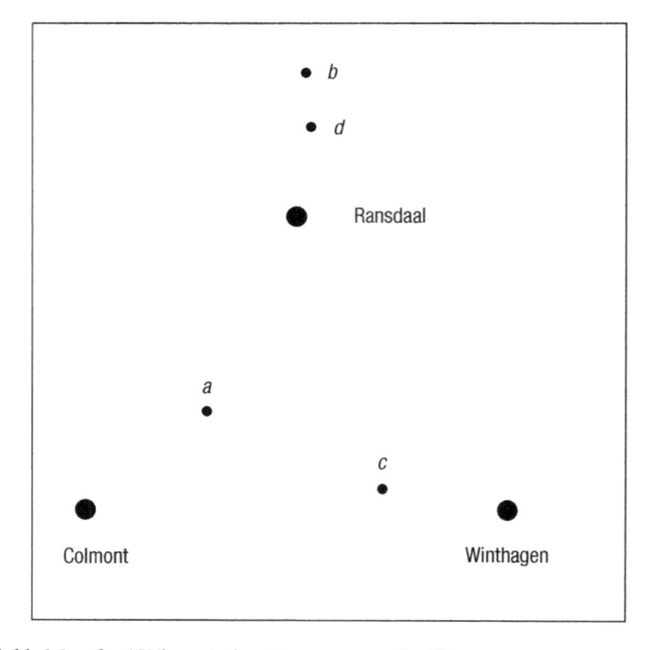

Figure 2.11 Map for "Where to locate a supermarket?"

inhabitant will always visit the supermarket that is closer to his village. If you happen to choose the same location, you will share the market equally with him.

(a) Compute, for every combination of locations that you and your competitor can choose, the number of customers for you and your competitor, and put these numbers in a table.

(b) Which locations are rational for you, and which are not? For every rational location, find a belief about the competitor's choice for which this location is optimal. For every irrational location, explain why there can be no belief for which this location is optimal.

(c) Make a beliefs diagram in which you depict beliefs that you may have about the competitor's choice and also beliefs that the competitor may have about your choice.

(d) Use the graphical method to determine those locations you can rationally choose while believing that the competitor chooses rationally as well.

(e) Apply the algorithm of two-fold elimination of strictly dominated choices to determine those locations you can rationally choose while believing that the competitor chooses rationally as well.

2.2 Preparing for a piano exam

Suppose you study piano. In two weeks you have an important exam, but you have not been studying too hard for it lately. There are three pieces that you may be asked to play in the exam: a relatively easy piece by Mozart, a somewhat more difficult piece by Chopin and a very tough piece by Rachmaninov. During the exam, the jury will select two out of these three pieces, but you do not know which. The jury will give you a grade for both pieces and your final grade will be the average of these two grades.

Since you have only two weeks left, you decide that you will focus on at most two pieces for the remaining time. So, you can dedicate the full two weeks to a single piece or you can dedicate one week to one of the pieces and one week to another. Suppose that your expected grade for the Mozart piece is given by

$$4 + 3\sqrt{x},$$

where x is the number of weeks you dedicate to this piece. Similarly, your expected grade for the Chopin piece is

$$4 + 2.5x,$$

whereas the expected grade for the piece by Rachmaninov is given by

$$4 + 1.5x^2.$$

Suppose now that the jury wants to see you perform well during the exam, but that they prefer listening to Chopin rather than listening to Rachmaninov, and they prefer Rachmaninov to Mozart. More precisely, the jury's utilities for listening to Chopin, Rachmaninov and Mozart are equal to 3, 2 and 1, respectively, and the jury's overall utility is given by the sum of your grade and the utilities they obtain from listening to the two pieces.

(a) Make a table which lists, for each of your possible practice schedules, and for each of the possible selections by the jury, the expected grade you would obtain for the exam, and the utility for the jury.
(b) Which are the rational practice schedules for you? For every rational practice schedule, find a belief about the jury's choice for which this schedule is optimal. For every irrational schedule, find another choice, or a randomized choice, that strictly dominates it.
(c) Which selections of pieces are rational for the jury? For every rational selection, determine a belief about your practice schedule for which this selection is optimal. For every irrational selection, determine some other selection, or randomized selection, for the jury that strictly dominates it.
(d) Make a beliefs diagram in which you depict beliefs for you and for the jury.
(e) Use the graphical method to determine those practice schedules you can rationally choose if you believe the jury to choose rationally as well.
(f) Use the algorithm of two-fold elimination of strictly dominated choices to determine those practice schedules you can rationally choose if you believe the jury to choose rationally as well.

2.3 Competition between two cinemas
In your village there are two cinemas that attract visitors from all over the neighborhood. You are the manager of one of these cinemas, Cinemax. Clearly, your number of visitors will not only depend on the price *you* choose for a ticket, but also on the price chosen by your competitor. More precisely, if you choose a price P_1 (in euros), and your competitor chooses a price P_2, then your total number of customers per week will be approximately

$$600 - 60P_1 + 30P_2.$$

So, if you increase your price by 1 euro, you will lose 60 customers per week, and if your rival increases its price by 1 euro, you will have 30 extra customers per week. The city council has decided that the maximum price for a ticket is 10 euros. Your goal is to maximize your weekly profit.
(a) Suppose you believe that the rival will choose a price of 4 euros. Describe your profit as a function of P_1, and draw this function as a graph. What is your optimal price in this case? Do the same exercise for the cases where you believe that the rival will choose a price of 0 euros, 7 euros, 9 euros and 10 euros. Draw all the functions on the same figure!
(b) Now, suppose that you believe that the rival will choose a price of P_2 euros, where P_2 can be any price between 0 and 10 euros. Describe your profit as a function of P_1 and P_2. What is your optimal choice in this case, and how does it depend on P_2?
(c) What are the rational prices that you can choose? For every rational price, find a belief about the rival's choice for which this price is optimal. For every irrational price, find another price that strictly dominates it.

Table 2.11. *Utilities for Barbara, Chris and you in "Going to a party"*

	white	red	blue	green	brown	black
You	3	6	2	5	1	4
Barbara	4	1	6	2	3	5
Chris	6	4	3	1	5	2

Suppose now that the number of customers per week attracted by your competitor is given by

$$400 - 50P_2 + 30P_1,$$

where P_1 denotes your price and P_2 the price chosen by the competitor.

(d) What are the rational prices for your competitor? For every rational price for the competitor, find a belief about your own choice for which this price is optimal. For every irrational price for the competitor, find another price that strictly dominates it.

(e) Take the lowest price P_2 that your competitor can rationally choose. For this P_2, draw your profit function as a function of P_1. Which price can you rationally choose if you believe that your competitor chooses exactly this lowest price P_2?

(f) Take the highest price P_2 that your competitor can rationally choose. For this P_2, draw your profit function as a function of P_1. Which price can you rationally choose if you believe that your competitor chooses exactly this highest price P_2?

(g) Determine all the prices that you can rationally choose if you believe that your competitor chooses rationally as well.

2.4 Going to a party

Tonight you are going to a party with Barbara and Chris. You must decide which color to wear this evening. Suppose that you and your friends have the choice between white, red, blue, green, brown and black, and that your utilities for these colors are given by Table 2.11. There is one problem, however: You do not know which colors they will wear this evening, and you *strongly dislike* wearing the same color as your friends. More precisely, if at least one of your friends is wearing the same color as you, your utility will be 0. If you wear a different color from both of your friends, your utility is given by the table. The same holds for your two friends.

(a) Show that the colors white, red, green and black are rational choices for you. For each of these colors, find a belief about your friends' choices for which this color is optimal.

(b) Show that the colors blue and brown are irrational. For both of these colors, find another choice or a randomized choice that strictly dominates it.

(c) Make a beliefs diagram for this situation, similar to the one in Figure 2.10.

(d) Use the graphical method to determine those colors you can rationally choose if you believe that your friends will choose rationally as well.

Table 2.12. *New utilities for Chris in "Going to a party"*

	white	red	blue	green	brown	black
Chris	1	4	3	6	5	2

(e) Use the algorithm of two-fold elimination of strictly dominated choices to determine those colors you can rationally choose if you believe that your friends will choose rationally as well.
(f) Suppose now that the preferences for Chris are given by Table 2.12, whereas the preferences for Barbara and you are still the same as before. Answer the questions (c) to (e) for these new utilities for Chris.

2.5 The three barkeepers

Suppose that you, Barbara and Chris all want to open a pub in your street. Figure 2.1, at the beginning of this chapter, shows a graphical illustration of this street. You and your friends can choose one of the locations $a, b, ..., g$ for the pub. As before, there are 100 thirsty fathers living between every two neighboring locations – equally distributed between these two locations – and every father will visit the pub that is nearest to his house. If two or more friends choose the same location, then these friends will equally share the group of fathers they attract.
(a) Which locations are rational for you? For every rational location, find a belief about your friends' choices for which this location is optimal. For every irrational location, find another location, or a randomization over locations, that strictly dominates it.
(b) Make a beliefs diagram for this game.
(c) Use the graphical method to determine those locations you can rationally choose if you believe that your friends will choose rationally too.
(d) Use the algorithm of two-fold elimination of strictly dominated choices to determine those locations you can rationally choose if you believe that your friends will choose rationally too.

2.6 A game of cards

Barbara, Chris and you are sitting in a bar, having a drink before the movie starts. You have brought a pack of playing cards with you, and tell your friends about a new card game you invented last night. The rules are easy: There are three piles of cards on the table with their faces down. One pile contains the 2, 5, 8 and jack of hearts, another pile contains the 3, 6, 9 and queen of hearts, and the last pile contains the 4, 7, 10 and king of hearts, and everybody knows this. The jack is worth 11 points, the queen is worth 12 points and the king 13 points. Each of the three players receives one of these decks, and everybody knows the decks that are given to the other two players. Then, all players

simultaneously choose one card from their deck, and put it on the table. The player putting the card whose value is the middle value on the table wins the game. Every losing player pays the value of his own card in euros to the winning player. Suppose that when you start playing the game, you hold the deck with the 3, 6, 9 and the queen.
(a) Which cards are rational for you? For every rational card, find a belief about your friends' choices for which this card is optimal. For every irrational card, find another card, or a randomization over cards, that strictly dominates it.
(b) Which cards are rational for your two friends? For every rational card, find a belief for which this card is optimal. For every irrational card, find another card, or a randomization over cards, that strictly dominates it.
(c) Make a beliefs diagram for this game.
(d) Use the graphical method to determine those cards you can rationally put on the table if you believe that your friends will play rationally too.
(e) Use the algorithm of two-fold elimination of strictly dominated choices to determine those cards you can rationally put on the table if you believe that your friends will play rationally too.

2.7 The big race

Today there will be a big race between a horse-drawn wagon, a truck and a Maserati. The horse wagon can reach a maximum speed of only 30 mph, the truck can go at 70 mph, whereas the Maserati is able to reach a speed of 120 mph. The racing circuit is depicted in Figure 2.12. The finish is at H, and the numbers along the roads represent the lengths of the roads in miles. To make the race more exciting, the vehicles will start from different places: The horse wagon will start at point D, the truck will start at C, and the Maserati will start at A. Every vehicle may choose any route it likes, as long as it leads to H. There is one problem, however: The roads are so narrow that it is impossible to overtake another vehicle! The objective for every vehicle is to finish

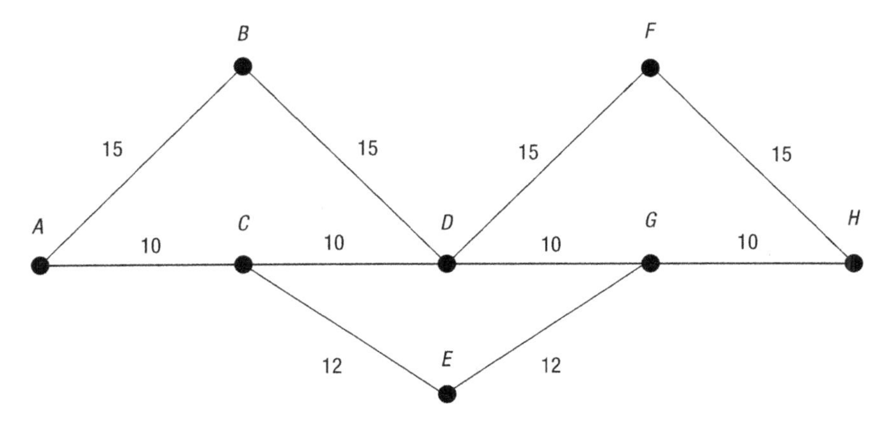

Figure 2.12 The big race

as quickly as possible, but the place is of no importance. So, for instance, finishing in second place in 30 minutes is better than finishing in first place in 35 minutes.
(a) Which routes are rational for the horse wagon, the truck and the Maserati? For every rational route, find a belief about the opponents' routes for which this route is optimal. For every irrational route, find another route, or a randomized route, that strictly dominates it.
(b) Make a beliefs diagram for this race.
(c) Suppose you drive the truck. Use the graphical method to determine those routes you can rationally choose if you believe that your opponents drive rationally too.
(d) Now, suppose you drive the Maserati. Using the algorithm of two-fold elimination of strictly dominated choices, find those routes that you can rationally choose as a Maserati driver if you believe that the horse wagon and truck will drive rationally too.

Theoretical Problems

2.8 Domination by undominated choices

Consider a choice c_i that is strictly dominated by a randomized choice. Show that c_i is strictly dominated by a randomized choice r_i that only assigns positive probability to choices c_i' that are *not* strictly dominated themselves.

2.9 Guaranteeing an expected utility level

(a) Let v be a number such that, for every belief b_i for player i about the opponents' choices, there is a choice c_i for player i with $u_i(c_i,b_i) > v$. Show that player i has a randomized choice r_i that guarantees player i strictly more than v, that is, $u_i(r_i,c_{-i}) > v$ for every opponents' choice-combination $c_{-i} \in C_{-i}$. (Here, we denote by

$$C_{-i} := C_1 \times ... \times C_{i-1} \times C_{i+1} \times ... \times C_n$$

the set of opponents' choice-combinations, and we denote an opponents' choice-combination by c_{-i}.)
Hint for (a): Use Theorem 2.5.3.
(b) Let v be a number such that, for every belief b_i for player i about the opponents' choices, there is a choice c_i for player i with $u_i(c_i,b_i) \geq v$. Show that player i has a randomized choice r_i that guarantees player i at least v, that is, $u_i(r_i,c_{-i}) \geq v$ for every opponents' choice-combination $c_{-i} \in C_{-i}$.
(c) Show that the converse statement in (b) is also true. So, show that, if player i has a randomized choice r_i that guarantees player i at least v, then for every belief b_i for player i about the opponents' choices, there is a choice c_i for player i with $u_i(c_i,b_i) \geq v$.

2.10 Zero-sum games

A *zero-sum game* is a static game with two players – player 1 and player 2 – such that $u_1(c_1,c_2) + u_2(c_1,c_2) = 0$ for all choice-combinations $(c_1,c_2) \in C_1 \times C_2$. That is, the sum of the utilities for the players is always equal to 0.

(a) Show that there is a number v such that

- player 1 has a randomized choice r_1 that guarantees him at least v, and
- player 2 has a randomized choice r_2 that guarantees him at least $-v$.

Hint for (a): Take v to be the highest expected utility level that player 1 can guarantee for himself. Use Problem 2.9, part (b).
(b) Show that the number v in part (a) is unique. This expected utility level v is called the *value* of the zero-sum game.
Hint for (b): Use Problem 2.9, part (c).

Literature

Early days of game theory
As with many scientific disciplines, it is difficult to say just when game theory really started. In the nineteenth century, Cournot (1838) and Bertrand (1883) analyzed a specific kind of game, namely the competition between firms producing homogeneous goods. In Cournot's model it is assumed that firms compete by choosing the quantity to produce, whereas Bertrand's model assumes that firms choose prices rather than quantities. Later, von Stackelberg (1934) extended Cournot's model to a dynamic setting, in which two firms choose their quantities *sequentially* rather than simultaneously. Hotelling (1929) studied a different form of competition between firms, where firms not only choose prices but also locations. Our example "Where to locate my pub?" can be seen as an instance of Hotelling's model in which the parties only choose locations.

At the beginning of the twentieth century, Zermelo (1913) and von Neumann (1928) investigated recreational games between two players. Zermelo (1913) proved that in the game of chess, either white has a strategy that guarantees a win, or black has a strategy that guarantees a win, or both white and black have strategies that guarantee at least a draw. This result is known as Zermelo's theorem. To this very day, we still do not know which of the three options above is true for the game of chess! Von Neumann (1928) studied general two-person zero-sum games, also called *strictly competitive games* (see Problem 2.10), which cover a broad spectrum of recreational games. He proved that for such games we can always find a unique expected utility level v such that player 1 has a randomized choice that guarantees him at least expected utility v, and player 2 has a randomized choice that guarantees him at least expected utility $-v$. This number v is called the *value* of the zero-sum game, and the corresponding randomized choices are called *maximin strategies* for the players. Some years before von Neumann's paper, Borel (1921, 1924, 1927) had made important contributions to the systematic study of recreational games. Borel was the first to formally define zero-sum games and the notion of a *randomized choice*, or *mixed strategy*, which he used for the analysis of recreational games. In a sense, von Neumann (1928) completed Borel's initial analysis of zero-sum games.

A major breakthrough in the development of game theory as a formal science was the book *Theory of Games and Economic Behavior* by von Neumann and Morgenstern in 1944. This book laid the foundation for modern game theory by providing a general definition of n-person games and offering a systematic analysis of this broad class of games. Often, this book is considered to mark the beginning of game theory, but we should not forget the important work by its predecessors on more restricted classes of games.

Choices

In the game theory literature, the players' choices are typically called *strategies*. In this book, we have reserved the term *strategy* for dynamic games (see Chapters 8 and 9), where it indicates a *plan* of choices.

Beliefs and expected utility

The central assumptions we make in this book are that a player (a) holds a probabilistic *belief* about the opponents' choices, (b) assigns a *utility* to each of the possible choice-combinations by the players, and (c) ranks his own choices by looking at the *expected utility* they generate, given his belief and utility function. In short, we assume that the players in a game conform to the *expected utility theory*. An axiomatic foundation for the expected utility theory was given in the books by von Neumann and Morgenstern (1944) and Savage (1954). In von Neumann and Morgenstern (1944), a person is supposed to have preferences over lotteries with objective probabilities. They then impose conditions on such preferences, the von Neumann–Morgenstern axioms, which guarantee that these preferences are generated by some utility function over the possible outcomes. That is, if the preferences satisfy the von Neumann–Morgenstern axioms, then we can find a utility function u on the possible outcomes such that the person prefers one lottery to another if the former induces a higher expected utility under u than the latter. Such a utility function u on the possible outcomes is often called a von Neumann–Morgenstern utility function. In Savage (1954), the objects of choice are not lotteries with objective probabilities, but rather *acts*. In Savage's model, the objects about which the person is uncertain are called *states of the world*, and an act is a mapping that assigns to every state of the world some possible outcome. The person is then supposed to have preferences over acts, rather than preferences over lotteries. Savage places a number of conditions on these preferences, the Savage axioms, which guarantee that these preferences are generated by a utility function over the possible outcomes and a *subjective probabilistic belief* about the states of the world. By the latter we mean that the person prefers one act to another, if the former induces a higher expected utility, given the utility function *and* the subjective probabilistic belief, than the latter. Anscombe and Aumann (1963) combine the approaches taken by von Neumann and Morgenstern (1944) and Savage (1954), by allowing for acts that assign to every state of the world a lottery with objective probabilities. Anscombe and Aumann's model contains both objective and subjective probabilities.

Players as decision makers under uncertainty

We can apply Savage's framework in particular to games. A player in a game is a decision maker who faces uncertainty about the choices to be made by his opponents. Hence, we can model the opponents' choice-combinations as the states of the world. Throughout this book we assume that a player in a game always conforms to Savage's expected utility theory, so that he makes his choice on the basis of the utility function he holds over the possible outcomes, and the subjective probabilistic belief he holds about the opponents' choice-combinations. However, in contrast to Savage's approach, we take the utility function and the probabilistic belief as the *primitives* of a player, on the basis of which he ranks his choices. Savage, in turn, takes the person's preferences over acts as the primitive, on the basis of which we can *derive* the utility function and the probabilistic belief under the Savage axioms. Although it seems entirely natural to model a player in a game explicitly as a decision maker under uncertainty, it took game theory a surprisingly long time to do so. The first to take this approach were Armbruster and Böge (1979) and Böge and Eisele (1979), later followed by, for instance, Aumann (1987) and Tan and Werlang (1988).

Choosing rationally

We say that a player chooses rationally if he makes a choice that maximizes his expected utility, given the subjective probabilistic belief he holds about the opponents' choices. In Aumann (1987) and Tan and Werlang (1988), such a player is called *Bayesian rational*. One should be careful with the name *rational* though, as a rational choice is not necessarily a *reasonable* choice! Rational just means that the choice is optimal for *some* belief, and does not put any restrictions on this belief. Obviously, a rational choice that can only be supported by unreasonable beliefs is itself unreasonable. In fact, the purpose of this book is to impose appealing restrictions on the beliefs that a player may hold about his opponents, thereby separating "reasonable" rational choices from "unreasonable" rational choices.

Randomized choices

In the literature, randomized choices are usually called *mixed strategies*. In many books and papers on game theory, mixed strategies are interpreted as conscious randomizations by the player himself, and are therefore treated as real objects of choice. In this book, however, we take the viewpoint that players do not randomize. In fact, both on a theoretical and a practical level there is nothing a player can gain from randomizing: If by randomizing over various choices he maximizes his expected utility, then he could achieve the same expected utility by simply committing to one of the choices he is randomizing over. But in that case, there is no point in randomizing at all. We therefore do not interpret a randomized choice as something a player would really choose, but rather treat it as an abstract mathematical tool to characterize rational choices in a game.

Characterization of rational choices

In Theorem 2.5.3 we showed that the rational choices in a static game are exactly those choices that are neither strictly dominated by another choice nor by a randomized choice. The first proof for this theorem that we are aware of was by Pearce (1984), who proved this result for two-player games in Lemma 3. His proof, however, can easily be extended to games with more than two players. An important difference between our proof and Pearce's is that Pearce's proof relies on the existence of a Nash equilibrium (see Chapter 4) in the game, whereas our proof does not. In fact, most proofs for this result in books and papers either rely on the existence of a Nash equilibrium, or use the duality theorem from linear programming or some variant of it, or use the existence of a value for zero-sum games. Our proof, in contrast, is completely elementary, and does not use any of these results.

Belief in the opponents' rationality

We say that a player believes in the opponents' rationality if he only assigns positive probability to the opponents' choices that are rational. This was defined in Tan and Werlang (1988) in their formulation of K_i^1 in Definition 5.2. In their Theorem 5.1, Tan and Werlang also show that if a player believes in the opponents' rationality, then every choice that is rational for him must survive the procedure of two-fold elimination of strictly dominated choices. This is obtained by choosing $l = 1$ in their theorem.

Belief and knowledge

In the literature, people often use the term "knowledge" instead of "belief" to express the players' epistemic state of mind. For instance, Tan and Werlang (1988) say that a player *knows* that his opponents choose rationally, rather than that he *believes* so. For our formal analysis of static games it does not make much of a difference whether we use the term "knowledge" or "belief" – it basically amounts to the same thing. However, on an intuitive level we think that the term "knowledge" is too strong to describe a player's state of mind in a game. Namely, if we say that player i *knows* that opponent j chooses rationally, then it suggests that player i is absolutely certain that player j indeed will choose rationally. But this seems too strong – player i can at best have a *belief* about the rationality of player j, and this belief may very well be wrong! Player i cannot look inside player j's mind, and can therefore not be absolutely certain that he will indeed choose rationally. The same applies to other mental characteristics of player j about which player i is uncertain. We therefore use the term "belief" rather than "knowledge" throughout this book. But, again, on a purely technical level it makes no difference – at least for the analysis of static games.

Examples and problems

The example "Where to locate my pub?" and the practical problems "Where to locate a supermarket?" and "The three barkeepers" are based on a model by Hotelling (1929) in which firms compete by choosing locations. The example "Going to a party" and the practical problem of the same name can be viewed as an instance of an *anti-coordination game*. The example "The traveler's dilemma" is due to Basu (1994), who used it to show that common belief in rationality (see Chapter 3) may sometimes lead to unexpected, and even counterintuitive, outcomes. The practical problem "Competition between two cinemas" is based on a model by Bertrand (1883) in which firms compete by choosing prices. The theoretical problem "Zero-sum games" is largely based on the work by von Neumann (1928).

3 Common belief in rationality

3.1 Beliefs about the opponents' beliefs

The main message from the previous chapter is that choosing wisely is more than just choosing rationally: A reasonable choice should not only be rational, in the sense that it can be supported by *some* belief about the opponents' choices, but it should be optimal for a *reasonable* belief about the opponents. The main question we will ask ourselves in this book is: When can a belief about the opponents be called *reasonable*?

The previous chapter was only a first step towards answering this question. In the second half of that chapter we concentrated on beliefs that only assign positive probability to the opponents' choices that are rational for them. Such beliefs are said to express "belief in the opponents' rationality." This is not the end of the story, however, since not every belief that expresses belief in the opponents' rationality will be entirely reasonable. As an illustration of this fact let us return to the example "Where to locate my pub?"

Example 3.1 Where to locate my pub?

Recall the story from Example 2.1. In Example 2.5 we saw that, if you believe that Barbara chooses rationally, then you can rationally choose the locations c, d and e yourself. The easiest way to see this is to look at the beliefs diagram in Figure 3.1, which is taken from the previous chapter. You can rationally choose c if you believe that Barbara rationally chooses b. Moreover, you can rationally choose d if you believe that Barbara rationally chooses d as well. Finally, you can rationally choose e if you believe that Barbara rationally chooses f. Does this mean that each of these choices c, d and e is also reasonable? The answer is "no."

Consider the choice c. The arrow from your choice c to Barbara's choice b means that c is optimal for you if you believe that Barbara chooses b. In turn, the arrow from Barbara's choice b to your choice a indicates that choice b is optimal for Barbara if she believes that you will *irrationally* choose a. If we put these two facts together, we can say that choice c is optimal for you if

- you believe that Barbara rationally chooses b, and
- you believe that Barbara believes that you *irrationally* choose a.

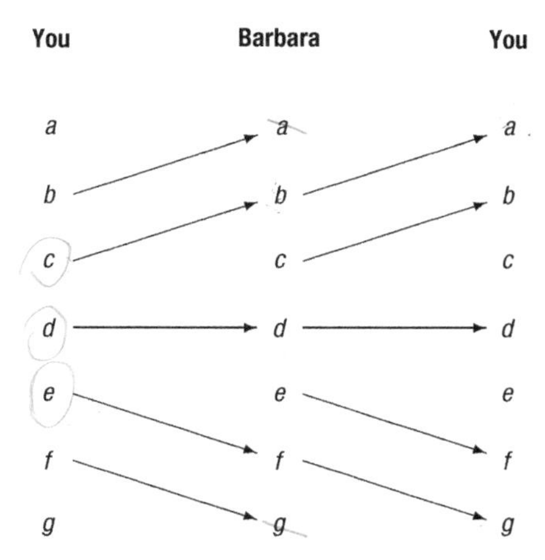

Figure 3.1 A beliefs diagram for "Where to locate my pub?" (III)

Such a belief seems unreasonable, however. If it is unreasonable to believe that Barbara chooses irrationally then, for the same reason, it should also be unreasonable to believe that Barbara believes that *you* choose irrationally. Or, to put it differently, if you take your opponent seriously, then it seems natural to also believe that your opponent will take *you* seriously.

If we follow this principle we must discard the belief above, as it states that "you believe that Barbara believes that *you* choose irrationally." Now, what are the consequences of following this principle in this example? So, which choices can you rationally make if

- you believe that Barbara chooses rationally, and
- you believe that Barbara believes that *you* choose rationally?

We have already seen in Chapter 2 that a and g are irrational choices for you. So, if you believe that Barbara believes that you choose rationally, you must believe that Barbara believes that you will not choose a and g.

In our analysis of Example 2.5 we have seen that, if you believe that Barbara does not choose a and g, then you can only rationally choose locations c,d and e. Namely, if you believe that Barbara will not choose a and g, then choosing c is always better than choosing b and choosing e is always better than choosing f. Since locations a and g are never optimal anyway, we may conclude that you can only rationally choose c,d and e if you believe that Barbara will not choose a and g. Because of the symmetry of the game, we can also turn the argument around: If Barbara believes that you will not choose a and g, then Barbara can only rationally choose locations c,d and e. So, if Barbara believes that you choose rationally, then Barbara can only rationally choose locations c,d and e. This, however, means that if

- you believe that Barbara chooses rationally, and
- you believe that Barbara believes that *you* choose rationally,

Table 3.1. *Number of customers you would obtain if you believe that Barbara will only choose from* $\{c,d,e\}$

		Barbara		
		c	d	e
You	c	300	250	300
	d	350	300	350
	e	300	250	300

then you must believe that Barbara will only choose c,d or e.

Suppose, now, you believe that Barbara will only choose from $\{c,d,e\}$. The outcomes of your choices are summarized in Table 3.1. Here, we have left out your choices a,b,f and g, since you cannot rationally make these choices anyway if you believe that Barbara chooses rationally. From this table it is clear that only location d can be optimal if you believe that Barbara will only choose c,d or e – any other location would always lead to fewer customers than location d.

Combining our insights so far, we see that if

- you believe that Barbara chooses rationally, and
- you believe that Barbara believes that *you* choose rationally,

then you can only rationally choose the location d.

Does this also mean that location d is a *reasonable* choice? In this example the answer is "yes." The key to this insight lies in the beliefs diagram of Figure 3.1. From this diagram, we see that you can rationally choose d if you believe that Barbara chooses d. Moreover, d is a rational choice for Barbara if she believes that you will choose d as well. In turn, d is a rational choice for yourself if you believe that Barbara will choose d, and so on. So, you can rationally choose d if

- you believe that Barbara will rationally choose d,
- you believe that Barbara believes that you will rationally choose d,
- you believe that Barbara believes that you believe that Barbara will rationally choose d,

and so on, ad infinitum.

The object we have constructed above is not just a *single* belief: it is a complete *hierarchy of beliefs*. The first belief in this hierarchy, which states that you believe that Barbara will choose d, is a belief that you hold about the opponent's choice. Such a belief is called a *first-order belief*. In fact, these are the kind of beliefs we discussed intensively in the first part of Chapter 2. The second belief, stating that you believe that Barbara believes that you will choose d, is more complex as it is a belief that you hold

about the opponent's belief about your own choice. Such a belief is called a *second-order belief*. Formally, it is a belief you hold about the opponent's first-order belief. The third belief in this hierarchy is even more involved, as it is a belief that you hold about what your opponent believes that you believe about the opponent's choice. So, it is a belief you hold about the opponent's second-order belief. Such a belief is called a *third-order belief*. Of course, we could continue this chain of arguments, but the idea should now be clear. In general, one can recursively define for every number k a kth-order belief, representing a belief you hold about the opponent's $(k-1)$th-order belief.

What can we say about the belief hierarchy above? Is it a reasonable belief hierarchy to hold or not? In our opinion it certainly is. Suppose that you hold this belief hierarchy. Then, you not only believe that your opponent chooses rationally, but you also believe that your opponent believes that you choose rationally as well. Not only this, you also believe that your opponent believes that you believe that your opponent chooses rationally also, and so on, ad infinitum. Or, to put it differently, the belief hierarchy does not contain any belief of any order in which the rationality of your opponent, or your own rationality, is questioned. Such belief hierarchies are said to express *common belief in rationality*.

In this example there is only one location you can rationally choose under common belief in rationality, namely location d. We have seen that if you believe that Barbara chooses rationally, and believe that Barbara believes that you choose rationally, then only d can be a possible rational choice for you. Moreover, we have convinced ourselves that location d *can* rationally be chosen under common belief in rationality. In summary, we may conclude that the idea of common belief in rationality points towards a unique choice for you, namely location d. So common belief in rationality tells us that you should locate your pub in the middle of the street! □

In the example above, we have informally introduced the idea of *common belief in rationality*, which states that you not only believe that your opponent chooses rationally, but also that he believes that you choose rationally, and that he believes that you believe that he chooses rationally, and so on. Let us now investigate the consequences of this idea for our example "Going to a party."

Example 3.2 Going to a party

Recall the story from Example 2.2 and the utilities for Barbara and you depicted in Table 2.9. We have reproduced these utilities in Table 3.2 for convenience. A possible beliefs diagram for this situation is the one in Figure 3.2, which is a copy of Figure 2.9. We have already seen in Chapter 2 that it is never optimal to choose your least preferred color *yellow*, and that it is never optimal to choose *red* if you believe that Barbara chooses rationally. In particular we may conclude that you cannot rationally choose *yellow* or *red* under common belief in rationality.

What about your choice *blue*? If we look at the beliefs diagram in Figure 3.2 and follow the arrows from your choice *blue*, we see that *blue* is optimal for you if you believe that Barbara will choose *red*, and *red* is optimal for her if she believes that

Table 3.2. *Utilities for you and Barbara in "Going to a party" (II)*

	blue	green	red	yellow	same color as friend
you	4	3	2	1	0
Barbara	2	1	4	3	0

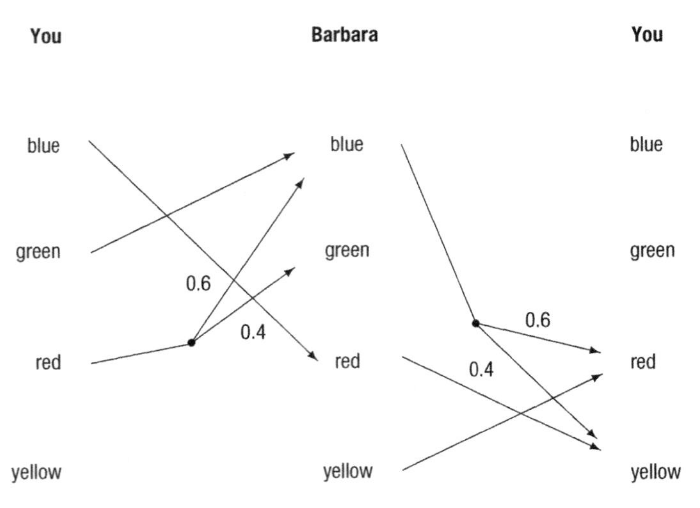

Figure 3.2 A beliefs diagram for "Going to a party" (III)

you will *irrationally* choose *yellow*. So, in this beliefs diagram your choice *blue* is supported by a belief hierarchy in which

- you believe that Barbara will rationally choose *red*, and
- you believe that Barbara believes that you will irrationally choose *yellow*.

This belief hierarchy certainly does not express common belief in rationality, as you believe that Barbara believes that you will make an irrational choice.

Does this mean that you cannot rationally choose *blue* under common belief in rationality? The answer is "no:" We could construct an alternative beliefs diagram, as in Figure 3.3, in which your choice *blue* would be supported by a belief hierarchy that expresses common belief in rationality. In the new beliefs diagram, we have replaced the arrow from Barbara's choice *red* to your choice *yellow* by an arrow from *red* to *blue*. This arrow is also valid, as it is optimal for Barbara to choose *red* if she believes that you will choose *blue*. If in the new beliefs diagram we start at your choice *blue*, and follow the arrows, then we obtain the belief hierarchy in which

- you believe that Barbara will rationally choose *red*,
- you believe that Barbara believes that you will rationally choose *blue*,

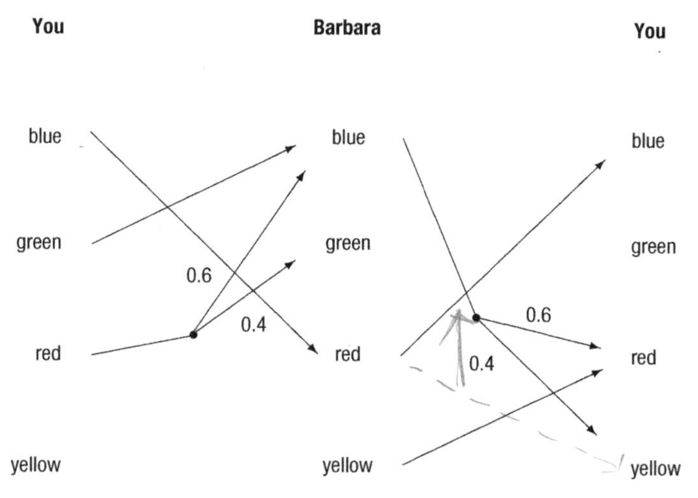

Figure 3.3 An alternative beliefs diagram for "Going to a party" (II)

- you believe that Barbara believes that you believe that Barbara will rationally choose *red*

and so on. This belief hierarchy does not question any player's rationality in any of his beliefs, and therefore expresses common belief in rationality. The first-order belief states that you believe that Barbara will choose *red*, which supports your own choice *blue*. You can thus rationally choose *blue* under common belief in rationality.

We finally turn to your choice *green*. Consider the beliefs diagram in Figure 3.3, and start following the arrows from your choice *green*. We first move to Barbara's choice *blue*, then, with probability 0.6 to your choice *red* and with probability 0.4 to your choice *yellow*. This chain of arrows represents a belief hierarchy in which

- you believe that Barbara will rationally choose *blue*,
- you believe that Barbara believes that, with probability 0.6, you will rationally choose *red*, and with probability 0.4 you will irrationally choose *yellow*.

So, this belief hierarchy does not express common belief in rationality, since you believe that Barbara assigns positive probability to the event that you make an irrational choice. However, as we have seen above, this does not automatically mean that you cannot rationally choose *green* under common belief in rationality. Perhaps there is a different beliefs diagram in which your choice *green* could be supported by a belief hierarchy that expresses common belief in rationality.

So, let us see whether we can indeed find such a beliefs diagram. This time you will notice, however, that this cannot be done. The reason is as follows: In Chapter 2 we saw that if you believe that Barbara chooses rationally, then you can only rationally choose your two most preferred colors *blue* and *green*. By symmetry of the game, it

Table 3.3. *New utilities for Barbara in "Going to a party"*

	blue	green	red	yellow	same color as friend
you	4	3	2	1	0
Barbara	4	3	2	1	5

can be shown in the same way that if Barbara believes that you choose rationally, then she can only rationally choose her two most preferred colors *red* and *yellow*. Hence, if you believe that Barbara chooses rationally, and believe that Barbara believes that you choose rationally, then you must believe that Barbara will not choose *blue* or *green*. But then, choosing *blue* yourself would give you an expected utility of 4, since you believe that, with probability 1, you will be wearing a different color from Barbara. On the other hand, choosing *green* would only give you an expected utility of 3. So, you can no longer rationally choose *green* if you believe that Barbara chooses rationally, and believe that Barbara believes that you choose rationally. In particular, you cannot rationally choose *green* under common belief in rationality.

Overall, we learn that under common belief in rationality, you should definitely go for your most preferred color *blue*. The other colors are no longer rational under common belief in rationality.

Let us finally look at a situation where Barbara's preferences have changed. Assume that Barbara now has the same preferences over colors as you but, unlike you, she *likes* it when she wears the same color as you. More specifically, if she happens to wear the same color as you, it would give her a utility of 5 instead of 0. Barbara's new preferences are given in Table 3.3. In this new situation, every color is rational for Barbara. Namely, each of the colors is optimal for her if she believes that you will be wearing exactly this color. Hence, we may construct the beliefs diagram shown in Figure 3.4. From this beliefs diagram it can immediately be seen that you can rationally choose *blue*, *green* and *red* under common belief in rationality, but not *yellow*. Let us start with your choice *blue*. By following the arrows in the beliefs diagram and translating these arrows into a belief hierarchy, we see that *blue* is optimal for you if

- you believe that Barbara will choose *green*,
- you believe that Barbara believes that you will choose *green*,
- you believe that Barbara believes that you believe that Barbara will choose *blue*,
- you believe that Barbara believes that you believe that Barbara believes that you will choose *blue*,
- you believe that Barbara believes that you believe that Barbara believes that you believe that Barbara will choose *green*,

and so on. In this way, we obtain a belief hierarchy in which you believe that Barbara chooses rationally, in which you believe that Barbara believes that you choose rationally,

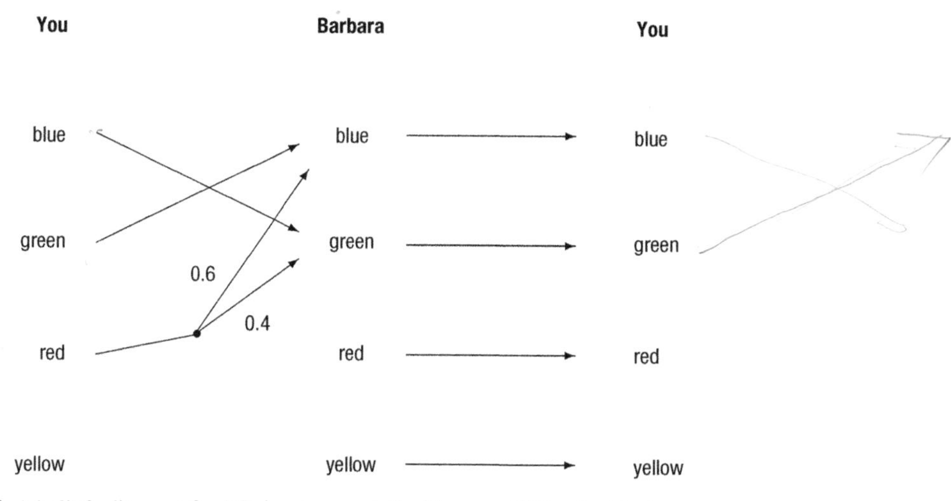

Figure 3.4 A beliefs diagram for "Going to a party" with new utilities for Barbara

and so on, ad infinitum. So, the belief hierarchy above, if completed, would express *common belief in rationality*. As your choice *blue* is optimal under this belief hierarchy, you can rationally choose *blue* under common belief in rationality.

The reason that this belief hierarchy expresses common belief in rationality is that, when you start at your choice *blue* and follow the arrows in the beliefs diagram, you only reach rational choices. In the same way, it can be verified that your choice *green* can also be chosen rationally under common belief in rationality.

Let us now turn to your choice *red*. If we start at your choice *red* and follow the arrows in the beliefs diagram, we obtain the belief hierarchy in which you believe that, *with probability 0.6*,

- Barbara will choose *blue*,
- Barbara believes that you will choose *blue*,
- Barbara believes that you believe that Barbara will choose *green*,
- Barbara believes that you believe that Barbara believes that you will choose *green*,
- Barbara believes that you believe that Barbara believes that you believe that Barbara will choose *blue*,

and so on, and you believe that, *with probability 0.4*,

- Barbara will choose *green*,
- Barbara believes that you will choose *green*,
- Barbara believes that you believe that Barbara will choose *blue*,
- Barbara believes that you believe that Barbara believes that you will choose *blue*,
- Barbara believes that you believe that Barbara believes that you believe that Barbara will choose *green*,

and so on. In this belief hierarchy, you believe with probability 1 that Barbara chooses rationally, you believe with probability 1 that Barbara believes with probability 1 that you choose rationally, and so on, ad infinitum. Or, graphically speaking, if you follow the arrows in the beliefs diagram, you only reach rational choices. So, this belief hierarchy expresses common belief in rationality. As your choice *red* is optimal if you hold this belief hierarchy, it follows that you can rationally choose *red* under common belief in rationality.

It is clear that *yellow* cannot be chosen rationally under common belief in rationality, as it is not a rational choice to begin with. Summarizing, we thus see that under the new utilities for Barbara, you can rationally choose *blue*, *green* and *red* under common belief in rationality, but not *yellow*. □

Let us now apply the idea of *common belief in rationality* to situations with more than two players. In such situations, common belief in rationality means that you believe that every opponent chooses rationally, that you believe that every opponent believes that each of his opponents chooses rationally, and so on. As an illustration, let us return to the example "Waiting for a friend."

Example 3.3 Waiting for a friend

Recall the story from Example 2.3. In Chapter 2 we saw that it is never optimal for you to arrive at 8.30 pm, since it is always better to arrive at 8.20 pm. It is also not optimal to arrive at 7.30 pm, as arriving at 7.40 pm is always better. Moreover, if you believe that both of your friends will arrive at rational times, then it is never optimal for you to arrive at 8.20. So you cannot rationally choose to arrive at 7.30, 8.20 or 8.30 under common belief in rationality. What about the other arrival times?

Let us first look at a beliefs diagram and see whether it can be of any help. A possible beliefs diagram for this situation is Figure 3.5, which is a copy of Figure 2.10. Let us start with your arrival time 8.10. As for games with two players, we can follow the chain of arrows that start at your choice 8.10, and translate this chain of arrows into a belief hierarchy. The forked arrow from your choice 8.10 to Barbara's choice 8.20 and Chris' choice 8.20 indicates that you can rationally choose 8.10 if you believe that Barbara will choose 8.20 and Chris will choose 8.20. This would be your first-order belief. The forked arrow from Barbara's choice 8.20 to your choice 8.30 and Chris' choice 8.30 means that Barbara can rationally choose 8.20 if she believes that you and Chris will both irrationally choose 8.30. The forked arrow from Chris' choice 8.20 to your choice 8.30 and Barbara's choice 8.30 means that Chris can rationally choose 8.20 if he believes that you and Barbara will both irrationally choose 8.30. So, this chain of arrows can be translated into a belief hierarchy in which

- you believe that Barbara and Chris will rationally arrive at 8.20,
- you believe that Barbara believes that you and Chris will irrationally arrive at 8.30,
- you believe that Chris believes that you and Barbara will irrationally arrive at 8.30.

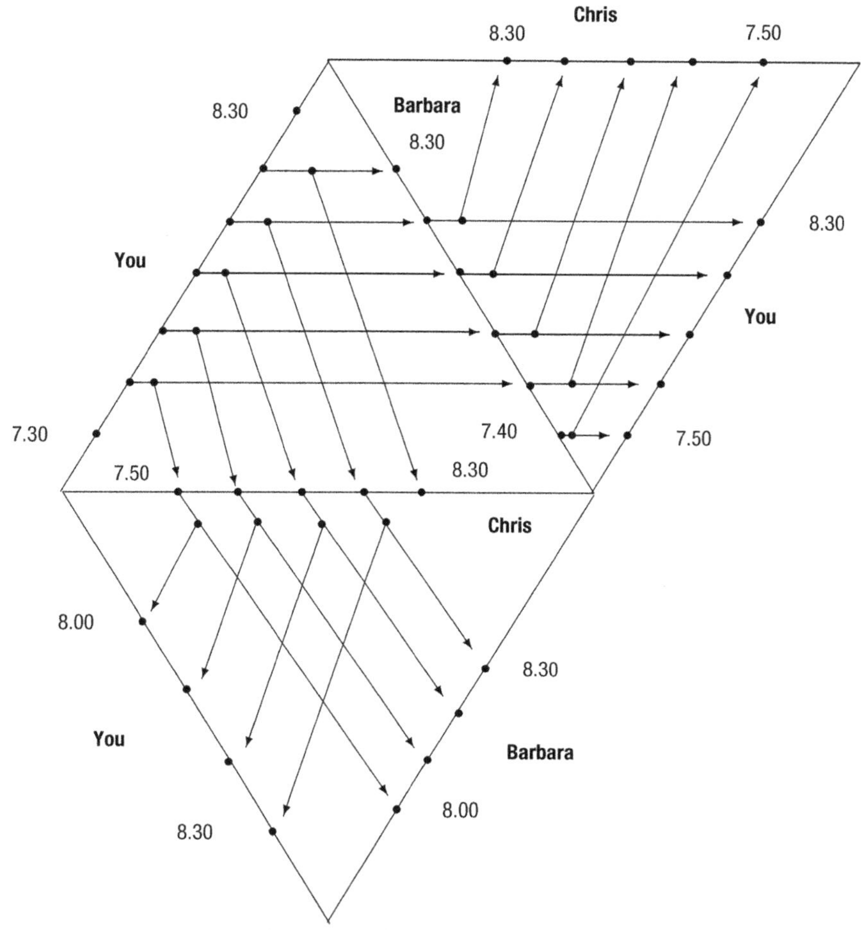

Figure 3.5 A beliefs diagram for "Waiting for a friend" (III)

The first belief, which specifies what you believe the others will do, is your first-order belief. The second and third belief, which specify what you believe the opponents believe that the other players will do, constitute your second-order belief. This belief hierarchy is not complete since we would also have to specify your third-order belief, that is, what you believe your opponents believe that the other players believe about the other players' choices. Similarly, we would have to define beliefs of higher order also. Importantly, however, we can already conclude that this belief hierarchy does not express common belief in rationality, since you believe that Barbara believes that the other players will choose irrationally. So, your own rationality and Chris' rationality are questioned in the second-order belief in the hierarchy. This belief hierarchy was intended to justify your choice to arrive at 8.10. But since this belief hierarchy does not express common belief in rationality, we cannot conclude that you can rationally choose 8.10 under common belief in rationality.

What should we do now? The beliefs diagram does not help us to verify whether you can rationally arrive at 8.10 under common belief in rationality, since the corresponding chain of arrows delivers a belief hierarchy that does not express common belief in rationality. But perhaps there is another belief hierarchy that justifies 8.10 and that *does* express common belief in rationality. So perhaps we can try another beliefs diagram, and see whether we are lucky there. The problem, however, is that there are many alternative beliefs diagrams we could try, and trying them all would simply take too much time. For instance, the choice to arrive at 8.10 is optimal not only if you believe that Barbara and Chris both arrive at 8.20, but it is also optimal if Barbara arrives at 8.10 and Chris arrives at 8.20, or if Barbara arrives at 7.40 and Chris at 8.20, or if Barbara arrives at 8.20 and Chris at 8.00, to name just a few options. Similarly for all other choices you can make.

A better idea in this case is to use some *logical reasoning* in order to verify whether you can rationally choose 8.10 under common belief in rationality. As an important first observation, note that it can only be optimal for you to arrive at 8.10 if at least one opponent arrives *later* than 8.10. Namely, if both opponents arrive at 8.10 or before, then it would be better for you to arrive *before* 8.10. So, you can only rationally choose 8.10 if some opponent arrives at 8.20 or at 8.30.

We have seen before that arriving at 8.30 is never optimal for you, and 8.20 is not optimal if you believe that your opponents choose rationally. The same applies to your opponents: For each opponent it cannot be optimal to arrive at 8.20 or 8.30 if he believes that the others choose rationally.

Altogether, we see that arriving at 8.10 can only be optimal if you believe that at least one opponent will arrive after 8.10, but for this opponent it cannot be optimal to arrive after 8.10 if he believes that the others choose rationally. This means, however, that you cannot rationally choose to arrive at 8.10 if

- you believe that both friends choose rationally, and
- you believe that both friends believe that the others choose rationally as well.

Therefore, you cannot rationally choose to arrive at 8.10 under common belief in rationality.

What about your choice to arrive at 8.00? A similar argument applies. First, note that arriving at 8.00 can only be optimal if you believe that at least one opponent will arrive *after* 8.00. We have seen above that you cannot rationally choose to arrive at 8.10, or later, if you believe that your friends choose rationally, and believe that your friends believe that their friends will choose rationally. The same applies to each of your friends: Barbara and Chris cannot rationally arrive at 8.10, or later, if they believe that their friends choose rationally, and believe that their friends believe that their friends choose rationally. Altogether, this means that you cannot rationally choose to arrive at 8.00 if

- you believe that both friends choose rationally,
- you believe that both friends believe that their friends choose rationally, and

- you believe that both friends believe that their friends believe that their friends choose rationally.

We may thus conclude that you cannot rationally choose to arrive at 8.00 under common belief in rationality. So far we have thus seen that arriving at 8.00 or later cannot be rational under common belief in rationality. In fact, the same holds for your friends: Also Barbara and Chris cannot rationally arrive at 8.00 or later under common belief in rationality.

Here is our final step: Suppose you hold a belief hierarchy that expresses common belief in rationality. Then,

- you believe that both friends choose rationally,
- you believe that both friends believe that their friends choose rationally,
- you believe that both friends believe that their friends believe that their friends choose rationally,

and so on. So, you believe that both friends choose rationally under common belief in rationality! In particular, you must believe that Chris chooses rationally under common belief in rationality. However, we have seen above that Chris cannot rationally arrive at 8.00 or later under common belief in rationality. Since, by the description of the game, he cannot arrive before 7.50, he can only rationally arrive at exactly 7.50 under common belief in rationality. So, if you hold a belief hierarchy that expresses common belief in rationality, you *must* believe that Chris arrives exactly at 7.50. Similarly, as you believe that Barbara chooses rationally under common belief in rationality, you must believe that Barbara will arrive before 8.00, so at 7.40 or at 7.50.

Altogether, if your belief hierarchy expresses common belief in rationality, you must believe that Chris will arrive at 7.50, and that Barbara will arrive at 7.40 or 7.50. But then, your unique optimal choice is to arrive exactly 10 minutes before your last friend, which is to arrive exactly at 7.40. So, there is only one possible rational choice you could make under common belief in rationality, which is to arrive exactly at 7.40.

Let us now construct a belief hierarchy for you that expresses common belief in rationality and that justifies your choice to arrive at 7.40. The beliefs diagram in Figure 3.5 will be of little help here. Namely, if we start at your choice 7.40 and follow the arrows in the diagram, we eventually end up at an irrational choice 8.30. So, the chain of arrows starting at your choice 7.40 translates into a belief hierarchy that does not express common belief in rationality. So, we must construct a belief hierarchy by logical reasoning again.

First note that it is rational for you to arrive at 7.40 if you believe that Barbara will arrive at 7.40 and Chris at 7.50. Similarly, for Barbara it is rational to arrive at 7.40 if she believes that you will arrive at 7.40 and Chris at 7.50. Finally, for Chris it is rational to arrive at 7.50 if he believes that you will arrive at 7.40 and Barbara at 7.40, as Chris cannot arrive earlier than 7.50 by the description of the game. So, we can construct a belief hierarchy in which

- you believe that Barbara will rationally arrive at 7.40 and Chris will rationally arrive at 7.50,
- you believe that Barbara believes that you will rationally arrive at 7.40 and Chris will rationally arrive at 7.50,
- you believe that Chris believes that you will rationally arrive at 7.40 and Barbara will rationally arrive at 7.40,
- you believe that Barbara believes that you believe that Barbara will rationally arrive at 7.40 and Chris will rationally arrive at 7.50,
- you believe that Barbara believes that Chris believes that you will rationally arrive at 7.40 and Barbara will rationally arrive at 7.40,
- you believe that Chris believes that you believe that Barbara will rationally arrive at 7.40 and Chris will rationally arrive at 7.50,
- you believe that Chris believes that Barbara believes that you will rationally arrive at 7.40 and Chris will rationally arrive at 7.50,

and so on, ad infinitum.

This belief hierarchy nowhere questions the rationality of any player, and hence expresses common belief in rationality. Given the first-order belief, that Barbara will arrive at 7.40 and Chris at 7.50, it is optimal to arrive at 7.40. This belief hierarchy can thus be used to justify that you can rationally arrive at 7.40 under common belief in rationality. Summarizing, we see that arriving at 7.40 is the only choice you can rationally make under common belief in rationality. Similarly, you can show that Barbara can only rationally choose to arrive at 7.40 under common belief in rationality, and Chris can only rationally arrive at 7.50 under common belief in rationality. So, if all friends reason in accordance with common belief in rationality, Chris will arrive exactly at 7.50, ten minutes later than Barbara and you. □

3.2 Belief hierarchies

In this section we will discuss the idea of a *belief hierarchy* on a more formal level. Let us first refresh our memories by summarizing the idea of a belief hierarchy. If there are only two players in the game, say you and your opponent, a belief hierarchy for you should specify

- the belief that you have about the opponent's choice,
- the belief that you have about the belief that the opponent has about your choice,
- the belief that you have about the belief that the opponent has about the belief that you have about the opponent's choice,

and so on, ad infinitum. The first belief in the hierarchy is called your *first-order belief*, the second belief is called your *second-order belief*, the third belief your *third-order belief*, and so on. In general, if there are n players in the game, then a belief hierarchy for player i specifies

- the belief that player i has about his opponents' choice-combinations,
- the belief that player i has about the beliefs that his opponents have about their opponents' choice-combinations,
- the belief that player i has about the beliefs that his opponents have about the beliefs that their opponents have about the other players' choice-combinations,

and so on, ad infinitum. Again, the first belief for player i is his *first-order belief*, the second belief is his *second-order belief*, and so on.

In the examples of the previous section we saw that we can use a beliefs diagram to construct belief hierarchies for a given player, say player i. The idea was as follows: Start at a choice for player i, say choice c_i, and follow the arrows in the beliefs diagram. If by following the arrows we only reach rational choices, then every choice that is reached will have an outgoing arrow. So, we can follow the arrows ad infinitum, and we can translate this infinite chain of arrows into a first-order belief, a second-order belief, a third-order belief, and so on, thereby constructing a complete belief hierarchy for player i. Moreover, this belief hierarchy would have the special property that it expresses common belief in rationality and justifies the choice c_i.

As an illustration, consider the example "Going to a party" with the utilities as specified in Table 3.3. A beliefs diagram for this situation was given in Figure 3.4. Let us start at your choice *red*. The first – forked – arrow, from your choice *red* to Barbara's choices *blue* and *green*, with probabilities 0.6 and 0.4, can be translated into the *first-order belief* in which you believe that, with probability 0.6, Barbara will choose *blue* and with probability 0.4 she will choose *green*. The second arrows, from Barbara's choice *blue* to your choice *blue*, and from Barbara's choice *green* to your choice *green*, correspond to the *second-order beliefs* in which you believe that, with probability 0.6, Barbara believes that you will choose *blue*, and in which you believe that, with probability 0.4, Barbara believes that you will choose *green*. The third arrows, from your choice *blue* to Barbara's choice *green*, and from your choice *green* to Barbara's choice *blue*, are the *third-order beliefs* in which you believe that, with probability 0.6, Barbara believes that you believe that Barbara chooses *green*, and in which you believe that, with probability 0.4, Barbara believes that you believe that Barbara chooses *blue*. And so on.

What happens, however, if we start at a choice in the beliefs diagram, follow the arrows, and notice that we reach an irrational choice? Consider, for instance, the example "Where to locate my pub?" for which we have given a beliefs diagram in Figure 3.1. Suppose we would like to construct a belief hierarchy that supports your location c. If we start at your choice c, then we go to Barbara's choice b, and from Barbara's choice b to your choice a, which is an irrational choice for you with no outgoing arrow. That is, we get stuck after two arrows. The arrows in this beliefs diagram are not enough to construct a complete belief hierarchy: The first arrow gives us the first-order belief, the second arrow gives us the second-order belief, but there is no third arrow from which we can build a third-order belief. Nevertheless, we want to make a complete belief hierarchy that supports your location c. How can this be done?

The solution to this problem is fairly simple: For every *irrational* choice in the beliefs diagram add a *dotted arrow*, which goes from this choice to an arbitrary choice of the other player. So, we could add a dotted arrow from your choice a to Barbara's choice a, a dotted arrow from Barbara's choice a to your choice a, a dotted arrow from your choice g to Barbara's choice g, and a dotted arrow from Barbara's choice g to your choice g. This would result in the *extended* beliefs diagram shown in Figure 3.6. The interpretation of a dotted arrow is different from that of a solid arrow: The dotted arrow from your choice a to Barbara's choice a only means that you believe that Barbara chooses a, but it does *not* mean that this belief *supports* your own choice a. In fact, your choice a is irrational, and therefore cannot be supported by any belief (or arrow). On the other hand, the solid arrow from your choice c to Barbara's choice b does not just mean that you believe that Barbara chooses b, but it also *supports* your choice c. Namely, it is optimal for you to choose location c if you believe that Barbara chooses b.

Although these dotted arrows do not provide any justification for the choices they start at, they serve a clear purpose: For every rational choice we can now construct a *complete* belief hierarchy that supports this choice! Let us return to our task of constructing a complete belief hierarchy that supports your location c. By starting at your choice c,

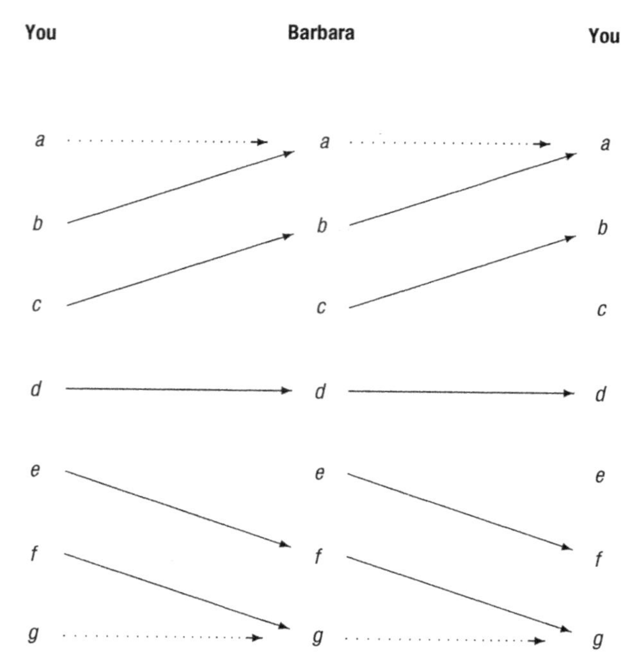

Figure 3.6 An extended beliefs diagram for "Where to locate my pub?"

and following the solid *and* dotted arrows, we first go to Barbara's choice b, then to your choice a, then to Barbara's choice a, then back to your choice a, and so on. So, we have an infinite chain of arrows. This enables us to construct the belief hierarchy in which

- you believe that Barbara will rationally choose b,
- you believe that Barbara believes that you will irrationally choose a,
- you believe that Barbara believes that you believe that Barbara will irrationally choose a,

and so on. This belief hierarchy clearly does not express common belief in rationality, as the second-order belief questions your own rationality. The graphical reason is that the chain of arrows used to construct the belief hierarchy passes through irrational choices. This automatically leads to belief hierarchies that question a player's rationality at some point.

The idea of adding dotted arrows also works for games with more than two players. Consider, for instance, the example "Waiting for a friend" for which we constructed the beliefs diagram shown in Figure 3.5. The irrational choices in this game are 8.30 and 7.30 for you, 8.30 for Barbara and 8.30 for Chris. By adding dotted arrows for these times we can make the extended beliefs diagram shown in Figure 3.7. With the extended beliefs diagram we can now, for instance, build a complete belief hierarchy that supports your choice to arrive at 8.10. If we start at your choice 8.10, we first go to Barbara's choice 8.20 and Chris' choice 8.20, which indicate that you believe that Barbara and Chris will arrive at 8.20. From Barbara's choice 8.20 we go to your choice 8.30 and Chris' choice 8.30, so you believe that Barbara believes that her two friends will both arrive at 8.30. From Chris' choice 8.20 we go to your choice 8.30 and Barbara's choice 8.30, hence you believe that Chris believes that his two friends will both arrive at 8.30. Once we arrive at a choice 8.30 (for any player) the dotted arrows point to the choices 8.30 for the other two players. This means you believe that your two friends believe that their two friends believe that their two friends will arrive at 8.30. Also, you believe that your two friends believe that their two friends believe that their two friends believe that their two friends will arrive at 8.30, and so on. In summary, the belief hierarchy we construct from starting at your choice 8.10 indicates that

- you believe that Barbara and Chris will rationally arrive at 8.20,
- you believe that Barbara believes that her two friends will both irrationally arrive at 8.30,
- you believe that Chris believes that his two friends will both irrationally arrive at 8.30,
- you believe that your two friends believe that their two friends believe that their two friends will irrationally arrive at 8.30,

and so on, ad infinitum. Clearly, this belief hierarchy does not express common belief in rationality, as the players' rationality is questioned in the second-order beliefs.

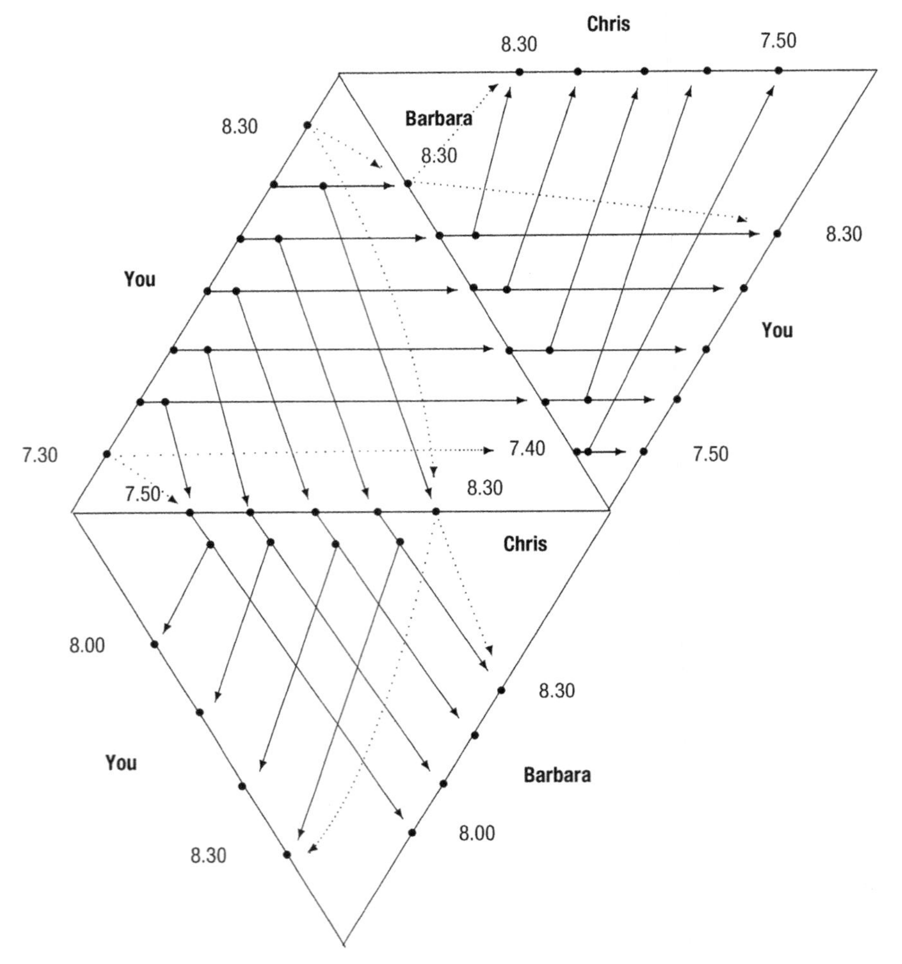

Figure 3.7 An extended beliefs diagram for "Waiting for a friend"

However, this belief hierarchy supports your choice to arrive at 8.10: Your first-order belief indicates that you believe that Barbara and Chris will both arrive at 8.20, and hence it is optimal to arrive at 8.10.

This method of adding dotted arrows to irrational choices works in fact for any game: Take an arbitrary game with an arbitrary number of players and a beliefs diagram for this game. We can always make an extended beliefs diagram by adding dotted arrows to all irrational choices. This would then guarantee that for every rational choice we can build a complete belief hierarchy that supports it. Start at a rational choice, and follow the (possibly dotted) arrows in the extended beliefs diagram. This chain of arrows will never stop, since at any point in the chain you are either at a rational choice, in which case you can continue by a solid arrow, or you are at an irrational choice, in which case you can continue by a dotted arrow. This infinite chain of arrows can be translated, as we have seen in a number of examples, into a complete belief hierarchy that supports the choice we started at.

3.3 Epistemic model

The idea of a belief hierarchy is certainly a central concept in this book. It has two disadvantages, however: a practical one and a theoretical one. The practical drawback is that writing down a complete belief hierarchy is often an impossible task, as we need to write down an infinite number of beliefs – first-order beliefs, second-order beliefs, third-order beliefs, and so on. The theoretical drawback is that providing a general mathematical definition of a belief hierarchy is far from easy. It would therefore be helpful if we could find a more compact way to represent a belief hierarchy, with an easier formal definition.

Such a compact representation can indeed be found. As an illustration, consider the example "Going to a party" with the utilities as stated in Table 3.3. An extended beliefs diagram for this situation can be found in Figure 3.8. The main idea will be to identify every choice in this diagram with the belief hierarchy we would obtain if we would start at the choice and follow the arrows. Take, for instance, your choice *red*. By following the arrows in the diagram we would obtain a belief hierarchy for you that supports your choice *red*. Let us call this belief hierarchy t_1^{red}. The subindex "1" indicates that this belief hierarchy belongs to player 1 – which is you. The superindex *red* means that this belief hierarchy is used to support your choice *red*. We use the letter t since in the literature a belief hierarchy for player i is often called an *epistemic type* – or simply *type* – for player i. So, instead of saying that player 1 holds belief hierarchy t_1^{red}

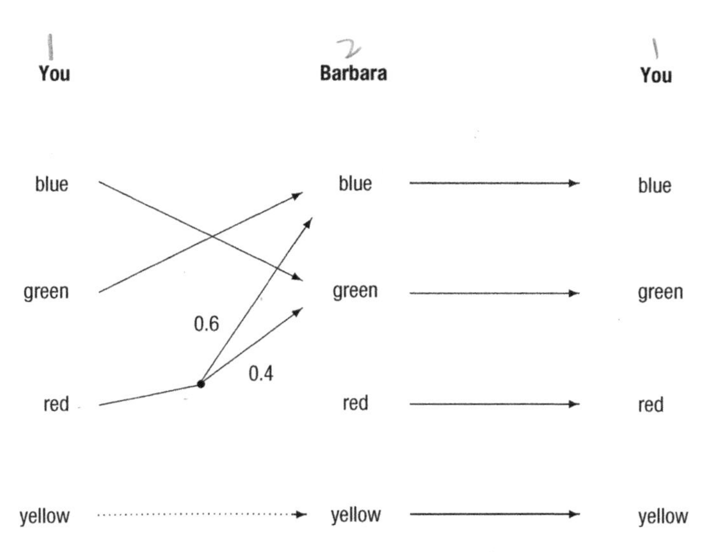

Figure 3.8 An extended beliefs diagram for "Going to a party" with utilities from Table 3.3

we could also say that *player 1 is of type* t_1^{red}. Similarly, we can define the type t_1^{blue} as the belief hierarchy we would obtain if we start at your choice *blue* and follow the arrows. In the same way we can define the types t_1^{green} and t_1^{yellow}. Note, however, that the belief hierarchy t_1^{yellow} cannot be used to *support* your choice *yellow*, since *yellow* is an irrational choice and therefore cannot be supported by any belief hierarchy. The type t_1^{yellow} simply corresponds to the belief hierarchy we obtain by starting at your choice *yellow*.

For player 2 – which is Barbara – we can also define types by this procedure. So, t_2^{blue} would be the belief hierarchy for Barbara which would result if we start at Barbara's choice *blue* and follow the arrows in the diagram. Similarly, we can define the types t_2^{green}, t_2^{red} and t_2^{yellow}.

Consider now the type t_1^{red} for you. For this type, your belief hierarchy is obtained by starting at your choice *red* and following the arrows. So, at type t_1^{red} you believe that, *with probability 0.6*,

- Barbara will choose *blue*,
- Barbara believes that you will choose *blue*,
- Barbara believes that you believe that Barbara will choose *green*,
- Barbara believes that you believe that Barbara believes that you will choose *green*,
- Barbara believes that you believe that Barbara believes that you believe that Barbara will choose *blue*,

and so on, and you believe that, *with probability 0.4*,

- Barbara will choose *green*,
- Barbara believes that you will choose *green*,
- Barbara believes that you believe that Barbara will choose *blue*,
- Barbara believes that you believe that Barbara believes that you will choose *blue*,
- Barbara believes that you believe that Barbara believes that you believe that Barbara will choose *green*,

and so on.

That is, type t_1^{red} believes that, with probability 0.6, Barbara will choose *blue* with the belief hierarchy that starts at her choice *blue*, and type t_1^{red} believes that, with probability 0.4, Barbara will choose *green* with the belief hierarchy that starts at her choice *green*. Remember that we have called the latter two belief hierarchies t_2^{blue} and t_2^{green}. Hence, if you are of type t_1^{red}, then you believe that, with probability 0.6, Barbara will choose *blue* and is of type t_2^{blue}, and you believe that, with probability 0.4, Barbara will choose *green* and is of type t_2^{green}.

In the same way, we can conclude that

- your type t_1^{blue} believes that, with probability 1, Barbara will choose *green* and is of type t_2^{green},

- your type t_1^{green} believes that, with probability 1, Barbara will choose *blue* and is of type t_2^{blue},
- your type t_1^{yellow} believes that, with probability 1, Barbara will choose *yellow* and is of type t_2^{yellow},
- Barbara's type t_2^{blue} believes that, with probability 1, you will choose *blue* and are of type t_1^{blue},
- Barbara's type t_2^{green} believes that, with probability 1, you will choose *green* and are of type t_1^{green},
- Barbara's type t_2^{red} believes that, with probability 1, you will choose *red* and are of type t_1^{red}, and
- Barbara's type t_2^{yellow} believes that, with probability 1, you will choose *yellow* and are of type t_1^{yellow}.

Let us now try to formulate these findings in a more formal way. By $T_1 = \{t_1^{blue}, t_1^{green}, t_1^{red}, t_1^{yellow}\}$ we denote the set of types for player 1 – you, and by $T_2 = \{t_2^{blue}, t_2^{green}, t_2^{red}, t_2^{yellow}\}$ the set of types for player 2 – Barbara. Recall that your type t_1^{red} believes that, with probability 0.6, Barbara will choose *blue* and is of type t_2^{blue}, and believes that, with probability 0.4, Barbara will choose *green* and is of type t_2^{green}. Mathematically, this sentence can be translated as

$$b_1(t_1^{red}) = (0.6) \cdot (blue,\ t_2^{blue}) + (0.4) \cdot (green,\ t_2^{green}).$$

Here, $b_1(t_1^{red})$ denotes the belief that type t_1^{red} has about the opponent's choice and the opponent's type. If we do the same for each of the other types, we obtain Table 3.4. We call this an *epistemic model* for the game.

From this epistemic model alone, we can construct for each type the complete belief hierarchy that belongs to it. Consider, for instance, your type t_1^{red}. The epistemic model tells us that t_1^{red} believes that, with probability 0.6, Barbara will choose *blue* and is of type t_2^{blue}, and believes that, with probability 0.4, Barbara will choose *green* and is of type t_2^{green}. So, in particular, t_1^{red} believes that, with probability 0.6, Barbara will choose *blue* and with probability 0.4 she will choose *green*. This is the first-order belief for t_1^{red}.

Since, according to the epistemic model, t_2^{blue} believes that you will choose *blue* and t_2^{green} believes that you will choose *green*, your type t_1^{red} also believes that, with probability 0.6, Barbara believes that you will choose *blue*, and believes that, with probability 0.4, Barbara believes that you will choose *green*. This is the second-order belief for t_1^{red}.

According to the epistemic model, t_2^{blue} believes that you are of type t_1^{blue}, and type t_1^{blue} believes that Barbara chooses *green*. So, from the epistemic model we can deduce that Barbara's type t_2^{blue} believes that you believe that Barbara chooses *green*.

Also, t_2^{green} believes that you are of type t_1^{green}, and type t_1^{green} believes that Barbara chooses *blue*. Hence, the epistemic model tells us that Barbara's type t_2^{green} believes that you believe that Barbara will choose *blue*.

Table 3.4. *An epistemic model for "Going to a party" with utilities from Table 3.3*

Types	$T_1 = \{t_1^{blue}, t_1^{green}, t_1^{red}, t_1^{yellow}\}$ $T_2 = \{t_2^{blue}, t_2^{green}, t_2^{red}, t_2^{yellow}\}$
Beliefs for player 1	$b_1(t_1^{blue}) = (green,\ t_2^{green})$ $b_1(t_1^{green}) = (blue,\ t_2^{blue})$ $b_1(t_1^{red}) = (0.6)\cdot(blue,\ t_2^{blue}) +$ $+ (0.4)\cdot(green,\ t_2^{green})$ $b_1(t_1^{yellow}) = (yellow,\ t_2^{yellow})$
Beliefs for player 2	$b_2(t_2^{blue}) = (blue,\ t_1^{blue})$ $b_2(t_2^{green}) = (green,\ t_1^{green})$ $b_2(t_2^{red}) = (red,\ t_1^{red})$ $b_2(t_2^{yellow}) = (yellow,\ t_1^{yellow})$

Recall that your type t_1^{red} believes that, with probability 0.6, Barbara is of type t_2^{blue}, and believes that, with probability 0.4, Barbara is of type t_2^{green}. Together with our insights above, we may thus conclude that your type t_1^{red} believes that, with probability 0.6, Barbara believes that you believe that Barbara chooses *green*, and your type t_1^{red} believes that, with probability 0.4, Barbara believes that you believe that Barbara chooses *blue*. This is the third-order belief for t_1^{red}.

By continuing in this fashion, we can deduce all higher-order beliefs for type t_1^{red} from the epistemic model. That is, by looking exclusively at the epistemic model, we can derive the complete belief hierarchy for t_1^{red}, and for all other types as well.

So, from the relatively small epistemic model in Table 3.4 we can construct rather complicated belief hierarchies as the one above by starting at a given type and following the beliefs in the epistemic model. The big practical advantage of this epistemic model is that writing it down requires considerably less effort and time than writing down complete belief hierarchies.

Of course, we can also construct epistemic models for games with more than two players. Take, for instance, the example "Waiting for a friend." In Figure 3.7 we showed an extended beliefs diagram. Our task is now to translate this extended beliefs diagram into an epistemic model. Similarly to above, we start by attaching to each of the choices *c* in the beliefs diagram some type. This type will represent the belief hierarchy we would obtain by starting at choice *c* and following the arrows in the diagram. Say you are player 1, Barbara is player 2 and Chris is player 3. Then, we attach to your choice 7.30 the type $t_1^{7.30}$, which is the belief hierarchy of player 1 – you – that results from starting at your choice 7.30 and following the dotted and solid arrows. In a similar way, we can define your other types $t_1^{7.40}, t_1^{7.50}, ..., t_1^{8.30}$. For Barbara we define, in the same fashion,

Table 3.5. *An epistemic model for "Waiting for a friend"*

Types	$T_1 = \{t_1^{7.30}, ..., t_1^{8.30}\}$ $T_2 = \{t_2^{7.40}, ..., t_2^{8.30}\}$ $T_3 = \{t_3^{7.50}, ..., t_3^{8.30}\}$
Beliefs for player 1	$b_1(t_1^{7.30}) = ((7.40, t_2^{7.40}), (7.50, t_3^{7.50}))$ $b_1(t_1^{7.40}) = ((7.50, t_2^{7.50}), (7.50, t_3^{7.50}))$ $\vdots$ $b_1(t_1^{8.20}) = ((8.30, t_2^{8.30}), (8.30, t_3^{8.30}))$ $b_1(t_1^{8.30}) = ((8.30, t_2^{8.30}), (8.30, t_3^{8.30}))$
Beliefs for player 2	$b_2(t_2^{7.40}) = ((7.50, t_1^{7.50}), (7.50, t_3^{7.50}))$ $\vdots$ $b_2(t_2^{8.20}) = ((8.30, t_1^{8.30}), (8.30, t_3^{8.30}))$ $b_2(t_2^{8.30}) = ((8.30, t_1^{8.30}), (8.30, t_3^{8.30}))$
Beliefs for player 3	$b_3(t_3^{7.50}) = ((8.00, t_1^{8.00}), (8.00, t_2^{8.00}))$ $\vdots$ $b_3(t_3^{8.20}) = ((8.30, t_1^{8.30}), (8.30, t_2^{8.30}))$ $b_3(t_3^{8.30}) = ((8.30, t_1^{8.30}), (8.30, t_2^{8.30}))$

the types $t_2^{7.40}, t_2^{7.50}, ..., t_2^{8.30}$ and for Chris we define the types $t_3^{7.50}, t_3^{8.00}, ..., t_3^{8.30}$. Let us begin with your type $t_1^{7.30}$. According to the extended beliefs diagram, type $t_1^{7.30}$ believes that Barbara chooses 7.40 and is of type $t_2^{7.40}$, and believes that Chris chooses 7.50 while being of type $t_3^{7.50}$. Or, written more formally,

$$b_1(t_1^{7.30}) = ((7.40, t_2^{7.40}), (7.50, t_3^{7.50})).$$

Here, $b_1(t_1^{7.30})$ denotes the belief that type $t_1^{7.30}$ has about the opponents' choices *and* the opponents' types. The expression above states that type $t_1^{7.30}$ assigns probability 1 to the opponents' choice-type combination in which Barbara will choose 7.40 and has type $t_2^{7.40}$ and in which Chris will choose 7.50 and has type $t_3^{7.50}$. In the same way, we can find the beliefs that the other types have about the opponents' choices and the opponents' types. This eventually leads to the epistemic model in Table 3.5.

From the two examples above it should be clear what an epistemic model looks like in general. Take an arbitrary game with n players. An epistemic model should first specify what the possible types are for every player. For every player i we denote by T_i the set of types that are considered for this player. A natural way to find T_i is to first draw an extended beliefs diagram and then identify every choice c_i for player i with a type $t_i^{c_i}$. The set T_i will then be the set of types $t_i^{c_i}$ obtained in this way. Secondly, every type t_i for player i holds a belief about the opponents' choices *and* types. How can we express such a belief mathematically? In Chapter 2 we have seen that a belief for player i about the opponents' choices can be expressed as a probability distribution b_i over the set $C_1 \times ... \times C_{i-1} \times C_{i+1} \times ... \times C_n$ of opponents' choice-combinations. Now, a type t_i should hold not only a belief about the opponents' choices, but also about the opponents' types. Hence, a type t_i should hold a belief about the opponents' choice-type combinations. The set of possible choice-type pairs for opponent j is given by $C_j \times T_j$, which denotes the set of possible pairs (c_j, t_j) we can make, with c_j being a choice in C_j and t_j being a type in T_j. The set of possible choice-type combinations for the opponents is therefore equal to

$$(C_1 \times T_1) \times ... \times (C_{i-1} \times T_{i-1}) \times (C_{i+1} \times T_{i+1}) \times ... \times (C_n \times T_n).$$

This set contains all possible combinations

$$((c_1, t_1), ..., (c_{i-1}, t_{i-1}), (c_{i+1}, t_{i+1}), ..., (c_n, t_n))$$

we can make, where (c_1, t_1) is a choice-type pair for opponent 1, ..., (c_{i-1}, t_{i-1}) is a choice-type pair for opponent $i-1$, and so on. The belief that type t_i for player i holds about the opponents' choices and types can therefore be formalized as a probability distribution $b_i(t_i)$ over the set

$$(C_1 \times T_1) \times ... \times (C_{i-1} \times T_{i-1}) \times (C_{i+1} \times T_{i+1}) \times ... \times (C_n \times T_n)$$

of opponents' choice-type combinations. We are now ready to provide a general definition of an epistemic model.

Definition 3.3.1 *(Epistemic model)*
*An **epistemic model** specifies for every player i a set T_i of possible types. Moreover, for every player i and every type $t_i \in T_i$ it specifies a probability distribution $b_i(t_i)$ over the set*

$$(C_1 \times T_1) \times ... \times (C_{i-1} \times T_{i-1}) \times (C_{i+1} \times T_{i+1}) \times ... \times (C_n \times T_n)$$

of opponents' choice-type combinations. The probability distribution $b_i(t_i)$ represents the belief that type t_i has about the opponents' choices and types.

As we have seen in the examples above, the information in an epistemic model allows us to construct for every type t_i the complete belief hierarchy that corresponds to it. The idea is that we start at t_i and then follow the beliefs in the epistemic model, just like we would follow the arrows in a beliefs diagram.

In general, it requires much less effort to write down an epistemic model than to write down complete belief hierarchies for a player in a game. This is an important practical advantage of the epistemic model. Moreover, as we have seen in the definition above, we can quite easily give a formal definition of an epistemic model, whereas it is much more difficult to provide a precise mathematical definition of a belief hierarchy. This is the main theoretical advantage of an epistemic model. In fact, in the remainder of this book we will always use the epistemic model when providing formal definitions of the concepts we will discuss. One should keep in mind, however, that types in an epistemic model are in fact *belief hierarchies*: Type t_i is just a short name for the complete belief hierarchy it represents.

To complete this section on epistemic models, we formally define what it means for a choice c_i to be optimal for type t_i.

Definition 3.3.2 *(Optimal choice for a type)*
*Consider a type t_i for player i within an epistemic model. Choice c_i is **optimal for type** t_i if it is optimal for the first-order belief that t_i holds about the opponents' choice-combinations.*

That is, to verify whether a certain choice c_i is optimal for type t_i, we only need to concentrate on the first-order belief that t_i holds – not on the higher-order beliefs.

3.4 Common belief in rationality

In this section we will formally define the idea of *common belief in rationality*. At the beginning of this chapter, when discussing the examples, we explained intuitively what we mean by common belief in rationality: It means that you not only believe that your opponents choose rationally, but you also believe that your opponents believe that their opponents will choose rationally, and that your opponents believe that their opponents believe that the other players will choose rationally, and so on. How can we express this idea in terms of an epistemic model?

Let us go back to the epistemic model in Table 3.4 for "Going to a party," where Barbara has the utilities as specified in Table 3.3. Let us concentrate on your type t_1^{red}. The epistemic model states that this type t_1^{red} believes that, with probability 0.6, Barbara will choose *blue* and is of type t_2^{blue}, and believes that, with probability 0.4, Barbara will choose *green* and is of type t_2^{green}. Moreover, following the beliefs in the epistemic model, we see that Barbara's type t_2^{blue} believes that you will choose *blue*. So it is optimal for Barbara's type t_2^{blue} to choose *blue*. Barbara's type t_2^{green} believes that you will choose *green*, and hence it is optimal for Barbara's type t_2^{green} to choose *green*. Altogether, we see that type t_1^{red} assigns positive probability to Barbara's choice-type pairs (*blue*, t_2^{blue}) and (*green*, t_2^{green}), where Barbara's choice *blue* is optimal for t_2^{blue} and Barbara's choice *green* is optimal for t_2^{green}. Hence, t_1^{red} only assigns positive probability to choice-type combinations for Barbara where the choice is optimal for the type. As such, we may conclude that your type t_1^{red} believes that Barbara chooses rationally.

Does your type t_1^{red} also believe that Barbara believes that you will choose rationally? Remember that t_1^{red} assigns probability 0.6 to Barbara's type t_2^{blue} and probability 0.4 to Barbara's type t_2^{green}. In turn, Barbara's type t_2^{blue} believes that you will choose *blue* while being of type t_1^{blue}. Since your type t_1^{blue} believes that Barbara will choose *green*, choosing *blue* is optimal for your type t_1^{blue}. So, Barbara's type t_2^{blue} believes that you choose rationally. In the same way, you can check also that Barbara's type t_2^{green} believes that you choose rationally. Hence, your type t_1^{red} only assigns positive probability to Barbara's types t_2^{blue} and t_2^{green}, and both types believe that you choose rationally. So, your type t_1^{red} indeed believes that Barbara believes that you will choose rationally.

Does your type t_1^{red} also believe that Barbara believes that you believe that Barbara chooses rationally? Remember that your type t_1^{red} assigns positive probability to Barbara's types t_2^{blue} and t_2^{green}. In turn, these types assign probability 1 to your types t_1^{blue} and t_1^{green}. Your type t_1^{blue} believes that Barbara will choose *green* while being of type t_2^{green}. Since choosing *green* is optimal for t_2^{green}, it follows that your type t_1^{blue} believes that Barbara will choose rationally. Your type t_1^{green} believes that Barbara chooses *blue* while being of type t_2^{blue}. As choosing *blue* is optimal for type t_2^{blue}, we may conclude that your type t_1^{green} also believes that Barbara chooses rationally. Summarizing, we see that your type t_1^{red} only assigns positive probability to Barbara's types t_2^{blue} and t_2^{green}, who believe that you are of type t_1^{blue} and t_1^{green}, respectively, who both believe that Barbara chooses rationally. So, your type t_1^{red} indeed believes that Barbara believes that you believe that Barbara chooses rationally.

By continuing this procedure we will eventually come to the conclusion that your type t_1^{red} expresses *common belief in rationality*. In a similar way, it can be verified that your types t_1^{blue} and t_1^{green} also express common belief in rationality within this epistemic model.

Suppose now we use the alternative extended beliefs diagram in Figure 3.9, and the associated epistemic model in Table 3.6. The only thing that has changed is that your type t_1^{blue} now believes that Barbara will choose *yellow* while being of type t_2^{yellow}. Previously, your type t_1^{blue} believed that Barbara would choose *green* while being of type t_2^{green}. Which of your types in this new epistemic model express common belief in rationality?

Let us start with your type t_1^{blue}. This type believes that Barbara is of type t_2^{yellow}, who believes that you irrationally will choose *yellow*. So, your type t_1^{blue} believes that

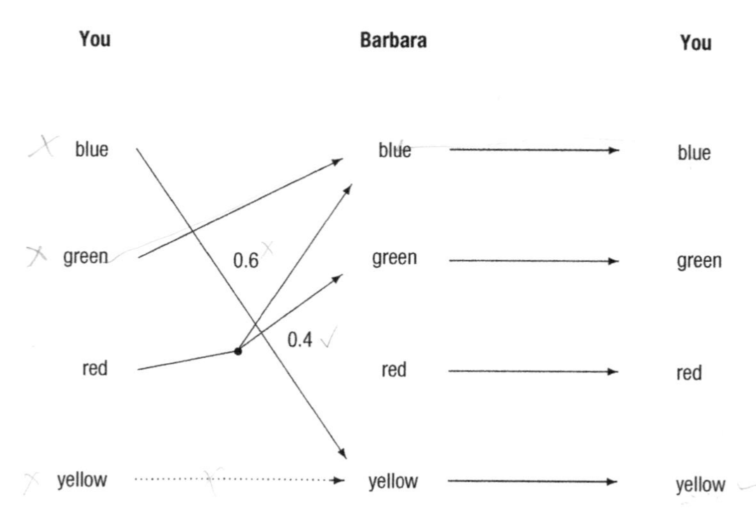

Figure 3.9 An alternative extended beliefs diagram for "Going to a party"

Table 3.6. *An alternative epistemic model for "Going to a party"*

Types	$T_1 = \{t_1^{blue}, t_1^{green}, t_1^{red}, t_1^{yellow}\}$ $T_2 = \{t_2^{blue}, t_2^{green}, t_2^{red}, t_2^{yellow}\}$
Beliefs for player 1	$b_1(t_1^{blue}) = (yellow,\ t_2^{yellow})$ $b_1(t_1^{green}) = (blue,\ t_2^{blue})$ $b_1(t_1^{red}) = (0.6)\cdot(blue,\ t_2^{blue}) + (0.4)\cdot(green,\ t_2^{green})$ $b_1(t_1^{yellow}) = (yellow,\ t_2^{yellow})$
Beliefs for player 2	$b_2(t_2^{blue}) = (blue,\ t_1^{blue})$ $b_2(t_2^{green}) = (green,\ t_1^{green})$ $b_2(t_2^{red}) = (red,\ t_1^{red})$ $b_2(t_2^{yellow}) = (yellow,\ t_1^{yellow})$

Barbara believes that you choose irrationally, and therefore does not express common belief in rationality.

What about your type t_1^{green}? This type believes that Barbara is of type t_2^{blue}, who believes that you are of type t_1^{blue} who, as we have seen, believes that Barbara believes that you choose irrationally. So, your type t_1^{green} believes that Barbara believes that you believe that Barbara believes that you choose irrationally, and hence does not express common belief in rationality.

And your type t_1^{red}? This type assigns positive probability to Barbara's type t_2^{blue}, who believes that you are of type t_1^{blue} who, as we have seen, believes that Barbara believes that you choose irrationally. So, your type t_1^{red} deems it possible that Barbara believes that you believe that Barbara believes that you choose irrationally, and hence does not express common belief in rationality.

Finally, your type t_1^{yellow} has the same belief hierarchy as your type t_1^{blue}. Namely, if you start at your choice *yellow* then you go to Barbara's choice *yellow*, so you follow the same chain of arrows as if you had started from your choice *blue*. Since we have seen that t_1^{blue} does not express common belief in rationality, we may immediately conclude that t_1^{yellow} does not express common belief in rationality either.

In the discussion above we have seen how we can formalize that your type t_1 believes in the opponent's rationality. It means that your type t_1 only assigns positive probability to choice-type pairs (c_2, t_2) for the opponent where the choice c_2 is optimal for the type t_2. If we extend this to games with more than two players, we obtain the following definition.

Definition 3.4.1 *(Type believes in the opponents' rationality)*
Consider an epistemic model, and a type t_i for player i within it. Type t_i ***believes***

in the opponents' rationality *if t_i only assigns positive probability to choice-type combinations*

$$((c_1,t_1),...,(c_{i-1},t_{i-1}),(c_{i+1},t_{i+1}),...,(c_n,t_n))$$

where choice c_1 is optimal for type t_1,..., choice c_n is optimal for type t_n.

With this definition at our disposal, it is now easy to define what it means for a type t_i to believe that his opponents believe in their opponents' rationality. It means that type t_i believes that each of his opponents is of a type who believes in his opponents' rationality. We say that type t_i expresses *2-fold belief in rationality*. In this fashion, we can also define what it means that type t_i believes that his opponents believe that their opponents believe in their opponents' rationality. This means, namely, that type t_i believes that each of his opponents is of a type who expresses 2-fold belief in rationality. In this case, we say that type t_i expresses 3-fold belief in rationality. And so on. This leads to the following definitions.

Definition 3.4.2 *(k-fold belief in rationality)*
Consider an epistemic model.
(1) Type t_i expresses ***1-fold belief in rationality*** *if t_i believes in the opponents' rationality.*
(2) Type t_i expresses ***2-fold belief in rationality*** *if t_i only assigns positive probability to the opponents' types that express 1-fold belief in rationality.*
(3) Type t_i expresses ***3-fold belief in rationality*** *if t_i only assigns positive probability to the opponents' types that express 2-fold belief in rationality.*
And so on.

In this way, we can recursively define k-fold belief in rationality for every number k. We say that a type t_i expresses *up to k-fold belief in rationality* if t_i expresses 1-fold, 2-fold, ..., and k-fold belief in rationality. We are now ready to provide a formal definition of common belief in rationality.

Definition 3.4.3 *(Common belief in rationality)*
Consider an epistemic model. Type t_i expresses ***common belief in rationality*** *if it expresses k-fold belief in rationality for every k.*

So, common belief in rationality means that the type believes in the opponents' rationality, believes that the opponents believe in their opponents' rationality, and so on, ad infinitum. Finally, what does it formally mean to say that player i can make a certain choice c_i under common belief in rationality? In terms of epistemic models, it means that somehow we should be able to construct *some* epistemic model in which there is *some* type for player i that expresses common belief in rationality, and that is willing to choose c_i.

Definition 3.4.4 *(Rational choice under common belief in rationality)*
A choice c_i for player i can ***rationally be made under common belief in rationality*** *if there is some epistemic model, and some type t_i within it, such that:*

(1) type t_i expresses common belief in rationality, and
(2) choice c_i is optimal for this type t_i.

In the same fashion, we can also define choices that can rationally be made when expressing up to k-fold belief in rationality. Say, namely, that a choice c_i can rationally be made when expressing up to k-fold belief in rationality if there is some epistemic model, and some type t_i within it, such that (1) type t_i expresses up to k-fold belief in rationality, and (2) choice c_i is optimal for this type t_i.

We should handle the definition above with some care, however. Suppose you construct an epistemic model with sets of types $T_1, ..., T_n$, and see that player i's choice c_i cannot be supported by any type in T_i that expresses common belief in rationality. This does not necessarily mean that c_i cannot rationally be chosen under common belief in rationality! Perhaps there is another epistemic model where the choice c_i *can* be supported by a type that expresses common belief in rationality. Consider, for instance, the example "Going to a party" where Barbara has the utilities as specified by Table 3.3. We have seen that within the epistemic model of Table 3.6, none of your types expresses common belief in rationality. Hence, within the epistemic model of Table 3.6, none of your choices can be supported by a type that expresses common belief in rationality. This does not mean, however, that none of your choices can rationally be made under common belief in rationality! Namely, within the other epistemic model of Table 3.4, your types t_1^{blue}, t_1^{green} and t_1^{red} all express common belief in rationality. Since *blue* is optimal for t_1^{blue}, *green* is optimal for t_1^{green} and *red* is optimal for t_1^{red}, we may conclude that, within the epistemic model of Table 3.4, your choices *blue*, *green* and *red* can all be supported by types that express common belief in rationality. In particular, your choices *blue*, *green* and *red* can all rationally be made under common belief in rationality.

3.5 Graphical method

The purpose of this section is to present an easy and intuitive way to verify whether a given choice can rationally be made under common belief in rationality. In fact, we have already implicitly introduced this method while discussing some examples earlier in this chapter. Namely, when trying to figure out which choices can be made under common belief in rationality, we often applied the following graphical method: Draw a beliefs diagram, start at some choice c_i for player i, and follow the arrows in the diagram. If the resulting chain of arrows only reaches *rational* choices, then we argued that player i can rationally make choice c_i under common belief in rationality. In these examples we also explained, somewhat informally, why this method works for the particular example.

We will now explain why this graphical method works for any game. Take an arbitrary game and create an extended beliefs diagram for it. In Section 3.3 we saw how we can translate this extended beliefs diagram into an epistemic model with sets of types $T_1, ..., T_n$. The key idea is as follows: If we start at a choice c_j and follow the arrows in the diagram, we obtain an infinite chain of arrows, which can be translated into a belief hierarchy, say $t_j^{c_j}$. Consider now the outgoing arrow – which may be a forked arrow – that leaves c_j. This arrow represents a belief, say $b_j^{c_j}$, about the opponents' choice-combinations. For every opponents' choice-combination $(c_1, ..., c_{j-1}, c_{j+1}, ..., c_n)$, the number $b_j^{c_j}(c_1, ..., c_{j-1}, c_{j+1}, ..., c_n)$ indicates the probability that $b_j^{c_j}$ assigns to the event that his opponents choose according to $(c_1, ..., c_{j-1}, c_{j+1}, ..., c_n)$. However, since all of the opponent's choices c_k are followed by the belief hierarchy $t_k^{c_k}$, the number $b_j^{c_j}(c_1, ..., c_{j-1}, c_{j+1}, ..., c_n)$ also indicates the probability that the belief hierarchy $t_j^{c_j}$ assigns to the event that

- his opponents choose according to $(c_1, ..., c_{j-1}, c_{j+1}, ..., c_n)$,
- his opponents' belief hierarchies are $(t_1^{c_1}, ..., t_{j-1}^{c_{j-1}}, t_{j+1}^{c_{j+1}}, ...t_n^{c_n})$.

So, every belief hierarchy $t_j^{c_j}$ that can be deduced from the beliefs diagram contains a belief about the opponents' choices and the opponents' belief hierarchies. Now, if we call every belief hierarchy $t_j^{c_j}$ a *type*, then every type $t_j^{c_j}$ that can be deduced from the beliefs diagram holds a belief about the opponents' possible choice-type combinations.

Summarizing, we see that every extended beliefs diagram can be translated into an epistemic model with sets of types $T_1, ..., T_n$, where

$$T_j = \{t_j^{c_j} \mid c_j \in C_j\}$$

for every player j. So, for every choice c_j we can construct a type $t_j^{c_j}$ that holds a belief about the opponents' possible choice-type combinations. These beliefs correspond exactly to the arrows, together with their associated probabilities, in the extended beliefs diagram.

Consider now a choice c_j whose outgoing arrow leads only to rational choices. Then, the corresponding type $t_j^{c_j}$ believes in the opponents' rationality. Namely, the belief $b_j^{c_j}$ only assigns positive probability to the opponents' choices c_k that are rational. Since every such rational choice c_k is supported by the corresponding belief hierarchy $t_k^{c_k}$, it follows that the belief hierarchy $t_j^{c_j}$ only assigns positive probability to the opponents' combinations $(c_k, t_k^{c_k})$ where the choice c_k is optimal under the associated belief hierarchy $t_k^{c_k}$. Or, in terms of types, the type $t_j^{c_j}$ only assigns positive probability to the opponents' choice-type combinations $(c_k, t_k^{c_k})$ where c_k is optimal for the type $t_k^{c_k}$. Hence, $t_j^{c_j}$ believes in the opponents' rationality.

Suppose now that we start at a rational choice c_i, follow the arrows in the diagram, and discover that we only reach rational choices by doing so. We will show that, in this case, c_i can be chosen rationally under common belief in rationality.

First, the arrow leaving c_i leads only to rational choices, and therefore, by the argument above, the type $t_i^{c_i}$ believes in the opponents' rationality.

Second, the arrow leaving c_i only leads to choices where the outgoing arrow leads only to rational choices. This means, however, that type $t_i^{c_i}$ only assigns positive probability to the opponents' choices c_j where the associated type $t_j^{c_j}$ believes in his opponents' rationality. That is, type $t_i^{c_i}$ only assigns positive probability to the opponents' types that believe in the opponents' rationality. So, type $t_i^{c_i}$ believes that his opponents believe in their opponents' rationality.

By continuing in this fashion, we come to the conclusion that type $t_i^{c_i}$ expresses common belief in rationality. Since c_i is rational, it is optimal for the associated type $t_i^{c_i}$. Therefore, the choice c_i we started with can be chosen rationally under common belief in rationality.

We thus obtain the following general result: If we have an extended beliefs diagram, start at a rational choice c_i, and observe that the induced chain of arrows only reaches rational choices, then we can conclude that the choice c_i can rationally be made under common belief in rationality.

What about the other direction? So, suppose that the choice c_i can be made rationally under common belief in rationality. Can we then always find an extended beliefs diagram in which the chain of arrows starting at c_i only reaches rational choices? The answer, as we will see, is "yes." Namely, take a choice c_i that can be made rationally under common belief in rationality. Then, we can find an epistemic model with sets of types $T_1, ..., T_n$, and a type t_i in T_i, such that type t_i expresses common belief in rationality within this epistemic model, and supports the choice c_i.

The idea is now to translate the epistemic model into (a part of) an extended beliefs diagram. Every type t_j in the epistemic model will correspond to a choice $c_j^{t_j}$ in the beliefs diagram, where the choice $c_j^{t_j}$ is chosen such that $c_j^{t_j}$ is optimal for type t_j. The beliefs that types have about the opponents' choices and types can be translated into solid or dotted arrows in the diagram. Now, consider the choice c_i we started with, which, by construction, is optimal for the type t_i that expresses common belief in rationality. Within the beliefs diagram, let us start at the choice c_i and consider the chain of arrows that results. Since type t_i believes in the opponents' rationality, the arrow that leaves c_i leads only to rational choices. Since t_i believes that his opponents' believe in their opponents' rationality, the arrow that leaves c_i leads only to choices where the outgoing arrow leads only to rational choices, and so on. By continuing in this fashion, we eventually conclude that the chain of arrows starting at c_i only reaches rational choices.

We thus have shown that for every choice c_i that can be made rationally under common belief in rationality, we can always find some extended beliefs diagram in which the chain of arrows starting at c_i only reaches rational choices. We had already shown that the opposite is also true: If in a certain extended beliefs diagram the chain of arrows starting at the rational choice c_i only reaches rational choices, then we can conclude that choice c_i can be made rationally under common belief in rationality. By putting these two insights together we obtain the following graphical method, which we call the *chain of arrows method*, to verify whether a choice c_i can rationally be made under common belief in rationality.

Graphical Method 3.5.1 *(Chain of arrows method)*
Consider a choice c_i for player i. If there is an extended beliefs diagram in which the chain of arrows starting at c_i begins with a solid arrow and only reaches rational choices, then player i can rationally choose c_i under common belief in rationality.

If no such extended beliefs diagram exists, then player i cannot rationally choose c_i under common belief in rationality.

One must be careful not to draw wrong conclusions when using this method. If you construct an extended beliefs diagram and find that the chain of arrows starting at c_i reaches an irrational choice, then this does not necessarily mean that c_i cannot be chosen rationally under common belief in rationality! Possibly there is some other beliefs diagram in which the chain of arrows starting at c_i *does* reach only rational choices. As an illustration, let us go back to the example "Going to a party" with the utilities as depicted in Table 3.3. Consider the extended beliefs diagram from Figure 3.9. If we start at any of your choices and follow the arrows, then the resulting chain of arrows will always reach your irrational choice *yellow*. So, in this extended beliefs diagram there is no chain of arrows that only reaches rational choices. This does not mean, however, that none of your choices can be made rationally under common belief in rationality! Consider, for instance, the other extended beliefs diagram in Figure 3.8. If we start at any of your choices *blue*, *green* or *red*, and follow the arrows in the diagram, then the resulting chain of arrows will only reach rational choices. So, from this extended beliefs diagram we can conclude that you can rationally choose any of the colors *blue*, *green* or *red* under common belief in rationality.

A similar warning applies to the example "Waiting for a friend," which has the extended beliefs diagram shown in Figure 3.7. If we start at your choice to arrive at 7.40 and follow the arrows in this diagram, then we first go to Barbara's choice 7.50 and Chris' choice 7.50, from which we go to your choice 8.00, Barbara's choice 8.00 and Chris' choice 8.00, from which we go to your choice 8.10, Barbara's choice 8.10 and Chris' choice 8.10, from which we go to your choice 8.20, Barbara's choice 8.20 and Chris' choice 8.20, from which we eventually reach your irrational choice 8.30, Barbara's irrational choice 8.30 and Chris' irrational choice 8.30. So, the chain of arrows starting at your choice 7.40 eventually reaches irrational choices. In the first section of this chapter we saw, however, that you can rationally choose to arrive at 7.40 under common belief in rationality. So, according to our graphical method, there must be an alternative extended beliefs diagram in which the chain of arrows starting at 7.40 only reaches rational choices. As an exercise, try to construct such an alternative extended beliefs diagram.

3.6 Existence

In this chapter we have been discussing the idea of common belief in rationality. But there is one important question that has not been answered yet, namely whether common belief in rationality is always possible. That is, for any given game in which you participate, does there always exist a belief hierarchy for you that expresses common belief in rationality? Or are there situations where common belief in rationality would simply be too much to ask?

In this section we will show that, if every player has finitely many possible choices, then we can *always* construct an epistemic model in which *all* types express common belief in rationality. In particular, in every such game there always exists at least one type for you that expresses common belief in rationality. The idea is best illustrated by means of an example.

Let us go back to the example "Going to a party" with the utilities as depicted in Table 3.3.

- We start at an arbitrary choice by you, say *yellow*. Subsequently, we ask the question: What choice can Barbara rationally make if she believes that you will choose *yellow*? The answer is *yellow*.
- We then ask the question: What choice can you rationally make if you believe that Barbara will choose *yellow*? The answer is *blue*.
- We subsequently ask the question: What choice can Barbara rationally make if she believes that you will choose *blue*? The answer is *blue*.
- Then we ask: What choice can you rationally make if you believe that Barbara will choose *blue*? The answer is *green*.
- Next, we ask: What choice can Barbara rationally make if she believes that you will choose *green*? The answer is *green*.
- We then ask: What choice can you rationally make if you believe that Barbara will choose *green*. The answer is *blue*.
- Again we ask: What choice can Barbara rationally make if she believes that you will choose *blue*? The answer is again *blue*.

And so on. From this moment on, we enter into a cycle of questions and answers that repeats itself after every four steps.

This chain of questions and answers can be translated into the diagram of Figure 3.10. The first answer corresponds to the arrow from Barbara's choice *yellow* to your choice *yellow*. According to the first answer, it is optimal for Barbara to choose *yellow* if she believes that you will choose *yellow*. Similarly, the second answer corresponds to the arrow from your choice *blue* to Barbara's choice *yellow*. The third answer corresponds to the arrow from Barbara's choice *blue* to your choice *blue*. The fourth answer corresponds to the arrow from your choice *green* to Barbara's choice *blue*. The fifth answer corresponds to the arrow from Barbara's choice *green* to your choice *green*. The sixth answer corresponds to the arrow from your choice *blue* to Barbara's choice *green*. The seventh answer corresponds to the arrow from Barbara's choice *blue* to your choice *blue*, and so on. So, we construct the chain of arrows *backwards*.

Suppose we delete the arrow that goes from your choice *blue* to Barbara's choice *yellow*. Then, the remaining diagram is (part of) a beliefs diagram. Moreover, if we start at either of your choices *blue* or *green* and follow the arrows, then we enter into a cycle

Table 3.7. *Epistemic model for "Going to a party," deduced from the cycle of arrows in Figure 3.10*

Types	$T_1 = \{t_1^{blue}, t_1^{green}\}$, $T_2 = \{t_2^{blue}, t_2^{green}\}$
Beliefs for player 1	$b_1(t_1^{blue}) = (green,\ t_2^{green})$ $b_1(t_1^{green}) = (blue,\ t_2^{blue})$
Beliefs for player 2	$b_2(t_2^{blue}) = (blue,\ t_1^{blue})$ $b_2(t_2^{green}) = (green,\ t_1^{green})$

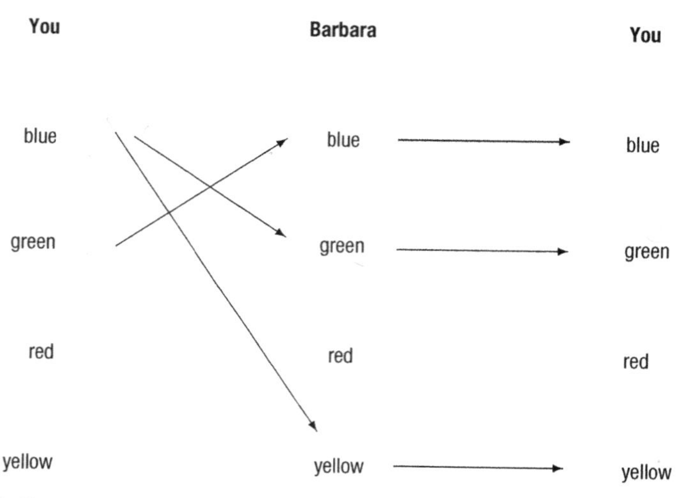

Figure 3.10 Common belief in rationality is always possible

of arrows that only reaches rational choices. This cycle of arrows can be translated into the epistemic model of Table 3.7. It is easy to see that all types in this epistemic model express common belief in rationality. The graphical argument is that, when starting at either of your choices *blue* or *green*, or when starting at either of Barbara's choices *blue* or *green*, by following the arrows in Figure 3.10 we enter into a cycle of arrows that only reaches rational choices. So, the four types that correspond to these chains of arrows all express common belief in rationality. In particular, this method shows that we *can* construct a type for you that expresses common belief in rationality.

Why exactly could we construct such an epistemic model in which all types express common belief in rationality? The reason is that the chain of questions and answers above starts cycling once we arrive at your choice *blue*. The arrows in the diagram are nothing other than translations of the answers to the questions above. So, the chain of arrows starting at your choice *blue* must eventually start to cycle. However, since every answer is, by construction, a rational choice, this means that the chain of arrows starting anywhere in this cycle only reaches rational choices. Now, if we translate the arrows in this cycle into an epistemic model, then all types will express common belief in rationality. The reason is that every type in the epistemic model corresponds to a belief hierarchy obtained by starting somewhere in this cycle. So, the belief hierarchy of every type is based on a chain of arrows that only reaches rational choices. Therefore, every type in the epistemic model expresses common belief in rationality.

This idea of questions and answers can in fact be applied to any game with two players: Start at an arbitrary choice for player 1 and then start a chain of questions and answers as described above. Since player 1 has only finitely many choices, there must be a place in this chain in which we reach the same choice for player 1 for a second time, after which the chain of questions and answers repeats itself. By translating this cycle of questions and answers into a chain of arrows in a beliefs diagram, we see that this chain of arrows enters into a cycle as well. If finally we translate this cycle of arrows into an epistemic model, then every type in this particular epistemic model will express common belief in rationality.

The idea can be extended to games with more than two players, although the construction will become more difficult. Take an arbitrary game with n players.

- Instead of starting at a choice, we now start at a *choice-combination* $(c_1^1, ..., c_n^1)$, where c_1^1 is a choice for player 1, ... , c_n^1 is a choice for player n.
- For every player i, let c_i^2 be a choice that player i can rationally make if he believes that his opponents choose according to $(c_1^1, ..., c_n^1)$. Then, the choice-combination $(c_1^1, ..., c_n^1)$ will lead to a new choice-combination $(c_1^2, ..., c_n^2)$.
- For every player i, let c_i^3 be a choice that player i can rationally make if he believes that his opponents choose according to $(c_1^2, ..., c_n^2)$. So, the choice-combination $(c_1^2, ..., c_n^2)$ will lead to a new choice-combination $(c_1^3, ..., c_n^3)$, and so on.

We thus obtain a chain of choice-combinations in this way. Since every player has only finitely many choices, there are also only finitely many choice-combinations. But then, the chain of choice-combinations constructed above must visit some choice-combination twice, after which it will repeat itself. That is, the chain of choice-combinations must run into a cycle.

In a similar way as with two players, the chain of choice-combinations constructed above can be translated into a (part of a) beliefs diagram with arrows. Since the chain of choice-combinations runs into a cycle, there is a choice c_i after which the chain of arrows runs into a cycle as well. But then, the cycle of arrows that starts at c_i can only reach rational choices. If we translate this cycle of arrows into an epistemic model, then every type within it will express common belief in rationality. So, by using this method we can always construct an epistemic model in which every type expresses common belief in rationality.

Moreover, the types we construct in this way have a very specific form – every type t_i so constructed always assigns probability 1 to one particular choice-type combination for the opponents. Namely, in the method above we only use "probability 1 beliefs" and "probability 1 arrows" – that is, beliefs and arrows that assign probability 1 to one specific choice-combination for the opponents. But then, every type we construct from the cycle of arrows always assigns, for every opponent, probability 1 to one specific choice and type for that opponent. In other words, every belief hierarchy we construct in this way always assigns probability 1 to one specific choice and one specific belief hierarchy for every opponent. Since such a construction is always possible in every static game, we arrive at the following general conclusion.

Theorem 3.6.1 *(Common belief in rationality is always possible)*
Consider a static game with finitely many choices for every player. Then, we can construct an epistemic model in which:
(1) every type expresses common belief in rationality, and
(2) every type assigns, for every opponent, probability 1 to one specific choice and one specific type for that opponent.

In particular, it is always possible to construct a belief hierarchy for every player that expresses common belief in rationality. That is, common belief in rationality never leads to logical contradictions in any static game with finitely many choices.

3.7 Algorithm

As with any method, the graphical method introduced in Section 3.5 has its strong and weak points. The strong point is that it not only verifies whether a given choice c_i can rationally be made under common belief in rationality, but, if it can be made, it also provides an epistemic model that explains *why* c_i can rationally be chosen under common belief in rationality. Suppose that we have constructed an extended beliefs diagram and see that the chain of arrows starting at c_i reaches rational choices only. Then, according to the chain of arrows method, we may conclude that c_i can rationally be chosen under common belief in rationality. The chain of arrows method does more than this, however. We can also translate the extended beliefs diagram into an epistemic model with sets of types $T_1, ..., T_n$, and use this epistemic model to justify, formally, that choice c_i can rationally be made under common belief in rationality.

The drawback of this method is that it does not provide a definite answer if, for a given extended beliefs diagram, the chain of arrows starting at choice c_i reaches an irrational choice. In this case, we are often not sure whether there is another beliefs diagram, not yet attempted, in which the chain of arrows starting at c_i *does* reach only rational choices, or whether there is simply *no* beliefs diagram in which the chain starting at c_i reaches only rational choices. For that reason, it would be very helpful if we could develop a method by which we can compute *all* choices that can rationally be made under common belief in rationality. This is exactly what we shall do in this section. As we will see, the algorithm we will develop here builds on the algorithm of two-fold elimination of strictly dominated choices from Chapter 2. In order to design the algorithm, we proceed recursively by the following steps.

Step 1: 1-fold belief in rationality
We start by identifying those choices you can rationally make under 1-fold belief in rationality. Remember that 1-fold belief in rationality simply means "belief in the opponents' rationality." Hence, the choices that can rationally be made under 1-fold belief in rationality are precisely those choices that can rationally be made if the player believes that his opponents choose rationally too. But we have seen in Chapter 2 that these are precisely the choices that survive the algorithm of two-fold elimination of strictly dominated choices.

Step 2: Up to 2-fold belief in rationality
We now wish to characterize those choices that can rationally be made by types that express 1-fold belief and 2-fold belief in rationality – that is, express up to 2-fold belief in rationality. Consider a type t_i for player i that expresses up to 2-fold belief in rationality. Then, type t_i only assigns positive probability to opponents' types t_j that express 1-fold belief in rationality. Moreover, as t_i expresses 1-fold belief in rationality himself, he only assigns positive probability to opponents' choice-type pairs (c_j, t_j) where c_j is optimal for t_j. Altogether, we see that type t_i only assigns positive probability to the opponents' choices c_j that are optimal for a type t_j that expresses 1-fold belief in rationality. But we know from step 1 that such choices c_j all survive the two-fold elimination of strictly dominated choices. Hence, type t_i only assigns positive probability to the opponents' choices that survive the procedure of two-fold elimination of strictly dominated choices.

Let Γ^2 be the reduced game that remains after applying two-fold elimination of strictly dominated choices. Then, we have learned from above that a type t_i that expresses up to 2-fold belief in rationality, will only assign positive probability to the opponents' choices in Γ^2. So, every choice c_i that is rational for type t_i is optimal for some belief about the opponents' choices in the reduced game Γ^2. But we know from Theorem 2.5.3 that every such choice c_i is not strictly dominated within the reduced game Γ^2. Hence, every choice c_i that can rationally be made by expressing up to 2-fold belief in rationality, is not strictly dominated within the reduced game Γ^2.

Now, let Γ^3 be the reduced game that remains if we eliminate from Γ^2 all strictly dominated choices. Then, we conclude that every choice c_i that can rationally be made by expressing up to 2-fold belief in rationality, must be in Γ^3. By construction, the reduced game Γ^3 is obtained by taking the original game Γ, applying two-fold elimination of strictly dominated choices, and then applying another round of elimination of strictly dominated choices. We call this procedure the *three-fold elimination of strictly dominated choices*. Summarizing, we see that every choice that can rationally be made by expressing up to 2-fold belief in rationality, must survive the procedure of three-fold elimination of strictly dominated choices.

Step 3: Up to 3-fold belief in rationality

We now ask: What are the choices that can rationally be made under expressing up to 3-fold belief in rationality? That is, which choices can be optimal for a type that expresses 1-fold belief, 2-fold belief and 3-fold belief in rationality? Consider a type t_i that expresses up to 3-fold belief in rationality. Then, type t_i only assigns positive probability to the opponents' types t_j that express 1-fold belief and 2-fold belief in rationality. That is, t_i only assigns positive probability to the opponents' types that express up to 2-fold belief in rationality. Moreover, as t_i also expresses 1-fold belief in rationality, he only assigns positive probability to the opponents' choice-type pairs (c_j, t_j) where c_j is optimal for t_j. Altogether, we see that type t_i only assigns positive probability to the opponents' choices c_j that are optimal for a type t_j that expresses up to 2-fold belief in rationality. But we know from step 2 that such choices c_j all survive the three-fold elimination of strictly dominated choices. Hence, type t_i only assigns positive probability to the opponents' choices that survive the procedure of three-fold elimination of strictly dominated choices.

Recall that Γ^3 is the reduced game that remains after applying three-fold elimination of strictly dominated choices. Then, we have learned from above that a type t_i that expresses up to 3-fold belief in rationality, will only assign positive probability to the opponents' choices in Γ^3. So, every choice c_i that is rational for type t_i is optimal for some belief about the opponents' choices in the reduced game Γ^3. But we know from Theorem 2.5.3 that every such choice c_i is not strictly dominated within the reduced game Γ^3. Hence, every choice c_i that can rationally be made by expressing up to 3-fold belief in rationality is not strictly dominated within the reduced game Γ^3.

Now, let Γ^4 be the reduced game that remains if we eliminate from Γ^3 all strictly dominated choices. Then, we conclude that every choice c_i that can rationally be made by expressing up to 3-fold belief in rationality, must be in Γ^4. By construction, the reduced game Γ^4 is obtained by taking the original game Γ, applying three-fold elimination of strictly dominated choices, and then applying another round of elimination of strictly dominated choices. We call this procedure *four-fold elimination of strictly dominated choices*. Summarizing, we see that every choice that can rationally be made by expressing up to 3-fold belief in rationality, must survive the procedure of four-fold elimination of strictly dominated choices.

By continuing in this fashion we conclude that, for every k, every choice that can rationally be made by a type that expresses up to k-fold belief in rationality, must survive the procedure of $(k+1)$-fold elimination of strictly dominated choices. Remember that a type is said to express common belief in rationality if it expresses k-fold belief in rationality for every k. So, all choices that can rationally be made by types that express common belief in rationality, must survive k-fold elimination of strictly dominated choices *for all* k! That is, every choice that can rationally be made under common belief in rationality must survive the procedure in which we keep eliminating strictly dominated choices until we can go no further. This procedure is called *iterated elimination of strictly dominated choices*, and is described formally as follows.

Algorithm 3.7.1 *(Iterated elimination of strictly dominated choices)*
Step 1: *Within the original game, eliminate all choices that are strictly dominated.*
Step 2: *Within the reduced game obtained after step 1, eliminate all choices that are strictly dominated.*
Step 3: *Within the reduced game obtained after step 2, eliminate all choices that are strictly dominated.*
Step 4: *Within the reduced game obtained after step 3, eliminate all choices that are strictly dominated.*

⋮

And so on.

This algorithm in defined through infinitely many steps, which may cause the impression that the procedure could go on forever, without ever reaching a final result. This, however, is not true. We can show that if the players have finitely many choices, then the algorithm will always stop after finitely many steps! At step 1 of the algorithm there are two possibilities: Either there are no strictly dominated choices in the original game or there are. If there are no strictly dominated choices, then there are no choices we can eliminate, and hence the reduced game obtained after step 1 will be the same as the original game. But then, step 2 will simply be a repetition of step 1 in which nothing happened. So, in step 2 we cannot eliminate any choices, and hence the reduced game after step 2 will still be the same as the original game, and so on. Hence, if there are no strictly dominated choices in the original game, then nothing will ever be eliminated in the algorithm, so the procedure would stop immediately. More precisely, step 2 and all further steps would be irrelevant, as nothing will ever happen in these steps. If, on the other hand, there *are* strictly dominated choices in the original game, then we would eliminate these, and the reduced game obtained after step 1 would contain less choices than the original game. Only in this case would step 2 be relevant.

So, we see that step 2 will only be relevant if we have eliminated *at least one choice* in the original game. In this case, the reduced game we start with in step 2 will have fewer choices than the original game. Again, there are two possibilities in this case: Either the reduced game at the start of step 2 has no strictly dominated choices or it has. If there are no strictly dominated choices, then, by the same argument as above, nothing will ever happen in the remaining steps, and so the algorithm would effectively stop at step 2. Step 3 would not be relevant in this case. If there *are* strictly dominated choices in the reduced game at the start of step 2, then we will eliminate these, and the reduced game at the start of step 3 would contain fewer choices than the reduced game at the start of step 2.

By repeating this argument, we see that any step k in the algorithm will only be relevant if at each of the previous steps $1,2,...,k-1$ *at least one choice* had been eliminated from the game. Or, to put it a little differently, the algorithm will only continue as long as we eliminated at least one choice in the preceding step. If at a given step no choice is eliminated, then nothing will ever happen in the remaining steps, and the algorithm would effectively stop. Since each player has only finitely many choices, there are only finitely many choices *we can possibly* eliminate in the game. But this means that the algorithm must effectively stop after finitely many steps!

In our discussion above we have seen that:

- every choice that can rationally be made by expressing up to k-fold belief in rationality, must survive the procedure of $(k+1)$-fold elimination of strictly dominated choices, and
- every choice that can rationally be made under *common* belief in rationality, survives the procedure of *iterated* elimination of strictly dominated choices.

It can be shown that the converse of these two statements is also true. That is, every choice that survives $(k+1)$-fold elimination of strictly dominated choices can rationally be made by a type that expresses up to k-fold belief in rationality. Moreover, every choice that survives iterated elimination of strictly dominated choices can rationally be made by a type that expresses common belief in rationality. We thus obtain the following general result.

Theorem 3.7.2 *(The algorithm works)*
Consider a static game with finitely many choices for each player.
(1) For every $k \geq 1$, *the choices that can rationally be made by expressing up to k-fold belief in rationality are exactly those choices that survive* $(k+1)$*-fold elimination of strictly dominated choices.*
(2) The choices that can rationally be made under common belief in rationality are exactly those choices that survive iterated elimination of strictly dominated choices.

The formal proof for this result can be found in the proofs section at the end of this chapter. To see how the algorithm of iterated elimination of strictly dominated choices works in practice, let us apply it to a number of examples in order to compute all choices the players can rationally make under common belief in rationality.

Example 3.4 Going to a party

Consider the scenario with the original utilities shown in Table 3.2. In Chapter 2 we saw that your choice *yellow* is strictly dominated by the randomized choice in which you choose *blue* and *green* with probability 0.5. Similarly, Barbara's choice *green* is strictly dominated by the randomized choice in which she chooses *red* and *yellow* with probability 0.5. Hence, in step 1 of the algorithm we eliminate the choice *yellow* for you, and the choice *green* for Barbara.

In the reduced game, your choice *green* will give you utility 3, since you expect Barbara not to choose *green*. So, your choice *red* is strictly dominated by your choice *green* in the reduced game. Similarly, Barbara's choice *blue* is strictly dominated by her choice *yellow* in the reduced game, since, within the reduced game, Barbara does not expect you to choose *yellow*. So, in step 2 we eliminate your choice *red* and Barbara's choice *blue*.

Within the reduced game we obtain, your choice *green* will then be strictly dominated by your choice *blue*, as, within the reduced game, you do not expect Barbara to choose *blue*. So, only your choice *blue* survives the algorithm of iterated elimination of strictly dominated choices. As such, you can only rationally choose *blue* under common belief in rationality. □

Example 3.5 Waiting for a friend

We have already seen in Chapter 2 that for you the choices 7.30 and 8.30 are strictly dominated, whereas for Barbara and Chris the choice 8.30 is strictly dominated. So, in step 1 we eliminate the choices 7.30 and 8.30 for you, the choice 8.30 for Barbara and the choice 8.30 for Chris.

In the reduced game we obtain, every player expects his friends to arrive no later than 8.20. So, in the reduced game to arrive at 8.20 is a strictly dominated choice, since it will always be better to arrive at 8.10. In step 2 we therefore eliminate the choice 8.20 for all three players.

In the reduced game we obtain, each player expects his friends to arrive no later than 8.10, and hence arriving at 8.10 is strictly dominated. We therefore eliminate 8.10 for all three players in step 3.

In the new reduced game, every player expects his friends to arrive no later than 8.00, so arriving at 8.00 is strictly dominated for all three players. In step 4 we thus eliminate the choice 8.00 for all players.

In the reduced game we obtain after step 4, you can choose between 7.40 and 7.50, Barbara can choose between 7.40 and 7.50, and Chris can only choose 7.50. Since you expect Chris to arrive exactly at 7.50 and expect Barbara to arrive no later than 7.50, your choice 7.50 is strictly dominated by 7.40. Similarly, for Barbara the choice 7.50 is also strictly dominated by 7.40. Of course, Chris' unique choice 7.50 in the reduced game cannot be strictly dominated within this reduced game. So, in step 5 we eliminate the choice 7.50 for you and Barbara.

At the beginning of step 6 we have a reduced game in which every player has a unique choice: You can only choose 7.40, Barbara can only choose 7.40 and Chris can only choose 7.50. Since, obviously, these choices cannot be strictly dominated within this reduced game, no choice can be eliminated here, and the algorithm stops after 5 steps. The conclusion is that Barbara and you can only rationally choose to arrive at 7.40 under common belief in rationality, whereas Chris can only rationally choose to arrive at 7.50 under common belief in rationality. Hence, if all three friends reason in accordance with common belief in rationality, Barbara and Chris would arrive as early as possible, whereas you would arrive at 7.40, ten minutes after your earliest possible arrival time. □

Example 3.6 The traveler's dilemma

Recall the story from Example 2.4. The compensation that you and Barbara receive for each combination of prices that can be written down is given by Table 3.8. The first number indicates the compensation you receive, and the second number is the amount that goes to Barbara. As we saw in Chapter 2, your choice 400 is strictly dominated by the randomized choice in which you choose price 100 with probability 0.45 and price

Table 3.8. *Compensation that you and Barbara receive in "The traveler's dilemma"*

		Barbara			
		100	200	300	400
You	100	100, 100	250, −75	300, −25	350, 25
	200	−75, 250	200, 200	350, 25	400, 75
	300	−25, 300	25, 350	300, 300	450, 125
	400	25, 350	75, 400	125, 450	400, 400

Table 3.9. *Reduced game after step 1 in "The traveler's dilemma" (II)*

		Barbara		
		100	200	300
You	100	100, 100	250, −75	300, −25
	200	−75, 250	200, 200	350, 25
	300	−25, 300	25, 350	300, 300

Table 3.10. *Reduced game obtained after step 2 in "The traveler's dilemma"*

		Barbara	
		100	200
You	100	100, 100	250, −75
	200	−75, 250	200, 200

300 with probability 0.55. By the symmetry of the game, the same applies to Barbara. So, in step 1 of the algorithm we eliminate the choices 400 for you and Barbara to give the reduced game shown in Table 3.9.

In the reduced game your choice 300 is strictly dominated by the randomized choice in which you choose the prices 100 and 200 both with probability 0.5. The same holds for Barbara. So, in step 2 we eliminate the choice 300 for you and Barbara, and arrive at the reduced game in Table 3.10.

In the new reduced game, choosing the price 200 is strictly dominated by choosing 100, and similarly for Barbara. So, in step 3 we eliminate the choice 200 for you and Barbara.

In the new reduced game, you and Barbara only have one choice left, which is choosing a price of 100. So, no choices can be eliminated anymore and the algorithm stops after 3 steps. We thus see that you can only rationally choose a price of 100 under common belief in rationality, and similarly for Barbara. Hence, if you and Barbara follow the idea of common belief in rationality, you would both write down the lowest possible price, which is 100. □

Example 3.7 The number machine

This last example will show that the idea of common belief in rationality may sometimes have very striking, and unexpected, consequences. Suppose you visit a casino in Las Vegas where your attention is drawn to a remarkable machine. It says: "Guess two-thirds of the average and you will be rich." The idea is simple: After putting 5 dollars in the machine you must enter a whole number between 2 and 100, and you will always receive some prize money. However, the amount of prize money will be bigger the closer your number is to two-thirds of the average of all the numbers that have been previously entered on this machine. If your number is very, very close to two-thirds of the average, you will win thousands of dollars. Which number should you choose?

Let us analyze this game by using the idea of common belief in rationality. So, which numbers can you rationally choose if you believe that all players before you chose rationally, believe that all players before you believed that all players before them chose rationally, and so on? It should be clear that two-thirds of the average of all previous numbers will always be less than 67, so by choosing 67 you will always be closer to two-thirds of the average than by choosing any number above 67. Hence, you will never choose any number above 67 since it is strictly dominated by the number 67. But the same applies to all players before you, so if you believe that all players before you chose rationally, you must believe that all players before you chose a number less than or equal to 67.

However, if that is true then two-thirds of the average cannot be bigger than two-thirds of 67, which is 44.67. So, if you believe that all previous players chose rationally it will never be optimal to choose any number above 45, since by choosing 45 instead you will always be closer to two-thirds of the average. But the same applies to the players before you. So, if you believe that all players before you chose rationally, and believe that all players before you believed that all players before them chose rationally, you must believe that all players before you chose at most 45.

But then, you expect that two-thirds of the average will not exceed two-thirds of 45, which is 30. So, choosing any number above 30 will now be strictly dominated by choosing 30, since by choosing 30 you expect to be closer to two-thirds of the average. However, the same applies to the previous players. So, if you reason in accordance with common belief in rationality, you must also expect that the players before you will have chosen at most 30.

As a consequence, you must expect that two-thirds of the average cannot be bigger than 20. So, you will not choose any number above 20, and believe your opponents did not do so either, and so on. By continuing this argument, we finally conclude that there is only one number that you can rationally choose under common belief in rationality, which is choosing the lowest possible number 2! Moreover, under common belief in rationality you believe all players before you also reasoned in this way, so you believe that also all players before you will have chosen 2!

The question remains whether you would really choose 2 if you were to visit this casino in reality. Probably not. The problem is that common belief in rationality imposes a lot in this example: It not only imposes that you are capable of making a rational choice yourself, but also that you believe that all players before you were capable of making a rational choice, and that you are capable of taking this into account. Not only this, you must also believe that all players before you believed that all players before them were capable of making a rational choice, and that all players before you were capable of taking this account, and so on. This does not seem entirely realistic. If you wish, you may play this game with your friends and family. Good luck! □

3.8 Order independence

In the previous section we presented an algorithm, iterated elimination of strictly dominated choices, which yields precisely those choices you can rationally make under common belief in rationality. In step 1 we start by eliminating *all* choices that are strictly dominated in the original game, which gives a reduced game. Subsequently, in step 2, we eliminate *all* choices that are strictly dominated within the reduced game obtained after step 1. And so on, until no further choices can be eliminated in this way. That is, at every step of the procedure we always eliminate *all* choices that are strictly dominated within the reduced game we are considering.

Suppose now that instead of always eliminating *all* strictly dominated choices, we eliminate at every step only *some* – but not necessarily *all* – choices that are strictly dominated. Does it alter the final result? As we will see, the answer is "no"! Hence, if we are only interested in those choices you can rationally make under common belief in rationality, then it is not necessary to always eliminate *all* strictly dominated choices in the algorithm, because we could instead eliminate only *some of the* strictly dominated choices at each step. The final set of choices does not depend on the specific order and speed of elimination!

To see why this is true, consider a specific game Γ, and two different elimination procedures – Procedure 1 and Procedure 2.

- **Procedure 1:** This is the original algorithm of iterated elimination of strictly dominated choices, in which we eliminate, at every round k, *all* choices that are strictly dominated within the reduced game Γ^{k-1}. This procedure yields a reduced game Γ^k at every round k.
- **Procedure 2:** This is a procedure in which we eliminate, at every round k, *some* – but not necessarily *all* – choices that are strictly dominated within the reduced game $\hat{\Gamma}^{k-1}$ obtained after round $k-1$. Suppose this procedure yields a reduced game $\hat{\Gamma}^k$ at every round k.

We will prove that Procedure 1 and Procedure 2 yield exactly the same final set of choices. We will first show that every choice that survives Procedure 1 must also survive Procedure 2. To show this, let us first compare round 1 for both procedures. Obviously, every choice which survives round 1 of Procedure 1 also survives round 1 of Procedure 2. That is, all choices in Γ^1 are also in $\hat{\Gamma}^1$.

Let us now compare round 2 for both procedures. Consider some choice c_i that is in the reduced game Γ^2 of Procedure 1. Then, choice c_i is not strictly dominated within the reduced game Γ^1. As all choices in Γ^1 are also in $\hat{\Gamma}^1$, it follows that c_i is also not strictly dominated within the reduced game $\hat{\Gamma}^1$ of Procedure 2. But then, c_i will definitely survive round 2 of Procedure 2, and hence c_i will be in $\hat{\Gamma}^2$. So, we see that every choice in Γ^2 will also be in $\hat{\Gamma}^2$.

By repeating this argument, we finally conclude that, in every round k, every choice in Γ^k will also be in $\hat{\Gamma}^k$. That is, in every round k, every choice that survives round k of Procedure 1 will also survive round k of Procedure 2. Therefore, every choice that survives Procedure 1 must also survive Procedure 2.

We now prove that the converse is also true: Every choice that survives Procedure 2 will also survive Procedure 1. In order to prove this, we show that, for every round k, every choice that is eliminated in round k of Procedure 1, will also *eventually* be eliminated in Procedure 2 – maybe at round k, but maybe at some later round.

Suppose that this were not true. So, suppose that there is some round k, and some choice c_i for player i, such that c_i is eliminated in round k of Procedure 1, but is never eliminated by Procedure 2. Then, let k^* be the *first* round where this happens. So, let k^* be the *first* round where there is some choice c_i, such that c_i is eliminated at round k^* of Procedure 1, but is never eliminated by Procedure 2.

Since c_i is eliminated at round k^* of Procedure 1, choice c_i is strictly dominated in the reduced game Γ^{k^*-1} of Procedure 1. By our choice of k^*, we know that all choices that are eliminated up to round $k^* - 1$ of Procedure 1, are also eventually eliminated by Procedure 2. This means that there is some round m, such that all choices that are eliminated up to round $k^* - 1$ of Procedure 1, are eliminated up to round m of Procedure 2. So, all choices that are in $\hat{\Gamma}^m$ are also in Γ^{k^*-1}. As choice c_i is strictly dominated in the reduced game Γ^{k^*-1} of Procedure 1, it must also be strictly dominated in the reduced game $\hat{\Gamma}^m$ of Procedure 2, because the set of choices in $\hat{\Gamma}^m$ is a subset of the set of choices in Γ^{k^*-1}. This means, however, that choice c_i must be eliminated sooner or later by Procedure 2. This contradicts our assumption that c_i is never eliminated by Procedure 2.

So, we conclude that every choice that is eliminated by Procedure 1 at some round k, must also eventually be eliminated by Procedure 2. Hence, every choice that survives Procedure 2 must also survive Procedure 1.

As we have already seen above that the converse is also true, we conclude that the set of choices that survive Procedure 1 is exactly the same as the set of choices that survive Procedure 2. So, the order and speed with which we eliminate choices does not matter for the eventual output of the algorithm of iterated elimination of strictly dominated choices.

Theorem 3.8.1 *(Order and speed of elimination does not matter)*
The set of choices that survive the algorithm of iterated elimination of strictly dominated choices does not change if we change the order and speed of elimination.

This theorem is not only of theoretical interest, but is also important for practical applications. In some games it may be difficult to always find *all* choices that are strictly dominated within the reduced game we are considering. But the above theorem says that this does not matter for the final result of the algorithm.

3.9 Proofs

In this section we will prove Theorem 3.7.2, which states that for every k, the choices that can rationally be chosen by expressing up to k-fold belief in rationality are precisely the choices that survive $(k+1)$-fold elimination of strictly dominated choices. Moreover, it states that the choices that can rationally be chosen under *common* belief in rationality are precisely the choices that survive the *full* procedure of iterated elimination of strictly dominated choices. Before we prove this result, we first investigate an important property of the algorithm of iterated elimination of strictly dominated choices, which we refer to as the "optimality principle." We will use the optimality principle to prove the theorem.

Optimality principle
For every player i and every $k \in \{0,1,2,...\}$, let C_i^k denote the set of choices for player i that survive k-fold elimination of strictly dominated choices. In particular, C_i^0 is the full set of choices C_i. By

$$C_{-i}^k := C_1^k \times ... \times C_{i-1}^k \times C_{i+1}^k \times ... \times C_n^k$$

we denote the set of choice combinations for i's opponents that survive k-fold elimination of strictly dominated choices. Let Γ^k denote the reduced game that remains after k-fold elimination of strictly dominated choices. By construction, C_i^k contains exactly those choices in C_i^{k-1} that are not strictly dominated in the reduced game Γ^{k-1}. Player i can only choose from C_i^{k-1} and the opponents can only choose from C_{-i}^{k-1}. So, by applying Theorem 2.5.3 to the reduced game Γ^{k-1}, we know that every $c_i \in C_i^k$ is optimal, among the choices in C_i^{k-1}, for some belief $b_i \in \Delta(C_{-i}^{k-1})$. That is,

$$u_i(c_i,b_i) \geq u_i(c_i',b_i) \text{ for all } c_i' \in C_i^{k-1}.$$

Here, by $\Delta(C_{-i}^{k-1})$ we mean the set of beliefs that assign positive probability only to choice combinations in C_{-i}^{k-1}.

Now, we can say a little more about choices $c_i \in C_i^k$. Not only is c_i optimal for some belief $b_i \in \Delta(C_{-i}^{k-1})$ among the *choices in* C_i^{k-1} but for this particular belief b_i, choice c_i is even optimal *among all choices in the original game*. That is,

$$u_i(c_i,b_i) \geq u_i(c_i',b_i) \text{ for all } c_i' \in C_i.$$

We refer to this result as the "optimality principle" for iterated elimination of strictly dominated choices.

Lemma 3.9.1 *(Optimality principle)*
Let $k \geq 1$. Consider a choice $c_i \in C_i^k$ that survives k-fold elimination of strictly dominated choices. Then, there is some belief $b_i \in \Delta(C_{-i}^{k-1})$ such that c_i is optimal for b_i among all choices in the original game.

Proof: Consider some choice $c_i \in C_i^k$. Then, we know from the above that there is some belief $b_i \in \Delta(C_{-i}^{k-1})$ such that

$$u_i(c_i, b_i) \geq u_i(c_i', b_i) \text{ for all } c_i' \in C_i^{k-1}.$$

We will prove that, in fact,

$$u_i(c_i, b_i) \geq u_i(c_i', b_i) \text{ for all } c_i' \in C_i.$$

Suppose, on the contrary, that there is some $c_i' \in C_i$ with $u_i(c_i', b_i) > u_i(c_i, b_i)$. Let c_i^* be an optimal choice for b_i from the choices in C_i. Then,

$$u_i(c_i^*, b_i) \geq u_i(c_i', b_i) > u_i(c_i, b_i).$$

Since c_i^* is optimal for the belief $b_i \in \Delta(C_{-i}^{k-1})$ for the choices in C_i, we can apply Theorem 2.5.3 to the reduced game Γ^{k-1} and conclude that c_i^* is not strictly dominated in Γ^{k-1}, and hence $c_i^* \in C_i^k$. In particular, $c_i^* \in C_i^{k-1}$. So, we have found a choice $c_i^* \in C_i^{k-1}$ with $u_i(c_i^*, b_i) > u_i(c_i, b_i)$. This, however, contradicts our assumption that

$$u_i(c_i, b_i) \geq u_i(c_i', b_i) \text{ for all } c_i' \in C_i^{k-1}.$$

Hence, we conclude that there is no $c_i' \in C_i$ with $u_i(c_i', b_i) > u_i(c_i, b_i)$. So, c_i is indeed optimal for b_i among the choices in the original game. This completes the proof. ◇

The algorithm works
We will now use the optimality principle to prove the following theorem.

Theorem 3.7.2 *(The algorithm works)*
Consider a static game with finitely many choices for each player.
(1) For every $k \geq 1$, the choices that can rationally be made by expressing up to k-fold belief in rationality are exactly those choices that survive $(k+1)$-fold elimination of strictly dominated choices.
(2) The choices that can rationally be made under common belief in rationality are exactly those choices that survive iterated elimination of strictly dominated choices.

Proof: Recall that, for every k, the set C_i^k denotes the set of choices for player i that survive k-fold elimination of strictly dominated choices. We denote by BR_i^k the set of choices that player i can rationally make by expressing up to k-fold belief in rationality. We show that $BR_i^k = C_i^{k+1}$ for every $k \geq 1$. We proceed by two parts. In part (a) we show, by induction on k, that $BR_i^k \subseteq C_i^{k+1}$ for all $k \geq 1$. In part (b) we prove that $C_i^{k+1} \subseteq BR_i^k$ for every $k \geq 1$.

(a) Show that $BR_i^k \subseteq C_i^{k+1}$.
We show this statement by induction on k.

Induction start: Start with $k = 1$. By definition, BR_i^1 contains those choices for player i that he can rationally make by expressing 1-fold belief in rationality – that is, while believing in the opponents' rationality. We know from Chapter 2 that every such choice survives two-fold elimination of strictly dominated choices and is therefore in C_i^2. It thus follows that $BR_i^1 \subseteq C_i^2$.

Induction step: Take some $k \geq 2$, and assume that $BR_i^{k-1} \subseteq C_i^k$ for all players i. We prove that $BR_i^k \subseteq C_i^{k+1}$ for all players i.

Consider a player i, and take some $c_i \in BR_i^k$. Then, c_i is optimal for some type t_i that expresses up to k-fold belief in rationality. Hence, t_i only assigns positive probability to the opponents' choice-type pairs (c_j, t_j) where c_j is optimal for t_j, and t_j expresses up to $(k-1)$-fold belief in rationality. As a consequence, t_i only assigns positive probability to the opponents' choices that can rationally be chosen by expressing up to $(k-1)$-fold belief in rationality. That is, t_i only assigns positive probability to the opponents' choices in BR_j^{k-1}. Since, by the induction assumption, $BR_j^{k-1} \subseteq C_j^k$ for all opponents j, it follows that t_i only assigns positive probability to the opponents' choice-combinations in C_{-i}^k. Since c_i is optimal for type t_i, it follows that choice c_i is optimal for some belief $b_i \in \Delta(C_{-i}^k)$. But then, by Theorem 2.5.3, choice c_i is not strictly dominated in the reduced game Γ^k that remains after k-fold elimination of strictly dominated choices. Hence, by construction, choice c_i must be in C_i^{k+1}. So, we have shown that every $c_i \in BR_i^k$ must be in C_i^{k+1}. Therefore, $BR_i^k \subseteq C_i^{k+1}$ for all players i. By induction, the proof for part (a) is complete.

(b) Show that $C_i^{k+1} \subseteq BR_i^k$.
We must show that every choice $c_i \in C_i^{k+1}$ is optimal for some type t_i that expresses up to k-fold belief in rationality. Suppose that the procedure of iterated elimination of strictly dominated choices terminates after K rounds – that is, $C_i^{K+1} = C_i^K$ for all players i. In this proof we will construct an epistemic model M with the following properties:

- For every choice $c_i \in C_i^1$, there is a type $t_i^{c_i}$ for which c_i is optimal.
- For every $k \geq 2$, if c_i is in C_i^k, then the associated type $t_i^{c_i}$ expresses up to $(k-1)$-fold belief in rationality.
- If c_i is in C_i^K, then the associated type $t_i^{c_i}$ expresses common belief in rationality.

For this construction we proceed by the following steps. In step 1 we construct, for every choice $c_i \in C_i^1$, some belief $b_i^{c_i}$ about the opponents' choices for which c_i is optimal. In step 2 we use these beliefs to construct our epistemic model M, where we define for every choice $c_i \in C_i^1$ some type $t_i^{c_i}$ for which c_i is optimal. In step 3 we show that, for every $k \geq 2$ and every $c_i \in C_i^k$, the associated type $t_i^{c_i}$ expresses up to $(k-1)$-fold belief in rationality. In step 4 we finally prove that, for every choice $c_i \in C_i^K$, the associated type $t_i^{c_i}$ expresses common belief in rationality.

Step 1: Construction of beliefs

We start by constructing, for every $c_i \in C_i^1$, a belief $b_i^{c_i}$ about the opponents' choices for which c_i is optimal. For every $k \in \{1, ..., K-1\}$, let D_i^k be the set of choices that survive k-fold elimination of strictly dominated choices, but not $(k+1)$-fold. So, $D_i^k = C_i^k \backslash C_i^{k+1}$. To define the beliefs $b_i^{c_i}$, we distinguish the following two cases:

(1) First take some $k \in \{1, ..., K-1\}$ and some choice $c_i \in D_i^k$. Then, by Lemma 3.9.1, choice c_i is optimal, among the choices in the original game, for some belief $b_i^{c_i} \in \Delta(C_{-i}^{k-1})$.

(2) Consider next some choice $c_i \in C_i^K$ that survives the full procedure of iterated elimination of strictly dominated choices. Then, $c_i \in C_i^{K+1}$. Hence, by Lemma 3.9.1 we can find some belief $b_i^{c_i} \in \Delta(C_{-i}^K)$ for which c_i is optimal, among the choices in the original game.

In this way we have defined, for every $k \geq 1$, and for every choice $c_i \in C_i^k$, a belief $b_i^{c_i}$ for which c_i is optimal.

Step 2: Construction of types

We will now use the beliefs $b_i^{c_i}$ from step 1 to construct, for every choice $c_i \in C_i^1$, some type $t_i^{c_i}$ for which c_i is optimal. For every player i, let the set of types be given by

$$T_i = \{t_i^{c_i} : c_i \in C_i^1\}.$$

For every $k \in \{1, ..., K\}$, let T_i^k be the set of types $t_i^{c_i} \in T_i$ with $c_i \in C_i^k$. Since $C_i^K \subseteq C_i^{K-1} \subseteq ... \subseteq C_i^1$, we have that $T_i^K \subseteq T_i^{K-1} \subseteq ... \subseteq T_i^1$, where T_i^1 is the full set of types T_i. We will now define the belief for each type, using the following three cases:

(1) Consider first the types $t_i^{c_i}$ with $c_i \in D_i^1$. That is, $t_i^{c_i} \in T_i^1 \backslash T_i^2$. We define the belief $b_i(t_i^{c_i})$ in the following way: Fix some arbitrary type $t_j \in T_j$ for each opponent j. For every opponents' combination of choices $(c_1, ..., c_{i-1}, c_{i+1}, ..., c_n)$ in C_{-i}, type t_i's belief $b_i(t_i^{c_i})$ assigns probability $b_i^{c_i}(c_1, ..., c_{i-1}, c_{i+1}, ..., c_n)$ to the choice-type combination

$$((c_1, t_1), ..., (c_{i-1}, t_{i-1}), (c_{i+1}, t_{i+1}), ..., (c_n, t_n)).$$

Type $t_i^{c_i}$ assigns probability zero to all other choice-type combinations for the opponents.

So, type $t_i^{c_i}$ holds the same belief about the opponents' choices as $b_i^{c_i}$. By construction in step 1, choice c_i is optimal for the belief $b_i^{c_i}$, so it follows that c_i is optimal for the type $t_i^{c_i}$.

(2) Consider next the types $t_i^{c_i}$ with $c_i \in D_i^k$ for some $k \in \{2, ..., K-1\}$. Hence, $t_i^{c_i} \in T_i^k \backslash T_i^{k+1}$ for some $k \in \{2, ..., K-1\}$. We define the belief $b_i(t_i^{c_i})$ as follows: For every opponents' combination of choices $(c_1, ..., c_{i-1}, c_{i+1}, ..., c_n)$ in C_{-i}, type t_i's belief $b_i(t_i^{c_i})$ assigns probability $b_i^{c_i}(c_1, ..., c_{i-1}, c_{i+1}, ..., c_n)$ to the choice-type combination

$$((c_1, t_1^{c_1}), ..., (c_{i-1}, t_{i-1}^{c_{i-1}}), (c_{i+1}, t_{i+1}^{c_{i+1}}), ..., (c_n, t_n^{c_n})).$$

Type $t_i^{c_i}$ assigns probability zero to all other choice-type combinations for the opponents.

So, type $t_i^{c_i}$ holds the same belief about the opponents' choices as $b_i^{c_i}$. By construction in step 1, choice c_i is optimal for the belief $b_i^{c_i}$, so it follows that c_i is optimal for type $t_i^{c_i}$. Since, by step 1, $b_i^{c_i} \in \Delta(C_{-i}^{k-1})$, and type $t_i^{c_i}$ only assigns positive probability to choice-type pairs (c_j, t_j) with $t_j = t_j^{c_j}$, it follows that $t_i^{c_i}$ only assigns positive probability to the opponents' types $t_j^{c_j}$ with $c_j \in C_j^{k-1}$. That is, $t_i^{c_i}$ only assigns positive probability to the opponents' types in T_j^{k-1}.

(3) Consider finally the types $t_i^{c_i}$ with $c_i \in C_i^K$. That is, $t_i^{c_i} \in T_i^K$. We define the belief $b_i(t_i^{c_i})$ as follows: For every opponents' combination of choices $(c_1, ..., c_{i-1}, c_{i+1}, ..., c_n)$ in C_{-i}, type t_i's belief $b_i(t_i^{c_i})$ assigns probability $b_i^{c_i}(c_1, ..., c_{i-1}, c_{i+1}, ..., c_n)$ to the choice-type combination

$$((c_1, t_1^{c_1}), ..., (c_{i-1}, t_{i-1}^{c_{i-1}}), (c_{i+1}, t_{i+1}^{c_{i+1}}), ..., (c_n, t_n^{c_n})).$$

Type $t_i^{c_i}$ assigns probability zero to all other choice-type combinations for the opponents.

So, type $t_i^{c_i}$ holds the same belief about the opponents' choices as $b_i^{c_i}$. By construction in step 1, choice c_i is optimal for the belief $b_i^{c_i}$, so it follows that c_i is optimal for type $t_i^{c_i}$. Since, by step 1, $b_i^{c_i} \in \Delta(C_{-i}^K)$, and type $t_i^{c_i}$ only assigns positive probability to choice-type pairs (c_j, t_j) with $t_j = t_j^{c_j}$, it follows that $t_i^{c_i}$ only assigns positive probability to the opponents' types $t_j^{c_j}$ with $c_j \in C_j^K$. That is, $t_i^{c_i}$ only assigns positive probability to the opponents' types in T_j^K.

Step 3: Every type $t_i \in T_i^k$ expresses up to $(k-1)$-fold belief in rationality
We prove this statement by induction on k.

Induction start: We start with $k = 2$. Take some type $t_i \in T_i^2$. That is, $t_i = t_i^{c_i}$ for some choice $c_i \in C_i^2$. By construction of step 2, type $t_i^{c_i}$ only assigns positive probability to the opponents' choice-type pairs $(c_j, t_j^{c_j})$ where c_j is optimal for $t_j^{c_j}$. This means that type $t_i^{c_i}$ believes in the opponents' rationality – that is, it expresses 1-fold belief in rationality. So, we have shown that every type $t_i \in T_i^2$ expresses 1-fold belief in rationality.

Induction step: Take some $k \geq 3$, and assume that for every player i, every type $t_i \in T_i^{k-1}$ expresses up to $(k-2)$-fold belief in rationality. Consider now some type $t_i \in T_i^k$. That is, $t_i = t_i^{c_i}$ for some choice $c_i \in C_i^k$. By construction of step 2, type $t_i^{c_i}$ only assigns positive probability to the opponents' types $t_j \in T_j^{k-1}$. By our induction assumption, every such type $t_j \in T_j^{k-1}$ expresses up to $(k-2)$-fold belief in rationality. Hence, type $t_i^{c_i}$ only assigns positive probability to opponents' types that express up to $(k-2)$-fold belief in rationality. This means, however, that type $t_i^{c_i}$ expresses up to $(k-1)$-fold belief in rationality. We have thus shown that every type $t_i \in T_i^k$

expresses up to $(k-1)$-fold belief in rationality. By induction on k, the proof of step 3 is complete.

Step 4: Every type $t_i \in T_i^K$ expresses common belief in rationality
In order to prove this, we use the following lemma.

Lemma 3.9.2 *For all $k \geq K-1$, every type $t_i \in T_i^K$ expresses up to k-fold belief in rationality.*

Proof of Lemma 3.9.2: We prove the statement by induction on k.

Induction start: Begin with $k = K-1$. Take some type $t_i \in T_i^K$. Then, we know from step 3 that t_i expresses up to $(K-1)$-fold belief in rationality.

Induction step: Consider some $k \geq K$, and assume that for every player i, every type $t_i \in T_i^K$ expresses up to $(k-1)$-fold belief in rationality. Now consider some player i and some type $t_i \in T_i^K$. So, $t_i = t_i^{c_i}$ for some $c_i \in C_i^K$. By construction of step 2, type $t_i^{c_i}$ only assigns positive probability to the opponents' types $t_j \in T_j^K$. By our induction assumption, every such type $t_j \in T_j^K$ expresses up to $(k-1)$-fold belief in rationality. Hence, type $t_i^{c_i}$ only assigns positive probability to opponents' types that express up to $(k-1)$-fold belief in rationality. This means, however, that type $t_i^{c_i}$ expresses up to k-fold belief in rationality. We have thus shown that every type $t_i \in T_i^K$ expresses up to k-fold belief in rationality. By induction on k, the proof of the lemma is complete. $\diamond$

By the lemma above we thus see that every type $t_i \in T_i^K$ expresses up to k-fold belief in rationality for every k. That is, every type $t_i \in T_i^K$ expresses common belief in rationality, which completes the proof of step 4.

With steps 1–4 we can now easily prove part (b), namely that $C_i^{k+1} \subseteq BR_i^k$ for every $k \geq 1$. Take some $k \geq 1$ and some choice $c_i \in C_i^{k+1}$. By step 3 we know that the associated type $t_i^{c_i} \in T_i^{k+1}$ expresses up to k-fold belief in rationality. By our construction in steps 1 and 2, the choice c_i is optimal for the type $t_i^{c_i}$, so it follows that c_i is optimal for a type that expresses up to k-fold belief in rationality. That is, $c_i \in BR_i^k$. So, we have shown that every choice $c_i \in C_i^{k+1}$ is also in BR_i^k, and therefore $C_i^{k+1} \subseteq BR_i^k$ for every $k \geq 1$. This completes the proof of part (b).

With parts (a) and (b) it is now easy to prove Theorem 3.7.2. Let us start with part (1) of the theorem. By parts (a) and (b) we know that $C_i^{k+1} = BR_i^k$ for every player i and every $k \geq 1$. In other words, the choices that can rationally be made by expressing up to k-fold belief in rationality are exactly those choices that survive $(k+1)$-fold elimination of strictly dominated choices for all $k \geq 1$. This completes the proof of part (1) of the theorem.

Consider next part (2) of the theorem. The choices that survive the full procedure of iterated elimination of strictly dominated choices are precisely the choices in C_i^K. From step 4 in part (b) above, we know that for every choice $c_i \in C_i^K$, the associated type

$t_i^{c_i} \in T_i^K$ expresses common belief in rationality. By construction, choice c_i is optimal for the associated type $t_i^{c_i}$, so it follows that every choice $c_i \in C_i^K$ is optimal for some type that expresses common belief in rationality. So, every choice that survives iterated elimination of strictly dominated choices can rationally be chosen under common belief in rationality.

Next, consider a choice c_i that can rationally be chosen under common belief in rationality. Then, $c_i \in BR_i^k$ for every k. So, by part (a) above, $c_i \in C_i^{k+1}$ for every k, which means that c_i survives the full procedure of iterated elimination of strictly dominated choices. Hence, we see that every choice that can rationally be chosen under common belief in rationality survives the full procedure of iterated elimination of strictly dominated choices. Altogether, we conclude that the choices that can rationally be chosen under common belief in rationality are exactly the choices that survive the full procedure of iterated elimination of strictly dominated choices. This completes the proof of part (2) of the theorem. ■

Practical problems

3.1 Where to locate a supermarket?

Recall the story from Problem 2.1 and have a look at your solutions to it.

(a) Consider the beliefs diagram you constructed in Problem 2.1. Create an extended beliefs diagram, and translate the extended beliefs diagram into an epistemic model for this situation.

(b) For each of the types for you in this model, describe its belief hierarchy.

(c) Which of your types in this epistemic model believe in the opponent's rationality? Which of these types believe, moreover, that the opponent believes in your rationality? Which of your types express common belief in rationality?

(d) Use the graphical method to determine those locations you and your competitor can rationally choose under common belief in rationality.

(e) Use the algorithm of iterated elimination of strictly dominated choices to determine the locations you and your competitor can rationally choose under common belief in rationality, and verify that it yields the same answer as (d). After how many rounds does the algorithm stop?

(f) Construct an epistemic model such that, for each of the locations c_i found in (d) and (e), there is a type $t_i^{c_i}$ such that:

- location c_i is optimal for $t_i^{c_i}$, and
- type $t_i^{c_i}$ expresses common belief in rationality.

3.2 Preparing for a piano exam

Recall the story from Problem 2.2.

(a) Consider the beliefs diagram you constructed in Problem 2.2. Create an extended beliefs diagram and translate the extended beliefs diagram into an epistemic model for this situation.

(b) Consider your type that supports dedicating two weeks to Mozart, and your type that supports dedicating two weeks to Chopin. For both types, describe the first three levels of their belief hierarchies.
(c) Which of your types in this epistemic model believe in the jury's rationality? Which of these types believe, moreover, that the jury believes in your rationality? Which of your types express common belief in rationality?
(d) Use the graphical method to find the choices that you and the jury can rationally make under common belief in rationality.
(e) Use the algorithm of iterated elimination of strictly dominated choices to find the choices that you and the jury can rationally make under common belief in rationality, and compare your answer with (d). After how many rounds does the algorithm stop?
(f) Construct an epistemic model such that, for each of the choices c_i found in (d) and (e), there is a type $t_i^{c_i}$ such that:

- choice c_i is optimal for $t_i^{c_i}$, and
- type $t_i^{c_i}$ expresses common belief in rationality.

3.3 Competition between two cinemas

Recall the story from Problem 2.3.
(a) In Problem 2.3, part (g), you found the prices that you can rationally choose if you believe that your competitor chooses rationally as well. Now, determine those prices that your competitor can rationally choose if he believes that you choose rationally.
(b) On the basis of your answer in (a), determine those prices you can rationally choose under expressing up to 2-fold belief in rationality.
(c) In Problem 2.3 and part (b) of this problem you have computed:

- the prices you can rationally choose,
- the prices you can rationally choose under 1-fold belief in rationality,
- the prices you can rationally choose under expressing up to 2-fold belief in rationality.

These three results can be interpreted as the first three steps of an algorithm. Develop a complete algorithm that eventually computes the prices you can rationally choose under common belief in rationality.
(d) Use this algorithm to compute the unique price you can rationally choose under common belief in rationality. Does the algorithm stop after finitely many rounds?
(e) Which price do you believe your competitor will choose under common belief in rationality? What profit do you expect to make under common belief in rationality?

3.4 Going to a party

Recall the story from Problem 2.4. Suppose first that the preferences over colors for the players are given by Table 2.11 in Problem 2.4.
(a) Consider the beliefs diagram you made in Problem 2.4. Create an extended beliefs diagram and translate the extended beliefs diagram into an epistemic model.

(b) Consider your type that supports wearing red and your type that supports wearing white. For both types, describe the first three levels of their belief hierarchies.
(c) Which of your types, if any, expresses common belief in rationality within this epistemic model?
(d) In this example, using the graphical method is difficult. (You may try if you wish and you will see that it is.) Use logical reasoning to find those colors you and your friends can rationally choose under common belief in rationality.
(e) Use the algorithm of iterated elimination of strictly dominated choices to find those colors you and your friends can rationally wear under common belief in rationality.
(f) Construct an epistemic model such that, for each of the choices c_i found in (d) and (e), there is a type $t_i^{c_i}$ such that:

- choice c_i is optimal for $t_i^{c_i}$, and
- type $t_i^{c_i}$ expresses common belief in rationality.

(g) Answer the same questions, but now for the case where Chris' utilities are given by Table 2.12 in Problem 2.4.

3.5 A game of cards

Recall the cards game from Problem 2.6.
(a) Consider the beliefs diagram you constructed in Problem 2.6. Create an extended beliefs diagram and translate the extended beliefs diagram into an epistemic model.
(b) Consider your type that supports playing the queen and your type that supports playing the 6. For both types, write down the first three levels of their belief hierarchies.
(c) Use the graphical method to determine those cards you and your friends can rationally play under common belief in rationality.
(d) Use the algorithm of iterated elimination of strictly dominated choices to find those cards you and your friends can rationally play under common belief in rationality, and compare your answer with (c). After how many rounds does the algorithm stop?
(e) Construct an epistemic model such that, for each of the choices c_i found in (c) and (d), there is a type $t_i^{c_i}$ such that:

- choice c_i is optimal for $t_i^{c_i}$, and
- type $t_i^{c_i}$ expresses common belief in rationality.

(f) Which player, or players, do you expect to be able to win the game under common belief in rationality? How much money do you expect the winner, or possible winners, to earn?

3.6 Snow White and the seven dwarfs

On a rainy evening, Snow White and the seven dwarfs feel like dancing. In order to decide who can dance with Snow White, they play a little game. The seven dwarfs line up in a row, one behind the other. The order of the dwarfs is: Doc at the back, then Grumpy, then Sleepy, then Happy, then Sneezy, then Bashful and finally Dopey at the front. Doc can see all the dwarfs in front of him, Grumpy can see the dwarfs in front of him, but not Doc, who is behind him, and so on. Finally, Dopey cannot see any dwarf, as he is at the front.

There is a box with seven red hats and six blue hats, and every dwarf knows this. Snow White randomly picks seven hats from the box and places them on the heads of the dwarfs. Every dwarf can only see the hats of the dwarfs in front of him, without knowing his own color or that of the dwarfs behind him. Then, Snow White asks Doc, the last dwarf in line: "Do you know the color of your hat?" If Doc answers with the correct color he is invited for a dance with Snow White, whereas the other dwarfs must dance with each other. If he says the wrong color then there will be no dancing and all dwarfs must do the dishes instead – an activity they all hate! If he says "I don't know," Snow White will turn to the next dwarf in line, Grumpy. Snow White will then ask Grumpy the same question "Do you know the color of your hat?" Grumpy can either give a color or say that he doesn't know. If he correctly guesses his color, he will dance with Snow White. If he is wrong about his color, they must all do the dishes. If he says he doesn't know, Snow White will turn to the next dwarf, Sleepy, and so on, until some dwarf gives a color, or the last dwarf, Dopey, has said that he doesn't know either. If all dwarfs say that they don't know, all the dwarfs must dance with each other and Snow White will go to bed.

Suppose that the dwarfs value a dance with Snow White at 10, that they value a dance with another dwarf at 2, and that they value doing the dishes at -1000!

(a) For every possible configuration of hats that Doc could see before him, what would be his rational answer?

(b) Suppose that Doc has said "I don't know,"and that Snow White now asks Grumpy, who believes that Doc has answered rationally. For every possible configuration of hats that he could see before him, what would be his rational answer?

(c) Suppose that Doc and Grumpy have said "I don't know." Snow White now asks Sleepy, who believes that Doc and Grumpy have answered rationally, and believes that Grumpy believes that Doc has answered rationally. For every possible configuration of hats that he could see before him, what would be his rational answer?

(d) Suppose that Snow White asks Happy, Sneezy and Bashful, who believe that all dwarfs before them have answered rationally, that all dwarfs before them believe that all dwarfs before them have answered rationally, and so on. For every possible configuration of hats that they could see before them, what would be their rational answers?

(e) Suppose finally that Doc, Grumpy, Sleepy, Happy, Sneezy and Bashful have all said "I don't know" and that Snow White now asks Dopey, who believes that all dwarfs have answered rationally, that all dwarfs before him believe that all dwarfs before them have answered rationally, and so on. Remember that Dopey cannot see any hats. What would be the rational answer for Dopey?

(f) Compute for every dwarf the chance that he will dance with Snow White.

3.7 The mother-in-law

Suppose that you and your partner are planning to move to another village, which has only three streets. There are seven houses for sale in that village, and their locations are depicted in Figure 3.11. The distance between two adjacent houses for sale in this village is 100 meters. So, the distance between house A and house B is 100 meters, the distance between house B and house C is 100 meters, and so on.

When your mother-in-law learned about your plans, she decided to move to the same village as well, and she can choose from the same seven houses. Tomorrow, you and your mother-in-law have to sign up for one of the seven houses. If you both choose different houses, you will both get the house of your choice. If you both choose the same house, the house will go to a third party, and you and your mother-in-law will not be able to move.

Suppose that your relationship with her is not quite optimal and that you wish to maximize the distance to her house. Your utility is equal to the distance between the houses if you both get a house in the village, and is equal to 0 if you cannot move to the village. Your mother-in-law, on the other hand, wishes to minimize the distance to your house, as she likes to visit her child every day, and check whether you have cleaned the house properly. More precisely, the utility for your mother-in-law is equal to

$$600 - \text{distance between the houses}$$

if you both get a house in the village, and is equal to 0 if she cannot move to the village.
(a) Show that location C is strictly dominated for you by a randomized choice in which you randomize over the locations A, D and G. That is, find probabilities α, β and γ with

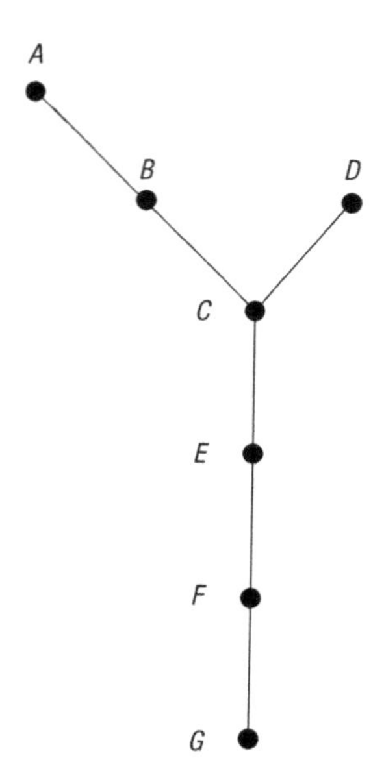

Figure 3.11 Houses for sale in "The mother-in-law"

$\alpha + \beta + \gamma = 1$ such that location C is strictly dominated by the randomized choice in which you choose location A with probability α, location D with probability β and location G with probability γ.
(b) Which are the rational locations for you? For every rational location, find a belief about your mother-in-law's choice for which this location is optimal. For every irrational location, find a randomized choice that strictly dominates it.
Hint for (b): Every irrational location is strictly dominated by a randomized choice in which you randomize over the locations A, D and G.
(c) Use the algorithm of iterated elimination of strictly dominated choices to find those locations you and your mother-in-law can rationally choose under common belief in rationality. After how many steps does the algorithm stop? Do you expect that both you and your mother-in-law will choose the same house?
(d) Construct an epistemic model such that, for each of the choices c_i found in (c), there is a type $t_i^{c_i}$ such that:

- choice c_i is optimal for $t_i^{c_i}$, and
- type $t_i^{c_i}$ expresses common belief in rationality.

Theoretical problems

3.8 Properties of common belief in rationality

Consider a game with n players and finitely many choices for every player.
(a) Suppose you construct an epistemic model with sets of types $T_1, ..., T_n$, such that every type in this epistemic model believes in the opponents' rationality. Show that every type in this epistemic model expresses common belief in rationality.
(b) Consider an epistemic model with sets of types $T_1, ..., T_n$, and a type $t_i \in T_i$ that expresses common belief in rationality. Show that t_i only assigns positive probability to opponents' types that express common belief in rationality too.
(c) Consider a choice c_i for player i that can rationally be chosen under common belief in rationality. Show that you can construct an epistemic model with sets of types $T_1, ..., T_n$ such that:

- every type in this epistemic model expresses common belief in rationality, and
- c_i is optimal for some type $t_i \in T_i$.

3.9 Best-response sets

Consider a game with n players and finitely many choices for every player. Let $D_i \subseteq C_i$ be a subset of choices for every player i. The collection $(D_1, D_2, ..., D_n)$ of subsets of choices is called a *best-response set* if for every player i, every choice $c_i \in D_i$ is optimal under some probabilistic belief

$$b_i^{c_i} \in \Delta(D_1 \times ... \times D_{i-1} \times D_{i+1} \times ... \times D_n).$$

That is, every choice in the collection $(D_1, D_2, ..., D_n)$ can be optimal if the player believes that his opponents will choose from the same collection $(D_1, D_2, ..., D_n)$.

Show that a choice c_i can rationally be chosen under common belief in rationality *if and only if* there is a best-response set $(D_1, D_2, ..., D_n)$ with $c_i \in D_i$.

3.10 Maximal number of steps in the algorithm

Consider a game with two players, 1 and 2, in which both players have k choices.
(a) Show that the algorithm of iterated elimination of strictly dominated choices needs at most $2k-2$ steps to finish.
(b) Show that the upper bound derived in (a) is sharp. That is, for any number k, construct a two-player game in which both players have k choices, and the algorithm needs exactly $2k-2$ steps to finish.

Literature

Belief hierarchies and types

A belief hierarchy specifies your belief about the opponents' choices, your belief about the opponents' beliefs about the other players' choices, and so on, ad infinitum. So, a belief hierarchy necessarily involves an *infinite* stream of beliefs, which makes it a rather complicated object to describe. Harsanyi (1967–1968) was the first to formally develop the idea of a belief hierarchy in games, although he did it for a different purpose. Harsanyi concentrated on games with *incomplete information* in which the players have uncertainty about certain parameters of the game, such as the opponents' utility functions. Instead of modeling the beliefs that players have about the opponents' choices, Harsanyi wanted to describe the beliefs that players have about the *parameters* of the game. He also wanted to describe the beliefs that players have about the beliefs the opponents have about these parameters, and so on. Thus, Harsanyi introduced the notion of an infinite belief hierarchy into game theory. However, the belief hierarchies in Harsanyi's work had a different content than the belief hierarchies in this book – Harsanyi's belief hierarchies are about the parameters of the game, such as the opponents' utility functions, but not about the opponents' choices, whereas the belief hierarchies in this book are exclusively about the players' choices in the game, not about the opponents' utility functions. In fact, throughout this book we assume that the players' utility functions are transparent to all players in the game. Formally speaking, we only consider games with *complete information* in this book.

A major problem that Harsanyi faced was how to formally describe these infinite belief hierarchies in a compact way. His elegant solution to this problem was the following: Call a player's belief hierarchy about the opponents' utility functions, in combination with his own utility function, a *type* – also called an *information vector* or *attribute vector* in other parts of his work – and assume that every type holds a probabilistic belief about the opponents' types. From this construction we can then *derive*, for every type, an infinite belief hierarchy. Because a type holds a belief about the opponents' types, and every type has a specific utility function, a type holds in particular a (first-order) belief about the opponents' utility functions. Moreover, as every type holds a (first-order) belief about the other players' utility functions, a type also holds a (second-order) belief about the opponents' (first-order) beliefs about the other players' utility functions, and so on. So, the types that Harsanyi constructed are an elegant and efficient way to encode the infinite belief hierarchies that players hold in a game with incomplete information.

Harsanyi's type-based model can easily be adapted to the needs of this chapter, and as such provides the basis for the epistemic model as we use it here. The only thing we need to change is to assume that the players' basic uncertainty concerns the opponents' *choices* – not the opponents' *utilities* – but the remainder of the construction basically resembles Harsanyi's. Indeed, in the epistemic model as we use it we also assume – as Harsanyi – that a player's type holds a probabilistic belief about the other players' types, but in addition we assume that a type also holds a belief about the opponents' choices. In fact, every type in our model holds a probabilistic belief about the possible choice-type combinations of the opponents. From this construction, we can then derive an infinite belief hierarchy for every type. An alternative way to do so – which would be even closer to Harsanyi's original model – would be to assume that a player's type not only specifies the player's beliefs but also his *choice*. In that case it would be sufficient to let every type hold a belief about the other players' types – precisely as Harsanyi does – in order to derive the full belief hierarchy for each of the types. In fact, some papers, such as Aumann and Brandenburger (1995), model belief hierarchies in precisely this way. In their model the type of a player also determines the choice he will make. The rationale behind this approach is that a combination of types – one for each player – provides a complete description of the world when seen from the outside. The reason we do *not* include a player's choice as part of his type is that we wish to distinguish clearly between a player's epistemic state of mind – or belief hierarchy – encoded by his type, and the choice he makes. In our view, a player first forms a belief hierarchy about his opponents, and subsequently makes his choice based on his beliefs.

Armbruster and Böge (1979) and Böge and Eisele (1979) were probably the first to use belief hierarchies to describe the beliefs that players have about the opponents' *choices*. To be precise, they consider a context in which players have beliefs about the opponents' choices *and* utility functions. Bernheim (1984) introduced the notion of *systems of beliefs*, which is closely related to belief hierarchies, and used them to express the beliefs that players may hold about the opponents' choices. Building upon their work and that of Harsanyi, Tan and Werlang (1988) subsequently give a definition of an epistemic model – see their Definition 3.10 – in which the players' belief hierarchies are modeled in essentially the same way as we do.

Alternative ways of describing belief hierarchies

Encoding belief hierarchies by means of types *à la* Harsanyi is just one possible way of doing so. In the game-theoretic literature there are at least two important alternative ways for describing a belief hierarchy. The first approach, which is based on the models by Aumann (1976) and Kripke (1963), assumes that there is a set of *states of the world*, and a function that assigns to every state of the world a choice for each of the players.

A player, however, has uncertainty about the true state of the world. This is modeled by assuming that for every state there is a set of states – typically containing more than one state – which the player deems possible. Moreover, for every state a player may hold a probabilistic belief about the states of the world he deems possible. Since a state specifies a choice for every opponent, a player holds in particular a (first-order) belief about the opponents' choices for every state. Moreover, as every state also specifies a belief for every opponent about the other players' choices, a player also holds for every state a (second-order) belief about the opponents' (first-order) beliefs about the other players' choices, and so on. Hence, in this model we can derive, for every state, a complete belief hierarchy for each of the players. We call this the *state-based approach*. The second approach, which is often used by scientists from logic and computer science, explicitly describes all levels in the belief hierarchy as formulae in some formal syntax. We call this the *syntactic* approach.

Among others, Brandenburger and Dekel (1993), Tan and Werlang (1992) and Tsakas (2011) explicitly compare the type-based approach, as used in this book, with the state-based approach described above. They show how to transform the encoding of a belief hierarchy in one model into an "epistemically equivalent" encoding in the other model, thereby establishing that the two approaches are essentially equivalent. For the presentation and analysis of epistemic concepts it does not really matter which language one uses for describing the belief hierarchies – the type-based language, the state-based language or the syntactic language. What matters is the content of the belief hierarchies and the conditions imposed on these. Hence, we could have written this book equally well by using one of the other two approaches described above. In a sense, all these models just provide different representations of the same primitive notion: the belief hierarchy of a player.

Common knowledge and common belief

Common belief in rationality states that a player believes that his opponents choose rationally, believes that his opponents believe that the other players choose rationally, and so on, ad infinitum. Or, equivalently, a player expresses common belief in the event that all players in a game will choose rationally. The idea of "common belief in an event" is based on the notion of *common knowledge* – a concept that was independently introduced by the philosopher Lewis (1969) and the game theorist Aumann (1976). We say that there is *common knowledge* of an event if all players know the event occurs, all players know that all players know this, and so on. Aumann (1976) formalizes this notion in the following way. Assume there is a set of states of the world, and for every state ω each player i knows that the true state of the world is in some set of states $P_i(\omega)$. An event E is a set of states. Player i then *knows* the event E at ω if the event E is true at every state $\omega' \in P_i(\omega)$ that player i deems possible at ω. That is, $\omega' \in E$ for every $\omega' \in P_i(\omega)$. Player i knows at ω that player j knows the event E if at every state $\omega' \in P_i(\omega)$ it is true that player j knows E. And so on. In this way, common knowledge of the event E can be defined for a state ω. Aumann's model is very close to Kripke's (1963) possible worlds model in logic, also referred to as a *Kripke structure*. The main difference between the two models is that Kripke's model is formulated in a logical language, whereas Aumann's model is set-theoretic. Moreover, Kripke's model is more general than Aumann's model as it uses an accessibility relation to describe the person's information for a given state, which is more general than the knowledge operator used by Aumann.

Common belief in rationality

Many of the ideas behind common belief in rationality can be found in the important papers by Bernheim (1984) and Pearce (1984). They independently developed the concept of *rationalizability* – which is very closely related to common belief in rationality as we will see – although they did not explicitly embed their analysis within a formal epistemic framework. Bernheim's key concept is that of a *consistent system of beliefs* – a notion which essentially mimics the requirement that a belief hierarchy must express common belief in rationality. An important difference with our model is, however, that Bernheim assumes that in games with three players or more, the belief that a player holds about his opponents' choices must be *independent* across the opponents. That is, the belief by player i about opponent j's choice must be stochastically independent from the belief about opponent k's choice. We do not make such an independence assumption in this chapter – nor in the remaining chapters of this book. In Bernheim's model, the choices that can rationally be chosen under a consistent system of beliefs are called *rationalizable*.

Pearce provides a different – yet equivalent – definition of rationalizable choices. He presents a recursive elimination procedure, similar to the procedure of iterated elimination of strictly dominated choices, that works as follows: At every round, we eliminate for each player those choices that are not optimal for any *independent* belief about the opponents' choices that have survived up to this round. The choices that survive this elimination procedure are called *rationalizable* by Pearce. By an "independent belief" we mean independent across the opponents, as discussed above for Bernheim's model. The only difference between Pearce's procedure and the algorithm of iterated elimination of strictly dominated choices in this chapter is that Pearce assumes that players hold *independent* beliefs about the opponents' choices, whereas we do not make that assumption in this chapter. Recall that a choice is strictly dominated if and only if it is not optimal for any belief – independent or not – about the opponents' choices. So, in the algorithm of iterated elimination of strictly dominated choices we eliminate, at every round, those choices for a player that are not optimal for any belief – independent or not – about the opponents' choices that have survived up to that round. Hence, it is immediately clear that Pearce's procedure leads to smaller sets of choices than the procedure of iterated elimination of strictly dominated choices. As the latter procedure characterizes the choices that can rationally be made under common belief in rationality, it follows that every rationalizable choice – in the sense of Bernheim (1984) and Pearce (1984) – can rationally be made under common belief in rationality. The opposite is not true, however: There are games in which some choices can rationally be made under common belief in rationality, which are not rationalizable. Obviously, the concepts of rationalizability and common belief in rationality coincide for games with two players, as in such games independence of beliefs is not an issue.

Both Bernheim (1984) and Pearce (1984) argue informally that rationalizability follows from imposing common belief in rationality – provided that the players hold independent beliefs about the opponents' choices. They do not provide a formal justification for this claim, however, as neither author formally describes common belief in rationality or rationalizability within an epistemic model.

Tan and Werlang (1988) do provide a formal definition of common belief in rationality within an epistemic model. Unlike Bernheim (1984) and Pearce (1984) they do not assume that players hold independent beliefs about the opponents' choices. They show, through Theorems 5.2 and 5.3, that the choices that can rationally be made under common belief in rationality are precisely those choices that survive the procedure of iterated elimination of strictly dominated choices – see Theorem 3.7.2 in this chapter. They also formally argue that rationalizability is obtained by additionally imposing common belief in the event that players hold independent beliefs about the opponents' choices. Similar results can be found in Brandenburger and Dekel (1987).

Independent beliefs

In a game with three or more players, player i is said to have independent beliefs about the opponents' choices if for every two opponents j and k, his belief about opponent j's choice is stochastically independent from his belief about opponent k's choice. We have seen above that if we take the concept of common belief in rationality and additionally impose that there is common belief in the event that players hold independent beliefs about the opponents' choices, then we arrive at the concept of rationalizability as defined by Bernheim (1984) and Pearce (1984). As all beliefs in two-player games are automatically independent, it follows that the concepts of common belief in rationality and rationalizability are equivalent for two-player games. For games with three or more players, rationalizability is a strict refinement of common belief in rationality, as there are games with three or more players where some choices are not rationalizable, yet can rationally be chosen under common belief in rationality.

Bernheim (1984) defends the independent beliefs assumption by arguing that the players typically make their choices *independently* of each other, without any possibility of communication, and that therefore a player's belief about the opponents' choices must be independent. In our view, this conclusion is not entirely correct: Even if player i believes that his opponents j and k choose independently, then it may still be that his belief about j's belief hierarchy is *correlated* with his belief about k's belief hierarchy. As a consequence, his belief about j's choice may well be correlated with his belief about k's choice, as these choices may arise as the optimal choices under different belief hierarchies, about which player i holds correlated beliefs.

This is precisely the viewpoint taken by Brandenburger and Friedenberg (2008), who call the above type of correlation between beliefs about different opponents' belief hierarchies *intrinsic*. They weaken the independence assumption in Bernheim (1984) and Pearce (1984) by stating that in a game with three or more players, the belief that player i has about opponent j's choice must be stochastically independent from his belief about opponent k's choice, once we *condition on a fixed belief hierarchy* for opponents j and k. They call this condition *conditional independence*. In other words, if we fix two belief hierarchies for opponents j and k, then conditional on these belief hierarchies being the actual belief hierarchies held by the opponents, the beliefs about the opponents' choices must be independent. This still reflects Bernheim's viewpoint that the two opponents choose independently, but recognizes that it does not automatically mean that the beliefs about the opponents' choices must be independent – which will be the case only if we condition on some fixed belief hierarchies for the opponents. In terms of our epistemic model, the conditional independence condition can be formalized as follows: Consider a type t_i for player i, and fix some types t_j and t_k for the opponents j and k, which are assigned positive probability by type t_i. Then, t_i's belief about t_j's choice should be stochastically independent from t_i's belief about t_k's choice. This condition is obviously weaker than the independence condition in Bernheim (1984) and Pearce (1984). Brandenburger and Friedenberg (2008) take common belief in rationality and additionally impose common belief in the event that types have conditionally independent beliefs. The concept obtained is, in terms of choices selected, somewhere between the concept of common belief in rationality and the concept of rationalizability.

Common prior

Harsanyi (1967–1968) devoted Part III of his work on games with incomplete information to a special case, in which the players' beliefs about the opponents' types are derived from a *common prior*. He calls this the *consistent priors* case. Formally, it means that there is a *unique* prior probability distribution p for the players' type combinations, such that the belief of a type t_i about the opponents' type combinations is obtained by conditioning the probability distribution p on his own type t_i. This imposes a severe restriction on the beliefs that the various types in the model may hold about the opponents' types, and hence seriously restricts the possible belief *hierarchies* that the various players may hold in a game.

Aumann (1987) takes the concept of common belief in rationality and restricts the players' belief hierarchies by imposing the common prior assumption. More precisely, Aumann considers a set of states of the world, assigns to every state ω a choice $c_i(\omega)$ for every player i, a set of states $P_i(\omega)$ that player i deems possible at ω, and a probabilistic belief $b_i(\omega) \in \Delta(P_i(\omega))$ that player i holds at ω about the states of the world. From this model we can then derive a complete belief hierarchy for the players at every state. Aumann then imposes that (1) for every state ω, the choice $c_i(\omega)$ for every player i is optimal, given his belief at ω about the opponents' choices, and (2) the beliefs $b_i(\omega)$ are derivable from a *common* prior probability distribution p on the set of states. By the latter, we mean that the belief $b_i(\omega)$ of player i at ω is obtained by conditioning the prior probability distribution p on the set of states $P_i(\omega)$ that player i deems possible at ω.

Aumann shows that under the conditions (1) and (2), the induced prior probability distribution for the players' choice-combinations constitutes a *correlated equilibrium*, as defined in Aumann (1974). Condition (1) is essentially equivalent to requiring common belief in rationality for every state of the world. Aumann (1987) shows that if we take common belief in rationality, and additionally require that the players' belief hierarchies are derivable from a common prior, then we obtain correlated equilibrium. Tan and Werlang (1988) also prove this result within their framework. Formally, a *correlated equilibrium* (Aumann, 1974) is a probability distribution q for the players' choice-combinations such that, for every player i and every choice c_i that has a positive probability under q, the choice c_i is optimal for player i given the belief about the opponents' choices obtained by conditioning q on the choice c_i. In terms of choices selected, a correlated equilibrium lies between common belief in rationality and Nash equilibrium (see Chapter 4).

Probability 1 beliefs

In Theorem 3.6.1 we have shown that we can always find belief hierarchies that (1) express common belief in rationality, and (2) assign probability 1 to a specific choice and belief hierarchy for every opponent. Such belief hierarchies correspond to what Bernheim (1984) calls *point-rationalizability* – that is, rationalizability if we restrict to beliefs that always assign probability 1 to one specific choice for every opponent. These beliefs are called *point beliefs* by Bernheim. Hence, in Theorem 3.6.1 we prove, in particular, the existence of point-rationalizable choices in the sense of Bernheim (1984). In Bernheim's paper, the existence of these choices is proved differently in Proposition 3.1.

Best-response sets

The concept of a *best-response set*, as in theoretical problem 3.9, was first defined by Pearce (1984) who called them "sets with the best response property." Pearce used best-response sets to characterize the rationalizable choices in a game: He showed that the rationalizable choices are precisely those choices that belong to some best-response set. This is analogous to the result found for our theoretical problem 3.9. The only difference is that we use the concept of common belief in rationality instead of rationalizability, and that, unlike Pearce, we do not assume that the players' beliefs about the opponents' choices are independent across the opponents. Brandenburger and Dekel (1987) provide a definition of a best-response set – which they call the *best-reply set* – which is identical to ours, in the sense that they do not require the players' beliefs to be independent. In fact, Brandenburger and Dekel (1987) use their notion of a best-reply set to define *correlated rationalizability*, which differs from the concept of rationalizability (Bernheim, 1984 and Pearce, 1984) in that beliefs are no longer required to be independent. In terms of choices, correlated rationalizability is equivalent to common belief in rationality, as the choices that can rationally be made under common belief in rationality are precisely the correlated rationalizable choices. Also Bernheim (1984), when proving the existence of rationalizable choices, discusses a notion that is similar to – but slightly different from – best-response sets. The difference is as follows: For every player i consider some subset $D_i \subseteq C_i$ of choices. Then, the collection $(D_1, D_2, ..., D_n)$ of subsets of choices is called a *best-response set* in the sense of Pearce (1984) if for every player i, every choice $c_i \in D_i$ is optimal under some independent probabilistic belief

$$b_i^{c_i} \in \Delta(D_1 \times ... \times D_{i-1} \times D_{i+1} \times ... \times D_n).$$

Bernheim (1984) requires, in addition, that every choice $c_i \in C_i$ that is optimal under some independent probabilistic belief

$$b_i^{c_i} \in \Delta(D_1 \times ... \times D_{i-1} \times D_{i+1} \times ... \times D_n)$$

must be in D_i. This is exactly what Basu and Weibull (1991) call a *tight CURB set*, where CURB stands for "closed under rational behavior." So, Bernheim's notion is more restrictive than Pearce's notion of a best-response set. Bernheim (1984) shows that the rationalizable choices in a game are precisely those choices in the largest tight CURB set of the game – a result that is similar to that for the theoretical problem 3.9.

Large epistemic models

Recall that an epistemic model M assigns to every player i a set of types T_i, and assigns to every type $t_i \in T_i$ some probabilistic belief $b_i(t_i)$ about the opponents' choice-type combinations. Within an epistemic model M we can derive for every type $t_i \in T_i$ the full belief hierarchy it induces. We say that the epistemic model M is *terminal* if every possible belief hierarchy we can think of is already contained in M. That is, for every player i, and every infinite belief hierarchy for player i that we could possibly construct, there is a type $t_i \in T_i$ that has precisely this belief hierarchy. Since there are obviously infinitely many – and in fact uncountably many – possible belief hierarchies, any terminal epistemic model must necessarily contain infinitely many – in fact, uncountably many – types for every player.

An important – but difficult – question that has been addressed in the literature is whether we can always construct a terminal epistemic model for every game. That is, can we always find an epistemic model that contains every possible belief hierarchy we can think of? Armbruster and Böge (1979), Böge and Eisele (1979) and Mertens and Zamir (1985) where the first to explicitly construct such terminal epistemic models. Later, Brandenburger and Dekel (1993), Heifetz (1993) and Heifetz and Samet (1998a) extended the above constructions by relaxing the topological assumptions made in the model. Epstein and Wang (1996) show that a similar construction also works in a more general framework in which the players hold hierarchies of *preferences over acts*, rather than hierarchies of beliefs, satisfying certain regularity conditions. In a nutshell, the above constructions of a terminal epistemic model proceed as follows:

We start by defining, for every player i, the basic space of uncertainty X_i^1, which in this chapter is the set of opponents' choice-combinations. So, we set $X_i^1 = C_{-i}$ for all players i. Define the set of all first-order beliefs for player i by taking all probabilistic beliefs on X_i^1. Hence, $B_i^1 = \Delta(X_i^1)$ is the set of all first-order beliefs for player i.

For the second step, we define the second-order space of uncertainty X_i^2 for player i by

$$X_i^2 = X_i^1 \times (\times_{j \neq i} B_j^1).$$

That is, X_i^2 contains all the opponents' choices and first-order beliefs. The set of all second-order beliefs B_i^2 for player i is then obtained by taking all probabilistic beliefs on X_i^2, so $B_i^2 = \Delta(X_i^2)$.

By continuing in this fashion, we recursively define for every k the kth-order space of uncertainty X_i^k by

$$X_i^k = X_i^{k-1} \times (\times_{j \neq i} B_j^{k-1}).$$

Hence, the kth-order space of uncertainty contains all elements from the $(k-1)$th-order space of uncertainty, together with the opponents' $(k-1)$th-order beliefs. The set B_i^k of all kth-order beliefs for player i is then $B_i^k = \Delta(X_i^k)$.

A *belief hierarchy* for player i within this model is then an infinite sequence $b_i = (b_i^1, b_i^2, ...)$ of beliefs where $b_i^k \in B_i^k$ for all k. So, a belief hierarchy consists of a first-order belief, a second-order belief, and so on, ad infinitum. We say that a belief hierarchy b_i is *coherent* if the various beliefs in this sequence do not contradict each other. That is, if we take two beliefs b_i^k and b_i^m in the sequence with $k < m$, then b_i^k and b_i^m should induce the same beliefs on X_i^k. It can then be shown that every coherent belief hierarchy holds a probabilistic belief about the opponents' choices and opponents' belief hierarchies.

Now, denote by T_i the set of belief hierarchies for player i that are coherent, and express common belief in the event that all players hold a coherent belief hierarchy. Call every belief hierarchy in T_i a *type*. Let T_{-i} be the set of opponents' type combinations. The crucial steps in the construction are to prove that (1) the sets of types so constructed are compact, (2) every type $t_i \in T_i$ holds a probabilistic belief $b_i(t_i)$ about the opponents' choices and the opponents' types in T_{-i}, (3) the mapping b_i that assigns to every type t_i this belief $b_i(t_i)$ is continuous, and (4) for every probabilistic belief about the opponents' choices and the opponents' types in T_{-i}, there is a type in T_i that has exactly this belief.

Epistemic models that satisfy condition (3) are called *continuous*, whereas epistemic models with property (4) are said to be *complete*. Friedenberg (2010) shows that, whenever the epistemic model is complete and continuous, and the sets of types T_i are compact, then the epistemic model is also terminal. Hence, the properties (1)–(4) above guarantee that the epistemic model M^* so constructed is *terminal*. In the literature, the specific epistemic model M^* constructed by this recursive procedure is often called the *canonical* epistemic model.

The canonical epistemic model M^* also has the special property that every epistemic model M we can possibly construct can be embedded inside M^*. So, in a sense, M^* is the "largest" epistemic model we can think of. We say that the epistemic model M^* is *universal*.

The epistemic models we use in this book all contain finitely many types for every player, and are therefore necessarily not terminal. The reason we do not use terminal epistemic models is that we do not really need them. Moreover, epistemic models with finitely many types have the advantage that they can more easily be represented in examples – which is very important for this book.

In each of the models discussed above, the belief that a player has about the opponents' choices – and about the opponents' beliefs – has always been represented by a probability distribution. Within such settings, it has been shown that terminal epistemic models generally exist. This is no longer true, however, if we represent the players' beliefs by *possibility sets* rather than probability distributions. By possibility sets we mean that every type deems possible some set of opponents' choice-type combinations, without specifying the probabilities for these combinations. In such settings, there may not be a single epistemic model that captures all possible belief hierarchies – see, for instance, Heifetz and Samet (1998b), Meier (2005), Brandenburger and Keisler (2006) and Brandenburger (2003). For more details on the possibility and impossibility of complete and terminal epistemic models the reader should consult Brandenburger (2003) and Siniscalchi (2008).

Finite belief hierarchies

Throughout this chapter, and in fact throughout this book, we assume that the belief hierarchies for the players consist of *infinitely* many levels. That is, a player holds a *first-order* belief about the opponents' choices, a *second-order* belief about the opponents' first-order beliefs about their opponents' choices, a *third-order* belief about the opponents' second-order beliefs and so on, ad infinitum. We thus assume that players can reason about beliefs about beliefs about beliefs – without restriction. Kets (2010) proposed an extended epistemic model in which certain types may not be able to reason all the way, but will in fact "stop reasoning" after finitely many steps in their belief hierarchy. Such types thus essentially possess a *finite* belief hierarchy, in which beliefs are only formed about opponents' first-order beliefs up to opponents' kth-order beliefs, for a given k.

The number machine

The example "The number machine" at the end of this chapter is based on a well-known game that is usually called "guessing two-thirds of the average" or "beauty contest."See Nagel (1995) for an experimental study of this game.

4 Simple belief hierarchies

4.1 Simple belief hierarchies

For many decades, game theory has been dominated by the concept of *Nash equilibrium*. In this chapter we will see that Nash equilibrium is obtained if we take *common belief in rationality* as the starting point and additionally impose conditions that go beyond the idea of common belief in rationality. We argue, however, that the additional conditions that are needed to arrive at Nash equilibrium are not very natural – in fact they are rather artificial. As a consequence, we believe that Nash equilibrium is not a very plausible concept to use when studying the reasoning processes of players in a game, even though Nash equilibrium has played a central role in game theory for many years. And this may in fact be the main message of this chapter. At the same time there is a very natural and basic alternative to Nash equilibrium – namely *common belief in rationality* – which does not suffer from these drawbacks.

In this chapter we will identify precisely those conditions that separate the concept of *common belief in rationality* – which we have studied in the previous two chapters – from the concept of *Nash equilibrium*. We will see that these conditions can be summarized by the concept of a *simple belief hierarchy*. As a first illustration of a simple belief hierarchy, let us consider the following example.

Example 4.1 Teaching a lesson

Suppose you are back at high school. During the biology class on Friday, the teacher tells you that you are very talented but that your grades could be much better if you would study more. For that reason, he announces that he will teach you a lesson! More precisely, he will give you a surprise exam someday next week, without telling you which day, and he wants to see you suffer and fail the exam, so that you will learn your lesson. When you go home on Friday afternoon, you have to decide on which day you will start preparing for the exam. Experience has taught you that you need to study at least the last two days before an exam in order to pass. To get full marks would require that you study at least six days. In this case, your father will say that he is proud of you. He has always been very tough with his children and he only gives compliments if somebody gets full marks in an exam at school.

Table 4.1. *Utilities for you and the teacher in "Teaching a lesson"*

		Teacher				
		Mon	Tue	Wed	Thu	Fri
You	Sat	3, 2	2, 3	1, 4	0, 5	3, 6
	Sun	−1, 6	3, 2	2, 3	1, 4	0, 5
	Mon	0, 5	−1, 6	3, 2	2, 3	1, 4
	Tue	0, 5	0, 5	−1, 6	3, 2	2, 3
	Wed	0, 5	0, 5	0, 5	−1, 6	3, 2

The utilities for you and the teacher are as follows: Passing the exam increases your utility by 5, whereas every day you study for it decreases your utility by 1. A compliment from your father is worth 4 to you. The teacher enjoys a utility of 5 if you fail the exam and every day you study for the exam increases his utility by 1.

The question is: On which day would you start studying for the exam? In order to analyze this question, we first put the utilities for you and the teacher for every possible combination of choices into a table. This results in Table 4.1. Here, your choice *Sat* means that you start studying on Saturday and that you keep studying every day until the day of the exam. Similarly for your other choices. Since you must study at least the last two days before the exam in order to pass, and the exam will be on Friday at the very latest, it does not make any sense to start studying on Thursday. The teacher, on the other hand, can choose any day on which to set the exam.

The utilities in the table are derived as follows: If you start studying on Saturday, and the teacher sets the exam on Wednesday, then you will pass the exam by studying during exactly four days. So, your utility would be $5-4=1$ and the teacher's utility would be 4. If you start studying on Monday and the teacher sets the exam on Tuesday, then you will fail by studying only one day. So, your utility would be -1 and the teacher's utility would be $5+1=6$. The only situation in which you will get full marks is when you start studying on Saturday and the teacher sets the exam on Friday. In that case you would study for exactly six days, pass the exam – with full marks – and receive a compliment from your father, so your utility would be $5+4-6=3$ and the teacher's utility would be 6. Similarly, you may verify the other utilities in the table.

For this situation, we can draw the beliefs diagram as depicted in Figure 4.1. If you believe that your teacher will set the exam on Friday, then starting studying on either Saturday or on Wednesday are optimal: By starting on Saturday you would expect to get full marks, enjoying the compliments from your father, whereas starting on Wednesday would reduce your expected working load to two days only, while still expecting to pass the exam. Starting studying on any other day would be optimal if you believe that the teacher will set the exam exactly two days later. On the other hand, if the teacher believes that you will start studying on Saturday, then he believes that he cannot fail

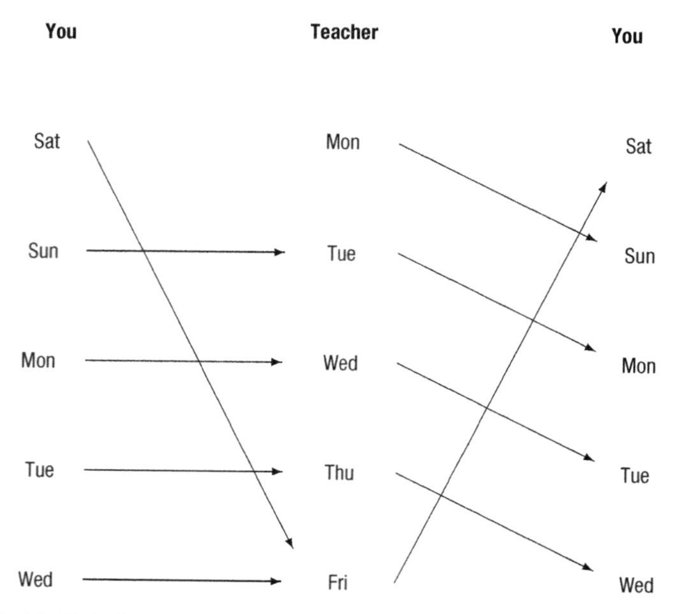

Figure 4.1 A beliefs diagram for "Teaching a lesson"

you, so the best he can do is to maximize your working load by setting the exam on Friday. Putting the exam on any other day of the week would be optimal for him if he believes that you will start studying exactly one day before. By doing so, he would expect you to study for one day and fail.

By starting at any of your choices and following the arrows, we only reach rational choices. Therefore, by using the graphical method from Chapter 3, we can conclude that each of your study schedules can rationally be chosen under common belief in rationality. Despite this, it is still possible to make a distinction between your various choices by looking at the *structure* of the belief hierarchies that support them.

Consider, for instance, your choice to start studying on Saturday. In the beliefs diagram, it is supported by a belief hierarchy in which

- you believe that the teacher will hold the exam on Friday,
- you believe that the teacher believes that you will start studying on Saturday,
- you believe that the teacher believes that you believe that the teacher will set the exam on Friday,
- you believe that the teacher believes that you believe that the teacher believes that you will start studying on Saturday,

and so on.

You could say that this belief hierarchy is entirely *generated* by the belief σ_2 that the teacher will hold the exam on Friday, and the belief σ_1 that you will start studying on Saturday. The belief hierarchy states that:

- your belief about the teacher's choice will be σ_2,
- you believe (with probability 1) that the teacher's belief about your choice is σ_1,
- you believe (with probability 1) that the teacher believes (with probability 1) that your belief about the teacher's choice is σ_2,
- you believe (with probability 1) that the teacher believes (with probability 1) that you believe (with probability 1) that the teacher's belief about your own choice is σ_1,

and so on.

Such a belief hierarchy will be called a *simple belief hierarchy* because it is entirely generated by the beliefs σ_1 and σ_2 above. A consequence of holding such a simple belief hierarchy is that you believe that the teacher is *correct* about the belief you hold about the teacher's choice. Namely, not only is your belief about the teacher's choice given by σ_2, but you also believe that the teacher believes that your belief about the teacher's choice is *indeed* σ_2.

Consider next the belief hierarchy that supports your choice to start studying on Sunday. In this belief hierarchy,

- you believe that the teacher will hold the exam on Tuesday,
- you believe that the teacher believes that you will start studying on Monday,
- you believe that the teacher believes that you believe that the teacher will hold the exam on Wednesday,
- you believe that the teacher believes that you believe that the teacher believes that you will start studying on Tuesday,

and so on.

This belief hierarchy is *not simple*. Your first-order belief states that you believe that the teacher will hold the exam on Tuesday, but your third-order belief states that you believe that the teacher believes that you believe that the teacher will hold the exam on *Wednesday*, and not on Tuesday. That is, you believe that the teacher is actually *wrong* about your belief about the teacher's choice. Also, your second-order belief states that you believe that the teacher believes that you will start studying on Monday, but your fourth-order belief states that you believe that the teacher believes that you believe that the teacher believes that you will start studying on *Tuesday*, and not on Monday. Hence, you believe that the teacher believes that you are *wrong* about the teacher's belief about your choice.

In a similar fashion, you may verify that the belief hierarchies that support your choices Monday and Tuesday are also not simple. Finally, the belief hierarchy that supports your choice to start studying on Wednesday is exactly the same as the belief hierarchy that supports your choice to start on Saturday, and is therefore simple.

Summarizing, we may conclude that within this beliefs diagram your choices to start studying on Saturday and Wednesday are supported by a simple belief hierarchy, whereas your choices to start on Sunday, Monday and Tuesday are not supported by simple belief hierarchies.

Now, does this render your choices to start studying on Saturday and Wednesday more plausible than your other choices? We believe not. The fact that a belief hierarchy is *simple* does not mean that this belief hierarchy is more plausible than some other belief hierarchy which is *not simple*. Remember that in a simple belief hierarchy, you believe that the teacher is *correct* about your belief about the teacher's choice. But in this game there are *many* plausible beliefs one can hold about the teacher's choice, so there is no good reason to expect that the teacher will be correct about your belief about the teacher's choice. In fact, it would be an absolute coincidence if the teacher were to be correct about your belief. As such, it is natural to allow for belief hierarchies in which you believe that the teacher is wrong about your belief about the teacher's choice, and hence there is no need to insist on simple belief hierarchies. □

In the example above, the simple belief hierarchy that supports your choices Saturday and Wednesday was completely determined by your first-order belief and your second-order belief. Moreover, your first- and second-order beliefs were both probability 1 beliefs: Your first-order belief states that you believe that, with probability 1, the teacher will set the exam on Friday, whereas the second-order belief states that you believe that the teacher believes that, with probability 1, you will start studying on Saturday.

There may be other situations in which you have a simple belief hierarchy, completely determined by your first- and second-order beliefs, but in which these two beliefs assign probabilities to *various* choices rather than assigning probability 1 to one particular choice. The following example illustrates this.

Example 4.2 Going to a party

Recall the story from the previous two chapters. Assume that the utilities are given by Table 4.2, which is a copy of Table 3.3 in the previous chapter. That is, Barbara likes it when you wear the same color as she does, whereas you strongly dislike this. For this situation we can make the beliefs diagram as depicted in Figure 4.2. Here, the arrows leaving your choices *green* and *red* should be read as follows: Both *green* and *red* are optimal for you if you believe that, with probability 2/3, Barbara will choose *blue* and with probability 1/3 she will choose *green*. So, your choices *green* and *red* can both be supported by this very same belief. Suppose that you hold this belief about Barbara's choice. Then, wearing *green* would give you an expected utility of

$$(2/3) \cdot 3 + (1/3) \cdot 0 = 2,$$

since you believe that, with probability 1/3, Barbara will wear the same color as you. Wearing *red* would also give you an expected utility of 2, since you believe that, with probability 1, Barbara will wear a different color. On the other hand, choosing *blue* would only give you an expected utility of

$$(2/3) \cdot 0 + (1/3) \cdot 4 = 4/3,$$

since you believe that, with probability 2/3, Barbara will wear the same color. Finally, choosing *yellow* would yield an expected utility of 1. So, if you believe that, with

Table 4.2. *Utilities for Barbara and you in "Going to a party"*

	blue	green	red	yellow	same color as friend
you	4	3	2	1	0
Barbara	4	3	2	1	5

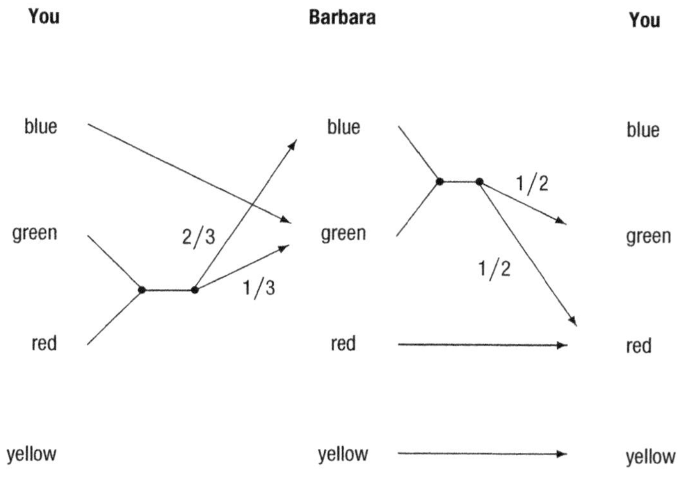

Figure 4.2 A beliefs diagram for "Going to a party" (IV)

probability 2/3, Barbara will choose *blue*, and with probability 1/3 she will choose *green*, then both *green* and *red* are optimal choices for you.

Similarly, the arrows leaving Barbara's choices *blue* and *green* indicate that both *blue* and *green* are optimal for her if she believes that, with probability 1/2, you will choose *green*, and with probability 1/2 you will choose *red*. Namely, if Barbara holds this belief then choosing *blue* would give her an expected utility of 4, since she believes that, with probability 1, you will wear a different color. Wearing *green* would give her an expected utility of

$$(1/2) \cdot 5 + (1/2) \cdot 3 = 4,$$

since she believes that, with probability 1/2, you will wear the same color. On the other hand, choosing *red* would yield her an expected utility of

$$(1/2) \cdot 2 + (1/2) \cdot 5 = 3.5,$$

since she believes that, with probability 1/2, you will wear the same color. Finally, choosing *yellow* would give her an expected utility of 1 only. So, both *blue* and *green* are optimal for Barbara if she believes that, with probability 1/2, you will choose *green*, and with probability 1/2 you will choose *red*.

From the beliefs diagram it can immediately be seen that you can rationally make your choices *blue*, *green* and *red* under common belief in rationality. If you start from any of these choices and follow the arrows in the diagram of Figure 4.2, then you only reach rational choices. So, by our graphical method from Chapter 3 we can conclude that any of these choices can rationally be made under common belief in rationality.

However, we can still distinguish between these choices by looking at the structure of the belief hierarchies that support them. Start, for instance, at your choice *blue*. Then, by following the arrows in the beliefs diagram, we obtain the belief hierarchy in which

- you believe that Barbara will choose *green*,
- you believe that Barbara believes that, with probability 1/2, you will choose *green* and with probability 1/2 you will choose *red*,
- you believe that Barbara believes that you believe that, with probability 2/3, she will choose *blue*, and with probability 1/3 she will choose *green*,

and so on.

Here, we have derived your third-order belief as follows: By following the arrows, we see that you believe that Barbara believes that, *with probability 1/2*,

- you believe that, with probability 2/3, Barbara will choose *blue*, and with probability 1/3 Barbara will choose *green*,

and you believe that Barbara believes that, *with probability 1/2*,

- you believe that, with probability 2/3, Barbara will choose *blue*, and with probability 1/3 Barbara will choose *green*.

So, in total, you believe that Barbara believes that, *with probability 1*,

- you believe that, with probability 2/3, Barbara will choose *blue*, and with probability 1/3 Barbara will choose *green*,

which is exactly the third-order belief we have written down in the belief hierarchy above.

This belief hierarchy is *not simple* since you believe that Barbara is incorrect about your first-order belief: On the one hand you believe that Barbara will choose *green*, but at the same time you believe that Barbara believes that you believe that, with probability 2/3, she will chooses *blue*, and not *green*.

Consider next the belief hierarchy that supports your choices *green* and *red*. By starting at either of these two choices, and following the arrows in the diagram, we see that:

- you believe that with probability 2/3, Barbara will choose *blue*, and with probability 1/3 Barbara will choose *green*,
- you believe that Barbara believes that, with probability 1/2, you will choose *green* and with probability 1/2 you will choose *red*,

- you believe that Barbara believes that you believe that, with probability 2/3, she will choose *blue*, and with probability 1/3 she will choose *green*,
- you believe that Barbara believes that you believe that Barbara believes that, with probability 1/2, you will choose *green* and with probability 1/2 you will choose *red*,

and so on, ad infinitum.

This belief hierarchy is completely determined by the belief σ_2 stating that, with probability 2/3, Barbara will choose *blue* and with probability 1/3 she will choose *green*, and the belief σ_1 stating that, with probability 1/2, you will choose *green* and with probability 1/2 you will choose *red*. Your belief hierarchy above states that:

- your belief about Barbara's choice is σ_2,
- you believe (with probability 1) that Barbara's belief about your choice is σ_1,
- you believe (with probability 1) that Barbara believes (with probability 1) that your belief about Barbara's choice is σ_2,
- you believe (with probability 1) that Barbara believes (with probability 1) that you believe (with probability 1) that Barbara's belief about your choice is σ_1,

and so on, ad infinitum.

In other words, this belief hierarchy is completely determined by your first-order and second-order beliefs, and is therefore a *simple* belief hierarchy. In particular, you believe within this belief hierarchy that Barbara is entirely correct about your precise probabilistic belief about Barbara's choice. Namely, you not only believe that, with probability 2/3 Barbara will choose *blue* and with probability 1/3 she will choose *green*, but you also believe that Barbara is correct in her belief about the probabilities you assign to Barbara's choices. This, however, seems a rather unnatural condition to impose! Why would you expect Barbara to be correct about these precise probabilities? Indeed, we see no good reason to impose such a condition.

Anyhow, we conclude that in this particular game, your choices *green* and *red* can be supported by a *simple* belief hierarchy that expresses common belief in rationality. □

The two examples above have illustrated the notion of a *simple belief hierarchy* in games with two players. For such games, we say that a belief hierarchy for player 1 is *simple* if it is completely generated by some probability distribution σ_2 over player 2's choices, and some probability distribution σ_1 over player 1's choices. More precisely, a belief hierarchy for player 1 is called *simple* if there is some probability distribution σ_2 over player 2's choices and some probability distribution σ_1 over player 1's choices such that:

- player 1's belief about player 2's choice is σ_2,
- player 1 believes that player 2's belief about player 1's choice is σ_1,
- player 1 believes that player 2 believes that player 1's belief about player 2's choice is σ_2,

- player 1 believes that player 2 believes that player 1 believes that player 2's belief about player 1's choice is σ_1,

and so on. In this case, we say that the belief hierarchy above is the simple belief hierarchy *generated* by the pair (σ_1, σ_2) of probability distributions.

Let us return to the example "Teaching a lesson." We have seen that the belief hierarchy supporting your choice to start studying on Saturday and Wednesday is a simple belief hierarchy. Consider the probability distributions σ_1 and σ_2 where

$\sigma_1 =$ the belief that, with probability 1, you will start studying on Saturday,
$\sigma_2 =$ the belief that, with probability 1, the teacher will hold the exam on Friday.

Then, the belief hierarchy is generated by the pair (σ_1, σ_2), and is therefore a simple belief hierarchy. The other three belief hierarchies for you from Figure 4.1 are not simple.

In the example "Going to a party" we have seen that the belief hierarchy that supports your choices *green* and *red* is a simple belief hierarchy. Consider the probability distributions σ_1 and σ_2 where

$\sigma_1 =$ the belief that, with probability 1/2, you will choose *green* and with probability 1/2, you will choose *red*;
$\sigma_2 =$ the belief that, with probability 2/3, Barbara will chooses *blue* and with probability 1/3 she will choose *green*.

Then, the belief hierarchy supporting your choices *green* and *red* is generated by the pair (σ_1, σ_2), and is thus a simple belief hierarchy. The other belief hierarchy for you from Figure 4.2 is not simple.

Let us now turn to games with more than two players. As we will see, the idea of a *simple belief hierarchy* can be extended to such games. The following example will illustrate this.

Example 4.3 Movie or party?

This evening you face a difficult decision. You have been invited to a party, together with your friends Barbara and Chris, but at the same time one of your favorite movies *Once upon a time in America* will be on TV. You prefer going to a party if you will have a good time there, but you prefer watching a movie to having a bad time at a party. More precisely, having a good time at a party gives you a utility of 3, watching a movie gives you a utility of 2, whereas having a bad time at a party gives you a utility of 0. Experience has taught you that you only have a good time at a party if both Barbara and Chris go. The problem, however, is that you are not sure whether both will come to the party, since they also like the movie.

In fact, the decision problems for Barbara and Chris are similar to yours: Watching the movie gives them both a utility of 2, having a good time at a party gives them a utility of 3, whereas having a bad time at a party gives them a utility of 0. However, Barbara and Chris were having a fierce discussion yesterday evening, and for that reason would like to avoid each other today. More precisely, assume that Barbara will only have a good time at the party if you will be there but not Chris, and that Chris will only have a good time if you will be there but not Barbara.

What should you do this evening: Go to the party or stay at home and watch the movie?

Let us first try to make a beliefs diagram for this situation. Note that it is optimal for you to stay at home if you believe that both friends will stay at home as well. On the other hand, it is optimal for you to go to the party if you believe that both friends will go as well.

For Barbara it is optimal to stay at home if she believes that you and Chris will stay at home as well. On the other hand, it would be optimal for her to go to the party if she believes that you will go but not Chris.

Finally, for Chris it is optimal to stay at home if he believes that you and Barbara will stay at home as well. On the other hand, it would be optimal for him to go to the party if he believes that you will go but not Barbara.

Based on these considerations, we can make the beliefs diagram in Figure 4.3. As you can see from the diagram, you can rationally make both choices under common

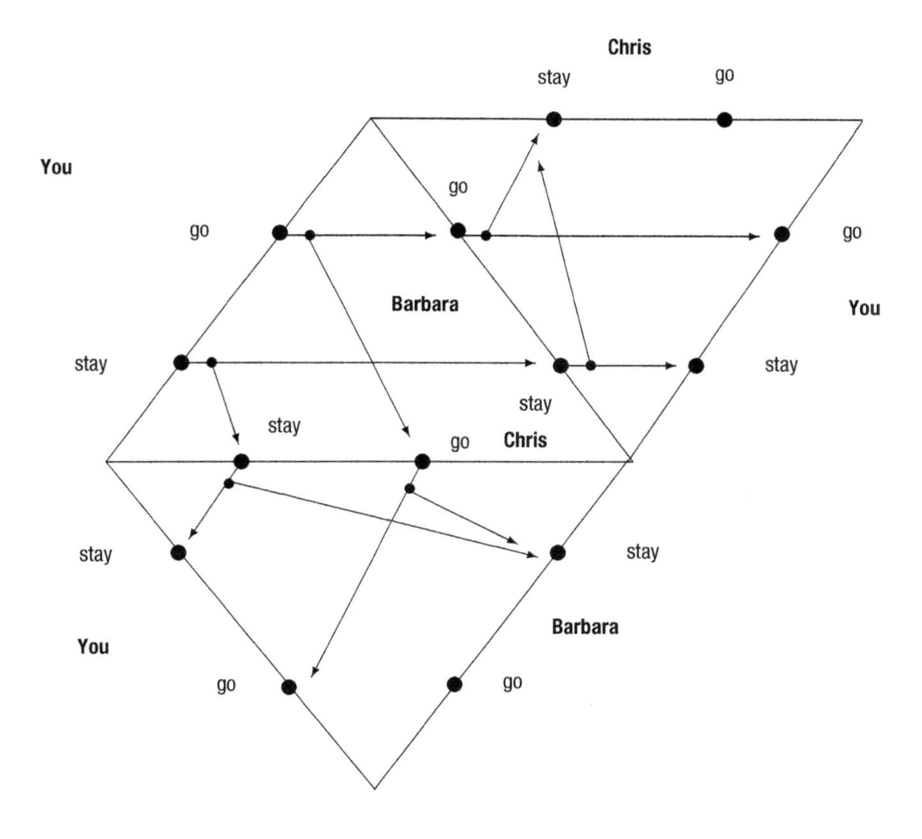

Figure 4.3 A beliefs diagram for "Movie or party?"

belief in rationality. If you start at your choice "stay" or at your choice "go" and follow the arrows in the diagram, you will only reach rational choices.

Still, we can distinguish between "going to the party" and "staying at home" by looking at the structure of the belief hierarchy that supports these choices. Consider first the belief hierarchy in the diagram that supports your choice to stay at home. It states that:

- you believe that both friends will stay at home,
- you believe that both friends believe that the others will stay at home,
- you believe that both friends believe that the others believe that everybody else will stay at home,

and so on. This belief hierarchy is completely generated by the beliefs σ_1, σ_2, and σ_3 where

$$\sigma_1 = \text{the belief that you will stay at home,}$$
$$\sigma_2 = \text{the belief that Barbara will stay at home,}$$
$$\sigma_3 = \text{the belief that Chris will stay at home.}$$

Your belief hierarchy that supports your choice to stay at home states that:

- your belief about Barbara's choice is σ_2,
- your belief about Chris' choice is σ_3,
- you believe that Barbara has belief σ_1 about your choice,
- you believe that Barbara has belief σ_3 about Chris' choice,
- you believe that Chris has belief σ_1 about your choice,
- you believe that Chris has belief σ_2 about Barbara's choice,

and so on. For this reason, we call it a *simple* belief hierarchy.

Compare this to the belief hierarchy in the diagram that supports your choice to go to the party. This belief hierarchy states that:

- you believe that Barbara and Chris will go to the party,
- you believe that Barbara believes that you will go to the party but not Chris,
- you believe that Chris believes that you will go to the party but not Barbara,

and so on. In particular, you believe that Chris will go the party, but you believe that Barbara believes that Chris will stay at home. Hence, you believe that Barbara's belief about Chris' choice differs from your own belief about Chris' choice. This means, however, that we cannot formulate a unique belief σ_3 about Chris' choice that can be used for the entire belief hierarchy, and therefore this belief hierarchy is *not simple*.

But again, we do not believe that the simple belief hierarchy above is more plausible than the other, non-simple, belief hierarchy. Why should you believe that Barbara has the same belief about Chris' choice as you do? Both of Chris' choices seem plausible, and hence there is no need to believe that Barbara will share your belief about Chris' choice. And as such, there is no good reason to prefer the simple belief hierarchy above, supporting your choice "staying at home," to the non-simple belief hierarchy that supports your choice "go to the party."

But we see that in this example, your choice "staying at home" is supported by a *simple* belief hierarchy in the beliefs diagram, whereas "going to the party" is not. □

The three examples above should give a good feel for what we mean by a *simple belief hierarchy*. In each of the examples, a belief hierarchy for player i was called *simple* if it is generated by some combination $(\sigma_1, ..., \sigma_n)$ of beliefs about the players' choices. That is, not only is player i's belief about player j's choice given by σ_j for every opponent j, but player i also believes that his opponents share these same beliefs, and he believes that his opponents believe that all other players share these same beliefs as well, and so on. More formally, we have the following definition.

Definition 4.1.1 *(Belief hierarchy generated by $(\sigma_1, ..., \sigma_n)$)*
Let $(\sigma_1, ..., \sigma_n)$ be a combination of beliefs, where σ_1 is a probabilistic belief about player 1's choice, ..., σ_n is a probabilistic belief about player n's choice. The belief hierarchy for player i that is ***generated by*** *$(\sigma_1, ..., \sigma_n)$ states that:*
(1) for every opponent j, player i has belief σ_j about player j's choice,
(2) for every opponent j and every player $k \neq j$, player i believes that player j has belief σ_k about player k's choice,
(3) for every opponent j, every player $k \neq j$, and every player $l \neq k$, player i believes that player j believes that player k has belief σ_l about player l's choice, and so on.

With this definition, it easy to define a simple belief hierarchy.

Definition 4.1.2 *(Simple belief hierarchy)*
Consider an epistemic model with sets of types $T_1, ..., T_n$. A type $t_i \in T_i$ for player i is said to have a ***simple belief hierarchy*** *if t_i's belief hierarchy is generated by some combination $(\sigma_1, ..., \sigma_n)$ of beliefs about the players' choices.*

In general it is possible – and often even likely – that a player has certain beliefs about his opponents' choices and beliefs, but believes that his opponents are *wrong* about his actual beliefs. For instance, you may very well believe that the teacher will hold the exam on Monday, but at the same time believe that the teacher believes that you believe that the teacher will hold the exam on Tuesday. Or, you may believe that Barbara believes that Chris will stay at home, and at the same time believe that Barbara believes that you believe that Barbara believes that Chris will go to the party.

However, if a type has a simple belief hierarchy, then it will always believe that his opponents are *correct* about his own beliefs! Consider a type t_i with a simple belief hierarchy. Then, t_i's belief hierarchy must be generated by some combination $(\sigma_1, ..., \sigma_n)$ of beliefs about the players' choices. Take an opponent j. Then, type t_i has belief σ_j about player j's choice, but t_i also believes that all opponents believe that he – player i – has belief σ_j about player j's choice. So, type t_i believes that his opponents are right about his belief about player j's choice. Similarly, type t_i believes that player j has belief σ_k about k's choice. At the same time, he believes that all opponents believe that he – player i – believes that player j has belief σ_k about k's choice. Hence, type t_i

believes that his opponents are right about his own belief about player j's belief about k's choice. In a similar way, you can verify that if type t_i has a certain belief about an opponent's choice or belief, then he believes that all other players are right about this belief.

This, however, is not a very natural condition to impose! In general, there are many plausible beliefs you can hold about your opponents, so there is no good reason to believe that your opponents will actually be correct about the beliefs you hold. In most games, it would be an absolute coincidence if your opponents were to be entirely correct about the beliefs you hold about the opponents' choices.

Another important characteristic of a simple belief hierarchy is that you believe that your opponents share your thoughts about other players. To see this, consider a type t_i with a simple belief hierarchy generated by a combination $(\sigma_1, ..., \sigma_n)$ of beliefs. Suppose that there are at least three players. Then, type t_i not only has belief σ_j about opponent j's choice, but also believes that every other player has this particular belief σ_j about j's choice. Similarly, type t_i not only believes that opponent j has belief σ_k about k's choice, but also believes that every other player believes that player j has belief σ_k about k's choice. In general, whenever type t_i holds a certain belief about opponent j, which may be a belief about j's choice or a belief about j's belief, then t_i believes that every other player shares the same belief about player j.

But this is also not a natural condition to impose! In general, there will be many plausible beliefs one can hold about player j, so why should you believe that every other player has the same belief about player j as you do? This seems rather artificial and overly restrictive.

Summarizing, we see that a type with a simple belief hierarchy will always believe that his opponents are correct about his own beliefs and that his opponents share his own beliefs about other players. Unlike common belief in rationality, this is certainly not a basic condition to impose on a player's reasoning process. In fact, there is nothing wrong with believing that some of your opponents may have incorrect beliefs about your own beliefs. After all, your opponents cannot look inside your head, so why should they be correct about your beliefs? And there is also nothing strange about believing that some opponent holds a different belief about a third player than you do. All people think differently, so why should your opponents hold exactly the same beliefs as you do? For these reasons we believe that *imposing* a simple belief hierarchy on a player is rather unnatural and overly restrictive.

4.2 Nash equilibrium

In Chapters 2 and 3 we introduced the idea of *common belief in rationality*, stating that you not only believe that your opponents choose rationally, but you also believe that your opponents believe that their opponents will choose rationally as well, and so on. In the previous section we introduced the idea of a *simple belief hierarchy*, which states that the whole belief hierarchy can be summarized by a combination $(\sigma_1, ..., \sigma_n)$ of beliefs about players' choices. In particular, a player with a simple belief hierarchy always believes that his opponents are *correct* about the beliefs he holds, and believes that every other player will hold the same belief about a given opponent as he does – if there are at least three players. We have argued that these are not very natural conditions to impose and that they are in fact overly restrictive.

In this section we will be interested in those belief hierarchies that not only express common belief in rationality but are also simple. As we will see, such belief hierarchies will lead to *Nash equilibrium*.

Let us consider an arbitrary game with n players and a simple belief hierarchy for player i generated by a combination of beliefs $(\sigma_1, ..., \sigma_n)$ about the players' choices. So, the belief hierarchy states that:

- player i's belief about j's choice is given by σ_j for every opponent j,
- player i believes that every opponent j has belief σ_k about player k, for every opponent j and every player $k \neq j$,
- player i believes that every opponent j believes that every player k has belief σ_l about player l, for every opponent j, every player $k \neq j$, and every player $l \neq k$,

and so on, ad infinitum.

There is a more compact way to write this belief hierarchy. Let us denote by σ_{-i} the combination $(\sigma_1, ..., \sigma_{i-1}, \sigma_{i+1}, ..., \sigma_n)$ of beliefs about all players' choices, except player i's. Then, player i's belief about his opponents' choices is given by σ_{-i}. Moreover, the simple belief hierarchy generated by $(\sigma_1, ..., \sigma_n)$ can be written as the belief hierarchy which states that:

- player i's belief about his opponents' choices is σ_{-i},
- player i believes that every opponent j has belief σ_{-j} about his opponents' choices,
- player i believes that every opponent j believes that each of his opponents k has belief σ_{-k} about his opponents' choices,

and so on.

Now, suppose that the simple belief hierarchy above expresses common belief in rationality. Consider an opponent j. By construction of the belief hierarchy, player i holds belief σ_j about player j's choice, and believes that player j holds belief σ_{-j} about his opponents' choices. Take a choice c_j for player j with $\sigma_j(c_j) > 0$. Then, player i assigns positive probability to choice c_j. Since player i's belief hierarchy expresses common belief in rationality, player i must, in particular, believe that player j chooses rationally. So, if player i assigns positive probability to choice c_j, then player i must believe that this choice c_j is optimal for player j. That is, choice c_j must be optimal for player j given the belief σ_{-j} he is believed to hold about his opponents' choices. We may thus conclude that if $\sigma_j(c_j) > 0$, then choice c_j must be optimal for player j if he holds belief σ_{-j} about his opponents' choices. This must hold for every opponent j.

Consider now a choice c_i for player i with $\sigma_i(c_i) > 0$. Choose an arbitrary opponent j. Remember that player i believes that player j holds belief σ_i about player i's choice. So, if $\sigma_i(c_i) > 0$, then player i believes that player j assigns positive probability to player

i's choice c_i. Since player i's belief hierarchy expresses common belief in rationality, player i must, in particular, believe that player j believes that player i chooses rationally. So, player i must believe that player j believes that choice c_i is optimal for player i. Since player i believes that player j believes that player i holds belief σ_{-i} about the opponents' choices, it must therefore be the case that choice c_i is optimal for player i given his belief σ_{-i} about the opponents' choices. We have thus seen that if $\sigma_i(c_i) > 0$, then choice c_i must be optimal for player i if he holds belief σ_{-i} about his opponents' choices.

Summarizing, we may conclude the following: If the simple belief hierarchy for player i generated by $(\sigma_1, ..., \sigma_n)$ expresses common belief in rationality, then for every player j, the belief σ_j only assigns positive probability to a choice c_j if c_j is optimal for player j if he had the belief σ_{-j} about his opponents' choices.

What about the opposite direction? Consider the simple belief hierarchy generated by $(\sigma_1, ..., \sigma_n)$, and suppose that for every player j the belief σ_j only assigns positive probability to choices c_j that are optimal for player j under the belief σ_{-j} about the opponents' choices. Can we then conclude that the belief hierarchy expresses common belief in rationality? As we will see, the answer is "yes."

We first show that player i believes in his opponents' rationality. Consider an arbitrary opponent j, and suppose that player i assigns positive probability to some choice c_j for player j. Then, $\sigma_j(c_j) > 0$, and hence, by our assumption above for $(\sigma_1, ..., \sigma_n)$, choice c_j is optimal for player j under the belief σ_{-j}. As player i believes that player j indeed has belief σ_{-j} about his opponents' choices, we may conclude that player i believes that choice c_j is optimal for player j. So, if player i assigns positive probability to some choice c_j, then he believes that c_j is optimal for player j. This, however, means that player i believes in opponent j's rationality. As this is true for every opponent j, we can conclude that player i believes in his opponents' rationality.

We next show that player i believes that his opponents believe in their opponents' rationality. Consider an arbitrary opponent j, and some other player $k \neq j$ (possibly $k = i$). Suppose that player i believes that player j assigns positive probability to choice c_k for player k. As player i believes that player j has belief σ_k about k's choice, it must be that $\sigma_k(c_k) > 0$. By our assumption for $(\sigma_1, ..., \sigma_n)$, choice c_k must be optimal for player k under the belief σ_{-k} about his opponents' choices. Since player i believes that player j believes that player k's belief about his opponents' choices is indeed σ_{-k}, player i believes that player j believes that choice c_k is optimal for player k. So, we see that player i believes that player j assigns positive probability to some choice c_k for player k only if player i believes that player j believes that choice c_k is optimal for player k. As such, player i believes that player j believes in player k's rationality. Since this is true for all opponents j and all players $k \neq j$, we may conclude that player i believes that his opponents believe in their opponents' rationality.

In a similar way, we can show that player i believes that his opponents believe that their opponents believe in their opponents' rationality, and so on, ad infinitum. We can therefore show that player i's simple belief hierarchy generated by $(\sigma_1, ..., \sigma_n)$ expresses common belief in rationality.

Together with our previous insight, we obtain the following characterization: The simple belief hierarchy for player i generated by $(\sigma_1, ..., \sigma_n)$ expresses common belief in rationality *if and only if* for every player j, the belief σ_j only assigns positive probability to choices c_j that are optimal for player j under the belief σ_{-j} about the opponents' choices.

The condition we have formulated above for $(\sigma_1, ..., \sigma_n)$ coincides exactly with that of *Nash equilibrium* – an idea that was first formulated by the American mathematician John Nash.

Definition 4.2.1 *(Nash equilibrium)*
Consider a game with n players. Let $(\sigma_1, ..., \sigma_n)$ be a combination of beliefs where σ_1 is a probabilistic belief about player 1's choice, ..., σ_n is a probabilistic belief about player n's choice. The combination of beliefs $(\sigma_1, ..., \sigma_n)$ is a ***Nash equilibrium*** *if for every player j, the belief σ_j only assigns positive probability to choices c_j that are optimal for player j under the belief σ_{-j} about the opponents' choices.*

In view of our characterization above, we can formulate the following relation between simple belief hierarchies expressing common belief in rationality and Nash equilibrium.

Theorem 4.2.2 *(Simple belief hierarchies versus Nash equilibrium)*
Consider the simple belief hierarchy for player i generated by the combination $(\sigma_1, ..., \sigma_n)$ of beliefs about players' choices. Then, this simple belief hierarchy expresses common belief in rationality if and only if the combination $(\sigma_1, ..., \sigma_n)$ of beliefs is a Nash equilibrium.

That is, Nash equilibrium is obtained if we take common belief in rationality and additionally assume that each player holds a *simple belief hierarchy*. We have argued, however, that assuming a simple belief hierarchy is rather unnatural and often overly restrictive. As a consequence, we believe that Nash equilibrium is not a very plausible concept for studying how people reason in games.

The result above shows that searching for simple belief hierarchies that express common belief in rationality can be reduced to searching for Nash equilibria. Every Nash equilibrium will generate a simple belief hierarchy expressing common belief in rationality, and, conversely, every simple belief hierarchy expressing common belief in rationality can be generated by a Nash equilibrium.

What about existence? Will we always be able to find a simple belief hierarchy that expresses common belief in rationality? In fact, John Nash solved this problem for us in his PhD dissertation. There, he showed that we can always find a Nash equilibrium $(\sigma_1, ..., \sigma_n)$ for every game with finitely many choices.

Theorem 4.2.3 *(Nash equilibrium always exists)*
For every game with finitely many choices there is at least one Nash equilibrium $(\sigma_1, ..., \sigma_n)$.

The proof of this result can be found in the proofs section at the end of this chapter. If we combine John Nash's result with Theorem 4.2.2, we come to the conclusion that for every player i there is always a simple belief hierarchy that expresses common belief in rationality.

Theorem 4.2.4 *(Common belief in rationality with simple belief hierarchies is always possible)*
Consider a game with finitely many choices. Then, for every player i there is at least one simple belief hierarchy that expresses common belief in rationality.

Although it may be difficult in some examples to find a simple belief hierarchy that expresses common belief in rationality, the theorem above guarantees that one always exists! From Theorem 3.6.1 in Chapter 3 we know that, for every game, we can always construct an epistemic model in which every type expresses common belief in rationality, and every type assigns probability 1 to one specific choice and type for every opponent. This, however, may no longer be possible if we additionally require simple belief hierarchies! Consider, for instance, the example "Going to a party" with the utilities from Table 4.2. We will see in the next section that for this game there is only one Nash equilibrium (σ_1,σ_2), where

$\sigma_1 =$ the belief that, with probability $1/2$, you will choose *green*, and, with probability $1/2$, you will choose *red*;
$\sigma_2 =$ the belief that, with probability $2/3$, Barbara will choose *blue*, and, with probability $1/3$, Barbara will choose *green*.

By Theorem 4.2.2 it follows that there is only one simple belief hierarchy for you that expresses common belief in rationality, namely the simple belief hierarchy generated by the beliefs (σ_1,σ_2) above. This simple belief hierarchy, however, does not assign probability 1 to one particular choice and type for Barbara. We may therefore conclude that for the example "Going to a party," there is no type for you that expresses common belief in rationality, has a simple belief hierarchy and assigns probability 1 to one particular choice and type for Barbara. These three requirements are thus in general incompatible with each other.

4.3 Computational method

In this section we will discuss those choices you can rationally make under common belief in rationality *with a simple belief hierarchy*. As we saw at the very end of the previous section, for every player i there is at least one simple belief hierarchy that expresses common belief in rationality. Therefore, every player i will always have at least one choice he can rationally make under common belief in rationality while having a simple belief hierarchy.

An important question that remains is: How can we compute such choices? In fact, we want to have a method that computes *all* choices you can rationally make with a simple belief hierarchy under common belief in rationality. Unfortunately, we will not end up with an easy elimination procedure as in Chapter 3 when studying common belief in rationality. Instead, we will have a computational method that is not as convenient as the iterated elimination of strictly dominated choices, but that will nevertheless greatly simplify our search for choices that can rationally be made with a simple belief hierarchy under common belief in rationality. The connection with Nash equilibrium, which we established in the previous section, will play a prominent role in our method.

Consider an arbitrary game with n players, and suppose we wish to find *all* choices that player i can rationally make with a simple belief hierarchy expressing common belief in rationality. In the previous section, we saw that a simple belief hierarchy for player i expressing common belief in rationality can be generated by a Nash equilibrium $(\sigma_1, ..., \sigma_n)$. So, a choice c_i can rationally be made with a simple belief hierarchy expressing common belief in rationality *if and only if* the choice c_i is optimal for a simple belief hierarchy generated by a Nash equilibrium $(\sigma_1, ..., \sigma_n)$.

What does it mean to say that c_i is optimal for a simple belief hierarchy generated by a Nash equilibrium $(\sigma_1, ..., \sigma_n)$? Remember that in this belief hierarchy, player i's belief about the opponents' choices is given by σ_{-i}. As such, the choice c_i is optimal for the simple belief hierarchy generated by the Nash equilibrium $(\sigma_1, ..., \sigma_n)$ *if and only if* the choice c_i is optimal for player i under the belief σ_{-i} about the opponents' choices. If we combine our two insights above, we reach the following conclusion:

A choice c_i can rationally be made with a simple belief hierarchy expressing common belief in rationality *if and only if* there is a Nash equilibrium $(\sigma_1, ..., \sigma_n)$ such that c_i is optimal for player i under the belief σ_{-i} about the opponents' choices. Such choices c_i are called *Nash choices*.

Definition 4.3.1 *(Nash choice)*
A choice c_i for player i is a **Nash choice** *if there is some Nash equilibrium $(\sigma_1, ..., \sigma_n)$ such that c_i is optimal for player i under the belief σ_{-i} about the opponents' choices.*

We thus obtain the following characterization.

Theorem 4.3.2 *(Simple belief hierarchies versus Nash choices)*
A choice c_i for player i can rationally be made with a simple belief hierarchy under common belief in rationality if and only if the choice c_i is a Nash choice.

Therefore, if you wish to find all choices you can rationally make with a simple belief hierarchy under common belief in rationality, you must compute all of the Nash choices in the game. This leads to the following computational method.

Algorithm 4.3.3 *(Nash equilibrium method)*
Step 1. *Compute all Nash equilibria $(\sigma_1, ..., \sigma_n)$ for the game.*
Step 2. *For every Nash equilibrium $(\sigma_1, ..., \sigma_n)$, determine all choices that are optimal for player i under the belief σ_{-i} about the opponents' choices. This gives you all Nash choices in the game.*

The choices found in step 2 are exactly the choices that player i can rationally make with a simple belief hierarchy under common belief in rationality.

Unfortunately, there is no simple algorithm that will compute all the Nash equilibria of a given game. For some games, computing all the Nash equilibria, or even finding a single Nash equilibrium, can in fact be a very difficult task. Despite this, the method above will still be of great help. As an illustration, let us go back to the examples we discussed in Section 4.1. For each of these examples, we will apply the method above to compute all the choices you can rationally make under common belief in rationality with a simple belief hierarchy.

Example 4.4 Teaching a lesson

Recall the story from Example 4.1. The choices and utilities for you and the teacher are reproduced in Table 4.3. In Section 4.1 we saw that you can rationally choose to start studying on Saturday or on Wednesday under common belief in rationality with a simple belief hierarchy. These two choices can be supported by the simple belief hierarchy in which

- you believe that the teacher will hold the exam on Friday,
- you believe that the teacher believes you will start studying on Saturday,
- you believe that the teacher believes that you believe that the teacher will hold the exam on Friday,

and so on. As this belief hierarchy expresses common belief in rationality, we may conclude that you can rationally choose to start studying on Saturday or on Wednesday under common belief in rationality with a simple belief hierarchy.

Are there other choices you can rationally make under common belief in rationality with a simple belief hierarchy? If we look at the beliefs diagram in Figure 4.1, we see that the choices *Sun*, *Mon* and *Tue* are supported by non-simple belief hierarchies that express common belief in rationality. This, however, does not mean that these choices cannot be made rationally under common belief in rationality with a simple belief hierarchy! Perhaps there is some other beliefs diagram in which some of these choices *are* supported by a simple belief hierarchy expressing common belief in rationality. We do not know yet! Let us use the computational method above in order to answer this question.

As a first step we will try to compute all Nash equilibria (σ_1, σ_2) in this game, where σ_1 is a probabilistic belief about your choices and σ_2 is a probabilistic belief about the teacher's choices. Suppose that (σ_1, σ_2) is a Nash equilibrium.

We first show that σ_2 must assign probability 0 to the teacher's choice *Thu*. Suppose, on the contrary, that $\sigma_2(Thu) > 0$. Then, by the definition of a Nash equilibrium, *Thu* must be optimal for the teacher under the belief σ_1 about your choice. This is only possible if $\sigma_1(Wed) > 0$, because otherwise *Fri* is always a better choice for the teacher than *Thu*. If the teacher believes that you will not choose *Wed*, then *Thu* always yields the teacher a lower utility than *Fri*. On the other hand, $\sigma_1(Wed) > 0$ implies that *Wed* must be optimal for you under the belief σ_2 about the teacher's choice. This is only

Table 4.3. *Utilities for you and the teacher in "Teaching a lesson" (II)*

		Teacher				
		Mon	Tue	Wed	Thu	Fri
You	Sat	3,2	2,3	1,4	0,5	3,6
	Sun	−1,6	3,2	2,3	1,4	0,5
	Mon	0,5	−1,6	3,2	2,3	1,4
	Tue	0,5	0,5	−1,6	3,2	2,3
	Wed	0,5	0,5	0,5	−1,6	3,2

possible if σ_2 assigns probability 1 to the teacher's choice *Fri*, because otherwise *Sat* is a better choice for you than *Wed*. This, however, contradicts our assumption that $\sigma_2(Thu) > 0$. So, in a Nash equilibrium it cannot be the case that $\sigma_2(Thu) > 0$. In other words, in every Nash equilibrium we must have that $\sigma_2(Thu) = 0$.

We now show that σ_2 must also assign probability 0 to the teacher's choice *Wed*. Suppose, on the contrary, that $\sigma_2(Wed) > 0$. Then, *Wed* must be optimal for the teacher under the belief σ_1 about your choice. This is only possible if $\sigma_1(Tue) > 0$, because otherwise *Thu* is a better choice for the teacher than *Wed*. However, $\sigma_1(Tue) > 0$ implies that *Tue* must be optimal for you under the belief σ_2 about the teacher's choice. This is only possible if $\sigma_2(Thu) > 0$, because otherwise *Sat* is a better choice for you than *Tue*. However, we have seen above that $\sigma_2(Thu) > 0$ is not possible in a Nash equilibrium. We must therefore conclude that our assumption, that $\sigma_2(Wed) > 0$, cannot hold in a Nash equilibrium. So, $\sigma_2(Wed) = 0$ in every Nash equilibrium.

We next show that σ_2 must assign probability 0 to the teacher's choice *Tue* as well. Suppose, on the contrary, that $\sigma_2(Tue) > 0$. Then, *Tue* must be optimal for the teacher under the belief σ_1 about your choice. This is only possible if $\sigma_1(Mon) > 0$. In order to see this, suppose that $\sigma_1(Mon) = 0$. Then, the teacher's choice *Tue* would always be worse than the randomized choice in which he chooses *Wed* with probability 0.9 and *Thu* with probability 0.1. So, *Tue* cannot be optimal for the teacher if he assigns probability 0 to your choice *Mon*. Therefore, it must be that $\sigma_1(Mon) > 0$. This implies that *Mon* must be optimal for you under the belief σ_2 about the teacher's choice. This is only possible if $\sigma_2(Wed) > 0$ or $\sigma_2(Thu) > 0$, since otherwise your choice *Mon* would be worse than your choice *Sat*. However, we have seen above that $\sigma_2(Wed) > 0$ and $\sigma_2(Thu) > 0$ are impossible in a Nash equilibrium. Therefore, our assumption that $\sigma_2(Tue) > 0$ cannot be true. So, $\sigma_2(Tue) = 0$ in all Nash equilibria.

We finally show that σ_2 must also assign probability 0 to the teacher's choice *Mon*. Assume, on the contrary, that $\sigma_2(Mon) > 0$. Then, *Mon* must be optimal for the teacher under the belief σ_1 about your choice. This is only possible if $\sigma_1(Sun) > 0$. In order to see this, suppose that $\sigma_1(Sun) = 0$. Then, the teacher's choice *Mon* would always be worse than the randomized choice in which he chooses *Tue* with probability 0.9, *Wed* with probability 0.09 and *Thu* with probability 0.01. This means that *Mon* cannot be optimal for the teacher if he assigns probability 0 to your choice *Sun*. So, we must have that $\sigma_1(Sun) > 0$. Hence, *Sun* must be optimal for you under the belief σ_2 about the teacher's choice. This, in turn, is only possible if $\sigma_2(Tue) > 0$, because otherwise *Sun* is always worse for you than *Mon*. However, as we have seen above, $\sigma_2(Tue) > 0$ is not possible in a Nash equilibrium. Hence, our assumption that $\sigma_2(Mon) > 0$ cannot be true. So, $\sigma_2(Mon) = 0$ in all Nash equilibria.

Summarizing, we have seen so far that σ_2 must assign probability 0 to *Mon*, *Tue*, *Wed* and *Thu* in every Nash equilibrium (σ_1, σ_2). This means, however, that $\sigma_2(Fri) = 1$ in every Nash equilibrium (σ_1, σ_2).

By definition of a Nash equilibrium, σ_1 only assigns positive probability to your choices that are optimal under the belief σ_2. Since σ_2 states that you believe that the teacher will, with probability 1, choose *Fri*, your optimal choices under the belief σ_2 are *Sat* and *Wed*. So, σ_1 can only assign positive probability to your choices *Sat* and *Wed*. Suppose that σ_1 assigns probability α to *Sat* and probability $1-\alpha$ to *Wed*.

Since $\sigma_2(Fri) = 1$, it must be the case that *Fri* is optimal for the teacher under the belief σ_1 about your choice. The expected utility for the teacher of choosing *Fri* under the belief σ_1 is

$$u_2(Fri) = \alpha \cdot 6 + (1-\alpha) \cdot 2 = 4\alpha + 2.$$

If the teacher were to choose *Thu* instead, his expected utility would be

$$u_2(Thu) = \alpha \cdot 5 + (1-\alpha) \cdot 6 = 6 - \alpha.$$

For all other choices, the teacher's expected utility is always less than $u_2(Thu)$. Since *Fri* must be optimal for the teacher, we must have that $u_2(Fri) \geq u_2(Thu)$, which amounts to $\alpha \geq 0.8$.

So, the Nash equilibria in this game are all pairs of beliefs (σ_1, σ_2) in which

σ_1 assigns probability α to your choice *Sat*,
and probability $1-\alpha$ to your choice *Wed*,

with $\alpha \geq 0.8$, and

σ_2 assigns probability 1 to the teacher's choice *Fri*.

In each of these Nash equilibria (σ_1, σ_2), your optimal choices under the belief σ_2 are *Sat* and *Wed*. So, *Sat* and *Wed* are your only Nash choices in this game.

We may thus conclude, from our computational method, that you can only rationally choose *Sat* and *Wed* under common belief in rationality with a simple belief hierarchy. □

Example 4.5 Going to a party

Assume that the utilities for you and Barbara are given by Table 4.4, which is a copy of Table 4.2. So, Barbara likes it when you wear the same color as she does, whereas

Table 4.4. *Utilities for Barbara and you in "Going to a party"*

	blue	green	red	yellow	same color as friend
you	4	3	2	1	0
Barbara	4	3	2	1	5

you strongly dislike this. In Example 4.2 we saw that you can rationally choose *green* and *red* under common belief in rationality with a simple belief hierarchy. From the beliefs diagram in Figure 4.2 we can see that your choices *green* and *red* are supported by the simple belief hierarchy where

- you believe that, with probability 2/3, Barbara will choose *blue*, and with probability 1/3 Barbara will choose *green*,
- you believe that Barbara believes that, with probability 1/2, you will choose *green* and with probability 1/2 you will choose *red*,
- you believe that Barbara believes that you believe that, with probability 2/3, she will choose *blue*, and with probability 1/3 she will choose *green*,

and so on, ad infinitum.

Since we have also seen that this belief hierarchy expresses common belief in rationality, it follows that you can rationally choose *green* and *red* under common belief in rationality with a simple belief hierarchy.

What about your other two choices? Can they also be chosen rationally under common belief in rationality with a simple belief hierarchy? Obviously not *yellow*, since *yellow* is not a rational choice to start with. But can you rationally choose your most preferred color *blue* under common belief in rationality with a simple belief hierarchy? Somewhat surprisingly, we will show that you cannot!

In order to prove this, we will use the computational method from the previous section. We first compute all of the Nash equilibria (σ_1,σ_2) for this situation, where σ_1 is a probability distribution over your own choices, and σ_2 is a probability distribution over Barbara's choices.

Suppose that (σ_1,σ_2) is a Nash equilibrium. Then, σ_1 should only assign positive probability to choices for you that are optimal under the belief σ_2 about Barbara's choice. Since your choice *yellow* is irrational it cannot be optimal under any belief, and hence σ_1 should assign probability 0 to your choice *yellow*.

But then, Barbara's choice *yellow* cannot be optimal for her under the belief σ_1 about your choice: Since σ_1 assigns probability 0 to your choice *yellow*, Barbara's expected utility from choosing *yellow* under the belief σ_1 will be 1, which is lower than the expected utility she gets from any of her other choices. So, *yellow* cannot be optimal for Barbara under σ_1. Since σ_2 should only assign positive probability to Barbara's choices that are optimal under σ_1, it follows that σ_2 should assign probability 0 to Barbara's choice *yellow*.

We now show that σ_1 and σ_2 must assign positive probability to at least two different choices. Assume that σ_1 assigns probability 1 to a single choice for you. We distinguish the following cases.

If σ_1 assigns probability 1 to your choice *blue*, then only *blue* can be optimal for Barbara under σ_1. Hence, σ_2 should assign probability 1 to Barbara's choice *blue*. But then, your choice *blue* is not optimal under σ_2, so σ_1 should have assigned probability 0 to your choice *blue*, which is a contradiction. So, σ_1 cannot assign probability 1 to your choice *blue*.

In exactly the same way, we can show that σ_1 cannot assign probability 1 to your choice *green* or *red* either. We also know that σ_1 cannot assign probability 1 to your choice *yellow*, since *yellow* is an irrational choice for you. Hence, we see that σ_1 cannot assign probability 1 to a single choice for you, and should therefore assign positive probability to at least two of your choices.

We now show that σ_2 cannot assign probability 1 to a single choice for Barbara. We distinguish the following cases.

If σ_2 assigns probability 1 to her choice *blue*, then only *green* can be optimal for you under σ_2. But then, σ_1 should assign probability 1 to your choice *green* which, as we have seen, is impossible in a Nash equilibrium.

If σ_2 assigns probability 1 to any of her other choices, then only *blue* can be optimal for you under σ_2. So, σ_1 should assign probability 1 to your choice *blue* which, as we have seen, is impossible in a Nash equilibrium.

Overall, we see that in a Nash equilibrium, both σ_1 and σ_2 should assign positive probability to at least two different choices.

Next, we show that σ_2 cannot assign positive probability to Barbara's choices *green* and *red* at the same time. Suppose that σ_2 assigns positive probability to Barbara's choices *green* and *red*. Then, both *green* and *red* should be optimal for Barbara under σ_1. In particular, *green* and *red* should give Barbara a higher expected utility than *blue* under the belief σ_1. By choosing *blue*, Barbara gets at least utility 4. So, Barbara's expected utilities for choosing *green* and *red* under the belief σ_1 should both exceed 4. Her expected utility from choosing *green* is

$$\sigma_1(green) \cdot 5 + (1 - \sigma_1(green)) \cdot 3$$

since she believes that, with probability $\sigma_1(green)$, you will choose the same color. This expected utility must be at least 4, so

$$\sigma_1(green) \geq 1/2.$$

On the other hand, her expected utility from choosing *red* is

$$\sigma_1(red) \cdot 5 + (1 - \sigma_1(red)) \cdot 2$$

since she believes that, with probability $\sigma_1(red)$, you will choose the same color. This expected utility must be at least 4, so

$$\sigma_1(red) \geq 2/3.$$

However, σ_1 cannot assign probability at least 1/2 to *green* and probability at least 2/3 to *red*, since the sum of σ_1's probabilities must be 1. So, we must conclude that σ_2 cannot assign positive probability to Barbara's choices *green* and *red* at the same time.

We next show that σ_2 cannot assign positive probability to Barbara's choices *blue* and *red* at the same time. Suppose, on the contrary, that σ_2 assigns positive probability to Barbara's choices *blue* and *red*. Since we have seen above that σ_2 cannot assign positive probability to both *green* and *red*, we must conclude that σ_2 assigns probability 0 to Barbara's choice *green*. Then, your expected utility from choosing *green* under the belief σ_2 will be 3, which means that your choice *red* cannot be optimal under σ_2. As such, σ_1 should assign probability 0 to *red*. But then, *red* cannot be optimal for Barbara under σ_1, as choosing *red* would only give her an expected utility of 2, which is less than the expected utility she would get from choosing *blue* or *green*. Therefore, σ_2 should have assigned probability 0 to *red*, which is a contradiction. So, σ_2 cannot assign positive probability to Barbara's choices *blue* and *red* at the same time.

This means, however, that in a Nash equilibrium the belief σ_2 should assign positive probability to Barbara's choices *blue* and *green*, and probability 0 to her choices *red* and *yellow*.

We now show that σ_1 cannot assign positive probability to your choice *blue*. Suppose, on the contrary, that σ_1 assigns positive probability to your choice *blue*. Then, your choice *blue* must be optimal under σ_2. Since σ_2 assigns probability 0 to Barbara's choice *red*, your choice *red* gives you an expected utility of 2 under σ_2. As your choice *blue* must be optimal, as we have seen, your expected utility from choosing *blue* must exceed 2. Now, your expected utility from choosing *blue* under σ_2 is

$$\sigma_2(blue) \cdot 0 + (1 - \sigma_2(blue)) \cdot 4,$$

since you believe that, with probability $\sigma_2(blue)$, Barbara will wear the same color. This expected utility can only exceed 2 if

$$\sigma_2(blue) \leq 1/2.$$

Since σ_2 can only assign positive probability to *blue* and *green*, this means that

$$\sigma_2(green) \geq 1/2.$$

But then, your expected utility from choosing *green* will be

$$\sigma_2(green) \cdot 0 + (1 - \sigma_2(green)) \cdot 3 \leq 3/2,$$

since $\sigma_2(green) \geq 1/2$. This means, however, that *green* cannot be optimal for you under σ_2, since we have seen that choosing *red* would give you an expected utility of 2.

So, σ_1 must assign probability 0 to your choice *green*. But then, Barbara's expected utility from choosing *green* under σ_1 would be only 3, which is less than what she would obtain from choosing *blue*. So, *green* would not be optimal for Barbara under σ_1, which means that σ_2 should assign probability 0 to Barbara's choice *green*. This, however, is a contradiction, since we have seen that σ_2 must assign positive probability to Barbara's choices *blue* and *green*. So, σ_1 cannot assign positive probability to your choice *blue*.

Since σ_1 assigns probability 0 to your choices *blue* and *yellow*, but on the other hand must assign positive probability to at least two choices, we must conclude that σ_1 assigns positive probability to your choices *green* and *red*, and probability 0 to your choices *blue* and *yellow*.

Until now we have seen that in a Nash equilibrium, σ_1 must assign positive probability to both of your choices *green* and *red*, and σ_2 should assign positive probability to both of Barbara's choices *blue* and *green*.

Since σ_1 assigns positive probability to both *green* and *red*, it should be the case that both *green* and *red* are optimal for you under σ_2. In particular, your expected utilities from choosing *green* and *red* should be equal. Your expected utility from choosing *green* is

$$\sigma_2(green) \cdot 0 + \sigma_2(blue) \cdot 3,$$

whereas your expected utility from choosing *red* is 2, since you believe that Barbara will not choose *red*. Hence, we must have that

$$\sigma_2(green) \cdot 0 + \sigma_2(blue) \cdot 3 = 2,$$

meaning that $\sigma_2(blue) = 2/3$, and $\sigma_2(green) = 1/3$. As you may verify, under this belief σ_2 your choices *green* and *red* are both optimal.

On the other hand, since σ_2 assigns positive probability to both *blue* and *green*, both *blue* and *green* should be optimal choices for Barbara under σ_1. In particular, *blue* and *green* should give Barbara the same expected utility under σ_1. Her expected utility from choosing *blue* is 4, since σ_1 assigns probability 0 to your choice *blue*. Her expected utility from choosing *green* is

$$\sigma_1(green) \cdot 5 + (1 - \sigma_1(green)) \cdot 3.$$

Since both expected utilities must be equal, we must have that

$$\sigma_1(green) \cdot 5 + (1 - \sigma_1(green)) \cdot 3 = 4,$$

which means that $\sigma_1(green) = 1/2$ and $\sigma_1(red) = 1/2$. As you may easily check, both *blue* and *green* are optimal choices for Barbara under this belief σ_1.

Summarizing, we see that there is only *one* Nash equilibrium for this situation, namely the pair (σ_1, σ_2) where

$$\sigma_1(green) = 1/2,\ \sigma_1(red) = 1/2,$$
$$\sigma_2(blue) = 2/3,\ \sigma_2(green) = 1/3.$$

Table 4.5. *Utilities for you, Barbara and Chris in "Movie or party?"*

You stay	Chris stays	Chris goes
Barbara stays	2,2,2	2,2,0
Barbara goes	2,0,2	2,0,0
You go	Chris stays	Chris goes
Barbara stays	0,2,2	0,2,3
Barbara goes	0,3,2	3,0,0

Under the belief σ_2, only your choices *green* and *red* are optimal. By the computational method, we may therefore conclude that you can only rationally choose *green* and *red* under common belief in rationality with a simple belief hierarchy, but not your most preferred color *blue*! □

Example 4.6 Movie or party?

Recall the story from Example 4.3. We have seen that you can rationally choose to stay at home under common belief in rationality with a simple belief hierarchy. There is a simple belief hierarchy in which

- you believe that your friends will stay at home,
- you believe that your friends believe that the others will stay at home,
- you believe that your friends believe that the others believe that the others will stay at home,

and so on. From the beliefs diagram in Figure 4.3 we can see that this simple belief hierarchy expresses common belief in rationality. As this belief hierarchy supports your choice to stay at home, we know that you can rationally choose to stay at home under common belief in rationality with a simple belief hierarchy. From the same beliefs diagram, we also see that there is a non-simple belief hierarchy that expresses common belief in rationality and that supports your other choice, which is to go to the party. So, you can also rationally choose to go to the party under common belief in rationality. But can you do so with a simple belief hierarchy?

In order to answer this question, we will use our computational method to compute all the choices you can rationally make under common belief in rationality with a simple belief hierarchy. We must first compute all Nash equilibria (σ_1, σ_2, σ_3) in the game, where σ_1 is a belief about your choice, σ_2 is a belief about Barbara's choice and σ_3 is a belief about Chris' choice. To make this computation a bit easier, we list in Table 4.5 the choices and the utilities for all three players in the game. Here, the first utility refers to you, the second utility to Barbara and the third utility to Chris.

We first show that in every Nash equilibrium ($\sigma_1,\sigma_2,\sigma_3$), the belief σ_1 must assign probability 0 to your choice *go*.

Suppose, on the contrary, that $\sigma_1(go) > 0$. Then, by the definition of a Nash equilibrium, the choice *go* must be optimal for you under the belief (σ_2, σ_3) about the opponents' choices. The expected utility for you of choosing *go* under the belief (σ_2, σ_3) is equal to

$$u_1(go) = 3 \cdot \sigma_2(go) \cdot \sigma_3(go),$$

whereas your utility of choosing *stay* is simply 2. Since *go* must be optimal for you under the belief (σ_2, σ_3), it must be the case that $\sigma_2(go) \cdot \sigma_3(go) \geq \frac{2}{3}$, and hence

$$\sigma_2(go) > 0 \text{ and } \sigma_3(go) > 0.$$

In fact, since $\sigma_2(go) \cdot \sigma_3(go) \geq \frac{2}{3}$, we must even have that

$$\sigma_2(go) \geq \frac{2}{3} \text{ and } \sigma_3(go) \geq \frac{2}{3}.$$

So, $\sigma_3(stay) \leq \frac{1}{3}$. Since $\sigma_2(go) > 0$, the choice *go* must be optimal for Barbara under the belief (σ_1, σ_3). However, her expected utility from choosing *go* under the belief (σ_1, σ_3) is equal to

$$u_2(go) = 3 \cdot \sigma_1(go) \cdot \sigma_3(stay),$$

which is less than or equal to 1 since $\sigma_3(stay) \leq \frac{1}{3}$. This means, however, that *stay* is better for Barbara than *go* under the belief (σ_1, σ_3). This is a contradiction, as we have seen that $\sigma_2(go) > 0$, and therefore *go* must be optimal for Barbara under the belief (σ_1, σ_3). So, our assumption that $\sigma_1(go) > 0$ cannot be correct. Hence, $\sigma_1(go) = 0$ in every Nash equilibrium. That is, $\sigma_1(stay) = 1$ in every Nash equilibrium.

Now, let $(\sigma_1, \sigma_2, \sigma_3)$ be a Nash equilibrium. Since $\sigma_1(stay) = 1$, for Barbara it is always better to *stay* than to *go* under the belief (σ_1, σ_3) about the opponents' choices. So, σ_2 can only assign positive probability to Barbara's choice *stay*, which means that $\sigma_2(stay) = 1$. Similarly, since $\sigma_1(stay) = 1$, for Chris it is always better to *stay* than to *go* under the belief (σ_1, σ_2). Therefore, σ_3 can only assign positive probability to Chris' choice *stay*, and hence $\sigma_3(stay) = 1$.

So, there is only one candidate for a Nash equilibrium, namely the combination $(\sigma_1, \sigma_2, \sigma_3)$ of beliefs with

$$\sigma_1(stay) = 1, \ \sigma_2(stay) = 1 \text{ and } \sigma_3(stay) = 1.$$

It is easily verified that this combination is indeed a Nash equilibrium. We therefore have a unique Nash equilibrium in this game, namely the combination of beliefs in which every player is believed to stay at home with probability 1.

For this unique Nash equilibrium $(\sigma_1, \sigma_2, \sigma_3)$, your optimal choice under the belief (σ_2, σ_3) is to stay at home. Hence, *stay* is your only Nash choice in this game.

By the computational method, we may therefore conclude that you can only rationally choose to stay at home under common belief in rationality with a simple belief hierarchy. □

4.4 Belief that opponents hold correct beliefs

In the previous sections we have explored the consequences of having a *simple belief hierarchy*. In this section we will be concerned with a different question, namely *when* do people have simple belief hierarchies about their opponents? More precisely, we will identify some conditions such that, *if* your way of reasoning about the opponents satisfies these conditions, *then* you are guaranteed to hold a simple belief hierarchy. In order to see what these conditions may look like, let us return to the example "Teaching a lesson."

Example 4.7 Teaching a lesson

Let us have a close look again at the beliefs diagram of Figure 4.1. We have seen that the belief hierarchy supporting your choices Saturday and Wednesday is simple, since you believe that the teacher will hold the exam on Friday, you believe that the teacher believes that you will start studying on Saturday, you believe that the teacher believes that you indeed believe that the teacher will hold the exam on Friday, you believe that the teacher believes that you indeed believe that the teacher believes that you will start studying on Saturday, and so on.

In particular, you believe that the teacher *is correct about your own beliefs*. Namely, you not only believe that the teacher will hold the exam on Friday, but you also believe that the teacher believes that you indeed believe so. Moreover, you not only believe that the teacher believes that you will start studying on Saturday, but you also believe that the teacher believes that you indeed believe so. In fact, for whatever you believe in this belief hierarchy, it is the case that you believe that the teacher believes that you indeed believe this. We say that you *believe that the teacher holds correct beliefs*.

Within your simple belief hierarchy described above, you also *believe that the teacher believes that you hold correct beliefs*. For instance, you not only believe that the teacher believes that you will start studying on Saturday, but you also believe that the teacher believes that you believe that the teacher indeed believes this. In general, within the simple belief hierarchy it is true that, whenever you believe that the teacher believes something, you believe that the teacher believes that you believe that the teacher indeed believes this. □

In the example above we have seen that in the simple belief hierarchy supporting your choices Saturday and Wednesday, you *believe that the teacher holds correct beliefs* and you *believe that the teacher believes that you hold correct beliefs*. We will show that in games with two players, these two conditions guarantee that you have a simple belief hierarchy. Before we come to this result, let us first formally define what we mean by *believing that your opponent holds correct beliefs*.

Definition 4.4.1 *(Believing that the opponents have correct beliefs)*
A type t_i for player i ***believes that his opponents have correct beliefs*** *if, whenever type t_i assigns probability p to an opponents' choice-type combination $((c_1,t_1),...,(c_{i-1},t_{i-1}),(c_{i+1},t_{i+1}),...,(c_n,t_n))$, then type t_i believes (with probability 1) that every opponent believes (with probability 1) that player*

i indeed assigns probability p to this opponents' choice-type combination $((c_1,t_1),...,(c_{i-1},t_{i-1}),(c_{i+1},t_{i+1}),...,(c_n,t_n))$.

For instance, if you believe that, with probability 0.7, the teacher will choose Monday, and you believe that the teacher has correct beliefs, then you must believe, with probability 1, that the teacher believes, with probability 1, that you believe, with probability 0.7, that the teacher will choose Monday.

In general, a type t_i that believes that the opponents have correct beliefs, must believe that everybody is correct about his entire belief hierarchy. Now, a type, as we have seen, is nothing else than a name that is given to a particular belief hierarchy. So, if t_i believes that everybody is correct about his belief hierarchy, then he believes that all opponents believe that his type is indeed t_i. As such, a type t_i that believes that his opponents have correct beliefs, must believe that everybody is correct about his type! Conversely, if type t_i believes that everybody is correct about his type, he believes that everybody is correct about his entire belief hierarchy, and hence t_i believes that his opponents hold correct beliefs. We thus obtain the following useful characterization.

Lemma 4.4.2 *(Characterization of types that believe that opponents hold correct beliefs)*
A type t_i for player i believes that his opponents hold correct beliefs if and only if type t_i believes that every opponent believes that i's type is t_i.

We will now show that in a game with two players, any type that believes that his opponent holds correct beliefs, and believes that his opponent believes that he holds correct beliefs himself, must have a *simple belief hierarchy*. In fact, we shall prove that the converse is also valid, that is, every type with a simple belief hierarchy necessarily believes that his opponent holds correct beliefs, and believes that his opponent believes that he holds correct beliefs himself.

Theorem 4.4.3 *(Characterization of types with a simple belief hierarchy in two-player games)*
Consider a game with two players. A type t_i for player i has a simple belief hierarchy if and only if t_i believes that his opponent holds correct beliefs, and believes that his opponent believes that he holds correct beliefs himself.

Proof: Suppose first that type t_i believes that opponent j holds correct beliefs, and believes that opponent j believes that i holds correct beliefs as well. Our goal is to show that t_i has a simple belief hierarchy.

We first prove that t_i assigns probability 1 to a single type t_j for player j. Suppose this is not the case, so suppose that t_i assigns positive probability to at least two different types t_j and t_j' for player j. See Figure 4.4 for an illustration. Since t_i believes that opponent j holds correct beliefs, t_i must believe – see Lemma 4.4.2 – that j believes that i's type is t_i. Type t_i considers both types t_j and t_j' possible, and therefore both types t_j and t_j' must believe that i's type is t_i. Now, consider type t_j for player j. Type t_j believes

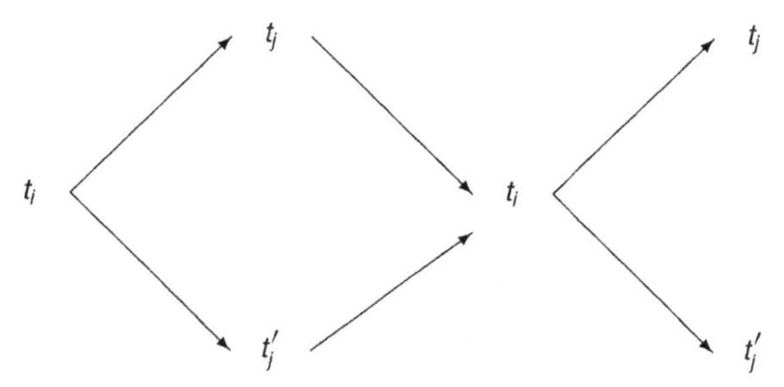

Figure 4.4 Suppose that t_i assigns positive probability to t_j and t_j'

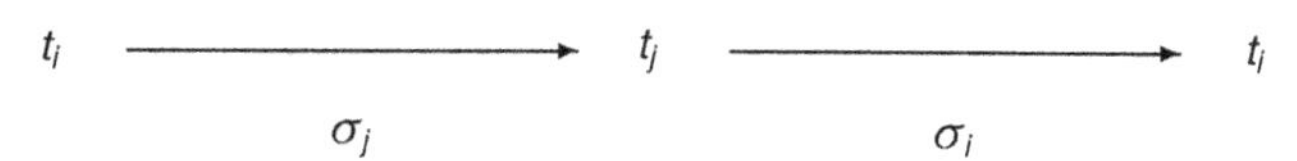

Figure 4.5 Type t_i must assign probability 1 to a single type t_j for player j

that i's type is t_i, who considers it possible that j's type is t_j'. Therefore, type t_j believes that i considers it possible that j's type is t_j', and not t_j. This, however, means that t_j does not believe that i has correct beliefs. Since t_i considers type t_j possible for player j, it would follow that type t_i does not believe with probability 1 that j believes that i has correct beliefs, which contradicts our assumption above. So, we must conclude that t_i assigns probability 1 to a single type t_j for player j who, in turn, assigns probability 1 to type t_i for player i. That is, the situation must be as in Figure 4.5.

Let σ_j be the belief that type t_i has about j's choice and let σ_i be the belief that t_j has about i's choice. We will show that t_i's belief hierarchy is generated by (σ_i, σ_j). First of all, t_i's belief about j's choice is σ_j. Moreover, since t_i believes that j has type t_j, it follows that t_i believes that j has belief σ_i about i's choice. As t_i believes that j believes that i's type is t_i, we have that t_i believes that j believes that i has belief σ_j about j's choice, and so on. So, indeed, t_i's belief hierarchy is generated by (σ_i, σ_j), and therefore we conclude that t_i has a *simple belief hierarchy*. This proves the first direction in the theorem.

For the converse, assume that type t_i has a simple belief hierarchy. Our goal is to show that t_i believes that opponent j has correct beliefs, and believes that j believes that i has correct beliefs as well.

Since t_i has a simple belief hierarchy, it must be generated by some pair (σ_i, σ_j) of beliefs, where σ_i is a belief about i's choice and σ_j is a belief about j's choice. So, type t_i not only has belief σ_j about j's choice, he also believes that j believes that i's belief about j's choice is indeed σ_j. Moreover, type t_i not only believes that j has belief σ_i about i's choice, he also believes that j believes that i believes that j's belief about i's choice is indeed σ_i. And so on. So, type t_i believes that j holds correct beliefs.

Now, suppose that t_i considers type t_j possible for player j. Then, type t_j not only has belief σ_i about i's choice, he also believes that i believes that j's belief about i's choice is indeed σ_i. Moreover, type t_j not only believes that i has belief σ_j about j's choice, he also believes that i believes that j believes that i's belief about j's choice is indeed σ_j. And so on. So, type t_j believes that i holds correct beliefs. Since this holds for every type t_j that t_i considers possible, type t_i believes that j believes that i has correct beliefs, as was to be shown.

We thus conclude that type t_i believes that j has correct beliefs and believes that j believes that i has correct beliefs. This completes the proof of the theorem. ■

In Theorem 4.2.2 we saw that, if you – being player i – hold a simple belief hierarchy generated by a combination of beliefs $(\sigma_1, ..., \sigma_n)$, and express common belief in rationality, then the combination of beliefs $(\sigma_1, ..., \sigma_n)$ must be a Nash equilibrium. In our Theorem 4.4.3 we have seen that, if in a two-player game you believe that the opponent has correct beliefs and believe that the opponent believes that you have correct beliefs, then you have a simple belief hierarchy. Moreover, this simple belief hierarchy is generated by the pair of beliefs (σ_i, σ_j), where σ_j is your belief about j's choice and σ_i is your belief about j's belief about your choice. So, if we combine the insights from both theorems, we can say the following for two-player games: If you believe that the opponent has correct beliefs, believe that the opponent believes that you have correct beliefs, and express common belief in rationality, then your belief σ_j about your opponent's choice together with your belief σ_i about your opponent's belief about your own choice, constitute a Nash equilibrium.

So, we have characterized a Nash equilibrium for two-player games by the following conditions: (1) you express common belief in rationality, (2) you believe that the opponent has correct beliefs, and (3) you believe that the opponent believes that you have correct beliefs. As we argued before, we believe that conditions (2) and (3) are rather unnatural as in general there is no good reason to expect that your opponent will be correct about your beliefs.

In the conditions above that lead to Nash equilibrium we can still weaken condition (1), though. It is not necessary to impose that you express *common* belief in rationality. It is sufficient to require that you believe in your opponent's rationality, and believe that your opponent believes in your own rationality.

Here is the reason: Suppose you are of type t_i, you believe that the opponent has correct beliefs, and you believe that the opponent believes that you have correct beliefs. From the proof of Theorem 4.4.3 we then know that you assign probability 1 to a single type t_j for player j, which assigns probability 1 to your true type t_i. Now, assume that you believe in j's rationality. Then, you only assign positive probability to choices c_j that are rational for t_j. Since you believe that j is correct about your beliefs, you believe that j believes that you only assign positive probability to choices c_j that are rational for t_j. On the other hand, you also believe that j believes that you believe that j's type is t_j. Taken together, you believe that j believes that you believe in j's rationality. So, from the assumption that you believe in j's rationality, it follows that you believe that j believes that you believe in j's rationality. In a similar fashion, this implies that you believe that j believes that you believe that j believes that you believe in j's rationality, and so on.

Assume also that you believe that j believes in your rationality. Then, type t_j can only assign positive probability to choices c_i that are rational for your type t_i. Since you believe that j believes that you are correct about his beliefs, t_j believes that you believe that j only assigns positive probability to choices c_i that are rational for t_i. Therefore, t_j believes that you believe that j believes in your rationality. Since you believe that j's type is t_j, it follows that you believe that j believes that you believe that j believes in your rationality. So, from the assumption that you believe that j believes in your rationality, it follows that you believe that j believes that you believe that j believes in your rationality. This, in turn, implies that you believe that j believes that you believe that j believes that you believe that j believes in your rationality, and so on.

Summarizing, we conclude that if you believe that the opponent has correct beliefs, you believe that the opponent believes that you have correct beliefs, you believe in your opponent's rationality and believe that your opponent believes in your rationality, then you express *common* belief in rationality. Above we saw that if you believe that the opponent has correct beliefs, you believe that the opponent believes that you have correct beliefs and you express common belief in rationality, then your belief σ_j about your opponent's choice together with your belief σ_i about your opponent's belief about your own choice, constitute a Nash equilibrium. By combining these two insights, we obtain the following result.

Theorem 4.4.4 *(Sufficient conditions for Nash equilibrium in a two-player game) Consider a game with two players. If type t_i believes that j has correct beliefs, believes that j believes that i has correct beliefs, believes in j's rationality and believes that j believes in i's rationality, then t_i's belief σ_j about j's choice together with t_i's belief σ_i about j's belief about i's choice, constitute a Nash equilibrium.*

The reader should be warned that this result only holds for games with two players. If we have three players or more, then the conditions above no longer lead to Nash equilibrium in general. Consider, for instance, the example "Movie or party?" with three players. Suppose:

- you believe that Barbara and Chris will go to the party,
- you believe that Barbara believes that you will go to the party but not Chris,
- you believe that Chris believes that you will go to the party but not Barbara,

and you believe that your friends believe that you hold these beliefs, you believe that your friends believe that you believe that your friends believe that you hold these beliefs, and so on. Assume, moreover, that you not only believe that Barbara believes that you will go to the party but not Chris, but you also believe that she believes that her friends believe that she believes so, and so on. Similarly, assume that you not only believe that Chris believes that you will go to the party but not Barbara, but you also believe that he believes that his friends believe that he believes so, and so on.

If you hold this belief hierarchy, then

- you believe that your friends have correct beliefs,

- you believe that Barbara believes that her friends have correct beliefs, and
- you believe that Chris believes that his friends have correct beliefs.

Moreover, it can be checked that you believe in your friends' rationality, and you believe that your friends believe in their friends' rationality. So, your belief hierarchy satisfies all the conditions of Theorem 4.4.4. However, your belief about your friends' choices, together with your belief about your friends' belief about your own choice, leads to the combination of beliefs (go, go, go), which is not a Nash equilibrium in the game. Namely, for Barbara it is not optimal to go to the party if her two friends go.

The problem is that the belief hierarchy described above is *not* a simple belief hierarchy: You believe that Barbara will go to the party but at the same time you believe that Chris believes that Barbara will stay at home. So, for games with three or more players, the conditions that you believe that the opponents have correct beliefs and believe that your opponents believe that their opponents have correct beliefs, no longer imply that your belief hierarchy is simple.

The conclusion is that for games with three or more players we need to add extra conditions in order to arrive at simple belief hierarchies. One extra condition we need is that you believe that opponent j has the same belief about opponent k as you do. In other words, you must project your belief about opponent k onto opponent j. We say that you hold *projective beliefs*. Another condition we must add is that your belief about opponent j's choice must be stochastically independent from your belief about opponent k's choice. We say that you have *independent beliefs*.

In a game with three or more players, it can be shown that if you believe that the opponents hold correct beliefs, your beliefs are *projective* and *independent*, you believe that every opponent believes that his opponents have correct beliefs and you believe that every opponent has *projective* and *independent* beliefs, then your belief hierarchy is guaranteed to be simple. In a similar way to above, we can then prove that if your belief hierarchy satisfies these conditions, and moreover you believe in the opponents' rationality and believe that your opponents believe in their opponents' rationality, then you express *common* belief in rationality. If we combine this insight with Theorem 4.2.2, we obtain the following result.

Theorem 4.4.5 *(Sufficient conditions for Nash equilibrium in a game with more than two players)*
Consider a game with more than two players. Let t_i be a type that believes in the opponents' rationality, holds projective and independent beliefs, believes that his opponents hold correct beliefs, and believes that every opponent has these four properties too. For every $j \neq i$, let σ_j be t_i's belief about j's choice and let σ_i be t_i's belief about any opponent's belief about i's choice. Then, $(\sigma_1, ..., \sigma_n)$ is a Nash equilibrium.

The proof for this result can be found in the proofs section at the end of this chapter. So we see that if we move from two players to three players or more, we need to impose extra conditions in order to arrive at Nash equilibrium. Moreover, the extra condition of *projective beliefs* is a rather problematic condition. In general there are many plausible beliefs one can hold about opponent j, so why should you believe that the other players share your beliefs about opponent j? Typically, there is no need for believing so, but Nash equilibrium nevertheless makes this assumption. This renders the concept of Nash equilibrium even more problematic when we consider three players or more.

4.5 Proofs

In the first two parts of this section we will prove Theorem 4.2.3, which shows that a Nash equilibrium always exists if all players have finitely many choices. In order to prove this result we will use *Kakutani's fixed point theorem* – a theorem in mathematics that shows that under certain conditions, a correspondence is guaranteed to have at least one fixed point. Before we present Kakutani's fixed point theorem in the first part of this section, we first explain what we mean by a correspondence and a fixed point, and introduce some other definitions that are needed to understand Kakutani's fixed point theorem. We then apply this theorem to prove the existence of Nash equilibria in the second part of this section. Finally, in the third part of this section we prove Theorem 4.4.5, which provides sufficient conditions for Nash equilibrium in games with more than two players.

Kakutani's fixed point theorem

By $\mathbb{R}^m$ we denote the set of all vectors $(x_1,...,x_m)$, where $x_k \in \mathbb{R}$ for all $k \in \{1,...,m\}$. Let X be some subset of $\mathbb{R}^m$. A *correspondence* from X to X is a mapping F that assigns to every point $x \in X$ a non-empty *set* of points $F(X) \subseteq X$. A *function* is a special case of a correspondence where $F(x)$ consists of a single point for every $x \in X$. A point $x^* \in X$ is called a *fixed point* of the correspondence F if $x^* \in F(x^*)$. That is, the image of the point x^* under F contains the point x^* itself. Not every correspondence F from X to X has a fixed point. The question we wish to answer is: Can we find conditions on the correspondence F that guarantee that F has at least one fixed point? Kakutani's fixed point theorem presents one such set of conditions. We will now present these conditions.

We say that a set $A \subseteq \mathbb{R}^m$ is *convex* if for every two points $a, b \in A$, and every number $\lambda \in [0,1]$, the convex combination $(1-\lambda)\cdot a + \lambda \cdot b$ is also in A. Geometrically, the convex combination $(1-\lambda)\cdot a + \lambda \cdot b$ is a point on the line between a and b. Hence, in geometric terms, a set A is convex if the line between any two points $a, b \in A$ is completely contained in A. We call the correspondence F from X to X *convex-valued* if $F(x)$ is a convex set for every $x \in X$. A set $A \subseteq \mathbb{R}^m$ is *compact* if it is both *closed* and *bounded*.

Finally, the correspondence F from X to X is *upper-semicontinuous* if for every two sequences $(x^k)_{k\in\mathbb{N}}$ and $(y^k)_{k\in\mathbb{N}}$ in X the following holds: if $y^k \in F(x^k)$ for all k, the sequence $(x^k)_{k\in\mathbb{N}}$ converges to $x \in X$, and the sequence $(y^k)_{k\in\mathbb{N}}$ converges to $y \in X$, then $y \in F(x)$. If the correspondence F is a function, then upper-semicontinuity is just the same as continuity. We are now ready to present Kakutani's fixed point theorem.

Theorem 4.5.1 *(Kakutani's fixed point theorem)*
Let X be a non-empty, compact and convex subset of $\mathbb{R}^m$. Moreover, let F be a correspondence from X to X that is convex-valued and upper-semicontinuous. Then, the correspondence F has at least one fixed point.

We will not prove this theorem here, as it would be too much of a diversion.

Existence of Nash equilibrium
We will now use Kakutani's fixed point theorem to prove Theorem 4.2.3, which states that a Nash equilibrium always exists if the players have finitely many choices.

Theorem 4.2.3 *(Nash equilibrium always exists)*
For every game with finitely many choices there is at least one Nash equilibrium $(\sigma_1, ..., \sigma_n)$.

Proof: For every player i, let $\Delta(C_i)$ denote the set of probability distributions on C_i. Remember that a probability distribution on C_i is a vector σ_i that assigns to every choice $c_i \in C_i$ some number $\sigma_i(c_i) \in [0, 1]$, such that

$$\sum_{c_i \in C_i} \sigma_i(c_i) = 1.$$

So, every combination of beliefs $(\sigma_1, ..., \sigma_n)$ belongs to the set $\Delta(C_1) \times ... \times \Delta(C_n)$. By

$$X := \Delta(C_1) \times ... \times \Delta(C_n)$$

we denote the set of all such belief combinations. Hence, X is a subset of some set $\mathbb{R}^m$. Moreover, it may easily be verified that the set X is non-empty, compact and convex.

For every $(\sigma_1, ..., \sigma_n) \in X$ and every player i, let $C_i^{\text{opt}}(\sigma_1, ..., \sigma_n)$ be the set of choices $c_i \in C_i$ that are optimal under the belief σ_{-i}. By $\Delta(C_i^{\text{opt}}(\sigma_1, ..., \sigma_n))$ we denote the set of probability distributions in $\Delta(C_i)$ that only assign positive probability to the choices in $C_i^{\text{opt}}(\sigma_1, ..., \sigma_n)$. Define the correspondence C^{opt} from X to X, which assigns to every belief combination $(\sigma_1, ..., \sigma_n) \in X$ the set of belief combinations

$$C^{\text{opt}}(\sigma_1, ..., \sigma_n) := \Delta(C_1^{\text{opt}}(\sigma_1, ..., \sigma_n)) \times ... \times \Delta(C_n^{\text{opt}}(\sigma_1, ..., \sigma_n)),$$

which is a subset of $\Delta(C_1) \times ... \times \Delta(C_n)$, and hence is a subset of X.

It may easily be verified that the set $C^{\text{opt}}(\sigma_1, ..., \sigma_n)$ is non-empty and convex for every $(\sigma_1, ..., \sigma_n)$. It thus follows that the correspondence C^{opt} is convex-valued. We now show that the correspondence C^{opt} is upper-semicontinuous. That is, we must show that for every sequence $(\sigma_1^k, ..., \sigma_n^k)_{k \in \mathbb{N}}$ converging to some $(\sigma_1, ..., \sigma_n)$, and every sequence $(\hat{\sigma}_1^k, ..., \hat{\sigma}_n^k)_{k \in \mathbb{N}}$ converging to some $(\hat{\sigma}_1, ..., \hat{\sigma}_n)$ with $(\hat{\sigma}_1^k, ..., \hat{\sigma}_n^k) \in C^{\text{opt}}(\sigma_1^k, ..., \sigma_n^k)$ for every k, then $(\hat{\sigma}_1, ..., \hat{\sigma}_n) \in C^{\text{opt}}(\sigma_1, ..., \sigma_n)$.

Suppose, contrary to what we want to prove, that $(\hat{\sigma}_1, ..., \hat{\sigma}_n) \notin C^{\text{opt}}(\sigma_1, ..., \sigma_n)$. Then, there is some player i such that $\hat{\sigma}_i$ assigns positive probability to some c_i, where c_i is not optimal under σ_{-i}. But then, if k is large enough, $\hat{\sigma}_i^k$ assigns positive probability to c_i, and c_i is not optimal under σ_{-i}^k. However, this contradicts the assumption that $(\hat{\sigma}_1^k, ..., \hat{\sigma}_n^k) \in C^{\text{opt}}(\sigma_1^k, ..., \sigma_n^k)$. So, we conclude that $(\hat{\sigma}_1, ..., \hat{\sigma}_n) \in C^{\text{opt}}(\sigma_1, ..., \sigma_n)$, and hence the correspondence C^{opt} is upper-semicontinuous.

Summarizing, we see that the set $X := \Delta(C_1) \times ... \times \Delta(C_n)$ is non-empty, compact and convex, and that the correspondence C^{opt} from X to X is upper-semicontinuous and convex-valued. By Kakutani's fixed point theorem, it then follows that C^{opt} has at least one fixed point $(\sigma_1^*, ..., \sigma_n^*) \in X$. That is, there is some $(\sigma_1^*, ..., \sigma_n^*) \in X$ with

$$(\sigma_1^*, ..., \sigma_n^*) \in C^{\text{opt}}(\sigma_1^*, ..., \sigma_n^*).$$

By the definition of C^{opt} this means that for every player i, we have that $\sigma_i^* \in \Delta(C_i^{\text{opt}}(\sigma_1^*, ..., \sigma_n^*))$. So, for every player i, the probability distribution σ_i^* only assigns positive probability to the choices c_i that are optimal under σ_{-i}^*. This means, however, that $(\sigma_1^*, ..., \sigma_n^*)$ is a Nash equilibrium. So, a Nash equilibrium always exists. ■

Sufficient conditions for Nash equilibrium with more than two players

We will finally prove the following theorem, which provides sufficient conditions for Nash equilibrium in a game with more than two players.

Theorem 4.4.5 *(Sufficient conditions for Nash equilibrium in a game with more than two players)*
Consider a game with more than two players. Let t_i be a type that believes in the opponents' rationality, holds projective and independent beliefs, believes that his opponents hold correct beliefs, and believes that every opponent has these four properties too. For every $j \neq i$, let σ_j be t_i's belief about j's choice and let σ_i be t_i's belief about any opponent's belief about i's choice. Then, $(\sigma_1, ..., \sigma_n)$ is a Nash equilibrium.

Proof: Consider an epistemic model and a type t_i for player i within it. Suppose that t_i believes in the opponents' rationality, holds projective and independent beliefs, believes that his opponents hold correct beliefs and believes that every opponent has these four properties too. We show that t_i's belief about the opponents' choices together with t_i's belief about the opponents' belief about i's choice, constitute a Nash equilibrium. As a first step towards this result, we prove the following lemma.

Lemma 4.5.2 *For every opponent j there is a type t_j such that:*
(1) type t_i believes (with probability 1) that opponent j is of type t_j,
(2) for every opponent j, type t_j believes (with probability 1) that i is of type t_i, and
(3) for every two opponents j and k, type t_j believes (with probability 1) that k is of type t_k.

Proof of Lemma 4.5.2 (1) and (2): Consider an opponent j. Suppose that type t_i assigns positive probability to some type t_j for opponent j. Since t_i believes that j holds correct beliefs about i, type t_i believes that player j believes that i's type is indeed t_i. Hence, type t_j must believe (with probability 1) that i is of type t_i.

Moreover, as type t_i believes that j believes that i has correct beliefs about j, it follows that type t_j believes that i has correct beliefs about j. As we have seen, type t_j believes that i is of type t_i, and hence type t_i must have correct beliefs about j, so type t_i must believe that, with probability 1, player j is of type t_j.

Summarizing, we see that for every opponent j there is a unique type t_j such that (1) type t_i believes (with probability 1) that opponent j is of type t_j, and (2) type t_j believes (with probability 1) that i is of type t_i. We have thus shown properties (1) and (2) of the lemma.

(3) Now consider two opponents j and k. We know from property (1) that there are types t_j and t_k such that type t_i believes (with probability 1) that j is of type t_j, and t_i believes (with probability 1) that k is of type t_k. As type t_i has projective beliefs, it follows that t_i believes that j believes that k is of type t_k. Since t_i believes that j is of type t_j, we have that type t_j believes (with probability 1) that k is of type t_k. This shows property (3) in the lemma, and hence the proof for the lemma is complete. ◇

So, from Lemma 4.5.2 it follows that type t_i believes that every opponent j is of type t_j, type t_i believes that every opponent j believes that every other player k is of type t_k, and so on. Hence, one type t_j for every opponent j is enough to describe t_i's entire belief hierarchy.

As a second step, we prove that type t_i's belief hierarchy is *simple*.

Lemma 4.5.3 *Type t_i has a simple belief hierarchy, generated by some combination of beliefs $(\sigma_1, ..., \sigma_n)$.*

Proof of Lemma 4.5.3: We first define the combination of beliefs $(\sigma_1, ..., \sigma_n)$, and then show that t_i's belief hierarchy is generated by $(\sigma_1, ..., \sigma_n)$.

As type t_i has independent beliefs, for every opponent j there is a belief σ_j about j's choice, such that for every opponents' choice-combination $(c_j)_{j \neq i}$, type t_i assigns probability $\prod_{j \neq i} \sigma_j(c_j)$ to it. This yields beliefs $\sigma_j \in \Delta(C_j)$ for every $j \neq i$. Now, consider an opponent j^*. We know from Lemma 4.5.2 that there is a type t_{j^*} such that t_i believes with probability 1 that j^* is of type t_{j^*}. Let $\sigma_i \in \Delta(C_i)$ be the belief that t_{j^*} holds about i's choice. In this way, we can construct a combination of beliefs $(\sigma_1, ..., \sigma_n)$.

We now show that t_i's belief hierarchy is generated by $(\sigma_1, ..., \sigma_n)$. We have already seen that t_i's belief about the opponents' choices is generated by $(\sigma_1, ..., \sigma_n)$.

Take an arbitrary opponent j. By Lemma 4.5.2 we know that there is a type t_j such that t_i believes (with probability 1) that j is of type t_j. As t_i believes that j has independent beliefs, it follows that t_j has independent beliefs about the opponents' choices. Consider an opponent $k \neq j, i$ of player j. Since t_i's belief about k's choice is σ_k, type t_i has projective beliefs, and type t_i believes that j is of type t_j, we must have that type t_j's belief about k's choice is also σ_k.

Above we chose a fixed opponent $j^* \neq i$ and defined σ_i as the belief that t_{j^*} holds about i's choice. Since type t_i believes that j^* is of type t_{j^*}, and t_i believes that j^* has projective beliefs, it follows that t_{j^*} must have projective beliefs. Moreover, by Lemma 4.5.2, property (3), we know that type t_{j^*} believes (with probability 1) that j is of type t_j. Since t_{j^*}'s belief about i's choice is σ_i and t_{j^*} holds projective beliefs, type t_{j^*} believes that j must also hold belief σ_i about i's choice. As t_{j^*} believes that j is of type t_j, it follows that t_j's belief about i's choice is σ_i. Summarizing, we see that type t_j's belief about the opponents' choices is independent and generated by $(\sigma_1, ..., \sigma_n)$.

Overall, we thus see that type t_i's belief about the opponents' choices is independent and generated by $(\sigma_1, ..., \sigma_n)$, and that every type t_j has independent beliefs about the opponents' choices generated by $(\sigma_1, ..., \sigma_n)$. Here, the opponents' types t_j are chosen as in Lemma 4.5.2. But then, together with properties (1), (2) and (3) of Lemma 4.5.2, it follows that t_i holds a simple belief hierarchy generated by the combination of beliefs $(\sigma_1, ..., \sigma_n)$. ◇

Hence, from Lemma 4.5.3 we know that type t_i has a *simple* belief hierarchy, generated by some combination of beliefs $(\sigma_1, ..., \sigma_n)$. We finally show that $(\sigma_1, ..., \sigma_n)$ is a Nash equilibrium.

Take an opponent j and assume that $\sigma_j(c_j) > 0$ for some choice c_j of player j. Hence, type t_i assigns positive probability to the choice c_j for player j. As type t_i believes in j's rationality, and believes that j's belief about the opponents' choices is given by σ_{-j}, then c_j must be optimal for player j under the belief σ_{-j}.

Consider now some choice c_i for player i with $\sigma_i(c_i) > 0$. Take an arbitrary opponent j. As t_i believes that j's belief about i's choice is given by σ_i, it follows that t_i believes that j assigns positive probability to c_i. Since type t_i believes that j believes in i's rationality, and t_i believes that j believes that i's belief about the opponents' choices is σ_{-i}, then c_i must be optimal under the belief σ_{-i}.

Altogether, we see that for every opponent j, the belief σ_j only assigns positive probability to a choice c_j if c_j is optimal under the belief σ_{-j}. Also, the belief σ_i only assigns positive probability to a choice c_i if c_i is optimal under the belief σ_{-i}. Hence, the combination of beliefs $(\sigma_1, ..., \sigma_n)$ is a Nash equilibrium. So, we conclude that type t_i has a simple belief hierarchy that is generated by a Nash equilibrium $(\sigma_1, ..., \sigma_n)$. This completes the proof of Theorem 4.4.5. ■

Practical problems

4.1 Black or white?

This evening there will be a party in the village and you and Barbara are invited. The problem is that you don't know whether to go or not, and if you go, which color to wear. Assume that you only have white and black suits in your wardrobe, and the same is true for Barbara. You and Barbara have conflicting interests when it comes to wearing clothes: You strongly dislike it when Barbara wears the same color as you, whereas Barbara prefers to wear the same color as you. Also, you know that you will only have a good time at the party if Barbara goes, and similarly for Barbara.

More precisely, your utilities are as follows: Staying at home gives you a utility of 2. If you go to the party and Barbara shows up wearing a different color to you, your utility will be 3. In all other cases, your utility is 0.

For Barbara the utilities are similar. The only difference is that she gets a utility of 3 if she goes to the party and you show up wearing the *same* color as her.

(a) Model this situation as a game between you and Barbara by listing the choices and utilities for you and Barbara.

(b) Draw an extended beliefs diagram. Translate this diagram into an epistemic model. Which choices can you rationally make under common belief in rationality?

(c) Which types t_i in your model believe that the opponent j has correct beliefs? Which of these types t_i believe that the opponent j believes that i has correct beliefs too? Which types have a simple belief hierarchy and which do not?

(d) Compute all Nash equilibria in this game. For every Nash equilibrium, write down the simple belief hierarchy it implies for you.

(e) Compute all choices you can rationally make under common belief in rationality with a simple belief hierarchy.

4.2 At the gambling table

You and Barbara are sitting at a gambling table and are playing the following game: Both players hold four cards in their hands: an ace, a two, a three and a four. They simultaneously lay one card on the table. If both players put down an ace, no payment is made. If one player puts down an ace and the opponent does not, then the player with the ace receives 1 euro from his opponent. If no player puts down an ace, then we add up the values of the two cards. If the sum is even Barbara pays this amount to you. If the sum is odd you pay this amount to Barbara.

(a) Specify for every combination of choices the utilities for you and Barbara.

(b) Create an extended beliefs diagram. For every irrational choice, specify a randomized choice that strictly dominates it. Which choices can you rationally make under common belief in rationality?

(c) Which belief hierarchies in your extended beliefs diagram are simple and which are not?

(d) Compute all Nash equilibria in this game. For every Nash equilibrium, write down the simple belief hierarchy it implies for you.

(e) Compute all the choices you can rationally make under common belief in rationality with a simple belief hierarchy.

4.3 To which pub shall I go?

In a village there are three pubs, say a, b and c. You and Barbara are both planning to go out tonight and you can choose between these three pubs. However, you have not told each other which pub you are going to. You both hold the same preferences over the pubs, represented by the utilities below:

Pub	Pub a	Pub b	Pub c
utility	6	5	4

There is one problem: Barbara likes to talk the whole evening without interruption and you have a headache. So, you would prefer to be in a pub *without* Barbara, whereas Barbara would prefer to be in a pub *with* you. More precisely, if Barbara goes to the same pub as you, your utility will be decreased by 4, whereas Barbara's utility would be increased by 4. If you visit different pubs, your utilities will be simply as specified in the table above.

(a) Build a table with the choices and utilities for you and Barbara.

(b) Draw an extended beliefs diagram for this situation. Translate this diagram into an epistemic model.

(c) Which pubs can you rationally visit under common belief in rationality?

(d) Which types t_i in your model believe that the opponent j has correct beliefs? Which of these types t_i believe that the opponent j believes that i has correct beliefs too? Which types have a simple belief hierarchy and which do not?

(e) Construct a Nash equilibrium (σ_1, σ_2) in which all choices have a positive probability. Write down the simple belief hierarchy it implies for you.

(f) Which pubs can you rationally visit under common belief in rationality with a simple belief hierarchy?

4.4 Summer holiday

You, Barbara and Chris must decide where to go on summer holiday. After a long discussion there seem to be two options left, *Spain* and *Iceland*, but there is no way to reach a unanimous agreement on either of the two destinations. You therefore agree on the following procedure: Each person will write a destination on a piece of paper. The holiday will be in the country chosen by the majority and only the people who actually voted for it will go. That is, somebody who votes for a country that receives the minority of votes will stay at home. The utilities assigned by the three friends to the three possible outcomes are given in Table 4.6.

(a) Create an extended beliefs diagram for this situation. Translate this extended beliefs diagram into an epistemic model. Which choices can you rationally make under common belief in rationality?

(b) Which belief hierarchies in your diagram are simple and which are not?

(c) Show that there is no Nash equilibrium $(\sigma_1, \sigma_2, \sigma_3)$ that consists solely of probability 1 beliefs.

(d) Show that there is a Nash equilibrium $(\sigma_1, \sigma_2, \sigma_3)$ in which σ_1 assigns probability 1 to your choice *Spain*.

(e) Show that the Nash equilibrium found in **(d)** is the only Nash equilibrium in the game.

Table 4.6. *Utilities in "Summer holiday"*

	You	Barbara	Chris
Spain	2	1	1
Iceland	1	2	0
Stay at home	0	0	2

Hint for (e): Show first that there is only one Nash equilibrium in which σ_1 assigns probability 1 to *Spain*. Suppose, next, that σ_1 assigns positive probability to *Iceland*. Show that this is only possible if σ_2 and σ_3 assign positive probability to *Iceland*. Show that σ_3 can only assign positive probability to *Iceland* if σ_1 and σ_2 assign positive probability to *Spain*. Show that σ_2 can only assign positive probability to *Spain* if σ_1 and σ_3 assign positive probability to *Spain*. Conclude that every player must be indifferent regarding the two countries. Show, finally, that in a Nash equilibrium it is not possible that all players are indifferent regarding the two countries.

(f) Which country, or countries, can you rationally choose under common belief in rationality with a simple belief hierarchy?

4.5 Playing hide-and-seek

Four children are playing hide-and-seek in the woods. For the three children that must hide, there are three particularly good hiding places: behind a tree, behind a big stone and in some long grass. If a child is the only one to hide behind the tree, it will be found with probability 0.4. If two or more children try to hide behind the tree they will all be found with probability 1, as there is only enough room for one child. If one or two children hide behind the stone or in the grass, they will be found with probability 0.1. If three children try to hide behind the stone or in the grass they will all be found with probability 1, as there is only room for two children at these two hiding places. The goal of every hiding child is to minimize the probability of being found.

(a) Create an extended beliefs diagram for this situation. Translate this diagram into an epistemic model. Show that all hiding places can rationally be chosen under common belief in rationality.

(b) Which belief hierarchies in your diagram are simple and which are not?

(c) Show that with a simple belief hierarchy it can never be optimal to hide behind the tree.

Hint for (c): Use the fact that for any two numbers $a, b \in [0,1]$, either $ab \leq 1/4$ or $(1-a)(1-b) \leq 1/4$.

(d) Which hiding places can rationally be chosen under common belief in rationality with a simple belief hierarchy? Support each of these hiding places by a simple belief hierarchy that expresses common belief in rationality.

4.6 Predicting the color of his suit

Suppose you are back at high school. It is commonly known that the French teacher always wears a blue suit or a brown suit. One day, Chris decides to organize the following bet for you and Barbara: You and Barbara must both give 2 euros to Chris today. If the teacher appears with his blue suit tomorrow, you will get 3 euros and Chris will keep the remaining euro. If the teacher shows up with his brown suit tomorrow, Barbara will get 3 euros and Chris will again keep the remaining euro. You and Barbara must decide whether to participate in the bet or not. If one person decides to participate and the other doesn't, the participating person will lose the 2 euros to Chris. Assume that the teacher is indifferent about whether he wears his blue or brown suit.

(a) Model this situation as a game between you, Barbara and the teacher. That is, for every combination of choices specify the utilities for you and Barbara.

(b) Create an extended beliefs diagram and translate it into an epistemic model. Which choices can you rationally make under common belief in rationality?

(c) Which types in your model have simple belief hierarchies and which types do not?

(d) Which choices can you rationally make under common belief in rationality with a simple belief hierarchy?

4.7 A high-school reunion

Tomorrow there will be a reunion of your former classmates from high school. Let us number the classmates $1, 2, 3, ..., 30$. Every person i had exactly one favorite classmate, namely $i + 1$, and one least preferred classmate, namely $i - 1$. If $i = 1$, then i's least preferred classmate is 30, and if $i = 30$, then i's favorite classmate is 1. For every classmate, the presence of his favorite classmate would increase his utility by 3, whereas the presence of his least preferred classmate would decrease his utility by 3. The presence of other classmates would not affect his utility. Every classmate must decide whether or not to go to the reunion. Assume that staying at home would yield a utility of 2.

(a) Explain why under common belief in rationality both going to the reunion and staying at home can be optimal. What would change if one of the classmates did *not* have a favorite classmate?

(b) For any number n between 0 and 29, construct a belief hierarchy expressing common belief in rationality in which you believe that exactly n fellow classmates will show up at the reunion.

(c) What choice, or choices, can you rationally make under common belief in rationality with a simple belief hierarchy?

(d) How many classmates do you think will show up under common belief in rationality with a simple belief hierarchy?

Theoretical problems

4.8 Games with two players and two choices

Consider a game with two players and two choices for both players. Show that every choice that can rationally be made under common belief in rationality, can also rationally be made under common belief in rationality *with a simple belief hierarchy*.

So, in these games, imposing a simple belief hierarchy puts no additional restrictions on the choices that the players can rationally make under common belief in rationality. Note that the example "Movie or party?" shows that this is not true for games with *three* players and two choices. Moreover, Problem 4.1 shows that it is not true for games with two players and *three* choices.

4.9 Zero-sum games

Remember from Problem 2.10 that a *zero-sum game* is a static game with two players – player 1 and player 2 – such that $u_1(c_1,c_2)+u_2(c_1,c_2)=0$ for all choice-combinations $(c_1,c_2)\in C_1\times C_2$. That is, the sum of the utilities for the players is always equal to 0. Remember also from that problem that for every zero-sum game there is a unique number v such that player 1 has a randomized choice r_1 that guarantees him at least v, and player 2 has a randomized choice r_2 that guarantees him at least $-v$. This number v is called the *value* of the zero-sum game.

For every belief $b_i\in\Delta(C_j)$ that player i can hold about j's choice, let $c_i^*(b_i)$ be an optimal choice for this belief. In other words, $u_i(c_i^*(b_i),b_i)\geq u_i(c_i,b_i)$ for every other choice $c_i\in C_i$. We call b_i a *most pessimistic* belief for player i if

$$u_i(c_i^*(b_i),b_i)\leq u_i(c_i^*(b_i'),b_i')$$

for every other belief $b_i'\in\Delta(C_j)$. So, the belief b_i yields player i the lowest expected utility among all possible beliefs, provided that player i chooses optimally given his belief.

(a) Consider a zero-sum game. Consider that player 1 has a type t_1 with a simple belief hierarchy, generated by a pair of beliefs (σ_1,σ_2) where $\sigma_1\in\Delta(C_1)$ and $\sigma_2\in\Delta(C_2)$. Suppose that type t_1 expresses common belief in rationality. Let c_1^* be an optimal choice for type t_1. Show that the expected utility for type t_1 by choosing c_1^* is equal to v – the value of the zero-sum game. Show that an analogous result holds for player 2.

Hint for (a): Use Problem 2.9, part (c).

(b) Consider a zero-sum game, and a type t_i for player i in some epistemic model. Suppose that type t_i has a simple belief hierarchy generated by a pair of beliefs (σ_1,σ_2), where $\sigma_1\in\Delta(C_1)$ and $\sigma_2\in\Delta(C_2)$. Show that type t_i expresses common belief in rationality if and only if σ_2 is a most pessimistic belief for player 1 and σ_1 is a most pessimistic belief for player 2.

Hint for (b): Use part (a) of this problem and part (c) in Problem 2.9.

(c) Show that parts (a) and (b) are not true if we drop the assumption that the type has a simple belief hierarchy. That is, provide an example of a zero-sum game in which player 1 has a type t_1 that expresses common belief in rationality, but where

- type t_1 can get an expected utility higher than v by choosing optimally, and
- type t_1's belief about 2's choice is not a most pessimistic belief.

4.10 Nash equilibrium versus Nash choice

(a) Consider an arbitrary game with n players. Suppose that $(\sigma_1, ..., \sigma_n)$ is a Nash equilibrium and that σ_i assigns positive probability to c_i. Show that c_i is a Nash choice. So, every choice that has positive probability in a Nash equilibrium is a Nash choice.

(b) Construct a two-player game in which player 1 has a Nash choice c_1, but there is no Nash equilibrium (σ_1, σ_2) in the game in which σ_1 assigns positive probability to c_1. So, not every Nash choice has positive probability in a Nash equilibrium!

Literature

Nash equilibrium

The concept of *Nash equilibrium* was developed by Nash (1950, 1951). For many decades it has played a dominant role in game theory and most textbooks on game theory still treat Nash equilibrium as the central concept. In part this is understandable, as the idea of Nash equilibrium has been extremely important for the development of game theory. But at the same time this chapter shows that Nash equilibrium is much less of a basic concept than common belief in rationality, for instance. Indeed, in order for a belief hierarchy to correspond to a Nash equilibrium it is not enough to just require common belief in rationality; we must impose additional restrictions, for example, that a player believes that his opponents are *correct* about the beliefs he holds about the opponents' choices. This is certainly *not* a *basic* condition to impose – and may even be an implausible restriction in many cases – and hence in our view Nash equilibrium is not at all a basic concept. For games with more than two players we need even more non-basic restrictions – beyond believing that the opponents are correct about your beliefs – to eventually arrive at Nash equilibrium.

Interpretation of Nash equilibrium

In this chapter we interpret a Nash equilibrium $(\sigma_1, ..., \sigma_n)$ as a combination of beliefs. More precisely, if player i holds a simple belief hierarchy generated by $(\sigma_1, ..., \sigma_n)$ then, for every opponent j, the belief σ_j represents the belief that player i has about j's choice, whereas σ_i is the belief that player i holds about the belief that his opponents hold about his own choice.

This interpretation of a Nash equilibrium differs considerably from the usual interpretations that are given in the literature. In many papers and books on game theory, a Nash equilibrium $(\sigma_1, ..., \sigma_n)$ is viewed as a combination of *randomized choices* – or *mixed strategies* – by the players. So, for every player i, the probability distribution $\sigma_i \in \Delta(C_i)$ is interpreted as the actual randomization carried out by player i. The idea of a Nash equilibrium is then motivated by stating that player i's randomization is optimal, *given* the randomizations that are carried out by his opponents. In our view, there are two serious problems to this interpretation. First, as we argued in Chapter 2, people typically do not randomize when making serious choices. Moreover, on a more theoretical level, there is nothing to gain for a player by randomizing, even if we seriously considered this option. Hence, interpreting σ_i as a conscious randomization by player i seems rather artificial. Secondly, we find it problematic to say that player i chooses optimally *given* the randomizations chosen by others. It suggests that player i somehow *knows*, or *is sure about*, the behavior of his opponents, and also this seems rather unrealistic. After all, player i cannot look into the minds of his opponents.

These problems – especially the first one – have also been noted by others, and this has led to another popular interpretation of Nash equilibrium, which we will now describe. In this alternative interpretation, the probability distributions $\sigma_1, ..., \sigma_n$ are no longer viewed as the randomizations carried out by the players, but rather as the *beliefs* that the various players hold about the opponents' choices. More precisely, σ_i is interpreted as the *common* belief that i's opponents hold about i's choice. The concept of Nash equilibrium is then motivated by saying that the common belief σ_i about i's choice should only assign positive probability to choices that are optimal for player i, *given* the beliefs that i holds about the opponents' choices. Although an improvement, we see major problems with this second interpretation as well. Again, there is a problem with the word "given" in the sentence above. If we say that σ_i must only assign positive probability to choices that are optimal for player i, *given* his beliefs about others' choices, then we suggest that i's opponents *know*, or *are sure about*, the beliefs that i holds about his opponents' choices, which again seems unrealistic.

In both interpretations above, the objects $\sigma_1, ..., \sigma_n$ in the Nash equilibrium are viewed as the *actual* randomized choices – or *actual* beliefs – of the various players in the game. Hence, if $\sigma_1, ..., \sigma_n$ are interpreted as randomized choices, then σ_1 represents the actual randomized choice by player 1, σ_2 the actual randomized choice by player 2, and so on. If the objects $\sigma_1, ..., \sigma_n$ are interpreted as beliefs, then σ_1 represents the actual common belief that 1's opponents hold about 1's choice, σ_2 the actual common belief that 2's opponents hold about 2's choice, and so on. Our interpretation of Nash equilibrium in this chapter is radically different, as we consider the beliefs $\sigma_1, ..., \sigma_n$ to be within the mind of a *single* player. More precisely, we analyze the game completely from the perspective of a single player – player i – and assume that he holds a simple belief hierarchy which is generated by the beliefs $\sigma_1, ..., \sigma_n$. That is, for every opponent j we assume that player i's belief about opponent j's choice is σ_j, and σ_i represents the belief that player i has about the opponents' belief about i's choice. So, the entire Nash equilibrium $(\sigma_1, ..., \sigma_n)$ is within the mind of a single player – player i! We actually make no statements at all about the *actual* beliefs of i's opponents – we only put restrictions on player i's actual belief hierarchy.

This is in line with the *one-person perspective* that we adopt throughout this book. In each of the chapters we always investigate the game from the perspective of a single player, and only put restrictions on the actual belief hierarchy of this player. No conditions are imposed on the actual belief hierarchies of the other players. Also Nash equilibrium is treated from this one-person perspective. In particular, in a two-player game it may be that player 1 has a simple belief hierarchy that is generated by a Nash equilibrium (σ_1, σ_2), whereas player 2 has a simple belief hierarchy that is generated by a different Nash equilibrium (σ_1', σ_2'). Within our framework this makes perfect sense, as a Nash equilibrium is viewed as an object that is in the mind of a single player, and hence different players may base their simple belief hierarchies on different Nash equilibria. But then, the *actual* belief of player 1 about 2's choice – σ_2 – together with the *actual* belief of player 2 about 1's choice – σ_1' – may not constitute a Nash equilibrium, although both players derive their belief hierarchy from a Nash equilibrium! This may seem strange at first sight, but it is perfectly in line with our interpretation of Nash equilibrium. If player 1's simple belief hierarchy is generated by the Nash equilibrium (σ_1, σ_2), then the only thing we are saying is that player 1's actual belief about 2's choice is σ_2, whereas 1's belief about 2's belief about 1's choice is σ_1. But player 1's belief about player 2's belief about 1's choice may very well be different from player 2's *actual* belief about 1's choice! Indeed, when we say that player 1's simple belief hierarchy is generated by (σ_1, σ_2), we are saying nothing about player 2's *actual* belief about 1's choice. So, it is no surprise that player 1's actual belief about 2's choice, together with player 2's actual belief about 1's choice, may not yield a Nash equilibrium, even if both players derive their belief hierarchy from a Nash equilibrium.

To illustrate this, consider a game between player 1 and player 2 in which both players must independently write down a number between 1 and 10. If they happen to write down the same number they both get 1 euro. Otherwise, they both get nothing. Then, player 1 may write down number 3 because he has a simple belief hierarchy that is generated by the Nash equilibrium $(3, 3)$. Similarly, player 2 may write down number 7 because he has a simple belief hierarchy that is generated by the Nash equilibrium $(7, 7)$. But then, player 1's actual belief about 2's choice – 3 – together with player 2's actual belief about 1's choice – 7 – do not constitute a Nash equilibrium! In fact, in this game it would be a complete surprise if the actual beliefs of players 1 and 2 about the opponent's choice were a Nash equilibrium.

Summarizing, we see that the main difference between our interpretation of Nash equilibrium and the usual interpretations found in the literature, is the fact that we view a Nash equilibrium as a "personal object" in the mind of a single person, whereas usually a Nash equilibrium is viewed as a combination of randomized choices – or beliefs – on which the various players in the game "agree." In fact, when we say that player i's belief hierarchy is generated by a Nash equilibrium, then we are only saying something about player i's belief hierarchy, and not about the actual belief hierarchies of the other players in the game. So, even if all players in the game derive their belief hierarchies from a Nash equilibrium, there is no reason to expect that the players' *actual* beliefs constitute a Nash equilibrium, provided there is more than one Nash equilibrium in the game.

Sufficient conditions for Nash equilibrium in two-player games

In this chapter we provided conditions on a belief hierarchy which guarantee that this belief hierarchy "corresponds" to a Nash equilibrium. Theorem 4.4.4 provides such a set of sufficient conditions for two-player games. It states that if player i believes that j is correct about i's beliefs, believes that j believes that i is correct about j's beliefs, believes in j's rationality, and believes that j believes in i's rationality, then i's belief about j's choice, together with i's belief about j's belief about i's choice, constitute a

Nash equilibrium. This theorem is based on Corollary 4.6 in Perea (2007a). There is a small difference in terminology, however. In Perea (2007a) a player is said to have *self-referential* beliefs if he believes that his opponents hold correct beliefs about his own beliefs.

In the literature we find alternative sufficient conditions for Nash equilibrium, which are similar to the ones above. Let us briefly discuss these alternative foundations for Nash equilibria here. Aumann and Brandenburger (1995) showed in their Theorem A the following result for two-player games: Suppose we have a pair of beliefs (σ_1, σ_2). If player 1 believes in 2's rationality, player 2 believes in 1's rationality, player 1's belief about 2's choice is σ_2, player 2's belief about 1's choice is σ_1, player 1 believes that player 2's belief about 1's choice is σ_1, and player 2 believes that player 1's belief about 2's choice is σ_2, then (σ_1, σ_2) constitutes a Nash equilibrium. An important difference with our Theorem 4.4.4 is that Aumann and Brandenburger (1995) impose conditions on the beliefs of *both* players simultaneously, whereas our theorem only imposes conditions on the beliefs of a *single* player. In that sense, one could say that Aumann and Brandenburger's model is more restrictive. On the other hand, the conditions that Aumann and Brandenburger impose on a *single* player's belief hierarchy are weaker than ours. For instance, Aumann and Brandenburger only require that player 1 believes in 2's rationality, but not that player 1 believes that player 2 believes in 1's rationality, as we require. Also, Aumann and Brandenburger do not require player 1 to believe that player 2 is correct about 1's beliefs. Indeed, in Aumann and Brandenburger's framework, player 1's actual belief about 2's choice is given by σ_2, and player 2's actual belief about player 1's belief about 2's choice is σ_2, but this does not imply that player 1 believes that player 2 believes that player 1's belief about 2's choice is σ_2.

Another important difference between Aumann and Brandenburger's conditions and our conditions in Theorem 4.4.4 is that Aumann and Brandenburger's conditions do *not* imply that there is common belief in rationality, whereas the conditions in Theorem 4.4.4 do. Polak (1999) has shown that, if Aumann and Brandenburger's conditions are strengthened by requiring *common* belief in the actual beliefs (σ_1, σ_2), then these new conditions *do* imply common belief in rationality.

Brandenburger and Dekel (1989) presented conditions that are more restrictive than the ones in Aumann and Brandenburger (1995) and which are related to Polak's (1999) conditions. They showed the following for two-player games: Consider a pair of beliefs (σ_1, σ_2). If both players express common belief in the beliefs (σ_1, σ_2), and both players express common belief in rationality, then (σ_1, σ_2) constitutes a Nash equilibrium. Here, by "common belief in the beliefs (σ_1, σ_2)" we mean that player 1's belief about 2's choice is σ_2, player 1 believes that player 2's belief about 1's choice is σ_1, player 1 believes that player 2 believes that 1's belief about 2's choice is σ_2, and so on, and similarly for player 2. Obviously, the conditions imposed by Brandenburger and Dekel (1989) are stronger than those imposed in Aumann and Brandenburger (1995). In fact, Brandenburger and Dekel (1989) showed that the converse of their result is also true: If both players express common belief in the beliefs (σ_1, σ_2) and (σ_1, σ_2) constitutes a Nash equilibrium, then both players also express common belief in rationality. Hence, they showed that, if both players express common belief in (σ_1,σ_2), then both players express common belief in rationality *if and only if* (σ_1,σ_2) is a Nash equilibrium. This result is very closely related to Theorem 4.2.2 in this chapter. Namely, stating that both players express common belief in (σ_1,σ_2) is equivalent to stating that both players have a simple belief hierarchy generated by (σ_1,σ_2). Hence, Brandenburger and Dekel (1989) prove that, if both players have a simple belief hierarchy generated by (σ_1,σ_2), then both players express common belief in rationality *if and only if* (σ_1,σ_2) is a Nash equilibrium. This, however, is exactly the message of Theorem 4.2.2 in this chapter, the only difference being that Theorem 4.2.2 focuses on the beliefs of one player only.

Asheim (2006, p. 5) considers a pair of *belief hierarchies* rather than a pair of beliefs about the opponent's choice. For a fixed pair (b_1,b_2) of belief hierarchies, Asheim (2006) proved the following result: If player 1 believes in 2's rationality, player 2 believes in 1's rationality, player 1 has belief hierarchy b_1, player 2 has belief hierarchy b_2, player 1 believes that player 2 has belief hierarchy b_2, and player 2 believes that player 1 has belief hierarchy b_1, then player 1's belief about 2's choice, together with player 2's belief about 1's choice, constitute a Nash equilibrium. It is clear that Asheim's conditions are stronger than those in Aumann and Brandenburger (1995). The crucial difference is that Asheim assumes that player i is correct about player j's *entire* belief hierarchy, whereas Aumann and Brandenburger only assume that player i is correct about j's belief about i's choice – not necessarily about the other levels in j's belief hierarchy. On a one-person level, Asheim's conditions imply our conditions in Theorem 4.4.4. Within Asheim's framework, player 1 has belief hierarchy b_1, player 2 has belief hierarchy b_2, player 2 believes that player 1 has belief hierarchy b_1, and player 1 believes that player 2 indeed has belief hierarchy b_2. This implies, however, that player 1 believes that player 2 believes that player 1 indeed has belief hierarchy b_1, and hence player 1 believes that player 2 holds correct beliefs about player 1's beliefs. In a similar way, it can be shown that Asheim's conditions imply that player 1 believes that player 2 believes that player 1 holds correct beliefs about player 2's beliefs. Moreover, as player 2 has belief hierarchy b_2, player 2 believes in 1's rationality, and player 1 believes that player 2 indeed has belief hierarchy b_2, it follows from Asheim's conditions that player 1 believes that player 2 believes in 1's rationality. Summarizing, we see that Asheim's conditions imply the conditions we impose in Theorem 4.4.4.

Tan and Werlang (1988), in their Theorem 6.2.1, provided a set of conditions that is similar to, but stronger than, Asheim's. Instead of requiring that player 1 believes in 2's rationality, and player 2 believes in 1's rationality, they require both players to express *common* belief in rationality, which is stronger. The other conditions they impose are the same as in Asheim (2006).

Sufficient conditions for Nash equilibrium with more than two players

The sufficient conditions for Nash equilibrium as presented in Theorem 4.4.4 will not lead to Nash equilibrium if we have more than two players. One reason is that these conditions do not imply that player i's belief about j's choice is the same as player i's belief about player k's belief about j's choice, if j and k are two different opponents.

Nor do these conditions guarantee that player i's belief about j's choice is independent from his belief about k's choice. But the latter two properties are needed for Nash equilibrium in a game with more than two players! Theorem 4.4.5 extends the conditions in Theorem 4.4.4 such that they lead to Nash equilibrium in a game with more than two players. More precisely, Theorem 4.4.5 shows the following for games with more than two players: If player i believes in the opponents' rationality, holds projective and independent beliefs, believes that his opponents hold correct beliefs about his own beliefs, and believes that his opponents have these properties too, then player i's belief about his opponents' choices, together with i's belief about the opponents' common belief about i's choice, constitute a Nash equilibrium. Remember that player i is said to have projective beliefs if his belief about k's belief about j's choice is the same as his own belief about j's choice. Moreover, having independent beliefs means that i's belief about j's choice is stochastically independent from his belief about k's choice. Theorem 4.4.5 follows from Corollary 4.6 in Perea (2007a). Actually, the conditions in Perea (2007a) are a bit weaker than the conditions we use in Theorem 4.4.5. Perea (2007a) assumes, in Corollary 4.6, *conditionally* independent beliefs instead of independent beliefs. Remember from the literature section in Chapter 3 that conditional independence is weaker than independence: It means that player i's belief about j's choice is independent from his belief about k's choice, once we condition on a fixed belief hierarchy for opponents j and k. Moreover, Corollary 4.6 in Perea (2007a) does not require player i to have projective beliefs himself, but only requires player i to believe that his opponents hold projective beliefs. It can be shown, however, that Perea's (2007a) conditions *imply* that player i has projective beliefs as well. We will now compare our conditions in Theorem 4.4.5 with other sufficient conditions for Nash equilibrium that have been proposed in the literature.

Aumann and Brandenburger (1995) showed, in their Theorem B, the following result for games with more than two players: For every player i consider a belief b_i about the opponents' choices. If (1) the players' belief hierarchies are derived from some common prior (see the literature section of Chapter 3), (2) every player believes in the opponents' rationality, and (3) there is common belief in the beliefs $(b_1, ..., b_n)$, then for every player i, each of i's opponents holds the same belief σ_i about i's choice, and $(\sigma_1, ..., \sigma_n)$ is a Nash equilibrium. Here, the conditions (1) and (3) imply that every player holds independent beliefs, and that every two players i and j hold the same belief about k's choice. Hence, the conditions (1) and (3) in Aumann and Brandenburger's (1995) Theorem B have a similar role to the conditions of projective and independent beliefs in our Theorem 4.4.5.

Brandenburger and Dekel (1987) also provided, in their Proposition 4.1, sufficient conditions for Nash equilibrium if there are more than two players. Their condition that the prior beliefs are *concordant* guarantees that two players i and j have the same belief about k's choice – a property which is similar to our condition of projective beliefs. They also assume that the players' information partitions are conditionally independent, which implies that player i's belief about j's choice is independent from his belief about k's choice. This corresponds to our condition of independence. Finally, their condition that the information partitions are informationally independent implies that player i believes that his opponents are correct about i's beliefs. Hence, the conditions imposed in Brandenburger and Dekel (1987) are quite similar to those used in Theorem 4.4.5.

Correct beliefs

In two-player games, what separates Nash equilibrium from common belief in rationality is the additional requirement that player i believes that opponent j is *correct* about i's beliefs. This condition is not needed for common belief in rationality, but some version of it is needed in order to obtain Nash equilibrium. Tan and Werlang (1988) provided a weaker version of this condition, which they call *conjectural consistency*. Formally, a player i is *conjecturally consistent* if all of the opponent's types that he deems possible assign *positive probability* to i's actual type. In other words, player i believes that every opponent deems i's actual type *possible*. Remember that in this chapter, requiring that player i believes that his opponents have correct beliefs is equivalent to stating that all of the opponent's types that i deems possible must assign *probability 1* to i's actual type. Hence, Tan and Werlang's (1988) notion of conjectural consistency is indeed weaker than our notion of belief in the opponents' correct beliefs. Tan and Werlang (1988) use the notion of conjectural consistency to provide an alternative set of sufficient conditions for Nash equilibria – for two-player games only – in their Theorem 6.3.1.

Existence of Nash equilibrium

In the proofs section of this chapter, we showed that for every game with finitely many choices there is at least one Nash equilibrium – see Theorem 4.2.3. For the proof we made use of Kakutani's fixed point theorem (Kakutani, 1941), which is a generalization of Brouwer's fixed point theorem. The main difference is that Brouwer's fixed point theorem refers to fixed points of *functions*, whereas Kakutani's fixed point theorem is about fixed points of *correspondences*. The correspondence C^{opt} we constructed in the proof of Theorem 4.2.3 is usually called the *best-response correspondence* in the literature.

Nash choice versus Nash equilibrium

It is important to clearly distinguish between Nash *choice* and Nash *equilibrium*. In this chapter, we called a choice c_i a Nash *choice* if it is optimal for player i under some simple belief hierarchy that is generated by a Nash equilibrium. Or, equivalently, choice c_i is optimal in some Nash equilibrium $(\sigma_1, ..., \sigma_n)$. One should always keep in mind that a Nash choice is an *object of choice*, whereas a Nash equilibrium is a *combination of beliefs* that generates a simple belief hierarchy for a player. So, a Nash equilibrium is within the mind of a player, whereas a Nash choice is something he chooses based on the Nash equilibrium he has in his mind. From theoretical problem 4.10 we know that every choice c_i that has positive probability in some Nash equilibrium $(\sigma_1, ..., \sigma_n)$ is always a Nash choice. However, the converse is not true – not every Nash choice c_i has positive probability in a Nash equilibrium $(\sigma_1, ..., \sigma_n)$! In theoretical problem 4.10, the reader is asked to provide an example of a game where this is indeed the case. So, there are choices that player i can rationally make if his belief hierarchy is generated by a Nash equilibrium, but that do not form part of a belief hierarchy generated by a Nash equilibrium!

In the literature one often talks about *Nash equilibrium strategies*, which are choices with positive probability in some Nash equilibrium. In view of the above, it is thus crucial to distinguish between Nash choices on the one hand, and Nash equilibrium strategies on the other. Every Nash equilibrium strategy is a Nash choice, but a Nash choice is not necessarily a Nash equilibrium strategy!

Part II

Lexicographic beliefs in static games

5 Primary belief in the opponent's rationality

In this chapter, and also the following two, we will restrict our attention exclusively to games with two players. The reason is that we want to keep our formal analysis as simple as possible, while still being able to convey the main ideas. In fact, all ideas, definitions and results in this chapter and the following two can be extended to games with more than two players.

5.1 Cautious reasoning about the opponent

In the chapters we have seen so far, your belief about the opponent's choice has always been modeled by a *single probability distribution* over the opponent's choices. In particular, you may assign probability *zero* to an opponent's choice a, which would reflect a state of mind in which you completely rule out the possibility that your opponent will choose a.

However, in many situations it may be wise to adapt a more *cautious* way of reasoning about your opponent, in which you could deem some of the opponent's choices much more likely than some of the other choices, but where you *never completely rule out any opponent's choice from consideration*. As an illustration, let us consider the following example.

Example 5.1 Should I call or not?

This evening, your friend Barbara will go to the cinema. She has told you that you can call her if you want to go, but that she will decide on the movie. In the village there is a small cinema with two screens that shows classic movies, and this evening there is the choice between *The Godfather* and *Casablanca*. You prefer watching *The Godfather* (utility 1) to *Casablanca* (utility 0), whereas Barbara has the opposite preference. If you don't call then you will stay at home, which would yield you a utility of 0. Barbara – as a real cinema addict – is only interested in the movie, and her utility of watching a movie does not depend on whether you go or not. The utilities for you and Barbara are thus given by Table 5.1. The question is: Should you call Barbara or not?

Intuitively, the best choice is definitely to call her – by not calling your utility will be 0 for sure, whereas by calling your utility will be at least 0, but there is a chance of getting

Table 5.1. *Utilities for you and Barbara in "Should I call or not?"*

		Barbara	
		The Godfather	Casablanca
You	Call	1, 0	0, 1
	Don't call	0, 0	0, 1

1 if she – surprisingly – were to choose *The Godfather*. So, there is nothing to lose by calling her: there is only something to gain, and therefore you should definitely call her.

This conclusion, however, cannot be reached if we model your belief about Barbara's choice the way we did in Part I of this book! Suppose that your belief about Barbara's choice is modeled by a *single* probability distribution on Barbara's choices, as we assumed in Part I of the book. Then, if you believe that Barbara chooses rationally, you must assign probability 0 to Barbara choosing *The Godfather*. But given this belief, you will be completely indifferent between calling and not calling, so you might as well not call her!

In order to conclude that you should call her, we must somehow model the following state of mind:

- since you believe in Barbara's rationality, you must deem her rational choice *Casablanca* much more likely (in fact, *infinitely more likely*) than her irrational choice *The Godfather*,
- at the same time, you should *not completely rule out* the possibility that she will irrationally choose *The Godfather*.

Only then will you have a clear preference for calling her. But how can we formally model such a state of mind? From our discussion, it should be clear that it is no longer enough to model your belief by a single probability distribution, so we should look for an alternative way of representing your belief.

A possible way to represent the state of mind above is to separate your belief about Barbara's choice into a *primary* belief and a *secondary* belief, as follows: Your *primary* belief is that you assign probability 1 to her choosing *Casablanca*, as you would expect, and your *secondary* belief is that you assign probability 1 to her choosing *The Godfather*, which you deem much less likely. Here, the primary belief is what you "believe most," whereas the secondary belief refers to an event that you deem much less likely – in fact, infinitely less likely – but that *you do not completely rule out*! If you hold this belief, then you do not completely exclude the possibility that Barbara will – surprisingly – choose *The Godfather*, and for that reason your unique best choice is indeed to call her. We call such a belief, consisting of a primary and a secondary belief, a *lexicographic belief* – a notion that we will describe more formally in the next section. □

As the example above shows, a *lexicographic belief* is a way to represent a state of mind in which you deem some choice a of the opponent much more likely than some other choice b, without completely discarding the latter choice from consideration. So, a lexicographic belief seems a suitable tool for modeling *cautious reasoning*, in which you do not exclude any choice by an opponent completely, but where, at the same time, you can deem some choices much more likely than others.

In some sense the previous example was a special case, as the lexicographic belief only consisted of a primary belief and a secondary belief. In general, however, a lexicographic belief may contain more levels than two. The following example will illustrate this.

Example 5.2 Where to read my book?

It is Saturday afternoon, and you want to go to a pub to read your book while enjoying a cup of coffee. A few minutes ago Barbara called, and she said she will go to a pub too, but you forgot to ask her which pub she is going to. In the village there are three pubs, which we call a, b and c, and you must decide which pub to go to. Your only objective is to avoid Barbara, since you want to read your book in silence. For simplicity, assume that your utility is 1 if you avoid Barbara and your utility is 0 if you happen to go to the same pub as her. On the other hand, Barbara prefers pub a to pub b, and prefers pub b to pub c, but she does not really care today whether you will be at the same pub or not. More precisely, going to pub a gives her a utility of 3, going to pub b yields her a utility of 2 and going to pub c gives her a utility of 1. So, the utilities for you and Barbara are as given in Table 5.2. Intuitively, your unique best choice is to go to pub c, since it seems least likely that Barbara will go to pub c, and your only objective is to avoid her. This choice, however, cannot be singled out if we model your belief about Barbara's choice by a *single* probability distribution, as we did in the previous chapters. In that case, you would assign probability 0 to Barbara going to pub b or pub c, since these choices are both worse for Barbara than pub a. Therefore, you would be indifferent between pubs b and c.

However, you will in general no longer be indifferent if you reason *cautiously* about Barbara's choice – that is, if you do not completely discard the possibility that she might go to pub b or c.

Consider the following lexicographic belief that you could hold about Barbara's choice: Your *primary* belief is that you assign probability 1 to Barbara going to pub a, your *secondary* belief is that you assign probability 1 to Barbara going to pub b, and your *tertiary* belief is that you assign probability 1 to her choosing pub c.

So, this is a lexicographic belief with three levels, representing a state of mind in which

- you deem it infinitely more likely that Barbara will go to pub a rather than to pub b, and
- you deem it infinitely more likely that Barbara will go to pub b rather than to pub c, but
- you do not discard the possibility that Barbara will go to pub b or pub c.

Table 5.2. *Utilities for you and Barbara in "Where to read my book?"*

		Barbara		
		Pub a	Pub b	Pub c
You	Pub a	0, 3	1, 2	1, 1
	Pub b	1, 3	0, 2	1, 1
	Pub c	1, 3	1, 2	0, 1

Given this lexicographic belief, you deem it least likely that Barbara will go to pub c. Since your only objective is to avoid Barbara, your unique best choice would therefore be to go to pub c!

Consider now the following alternative lexicographic belief: Your *primary* belief is that you assign probability 1 to Barbara going to pub a, your *secondary* belief is that you assign probability 1 to Barbara going to pub c, and your *tertiary* belief is that you assign probability 1 to her choosing pub b.

In this case, your deem it least likely that Barbara will go to pub b, so your unique best choice is now to go to pub b! So, although both lexicographic beliefs have the same primary belief, namely that Barbara will go to pub a, they yield different optimal choices for you. Hence, your secondary and tertiary beliefs may play an important role when deciding what to do in the game.

Another possible lexicographic belief you could hold is the following: Your *primary* belief is that you assign probability 1 to Barbara going to pub a, and your *secondary* belief is that you assign probability $1/3$ to her choosing pub b, and probability $2/3$ to her choosing pub c. This would model a state of mind in which you deem it infinitely more likely that Barbara will go to pub a rather than to pub b or c, and you deem it twice as likely that she will go to pub c rather than to pub b. So, this is a lexicographic belief with two levels, where the secondary belief assigns positive probability to various choices. Given this belief, your unique best choice would be to visit pub b, since this is the pub that you deem least likely for Barbara. □

5.2 Lexicographic beliefs

In this section we will formally introduce a *lexicographic belief*. The crucial difference between this belief and the one we discussed in Part I of this book is that a lexicographic belief may consist of various *levels* – a *primary* belief, a *secondary* belief, a *tertiary* belief, and so on. The interpretation is that the *primary* belief represents what you believe most, the *secondary* belief refers to events that you deem infinitely less likely but still possible, the *tertiary* belief refers to events that, in turn, you deem infinitely less likely than the events in the secondary belief, but still possible, and so on.

The examples we saw in the previous section contained lexicographic beliefs with two or three levels. These were special cases, as in general a lexicographic belief may contain *any* number of levels. Formally, a lexicographic belief can be defined as follows.

Definition 5.2.1 *(Lexicographic belief)*
Consider a game with two players, i and j. A ***lexicographic belief*** *for player i about j's choice is a finite sequence*

$$b_i = (b_i^1; b_i^2; ...; b_i^K),$$

where $b_i^1, b_i^2, ..., b_i^K$ are probability distributions on the set C_j of j's choices. Here, b_i^1 is called the primary belief, b_i^2 is called the secondary belief, and so on. In general, b_i^k is called the level k belief.

An important property of a lexicographic belief is that it allows a player to deem one of the opponent's choices *infinitely more likely* than another, without completely discarding the latter choice. But what do we mean precisely by deeming some choice *infinitely more likely* than another? Take a lexicographic belief $b_i = (b_i^1; b_i^2; ...; b_i^K)$ for player i, and compare two choices for an opponent c_j and c_j'. Suppose that the primary belief b_i^1 assigns positive probability to c_j but not to c_j'. Then, player i deems c_j infinitely more likely than c_j', because c_j is considered in the most important belief – the primary belief – whereas c_j' is only considered in some less important belief in b_i, or in no belief at all. Suppose now that the primary belief b_i^1 assigns probability zero to both choices c_j and c_j', and that the secondary belief b_i^2 assigns positive probability to c_j but not to c_j'. Then, we also say that player i deems c_j infinitely more likely than c_j'. Namely, the first belief that assigns positive probability to c_j is the secondary belief, whereas c_j' is only considered in some less important belief, or in no belief at all. By iterating this argument, we arrive at the following definition.

Definition 5.2.2 *(Infinitely more likely choices)*
Consider a lexicographic belief $b_i = (b_i^1; b_i^2; ...; b_i^K)$ for player i about j's choice. Take two choices c_j and c_j' for player j. Then, the lexicographic belief b_i deems choice c_j ***infinitely more likely*** *than choice c_j' if there is some level k such that*
(1) the beliefs $b_i^1, ..., b_i^{k-1}$ assign probability zero to both c_j and c_j', and
(2) belief b_i^k assigns positive probability to c_j but not to c_j'.

Of course, condition (1) only applies if both c_j and c_j' have probability zero under the primary belief b_i^1. Otherwise, condition (1) is an empty statement. To illustrate the concept of a lexicographic belief, and the notion of *infinitely more likely*, let us return to the example "Where to read my book?" from the previous section. Table 5.3 contains a list of some possible lexicographic beliefs you could hold about Barbara's choice. Here, b_1 represents the lexicographic belief with three levels, in which your primary belief assigns probability 1 to a, your secondary belief assigns probability 1 to b and your tertiary belief assigns probability 1 to c. So, you deem the event that Barbara will

Table 5.3. *Some lexicographic beliefs you can hold about Barbara's choice in "Where to read my book?"*

$b_1 = (a;b;c)$,	$b_1' = (a;c;b)$,	$b_1'' = (a;\frac{1}{3}b+\frac{2}{3}c)$,
$b_1''' = (\frac{3}{4}a+\frac{1}{4}c;b)$,	$\hat{b}_1 = (\frac{1}{2}a+\frac{1}{2}b;\frac{1}{3}b+\frac{2}{3}c)$,	$\tilde{b}_1 = (\frac{1}{2}a+\frac{1}{2}b)$.

go to pub a infinitely more likely than the event that she will go to pub b, and you deem the event that she will go to pub b infinitely more likely than the event that she will go to pub c.

The lexicographic belief b_1' also has three levels. Your primary belief assigns probability 1 to a, your secondary belief assigns probability 1 to c and your tertiary belief assigns probability 1 to b. So, you deem the event that Barbara will go to pub a infinitely more likely than the event that she will go to pub c, and you deem the event that she will go to pub c infinitely more likely than the event that she will go to pub b.

The lexicographic belief b_1'' has two levels only. The primary belief assigns probability 1 to a, and the secondary belief $\frac{1}{3}b+\frac{2}{3}c$ assigns probability $\frac{1}{3}$ to b and probability $\frac{2}{3}$ to c. So, you deem the event that Barbara will go to pub a infinitely more likely than the event that she will go either to pub b or c, and you deem the event that she will go to pub c twice as likely as the event that she will go to pub b.

The lexicographic belief b_1''' also has two levels. The primary belief assigns probability $\frac{3}{4}$ to a and probability $\frac{1}{4}$ to c, whereas the secondary belief assigns probability 1 to b. So, you deem the event that Barbara will go to pub a three times as likely as the event that she will go to pub c, but you deem both events infinitely more likely than the event that she will go to pub b.

The lexicographic belief $\hat{b}_1$ also has two levels. Note that pub b has a positive probability both in the primary belief and in the secondary belief. According to the secondary belief, one might be tempted to say that you deem the event that Barbara will go to pub c twice as likely as the event that she will go to pub b. This, however, is not true! Since b has a positive probability in the primary belief, which is what you believe most, whereas c has a positive probability for the first time in the secondary belief, you deem the event that Barbara will go to pub b infinitely more likely than the event that she will go to pub c!

Finally, the lexicographic belief $\tilde{b}_1$ has only one level. According to the primary belief, you deem the event that Barbara will go to pub a equally as likely as the event that she will go to pub b. However, this belief completely rules out the possibility that Barbara will go to pub c. For this reason, we say that this lexicographic belief is *not cautious*.

In general, a lexicographic belief is called *cautious* if it does not discard any opponent's choice from consideration. Formally, we have the following definition.

Definition 5.2.3 *(Cautious lexicographic belief)*
Consider a lexicographic belief $b_i = (b_i^1;b_i^2;...;b_i^K)$ *about* j*'s choice. We say that the*

*lexicographic belief b_i is **cautious** if for all of the opponent's choices $c_j \in C_j$ there is some $k \in \{1,...,K\}$ such that b_i^k assigns positive probability to c_j.*

A lexicographic belief is cautious if all of the opponent's choices receive positive probability in at least *some* of its levels. For instance, the lexicographic belief $\hat{b}_1 = (\frac{1}{2}a + \frac{1}{2}b; \frac{1}{3}b + \frac{2}{3}c)$ in Table 5.3 is cautious. Namely, a and b receive positive probability in the primary belief, whereas c receives positive probability in the secondary belief.

Suppose now you hold a lexicographic belief $b_i = (b_i^1; b_i^2; ...; b_i^K)$ about your opponent's choice, and that you must decide between two of your choices, say a and b. How do we determine which of these two choices is best for you? The idea is first to focus on the primary belief b_i^1. If, under the primary belief b_i^1, choice a gives a higher expected utility than b, then a is better for you than b. So, in this case we do not need to consider the secondary belief, the tertiary belief, and so on, since the primary belief discriminates between a and b. Of course, if under the primary belief b gives a higher expected utility than a, then b is better than a.

Suppose, however, that under the primary belief a and b give the same expected utility. Only in that case do we turn to the secondary belief b_i^2 – if there is a secondary belief, of course. If, under the secondary belief b_i^2, choice a gives a higher expected utility than b, then a is better than b. Hence, if a and b are equally good under the primary belief, but a is better than b under the secondary belief, then, under the full lexicographic belief, choice a is better than b.

Suppose now that the lexicographic belief has at least three levels, and that a and b yield the same expected utility under the primary belief, and also yield the same expected utility under the secondary belief. Only in that case do we turn to the tertiary belief. If, under the tertiary belief, a gives a higher expected utility than b, then, under the full lexicographic belief, choice a is better than b.

As an illustration, let us apply this principle to some lexicographic beliefs from Table 5.3. The utilities for both players in "Where to read my book?" can be found in Table 5.2. Start with the lexicographic belief $b_1 = (a; b; c)$. Which of your choices is best under this belief? Under the primary belief, assigning probability 1 to Barbara visiting pub a, your choice a gives utility 0, whereas your choices b and c give utility 1. So, by looking at the primary belief only, we can already conclude that you prefer b and c to a, but it does not tell us yet whether you prefer b to c, or vice versa. In order to decide between b and c, we must turn to the secondary belief. Under the secondary belief, assigning probability 1 to Barbara visiting pub b, your choice b gives utility 0, whereas your choice c gives utility 1. So, we can conclude that, under the lexicographic belief b_1, your choice c is better than b, and your choice b is better than a.

Let us now look at the lexicographic belief $b_1'' = (a; \frac{1}{3}b + \frac{2}{3}c)$. Which of your choices is best under b_1''? Under the primary belief your choice a gives a utility of 0, whereas your choices b and c give a utility of 1. We write

$$u_1^1(a, b_1'') = 0, \; u_1^1(b, b_1'') = 1, \text{ and } u_1^1(c, b_1'') = 1.$$

Here, $u_1^1(a, b_1'')$ denotes the expected utility you (player 1, therefore the subscript 1) obtain by choosing a under the primary belief (therefore the superscript 1) of b_1''. Similarly for the expressions $u_1^1(b, b_1'')$ and $u_1^1(c, b_1'')$. So, by looking at the primary belief we can conclude that you prefer your choices b and c to a, but we cannot yet decide between b and c. So, we must turn to the secondary belief. Under the secondary belief $\frac{1}{3}b + \frac{2}{3}c$, the expected utility you obtain from choosing b is

$$u_1^2(b, b_1'') = \tfrac{1}{3} \cdot 0 + \tfrac{2}{3} \cdot 1 = \tfrac{2}{3},$$

whereas the expected utility you obtain from choosing c under the secondary belief is

$$u_1^2(c, b_1'') = \tfrac{1}{3} \cdot 1 + \tfrac{2}{3} \cdot 0 = \tfrac{1}{3}.$$

Here, the superscript 2 in $u_1^2(b, b_1'')$ indicates that this is the expected utility from choosing b under the *secondary belief*. Since b and c yield the same expected utility under the primary belief, but b gives a higher expected utility under the secondary belief, we may conclude that b is better than c under the lexicographic belief b_1''. Overall, you prefer your choice b to c, and your choice c to a under the lexicographic belief b_1''.

Finally, consider the lexicographic belief $\hat{b}_1 = (\frac{1}{2}a + \frac{1}{2}b; \frac{1}{3}b + \frac{2}{3}c)$. Which choice is best for you under $\hat{b}_1$? Under the primary belief $\frac{1}{2}a + \frac{1}{2}b$, the expected utilities from choosing a, b and c are given by

$$u_1^1(a, \hat{b}_1) = \tfrac{1}{2} \cdot 0 + \tfrac{1}{2} \cdot 1 = \tfrac{1}{2},$$
$$u_1^1(b, \hat{b}_1) = \tfrac{1}{2} \cdot 1 + \tfrac{1}{2} \cdot 0 = \tfrac{1}{2},$$
$$u_1^1(c, \hat{b}_1) = \tfrac{1}{2} \cdot 1 + \tfrac{1}{2} \cdot 1 = 1.$$

So, by looking at the primary belief only, we can already conclude that c is your best choice.

These illustrations should give the reader a clear indication how we can decide, on the basis of a lexicographic belief, which of your choices is best. The key idea is that, for any two of your choices a and b, we compare the expected utilities that these two choices have under your primary belief, under your secondary belief, and so on. Next we give the formal definition of the expected utility that a choice has under the primary belief, secondary belief, and so on.

Definition 5.2.4 *(Expected utility under a lexicographic belief)*
Consider a lexicographic belief $b_i = (b_i^1; b_i^2; ...; b_i^K)$ for player i about j's choice, and a choice $c_i \in C_i$ for player i. For every level $k \in \{1, ..., K\}$, the ***expected utility*** *of choosing c_i under the level k belief b_i^k is equal to*

$$u_i^k(c_i, b_i) = \sum_{c_j \in C_j} b_i^k(c_j) \cdot u_i(c_i, c_j).$$

So, every choice c_i has, under the lexicographic belief b_i, a list of expected utilities

$$(u_i^1(c_i, b_i); u_i^2(c_i, b_i); ...; u_i^K(c_i, b_i)).$$

In this list, $u_i^1(c_i, b_i)$ is the expected utility induced under the primary belief, $u_i^2(c_i, b_i)$ is the expected utility induced under the secondary belief (if there is any), and so on.

Suppose now we compare two of your choices c_i and c_i', and want to decide which of these choices is best under the lexicographic belief b_i. The idea is first to focus on the expected utilities under the primary belief, that is, to compare $u_i^1(c_i, b_i)$ and $u_i^1(c_i', b_i)$. If $u_i^1(c_i, b_i) > u_i^1(c_i', b_i)$, then we may already conclude that c_i is better, and we do not have to look at the other belief levels. If $u_i^1(c_i, b_i) < u_i^1(c_i', b_i)$, we know that c_i' is better, and we are done too. If, however, $u_i^1(c_i, b_i) = u_i^1(c_i', b_i)$, then the expected utilities under the primary belief do not give us an indication yet which choice is better. In this case we turn to the expected utilities under the secondary belief – if there is any – that is, to $u_i^2(c_i, b_i)$ and $u_i^2(c_i', b_i)$. If $u_i^2(c_i, b_i) > u_i^2(c_i', b_i)$, then we know that c_i is better. If $u_i^2(c_i, b_i) < u_i^2(c_i', b_i)$, our conclusion is that c_i' is better. If, however, $u_i^2(c_i, b_i) = u_i^2(c_i', b_i)$, then the expected utilities under the primary belief and the secondary belief do not tell us yet which choice is better. In that case we would turn to the expected utilities under the tertiary belief – if there is any – and so on. So, in general, every lexicographic belief about the opponent's choice induces a preference relation on your own choices.

Definition 5.2.5 *(Preference relation induced by a lexicographic belief)*
*Consider a lexicographic belief $b_i = (b_i^1; b_i^2; ...; b_i^K)$ for player i about j's choice, and two choices c_i and c_i' for player i. Player i **prefers** choice c_i to choice c_i' under the lexicographic belief b_i if there is some $k \in \{1, ..., K\}$ such that*
(1) $u_i^k(c_i, b_i) > u_i^k(c_i', b_i)$, and
(2) for every $l \in \{1, ..., k-1\}$ we have $u_i^l(c_i, b_i) = u_i^l(c_i', b_i)$.

So, you prefer choice c_i to c_i' if

- either c_i is better than c_i' under the primary belief, or
- there is some k such that c_i and c_i' are equally good under the first $k-1$ levels of belief, but c_i is better under the level k belief.

With this definition, it is now easy to define what an *optimal choice* under a lexicographic belief is.

Definition 5.2.6 *(Optimal choice under a lexicographic belief)*
*Consider a lexicographic belief $b_i = (b_i^1; b_i^2; ...; b_i^K)$ for player i about j's choice. A choice c_i for player i is **optimal** under the lexicographic belief b_i if there is no other choice c_i' that player i prefers to c_i under b_i.*

In the special case where a lexicographic belief consists of a single level only, the definition above coincides exactly with that of an optimal choice as defined in Chapter 2.

5.3 Belief hierarchies and types

So far in this chapter we have seen that lexicographic beliefs are useful for modeling *cautious reasoning*. More precisely, if we want to model a state of mind in which you deem some of the opponent's choices infinitely more likely than some other choices, without completely discarding any of these choices, then this can be done by means of lexicographic beliefs.

In this section we will incorporate the idea of cautious reasoning into a *belief hierarchy*, in which you not only hold a belief about the opponent's choice, but also a belief about what the opponent believes about your own choice, and a belief about your opponent's belief about what you believe about the opponent's choice, and so on. The way to proceed will be pretty much the same as in Chapter 3. There, we modeled a player's belief hierarchy by constructing an *epistemic model*. Remember that an epistemic model in Chapter 3 defined for every player a set of types, and for every type a *standard* belief – that is, a single probability distribution – about the opponents' possible choice-type combinations.

In particular, every type holds a belief about his opponents' choices. Moreover, since every type holds a belief about his opponents' types, and every type holds a belief about the other players' choices, every type holds a belief about what others believe about their opponents' choices, and so on. Hence, for every type in the epistemic model we can derive a complete belief hierarchy.

In this section we shall do exactly the same. The only difference is that within the epistemic model we will replace "standard belief" by "lexicographic belief." So, a type within our new epistemic model will now have a *lexicographic* belief about the opponent's possible choice-type pairs. In particular, a type will hold a lexicographic belief about the opponent's choice, which is exactly the object we have been discussing in the previous two sections. Moreover, since a type holds a lexicographic belief about the opponent's possible types, and every type holds a lexicographic belief about your own choice, each of your types will have a lexicographic belief about the possible lexicographic beliefs that your opponent can hold about your own choice, and so on.

In order to define such an epistemic model with lexicographic beliefs, we must first formalize what we mean by a lexicographic belief about the opponent's choice-type pairs.

Definition 5.3.1 *(Lexicographic belief about opponent's choice-type pairs)*
Let T_j be a finite set of types for player j. A ***lexicographic belief*** *for player i about the opponent's choice-type pairs is a finite sequence*

$$b_i = (b_i^1; b_i^2; ...; b_i^K)$$

where, for every $k \in \{1, ..., K\}$, b_i^k is a probability distribution on the set $C_j \times T_j$ of the opponent's choice-type pairs.

As in the previous two sections, we refer to b_i^1 as the *primary* belief, to b_i^2 as the *secondary* belief, and so on. To illustrate this concept, let us return to the example "Where to read my book?" Barbara's set of choices is $C_2 = \{a, b, c\}$. Suppose we consider two different types for Barbara, t_2 and t_2', so $T_2 = \{t_2, t_2'\}$. Consider the lexicographic belief

$$b_1 = ((a, t_2); \tfrac{1}{3}(b, t_2) + \tfrac{2}{3}(c, t_2'))$$

for you about Barbara's choice-type combinations. So, your primary belief assigns probability 1 to the event that Barbara is of type t_2 and chooses a, whereas the secondary belief assigns probability $\frac{1}{3}$ to the event that Barbara is also of type t_2 but chooses b, and assigns probability $\frac{2}{3}$ to the event that Barbara is of the different type t_2' and chooses c.

The interpretation is that:

- you deem the event that Barbara is of type t_2 and chooses a infinitely more likely than the event that she is of type t_2 and chooses b, but you do not discard the latter event,
- you deem the event that Barbara is of type t_2 and chooses a infinitely more likely than the event that she is of type t_2' and chooses c, but you do not discard the latter event,
- you deem the event that she is of type t_2' and chooses c twice as likely as the event that she is of type t_2 and chooses b.

In particular, you deem Barbara's choice a infinitely more likely than her two other choices, and you deem Barbara's type t_2 infinitely more likely than the other type t_2'. Namely, your primary belief assigns positive probability to type t_2, whereas type t_2' only receives positive probability in your secondary belief. Formally, the notion of *infinitely more likely* for a lexicographic belief about the opponent's choice-type pairs is defined as follows.

Definition 5.3.2 *(Infinitely more likely choice-type pairs)*
Consider a lexicographic belief $b_i = (b_i^1; b_i^2; ...; b_i^K)$ for player i about j's choice-type pairs. Take two choice-type pairs (c_j, t_j) and (c_j', t_j') for player j. Then, the lexicographic belief b_i deems the choice-type pair (c_j, t_j) ***infinitely more likely*** *than the choice-type pair (c_j', t_j') if there is some level k such that*
(1) the beliefs $b_i^1, ..., b_i^{k-1}$ assign probability zero to both (c_j, t_j) and (c_j', t_j'), and
(2) belief b_i^k assigns positive probability to (c_j, t_j) but not to (c_j', t_j').

We are now ready to formally define an epistemic model with lexicographic beliefs.

Definition 5.3.3 *(Epistemic model with lexicographic beliefs)*
Consider a game with two players. An ***epistemic model with lexicographic beliefs*** *defines, for both players i, a finite set of types T_i. Moreover, for both players i, and every type $t_i \in T_i$, it specifies a lexicographic belief $b_i(t_i) = (b_i^1(t_i); b_i^2(t_i); ...; b_i^K(t_i))$ for player i about the opponent's choice-type pairs.*

Here, $b_i^1(t_i)$ is the primary belief that type t_i holds about the opponent's choice-type pairs, $b_i^2(t_i)$ the secondary belief, and so on.

As we argued above, although somewhat informally, we can deduce for every type t_i within the epistemic model the complete belief hierarchy it has about the opponent's choice, about the opponent's belief about his own choice, and so on. In order to see how this works, let us return again to the example "Where to read my book?" Consider the epistemic model as given by Table 5.4. Let us try to derive a part of the belief

Table 5.4. *An epistemic model with lexicographic beliefs for "Where to read my book?"*

Types	$T_1 = \{t_1, \hat{t}_1, t_1'\}$, $T_2 = \{t_2, \hat{t}_2, t_2'\}$
Beliefs for you	$b_1(t_1) = ((a,t_2); \frac{1}{3}(b,t_2) + \frac{2}{3}(c,\hat{t}_2))$ $b_1(\hat{t}_1) = (\frac{1}{2}(a,t_2) + \frac{1}{2}(b,\hat{t}_2); (c,\hat{t}_2))$ $b_1(t_1') = ((a,t_2'); \frac{1}{3}(b,t_2') + \frac{2}{3}(c,t_2'))$
Beliefs for Barbara	$b_2(t_2) = ((b,t_1); \frac{3}{4}(a,\hat{t}_1) + \frac{1}{4}((c,t_1))$ $b_2(\hat{t}_2) = ((a,\hat{t}_1); (b,t_1); (c,\hat{t}_1))$ $b_2(t_2') = ((b,t_1'); \frac{3}{4}(a,t_1') + \frac{1}{4}(c,t_1')))$

hierarchy for your type t_1 from this epistemic model. First of all, for this type your lexicographic belief about Barbara's choice is $(a; \frac{1}{3}b + \frac{2}{3}c)$. So, you deem her choice a infinitely more likely than her choices b and c, and you deem her choice c twice as likely as her choice b.

Next, you deem her type t_2 infinitely more likely than her type $\hat{t}_2$. Her type t_2 has lexicographic belief $(b; \frac{3}{4}a + \frac{1}{4}c)$ about your choice, whereas her type $\hat{t}_2$ has lexicographic belief $(a; b; c)$ about your choice. So, you deem the event that she has lexicographic belief $(b; \frac{3}{4}a + \frac{1}{4}c)$ about your choice infinitely more likely than the event that she has lexicographic belief $(a; b; c)$ about your choice, without discarding the latter event.

Note that her type t_2 deems your type t_1 infinitely more likely than your type $\hat{t}_1$. In turn, your type t_1 has lexicographic belief $(a; \frac{1}{3}b + \frac{2}{3}c)$ about her choice, and your type $\hat{t}_1$ has lexicographic belief $(\frac{1}{2}a + \frac{1}{2}b; c)$ about her choice. So, her type t_2 deems the event that you have lexicographic belief $(a; \frac{1}{3}b + \frac{2}{3}c)$ about her choice infinitely more likely than the event that you have lexicographic belief $(\frac{1}{2}a + \frac{1}{2}b; c)$ about her choice.

On the other hand, her type $\hat{t}_2$ deems your type $\hat{t}_1$ infinitely more likely than your type t_1. Hence, her type $\hat{t}_2$ deems the event that you have lexicographic belief $(\frac{1}{2}a + \frac{1}{2}b; c)$ about her choice infinitely more likely than the event that you have lexicographic belief $(a; \frac{1}{3}b + \frac{2}{3}c)$ about her choice.

Since your type t_1 deems her type t_2 infinitely more likely than her type $\hat{t}_2$, we may conclude the following: Your type t_1 deems the event that "she deems your lexicographic belief $(a; \frac{1}{3}b + \frac{2}{3}c)$ about her choice infinitely more likely than your lexicographic belief $(\frac{1}{2}a + \frac{1}{2}b; c)$" *infinitely more likely than* the event that "she deems your lexicographic belief $(\frac{1}{2}a + \frac{1}{2}b; c)$ about her choice infinitely more likely than your lexicographic belief $(a; \frac{1}{3}b + \frac{2}{3}c)$."

By continuing like this, we may derive the full belief hierarchy for type t_1. As you can see, writing down the full belief hierarchy *explicitly* is an even more difficult task than with standard beliefs. For that reason, the concept of an *epistemic model* is extremely valuable for lexicographic beliefs, as it allows us to represent the belief hierarchies with lexicographic beliefs in a *much more compact way* than by writing them out explicitly.

5.4 Cautious types

The reason that we focus on lexicographic beliefs in this part of the book is that we would like to model *cautious reasoning* by players, that is, patterns of reasoning in which you deem all of the opponent's choices possible, but may still deem some choice of the opponent much more likely than some other choice. With the definition of an epistemic model, we can now define formally what we mean by a *cautious type*.

Recall that a type t_i holds a lexicographic belief about the opponent's choice-type pairs. Now, let t_j be an opponent's type that t_i deems possible, that is, some level in t_i's lexicographic belief assigns positive probability to this particular type t_j. If type t_i is *cautious* then, for every opponent's choice c_j, he should deem possible the event that the opponent is of type t_j *and* chooses c_j. In other words, if you deem a certain type t_j possible for your opponent, then you should also deem every choice possible for type t_j.

In order to formally define a cautious type, let us first define what it means to deem an opponent's type, or an opponent's choice-type pair, *possible*.

Definition 5.4.1 *(Deeming an event possible)*
Consider an epistemic model with lexicographic beliefs, and a type t_i for player i with lexicographic belief $b_i(t_i) = (b_i^1(t_i); b_i^2(t_i); ...; b_i^K(t_i))$ about j's choice-type pairs.
Type t_i deems an opponent's choice-type pair (c_j, t_j) ***possible*** *if there is some level $k \in \{1, ..., K\}$ such that $b_i^k(t_i)$ assigns positive probability to (c_j, t_j).*
Type t_i deems an opponent's type t_j ***possible*** *if there is some level $k \in \{1, ..., K\}$ and some choice $c_j \in C_j$ such that $b_i^k(t_i)$ assigns positive probability to (c_j, t_j).*

We can now formally define what we mean by a cautious type.

Definition 5.4.2 *(Cautious type)*
Consider an epistemic model with lexicographic beliefs, and a type t_i for player i with lexicographic belief $b_i(t_i) = (b_i^1(t_i); b_i^2(t_i); ...; b_i^K(t_i))$ about j's choice-type pairs. Type t_i is ***cautious*** *if, whenever it deems an opponent's type t_j possible, then for every choice $c_j \in C_j$ it deems the choice-type pair (c_j, t_j) possible.*

In particular, a cautious type t_i deems all of the opponent's choices possible. However, a cautious type does more than this – for all of the opponent's types t_j that are taken into account, and for every possible choice c_j for an opponent, type t_i deems possible the event that his opponent is of type t_j *and* chooses c_j.

As an illustration, let us have a look at the types in Table 5.4 of the previous section. Type t_1 deems possible all of the opponent's choices a, b and c, but t_1 is *not cautious*. Namely, type t_1 deems possible the opponent's type t_2, but does *not* deem possible the event that his opponent is of type t_2 *and* chooses c. Suppose we construct a new type t_1'' with lexicographic belief

$$b_1(t_1'') = ((a, t_2); (a, \hat{t}_2); \tfrac{1}{3}(b, t_2) + \tfrac{2}{3}(c, \hat{t}_2); \tfrac{1}{3}(b, \hat{t}_2) + \tfrac{2}{3}(c, t_2)),$$

consisting of four levels. In a sense, we have "extended" the lexicographic belief $b_1(t_1)$. The new type t_1'' *is* cautious: Type t_1'' deems possible the opponent's types t_2 and $\hat{t}_2$. For the opponent's type t_2, type t_1'' deems possible all corresponding choice-type pairs $(a,t_2),(b,t_2)$ and (c,t_2). Similarly, for the opponent's type $\hat{t}_2$, type t_1'' deems possible all corresponding choice-type pairs $(a,\hat{t}_2),(b,\hat{t}_2)$ and $(c,\hat{t}_2)$. Therefore, t_1'' is cautious.

In the same way, it can be verified that the types $\hat{t}_1, t_2$ and $\hat{t}_2$ in Table 5.4 are also *not cautious*, although these types deem possible all of the opponent's choices.

In turn, the type t_1' in Table 5.4 *is* cautious: It only deems possible the opponent's type t_2', and for this opponent's type t_2' it deems possible all of the opponent's choices a,b and c. Similarly, it can be seen that the type t_2' is cautious as well.

5.5 Primary belief in the opponent's rationality

In the first part of this book, where we focused on standard beliefs, we introduced the idea that you not only choose rationally yourself, but also believe that your opponent chooses rationally. In fact, this was the key idea that finally led to the concept of *common belief in rationality*, which plays a central role in this book.

We would like to introduce this same idea into a model with lexicographic beliefs. Let us first formally define what we mean by a rational choice for a type with lexicographic beliefs.

Remember that a type t_i has a lexicographic belief $b_i(t_i)$ about the opponent's possible choice-type pairs. In particular, this induces a lexicographic belief about the opponent's choices. Consider, for instance, the type t_1 in the epistemic model of Table 5.4 above. The lexicographic belief about the opponent's choice-type pairs is

$$b_1(t_1) = ((a,t_2); \tfrac{1}{3}(b,t_2) + \tfrac{2}{3}(c,\hat{t}_2)).$$

So, t_1's lexicographic belief about the opponent's choice is

$$(a; \tfrac{1}{3}b + \tfrac{2}{3}c).$$

As a further example, suppose we add a new type t_1'' to this epistemic model with lexicographic belief

$$b_1(t_1'') = ((a,t_2); \tfrac{1}{4}(b,t_2) + \tfrac{1}{2}(b,\hat{t}_2) + \tfrac{1}{4}(c,\hat{t}_2)).$$

Then, the lexicographic belief that t_1'' has about the opponent's choice is

$$(a; \tfrac{3}{4}b + \tfrac{1}{4}c).$$

In the secondary belief $\frac{1}{4}(b,t_2) + \frac{1}{2}(b,\hat{t}_2) + \frac{1}{4}(c,\hat{t}_2)$ type t_1'' assigns probability $\frac{1}{4}$ to the event that Barbara is of type t_2 and chooses b, and assigns probability $\frac{1}{2}$ to the event that she is of a different type $\hat{t}_2$ but still chooses b. So, in total, the secondary belief assigns probability $\frac{1}{4} + \frac{1}{2} = \frac{3}{4}$ to the event that Barbara chooses b.

So, for every type t_i we can derive in this way the lexicographic belief that t_i has about the opponent's choice. Earlier in this chapter, we saw what it means for a choice

c_i to be optimal under a lexicographic belief about the opponent's choice. This naturally leads to the following definition.

Definition 5.5.1 *(Rational choice for a type with a lexicographic belief)*
*Consider an epistemic model with lexicographic beliefs and a type t_i for player i. A choice c_i is **rational** for type t_i if c_i is optimal under the lexicographic belief that t_i has about the opponent's choice.*

Let us return to the epistemic model of Table 5.4. The utilities for the corresponding game can be found in Table 5.2. Your type t_1 has lexicographic belief $((a,t_2); \frac{1}{3}(b,t_2)+\frac{2}{3}(c,\hat{t}_2))$ about Barbara's choice-type pairs. Which choice is rational for type t_1? Note that t_1 has lexicographic belief $(a; \frac{1}{3}b+\frac{2}{3}c)$ about Barbara's choice. Your choice a has utility 0 under the primary belief, whereas your choices b and c have utility 1 under the primary belief. So, you prefer both b and c to a. Under the secondary belief $\frac{1}{3}b+\frac{2}{3}c$, your choice b has expected utility $\frac{2}{3}$, whereas your choice c has expected utility $\frac{1}{3}$. So, you prefer b to c. We may therefore conclude that only choice b is rational for t_1.

Now that we have defined rational choices for a type, our next step will be to formalize the idea that a type believes that his opponent chooses rationally. Let us first recall how we formalized this idea in an epistemic model with standard beliefs. There, a type t_i held a single probability distribution $b_i(t_i)$ on the opponent's choice-type pairs. In Chapter 3, we said that such a type *believes in the opponent's rationality* if the belief $b_i(t_i)$ only assigns positive probability to the opponent's choice-type pairs (c_j,t_j) where choice c_j is rational for type t_j.

Next, take a type t_i with a *lexicographic* belief $b_i(t_i)=(b_i^1(t_i);b_i^2(t_i);...;b_i^K(t_i))$ about the opponent's choice-type pairs, as we assume in this chapter. Remember that $b_i^1(t_i),...,b_i^K(t_i)$ are all probability distributions on the opponent's choice-type pairs. Let us also assume that type t_i is cautious. How do we now define the event that t_i believes in the opponent's rationality?

A first attempt might be to say that each of the probability distributions $b_i^1(t_i),...,b_i^K(t_i)$ should only assign positive probability to the opponent's choice-type pairs (c_j,t_j) where choice c_j is rational for type t_j. However, this may be too much to ask if we also want type t_i to be cautious! Consider the example "Where to read my book?" Suppose we would like to construct a type t_1 for you that at all levels only assigns positive probability to the choice-type pairs (c_2,t_2) for Barbara where c_2 is rational for t_2. For Barbara, only choice a can be rational, so the lexicographic belief $b_1(t_1)$ should, at all levels, assign probability 1 to Barbara's choice a. But this means that t_1 should completely ignore Barbara's inferior choices b and c, and hence t_1 cannot be cautious. So, it is impossible to construct a type for you that is both cautious, and at all levels only assigns positive probability to the choice-type pairs (c_2,t_2) for Barbara where c_2 is rational for t_2!

So, we must be more modest with our definition of *belief in the opponent's rationality*. The problem is that a cautious type should deem *all* of the opponent's choices possible – also the *irrational* ones – and hence it may not be possible to believe, at each of your levels in the lexicographic belief, that your opponent makes optimal choices.

However, the minimum we should require is that at least in your *primary belief* you should be convinced of the opponent's rationality. That is, at least in your primary belief you must assign positive probability only to opponent's choice-type pairs where the choice is optimal for the type. The following definition formalizes this idea.

Definition 5.5.2 *(Primary belief in the opponent's rationality)*
Consider a type t_i with lexicographic belief $b_i(t_i) = (b_i^1(t_i); b_i^2(t_i); ...; b_i^K(t_i))$ about j's choice-type pairs. Type t_i ***primarily believes in the opponent's rationality*** *if t_i's primary belief $b_i^1(t_i)$ only assigns positive probability to choice-type pairs (c_j, t_j) where choice c_j is rational for type t_j.*

Consider again the epistemic model from Table 5.4 for the game "Where to read my book?" Does type t_1 primarily believe in Barbara's rationality? Type t_1's primary belief assigns probability 1 to Barbara's choice-type pair (a, t_2). Since, obviously, a is rational for type t_2, type t_1 indeed primarily believes in Barbara's rationality. In the same way, it can be checked that your type t_1' also primarily believes in Barbara's rationality. On the other hand, your other type $\hat{t}_1$ does not primarily believe in Barbara's rationality, as the primary belief in $\hat{t}_1$ assigns probability $\frac{1}{2}$ to Barbara's choice-type pair $(b, \hat{t}_2)$ and b is not rational for $\hat{t}_2$.

5.6 Common full belief in "primary belief in rationality"

In the previous two sections we defined what it means for a type to be cautious and to primarily believe in the opponent's rationality. Since these are natural properties for a type, it seems reasonable to require that you also believe that *your opponent* is cautious and primarily believes in your own rationality, and that you also believe that your opponent believes that *you* have these properties, and so on. That is, we would like to define a version of *common belief in rationality* for a model with lexicographic beliefs.

Consider a type t_i with lexicographic belief $b_i(t_i) = (b_i^1(t_i); b_i^2(t_i); ...; b_i^K(t_i))$ about j's choice-type pairs. How do we define the event that t_i believes that j is cautious? We have two choices: We could adapt a *weak* version, stating that t_i's *primary* belief $b_i^1(t_i)$ should only assign positive probability to j's types that are cautious. Such a weak version would allow t_i to assign positive probability to non-cautious types in the secondary belief and further levels. Or we could go for a *strong* version, which states that *all* levels $b_i^1(t_i), ..., b_i^K(t_i)$ should only assign positive probability to j's types that are cautious. Or, in other words, type t_i should only deem possible opponent's types t_j that are cautious.

In fact, both choices are feasible. In this chapter we adapt the strong version above which states that you should only deem possible opponent's types that are cautious. That is, no level in your lexicographic belief should assign positive probability to opponent's types that are not cautious. We say that you *fully believe* in the opponent's caution. Or, equivalently, you express 1-fold full belief in caution. As a next step, we say that you express 2-fold full belief in caution if you only deem possible opponent's types that express 1-fold full belief in caution. That is, you only deem possible opponent's types that fully believe that you are cautious. And so on. By repeating this procedure, we obtain *common full belief in caution.*

Definition 5.6.1 *(Common full belief in caution)*
*(1) Type t_i expresses **1-fold full belief in caution** if t_i only deems possible j's types that are cautious.*
*(2) Type t_i expresses **2-fold full belief in caution** if t_i only deems possible j's types that express 1-fold full belief in caution.*
*(3) Type t_i expresses **3-fold full belief in caution** if t_i only deems possible j's types that express 2-fold full belief in caution.*
And so on.
*Type t_i expresses **common full belief in caution** if t_i expresses k-fold full belief in caution for every k.*

Let us return again to the types in Table 5.4. As we have already seen, the types $t_1, \hat{t}_1, t_2$ and $\hat{t}_2$ are not cautious. Since t_1 deems possible the types t_2 and $\hat{t}_2$, type t_1 does not express common full belief in caution. In the same way, it can be verified that also $\hat{t}_1, t_2$ and $\hat{t}_2$ do not express common full belief in caution.

On the other hand, the types t_1' and t_2' are cautious. Since t_1' only deems possible the opponent's type t_2', and t_2' only deems possible the opponent's type t_1', it follows that t_1' and t_2' *do* express common full belief in caution.

We next formally define a version of common belief in rationality. In the previous section we defined *primary belief in the opponent's rationality*, which means that a type, in his primary belief, only assigns positive probability to opponent's choice-type pairs where the choice is rational for the type. How do we define the event that type t_i believes that j primarily believes in i's rationality? Again, there are two possible roads to take. One can adopt a *weak* version, stating that type t_i, in his *primary* belief, should only assign positive probability to j's types that primarily believe in i's rationality. One could also adopt a *strong* version, which would require that t_i, in *each of the levels* of his lexicographic belief, only assigns positive probability to j's types that primarily believe in i's rationality. That is, t_i should only deem possible j's types t_j that primarily believe in i's rationality. We say that t_i *fully believes* that j primarily believes in i's rationality. Again, we choose the strong version here, although both versions are possible. This leads eventually to common full belief in primary belief in the opponent's rationality, which we formalize below. As we did for common full belief in caution, we first define k-fold full belief in primary belief in the opponent's rationality for every k.

Definition 5.6.2 *(Common full belief in primary belief in rationality)*
*(1) Type t_i expresses **1-fold full belief in primary belief in the opponent's rationality** if t_i primarily believes in j's rationality.*

*(2) Type t_i expresses **2-fold full belief in primary belief in the opponent's rationality** if t_i only deems possible j's types that express 1-fold full belief in primary belief in the opponent's rationality.*
*(3) Type t_i expresses **3-fold full belief in primary belief in the opponent's rationality** if t_i only deems possible j's types that express 2-fold full belief in primary belief in the opponent's rationality.*
And so on.
*Type t_i expresses **common full belief in primary belief in the opponent's rationality** if t_i expresses k-fold full belief in primary belief in the opponent's rationality for every k.*

In this chapter we will be interested in those choices you can rationally make if you are cautious and express common full belief in "caution and primary belief in the opponent's rationality." The following definition formalizes this concept.

Definition 5.6.3 *(Rational choice under common full belief in "caution and primary belief in rationality")*
*Let c_i be a choice for player i. Player i can rationally choose c_i under **common full belief in "caution and primary belief in rationality"** if there is some epistemic model with lexicographic beliefs and some type t_i within it, such that:*
(1) t_i is cautious and expresses common full belief in "caution and primary belief in rationality," and
(2) choice c_i is rational for t_i.

Obviously, when we say that type t_i expresses common full belief in "caution and primary belief in rationality," we mean that t_i expresses common full belief in caution and expresses common full belief in primary belief in the opponent's rationality. To explore the consequences of this concept, let us review the two examples at the beginning of this chapter.

Example 5.3 Should I call or not?

Recall the story from Example 5.1 and the utilities from Table 5.1. If you are cautious and primarily believe in Barbara's rationality, you must hold a lexicographic belief in which you deem Barbara choosing *Casablanca* as infinitely more likely than Barbara choosing *The Godfather*, but in which you still deem possible the latter event. Therefore, your only rational choice is to call her. In particular, the only choice you can possibly rationally make under common full belief in "caution and primary belief in rationality" is to *call*.

In order to show that you can indeed choose to *call* under common full belief in "caution and primary belief in rationality," consider the epistemic model in Table 5.5. It has only one type for you and for Barbara. Note that *call* is rational for your type t_1 and that *Casablanca* is rational for Barbara's type t_2. Since t_1's primary belief assigns probability 1 to the choice-type pair $(Casablanca, t_2)$, and t_2's primary belief assigns probability 1 to the choice-type pair $(call, t_1)$, we can conclude that

Table 5.5. *An epistemic model with lexicographic beliefs for "Should I call or not?"*

Types	$T_1 = \{t_1\}, \quad T_2 = \{t_2\}$
Beliefs for you	$b_1(t_1) = ((Casablanca, t_2); (Godfather, t_2))$
Beliefs for Barbara	$b_2(t_2) = ((call, t_1); (not\ call, t_1))$

t_1 and t_2 primarily believe in the opponent's rationality. Moreover, t_1 and t_2 are both cautious. As t_1 only deems possible the opponent's type t_2, and type t_2 only deems possible the opponent's type t_1, it follows that t_1 expresses common full belief in "caution and primary belief in rationality." As *call* is rational for t_1, it follows that you can rationally choose *call* under common full belief in "caution and primary belief in rationality." Hence, *call* is the *only* choice you can rationally make under common full belief in "caution and primary belief in rationality." □

Example 5.4 Where to read my book?

Recall the story from Example 5.2 and the utilities from Table 5.2. If you primarily believe in Barbara's rationality, your primary belief should assign probability 1 to Barbara choosing pub a. Therefore, going to pub a cannot be optimal for you. So, the only pubs you could possibly rationally choose under common full belief in "caution and primary belief in rationality" are pubs b and c. We will show that you can indeed rationally choose both pubs under common full belief in "caution and primary belief in rationality."

Consider first the epistemic model in Table 5.4. As we have seen, your type t_1' expresses common full belief in caution. It can be verified that choice b is rational for your type t_1'. Moreover, choice a is rational for Barbara's type t_2'. Since t_1''s primary belief assigns probability 1 to the rational choice-type pair (a, t_2'), and t_2''s primary belief assigns probability 1 to the rational choice-type pair (b, t_1'), it follows that t_1' and t_2' primarily believe in the opponent's rationality. Note that type t_1' only deems possible the opponent's type t_2', and that t_2' only deems possible the opponent's type t_1'. Therefore, t_1' expresses common full belief in primary belief in the opponent's rationality. Hence, t_1' is cautious and expresses common full belief in "caution and primary belief in rationality." As going to pub b is rational for t_1', it follows that you can rationally choose pub b under common full belief in "caution and primary belief in rationality."

What about going to pub c? Consider your type $\hat{t}_1$ in the epistemic model of Table 5.4. This type deems it infinitely more likely that Barbara will go to pub a or pub b than that she will go to pub c. Since your only objective is to avoid Barbara, going to pub c is rational for type $\hat{t}_1$. However, type $\hat{t}_1$ is not cautious, nor does it primarily believe in Barbara's rationality! Namely, $\hat{t}_1$'s primary belief assigns probability $\frac{1}{2}$ to Barbara's choice-type pair $(b, \hat{t}_2)$, but obviously b is not rational for type $\hat{t}_2$. In particular, $\hat{t}_1$ does not express common full belief in "caution and primary belief in rationality."

Table 5.6. *An epistemic model with lexicographic beliefs for "Where to read my book?" (II)*

Types	$T_1 = \{t_1\}, \quad T_2 = \{t_2\}$
Beliefs for you	$b_1(t_1) = ((a,t_2); \frac{2}{3}(b,t_2) + \frac{1}{3}(c,t_2))$
Beliefs for Barbara	$b_2(t_2) = ((c,t_1); \frac{1}{2}(a,t_1) + \frac{1}{2}(b,t_1))$

However, we can construct an alternative epistemic model in which there is a cautious type for you, expressing common full belief in "caution and primary belief in rationality," for which going to pub c is optimal. Consider the epistemic model in Table 5.6, which has only one type for you and Barbara. First of all, both types are cautious, so your type t_1 expresses common full belief in caution. It may be verified that c is rational for your type t_1, since type t_1 deems it least likely that Barbara will go to pub c. Moreover, a is rational for Barbara's type t_2. Since t_1's primary belief assigns probability 1 to Barbara's rational choice-type pair (a,t_2), and t_2's primary belief assigns probability 1 to your rational choice-type pair (c,t_1), it follows that both types t_1 and t_2 primarily believe in the opponent's rationality. Therefore, your type t_1 expresses common full belief in "caution and primary belief in rationality." As going to pub c is rational for t_1, it follows that you can rationally choose pub c under common full belief in "caution and primary belief in rationality."

Summarizing, we see that you can rationally choose pubs b and c under common full belief in "caution and primary belief in rationality," but not pub a. □

The idea of *common full belief in "caution and primary belief in rationality"* can be viewed as the counterpart of *common belief in rationality* that we discussed in the first part of this book. Both concepts are based on the idea that you not only choose rationally yourself, but also believe that your opponent chooses rationally, and believe that your opponent believes that you choose rationally, and so on.

The crucial difference between both ideas is that *common belief in rationality* allows you to completely ignore some choice of the opponent. This concept is defined for a model with standard beliefs, in which your belief is represented by a *single* probability distribution on the opponent's choices, and this belief may well assign probability zero to some of the opponent's choices. The concept of common full belief in "caution and primary belief in rationality," on the other hand, requires you to deem all of the opponent's choices possible. So, it is no longer possible to completely ignore one of the opponent's choices.

This difference can have some very striking consequences for your choices, as will be made clear by the following example.

Example 5.5 Teaching a lesson

Remember the story from Example 4.1 at the beginning of Chapter 4. The utilities for you and the teacher are reproduced in Table 5.7. In the discussion of this example in

Table 5.7. *Utilities for you and the teacher in "Teaching a lesson" (III)*

		Teacher				
		Mon	Tue	Wed	Thu	Fri
You	Sat	3,2	2,3	1,4	0,5	3,6
	Sun	−1,6	3,2	2,3	1,4	0,5
	Mon	0,5	−1,6	3,2	2,3	1,4
	Tue	0,5	0,5	−1,6	3,2	2,3
	Wed	0,5	0,5	0,5	−1,6	3,2

Table 5.8. *Reduced game after eliminating your irrational choice Wed*

		Teacher				
		Mon	Tue	Wed	Thu	Fri
You	Sat	3,2	2,3	1,4	0,5	3,6
	Sun	−1,6	3,2	2,3	1,4	0,5
	Mon	0,5	−1,6	3,2	2,3	1,4
	Tue	0,5	0,5	−1,6	3,2	2,3

Chapter 4, we saw that, within a framework of standard beliefs, you can rationally make *each* of your choices under common belief in rationality. Figure 4.1 is a beliefs diagram in which every choice for you and the teacher has an outgoing arrow. So, every choice can rationally be made under common belief in rationality.

Things change drastically, however, if we turn to a model with lexicographic beliefs, and require you to reason *cautiously* about the teacher. Suppose that you hold a cautious lexicographic belief about the teacher's choice. Then, *it can no longer be optimal for you to choose Wed*! In order to see this, compare your choice *Wed* with your choice *Sat*. As can be seen from Table 5.7, *Sat* is always strictly better for you than *Wed*, except when the teacher chooses *Fri*. In the latter case, *Sat* and *Wed* are equally good. So, if you deem all of the teacher's choices possible, *Sat* is a strictly better choice for you than *Wed*. Hence, *Wed* cannot be rational for you if you are cautious. For that reason, we may "eliminate" your choice *Wed* from the game, which results in the reduced game of Table 5.8.

If the teacher primarily believes in your caution and rationality, his primary belief should assign probability 0 to *Wed*. But then, it will always be better for the teacher to choose *Fri* rather than *Thu*. So, if the teacher is rational and primarily believes in your caution and rationality, he will not choose *Thu*. We can thus "eliminate" the teacher's choice *Thu* from the game, resulting in the reduced game of Table 5.9.

Table 5.9. *Reduced game after eliminating the teacher's choice Thu*

		Teacher			
		Mon	Tue	Wed	Fri
You	Sat	3,2	2,3	1,4	3,6
	Sun	−1,6	3,2	2,3	0,5
	Mon	0,5	−1,6	3,2	1,4
	Tue	0,5	0,5	−1,6	2,3

Table 5.10. *Reduced game after eliminating your choice Tue*

		Teacher			
		Mon	Tue	Wed	Fri
You	Sat	3,2	2,3	1,4	3,6
	Sun	−1,6	3,2	2,3	0,5
	Mon	0,5	−1,6	3,2	1,4

If you primarily believe in the teacher's rationality and primarily believe that the teacher reasons in the way above, then your primary belief should assign probability 0 to the teacher's choice *Thu*. But then, choosing *Sat* is always better for you than choosing *Tue*. So, if you are rational and primarily believe that the teacher is rational and reasons in the way above, then you will not choose *Tue*. So, we can "eliminate" your choice *Tue* from the game, leading to the reduced game in Table 5.10.

If the teacher primarily believes in your rationality and primarily believes that you reason in the way above, then his primary belief should assign probability 0 to your choices *Tue* and *Wed*. But then, choosing *Fri* will always be better for the teacher than choosing *Wed*. So, the teacher will not choose *Wed* if he is rational and primarily believes that you are rational and reason in the way above. We can therefore "eliminate" the teacher's choice *Wed* from the game. This leads to the reduced game of Table 5.11.

If you primarily believe in the teacher's rationality and primarily believe that he reasons in the way above, then your primary belief should assign probability 0 to the teacher's choices *Wed* and *Thu*. As a consequence, choosing *Sat* is always better for you than choosing *Mon*. Hence, *Mon* can no longer be rational for you if you primarily believe that the teacher is rational and reasons in the way above. We can therefore "eliminate" your choice *Mon* from the game. This results in the reduced game of Table 5.12.

If the teacher primarily believes that you are rational and primarily believes that you reason in the way above, then his primary belief should assign probability 0 to

Table 5.11. *Reduced game after eliminating the teacher's choice Wed*

		Teacher		
		Mon	Tue	Fri
You	Sat	3,2	2,3	3,6
	Sun	−1,6	3,2	0,5
	Mon	0,5	−1,6	1,4

Table 5.12. *Reduced game after eliminating your choice Mon*

		Teacher		
		Mon	Tue	Fri
You	Sat	3,2	2,3	3,6
	Sun	−1,6	3,2	0,5

Table 5.13. *Reduced game after eliminating the teacher's choice Tue*

		Teacher	
		Mon	Fri
You	Sat	3,2	3,6
	Sun	−1,6	0,5

your choices *Mon*, *Tue* and *Wed*. But then, choosing *Fri* will always be better for the teacher than choosing *Tue*. So, *Tue* can no longer be rational for the teacher if he primarily believes that you are rational and reason in the way above. We may therefore "eliminate" the teacher's choice *Tue* from the game. This leads to the reduced game in Table 5.13.

If you primarily believe in the teacher's rationality and primarily believe that the teacher reasons in the way above, then your primary belief should assign probability 0 to the teacher's choices *Tue*, *Wed* and *Thu*. In that case, choosing *Sat* is always better for you than choosing *Sun*. We may therefore conclude that if you primarily believe that the teacher is rational and reasons in the way above, then your only possible rational choice is to start studying on *Sat*. In particular, the only choice you can possibly rationally make under common full belief in "caution and primary belief in rationality" is to start studying on *Sat*!

Table 5.14. *An epistemic model with lexicographic beliefs for "Teaching a lesson"*

Types	$T_1 = \{t_1\}, \quad T_2 = \{t_2\}$
Beliefs for you	$b_1(t_1) = ((Fri, t_2);$ $\frac{1}{4}(Mon, t_2) + \frac{1}{4}(Tue, t_2) + \frac{1}{4}(Wed, t_2) + \frac{1}{4}(Thu, t_2))$
Beliefs for teacher	$b_2(t_2) = ((Sat, t_1);$ $\frac{1}{4}(Sun, t_1) + \frac{1}{4}(Mon, t_1) + \frac{1}{4}(Tue, t_1) + \frac{1}{4}(Wed, t_1))$

In order to show that you *can* indeed choose *Sat* rationally under common full belief in "caution and primary belief in rationality," consider the epistemic model in Table 5.14. There is only one type for you and the teacher. Both types are cautious, as can easily be seen. Moreover, *Sat* is rational for your type t_1, whereas *Fri* is rational for the teacher's type t_2. Since t_1's primary belief assigns probability 1 to the rational teacher's choice-type pair (Fri, t_2) and t_2's primary belief assigns probability 1 to your rational choice-type pair (Sat, t_1), it follows that both t_1 and t_2 primarily believe in the opponent's rationality. Therefore, t_1 expresses common full belief in primary belief in the opponent's rationality. So, we may conclude that t_1 is cautious and expresses common full belief in "caution and primary belief in rationality." Since *Sat* is rational for your type t_1, we may conclude that you can rationally choose *Sat* under common full belief in "caution and primary belief in rationality."

Summarizing, the conclusion is that under common full belief in "caution and primary belief in rationality," there is only one choice you can rationally make, namely to start studying right away on *Sat*. Compare this to *common belief in rationality* with standard beliefs under which you could rationally start studying on *any* day. So, requiring cautious reasoning can have an enormous impact on the possible choices you can rationally make! □

5.7 Existence

In the previous section we developed the central idea of this chapter, namely *common full belief in "caution and primary belief in rationality"*. For some specific examples, we identified those choices you can rationally make under common full belief in "caution and primary belief in rationality." An important question is: Does there always exist a belief hierarchy for you that expresses common full belief in "caution and primary belief in rationality"? In other words, is it always possible to satisfy all the requirements imposed by common full belief in "caution and primary belief in rationality"? In this section we will show that the answer is "yes."

As in Chapter 3, we will present an iterative procedure that will always lead to a belief hierarchy that expresses common full belief in "caution and primary belief in rationality." We will first illustrate this procedure by means of an example.

Table 5.15. *Utilities for you and Barbara in "Hide-and-seek"*

		Barbara		
		Pub *a*	Pub *b*	Pub *c*
You	Pub *a*	0,5	1,2	1,1
	Pub *b*	1,3	0,4	1,1
	Pub *c*	1,3	1,2	0,3

Example 5.6 Hide-and-seek

The story is largely the same as in "Where to read my book?" The difference is that Barbara would now very much like to talk to you. More precisely, if you both go to the same pub, then her utility will be increased by 2. So, Barbara is seeking you, whereas you are trying to hide from her. The new utilities are as depicted in Table 5.15. For convenience, let us denote your choices by a_1, b_1 and c_1, and let us denote Barbara's choices by a_2, b_2 and c_2.

Start with an arbitrary cautious lexicographic belief for you about Barbara's choice, say $(a_2; b_2; c_2)$. Under this belief, your optimal choice is c_1.

Then take a cautious belief for Barbara about your choice that deems this optimal choice c_1 infinitely more likely than your other choices, say $(c_1; a_1; b_1)$. Under this belief, Barbara's optimal choice is a_2.

Then take a cautious belief for you about Barbara's choice that deems this optimal choice a_2 infinitely more likely than her other choices, say $(a_2; c_2; b_2)$. Under this belief, your optimal choice is b_1.

Then take a cautious belief for Barbara about your choice that deems this optimal choice b_1 infinitely more likely than your other choices, say $(b_1; a_1; c_1)$. Under this belief, Barbara's optimal choice is b_2.

Then take a cautious belief for you about Barbara's choice that deems this optimal choice b_2 infinitely more likely than her other choices, say $(b_2; a_2; c_2)$. Under this belief, your optimal choice is c_1.

However, we have already seen c_1 as an optimal choice before. So, we can repeat the steps above over and over again, with the following cycle of lexicographic beliefs:

$$(c_1; a_1; b_1) \to (a_2; c_2; b_2) \to (b_1; a_1; c_1) \to (b_2; a_2; c_2) \to (c_1; a_1; b_1).$$

Here, choice c_1 is optimal under the belief $(b_2; a_2; c_2)$, choice b_2 is optimal under the belief $(b_1; a_1; c_1)$, and so on.

We will now transform the beliefs in this cycle into belief hierarchies that are cautious and express common full belief in "caution and primary belief in rationality." Define types $t_1^{a_2c_2b_2}$ and $t_1^{b_2a_2c_2}$ for you, and types $t_2^{c_1a_1b_1}$ and $t_2^{b_1a_1c_1}$ for Barbara, where

$$b_1(t_1^{a_2c_2b_2}) = ((a_2, t_2^{c_1a_1b_1}); (c_2, t_2^{c_1a_1b_1}); (b_2, t_2^{c_1a_1b_1})),$$

$$b_1(t_1^{b_2a_2c_2}) = ((b_2, t_2^{b_1a_1c_1}); (a_2, t_2^{b_1a_1c_1}); (c_2, t_2^{b_1a_1c_1})),$$
$$b_2(t_2^{c_1a_1b_1}) = ((c_1, t_1^{b_2a_2c_2}); (a_1, t_1^{b_2a_2c_2}); (b_1, t_1^{b_2a_2c_2})),$$
$$b_2(t_2^{b_1a_1c_1}) = ((b_1, t_1^{a_2c_2b_2}); (a_1, t_1^{a_2c_2b_2}); (c_1, t_1^{a_2c_2b_2})).$$

Obviously, all types here are cautious. Note that your type $t_1^{a_2c_2b_2}$ holds lexicographic belief $(a_2;c_2;b_2)$ about Barbara's choice, and believes that Barbara is of type $t_2^{c_1a_1b_1}$, who holds lexicographic belief $(c_1;a_1;b_1)$ about your choice. By construction, Barbara's choice a_2 is optimal under her belief $(c_1;a_1;b_1)$, and hence your type $t_1^{a_2c_2b_2}$ primarily believes in Barbara's rationality.

In the same way, it can be verified that the other types above also primarily believe in the opponent's rationality. But then, it follows that all types express common full belief in "caution and primary belief in rationality." Hence, all four types above are cautious and express common full belief in "caution and primary belief in rationality." In particular, there exists a belief hierarchy for you that is cautious and expresses common full belief in "caution and primary belief in rationality," namely the belief hierarchy of your type $t_1^{a_2c_2b_2}$ (or of your type $t_1^{b_2a_2c_2}$). □

In fact, the procedure above can be applied to *every* game with two players. In general, this procedure would look as follows:

- Start with an arbitrary cautious lexicographic belief for player i about j's choices, which assigns at every level probability 1 to one of j's choices. Let c_i^1 be an optimal choice for player i under this belief.
- Construct a cautious lexicographic belief for player j about i's choices where the primary belief assigns probability 1 to c_i^1, and the other levels always assign probability 1 to one of i's choices. Let c_j^2 be an optimal choice for player j under this belief.
- Construct a cautious lexicographic belief for player i about j's choices where the primary belief assigns probability 1 to c_j^2, and the other levels always assign probability 1 to one of j's choices. Let c_i^3 be an optimal choice for player i under this belief.

And so on. Since there are only finitely many choices for both players, there must be a round k in which the optimal choice c_i^k or c_j^k has already been an optimal choice at some earlier round l. But then, we can simply repeat the procedure between round l and round k over and over again, and hence enter into a cycle of beliefs.

So, this procedure is guaranteed to find a cycle of beliefs. In exactly the same way as we did for the example above, we can then transform the beliefs in this cycle into types that are cautious and express common full belief in "caution and primary belief in rationality." Since this is always possible, we conclude that we can always construct an epistemic model in which all types are cautious and express common full belief in "caution and primary belief in rationality." Moreover, the types we generate in this way have two very special properties: (1) they only deem possible one belief hierarchy – and hence one type – for the opponent, and (2) they assign at every level of the lexicographic belief probability 1 to one of the opponent's choices. We thus obtain the following general existence result.

Theorem 5.7.1 *(Common full belief in "caution and primary belief in rationality" is always possible)*
Consider a static game with two players and finitely many choices for both players. Then, we can construct an epistemic model with lexicographic beliefs in which
(1) every type is cautious and expresses common full belief in "caution and primary belief in rationality," and
(2) every type deems possible only one of the opponent's types, and assigns at each level of the lexicographic belief, probability 1 to one of the opponent's choices.

In particular, there always exists for every player at least one type that is cautious and expresses common full belief in "caution and primary belief in rationality."

5.8 Weakly dominated choices

So far, we have introduced the ideas of caution and primary belief in the opponent's rationality, have incorporated them into the concept of common full belief in "caution and primary belief in rationality," and have shown that the latter concept is always possible in every game. The purpose of this and the following section is to see whether we can construct an algorithm that always computes all choices you can rationally make under common full belief in "caution and primary belief in rationality."

Remember that in the first part of the book, we developed an algorithm for *common belief in rationality* with standard beliefs. The algorithm, called *iterated elimination of strictly dominated choices*, first eliminates all strictly dominated choices from the game, then eliminates from the reduced game all choices that are strictly dominated within the reduced game, then eliminates from the smaller reduced game all choices that are strictly dominated within this smaller reduced game, and so on, until no further choices can be eliminated. We saw that this algorithm selects exactly those choices that can rationally be made under common belief in rationality with standard beliefs. As we will see, the algorithm we will develop for common full belief in "caution and primary belief in rationality" is very similar to the algorithm above, with a little – but important – twist in the first step.

As a first step towards the algorithm, let us try to characterize those choices you can rationally make under a cautious lexicographic belief. After all, the basic requirement in *common full belief in "caution and primary belief in rationality"* is that a player holds a cautious lexicographic belief about his opponent's choice, and chooses optimally given this belief. Let us return to the example "Teaching a lesson" that we discussed earlier in this chapter. The utilities can be found in Table 5.7. We argued that, if you are cautious, then you will never choose *Wed*. The reason is that *Sat* is always at least as good for you as *Wed*, and is strictly better whenever the teacher does not choose *Fri*. Since you will never completely exclude any of the teacher's choices if you are cautious, you should definitely prefer *Sat* to *Wed*. In particular, you should never choose *Wed* if you are cautious.

The argument above can, in fact, be generalized: Consider a game with two players and two choices, c_i and c_i', for player i. Suppose that choice c_i is always at least as good for player i as c_i', no matter what the opponent chooses, and that for at least one of the opponent's choices c_j the choice c_i is strictly better than c_i'. If player i is cautious, then he will hold a lexicographic belief about j's choice that does not rule out any of the opponent's choices, so in particular deems it possible that player j will choose c_j. But then, since c_i is strictly better than c_i' against c_j, and as least as good as c_i' against any other of the opponent's choices, player i should definitely prefer c_i to c_i'. In particular, c_i' cannot be optimal for player i if he is cautious.

In the argument above we say that choice c_i' is *weakly dominated* by choice c_i, meaning that c_i is always at least as good as c_i', and sometimes strictly better. Or, more formally:

Definition 5.8.1 *(Weakly dominated by another choice)*
*A choice c_i' is **weakly dominated** by another choice c_i if for all of the opponent's choices c_j we have that $u_i(c_i,c_j) \geq u_i(c_i',c_j)$, and for at least one of the opponent's choices c_j that $u_i(c_i,c_j) > u_i(c_i',c_j)$.*

So, from our argument above we know that if a choice c_i' is weakly dominated by some other choice c_i, then c_i' can never be optimal if player i holds a cautious lexicographic belief about j's choice. This insight is analogous to what we saw for standard beliefs in Chapter 2: There, we saw that a choice c_i' that is *strictly* dominated by some other choice can never be optimal if player i holds a standard belief about j's choice. In fact, there was a little bit more there. A choice c_i' that is strictly dominated by another choice, or strictly dominated by a *randomized choice*, can never be optimal.

A similar result holds for cautious lexicographic beliefs: If a choice c_i' is weakly dominated by some other choice, or weakly dominated by a randomized choice, then it cannot be optimal under a cautious lexicographic belief. Let us first formally define what it means to be weakly dominated by a randomized choice. Recall that a randomized choice r_i for player i is a probability distribution on the set of i's choices. That is, you choose each of your choices c_i in C_i with a probability, $r_i(c_i)$. Recall, from Chapter 2, that $u_i(r_i,c_j)$ denotes the expected utility for player i if he uses the randomized choice r_i and his opponent chooses c_j.

Definition 5.8.2 *(Weakly dominated by a randomized choice)*
*A choice c_i is **weakly dominated** by a randomized choice r_i if for all of the opponent's choices c_j we have that $u_i(r_i,c_j) \geq u_i(c_i,c_j)$, and for at least one of the opponent's choices c_j that $u_i(r_i,c_j) > u_i(c_i,c_j)$.*

Hence, the randomized choice r_i is always at least as good as c_i, and sometimes strictly better. So, a choice c_i that is weakly dominated by another choice, or by a randomized choice, cannot be optimal under a cautious lexicographic belief. As in Chapter 2 for standard beliefs, we can show that the converse is also true: Every choice that is *not* weakly dominated by another choice, nor by a randomized choice, *is* optimal for some cautious lexicographic belief about the opponent's choice. In fact, we can show something even stronger: If a choice is not weakly dominated, then it is optimal for a cautious lexicographic belief that consists of *one level only*. We thus obtain the following characterization.

Theorem 5.8.3 *(Characterization of optimal choices with cautious lexicographic beliefs)*
A choice c_i can be made optimally under a cautious lexicographic belief about j's choice, if and only if, c_i is not weakly dominated by another choice, nor weakly dominated by a randomized choice. Moreover, every choice c_i that is not weakly dominated by another choice, nor by a randomized choice, is optimal for a cautious lexicographic belief with one level only.

The proof of this result can be found in the proofs section at the end of this chapter. This characterization will play a crucial role in the algorithm we will develop in the following section.

5.9 Algorithm

We want an algorithm that computes exactly those choices you can rationally make under common full belief in "caution and primary belief in rationality." As in Chapter 3, we proceed in steps. We start by characterizing those choices you can rationally make if you are cautious and express 1-fold full belief in "caution and primary belief in rationality." We then characterize those choices you can rationally make if you are cautious and express up to 2-fold full belief in "caution and primary belief in rationality." And so on. By doing so, we eventually generate those choices you can rationally make if you are cautious and express *common* full belief in "caution and primary belief in rationality."

Step 1: 1-fold full belief in "caution and primary belief in rationality"
We start by identifying those choices you can rationally make if you are cautious and express 1-fold full belief in "caution and primary belief in rationality." Remember that 1-fold full belief in "caution and primary belief in rationality" means that you fully believe that the opponent is cautious and primarily believe in the opponent's rationality. Consider a type t_i with lexicographic belief $b_i = (b_i^1; ...; b_i^K)$ about the opponent's choice-type pairs, and assume that t_i is cautious and expresses 1-fold full belief in "caution and primary belief in rationality." Since t_i is cautious, we know from Theorem 5.8.3 of the previous section that every optimal choice for t_i cannot be weakly dominated. Moreover, as t_i expresses 1-fold full belief in "caution and primary belief in rationality," we have that t_i fully believes that j is cautious and primarily believes in j's rationality. As a consequence, the primary belief b_i^1 only assigns positive probability to opponent's choice-type pairs (c_j, t_j) where t_j is cautious and c_j is optimal for t_j. From Theorem 5.8.3 we know that each of these choices c_j is not weakly dominated.

That is, t_i's primary belief b_i^1 only assigns positive probability to j's choices that are not weakly dominated. Let Γ^1 be the reduced game that remains after eliminating all weakly dominated choices from the game. Then, t_i's primary belief b_i^1 only assigns positive probability to j's choices in Γ^1. Since every optimal choice for t_i must in particular be optimal for the primary belief b_i^1, it follows that every optimal choice c_i for t_i must be optimal for some probabilistic belief in the reduced game Γ^1. From Theorem 2.5.3 in Chapter 2 we know that every such choice c_i is not strictly dominated in the reduced game Γ^1. Hence, we conclude that every optimal choice for t_i is not strictly dominated in the reduced game Γ^1.

Summarizing, we see that every optimal choice for t_i is (1) not *weakly* dominated in the original game Γ, and (2) not *strictly* dominated in the reduced game Γ^1 obtained by eliminating all weakly dominated choices from the original game Γ.

Now, let Γ^2 be the reduced game that remains if we eliminate from Γ^1 all strictly dominated choices. That is, Γ^2 is obtained from the original game by first eliminating all *weakly* dominated choices and then eliminating all *strictly* dominated choices from the reduced game that remains. Then, based on our findings above, we conclude that every choice that can rationally be made by a cautious type that expresses 1-fold full belief in "caution and primary belief in rationality," must be in Γ^2.

Step 2: Up to 2-fold full belief in "caution and primary belief in rationality"
We now wish to identify those choices that can rationally be made by a type that is cautious and expresses 1-fold and 2-fold full belief in "caution and primary belief in rationality" – that is, expresses *up to* 2-fold full belief in "caution and primary belief in rationality." Consider a type t_i with lexicographic belief $b_i = (b_i^1; ...; b_i^K)$ about the opponent's choice-type pairs, and assume that t_i is cautious and expresses up to 2-fold full belief in "caution and primary belief in rationality." Since t_i expresses 1-fold and 2-fold full belief in "caution and primary belief in rationality," it follows in particular that t_i's primary belief b_i^1 only assigns positive probability to the opponent's choice-type pairs (c_j, t_j) where t_j is cautious, t_j expresses 1-fold full belief in "caution and primary belief in rationality," and c_j is optimal for t_j. That is, the primary belief b_i^1 only assigns positive probability to the opponent's choices c_j that are optimal for a cautious type that expresses 1-fold full belief in "caution and primary belief in rationality." However, from step 1 above we know that all these choices c_j are in the reduced game Γ^2. So, the primary belief b_i^1 only assigns positive probability to j's choices in the reduced game Γ^2. As an optimal choice for t_i must in particular be optimal for the primary belief b_i^1, it follows that every optimal choice for t_i must be optimal for some belief in Γ^2. Hence, by Theorem 2.5.3 in Chapter 2, every optimal choice for t_i cannot be strictly dominated in Γ^2. Now, let us denote by Γ^3 the reduced game that is obtained from Γ^2 by eliminating all choices that are strictly dominated within Γ^2. Then, we conclude that all choices that are optimal for t_i must be in Γ^3.

Hence, we see that every choice that can rationally be made by a type that is cautious and expresses up to 2-fold full belief in "caution and primary belief in rationality," must be in the reduced game Γ^3. Here, Γ^3 is the reduced game obtained from the original game Γ by first eliminating all *weakly* dominated choices, and then performing two-fold elimination of *strictly* dominated choices.

By continuing in this fashion we arrive at the following conclusion: For every k, let Γ^k be the reduced game obtained from the original game by first eliminating all *weakly* dominated choices, and then performing $(k-1)$-fold elimination of *strictly* dominated choices. Then, for every k, all choices that are optimal for a cautious type that expresses up to k-fold full belief in "caution and primary belief in rationality," must be in the game Γ^{k+1}.

As a consequence, all choices that can rationally be made under *common* full belief in "caution and primary belief in rationality" must survive the algorithm in which we first eliminate all *weakly* dominated choices, and then apply the process of *iterated* elimination of *strictly* dominated choices. This algorithm is known as the *Dekel–Fudenberg procedure*. We now offer a formal definition of this procedure.

Algorithm 5.9.1 *(Dekel–Fudenberg procedure)*
Step 1: *Within the original game, eliminate all choices that are* ***weakly*** *dominated.*
Step 2: *Within the reduced game obtained after step 1, eliminate all choices that are* ***strictly*** *dominated.*
Step 3: *Within the reduced game obtained after step 2, eliminate all choices that are* ***strictly*** *dominated.*
Step 4: *Within the reduced game obtained after step 3, eliminate all choices that are* ***strictly*** *dominated.*
⋮
And so on.

This algorithm stops after finitely many steps because there are only finitely many choices for both players, so after finitely many steps we will reach a reduced game in which none of the choices will be strictly dominated.

Above we argued that every choice that can rationally be made when expressing up to k-fold full belief in "caution and primary belief in rationality" must survive the first $k+1$ steps of the Dekel–Fudenberg procedure. In fact, the converse is also true: Every choice that survives the first $k+1$ steps of the Dekel–Fudenberg procedure can rationally be made when expressing up to k-fold full belief in "caution and primary belief in rationality." Consequently, every choice that survives the *complete* Dekel–Fudenberg procedure can rationally be chosen under *common* full belief in "caution and primary belief in rationality." We thus obtain the following general result.

Theorem 5.9.2 *(The algorithm works)*
Consider a two-player static game with finitely many choices for both players.
(1) For every $k \geq 1$, the choices that can rationally be made by a cautious type that expresses up to k-fold full belief in "caution and primary belief in rationality" are exactly those choices that survive the first $k+1$ steps of the Dekel–Fudenberg procedure.

(2) The choices that can rationally be made by a cautious type that expresses common full belief in "caution and primary belief in rationality" are exactly those choices that survive the Dekel–Fudenberg procedure.

The proof of this result can be found in the proofs section at the end of this chapter. The same result would hold if, instead of common *full* belief in "caution and primary belief in rationality," we require common *primary* belief in "caution and primary belief in rationality." That is, a type only *primarily* believes that his opponent is cautious and primarily believes in rationality, and only *primarily* believes that his opponent primarily believes that he is cautious and primarily believes in rationality, and so on. As above, we can then show the following result: A choice c_i can rationally be made by a cautious player who expresses common *primary* belief in "caution and primary belief in rationality," *if and only if*, the choice c_i survives the Dekel–Fudenberg procedure. That is, common *primary* belief in "caution and primary belief in rationality" leads to exactly the same choices as common *full* belief in "caution and primary belief in rationality."

In order to illustrate the Dekel–Fudenberg procedure, let us return to the example "Teaching a lesson."

Example 5.7 Teaching a lesson

Earlier in this chapter we saw that under common full belief in "caution and primary belief in rationality" there is only one choice you can rationally make: to start studying on *Sat*. According to our theorem above, the same conclusion should be reached by applying the Dekel–Fudenberg procedure.

Consider the utilities for you and the teacher given in Table 5.7. Note that your choice *Wed* is weakly dominated by your choice *Sat*, and that no other choice for you or the teacher is weakly dominated in the original game. So, in step 1 of the Dekel–Fudenberg procedure we eliminate your weakly dominated choice *Wed* from the game, which results in the reduced game of Table 5.8.

In this first reduced game, the teacher's choice *Thu* is strictly dominated by his choice *Fri*, and there is no other strictly dominated choice. So, in step 2 we eliminate the teacher's choice *Thu* from the game, which leads to the reduced game of Table 5.9.

In this second reduced game, your choice *Tue* is strictly dominated by your choice *Sat*, and there is no other strictly dominated choice. Hence, in step 3 we eliminate your choice *Tue* from the game, resulting in the reduced game of Table 5.10.

In this third reduced game, the teacher's choice *Wed* is strictly dominated by his choice *Fri*, and there are no other strictly dominated choices. Therefore, in step 4 we eliminate the teacher's choice *Wed* from the game, which results in the reduced game of Table 5.11.

In this fourth reduced game, your choice *Mon* is strictly dominated by your choice *Sat*, and there are no other strictly dominated choices. So, in step 5 we eliminate your choice *Mon* from the game, which leads to the reduced game of Table 5.12.

Table 5.16. *"Teaching a lesson" with an additional choice Thu for you*

		Teacher				
		Mon	Tue	Wed	Thu	Fri
You	Sat	3,2	2,3	1,4	0,5	3,6
	Sun	−1,6	3,2	2,3	1,4	0,5
	Mon	0,5	−1,6	3,2	2,3	1,4
	Tue	0,5	0,5	−1,6	3,2	2,3
	Wed	0,5	0,5	0,5	−1,6	3,2
	Thu	0,5	0,5	0,5	0,5	−1,6

In this fifth reduced game, the teacher's choice *Tue* is strictly dominated by his choice *Fri*, and there are no other strictly dominated choices. Hence, in step 6, we eliminate the teacher's choice *Tue* from the game, which leads to the reduced game of Table 5.13.

In this sixth reduced game, your choice *Sun* is strictly dominated by your choice *Sat*. So, we eliminate your choice *Sun* from the game, which leaves only your choice *Sat*.

Hence, only your choice *Sat* survives the Dekel–Fudenberg procedure. By our theorem above, we may thus conclude that you can only rationally choose *Sat* under common full belief in "caution and primary belief in rationality." □

To conclude this chapter, let us investigate whether the order and speed of elimination is important for the eventual output of the Dekel–Fudenberg procedure. In Chapter 3, Theorem 3.8.1, we saw that for the algorithm of iterated elimination of strictly dominated choices, it does not matter in which order and with which speed we eliminate strictly dominated choices – the eventual output of the algorithm will be the same. That is, even if we eliminate at every round only *some* – but not necessarily *all* – strictly dominated choices, then the final sets of choices that survive will be exactly the same as under the original algorithm.

Now, does the same hold for the Dekel–Fudenberg procedure? The answer is "no." Consider the example "Teaching a lesson," and suppose we add a sixth choice for you: to start studying on *Thu*. Then, this gives rise to a new game with the utilities as given by Table 5.16. In this modified game there are two weakly dominated choices: the choices *Wed* and *Thu* for you. Suppose now that in the Dekel–Fudenberg procedure, we only eliminate the weakly dominated choice *Thu* for you in the first round, instead of eliminating both weakly dominated choices *Wed* and *Thu*. Then, by the definition of the Dekel–Fudenberg procedure, we can only eliminate *strictly* dominated choices after this elimination. But if we only eliminate your choice *Thu*, then in the reduced game that remains there are *no* strictly dominated choices. That is, the procedure would stop after round 1. So, if in the Dekel–Fudenberg procedure we only eliminate the weakly dominated choice *Thu* at round 1, then no further choices would be eliminated, and hence all choices except for your choice *Thu* would survive. But then, the set of choices that survive will be different from the choices that survive the *original* Dekel–Fudenberg procedure, which are your choice *Sat* and the teacher's choice *Fri*. Hence, for the eventual output of the Dekel–Fudenberg procedure it is important that at the first round we eliminate *all* weakly dominated choices. Otherwise, the final result can be very different.

But it is easily seen that *after* step 1 of the Dekel–Fudenberg procedure, it does not matter in which order, or how quickly, we eliminate choices. After step 1, in which we eliminate *all* weakly dominated choices from the original game, we apply iterated elimination of strictly dominated choices to the reduced game that remains. In Chapter 3 we saw, however, that for the eventual output of iterated elimination of strictly dominated choices the order and speed of elimination is not relevant. Hence, for the eventual output of the Dekel–Fudenberg procedure the order and speed of elimination *after step 1* does not matter – as long as we eliminate *all* weakly dominated choices at step 1. We thus reach the following conclusion.

Theorem 5.9.3 *(Order of elimination in the Dekel–Fudenberg procedure)*
The set of choices that survive the Dekel–Fudenberg procedure does not change if we change the order and speed of elimination ***after*** *step 1 of the procedure. However, the set of choices that survive may change if in step 1 we do not eliminate* ***all*** *weakly dominated choices.*

So, if we wish to find all choices you can rationally make under common full belief in "caution and primary belief in rationality," then it is absolutely crucial that in the first step of the Dekel–Fudenberg procedure we eliminate *all* weakly dominated choices from the game. From that step on, it is no longer necessary to always find *all* strictly dominated choices in the reduced games. If at every subsequent step we eliminate *some* – but not necessarily *all* – strictly dominated choices, we will still end up with the "correct" sets of choices for the players.

5.10 Proofs

This proofs section consists of three parts. In the first part we prove Theorem 5.8.3, which characterizes those choices that can rationally be made under a cautious lexicographic belief. In the other two parts we prove Theorem 5.9.2, which states that the choices that can rationally be made under common full belief in "caution and primary belief in rationality" are precisely those choices that survive the Dekel–Fudenberg procedure. Before we prove this result, we first derive in the second part an optimality principle for the Dekel–Fudenberg procedure, which is similar to the optimality principle for iterated elimination of strictly dominated choices in the proofs section of Chapter 3. In the third part we finally apply this optimality principle to prove Theorem 5.9.2.

Characterization of rational choices under cautious lexicographic beliefs

In this part we prove Theorem 5.8.3, which identifies those choices that can be optimal for a type with a cautious lexicographic belief.

Theorem 5.8.3 *(Characterization of optimal choices with cautious lexicographic beliefs)*
A choice c_i can be made optimally under a cautious lexicographic belief about j's choice, if and only if, c_i is not weakly dominated by another choice, nor weakly dominated by a randomized choice. Moreover, every choice c_i that is not weakly dominated by another choice, nor by a randomized choice, is optimal for a cautious lexicographic belief with one level only.

Proof: To prove this result, we proceed by two parts. In part (a) we show that every choice that is weakly dominated by another choice, or by a randomized choice, cannot be optimal for any cautious lexicographic belief. In part (b) we show that every choice that is not optimal for any cautious lexicographic belief with one level, is weakly dominated. Part (b) implies that every choice that is not optimal for *any* cautious lexicographic belief – including those with more than one level – is weakly dominated. Together with part (a), it follows that a choice is not optimal for any cautious lexicographic belief, if and only if, the choice is weakly dominated. Or, equivalently, a choice *is* optimal for some cautious lexicographic belief, if and only if, it is *not* weakly dominated. Moreover, part (b) implies that every choice that is *not* weakly dominated is optimal for a cautious lexicographic belief with one level only. Hence, parts (a) and (b) suffice to prove the theorem.

(a) We show that every choice that is weakly dominated by another choice, or by a randomized choice, cannot be optimal for any cautious lexicographic belief. Suppose that choice c_i is weakly dominated by some randomized choice r_i. In the special case where r_i assigns probability 1 to one choice c_i', we would have that c_i is weakly dominated by that choice c_i'. So, we do not have to consider this latter case separately. We show that c_i cannot be optimal for any cautious lexicographic belief.

Since c_i is weakly dominated by the randomized choice r_i, we have that $u_i(r_i,c_j) \geq u_i(c_i,c_j)$ for all of the opponent's choices c_j, and $u_i(r_i,c_j) > u_i(c_i,c_j)$ for at least one of the opponent's choices c_j. Take an arbitrary cautious lexicographic belief

$$b_i = (b_i^1; b_i^2; \ldots; b_i^K)$$

about j's choice. We show that there is some other choice c_i' that is better than c_i for player i under the lexicographic belief b_i.

For every level $k \in \{1, \ldots, K\}$, let $u_i^k(c_i,b_i)$ be the expected utility induced by c_i under the level k belief b_i^k. Similarly, let $u_i^k(r_i,b_i)$ be the expected utility induced by the randomized choice r_i under the level k belief b_i^k. By definition,

$$u_i^k(c_i,b_i) = \sum_{c_j \in C_j} b_i^k(c_j) \cdot u_i(c_i,c_j),$$

and

$$u_i^k(r_i, b_i) = \sum_{c_j \in C_j} b_i^k(c_j) \cdot u_i(r_i, c_j).$$

As the lexicographic belief b_i is cautious, all of the opponent's choices c_j have a positive probability in some level of b_i. Since $u_i(r_i, c_j) \geq u_i(c_i, c_j)$ for all of the opponent's choices c_j, and $u_i(r_i, c_j) > u_i(c_i, c_j)$ for at least one of the opponent's choices c_j, we have that $u_i^k(r_i, b_i) \geq u_i^k(c_i, b_i)$ for all levels k, and $u_i^k(r_i, b_i) > u_i^k(c_i, b_i)$ for some level k. Let l be the first level for which $u_i^l(r_i, b_i) > u_i^l(c_i, b_i)$. Then,

$$\begin{aligned} u_i^1(r_i, b_i) &= u_i^1(c_i, b_i), \\ &\vdots \\ u_i^{l-1}(r_i, b_i) &= u_i^{l-1}(c_i, b_i), \\ u_i^l(r_i, b_i) &> u_i^l(c_i, b_i). \end{aligned}$$

Suppose that $D_i \subseteq C_i$ is the set of choices to which r_i assigns positive probability. By definition,

$$u_i^k(r_i, b_i) = \sum_{c_i \in D_i} r_i(c_i) \cdot u_i^k(c_i, b_i)$$

for all levels k.

Suppose first that $l = 1$, that is, $u_i^1(r_i, b_i) > u_i^1(c_i, b_i)$. Then, there is some $c_i' \in D_i$ with $u_i^1(c_i', b_i) > u_i^1(c_i, b_i)$, which means that c_i' is better than c_i under the lexicographic belief b_i. Hence, c_i would not be optimal under b_i.

Suppose next that $l = 2$. That is, $u_i^1(r_i, b_i) = u_i^1(c_i, b_i)$ and $u_i^2(r_i, b_i) > u_i^2(c_i, b_i)$. Since $u_i^1(r_i, b_i) = u_i^1(c_i, b_i)$, there are two possibilities: Either there is some $c_i' \in D_i$ with $u_i^1(c_i', b_i) > u_i^1(c_i, b_i)$, or $u_i^1(c_i', b_i) = u_i^1(c_i, b_i)$ for every $c_i' \in D_i$. If there is some $c_i' \in D_i$ with $u_i^1(c_i', b_i) > u_i^1(c_i, b_i)$, then, by definition, c_i' is better than c_i under the lexicographic belief b_i, and we are done. Suppose, alternatively, that $u_i^1(c_i', b_i) = u_i^1(c_i, b_i)$ for every $c_i' \in D_i$. Then, we move to level 2 in the lexicographic belief. Since $u_i^2(r_i, b_i) > u_i^2(c_i, b_i)$, there must be some $c_i' \in D_i$ with $u_i^2(c_i', b_i) > u_i^2(c_i, b_i)$. As $u_i^1(c_i', b_i) = u_i^1(c_i, b_i)$, it follows that c_i' is better than c_i under the lexicographic belief b_i. Hence, c_i would not be optimal under b_i.

Suppose now that $l = 3$, that is, $u_i^1(r_i, b_i) = u_i^1(c_i, b_i)$ and $u_i^2(r_i, b_i) = u_i^2(c_i, b_i)$, but $u_i^3(r_i, b_i) > u_i^3(c_i, b_i)$. Similarly to above, there are two possibilities: Either there is some $c_i' \in D_i$ with $u_i^1(c_i', b_i) > u_i^1(c_i, b_i)$, or $u_i^1(c_i', b_i) = u_i^1(c_i, b_i)$ for every $c_i' \in D_i$. If there is some $c_i' \in D_i$ with $u_i^1(c_i', b_i) > u_i^1(c_i, b_i)$, then c_i' is better than c_i under the lexicographic belief b_i. So, c_i would not be optimal under b_i. Suppose, alternatively, that $u_i^1(c_i', b_i) = u_i^1(c_i, b_i)$ for every $c_i' \in D_i$. Then, we move to level 2 in the lexicographic belief. Since $u_i^2(r_i, b_i) = u_i^2(c_i, b_i)$, there are again two possibilities: Either there is some $c_i' \in D_i$ with $u_i^2(c_i', b_i) > u_i^2(c_i, b_i)$, or $u_i^2(c_i', b_i) = u_i^2(c_i, b_i)$ for every $c_i' \in D_i$. Suppose there is some $c_i' \in D_i$ with $u_i^2(c_i', b_i) > u_i^2(c_i, b_i)$. As $u_i^1(c_i', b_i) = u_i^1(c_i, b_i)$, it follows c_i' is better than c_i under the lexicographic belief b_i. Hence, c_i would not be optimal under b_i.

Suppose, on the other hand, that $u_i^2(c_i', b_i) = u_i^2(c_i, b_i)$ for every $c_i' \in D_i$. Then, we move to level 3 in the lexicographic belief. Since $u_i^3(r_i, b_i) > u_i^3(c_i, b_i)$, there must be some $c_i' \in D_i$ with $u_i^3(c_i', b_i) > u_i^3(c_i, b_i)$. As $u_i^1(c_i', b_i) = u_i^1(c_i, b_i)$ and $u_i^2(c_i', b_i) = u_i^2(c_i, b_i)$, it follows that c_i' is better than c_i under the lexicographic belief b_i. Hence, c_i would not be optimal under b_i.

By proceeding like this, we will eventually conclude that for every possible l, the choice c_i cannot be optimal for the cautious lexicographic belief b_i. Since this is true for *every* cautious lexicographic belief b_i about j's choice, we see that c_i cannot be optimal under any cautious lexicographic belief about j's choice. This completes the proof of part (a).

(b) We show that every choice that is not optimal for any cautious lexicographic belief with one level, is weakly dominated. Take a choice c_i^* that is not optimal for any cautious lexicographic belief with one level only. Clearly, a lexicographic belief with only one level corresponds to a standard belief, that is, a single probability distribution b_i on j's choices, as in Part I of this book. We call such a standard belief b_i *cautious* if it assigns positive probability to each of j's choices. Since c_i^* is not optimal for any cautious lexicographic belief with one level only, we conclude that choice c_i^* is not optimal for any cautious standard belief b_i about j's choices. For every $n \geq 2$, and all of the opponent's choices c_j, define the cautious standard belief $b_i^{c_j,n} \in \Delta(C_j)$ about j's choice by

$$b_i^{c_j,n}(c_j') = \begin{cases} 1 - \frac{1}{n}, & \text{if } c_j' = c_j, \\ \frac{1}{n(|C_j|-1)}, & \text{if } c_j' \neq c_j. \end{cases}$$

Here, $|C_j|$ denotes the number of choices in C_j. Clearly, the belief $b_i^{c_j,n}$ assigns positive probability to each of j's choices, and is thus cautious. Moreover, if n is large, then $b_i^{c_j,n}$ assigns a probability of almost 1 to choice c_j. For every $n \geq 2$ we define a modified game Γ^n in which player i's modified utility function u_i^n is given by

$$u_i^n(c_i, c_j) := u_i(c_i, b_i^{c_j,n})$$

for all $c_i \in C_i$ and $c_j \in C_j$, whereas player j's utility function in Γ^n is simply u_j – player j's utility function in the original game Γ. Hence, if n tends to infinity, then the utility $u_i^n(c_i, c_j)$ in the modified game Γ^n converges to the utility $u_i(c_i, c_j)$ in the original game Γ, as $b_i^{c_j,n}$ converges to the belief that assigns probability 1 to c_j.

We will now show that in every modified game Γ^n, choice c_i^* is not optimal for any standard belief b_i about j's choices. Suppose, contrary to what we want to prove, that c_i^* is optimal in some modified game Γ^n for some standard belief b_i. That is, $u_i^n(c_i^*, b_i) \geq u_i^n(c_i, b_i)$ for all $c_i \in C_i$. By definition of the modified utility function u_i^n in Γ^n we have that

$$u_i^n(c_i^*, b_i) = \sum_{c_j \in C_j} b_i(c_j) \cdot u_i^n(c_i^*, c_j)$$

$$= \sum_{c_j \in C_j} b_i(c_j) \cdot u_i(c_i^*, b_i^{c_j,n})$$

$$= u_i(c_i^*, \hat{b}_i),$$

where $\hat{b}_i \in \Delta(C_j)$ is the belief given by

$$\hat{b}_i(c_j) := \sum_{c_j' \in C_j} b_i(c_j') \cdot b_i^{c_j',n}(c_j)$$

for all $c_j \in C_j$. As each of the beliefs $b_i^{c_j',n}$ assigns positive probability to each of j's choices, it follows that $\hat{b}_i$ also assigns positive probability to each of j's choices, and hence $\hat{b}_i$ is a cautious belief. So, we conclude that $u_i^n(c_i^*, b_i) = u_i(c_i^*, \hat{b}_i)$ for some cautious belief $\hat{b}_i$. Similarly, it can be shown that $u_i^n(c_i, b_i) = u_i(c_i, \hat{b}_i)$ for every other choice c_i. As $u_i^n(c_i^*, b_i) \geq u_i^n(c_i, b_i)$ for all $c_i \in C_i$, it follows that $u_i(c_i^*, \hat{b}_i) \geq u_i(c_i, \hat{b}_i)$ for every choice $c_i \in C_i$, which would imply that choice c_i^* is optimal in the original game Γ for the cautious belief $\hat{b}_i$. This, however, contradicts our assumption that c_i^* is not optimal within the original game Γ for any cautious standard belief. Hence, we conclude that in every modified game Γ^n, choice c_i^* is not optimal for any standard belief b_i about j's choices.

By applying Theorem 2.5.3 from Chapter 2 to each of the modified games Γ^n, we find that within every modified game Γ^n, choice c_i^* is strictly dominated by some randomized choice. For every modified game Γ^n, let r_i^n be a randomized choice for which

$$\min_{c_j \in C_j} (u_i^n(r_i^n, c_j) - u_i^n(c_i^*, c_j)) \geq \min_{c_j \in C_j} (u_i^n(r_i, c_j) - u_i^n(c_i^*, c_j)) \tag{5.1}$$

for all randomized choices r_i. As in every modified game Γ^n, choice c_i^* is strictly dominated by some randomized choice, we know that in every Γ^n there is some randomized choice r_i for which

$$\min_{c_j \in C_j} (u_i^n(r_i, c_j) - u_i^n(c_i^*, c_j)) > 0.$$

But then, it follows that

$$\min_{c_j \in C_j} (u_i^n(r_i^n, c_j) - u_i^n(c_i^*, c_j)) > 0$$

for every modified game Γ^n.

In this way, we construct for every $n \geq 2$ some randomized choice r_i^n satisfying (5.1). That is, we obtain a sequence $(r_i^n)_{n \geq 2}$ of randomized choices in $\Delta(C_i)$. As the set $\Delta(C_i)$ is compact – that is, closed and bounded – we know there is a subsequence of the sequence $(r_i^n)_{n \geq 2}$ that converges to some randomized choice $r_i^* \in \Delta(C_i)$. We may therefore assume – without loss of generality – that the original sequence $(r_i^n)_{n \geq 2}$ converges to some randomized choice r_i^*.

We will finally prove that this randomized choice r_i^* *weakly* dominates the choice c_i^* in the original game Γ. By construction, every randomized choice r_i^n in the sequence strictly dominates the choice c_i^* within the modified game Γ^n. That is,

$$u_i^n(r_i^n, c_j) > u_i^n(c_i^*, c_j)$$

for every $c_j \in C_j$ and every $n \geq 2$. If n tends to infinity, then $u_i^n(r_i^n, c_j)$ converges to $u_i(r_i^*, c_j)$ and $u_i^n(c_i^*, c_j)$ converges to $u_i(c_i^*, c_j)$, which immediately implies that

$$u_i(r_i^*, c_j) \geq u_i(c_i^*, c_j)$$

for all $c_j \in C_j$.

It remains to show that $u_i(r_i^*, c_j) > u_i(c_i^*, c_j)$ for at least some $c_j \in C_j$. Suppose, contrary to what we want to prove, that $u_i(r_i^*, c_j) = u_i(c_i^*, c_j)$ for all $c_j \in C_j$. As the sequence $(r_i^n)_{n \geq 2}$ converges to r_i^*, there must be some $n \geq 2$ such that every choice c_i that has positive probability in r_i^*, also has positive probability in r_i^n. Take such an $n \geq 2$. Then, there is some $\lambda \in (0,1)$ small enough such that

$$r_i^n(c_i) \geq \lambda \cdot r_i^*(c_i)$$

for all $c_i \in C_i$. We may thus construct a new randomized choice $\hat{r}_i^n$ by

$$\hat{r}_i^n(c_i) := \frac{1}{1-\lambda} \cdot r_i^n(c_i) - \frac{\lambda}{1-\lambda} \cdot r_i^*(c_i) \tag{5.2}$$

for all $c_i \in C_i$. As $r_i^n(c_i) \geq \lambda \cdot r_i^*(c_i)$ for all c_i, we have that $\hat{r}_i^n(c_i) \geq 0$ for all c_i. Moreover,

$$\sum_{c_i \in C_i} \hat{r}_i^n(c_i) = \frac{1}{1-\lambda} \sum_{c_i \in C_i} r_i^n(c_i) - \frac{\lambda}{1-\lambda} \sum_{c_i \in C_i} r_i^*(c_i) = 1,$$

since

$$\sum_{c_i \in C_i} r_i^n(c_i) = \sum_{c_i \in C_i} r_i^*(c_i) = 1.$$

So, $\hat{r}_i^n$ is indeed a randomized choice. Then, we have for every choice $c_j \in C_j$ that

$$\begin{aligned} u_i^n(\hat{r}_i^n, c_j) &= \frac{1}{1-\lambda} u_i^n(r_i^n, c_j) - \frac{\lambda}{1-\lambda} u_i^n(r_i^*, c_j) \\ &= \frac{1}{1-\lambda} u_i^n(r_i^n, c_j) - \frac{\lambda}{1-\lambda} u_i(r_i^*, b_i^{c_j,n}) \\ &= \frac{1}{1-\lambda} u_i^n(r_i^n, c_j) - \frac{\lambda}{1-\lambda} u_i(c_i^*, b_i^{c_j,n}) \\ &= \frac{1}{1-\lambda} u_i^n(r_i^n, c_j) - \frac{\lambda}{1-\lambda} u_i^n(c_i^*, c_j). \end{aligned}$$

Here, the first equality follows from the definition of $\hat{r}_i^n$ in (5.2), the second and fourth equalities follow from the definition of the modified utility function u_i^n, and the third equality follows from the assumption that $u_i(r_i^*, c_j') = u_i(c_i^*, c_j')$ for all $c_j' \in C_j$. So, we conclude that

$$\begin{aligned} u_i^n(\hat{r}_i^n, c_j) - u_i^n(c_i^*, c_j) &= \frac{1}{1-\lambda} u_i^n(r_i^n, c_j) - \frac{\lambda}{1-\lambda} u_i^n(c_i^*, c_j) \\ &\quad - u_i^n(c_i^*, c_j) \\ &= \frac{1}{1-\lambda}(u_i^n(r_i^n, c_j) - u_i^n(c_i^*, c_j)) \\ &> u_i^n(r_i^n, c_j) - u_i^n(c_i^*, c_j) \end{aligned}$$

for all $c_j \in C_j$. Here, the inequality follows from the fact that $\lambda \in (0,1)$ – and hence $1/(1-\lambda) > 1$ – and the fact that $u_i^n(r_i^n, c_j) - u_i^n(c_i^*, c_j) > 0$. So, we see that

$$u_i^n(\hat{r}_i^n, c_j) - u_i^n(c_i^*, c_j) > u_i^n(r_i^n, c_j) - u_i^n(c_i^*, c_j)$$

for all $c_j \in C_j$, which implies that

$$\min_{c_j \in C_j} (u_i^n(\hat{r}_i^n, c_j) - u_i^n(c_i^*, c_j)) > \min_{c_j \in C_j} (u_i^n(r_i^n, c_j) - u_i^n(c_i^*, c_j)).$$

This, however, contradicts the choice of r_i^n according to (5.1). Hence, the assumption that $u_i(r_i^*, c_j) = u_i(c_i^*, c_j)$ for all $c_j \in C_j$ cannot be true. So, there must be some $c_j \in C_j$ with $u_i(r_i^*, c_j) > u_i(c_i^*, c_j)$. Altogether, we see that $u_i(r_i^*, c_j) \geq u_i(c_i^*, c_j)$ for all c_j, and $u_i(r_i^*, c_j) > u_i(c_i^*, c_j)$ for at least some c_j. This, however, implies that c_i^* is weakly dominated by the randomized choice r_i^*. The proof of part (b) is hereby complete.

Hence, we have shown parts (a) and (b). These two parts together imply the theorem. Hence, the proof of Theorem 5.8.3 is complete. ■

Optimality principle

In this subsection we show an *optimality principle* for the Dekel–Fudenberg procedure, which is similar to the optimality principle for iterated elimination of strictly dominated choices in the proofs section of Chapter 3. For every $k \in \{0, 1, 2, ...\}$, let C_i^k be the set of choices for player i that survive the first k steps in the Dekel–Fudenberg procedure. In particular, C_i^0 is the full set of choices. Let Γ^k denote the reduced game that remains after the first k steps of the Dekel–Fudenberg procedure.

Now, consider some $c_i \in C_i^k$ where $k \geq 1$. Then, in particular, $c_i \in C_i^1$, which means that c_i is not weakly dominated in the original game. By Theorem 5.8.3, we then know that c_i is optimal for some cautious lexicographic belief that consists of one level only. That is, c_i is optimal for some cautious standard belief b_i^2 that assigns positive probability to all choices in C_j.

Moreover, since $c_i \in C_i^k$, we know by the definition of the Dekel–Fudenberg procedure that c_i is not strictly dominated in the reduced game Γ^{k-1} that remains after step $k-1$ of the algorithm. In Γ^{k-1}, player i can choose from C_i^{k-1} and player j can choose from C_j^{k-1}. So, by applying Theorem 2.5.3 to the reduced game Γ^{k-1}, we know that every $c_i \in C_i^k$ is optimal, among the choices in C_i^{k-1}, for some belief $b_i^1 \in \Delta(C_j^{k-1})$.

That is,

$$u_i(c_i, b_i^1) \geq u_i(c_i', b_i^1) \text{ for all } c_i' \in C_i^{k-1}.$$

Together with our insight above, we may then conclude that c_i is optimal, among the choices in C_i^{k-1}, for the cautious lexicographic belief

$$b_i = (b_i^1; b_i^2)$$

where $b_i^1 \in \Delta(C_j^{k-1})$ and b_i^2 assigns positive probability to all choices in C_j.

In fact, we can say more about choice c_i. Not only is c_i optimal for the lexicographic belief b_i above *among the choices in* C_i^{k-1}. But for this particular lexicographic belief b_i, choice c_i is even optimal *among all choices in the original game*. We refer to this result as the "optimality principle for the Dekel–Fudenberg procedure."

Lemma 5.10.1 *(Optimality principle)*
Consider a choice $c_i \in C_i^k$ for player i that survives the first k steps in the Dekel–Fudenberg procedure. Then, c_i is optimal, among the choices in the original game, for some cautious lexicographic belief $b_i = (b_i^1; b_i^2)$, where $b_i^1 \in \Delta(C_j^{k-1})$ and b_i^2 assigns positive probability to all choices in C_j.

Proof. Consider some choice $c_i \in C_i^k$. Then, we know from above that there is some belief b_i^2, assigning positive probability to all choices in C_j, such that

$$u_i(c_i, b_i^2) \geq u_i(c_i', b_i^2) \text{ for all } c_i' \in C_i,$$

and a belief $b_i^1 \in \Delta(C_j^{k-1})$ such that

$$u_i(c_i, b_i^1) \geq u_i(c_i', b_i^1) \text{ for all } c_i' \in C_i^{k-1}.$$

We will prove that

$$u_i(c_i, b_i^1) \geq u_i(c_i', b_i^1) \text{ for all } c_i' \in C_i.$$

Suppose, on the contrary, that there is some $c_i' \in C_i$ with $u_i(c_i', b_i^1) > u_i(c_i, b_i^1)$. Let c_i^* be an optimal choice for b_i^1 among the choices in C_i. Then,

$$u_i(c_i^*, b_i^1) \geq u_i(c_i', b_i^1) > u_i(c_i, b_i^1).$$

Since c_i^* is optimal for belief $b_i^1 \in \Delta(C_j^{k-1})$ among the choices in C_i, we can apply Theorem 2.5.3 to the reduced game Γ^{k-1} and conclude that c_i^* is not strictly dominated in Γ^{k-1}, and hence $c_i^* \in C_i^k$. In particular, $c_i^* \in C_i^{k-1}$. So, we have found a choice $c_i^* \in C_i^{k-1}$ with $u_i(c_i^*, b_i^1) > u_i(c_i, b_i^1)$. This, however, contradicts our assumption that

$$u_i(c_i, b_i^1) \geq u_i(c_i', b_i^1) \text{ for all } c_i' \in C_i^{k-1}.$$

Hence, we conclude that there is no $c_i' \in C_i$ with $u_i(c_i', b_i^1) > u_i(c_i, b_i^1)$. So, c_i is indeed optimal for b_i^1 among the choices in the original game. Since c_i is also optimal for b_i^2 among the choices in the original game, we conclude that c_i is optimal, among all choices in the original game, for the cautious lexicographic belief $b_i = (b_i^1; b_i^2)$. This completes the proof. ◇

The algorithm works

We will now use the characterization of optimal choices in Theorem 5.8.3 and the optimality principle for the Dekel–Fudenberg procedure in Lemma 5.10.1, to prove that the Dekel–Fudenberg procedure yields exactly those choices that can rationally be made under common full belief in "caution and primary belief in rationality."

Theorem 5.9.2. *(The algorithm works)*
Consider a static game with finitely many choices for every player.
(1) For every $k \geq 1$, the choices that can rationally be made by a cautious type that expresses up to k-fold full belief in "caution and primary belief in rationality" are exactly those choices that survive the first $k+1$ steps of the Dekel–Fudenberg procedure.
(2) The choices that can rationally be made by a cautious type that expresses common full belief in "caution and primary belief in rationality" are exactly those choices that survive the Dekel–Fudenberg procedure.

Proof: Remember that, for every k, the set C_i^k denotes the set of choices for player i that survive the first k steps of the Dekel–Fudenberg procedure. We denote by BR_i^k the set of choices that a cautious player i can rationally make when expressing up to k-fold full belief in "caution and primary belief in rationality." We show that $BR_i^k = C_i^{k+1}$ for every $k \geq 1$. We proceed by two parts. In part (a), we show by induction on k that $BR_i^k \subseteq C_i^{k+1}$ for all $k \geq 1$. In part (b) we prove that $C_i^{k+1} \subseteq BR_i^k$ for every $k \geq 1$.

(a) Show that $BR_i^k \subseteq C_i^{k+1}$.

We show this statement by induction on k.

Induction start: Start with $k = 1$. By definition, BR_i^1 are those choices that a cautious player i can rationally make if he expresses 1-fold full belief in "caution and primary belief in rationality." That is, BR_i^1 contains those choices that are optimal for some cautious player i who fully believes that j is cautious and primarily believes in j's rationality. Now, take some cautious type t_i for player i that fully believes that j is cautious and primarily believes in j's rationality. Suppose that t_i holds the lexicographic belief $b_i(t_i) = (b_i^1; ...; b_i^K)$ about j's choice-type pairs. Then, the primary belief b_i^1 assigns positive probability only to choice-type pairs (c_j, t_j) where t_j is cautious and c_j is optimal for t_j. In particular, b_i^1 only assigns positive probability to the choices c_j that are optimal for some cautious lexicographic belief. From Theorem 5.8.3 we know that every such choice c_j is not weakly dominated in the original game Γ. Hence, the primary belief b_i^1 only assigns positive probability to j's choices that are not weakly dominated in the original game. Remember that Γ^1 denotes the reduced game that remains after the first step in the Dekel–Fudenberg procedure. So, the primary belief b_i^1 only assigns positive probability to j's choices in Γ^1.

Now, take an arbitrary choice c_i that is optimal for type t_i. As t_i is cautious, it follows from Theorem 5.8.3 that c_i is not weakly dominated in the original game Γ, and hence c_i is in Γ^1. Moreover, choice c_i must in particular be optimal for t_i's primary belief b_i^1, which – as we have seen – only assigns positive probability to the opponent's choices in Γ^1. But then, it follows from Theorem 2.5.3 in Chapter 2 that choice c_i cannot be strictly dominated in Γ^1, and hence must be in Γ^2 – the reduced game that remains after the first two steps of the Dekel–Fudenberg procedure. Summarizing, we see that every choice c_i that is optimal for a cautious type t_i that fully believes that j is cautious and primarily believes in j's rationality, must be in Γ^2. As such, every choice c_i that can rationally be made when expressing 1-fold full belief in "caution and primary belief in rationality" must be in C_i^2. That is, $BR_i^1 \subseteq C_i^2$, as was to be shown.

Induction step: Take some $k \geq 2$, and assume that $BR_i^{k-1} \subseteq C_i^k$ for both players i. We prove that $BR_i^k \subseteq C_i^{k+1}$ for both players i.

Consider a player i, and take some $c_i \in BR_i^k$. Then, c_i is optimal for some cautious type t_i that expresses up to k-fold full belief in "caution and primary belief in rationality." Let $b_i(t_i) = (b_i^1; ...; b_i^K)$ be the lexicographic belief that t_i holds about j's choice-type pairs. Then, the primary belief b_i^1 only assigns positive probability to the opponent's choice-type pairs (c_j, t_j) where c_j is optimal for t_j, type t_j is cautious and t_j expresses up to $(k-1)$-fold full belief in "caution and primary belief in rationality." As a consequence, b_i^1 only assigns positive probability to the opponent's choices that can rationally be made when expressing up to $(k-1)$-fold full belief in "caution and primary belief in rationality." That is, the primary belief b_i^1 only assigns positive probability to j's choices in BR_j^{k-1}. Since, by the induction assumption, $BR_j^{k-1} \subseteq C_j^k$, it follows that b_i^1 only assigns positive probability to j's choices in C_j^k.

Since c_i is optimal for type t_i, choice c_i must in particular be optimal for the primary belief b_i^1. As b_i^1 only assigns positive probability to j's choices in C_j^k, it follows that choice c_i is optimal for some belief $b_i^1 \in \Delta(C_j^k)$. But then, by Theorem 2.5.3 in Chapter 2, choice c_i is not strictly dominated in the reduced game Γ^k that remains after the first k steps in the Dekel–Fudenberg procedure. Hence, by construction, choice c_i must be in C_i^{k+1}. So, we have shown that every $c_i \in BR_i^k$ must be in C_i^{k+1}. Therefore, $BR_i^k \subseteq C_i^{k+1}$ for both players i. By induction, the proof for part (a) is complete.

(b) Show that $C_i^{k+1} \subseteq BR_i^k$.

So, we must show that every choice $c_i \in C_i^{k+1}$ is optimal for some cautious type t_i that expresses up to k-fold full belief in "caution and primary belief in rationality." Suppose that the Dekel–Fudenberg procedure terminates after K rounds – that is, $C_i^{K+1} = C_i^K$ for both players i. In this proof we will construct an epistemic model M with the following properties:

- For every choice $c_i \in C_i^1$, there is a cautious type $t_i^{c_i}$ for which c_i is optimal.
- For every $k \geq 2$, if c_i is in C_i^k, then the associated type $t_i^{c_i}$ is cautious and expresses up to $(k-1)$-fold full belief in "caution and primary belief in rationality."
- If c_i is in C_i^K, then the associated type $t_i^{c_i}$ is cautious and expresses common full belief in "caution and primary belief in rationality."

For this construction we proceed by the following steps. In step 1 we will construct, for every choice $c_i \in C_i^1$, a cautious lexicographic belief $b_i^{c_i}$, with two levels, about the opponent's choices for which c_i is optimal. In step 2 we will use these beliefs to construct our epistemic model M, where we define for every choice $c_i \in C_i^1$ a cautious type $t_i^{c_i}$ for which c_i is optimal. In step 3 we will show that, for every $k \geq 2$ and every $c_i \in C_i^k$, the associated type $t_i^{c_i}$ expresses up to $(k-1)$-fold full belief in "caution and primary belief in rationality." In step 4 we will finally prove that, for every choice $c_i \in C_i^K$, the associated type $t_i^{c_i}$ expresses common full belief in "caution and primary belief in rationality."

Step 1: Construction of beliefs

We start by constructing, for every $c_i \in C_i^1$, a cautious lexicographic belief $b_i^{c_i}$ about the opponent's choices, with two levels, for which c_i is optimal. For every $k \in \{1, ..., K-1\}$, let D_i^k be the set of choices that survive the first k steps of the Dekel–Fudenberg procedure, but not step $k+1$. So, $D_i^k = C_i^k \backslash C_i^{k+1}$. For defining the beliefs, we distinguish between the following two cases:

(1) First consider some $k \in \{1, ..., K-1\}$ and a choice $c_i \in D_i^k$. Then, by Lemma 5.10.1, choice c_i is optimal, among the choices in the original game, for a cautious, two-level lexicographic belief $b_i^{c_i} = (b_i^{c_i,1}; b_i^{c_i,2})$, where $b_i^{c_i,1} \in \Delta(C_j^{k-1})$ and $b_i^{c_i,2}$ assigns positive probability to all choices in C_j.

(2) Consider next some choice $c_i \in C_i^K$ that survives the full Dekel–Fudenberg procedure. Then, also $c_i \in C_i^{K+1}$. Hence, by Lemma 5.10.1, choice c_i is optimal, among the choices in the original game, for a cautious, two-level lexicographic belief $b_i^{c_i} = (b_i^{c_i,1}; b_i^{c_i,2})$, where $b_i^{c_i,1} \in \Delta(C_j^K)$ and $b_i^{c_i,2}$ assigns positive probability to all choices in C_j.

In this way we have defined, for every choice $c_i \in C_i^1$, a two-level cautious lexicographic belief $b_i^{c_i}$ for which c_i is optimal.

Step 2: Construction of types

We will now use the beliefs $b_i^{c_i} = (b_i^{c_i,1}; b_i^{c_i,2})$ from step 1 to construct, for every choice $c_i \in C_i^1$, a cautious type $t_i^{c_i}$ for which c_i is optimal. For every player i, let the set of types be given by

$$T_i = \{t_i^{c_i} : c_i \in C_i^1\}.$$

For every $k \in \{1, ..., K\}$, let T_i^k be the set of types $t_i^{c_i} \in T_i$ where $c_i \in C_i^k$. As $C_i^K \subseteq C_i^{K-1} \subseteq ... \subseteq C_i^1$, it follows that $T_i^K \subseteq T_i^{K-1} \subseteq ... \subseteq T_i^1$, where $T_i^1 = T_i$. We will now define the beliefs of these types and distinguish the following three cases:

(1) Consider first the types $t_i^{c_i} \in T_i$ with $c_i \in D_i^1$. That is, $t_i \in T_i^1 \backslash T_i^2$. We define the lexicographic belief $b_i(t_i^{c_i}) = (b_i^1(t_i^{c_i}); b_i^2(t_i^{c_i}))$ in the following way: Consider some arbitrary type $t_j \in T_j$ for opponent j.

For all of the opponent's choices c_j, the primary belief $b_i^1(t_i^{c_i})$ assigns probability $b_i^{c_i,1}(c_j)$ to the choice-type pair (c_j, t_j). The primary belief $b_i^1(t_i^c)$ assigns probability zero to all other choice-type pairs for player j. Hence, the primary belief $b_i^1(t_i^{c_i})$ has the same probability distribution over j's choices as $b_i^{c_i,1}$.

The secondary belief $b_i^2(t_i^{c_i})$ is defined as follows: For all of the opponent's choices $c_j \in C_j$, the secondary belief $b_i^2(t_i^{c_i})$ assigns probability $b_i^{c_i,2}(c_j)$ to the choice-type pair (c_j,t_j), where t_j is the same type as for the primary belief. So, the secondary belief $b_i^2(t_i^{c_i})$ has the same probability distribution over j's choices as $b_i^{c_i,2}$.

As type $t_i^{c_i}$ only deems possible type t_j for the opponent, and assigns, in its secondary belief, positive probability to all of the opponent's choices, it follows that $t_i^{c_i}$ is cautious. Moreover, the type $t_i^{c_i}$ holds the same lexicographic belief about j's choices as the belief $b_i^{c_i}$ from step 1. Since, by construction in step 1, the choice c_i is optimal for the lexicographic belief $b_i^{c_i}$, it follows that c_i is optimal for type $t_i^{c_i}$.

(2) Consider next the types $t_i^{c_i} \in T_i$ with $c_i \in D_i^k$ for some $k \geq 2$. That is, $t_i \in T_i^k \backslash T_i^{k+1}$ for some $k \geq 2$. We define the lexicographic belief $b_i(t_i^{c_i}) = (b_i^1(t_i^{c_i}); b_i^2(t_i^{c_i}))$ as follows:

For all of the opponent's choices c_j, the primary belief $b_i^1(t_i^{c_i})$ assigns probability $b_i^{c_i,1}(c_j)$ to the choice-type pair $(c_j, t_j^{c_j})$. It assigns probability zero to all other choice-type pairs for the opponent. Hence, $b_i^1(t_i^{c_i})$ has the same probability distribution over j's choices as $b_i^{c_i,1}$. As, by construction in step 1, $b_i^{c_i,1} \in \Delta(C_j^{k-1})$, the primary belief $b_i^1(t_i^{c_i})$ only assigns positive probability to opponent's types $t_j^{c_j}$ with $c_j \in C_j^{k-1}$. That is, $b_i^1(t_i^{c_i})$ only assigns positive probability the opponent's types in T_j^{k-1}.

The secondary belief $b_i^2(t_i^{c_i})$ is defined as follows: For all of the opponent's choices c_j and all of the opponent's types $t_j \in T_j^{k-1}$, the secondary belief $b_i^2(t_i^{c_i})$ assigns probability $b_i^{c_i,2}(c_j)/|T_j^{k-1}|$ to the choice-type pair (c_j,t_j). It assigns probability zero to all other choice-type pairs. Here, $|T_j^{k-1}|$ denotes the number of types in T_j^{k-1}. Hence, the secondary belief $b_i^2(t_i^{c_i})$ has the same probability distribution over j's choices as $b_i^{c_i,2}$, and assigns positive probability only to opponent's types in T_j^{k-1}.

By construction, type $t_i^{c_i}$ only deems possible opponent's types in T_j^{k-1}. Since the secondary belief $b_i^2(t_i^{c_i})$ assigns positive probability to every choice-type pair (c_j,t_j) with $c_j \in C_j$ and $t_j \in T_j^{k-1}$, it follows that type $t_i^{c_i}$ is cautious. Moreover, the type $t_i^{c_i}$ holds the same lexicographic belief about j's choices as the belief $b_i^{c_i}$ from step 1. Since, by construction in step 1, the choice c_i is optimal for the lexicographic belief $b_i^{c_i}$, it follows that c_i is optimal for type $t_i^{c_i}$.

(3) Finally, consider the types $t_i^{c_i} \in T_i$ with $c_i \in C_i^K$. That is, $t_i \in T_i^K$. We define the lexicographic belief $b_i(t_i^{c_i}) = (b_i^1(t_i^{c_i}); b_i^2(t_i^{c_i}))$ as follows:

For all of the opponent's choices c_j, the primary belief $b_i^1(t_i^{c_i})$ assigns probability $b_i^{c_i,1}(c_j)$ to the choice-type pair $(c_j, t_j^{c_j})$. It assigns probability zero to all other choice-type pairs for the opponent. Hence, $b_i^1(t_i^{c_i})$ has the same probability distribution over j's choices as $b_i^{c_i,1}$. As, by construction in step 1, $b_i^{c_i,1} \in \Delta(C_j^K)$, the primary belief $b_i^1(t_i^{c_i})$ only assigns positive probability to the opponent's types $t_j^{c_j}$ with $c_j \in C_j^K$. That is, $b_i^1(t_i^{c_i})$ only assigns positive probability to the opponent's types in T_j^K.

The secondary belief $b_i^2(t_i^{c_i})$ is defined as follows: For all of the opponent's choices c_j, and all of the opponent's types $t_j \in T_j^K$, the secondary belief $b_i^2(t_i^{c_i})$ assigns probability $b_i^{c_i,2}(c_j)/|T_j^K|$ to the choice-type pair (c_j, t_j). It assigns probability zero to all other choice-type pairs. Here, $|T_j^K|$ denotes the number of types in T_j^K. Hence, the secondary belief $b_i^2(t_i^{c_i})$ has the same probability distribution over j's choices as $b_i^{c_i,2}$, and assigns positive probability only to opponent's types in T_j^K.

By construction, type $t_i^{c_i}$ only deems possible opponent's types in T_j^K. Since the secondary belief $b_i^2(t_i^{c_i})$ assigns positive probability to every choice-type pair (c_j, t_j) with $c_j \in C_j$ and $t_j \in T_j^K$, it follows that type $t_i^{c_i}$ is cautious. Moreover, the type $t_i^{c_i}$ holds the same lexicographic belief about j's choices as the belief $b_i^{c_i}$ from step 1. Since, by construction in step 1, the choice c_i is optimal for the lexicographic belief $b_i^{c_i}$, it follows that c_i is optimal for type $t_i^{c_i}$.

Step 3: Every type $t_i \in T_i^k$ expresses up to $(k-1)$-fold full belief in "caution and primary belief in rationality"
We prove this statement by induction on k.

Induction start: We start with $k = 2$. Take some type $t_i \in T_i^2$. That is, $t_i = t_i^{c_i}$ for some choice $c_i \in C_i^2$. By construction in step 2, the primary belief $b_i^1(t_i^{c_i})$ of type $t_i^{c_i}$ only assigns positive probability to opponent's choice-type pairs $(c_j, t_j^{c_j})$ where $c_j \in C_j^1$ and choice c_j is optimal for type $t_j^{c_j}$. This means, in particular, that type $t_i^{c_i}$ primarily believes in the opponent's rationality. We also know, by construction in step 2, that type $t_i^{c_i}$ only deems possible the opponent's types in T_j^1. As each of these types is cautious, it follows that $t_i^{c_i}$ fully believes that j is cautious. Summarizing, we see that $t_i^{c_i}$ primarily believes in j's rationality, and fully believes that j is cautious – that is, it expresses 1-fold full belief in "caution and primary belief in rationality." So, we have shown that every type $t_i \in T_i^2$ expresses 1-fold full belief in "caution and primary belief in rationality."

Induction step: Take some $k \geq 3$, and assume that for both players i, every type $t_i \in T_i^{k-1}$ expresses up to $(k-2)$-fold full belief in "caution and primary belief in rationality." Consider now some type $t_i \in T_i^k$. That is, $t_i = t_i^{c_i}$ for some choice $c_i \in C_i^k$. By construction in step 2, type $t_i^{c_i}$ only deems possible opponent's types $t_j \in T_j^{k-1}$. By our induction assumption, every such type $t_j \in T_j^{k-1}$ expresses up to $(k-2)$-fold full belief in "caution and primary belief in rationality." Hence, type $t_i^{c_i}$ only deems possible opponent's types that express up to $(k-2)$-fold full belief in "caution and primary belief in rationality." This means, however, that type $t_i^{c_i}$ expresses up to $(k-1)$-fold full belief in "caution and primary belief in rationality."

We have thus shown that every type $t_i \in T_i^k$ expresses up to $(k-1)$-fold full belief in "caution and primary belief in rationality." By induction on k, the proof of step 3 is complete.

Step 4: Every type $t_i \in T_i^K$ expresses common full belief in "caution and primary belief in rationality"
In order to prove this, we show the following lemma.

Lemma 5.10.2 *For all $k \geq K-1$, every type $t_i \in T_i^K$ expresses up to k-fold full belief in "caution and primary belief in rationality."*

Proof of Lemma 5.10.2: We prove the statement by induction on k.

Induction start: Begin with $k = K-1$. Take some type $t_i \in T_i^K$. Then, we know from step 3 that t_i expresses up to $(K-1)$-fold full belief in "caution and primary belief in rationality."

Induction step: Consider some $k \geq K$, and assume that for both players i, every type $t_i \in T_i^K$ expresses up to $(k-1)$-fold full belief in "caution and primary belief in rationality." Consider player i and some type $t_i \in T_i^K$. That is, $t_i = t_i^{c_i}$ for some choice $c_i \in C_i^K$. By construction in step 2, the type $t_i^{c_i}$ only deems possible opponent's types $t_j \in T_j^K$. By our induction assumption, every such type $t_j \in T_j^K$ expresses up to $(k-1)$-fold full belief in "caution and primary belief in rationality." Hence, $t_i^{c_i}$ expresses up to k-fold full belief in "caution and primary belief in rationality." We have thus shown that every type $t_i \in T_i^K$ expresses up to k-fold full belief in "caution and primary belief in rationality." By induction on k, the proof of the lemma is complete. ◇

By step 3 and the lemma above we thus see that every type $t_i \in T_i^K$ expresses k-fold full belief in "caution and primary belief in rationality" for every k. That is, every type $t_i \in T_i^K$ expresses common full belief in "caution and primary belief in rationality," which completes the proof of step 4.

With steps 1–4, we can now easily prove part (b), namely that $C_i^{k+1} \subseteq BR_i^k$ for every $k \geq 1$. Take some $k \geq 1$ and some choice $c_i \in C_i^{k+1}$. So, the associated type $t_i^{c_i}$ is in T_i^{k+1}. By step 3 we then know that the associated type $t_i^{c_i}$ expresses up to k-fold full belief in "caution and primary belief in rationality." As, by construction in steps 1 and 2, the choice c_i is optimal for the type $t_i^{c_i}$ and type $t_i^{c_i}$ is cautious, it follows that c_i is optimal for a cautious type that expresses up to k-fold full belief in "caution and primary belief in rationality." That is, $c_i \in BR_i^k$. So, we have shown that every choice $c_i \in C_i^{k+1}$ is also in BR_i^k, and therefore $C_i^{k+1} \subseteq BR_i^k$ for every $k \geq 1$. This completes the proof of part (b).

With parts (a) and (b), it is now easy to prove Theorem 5.9.2. Let us start with part (1) of the theorem. By parts (a) and (b) we know that $C_i^{k+1} = BR_i^k$ for every player i and every $k \geq 1$. In other words, the choices that can rationally be made by a cautious type that expresses up to k-fold full belief in "caution and primary belief in rationality," are exactly those choices that survive the first $k+1$ steps of the Dekel–Fudenberg procedure, for all $k \geq 1$. This completes the proof of part (1) in the theorem.

Consider next part (2) of the theorem. The choices that survive the full Dekel–Fudenberg procedure are precisely the choices in C_i^K. So, the associated types are in

T_i^K. From step 4 in part (b) above, we then know that for every choice $c_i \in C_i^K$, the associated type $t_i^{c_i} \in T_i^K$ expresses common full belief in "caution and primary belief in rationality." As, by construction, choice c_i is optimal for the associated type $t_i^{c_i}$ and type $t_i^{c_i}$ is cautious, it follows that every choice $c_i \in C_i^K$ is optimal for some cautious type that expresses common full belief in "caution and primary belief in rationality." So, every choice that survives the Dekel–Fudenberg procedure can rationally be chosen under common full belief in "caution and primary belief in rationality."

Next, consider a choice c_i that can rationally be chosen under common full belief in "caution and primary belief in rationality." Then, $c_i \in BR_i^k$ for every k. So, by part (a) above, $c_i \in C_i^{k+1}$ for every k, which means that c_i survives the full Dekel–Fudenberg procedure. Hence, we see that every choice that can rationally be made under common full belief in "caution and primary belief in rationality," survives the Dekel–Fudenberg procedure. Altogether, we conclude that the choices that can rationally be made under common full belief in "caution and primary belief in rationality," are exactly the choices that survive the Dekel–Fudenberg procedure. This completes the proof of part (2) of the theorem. ■

Practical problems

5.1 Painting a room

A colleague at work wants to paint his living room, but he has two left hands. Therefore, he wants to ask somebody else to do the job. Both you and your friend Barbara have expressed interest in the task. In order to decide who will do the job, he invites you and Barbara to his favorite pub, where he proposes the following procedure: Barbara and you must simultaneously write down a price on a piece of paper. The price must be either 100 euros, 200 euros or 300 euros. The person who has written down the lowest price will get the job, and will be paid exactly that amount. If you both write down the same price, your colleague will choose Barbara, since she has known the colleague longer than you have.

(a) Formulate this situation as a game between Barbara and you.

(b) Which prices can you rationally write down under common belief in rationality with standard beliefs?

(c) Now, turn to a setting with lexicographic beliefs. Consider the following four lexicographic beliefs that you could hold about Barbara's choice.

$$b_1 = (100; \tfrac{2}{3} \cdot 200 + \tfrac{1}{3} \cdot 300), \qquad b_1' = (100; 300; 200),$$
$$b_1'' = (\tfrac{3}{4} \cdot 100 + \tfrac{1}{4} \cdot 200; 300), \qquad b_1''' = (\tfrac{1}{4} \cdot 100 + \tfrac{1}{4} \cdot 200 + \tfrac{1}{2} \cdot 300).$$

For each of these lexicographic beliefs, state which is your best choice, your second-best choice and your worst choice.

(d) Which prices can you rationally write down under common full belief in "caution and primary belief in rationality"?

(e) Construct an epistemic model with lexicographic beliefs such that:

- every type in this model is cautious and expresses common full belief in "caution and primary belief in rationality," and
- for each of the prices found in (d) there is a type in this model for which this price is rational.

5.2 A second-price auction

Today, a beautiful record by The Searchers, with their autographs on the cover, will be auctioned. You and Barbara are the only participants in this auction. The auction works as follows: You and Barbara write down a bid on a piece of paper, and hand it over to the auctioneer. He will compare the two bids and give the album to the higher bidder. However, the price that has to be paid by the higher bidder is the bid of the other person. So, the winner pays the second-highest bid. We call this a *second-price auction*. If you and Barbara happen to choose the same bid, the auctioneer will toss a coin and this will determine the person who gets the album. For simplicity, assume that you and Barbara cannot bid higher than 100 euros and that your bids must be multiples of one euro.

Suppose that the personal value you attach to the album, expressed in euros, is 50. Similarly, Barbara attaches a value of 40 to the album. If you win the auction and pay a price of P, your utility would be $50 - P$ and Barbara's utility would be 0. If Barbara wins the auction and pays a price of P, her utility would be $40 - P$, and your utility would be 0.

(a) Show that for you, every bid other than 50 is weakly dominated. Which bid, or bids, can you rationally make under a cautious lexicographic belief about Barbara's bid?

(b) Similarly, show that for Barbara, every bid other than 40 is weakly dominated. Which bid, or bids, can Barbara rationally make under a cautious lexicographic belief about your bid?

(c) Which bid, or bids, can you rationally make under common full belief in "caution and primary belief in rationality"?

(d) Construct an epistemic model with lexicographic beliefs such that:

- every type in this model is cautious and expresses common full belief in "caution and primary belief in rationality," and
- for each of the bids found in (c) there is a type in this model for which this bid is rational.

5.3 The closer the better

You and your friend Barbara participate as a team in the TV show *The Closer the Better*. In the final round, you must give an odd number between 1 and 9. Afterwards, Barbara must give an even number between 1 and 9, without knowing which number you have already chosen. The closer your numbers are, the better. More precisely, the prize money you earn as a team is equal to 10 minus the (absolute) difference between the two numbers.

(a) Model this situation as a game between you and Barbara.

(b) Show that, under common belief in rationality with standard beliefs, you can rationally choose any number.

(c) Which numbers can you rationally choose under a cautious lexicographic belief? For each of these numbers, determine a cautious lexicographic belief about Barbara's choice for which this number is optimal. For every other number, find another choice, or a randomized choice, that weakly dominates it.

(d) Which numbers can Barbara rationally choose under a cautious lexicographic belief? For each of these numbers, determine a cautious lexicographic belief for Barbara about your choice for which this number is optimal. For every other number, find another choice, or a randomized choice, for Barbara that weakly dominates it.

(e) Which numbers can you choose rationally under common full belief in "caution and primary belief in rationality"?

(f) Construct an epistemic model with lexicographic beliefs such that:

- every type in this model is cautious and expresses common full belief in "caution and primary belief in rationality," and
- for each of your numbers found in (e) there is a type in this model for which this number is rational.

5.4 A walk through the forest

You and Barbara are enjoying a beautiful walk through the forest. However, while you were studying some flowers, Barbara kept on walking, and now you have lost her. The question is: Where should you wait for her? Figure 5.1 is a map of the forest with its paths. There are eight open spots where you could wait for Barbara, and we call these spots $a, b, c, ..., h$. From each open spot you can see as far as the next open spot(s). Similarly, Barbara can also wait at any of these open spots, and can see as far as the next open spot(s). The objective for you and Barbara is, of course, to find each other. More precisely, if you and Barbara wait at spots where you can see each other, then you both get a utility of 5. At spot c there is a small bench where you can sit, and sitting

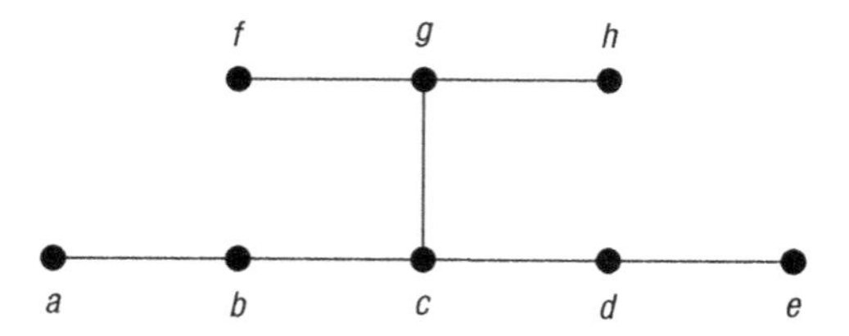

Figure 5.1 A walk through the forest

on this bench would give you, or Barbara, an additional utility of 1, since you are both tired from the long walk.
(a) Create a beliefs diagram with standard beliefs. Which spots can you rationally choose under common belief in rationality with standard beliefs?
(b) Which spots can you choose rationally under a cautious lexicographic belief about Barbara's choice? For each of these spots, find a cautious lexicographic belief about Barbara's choice for which this spot is optimal. For every other spot, find another choice, or a randomized choice, that weakly dominates it.
(c) Which spots can you choose rationally under common full belief in "caution and primary belief in rationality"?
(d) Construct an epistemic model with lexicographic beliefs such that:

- every type in this model is cautious and expresses common full belief in "caution and primary belief in rationality," and
- for each of the spots found in (c) there is a type in this model for which this spot is rational.

5.5 Dinner for two

This evening Barbara has invited you for dinner. You promised to bring something to drink, and as usual you either bring some bottles of beer, a bottle of white wine or a bottle of red wine. Barbara's favorite dishes are salmon, souvlaki and nasi goreng. Of course, you want to bring a drink that will go well with the dish that Barbara has prepared. Both you and Barbara agree that salmon goes reasonably well with beer, goes badly with red wine, but is excellent with white wine. You also agree that souvlaki goes reasonably well with beer, badly with white wine, but is excellent with red wine. Finally, you agree that nasi goreng is reasonable with white wine and red wine. However, you find that nasi goreng goes excellently with beer, whereas according to Barbara nasi goreng is only reasonable with beer. Suppose that a bad combination gives a utility of 0, a reasonable combination yields a utility of 1 and an excellent combination gives a utility of 3.
(a) Model this situation as a game between you and Barbara.
(b) Create a beliefs diagram with standard beliefs. Which choice(s) can you rationally make under common belief in rationality with standard beliefs?
(c) Which drinks can you rationally bring under a cautious lexicographic belief about Barbara's choice? For each of these drinks, determine a cautious lexicographic belief about Barbara's choice for which this drink is optimal. For every other drink, find another choice, or a randomized choice, that weakly dominates it.
(d) Which dishes can Barbara rationally prepare under a cautious lexicographic belief about your choice? For each of these dishes, determine a cautious lexicographic belief about your choice for which this dish is optimal. For every other dish, find another choice, or a randomized choice, that weakly dominates it.
(e) Which drink, or drinks, can you bring rationally under common full belief in "caution and primary belief in rationality"?
(f) Construct an epistemic model with lexicographic beliefs such that:

- every type in this model is cautious and expresses common full belief in "caution and primary belief in rationality," and
- for each of the drinks found in (e) there is a type in this model for which this drink is rational.

5.6 Bicycle race

In ten days there will be a bicycle race between you and Barbara. The race consists of twenty laps at the indoor cycling arena. You and Barbara must decide how many days to practice for this race. So, the maximum number of days you can practice is ten, and the minimum is zero. You are optimistic, and you believe you can beat Barbara even if you practice one day less than she does. Of course, if you practice the same number of days as she does, or more, then you expect to win. However, if you practice at least two days less than she does, you think you will lose. Finally, if you practice for the full ten days and Barbara does not practice at all, then you expect to beat her by at least one lap. Barbara, on the other hand, reasons in exactly the same way about *her* winning possibilities.

Assume that, for you and Barbara, a win would have a utility of 10, whereas every day of practice would decrease the utility by 1. A win by at least one lap would increase the utility by an extra 10.

(a) Create a beliefs diagram with standard beliefs. Which choices can you rationally make under common belief in rationality with standard beliefs?

(b) Which choices can you make rationally under a cautious lexicographic belief about Barbara's choice? For each of these choices find a cautious lexicographic belief about Barbara's choice for which this choice is rational. For every other choice, find another choice, or a randomized choice, that weakly dominates it.

(c) Which choices can you make rationally under common full belief in "caution and primary belief in rationality"?

5.7 Stealing an apple

Five minutes ago you stole an apple, and since then you have been followed by an angry farmer. Right now you are standing in front of a castle with 25 chambers, which seems like the perfect hiding place. Figure 5.2 is a map of these chambers. In every chamber there are eight doors – there is a door in each of the four walls and a door in each of the four corners. For instance, room 7 is connected to rooms 1, 2, 3, 6, 8, 11, 12 and 13. You must decide in which of these chambers to hide. The farmer, who will see you enter the castle, must decide in which of these chambers to look for you. Suppose that the farmer will find you whenever he enters the chamber where you are hiding or one of the chambers that is connected to it by a door.

If the farmer finds you, your utility will be 0 and the farmer's utility will be 1. If you are not found, your utility will be 1 whereas the farmer's utility will be 0.

1	2	3	4	5
6	7	8	9	10
11	12	13	14	15
16	17	18	19	20
21	22	23	24	25

Figure 5.2 Stealing an apple

(a) In which of the chambers can you hide rationally under common belief in rationality with standard beliefs?
(b) In which of the chambers can you rationally hide under a cautious lexicographic belief about the farmer's choice? For each of these chambers, find a cautious lexicographic belief about the farmer's choice for which this chamber is optimal. For every other chamber, find another choice, or a randomized choice, that weakly dominates it.
(c) Which chambers can the farmer rationally choose under a cautious lexicographic belief about your choice? For each of these chambers, find a cautious lexicographic belief about your choice for which this chamber is optimal. For every other chamber, find another choice, or a randomized choice, that weakly dominates it.
(d) In which of the chambers can you hide rationally under common full belief in "caution and primary belief in rationality"?

Theoretical problems

5.8 Permissible sets

Consider a game with two players and finitely many choices for both players. Let $D_i \subseteq C_i$ be a subset of choices for both players i. The collection (D_1, D_2) of subsets of choices is called a *permissible set* if for both players i, every choice $c_i \in D_i$ is optimal under some cautious lexicographic belief $b_i^{c_i}$ about j's choice where the primary belief assigns positive probability only to choices in D_j. That is, every choice in the collection (D_1, D_2) can be optimal if the player is cautious and primarily believes that his opponent will choose from (D_1, D_2).

Show that a choice c_i can be made rationally under common full belief in "caution and primary belief in rationality," *if and only if*, there is a permissible set (D_1, D_2) with $c_i \in D_i$.

5.9 Perfect equilibrium

Consider a game with two players and finitely many choices for both players. Let b_i be a lexicographic belief about j's choice, for both players i. The pair (b_1, b_2) of lexicographic beliefs is called a *perfect equilibrium* if, for both players i:

- the lexicographic belief b_i is cautious, and
- b_i's primary belief only assigns positive probability to j's choices that are optimal under b_j.

A choice c_i is called a *perfect equilibrium choice* if there is a perfect equilibrium (b_1, b_2) such that c_i is optimal under the lexicographic belief b_i.

(a) Show that every perfect equilibrium choice is a Nash choice but not *vice versa*.

Consider an epistemic model with lexicographic beliefs. A type t_i *fully believes that j is of type* t_j if the lexicographic belief $b_i(t_i)$ assigns, at *each* of its levels, probability 1 to type t_j. As in Chapter 4, say that type t_i *fully believes that j has correct beliefs* if the lexicographic belief $b_i(t_i)$ assigns, at *each* of its levels, only positive probability to j's types that fully believe that i is of type t_i.

Consider an epistemic model with lexicographic beliefs, and a type t_i that:

- is cautious,
- expresses common full belief in "caution and primary belief in rationality,"
- fully believes that j has correct beliefs,
- fully believes that j fully believes that i has correct beliefs.

(By the latter we mean that $b_i(t_i)$, at *each* of its levels, only assigns positive probability to j's types that fully believe that i has correct beliefs.)

(b) Show that there is a single type t_j such that t_i fully believes that j is of type t_j.

(c) Show that t_i's lexicographic belief about j's choice, together with t_j's lexicographic belief about i's choice, is a perfect equilibrium.

(d) Show that c_i is a perfect equilibrium choice, *if and only if*, it can be chosen rationally by a type t_i that:

- is cautious,
- expresses common full belief in "caution and primary belief in rationality,"
- fully believes that j has correct beliefs,
- fully believes that j fully believes that i has correct beliefs.

5.10 Full belief in the opponent's rationality

Consider a game with two players, and finitely many choices for both players. Within an epistemic model with lexicographic beliefs, consider a type t_i. We say that type t_i *fully believes in j's rationality* if the lexicographic belief $b_i(t_i)$, at *each* of its levels, only assigns positive probability to choice-type pairs (c_j, t_j) where c_j is rational for t_j.

In this way, we can also define *common full belief in caution and rationality*. Namely, a type t_i expresses *common full belief in caution and rationality* if

- t_i fully believes in j's rationality and j's caution,
- $b_i(t_i)$, at *each* of its levels, only assigns positive probability to j's types that fully believe in i's rationality and i's caution,

and so on. We say that common *full* belief in caution and rationality is *possible* for a specific game if there is an epistemic model with lexicographic beliefs, and a type t_i inside this epistemic model, that expresses common *full* belief in caution and rationality. In Section 5.5 we saw that *full* belief in the opponent's rationality may not be possible if we require the player to be cautious. So, as a consequence, common *full* belief in caution and rationality is often impossible.

Show that, for a specific game, common *full* belief in caution and rationality is possible, *if and only if*, no choice in this game is weakly dominated.

Literature

Cautious reasoning

Part II of this book – starting from Chapter 5 – is built upon the assumption that you reason *cautiously* about your opponent. That is, you may deem some of the opponent's choices much more likely than others, but you never discard any opponent's choice completely from consideration. A consequence of cautious reasoning is that you will never rationally make a choice that is *weakly dominated* in the game. Luce and Raiffa (1957) argued in favor of concepts that rule out weakly dominated choices. In spite of their observation, it took game theory a long time before it developed concepts that rule out weakly dominated choices by explicitly incorporating cautious reasoning. The first such concept was probably *perfect equilibrium*, proposed by Selten (1975). We will discuss this concept later in this section.

Tension between caution and belief in the opponent's rationality

As we saw at the beginning of this chapter in the example "Should I call or not?," there is a tension between cautious reasoning on the one hand, and belief in the opponent's rationality on the other hand. Within a model of standard beliefs, cautious reasoning implies that you must assign a positive probability to each of the opponent's choices, whereas belief in the opponent's rationality requires you to assign probability zero to each of the opponent's irrational choices. So how can we possibly combine these two conditions? This tension between cautious reasoning and belief in the opponent's rationality is discussed extensively in the literature – see, for instance, Samuelson (1992), Börgers and Samuelson (1992) and Börgers (1994). A possible solution to this problem – which is in fact the road we take in Part II of this book – is to move from a model of standard beliefs to a model of *lexicographic* beliefs. Within a lexicographic belief it is possible to simultaneously (1) take all of the opponent's choices into account, by giving all of the opponent's choices a positive probability in some of the levels –

not necessarily the first one, and (2) believe in the opponent's rationality, by assigning *in the first level* of the lexicographic belief only positive probability to rational choices for the opponent.

Lexicographic beliefs
In this book we model cautious reasoning of a player by means of a *lexicographic belief*. In contrast to a standard belief, which we used for Part I of the book, a lexicographic belief consists of various probability distributions – a primary belief, a secondary belief, and so on. Cautious reasoning then requires that all of the opponent's choices have a positive probability in at least some level of the lexicographic belief. The concept of a lexicographic belief was introduced by Blume, Brandenburger and Dekel (1991a, 1991b), who use the term *lexicographic probability system* instead. In Blume, Brandenburger and Dekel (1991a), the authors provide a decision theoretic foundation for the use of lexicographic beliefs. As with the approaches by Savage (1954) and Anscombe and Aumann (1963) discussed in the literature section of Chapter 2, the authors assume that a decision maker under uncertainty holds a preference relation over acts. They then impose some axioms on this preference relation, and show that under these axioms the preference relation can be represented by a utility function over the possible outcomes and a lexicographic belief on the possible states of the world. Hence, the concept of a lexicographic belief can be used for any context in which a person must choose under uncertainty. In particular it applies to games in which the states of the world contain the possible choices that can be made by the opponents.

Characterization of optimal choices with cautious lexicographic beliefs
In Theorem 5.8.3 we saw that a choice is optimal under a cautious lexicographic belief if and only if it is not weakly dominated. Moreover, every choice that is not weakly dominated is optimal for a cautious lexicographic belief with one level only. This result is based on Lemma 4 in Pearce (1984), which states that in two-player games a choice is optimal for a cautious *standard* belief – that is, a cautious lexicographic belief with one level only – if and only if, it is not weakly dominated. The last part of our proof of Theorem 5.8.3 is based on the last part of Pearce's proof.

Alternative ways of modeling cautious reasoning
In the literature there are several alternative ways of modeling cautious reasoning by a player. One way of doing so is to assume that player i's belief is not given by a single probability distribution on j's choices, but rather by a whole sequence $(b_i^n)_{n\in\mathbb{N}}$ of probability distributions on j's choices, where every b_i^n assigns positive probability to each of j's choices. That is, every b_i^n is *cautious*. This is essentially the approach taken by Selten (1975) when he introduced the concept of *perfect equilibrium* – a notion which we will discuss in more detail later in this section. What matters in this sequence $(b_i^n)_{n\in\mathbb{N}}$ are the likelihood ratios $b_i^n(c_j)/b_i^n(c_j')$ between two of the opponent's choices c_j and c_j' when n tends to infinity. If this likelihood ratio tends to zero,

then it means that player i deems c_j' *infinitely more likely* than c_j, even when both c_j and c_j' have probability zero when n tends to infinity. Hence, the idea of deeming one of the opponent's choices infinitely more likely than another can also be modelled by taking such a sequence $(b_i^n)_{n \in \mathbb{N}}$ of probability distributions. Moreover, if each of the probability distributions b_i^n is cautious, then this is a way to model cautious reasoning. We thus see that cautious reasoning by a player can also be modeled by taking a sequence $(b_i^n)_{n \in \mathbb{N}}$ of cautious probability distributions on j's choices, rather than considering a cautious lexicographic belief. In fact, Blume, Brandenburger and Dekel (1991b) show in their Propositions 1 and 2 that both approaches are in a sense equivalent.

Another approach is to model your belief about the opponent not by a standard probability distribution, but by a system of *conditional probabilities*. The concept of conditional probabilities was developed by Popper (1934, 1968), de Finetti (1936) and Rényi (1955). The main idea is that, for all of the opponent's choices c_j, and every subset of the opponent's choices $D_j \subseteq C_j$ containing c_j, you specify the conditional probability $p(c_j \mid D_j)$ assigned to c_j, conditional on the event that j chooses from the subset of choices D_j. So, even if the choices c_j and c_j' have probability zero conditional on the whole set C_j – that is, $p(c_j \mid C_j) = p(c_j' \mid C_j) = 0$ – you can still specify the conditional probability $p(c_j \mid \{c_j, c_j'\})$, conditional on player j choosing from $\{c_j, c_j'\}$. Within this model, you deem the opponent's choice c_j' *infinitely more likely* than another choice c_j if $p(c_j \mid \{c_j, c_j'\}) = 0$. Hence, the notion of conditional probabilities can very well be used to model cautious reasoning by a player.

The final alternative we discuss here is the representation of your belief by a *non-standard* probability distribution. That is, the probability you assign to an opponent's choice may involve an *infinitesimal* ϵ, where ϵ can be thought of as an "artificial" number that is smaller than any positive real number. Hence, as an approximation, one may think of ϵ as a very small positive number. Intuitively, when we say that you assign probability ϵ to an opponent's choice c_j, then you are assigning a very small probability to c_j, but you still deem c_j possible. Moreover, if you assign probability ϵ to c_j and probability ϵ^2 to c_j', then you deem c_j and c_j' both very unlikely, but still you deem c_j infinitely more likely than c_j', as ϵ is "infinitely much larger" than ϵ^2. In general, within this framework you deem an opponent's choice c_j' *infinitely more likely* than another choice c_j if the ratio between the probability assigned to c_j and the probability assigned to c_j' is *infinitesimally small*. This model of non-standard analysis with an infinitesimal ϵ is due to Robinson (1973).

Hammond (1994) and Halpern (2010) showed that the latter two approaches are essentially equivalent to the model of lexicographic beliefs. Therefore, we can model the cautious reasoning of players equivalently by lexicographic beliefs, sequences of standard beliefs, conditional probabilities or non-standard probability distributions. For this book we chose lexicographic beliefs, as in our view it provides a compact and transparent representation of cautious reasoning. But we could as well have chosen any of the other three models instead.

Primary belief in the opponent's rationality

The key concept in this chapter is primary belief in the opponent's rationality, which states that the first level of your lexicographic belief only assigns positive probability to opponent's choice-type pairs (c_j, t_j), where the choice c_j is optimal for the type t_j. Together with "caution," it provides the basis for the concept of common full belief in "caution and primary belief in rationality." This concept is essentially due to Brandenburger (1992b) and Börgers (1994).

Brandenburger (1992b) – as we do – uses a model with lexicographic beliefs and discusses the idea of "common first-order knowledge that players choose admissible actions." Here, first-order knowledge corresponds to our notion of primary belief, and admissible actions are choices that are not weakly dominated in the original game. From Theorem 5.8.3 we know that the latter choices are exactly those choices that are optimal for some cautious type. Hence, "first-order knowledge that players choose admissible actions" is essentially equivalent to imposing "primary belief in the opponent's caution and the opponent's rationality" within our set-up. Hence, Brandenburger's notion of "common first-order knowledge that players choose admissible actions" corresponds to common primary belief in the opponent's caution and the opponent's rationality within our model. That is, player i assigns, in his primary belief, only positive probability to opponent's choice-type pairs (c_j, t_j) where c_j is optimal for t_j and t_j is cautious, player i assigns in his primary belief only positive probability to j's types that assign, in their primary belief, only positive probability to i's choice-type pairs (c_i, t_i) where c_i is optimal for t_i and t_i is cautious, and so on.

Note that this concept is *weaker* than the notion of common full belief in "caution and primary belief in rationality" which we use in this chapter. Our notion requires that player i *fully* believes that "j is cautious and primarily believes in i's rationality." That is, player i must assign, in *each* of the levels of his lexicographic belief, only positive probability to j's types that are cautious and primarily believe in i's rationality. On the other hand, Brandenburger's concept only requires that player i's *primary* belief assigns positive probability exclusively to j's types that are cautious and primarily believe in i's rationality. And so on. Hence, our notion puts more restrictions on a player's belief hierarchy than Brandenburger's concept. However, in terms of choices selected, both concepts are equivalent. Brandenburger (1992b) calls a choice *permissible* if it can rationally be chosen by a cautious player who expresses common first-order knowledge that players choose admissible actions. He then shows that the permissible choices are exactly those choices that survive the Dekel–Fudenberg procedure. However, we know from Theorem 5.9.2 that these are precisely the choices that can rationally be chosen by a cautious player who expresses common full belief in "caution and primary belief in rationality." Hence, Brandenburger's concept leads to exactly the same choices as common full belief in "caution and primary belief in rationality," although the latter concept is more restrictive in terms of belief hierarchies selected!

Börgers (1994) used a model with *standard* beliefs rather than lexicographic beliefs. He called a player cautious if he assigns positive probability to each of the opponent's choices. For every number $p < 1$ he introduced "common p-belief in caution and the opponent's rationality." That is, player i believes, with probability at least p, that j is cautious and chooses optimally, player i believes, with probability at least p, that player j believes, with probability at least p, that i is cautious and chooses optimally, and so on. Hence, if p is close to 1, then player i believes, with probability almost 1, that j is cautious and chooses optimally, player i believes, with probability almost 1, that j believes, with probability almost 1, then i is cautious and chooses optimally, and so on. In terms of a lexicographic belief, this corresponds to saying that player i primarily believes that j is cautious and chooses optimally, that player i primarily believes that j primarily believes that i is cautious and chooses optimally, and so on. This, however, is precisely Brandenburger's concept above. In fact, Börgers (1994) showed that if we let p tend to 1, then the choices that player i can rationally make if he is cautious, and expresses common p-belief in caution and the opponent's rationality, are exactly the choices that survive the Dekel–Fudenberg procedure. Hence, in terms of choices selected, the concepts of Brandenburger (1992b) and Börgers (1994) are both equivalent to common full belief in "caution and primary belief in rationality" as used in this chapter. The notion of p-belief – or approximate knowledge/belief – used by Börgers (1994) in his model, was introduced by Monderer and Samet (1989) and Stinchcombe (1988).

Börgers' model is essentially equivalent to Gul's (1997) *perfect τ-theory*, and to *weak perfect rationalizability* proposed by Herings and Vannetelbosch (1999). Gul (1997) considered for both players i some set of choices $D_i \subseteq C_i$. Basically, Gul called the pair of choice sets (D_1, D_2) a *perfect τ-theory* if for every large enough $p < 1$, the set D_i contains exactly those choices that are optimal for some cautious belief that assigns probability at least p to the event that j chooses from D_j. As in theoretical problem 5.8, it can be shown that the choices that are selected by Börgers' common p-belief in caution and the opponent's rationality – when p is large enough – are exactly the choices that appear in a perfect τ-theory. Indeed, Gul showed in his Propositions 3 and 4 that the choices that appear in some perfect τ-theory are exactly the choices that survive the Dekel–Fudenberg procedure. Hence, in terms of choices selected, the models by Börgers (1994) and Gul (1997) are equivalent. Herings and Vannetelbosch (1999) define the concept of *weak perfect rationalizability* in essentially the same way as Gul defined a perfect τ-theory, but do so by means of an algorithm. Herings and Vannetelbosch (2000) showed in their Theorem 2 that the weakly perfectly rationalizable choices are exactly the choices that survive the Dekel–Fudenberg procedure, and hence their notion is equivalent, in terms of choices, to Gul's perfect τ-theory.

Summarizing, we thus see that in two-player games the models by Brandenburger (1992b) and Börgers (1994), the concept of perfect τ-theory by Gul (1997) and the concept of weak perfect rationalizability by Herings and Vannetelbosch (1999) are all equivalent, in terms of choices selected, to common full belief in "caution and primary belief in rationality." We finally note that Asheim and Dufwenberg (2003) defined common full belief in "caution and primary belief in rationality" in exactly the same way as we do. In their Proposition 5.2 they showed that the choices that can rationally be made under common full belief in "caution and primary belief in rationality" are exactly the choices that survive the Dekel–Fudenberg procedure. Hence, their Proposition 5.2 corresponds to our Theorem 5.9.2.

Dekel–Fudenberg procedure

The Dekel–Fudenberg procedure was first considered by Dekel and Fudenberg (1990), although they did so in a different setting. Dekel and Fudenberg considered games in which the players have a small amount of uncertainty about the opponent's utility function. Such games are called *elaborations* of the original game Γ. By "small amount of uncertainty" we mean that the players assign probability almost 1 to the opponent having the utility function specified by the original game, and attach a small probability to the opponent having different utilities. To every such elaboration $\tilde{\Gamma}$ of the original game Γ, the authors apply the algorithm of *iterated elimination of weakly dominated choices* – that is, first eliminate all weakly dominated choices from the entire game $\tilde{\Gamma}$, then eliminate from the reduced game that remains all choices that are *weakly* dominated within the reduced game, and so on, until no choices can be eliminated. This algorithm will be discussed in more detail in Chapter 7 of this book. Dekel and Fudenberg show that, if we take a sequence of elaborations $\tilde{\Gamma}^n$ in which the amount of uncertainty about the opponent's utilities tends to zero, and apply iterated elimination of weakly dominated choices to every elaboration $\tilde{\Gamma}^n$ in this sequence, then in the limit we obtain exactly those choices that survive one round of elimination of weakly dominated choices, followed by iterated elimination of strictly dominated choices, when applied to the original game Γ. The latter procedure was later called the *Dekel–Fudenberg procedure*.

Perfect equilibrium

The concept of *perfect equilibrium*, which was proposed by Selten (1975), is probably the first game-theoretic concept that explicitly incorporates the requirement that players are cautious. Although it was originally defined for dynamic games, it is most often applied to static games. Selten provided two equivalent definitions of perfect equilibrium – one using sequences of so-called "perturbed games," and the other using sequences of cautious standard beliefs. We will present the latter definition here. Take a probability distribution σ_1 on player 1's choices, and a probability distribution σ_2 on player 2's choices. From 1's perspective, σ_2 can be interpreted as his belief about 2's choice, whereas σ_1 can be understood as 1's belief about 2's belief about 1's choice. The pair (σ_1, σ_2) is a *perfect equilibrium* if there is a sequence $(\sigma_1^n, \sigma_2^n)_{n \in \mathbb{N}}$ converging to (σ_1, σ_2) such that, for both players i, (1) every σ_i^n assigns positive probability to each of i's choices, and (2) σ_i only assigns positive probability to choices c_i that are optimal for each of the beliefs σ_j^n. It can be shown that every perfect equilibrium (σ_1, σ_2) is also a Nash equilibrium, but not vice versa.

Blume, Brandenburger and Dekel (1991b) provided a characterization of perfect equilibrium in terms of lexicographic beliefs – at least for two-player games – which is the definition we used in the theoretical problem 5.9. In that problem, the reader is asked to show that there is a very close relation between common full belief in "caution and primary belief in rationality" on the one hand, and perfect equilibrium on the other. If we take common full belief in "caution and primary belief in rationality," and additionally require that player i believes that j has correct beliefs, and player i believes that j believes that i has correct beliefs, then we obtain perfect equilibrium. See theoretical problem 5.9 for the precise details. Hence, one could say that perfect equilibrium relates in the same way to common full belief in "caution and primary belief in rationality" as Nash equilibrium relates to common belief in rationality – in both cases, the former concept is obtained from the latter by additionally assuming that i believes that j has correct beliefs, and i believes that j believes that i has correct beliefs.

Related concepts

We finally discuss some other concepts proposed in the literature, which are related – at least in spirit – to the concept of common full belief in "caution and primary belief in rationality."

Bernheim (1984) proposed *perfect rationalizability*, which is somewhat similar to the notions of perfect τ-theory by Gul (1997) and weakly perfect rationalizability by Herings and Vannetelbosch (1999) discussed above. Bernheim defined the concept using an algorithm, but it can equivalently be defined as follows: For every choice c_i in the game take a small positive number $\epsilon(c_i)$, and let ϵ denote the vector of all these numbers. Consider for both players i a set of standard beliefs $B_i \subseteq \Delta(C_i)$ about i's choice. Say that the pair (B_1, B_2) of belief sets is ϵ-perfectly rationalizable if for both players i, and every belief $b_i \in B_i$, we have that (1) $b_i(c_i) \geq \epsilon(c_i)$ for all $c_i \in C_i$, and (2) $b_i(c_i) > \epsilon(c_i)$ only if c_i is optimal for some belief $b_j \in B_j$. A choice c_i is ϵ-perfectly rationalizable if there is an ϵ-perfectly rationalizable pair of belief sets (B_1, B_2), such that c_i is optimal for some belief $b_j \in B_j$. Finally, a choice c_i is *perfectly rationalizable* if there is a sequence $(\epsilon^n)_{n \in \mathbb{N}}$ converging to the zero vector, such that c_i is ϵ^n-perfectly rationalizable for large enough n. As shown in Börgers (1994) and Herings and Vannetelbosch (1999, 2000), every perfectly rationalizable choice survives the Dekel–Fudenberg procedure, but not vice versa. Hence, in terms of choices selected, perfect rationalizability is more restrictive than common full belief in "caution and primary belief in rationality."

Pearce (1984) introduced *cautious rationalizability*, which he defined by means of the following procedure. For a given game Γ, first take those choices that are rationalizable in Γ – see the literature section in Chapter 3. This leads to a reduced game Γ^1. Then, assume that the players assign positive probability to all of the opponent's choices in Γ^1, and take those choices in Γ^1 that are optimal for such a "cautious belief within Γ^1." This leads to a new reduced game Γ^2. Next, look at those choices that are rationalizable within Γ^2, leading to a further reduced game Γ^3. Then, assume that the players assign positive probability to all of the opponent's choices in Γ^3, and focus on those choices that are optimal for such a cautious belief within Γ^3, and so on, until no further choices can be eliminated. The choices that remain at the end are called *cautiously rationalizable*. We have seen in the literature section of Chapter 3 that in two-player games, the rationalizable choices are exactly those choices that can rationally be chosen under common belief in rationality, which – as we know from Theorem 3.7.2 in Chapter 3 – are precisely those choices that survive iterated elimination of strictly dominated choices. Moreover, as we learned in this chapter, the choices that are optimal for a cautious belief within a reduced game $\hat{\Gamma}$ are exactly those choices that are not weakly dominated within $\hat{\Gamma}$. Hence, Pearce's procedure for cautious rationalizability coincides in two-player games with the following algorithm: First, carry out iterated elimination of *strictly* dominated choices. Then, remove all choices that are *weakly* dominated within the reduced game that remains. Then, perform again iterated elimination of *strictly* dominated choices on the new reduced game, after which we again eliminate all *weakly* dominated choices. And so on, until no further choices can be removed. In terms of choices selected, cautious rationalizability is more restrictive than common belief in rationality within a framework of standard beliefs, whereas it bears no general logical relation with common full belief in "caution and primary belief in rationality": There are games in which cautious rationalizability is more restrictive than common full belief in "caution and primary belief in rationality," but there are other games where the opposite is true. However, it is not clear which epistemic conditions on a player's reasoning process would lead him to choose in accordance with cautious rationalizability.

Samuelson (1992) and Börgers and Samuelson (1992) considered *consistent pairs of choice sets*, which are somewhat related to *permissible sets* studied in theoretical problem 5.8. Formally, take for both players i a subset of choices $D_i \subseteq C_i$. Then, (D_1, D_2) is called a *consistent pair of choice sets* if for both players i, the set D_i contains *exactly* those choices that are not weakly dominated on D_j. The motivation for this concept is that D_i may be viewed as a candidate set of "reasonable choices" for player i, on which both players agree. If, in addition, both players i deem all of the opponent's choices in D_j possible, by holding a "cautious belief on D_j," then the reasonable choices for player i – that is, D_i – must contain all the choices that are optimal for such a cautious belief on D_j. These, in turn, are precisely the choices for player i that are not weakly dominated on D_j. Compare this to a permissible set: (D_1, D_2) is called a *permissible set* if for both players i, every choice $c_i \in D_i$ is optimal for some cautious lexicographic belief about j's choice in which primary belief only assigns positive probability to choices in D_j. Or, equivalently, every choice $c_i \in D_i$ is not weakly dominated on C_j and not strictly dominated on D_j. Although both concepts are related in spirit, there is no general logical relationship between the two. In fact, as shown in Samuelson (1992) and Börgers and Samuelson (1992), there are games in which a consistent pair of choice sets does not exist. Hence, the requirements imposed by a consistent pair of choice sets may sometimes lead to logical contradictions.

Asheim and Dufwenberg (2003) introduced *fully permissible choice sets*, which are somewhat related to *consistent pairs of choice sets*. Although Asheim and Dufwenberg used an algorithm to define fully permissible choice sets, we will use a different – yet equivalent – definition, which makes it easier to compare it to consistent pairs of choice sets. Consider for both players i a collection of choice sets $\mathcal{C}_i = \{D_i^1, ..., D_i^{K_i}\}$, where $D_i^k \subseteq C_i$ for every $k \in \{1, ..., K_i\}$. The pair $(\mathcal{C}_1, \mathcal{C}_2)$ of collections of choice sets is *fully permissible* if for both players i, and every choice set D_i^k in $\mathcal{C}_i$, there is some $M_j \subseteq \{1,...,K_j\}$ such that D_i^k contains *exactly* those choices that are not weakly dominated on C_j and not weakly dominated on the union of choice sets $\cup_{m \in M_j} D_j^m$. The intuition is that every choice set D_i^k in $\mathcal{C}_i$ is regarded as a "reasonable" choice set for player i. If player i is cautious and deems all of the opponent's choices in the subcollection of reasonable choice sets $\cup_{m \in M_j} D_j^m$ infinitely more likely than all of the opponent's other choices, then the choice set for player i will only contain choices that are not weakly dominated on C_j and not weakly dominated on the union of choice sets $\cup_{m \in M_j} D_j^m$. Moreover, Asheim and Dufwenberg require that the choice set D_i^k contains *all* of these choices. This is similar to the requirement in a consistent pair of choice sets (D_1, D_2), where D_i is assumed to contain *all* choices that are not weakly dominated on D_j. Now, say that a choice c_i for player i is *fully permissible* if there is a fully permissible pair $(\mathcal{C}_1, \mathcal{C}_2)$ of collections of choice sets, such that c_i is in at least one of the sets in $\mathcal{C}_i$. Asheim and Dufwenberg show, in their Proposition 3.2, that every fully permissible choice survives the Dekel–Fudenberg procedure, but the converse need not be true. Hence, in terms of choices selected, full permissibility is more restrictive than common full belief in "caution and primary belief in rationality."

We finally note that Herings and Vannetelbosch (1999) investigated the formal relations between various concepts discussed above, including perfect rationalizability, weakly perfect rationalizability and cautious rationalizability. The reader is referred to their paper for more details about these comparisons.

6 Respecting the opponent's preferences

6.1 Respecting the opponent's preferences

In the previous chapter we introduced the idea of reasoning *cautiously* about your opponent. By this we mean that you do not completely discard any of the opponent's choices from consideration, even though you may deem some of the opponent's choices much more likely than others. We have seen that cautious reasoning can be modeled by using *lexicographic* beliefs to express your thoughts about the opponent's choice. So, instead of representing your belief by a *single* probability distribution on the opponent's choices – as we did in Part I of this book – we model your belief by *various* probability distributions – your *primary* belief, your *secondary* belief, and so on. To reason cautiously about your opponent then means that all of the opponent's choices should enter *somewhere* in your lexicographic belief, maybe in your primary belief, or perhaps only in your tertiary belief, or even later.

The key problem we faced in the previous chapter was how to define "belief in the opponent's rationality." As we discovered, we cannot require you to believe in the opponent's rationality at *each level* of your lexicographic belief, since this is not possible in general when you reason cautiously. We then opted for a rather weak version of belief in the opponent's rationality, stating that you believe in the opponent's rationality at least in your *primary* belief. We called this condition *primary belief in the opponent's rationality*. This condition thus only puts restrictions on your *primary* belief, but not on your secondary belief and further. So, in a sense, it ignores your secondary and higher beliefs. Does this mean that these higher-level beliefs are not important?

The answer is that in many examples these higher-level beliefs *are* important, and should be taken seriously when thinking about your eventual choice. In fact, in many examples imposing only *primary belief in the opponent's rationality* may lead to unreasonable choices. The following example illustrates this fact.

Example 6.1 Where to read my book?

Recall the story from Example 5.2 in the previous chapter. The utilities for you and Barbara in this situation are given in Table 6.1. We saw in the previous chapter that under common full belief in "caution and primary belief in rationality," you can either

Table 6.1. *Utilities for you and Barbara in "Where to read my book?" (II)*

		Barbara		
		Pub a	Pub b	Pub c
You	Pub a	0,3	1,2	1,1
	Pub b	1,3	0,2	1,1
	Pub c	1,3	1,2	0,1

Table 6.2. *An epistemic model for "Where to read my book?"*

Types	$T_1 = \{t_1^b, t_1^c\}$, $T_2 = \{t_2^a\}$
Beliefs for you	$b_1(t_1^b) = ((a,t_2^a);(c,t_2^a);(b,t_2^a))$
	$b_1(t_1^c) = ((a,t_2^a);(b,t_2^a);(c,t_2^a))$
Beliefs for Barbara	$b_2(t_2^a) = ((c,t_1^c);(b,t_1^c);(a,t_1^c))$

rationally choose to read your book in pub b or pub c. Consider the epistemic model depicted in Table 6.2. It can easily be verified that every type in this epistemic model is cautious, and primarily believes in the opponent's rationality. Namely, the primary beliefs of your types t_1^b and t_1^c both assign probability one to Barbara's choice-type pair (a,t_2^a), and a is indeed optimal for Barbara's type t_2^a. On the other hand, Barbara's type t_2^a assigns, in her primary belief, probability one to your choice-type pair (c,t_1^c), and c is indeed optimal for your type t_1^c. But then, we can conclude that every type in this epistemic model expresses common full belief in "caution and primary belief in rationality." Since choosing b is optimal for your type t_1^b, and choosing c is optimal for your type t_1^c, we see that you can rationally go to pub b or to pub c under common full belief in "caution and primary belief in rationality."

However, the secondary and tertiary beliefs of your type t_1^b do not seem very reasonable here! Barbara clearly prefers going to pub b rather than going to pub c. Therefore, it seems reasonable to deem the event "Barbara will go to pub b" infinitely more likely than "Barbara will go to pub c," even if both events are much less likely than "Barbara will go to pub a." Your type t_1^b, however, deems in its secondary and tertiary beliefs the event "Barbara will go to pub c" infinitely more likely than "Barbara will go to pub b," which seems counter-intuitive.

So, in this example it seems natural to deem the event "Barbara will go to pub a" infinitely more likely than "Barbara will go to pub b," and to deem the latter event infinitely more likely than "Barbara will go to pub c." But then, the only reasonable choice that would remain for you would be to go to pub c! □

The example above thus shows that imposing only *primary belief in the opponent's rationality* may not be enough, as it may allow for unreasonable secondary, and higher-level, beliefs. The problem was that Barbara not only prefers pub a to the other two pubs, but she also prefers pub b to pub c. Therefore, it seems reasonable to deem Barbara's choice a infinitely more likely than her other two choices, which in fact is guaranteed by *primary belief in the opponent's rationality,* but also to deem her inferior choice b infinitely more likely than her even more inferior choice c.

We can actually apply this argument to any arbitrary game: If your opponent prefers a choice a to some other choice b, then it seems natural that your lexicographic belief deems the better choice a for the opponent infinitely more likely than the inferior choice b, even when both choices are suboptimal for the opponent. This condition is called *respecting the opponent's preferences,* and may be formalized as follows.

Definition 6.1.1 *(Respecting the opponent's preferences)*
Consider an epistemic model with lexicographic beliefs and a type t_i for player i. Type t_i ***respects the opponent's preferences*** *if for all of the opponent's types t_j deemed possible by t_i, and every two choices c_j, c_j' where t_j prefers c_j to c_j', type t_i deems the opponent's choice-type pair (c_j, t_j) infinitely more likely than the opponent's choice-type pair (c_j', t_j).*

Recall that t_i deems possible the opponent's type t_j if, somewhere in its lexicographic belief, type t_i assigns positive probability to type t_j. Respecting the opponent's preferences thus states that, if choice c_j is better for the opponent's type t_j than choice c_j', then you deem the event that the opponent is of type t_j and makes the *better* choice c_j *infinitely more likely* than the event that the opponent is of the same type t_j and makes the *inferior* choice c_j'.

As an illustration, reconsider the epistemic model in Table 6.2. Your type t_1^b does not respect Barbara's preferences, as it deems Barbara's choice b infinitely less likely than her choice c, although b is better for her than c. On the other hand, your type t_1^c *does* respect Barbara's preferences. Let us finally consider Barbara's type t_2^a. Her type t_2^a only deems possible your type t_1^c, for which your choice c is better than your choice b, and your choice b is better than your choice a. Barbara's type t_2^a deems your choice-type pair (c, t_1^c) infinitely more likely than your choice-type pair (b, t_1^c), which in turn it deems infinitely more likely than your choice-type pair (a, t_1^c). So, we see that Barbara's type t_2^a respects your preferences.

It is very easily seen that *respecting the opponent's preferences* is a stronger condition than *primary belief in the opponent's rationality.* More precisely, if you respect the opponent's preferences, than you automatically primarily believe in his rationality. Suppose that type t_i respects opponent j's preferences. Consider some opponent's type t_j deemed possible by t_i, and some choice c_j that is not optimal for t_j. Then, there is some choice c_j' that is better for t_j than c_j. As t_i respects j's preferences, type t_i must deem (c_j', t_j) infinitely more likely than (c_j, t_j). This means, in particular, that t_i cannot assign positive probability to (c_j, t_j) in his primary belief. So, we see that type t_i, in his primary belief, can only assign positive probability to pairs (c_j, t_j) where c_j is actually optimal for t_j. Hence, t_i primarily believes in j's rationality. As such, every type t_i that respects j's preferences also primarily believes in j's rationality.

The converse is not true, however. Consider again the epistemic model in Table 6.2. There, your type t_1^b primarily believes in Barbara's rationality but does not respect Barbara's preferences. Our conclusion is thus that *respecting the opponent's preferences* implies *primary belief in the opponent's rationality,* but not vice versa.

6.2 Common full belief in "respect of preferences"

In the previous section we introduced the idea of *respecting the opponent's preferences.* Intuitively, it states that you deem better choices for the opponent infinitely more likely than inferior choices. However, if this a reasonable line of reasoning for you, then it also makes sense to believe that your opponent will follow this line of reasoning as well. That is, you also believe that your opponent will respect *your* preferences. Additionally, it is also plausible to believe that your opponent believes that you will respect his preferences, and so on. As in the previous chapter, we will use *full belief* here to formalize these conditions. This will eventually lead to the idea of *common full belief in "respect of preferences."* To describe this idea formally, we first recursively define k-fold full belief in "respect of preferences" for all $k \in \{1, 2, ...\}$, as in the previous chapters.

Definition 6.2.1 *(Common full belief in "respect of preferences")*
(1) Type t_i expresses ***1-fold full belief in "respect of preferences"*** *if t_i respects the opponent's preferences.*
(2) Type t_i expresses ***2-fold full belief in "respect of preferences"*** *if t_i only deems possible j's types that express 1-fold full belief in "respect of preferences."*
(3) Type t_i expresses ***3-fold full belief in "respect of preferences"*** *if t_i only deems possible j's types that express 2-fold full belief in "respect of preferences."*
And so on.
Type t_i expresses ***common full belief in "respect of preferences"*** *if t_i expresses k-fold full belief in "respect of preferences" for all k.*

Of course, we still want to maintain the assumption that you are cautious, that is, you do not exclude any of your opponent's choices completely from consideration. More than this, we will still assume that you express *common full belief in caution,* as we defined it in the previous chapter. So, you only deem possible opponent's types that are cautious, you only deem possible opponent's types that only deem possible types for you that are cautious, and so on.

The choices we will be interested in throughout this chapter are those you can rationally make under common full belief in "caution and respect of preferences." These choices are formalized by the following definition.

Definition 6.2.2 *(Rational choice under common full belief in "caution and respect of preferences")*
Let c_i be a choice for player i. Player i can rationally choose c_i under ***common full belief in "caution and respect of preferences"*** *if there is some epistemic model with lexicographic beliefs, and some type t_i within it, such that:*
(1) t_i is cautious and expresses common full belief in "caution and respect of preferences," and
(2) choice c_i is rational for t_i.

In the previous section we saw that *respecting the opponent's preferences* is a stronger requirement than *primary belief in the opponent's rationality.* That is, every type that respects the opponent's preferences also primarily believes in the opponent's rationality, but not vice versa. A direct consequence of this observation is that *common full belief in "caution and respect of preferences"* is a stronger condition than *common full belief in "caution and primary belief in rationality."* In other words, the concept we explore in this chapter is more restrictive than the concept we investigated in the previous chapter. We thus obtain the following general relation between the two concepts.

Theorem 6.2.3 *(Relation with common full belief in "caution and primary belief in rationality")*
Suppose that player i can rationally choose c_i under common full belief in "caution and respect of preferences." Then, he can also rationally choose c_i under common full belief in "caution and primary belief in rationality."

The converse is not true. In Example 6.1, for instance, we saw that you can rationally choose pub b under common full belief in "caution and primary belief in rationality." However, if you are cautious and respect Barbara's preferences, you can no longer rationally choose pub b. Hence, you can certainly not rationally choose pub b under common full belief in "caution and respect of preferences."

We will now illustrate the consequences of *common full belief in "caution and respect of preferences"* by means of two examples.

Example 6.2 Where to read my book?

Recall the story from Example 5.2 and the utilities from Table 6.1. Which choices can you rationally make under common full belief in "caution and respect of preferences." In Example 6.1 we saw that, if you respect Barbara's preferences, then you must deem Barbara's choice a infinitely more likely than her choice b, and you must deem her choice b infinitely more likely than her choice c. So you should go to pub c. Hence, the only choice you can possibly rationally make under common full belief in "caution and respect of preferences" is c.

Can you indeed rationally choose c under common full belief in "caution and respect of preferences"? To see this, reconsider the epistemic model in Table 6.2. There, your type t_1^c and Barbara's type t_2^a are both cautious and respect the opponent's preferences. As your type t_1^c only deems possible Barbara's type t_2^a, and Barbara's type t_2^a only deems possible your type t_1^c, it follows that your type t_1^c expresses common full belief in "caution and respect of preferences." Since your choice c is optimal for your type t_1^c, we conclude that you can indeed rationally choose c under common full belief in "caution and respect of preferences." So, summarizing, we see that you can only rationally visit pub c under common full belief in "caution and respect of preferences." □

As you will have noted, the analysis in Example 6.2 was rather easy. The reason is that we did not really need all the requirements in common full belief in "caution and respect of preferences" to conclude that you can only choose pub c. By merely requiring that you are cautious and respect Barbara's preferences, we could deduce that you should go to pub c. In general, however, we will need some of the other requirements as well to see which choices you can rationally make under common full belief in "caution and respect of preferences." This will be illustrated by the following example.

Example 6.3 Dividing a pizza

You and Barbara have together ordered one pizza, which comes in four slices. In order to divide the pizza, you have agreed on the following procedure. You both write down on a piece of paper the number of slices you want. You have the option of writing down instead "I want the rest," which means that you will get the remaining number of slices if the other person writes down a number. So, if you write down "I want the rest," and Barbara writes down "I want three slices," then Barbara will get the three slices she wants, and you will receive the rest, which is one slice.

If you both write down a number, and the two numbers add up to at most four, then both you and Barbara will simply receive the number of slices you want. If the two numbers add up to more than four, then there is a problem since it is impossible to meet both persons' wishes. In this case you will enter into a fierce discussion, and at the end nobody will eat anything. If you write down a number and Barbara writes "I want the rest," then you will receive the number of slices you want and Barbara gets the rest. Similarly if Barbara writes down a number and you write "I want the rest." If you both write down "I want the rest," then you divide the pizza equally between the two of you.

So, both you and Barbara have six possible choices: to claim any number of slices between 0 and 4, or to claim the rest. The utilities for you and Barbara are as depicted in Table 6.3. What choices can you rationally make under common full belief in "caution and respect of preferences."

Note first that your choices 0, 1 and 2 are weakly dominated by "claim the rest." Namely, whatever Barbara decides, claiming the rest is always at least as good for you as claiming 0, 1 or 2 slices, and for some of Barbara's choices even strictly better. Since you are cautious, you do not exclude any of Barbara's choices. Therefore, you will definitely not claim 0, 1 or 2 slices, as "claim the rest" is better.

The same applies to Barbara. Her choices 0, 1 and 2 are also weakly dominated by "claim the rest." Since you believe that Barbara is cautious, you believe that Barbara will also prefer "claim the rest" to claiming 0, 1 or 2 slices. But then, since you respect Barbara's preferences, you must deem her choice "claim the rest" infinitely more likely than her choices 0, 1 and 2.

Table 6.3. *Utilities for you and Barbara in "Dividing a pizza"*

		Barbara					
		0	1	2	3	4	rest
	0	0,0	0,1	0,2	0,3	0,4	0,4
	1	1,0	1,1	1,2	1,3	0,0	1,3
You	2	2,0	2,1	2,2	0,0	0,0	2,2
	3	3,0	3,1	0,0	0,0	0,0	3,1
	4	4,0	0,0	0,0	0,0	0,0	4,0
	rest	4,0	3,1	2,2	1,3	0,4	2,2

However, in this case claiming 4 slices is always better for you than claiming 3 slices! As you deem Barbara's choice "claim the rest" infinitely more likely than her choices 0, 1 and 2, one of the following situations must occur:

- You deem Barbara's choice "claim the rest" infinitely more likely than her other choices. So, your primary belief assigns probability one to Barbara's choice "claim the rest." Under this primary belief, claiming 4 slices is better for you than claiming only 3. But then, under your full lexicographic belief, it is better for you to claim 4 slices rather than 3.
- You deem Barbara's choices "claim the rest" and "claim 4 slices" infinitely more likely than her other choices. Since claiming 4 slices is better for you than claiming 3 slices if Barbara claims the rest, and equally good if Barbara claims 4 slices, you will prefer claiming 4 slices to claiming only 3.
- You deem Barbara's choices "claim the rest" and "claim 3 slices" infinitely more likely than her other choices. Similarly to above, since claiming 4 slices is better for you than claiming 3 slices if Barbara claims the rest, and equally good if Barbara claims 3 slices, you will prefer claiming 4 slices to claiming only 3.
- The last possible situation is that you deem Barbara's choices "claim the rest," "claim 4 slices" and "claim 3 slices" infinitely more likely than her other choices. Since claiming 4 slices is better for you than claiming 3 slices if Barbara claims the rest, and equally good if Barbara claims 3 or 4 slices, you will prefer claiming 4 slices to claiming only 3.

Hence, we see that you will indeed prefer claiming 4 slices to claiming only 3, so you will definitely not claim 3 slices. Above, we have seen that you will also not claim 0, 1 or 2 slices. Hence, under common full belief in "caution and respect of preferences," the only choices you could possibly rationally make are claiming 4 slices and claiming the rest.

We will now show that, under common full belief in "caution and respect of preferences," both of these choices are possible. Consider the epistemic model in Table 6.4. Clearly, all types in this epistemic model are cautious. Moreover, each of the types

Table 6.4. *An epistemic model for "Dividing a pizza"*

Types	$T_1 = \{t_1^4, t_1^r\}, \quad T_2 = \{t_2^4, t_2^r\}$
Beliefs for you	$b_1(t_1^4) = ((rest, t_2^r); (1, t_2^r); (4, t_2^r); (3, t_2^r); (2, t_2^r); (0, t_2^r))$ $b_1(t_1^r) = ((4, t_2^4); (3, t_2^4); (rest, t_2^4); (2, t_2^4); (1, t_2^4); (0, t_2^4))$
Beliefs for Barbara	$b_2(t_2^4) = ((rest, t_1^r); (1, t_1^r); (4, t_1^r); (3, t_1^r); (2, t_1^r); (0, t_1^r))$ $b_2(t_2^r) = ((4, t_1^4); (3, t_1^4); (rest, t_1^4); (2, t_1^4); (1, t_1^4); (0, t_1^4))$

also respects the opponent's preferences. Consider, for instance, your type t_1^r, who fully believes that Barbara is of type t_2^4. Barbara's type t_2^4, in turn, primarily believes that you will claim the rest. Under this primary belief, Barbara prefers 4 to 3, 3 to 2 and "claim the rest," 2 and "claim the rest" to 1, and 1 to 0. So, Barbara's secondary belief in t_2^4 will only be relevant for deciding whether she prefers 2 to "claim the rest," or vice versa. Her other choices can be ranked on the basis of her primary belief alone. Barbara's secondary belief in t_2^4 assigns probability 1 to your choice 1, in which case claiming the rest is better for Barbara than claiming 2 slices. So, Barbara's type t_2^4 prefers "claim the rest" to 2. Summarizing, we see that Barbara's type t_2^4 prefers 4 to 3, 3 to "claim the rest," "claim the rest" to 2, 2 to 1 and 1 to 0. At the same time, your type t_1^r deems Barbara's choice 4 infinitely more likely than her choice 3, deems 3 infinitely more likely than "claim the rest," deems "claim the rest" infinitely more likely than 2, deems 2 infinitely more likely than 1 and deems 1 infinitely more likely than 0. Hence, we conclude that your type t_1^r respects Barbara's preferences.

Similarly we can show that your type t_1^4 also respects Barbara's preferences, although this is more difficult. Your type t_1^4 fully believes that Barbara is of type t_2^r. Barbara's type t_2^r, in turn, primarily believes that you will claim 4 slices. However, under this primary belief each of Barbara's choices would be equally good – or actually equally bad – and hence we must turn to the secondary and higher-level beliefs to determine the ranking of choices under Barbara's type t_2^r. The secondary belief of t_2^r assigns probability 1 to your choice 3, under which Barbara prefers "claim the rest" and 1 to each of her other choices. The tertiary belief of t_2^r assigns probability 1 to your choice "claim the rest." Under this tertiary belief, Barbara prefers "claim the rest" to 1 so that, overall, type t_2^r will prefer "claim the rest" to 1, and 1 to each of her other choices. Moreover, under the tertiary belief, Barbara prefers 4 to 3, 3 to 2 and 2 to 0. So, we see that, overall, type t_2^r will prefer 4 to 3, 3 to 2 and 2 to 0. Summarizing, we conclude that Barbara's type t_2^r prefers "claim the rest" to 1, 1 to 4, 4 to 3, 3 to 2 and 2 to 0. At the same time, your type t_1^4 deems Barbara's choice "claim the rest" infinitely more likely than her choice 1, deems 1 infinitely more likely than 4, deems 4 infinitely more likely than 3, deems 3 infinitely more likely than 2 and deems 2 infinitely more likely than 0. Hence, we conclude that your type t_1^4 also respects Barbara's preferences.

In exactly the same way, it can be verified that Barbara's types t_2^4 and t_2^r respect your preferences as well. So, all the types in the epistemic model are cautious and respect the opponent's preferences. So, we may conclude that all types in the epistemic model express common full belief in "caution and respect of preferences." Finally, since your choice 4 is rational for your type t_1^4 and your choice "claim the rest" is rational for your type t_1^r, it follows that you can both rationally claim 4 slices and claim the rest under common full belief in "caution and respect of preferences." □

6.3 Existence

So far in this chapter, we have introduced the idea of respecting the opponent's preferences and have incorporated it into the concept of common full belief in "caution and respect of preferences." By means of some examples we have illustrated the concept and shown how it restricts the possible choices you can make. A natural question that arises is: Does there always exist a belief hierarchy for you that expresses common full belief in "caution and respect of preferences"? So, can we always meet the conditions that common full belief in "caution and respect of preferences" imposes or are these sometimes too strong? In this section, we will show that we can indeed always construct a belief hierarchy that meets all of these conditions.

As in Chapters 3 and 5, we will give an easy iterative procedure that always yields a belief hierarchy that expresses common full belief in "caution and respect of preferences." We first illustrate this procedure by means of an example, and then show how the procedure works in general.

Example 6.4 Hide-and-seek

Recall the story from Example 5.6. For convenience, we have reproduced the utilities in Table 6.5. As in Chapter 5, we denote your choices by a_1, b_1 and c_1, and denote Barbara's choices by a_2, b_2 and c_2.

We start with an arbitrary cautious lexicographic belief for you about Barbara's choice, say $(a_2; b_2; c_2)$. Under this belief, your preferences on your own choices would be (c_1, b_1, a_1); that is, you would prefer c_1 to b_1 and b_1 to a_1.

Next, take a cautious lexicographic belief for Barbara about your choice that respects these preferences, say $(c_1; b_1; a_1)$. Under this belief, Barbara's preferences on her own choices would be (a_2, c_2, b_2).

Next, take a belief for you about Barbara's choice that respects these preferences, say $(a_2; c_2; b_2)$. Under this belief, your preferences on your own choices would be (b_1, c_1, a_1).

Next, take a belief for Barbara about your choice that respects these preferences, say $(b_1; c_1; a_1)$. Under this belief, Barbara's preferences on her own choices would be (b_2, a_2, c_2).

Next, take a belief for you about Barbara's choice that respects these preferences, say $(b_2; a_2; c_2)$. Under this belief, your preferences on your own choices would be (c_1, a_1, b_1).

Table 6.5. *Utilities for you and Barbara in "Hide-and-seek" (II)*

		Barbara		
		Pub a	Pub b	Pub c
You	Pub a	0,5	1,2	1,1
	Pub b	1,3	0,4	1,1
	Pub c	1,3	1,2	0,3

Next, take a belief for Barbara about your choice that respects these preferences, say $(c_1;a_1;b_1)$. Under this belief, Barbara's preferences on her own choices would be (a_2,c_2,b_2).

However, (a_2,c_2,b_2) has already been Barbara's induced preference relation. So, we can repeat the last four steps over and over again, with the following cycle of beliefs:

$$(a_2;c_2;b_2) \to (b_1;c_1;a_1) \to (b_2;a_2;c_2) \to (c_1;a_1;b_1) \to (a_2;c_2;b_2).$$

So, belief $(a_2;c_2;b_2)$ induces preference relation (b_1,c_1,a_1), belief $(b_1;c_1;a_1)$ induces preference relation $(b_2;a_2;c_2)$, and so on.

As in Chapter 5, we will now transform the beliefs in this cycle into belief hierarchies that are cautious and express common full belief in "caution and respect of preferences." We define types $t_1^{a_2c_2b_2}$ and $t_1^{b_2a_2c_2}$ for you, and types $t_2^{b_1c_1a_1}$ and $t_2^{c_1a_1b_1}$ for Barbara, where

$$b_1(t_1^{a_2c_2b_2}) = ((a_2,t_2^{c_1a_1b_1});(c_2,t_2^{c_1a_1b_1});(b_2,t_2^{c_1a_1b_1})),$$
$$b_1(t_1^{b_2a_2c_2}) = ((b_2,t_2^{b_1c_1a_1});(a_2,t_2^{b_1c_1a_1});(c_2,t_2^{b_1c_1a_1})),$$
$$b_2(t_2^{b_1c_1a_1}) = ((b_1,t_1^{a_2c_2b_2});(c_1,t_1^{a_2c_2b_2});(a_1,t_1^{a_2c_2b_2})),$$
$$b_2(t_2^{c_1a_1b_1}) = ((c_1,t_1^{b_2a_2c_2});(a_1,t_1^{b_2a_2c_2});(b_1,t_1^{b_2a_2c_2})).$$

Clearly, all types above are cautious. We show that all the types also respect the opponent's preferences. Consider, for instance, your type $t_1^{a_2c_2b_2}$. It holds lexicographic belief $(a_2;c_2;b_2)$ about Barbara's choice, and believes that Barbara is of type $t_2^{c_1a_1b_1}$, who holds lexicographic belief $(c_1;a_1;b_1)$ about your choice. By construction of the procedure, Barbara's preferences under the belief $(c_1;a_1;b_1)$ are (a_2,c_2,b_2), and hence your type $t_1^{a_2c_2b_2}$ respects Barbara's preferences. In the same way, it can be verified that the other three types respect the opponent's preferences as well.

This, however, implies that every type expresses common full belief in "caution and respect of preferences." Hence, we may conclude that every type above is cautious and expresses common full belief in "caution and respect of preferences." In particular, there exists a belief hierarchy for you that is cautious and expresses common full belief in "caution and respect of preferences." □

The iterative procedure we used in the example can actually be applied to *every* game with two players. In general, the procedure would work as follows:

- Start with an arbitrary cautious lexicographic belief for player i about j's choices, which assigns at every level probability 1 to one of j's choices. This belief induces a preference relation R_i^1 for player i on his own choices.
- Construct a cautious lexicographic belief for player j about i's choices that respects preference relation R_i^1, and which assigns at every level probability 1 to one of i's choices. This belief induces a preference relation R_j^2 for player j on his own choices.
- Construct a cautious lexicographic belief for player i about j's choices that respects preference relation R_j^2, and which assigns at every level probability 1 to one of j's choices. This belief induces a preference relation R_i^3 for player i on his own choices.

And so on. Note that there are only finitely many possible preference relations for both players, since every player has only finitely many choices. But then, there must be some round k where the induced preference relation R_i^k or R_j^k has already appeared as the induced preference relation at an earlier round l. So, we would then simply repeat the procedure between round l and round k over and over again, and would thus start a cycle of beliefs.

We may thus conclude that this procedure will always find a cycle of beliefs. In the same way as in the example, we can transform the beliefs in this cycle into types that are cautious and express common full belief in "caution and respect of preferences." This is always possible, and hence we can always construct in this way an epistemic model in which all types are cautious and express common full belief in "caution and respect of preferences." Moreover, the procedure above guarantees that the types we construct have the following two properties: (1) every type only deems possible one belief hierarchy – and hence one type – for the opponent, and (2) every type holds a lexicographic belief that assigns, at every level, probability 1 to one of the opponent's choices. We thus obtain the following general existence result.

Theorem 6.3.1 *(Common full belief in "caution and respect of preferences" is always possible)*
Consider a static game with two players and finitely many choices for both players. Then, we can construct an epistemic model with lexicographic beliefs in which:
(1) every type is cautious and expresses common full belief in "caution and respect of preferences," and
(2) every type deems possible only one of the opponent's types, and assigns at each level of the lexicographic belief probability 1 to one of the opponent's choices.

As a consequence, there always exists a belief hierarchy for you that is cautious and expresses common full belief in "caution and respect of preferences." In other words, in every game it is possible to meet all the requirements imposed by common full belief in "caution and respect of preferences."

6.4 Why elimination of choices does not work

Example 6.3 showed that it may be rather difficult to determine those choices you can rationally make under common full belief in "caution and respect of preferences." So, it would be very useful if we could find an algorithm that helps us to find those choices. But what would such an algorithm look like?

Up to this stage, we have seen two algorithms: *iterated elimination of strictly dominated choices* in Chapter 3, and the *Dekel–Fudenberg procedure* in Chapter 5. In the first algorithm, we started by eliminating all strictly dominated choices in the game. Within the reduced game we obtain, we then again eliminate all strictly dominated choices, and we continue in this fashion until no further choices can be eliminated. We showed that this algorithm yields all choices you can rationally make under common belief in rationality (with standard beliefs). In the second algorithm, we started by eliminating all *weakly* dominated choices in the game, and in the reduced game we obtain we then apply iterated elimination of *strictly* dominated choices. In Chapter 5 we showed that this algorithm yields all choices you can rationally make under common full belief in "caution and primary belief in rationality" (with lexicographic beliefs).

Both algorithms thus proceed by iteratively eliminating strictly or weakly dominated choices from the game. It therefore seems natural to see whether a similar algorithm would work for common full belief in "caution and respect of preferences," especially because elimination of choices is such an easy procedure. Unfortunately, it turns out that iterative elimination of choices will not work for common full belief in "caution and respect of preferences." The following example shows why.

Example 6.5 Spy game

The story is largely the same as in "Where to read my book?" You and Barbara must decide whether to visit pub a, pub b or pub c this afternoon. Since you want to read your book in peace, your only objective is to avoid Barbara. That is, if you visit the same pub as Barbara your utility will be zero, whereas your utility will be 1 if you go to a different pub than Barbara. On the other hand, Barbara prefers pub a to pub b and pub b to pub c. More precisely, her utility at pub a would be 3, her utility at pub b would be 2 and her utility at pub c would be 1.

So far, the story is exactly the same as in "Where to read my book?" Here comes the new part. Barbara now suspects that you are having an affair, and she would very much like to spy on you this afternoon. Since pubs a and c are in the same street, exactly opposite to each other, you can perfectly watch pub a from pub c, and vice versa. Pub b, however, is on the other side of town. Hence, if you go to pub a and Barbara goes to pub c, or vice versa, then Barbara can spy on you; otherwise not. Suppose now that spying would give Barbara an additional utility of 3. Then, the new utilities would be as in Table 6.6.

What choices can you rationally make under common full belief in "caution and respect of preferences"? Note that Barbara clearly prefers pub a to pub b. Therefore, if you respect Barbara's preferences, you must deem the event that Barbara will go to pub a infinitely more likely than the event that she will go to pub b. Since your only

Table 6.6. *Utilities for you and Barbara in "Spy game"*

		Barbara		
		Pub a	Pub b	Pub c
You	Pub a	0,3	1,2	1,4
	Pub b	1,3	0,2	1,1
	Pub c	1,6	1,2	0,1

Table 6.7. *An epistemic model for "Spy game"*

Types	$T_1 = \{t_1^c\}$, $T_2 = \{t_2^a\}$
Beliefs for you	$b_1(t_1^c) = ((a,t_2^a);(b,t_2^a);(c,t_2^a))$
Beliefs for Barbara	$b_2(t_2^a) = ((c,t_1^c);(b,t_1^c);(a,t_1^c))$

objective is to avoid Barbara, you will then prefer going to pub b rather than to pub a. So, you must believe that Barbara deems the event that you will go to pub b infinitely more likely than the event that you will go to pub a. In that case, Barbara would actually prefer going to pub b rather than going to pub c. Hence, you must deem the event that Barbara will go to pub b infinitely more likely than the event that she will go to pub c. Summarizing, we see that you must deem the event that Barbara will go to pub a infinitely more likely than the event that she will go to pub b, and you must deem the event that she will go to pub b infinitely more likely than the event that she will go to pub c. But then, you would go to pub c.

So, the only possible choice you could rationally make under common full belief in "caution and respect of preferences" is pub c. We now show that you can indeed rationally make this choice under common full belief in "caution and respect of preferences." Consider the epistemic model in Table 6.7. It can easily be verified that both types are cautious and respect the opponent's preferences. Therefore, both types express common full belief in "caution and respect of preferences." Since pub c is optimal for your type t_1^c, you can rationally visit pub c under common full belief in "caution and respect of preferences." We thus conclude that pub c is your only rational choice under common full belief in "caution and respect of preferences."

Can we obtain this result by some algorithm that iteratively eliminates strictly or weakly dominated choices? As we will see, the answer is "no." The reason is the following.

From the utilities in Table 6.6, it is clear that Barbara's choice pub b is the only weakly dominated choice in the game. Therefore, an algorithm that eliminates strictly or weakly dominated choices could only eliminate Barbara's choice pub b at the beginning. However, if we remove Barbara's choice pub b, then your choice pub b would be optimal for *every possible* belief you could hold in the reduced game. So, we would then never eliminate your choice pub b.

Hence, we see that an algorithm that iteratively eliminates strictly or weakly dominated choices will never be able to eliminate your choice pub b from the game. We have seen, however, that you cannot rationally choose pub b under common full belief in "caution and respect of preferences." So, such an algorithm cannot work for common full belief in "caution and respect of preferences." □

6.5 Preference restrictions and likelihood orderings

The previous example has taught us that an algorithm for common full belief in "caution and respect of preferences" cannot be based on eliminating strictly or weakly dominated choices. The key argument in this example was the following: At the beginning of such an algorithm we can eliminate only Barbara's choice pub b, as it is the only choice in the game that is weakly dominated. Eliminating Barbara's choice pub b from the game would actually require that you deem her choice pub b infinitely less likely than her two remaining choices pub a and pub c. In particular, you should then deem her choice pub b infinitely *less* likely than her choice pub c.

From our arguments in the previous section, we know that common full belief in "caution and respect of preferences" requires that you should deem Barbara's choice pub b infinitely *more* likely than her choice pub c. Obviously, these two requirements are in conflict with each other. So, an algorithm based on eliminating choices cannot work here.

But what type of algorithm *would* then work for common full belief in "caution and respect of preferences"? To answer this question, let us first return to the example "Spy game."

Example 6.6 Spy game

For convenience, let us denote your choices by a_1, b_1 and c_1, and denote Barbara's choices by a_2, b_2 and c_2. We have seen that eliminating Barbara's choice b_2 is not a good idea, as it would mean that you deem her eliminated choice b_2 infinitely less likely than her remaining choice c_2, which is not what we want.

What we *can* conclude, however, is that Barbara prefers her choice a_2 to her choice b_2, since a_2 weakly dominates b_2. We formally say that the pair (b_2, a_2) is a *preference restriction* for Barbara. This preference restriction should be read as follows: Barbara prefers the second choice in the pair, a_2, to the first choice in the pair, b_2.

If you respect Barbara's preferences, then your lexicographic belief about Barbara's choices should respect the preference restriction (b_2, a_2); that is, it should deem her choice a_2 infinitely more likely than her choice b_2. This means that your lexicographic belief will order Barbara's choices, in terms of likelihood, in one of the following ways:

- $(\{a_2\}; \{b_2\}; \{c_2\})$, or
- $(\{a_2\}; \{c_2\}; \{b_2\})$, or

- $(\{c_2\};\{a_2\};\{b_2\})$, or
- $(\{a_2\};\{b_2,c_2\})$, or
- $(\{a_2,c_2\};\{b_2\})$.

We call these the possible *likelihood orderings* that your lexicographic belief can induce on Barbara's choices. These likelihood orderings should be read as follows. Take, for instance, the first likelihood ordering $(\{a_2\};\{b_2\};\{c_2\})$. It states that you deem Barbara's choice a_2 infinitely more likely than her choice b_2, and that you deem her choice b_2 infinitely more likely than her choice c_2. The likelihood ordering $(\{a_2\};\{b_2,c_2\})$, for instance, states that you deem her choice a_2 infinitely more likely than her choices b_2 and c_2, but that you do not deem b_2 infinitely more likely than c_2 or vice versa. Note that the likelihood orderings above are all the possible likelihood orderings that deem Barbara's choice a_2 infinitely more likely than her choice b_2. So, if you respect Barbara's preference restriction (b_2,a_2), then your likelihood ordering on Barbara's choices should be one of the orderings above.

If your likelihood ordering is $(\{a_2\};\{b_2\};\{c_2\})$, $(\{a_2\};\{c_2\};\{b_2\})$ or $(\{a_2\};\{b_2,c_2\})$, then you deem Barbara's choice a_2 infinitely more likely than her other choices. So, you assign positive probability to a_2 before assigning positive probability to her other choices. We say that you *assume* Barbara's choice a_2. Since choosing b_1 is better for you than choosing a_1 if Barbara chooses a_2, we conclude that in this case you will prefer b_1 to a_1.

If your likelihood ordering is $(\{c_2\};\{a_2\};\{b_2\})$ or $(\{a_2,c_2\};\{b_2\})$, then you deem Barbara's choices a_2 and c_2 infinitely more likely than her remaining choice b_2. That is, you assign positive probability to a_2 and c_2 *before* you assign positive probability to b_2. We say that you *assume* Barbara's set of choices $\{a_2,c_2\}$. Note that your choice b_1 weakly dominates your choice a_1 on $\{a_2,c_2\}$, since b_1 is better than a_1 if Barbara chooses a_2, and they are equally good if Barbara chooses c_2. Since you assign positive probability to a_2 and c_2 *before* you assign positive probability to b_2, you must prefer b_1 to a_1.

So, overall we see that if your likelihood ordering respects Barbara's preference restriction (b_2,a_2), then you will prefer b_1 to a_1. The reason is that every likelihood ordering that respects the preference restriction (b_2,a_2) will either assume Barbara's choice $\{a_2\}$ or assume Barbara's set of choices $\{a_2,c_2\}$. Since b_1 weakly dominates a_1 on $\{a_2\}$ and on $\{a_2,c_2\}$, it follows that you will prefer b_1 to a_1.

Hence, we have (a_1,b_1) as a new preference restriction for you. If Barbara respects your preferences, then her likelihood ordering on your choices must deem your choice b_1 infinitely more likely than your choice a_1. That is, Barbara's likelihood ordering on your choices must be one of:

- $(\{b_1\};\{a_1\};\{c_1\})$, or
- $(\{b_1\};\{c_1\};\{a_1\})$, or
- $(\{c_1\};\{b_1\};\{a_1\})$, or
- $(\{b_1\};\{a_1,c_1\})$, or
- $(\{b_1,c_1\};\{a_1\})$.

Note that each of these likelihood orderings either assumes your choice $\{b_1\}$, or assumes your set of choices $\{b_1, c_1\}$. So, Barbara either assigns positive probability to b_1 before she assigns positive probability to your other choices, or she assigns positive probability to b_1 and c_1 before she assigns positive probability to a_1. Since Barbara's choice b_2 weakly dominates her choice c_2 on $\{b_1\}$ and on $\{b_1, c_1\}$ – in fact, strictly dominates c_2 on these sets – we conclude that Barbara must prefer b_2 to c_2.

So, we add (c_2, b_2) as a new preference restriction for Barbara. If you respect Barbara's preferences, you must respect her "old" preference restriction (b_2, a_2) and her "new" preference restriction (c_2, b_2). Hence, your likelihood ordering on Barbara's choices must deem a_2 infinitely more likely than b_2, and must deem b_2 infinitely more likely than c_2. So, your likelihood ordering on Barbara's choices must be

$$(\{a_2\}; \{b_2\}; \{c_2\}).$$

But then, there is only one optimal choice for you, namely c_1. So, this procedure yields a unique choice for you, c_1, which is the only choice you can rationally make under common full belief in "caution and respect of preferences."

In a nutshell, the procedure above can be summarized as follows: Since a_2 weakly dominates b_2 in the full game, Barbara will prefer a_2 to b_2, so (b_2, a_2) is a preference restriction for Barbara. Every likelihood ordering for you that respects Barbara's preference restriction (b_2, a_2) assumes either $\{a_2\}$ or $\{a_2, c_2\}$. Since your choice b_1 weakly dominates a_1 on $\{a_2\}$ and on $\{a_2, c_2\}$, you will prefer b_1 to a_1, and hence (a_1, b_1) is a new preference restriction for you. Every likelihood ordering for Barbara that respects your preference restriction (a_1, b_1) assumes either $\{b_1\}$ or $\{b_1, c_1\}$. Since Barbara's choice b_2 weakly dominates her choice c_2 on $\{b_1\}$ and on $\{b_1, c_1\}$, Barbara will prefer b_2 to c_2. So, (c_2, b_2) is a new preference restriction for Barbara. There is only one likelihood ordering for you that respects both of Barbara's preference restrictions (b_2, a_2) and (c_2, b_2), namely the likelihood ordering $(a_2; b_2; c_2)$. But then, the only possible optimal choice for you is c_1. □

In the procedure above, we have used the following important insight: If a likelihood ordering for player i *assumes* a set of choices D_j for player j, and choice a_i weakly dominates choice c_i for player i on D_j, then player i will prefer a_i to c_i.

Here is the reason: If player i assumes the set of choices D_j, then he assigns positive probability to each of the choices in D_j *before* assigning positive probability to any choice outside D_j. That is, in i's lexicographic belief the first few levels, say the first k levels, assign positive probability only to j's choices in D_j, and every choice in D_j has a positive probability in one of the levels $1, 2, ..., k$. As a_i weakly dominates c_i on D_j, choice a_i is always at least as good as c_i, and sometimes strictly better than c_i, whenever j chooses from D_j. But then, under the first k levels of the lexicographic belief, a_i always yields at least as high a utility as c_i, and sometimes a strictly higher utility. Therefore, player i will prefer a_i to c_i.

Suppose now that player i assumes a set of choices D_j for player j, and that his choice c_i is not weakly dominated by another choice a_i on D_j, but rather is weakly dominated on D_j by some *randomized* choice. What can we conclude then?

Recall that a randomized choice for player i is a probability distribution r_i on his set of choices C_i. Let us assume that r_i only assigns positive probability to choices in A_i, where A_i is some subset of choices for player i. We say that r_i is a *randomized choice on* A_i. So, consider the situation where player i assumes a set of choices D_j for player j, and where his choice c_i is weakly dominated on D_j by some randomized choice r_i on A_i. What can we say about player i's preferences? The following result tells us that in this case, you will prefer at least some choice in A_i to your choice c_i.

Lemma 6.5.1 *(Preferences if you assume an opponent's set of choices)*
Suppose player i holds a cautious lexicographic belief and assumes a set of choices D_j for player j. Let A_i be some set of choices for player i. If his choice c_i is weakly dominated on D_j by some randomized choice r_i on A_i, then player i will prefer some choice in A_i to c_i.

The proof of this result can be found in the proofs section at the end of this chapter.

Remember that a preference restriction (c_i,a_i) for player i means that player i will prefer choice a_i to c_i. Similarly, we can take a *set* of choices A_i for player i, and say that (c_i,A_i) is a preference restriction for player i if player i prefers at least one choice in A_i to c_i. The lemma above thus states that, if player i assumes some set of choices D_j for player j, and his choice c_i is weakly dominated on D_j by some randomized choice on A_i, then (c_i,A_i) will be a preference restriction for player i.

In the following example we use this insight to find those choices you can rationally make under common full belief in "caution and respect of preferences."

Example 6.7 Runaway bride

Today you are going to Barbara's wedding. Just when she was supposed to say "yes," she changed her mind and ran away. The groom, who is currently in shock, asked you to find her, and of course you want to help him. You have known Barbara for a long time now, and you can therefore guess the places where she might be hiding: the houses a,b,c,d and e depicted in Figure 6.1. Houses a and e are where her mother and grandmother live, and you know that in these circumstances they will simply not open the door to you. Houses b,c and d belong to friends of Barbara, who would probably open the door to you. The distance between two neighboring houses is one kilometer. Since your objective is to find Barbara, your utility will be 1 if you find her and your utility will be zero otherwise. Barbara, who suspects that you will follow her, has tried to be as far away from you as possible. More precisely, her utility is the distance between

Figure 6.1 Possible hiding places in "Runaway bride"

Table 6.8. *Utilities for you and Barbara in "Runaway bride"*

		Barbara				
		a	b	c	d	e
You	a	0,0	0,1	0,2	0,3	0,4
	b	0,1	1,0	0,1	0,2	0,3
	c	0,2	0,1	1,0	0,1	0,2
	d	0,3	0,2	0,1	1,0	0,1
	e	0,4	0,3	0,2	0,1	0,0

the place where she is hiding and the place where you will look for her. Hence, the utilities for you and Barbara are given by Table 6.8. For convenience, let us denote your choices by $a_1,...,e_1$ and denote Barbara's choices by $a_2,...,e_2$.

Which choices can you rationally make under common full belief in "caution and respect of preferences"? As for the previous example, we will use likelihood orderings and preference restrictions to determine these choices.

First, note that none of Barbara's choices is weakly dominated by some other choice. However, her choice c_2 is weakly dominated by her randomized choice $\frac{1}{2}b_2 + \frac{1}{2}d_2$. If you choose c_1, then the randomized choice is strictly better for Barbara than her choice c_2, and they are equally good for Barbara if you make any other choice. Since Barbara of course assumes your full set of choices $\{a_1,b_1,c_1,d_1,e_1\}$, we know from Lemma 6.5.1 that Barbara will prefer either b_2 or d_2 to c_2. Hence, $(c_2,\{b_2,d_2\})$ is a preference restriction for Barbara.

If you respect Barbara's preferences, then you must deem either b_2 or d_2 infinitely more likely than c_2. But then, there must be some set of choices D_2 for Barbara that contains b_2 or d_2 but not c_2, such that you deem all choices in D_2 infinitely more likely than all choices outside D_2. That is, your likelihood ordering must assume some set D_2 of Barbara's choices that contains b_2 or d_2 but not c_2. On every such set D_2, your choice c_1 is weakly dominated by your randomized choice $\frac{1}{2}b_1 + \frac{1}{2}d_1$. By Lemma 6.5.1 we then know that you will prefer either b_1 or d_1 to c_1, so $(c_1,\{b_1,d_1\})$ is a preference restriction for you.

Moreover, your choices a_1 and e_1 are weakly dominated in the full game by, for instance, c_1. Since, of course, you assume Barbara's full set of choices $\{a_2,b_2,c_2,d_2,e_2\}$, we know from Lemma 6.5.1 that you will prefer c_1 to a_1 and e_1. Hence, also $(a_1,\{c_1\})$ and $(e_1,\{c_1\})$ are preference restrictions.

Summarizing, we see that you will have the following preference restrictions:

$$(c_1,\{b_1,d_1\}),\ (a_1,\{c_1\}) \text{ and } (e_1,\{c_1\}).$$

So, you prefer either b_1 or d_1 to c_1, and prefer c_1 to both a_1 and e_1. So, your only possible optimal choices are b_1 and d_1.

Table 6.9. *An epistemic model for "Runaway bride"*

Types	$T_1 = \{t_1^b, t_1^d\}, \quad T_2 = \{t_2^a, t_2^e\}$
Beliefs for you	$b_1(t_1^b) = ((a, t_2^a); (b, t_2^a); (e, t_2^a); (c, t_2^a); (d, t_2^a))$
	$b_1(t_1^d) = ((e, t_2^e); (d, t_2^e); (a, t_2^e); (c, t_2^e); (b, t_2^e))$
Beliefs for Barbara	$b_2(t_2^a) = ((d, t_1^d); (c, t_1^d); (b, t_1^d); (a, t_1^d); (e, t_1^d))$
	$b_2(t_2^e) = ((b, t_1^b); (c, t_1^b); (d, t_1^b); (a, t_1^b); (e, t_1^b))$

What about Barbara? We have seen above that $(c_2, \{b_2, d_2\})$ is a preference restriction for Barbara. Moreover, in the full game her choice b_2 is weakly dominated by her randomized choice $\frac{3}{4}a_2 + \frac{1}{4}e_2$, which means that also $(b_2, \{a_2, e_2\})$ is a preference restriction for Barbara. Similarly, in the full game her choice d_2 is weakly dominated by her randomized choice $\frac{1}{4}a_2 + \frac{3}{4}e_2$, and hence $(d_2, \{a_2, e_2\})$ is a preference restriction for Barbara as well. So we see that Barbara will have the following preference restrictions:

$$(c_2, \{b_2, d_2\}), \ (b_2, \{a_2, e_2\}) \text{ and } (d_2, \{a_2, e_2\}).$$

So, Barbara prefers either b_2 or d_2 to c_2, prefers either a_2 or e_2 to b_2, and prefers either a_2 or e_2 to d_2. So, her only possible optimal choices are a_2 and e_2.

Altogether, the procedure above leads to choices b_1 and d_1 for you, and choices a_2 and e_2 for Barbara. We will now show that these are exactly the choices that can rationally be made under common full belief in "caution and respect of preferences."

Consider the epistemic model in Table 6.9. The reader may verify that every type in this model is cautious and respects the opponent's preferences. Therefore, every type here expresses common full belief in "caution and respect of preferences." Since choice b_1 is optimal for type t_1^b, choice d_1 is optimal for type t_1^d, choice a_2 is optimal for type t_2^a and e_2 is optimal for type t_2^e, it follows that indeed the choices b_1, d_1, a_2 and e_2 can rationally be made under common full belief in "caution and respect of preferences." But then, by our procedure above, these are the only choices that can rationally be made under common full belief in "caution and respect of preferences." □

To conclude this section, we will formally define *preference restrictions* and *likelihood orderings* as we have used them above.

Definition 6.5.2 *(Preference restrictions)*
A ***preference restriction*** *for player i is a pair (c_i, A_i), where c_i is a choice and A_i is a set of choices for player i. The interpretation is that player i prefers at least one choice in A_i to c_i.*

Here, A_i can contain one choice for player i or several choices. If A_i contains only one choice, say a_i, then the preference restriction states that player i prefers a_i to c_i.

Definition 6.5.3 *(Likelihood orderings)*
A ***likelihood ordering*** *for player i on j's choices is a sequence* $L_i = (L_i^1; L_i^2; ...; L_i^K)$, *where* $L_i^1, ..., L_i^K$ *are disjoint sets of choices for player j whose union is equal to the full choice set* C_j. *The interpretation is that player i deems all choices in* L_i^1 *infinitely more likely than all choices in* L_i^2, *deems all choices in* L_i^2 *infinitely more likely than all choices in* L_i^3 *and so on.*

Here, by "disjoint" we mean that the sets $L_i^1, ..., L_i^K$ have no overlap. That is, every choice for player j is in exactly one of the sets from $L_i^1, ..., L_i^K$. We now formally define what it means to say that a likelihood ordering respects a preference restriction or assumes a set of choices.

Definition 6.5.4 *(Respecting a preference restriction and assuming a set of choices)*
A likelihood ordering L_i *for player i* ***respects a preference restriction*** (c_j, A_j) *for player j, if* L_i *deems at least one choice in* A_j *infinitely more likely than* c_j.
A likelihood ordering L_i *for player i* ***assumes a set of choices*** D_j *for player j, if* L_i *deems all choices in* D_j *infinitely more likely than all choices outside* D_j.

In the following section we will use these definitions for our algorithm, which computes for every game those choices the players can rationally make under common full belief in "caution and respect of preferences."

6.6 Algorithm

In the previous two examples we have used a procedure for computing those choices a player can rationally make under common full belief in "caution and respect of preferences." This procedure was based on likelihood orderings and preference restrictions, and was designed as follows:

At the beginning we look, for both players i, if in the full game some choice c_i is weakly dominated by some other choice a_i, or by some randomized choice r_i. If c_i is weakly dominated by some other choice a_i, then we add $(c_i, \{a_i\})$ as a preference restriction for player i. If c_i is weakly dominated by some randomized choice r_i, assigning positive probability to choices in A_i, then we add (c_i, A_i) as a preference restriction for player i.

We then concentrate, for both players i, on those likelihood orderings that respect all of j's preference restrictions. If every such likelihood ordering for player i assumes some set of choices D_j for player j on which his choice c_i is weakly dominated by some other choice a_i, then we add $(c_i, \{a_i\})$ as a preference restriction for player i. If every such likelihood ordering for player i assumes some set of choices D_j for player j on which his choice c_i is weakly dominated by some randomized choice r_i on A_i, then we add (c_i, A_i) as a preference restriction for player i.

We then concentrate, for both players i, on those likelihood orderings that respect all of j's preference restrictions above. And so on.

We call this procedure *iterated addition of preference restrictions*. So, at the beginning we add – if possible – preference restrictions for both players. We then focus on likelihood orderings that respect these preference restrictions, which – possibly – leads to new preference restrictions for both players. We then focus on those likelihood orderings that respect all of these new preference restrictions as well. This – possibly – leads to new preference restrictions for both players, and so on, until no new preference restrictions can be added. Hence, throughout the procedure we add new preference restrictions for the players, and restrict further and further the possible likelihood orderings that players can have.

Note that, formally speaking, there is no need to distinguish between being weakly dominated by another choice a_i and being weakly dominated by a randomized choice r_i on A_i. If the randomized choice r_i assigns probability 1 to choice a_i, then being weakly dominated by the randomized choice r_i is the same as being weakly dominated by the choice a_i. So, in a sense, the randomized choices cover all "usual" choices as a special case. We can formally describe the algorithm of *iterated addition of preference restrictions* in the following way.

Algorithm 6.6.1 *(Iterated addition of preference restrictions)*
Step 1. *For both players i, add a preference restriction* (c_i, A_i) *for player i if in the full game* c_i *is weakly dominated by some randomized choice on* A_i.
Step 2. *For both players i, restrict to likelihood orderings* L_i *that respect all preference restrictions for player j in step 1. Add a preference restriction* (c_i, A_i) *for player i if every such likelihood ordering* L_i *assumes a set of choices* D_j *on which* c_i *is weakly dominated by some randomized choice on* A_i.
Step 3. *For both players i, restrict to likelihood orderings* L_i *that respect all preference restrictions for player j in steps 1 and 2. Add a preference restriction* (c_i, A_i) *for player i if every such likelihood ordering* L_i *assumes a set of choices* D_j *on which* c_i *is weakly dominated by some randomized choice on* A_i.
And so on, until no further preference restrictions can be added.

So, at every round the set of preference restrictions for both players increases, whereas the set of likelihood orderings we restrict to becomes smaller. Since there are only finitely many possible preference restrictions in the game, the algorithm must stop after finitely many rounds.

At the end of the algorithm we obtain, for both players i, a final set of preference restrictions. What then are the choices that can be optimal for player i after applying this algorithm? These are exactly those choices c_i that are not part of some final preference restriction (c_i, A_i). Namely, if c_i is in a final preference restriction (c_i, A_i), then player i should prefer some choice in A_i to c_i, and hence c_i can never be optimal. If, on the other hand, c_i is not in any final preference restriction (c_i, A_i), then there is no restriction which says that player i should prefer some other choice to c_i, and hence c_i can be an optimal choice. In this case, we say that choice c_i is *optimal after iterated addition of preference restrictions*. Similarly, after every step k of the algorithm, we can define the choices that are optimal after k-fold addition of preference restrictions. By the latter procedure, we mean the first k steps of iterated addition of preference restrictions.

Definition 6.6.2 *(Optimal choices after iterated addition of preference restrictions)*
Choice c_i is ***optimal after k-fold addition of preference restrictions,*** *if c_i is not in any preference restriction (c_i, A_i) generated during the first k steps of the algorithm.*
Choice c_i is ***optimal after iterated addition of preference restrictions*** *if c_i is not in any preference restriction (c_i, A_i) generated during the complete algorithm.*

We have seen in the two previous examples that the algorithm of iterated addition of preference restrictions yields exactly those choices you can rationally make under common full belief in "caution and respect of preferences." This is not only true for these examples; it is in fact true for *all* games.

We now give a partial argument for why the algorithm works in general. Take an arbitrary static game with two players. We will first focus on the choices that player i can rationally make if he is cautious and expresses 1-fold full belief in "caution and respect of preferences," then concentrate on the choices he can rationally make if he expresses up to 2-fold full belief in "caution and respect of preferences," and so on.

1-fold full belief in "caution and respect of preferences": Suppose that player i is of type t_i, which is cautious and expresses 1-fold full belief in "caution and respect of preferences." That is, t_i is cautious, respects j's preferences, and fully believes that j is cautious. Suppose that the opponent's choice c_j is weakly dominated by some randomized choice r_j, which assigns positive probability only to choices in $A_j \subseteq C_j$. Then, by Lemma 6.5.1, player j prefers some choice in A_j to c_j. Since t_i respects j's preferences, it must be that t_i deems some choice in A_j infinitely more likely than c_j, and hence t_i respects the preference restriction (c_j, A_j). So, we see that t_i must respect every preference restriction (c_j, A_j) where c_j is weakly dominated by some randomized choice on A_j. But these are all preference restrictions for player j generated in step 1 of the algorithm. Hence, type t_i's likelihood ordering must respect all of the opponent's preference restrictions (c_j, A_j) generated in step 1 of the algorithm.

Consider now a preference restriction (c_i, A_i) generated for player i up to step 2. We show that choice c_i cannot be optimal for t_i. Since the preference restriction (c_i, A_i) is generated in step 1 or step 2, we know that every likelihood ordering L_i which respects all of the opponent's preference restrictions from step 1, assumes a set of choices D_j on which choice c_i is weakly dominated by some randomized choice on A_i. We have seen above that t_i indeed respects all of the opponent's preference restrictions from step 1. Therefore, t_i assumes a set of choices D_j on which c_i is weakly dominated by some randomized choice on A_i. But then, by Lemma 6.5.1, type t_i prefers some choice in A_i to c_i. In particular, choice c_i cannot be optimal for type t_i.

Summarizing, we see that every choice c_i that is in some preference restriction (c_i, A_i) generated during steps 1 and 2 of the algorithm, cannot be optimal for type t_i. Or, equivalently, every choice c_i that is optimal for type t_i will not be in any preference restriction (c_i,A_i) generated during the first two steps. That is, every choice that is optimal for type t_i must be optimal after 2-fold addition of preference restrictions. So, we see that every choice that can rationally be made under 1-fold full belief in "caution and respect of preferences" must be optimal after 2-fold addition of preference restrictions.

Up to 2-fold full belief in "caution and respect of preferences": Consider a type t_i for player i which is cautious and expresses up to 2-fold full belief in "caution and respect of preferences." That is, type t_i is cautious, respects j's preferences, fully believes that j is cautious and only deems possible j's types t_j that express 1-fold full belief in "caution and respect of preferences." Take a preference restriction (c_i,A_i) generated during the first three steps. We show that type t_i prefers some choice in A_i to c_i, and hence c_i cannot be optimal for type t_i.

Above we have seen that for every type t_i' that expresses 1-fold full belief in "caution and respect of preferences," the following holds: If (c_i,A_i) is a preference restriction generated during the first two steps, then t_i' prefers some choice in A_i to c_i. The same applies to player j. Hence, for every type t_j that expresses 1-fold full belief in "caution and respect of preferences," and every preference restriction (c_j,A_j) generated during the first two steps, type t_j prefers some choice in A_j to c_j. As type t_i only deems possible opponent's types t_j that express 1-fold full belief in "caution and respect of preferences," and type t_i respects j's preferences, we can conclude that for every preference restriction (c_j,A_j) generated during the first two steps, type t_i will deem some choice in A_j infinitely more likely than c_j. That is, type t_i respects every preference restriction (c_j,A_j) generated during the first two steps.

Now, take some preference restriction (c_i,A_i) generated during the first three steps. Then, every likelihood ordering L_i that respects all preference restrictions (c_j,A_j) generated during the first two steps, will assume a set of choices D_j on which c_i is weakly dominated by some randomization on A_i. As type t_i indeed respects every preference restriction (c_j,A_j) generated during the first two steps, we conclude that type t_i assumes a set of choices D_j on which c_i is weakly dominated by some randomization on A_i. But then, by Lemma 6.5.1, type t_i prefers some choice in A_i to c_i, and hence c_i cannot be optimal for t_i.

Summarizing, we see that every choice c_i that is in some preference restriction (c_i,A_i) generated during the first three steps of the algorithm, cannot be optimal for type t_i. Or, equivalently, every choice c_i that is optimal for type t_i will not be in any preference restriction (c_i,A_i) generated during the first three steps. That is, every choice that is optimal for type t_i must be optimal after 3-fold addition of preference restrictions. So, we see that every choice that can rationally be chosen when expressing up to 2-fold full belief in "caution and respect of preferences" must be optimal after 3-fold addition of preference restrictions.

By continuing in this fashion, we conclude that for every $k \in \{1,2,...\}$, every choice c_i that can rationally be chosen when expressing up to k-fold full belief in "caution and respect of preferences" must be optimal after $(k+1)$-fold addition of preference restrictions. It can be shown that the converse is also true: Every choice c_i that is optimal after $(k+1)$-fold addition of preference restrictions, is optimal for a type t_i that is cautious and expresses up to k-fold full belief in "caution and respect of preferences." Hence, the choices that can rationally be chosen when expressing up to k-fold full belief in "caution and respect of preferences" are precisely those choices that are optimal after $(k+1)$-fold addition of preference restrictions. As a consequence, the choices that can rationally be chosen under *common* full belief in "caution and respect of preferences" are exactly the choices that are optimal after the *full* procedure of iterated addition of preference restrictions. We thus obtain the following general result.

Theorem 6.6.3 *(The algorithm works)*
Consider a two-player static game with finitely many choices for both players.
(1) For every $k \geq 1$, the choices that can rationally be made by a cautious type that expresses up to k-fold full belief in "caution and respect of preferences" are exactly those choices that are optimal after $(k+1)$-fold addition of preference restrictions.
(2) The choices that can rationally be made by a cautious type that expresses common full belief in "caution and respect of preferences" are exactly those choices that are optimal after the full procedure of iterated addition of preference restrictions.

So, indeed this algorithm always delivers all choices the players can rationally make under common full belief in "caution and respect of preferences." The proof of this result can be found in the proofs section at the end of this chapter. To conclude, we illustrate the algorithm by means of a final example.

Example 6.8 Take a seat

You and Barbara are the only ones who have to sit a mathematics exam. The problem is that you have concentrated exclusively on algebra, whereas Barbara has only studied geometry. In order to pass you must be able to copy from Barbara, who faces a similar problem. When you enter the room you discover, much to your disappointment, that the teacher has rearranged the tables. Figure 6.2 shows the set-up of the tables. There are eight possible seats that you can choose, which we have labeled a to h. The teacher will ask you and Barbara where you want to sit. If you happen to name the same seat, then the teacher will toss a coin to decide who will get that seat and the other person will get the seat horizontally next to it. Otherwise, you will both get the seat you have named. In order to copy from Barbara you must either sit horizontally next to her or diagonally behind her. The same is true for Barbara. So, for instance, if you sit at c and Barbara sits at b, then you can copy from Barbara but Barbara cannot copy from you. If you sit at e and Barbara at b, then nobody can copy. If you sit at c and Barbara at d, you can both copy from each other.

During the exam the teacher will read the *New York Times*, but will watch from time to time to see whether you are cheating. If you are sitting at table a, just in front of the teacher's table, there is no chance of copying without being caught. At table b you will be caught with probability 90% if you try to copy. The further you go to the back, the smaller the chance of being caught. More precisely, at tables c and d you would

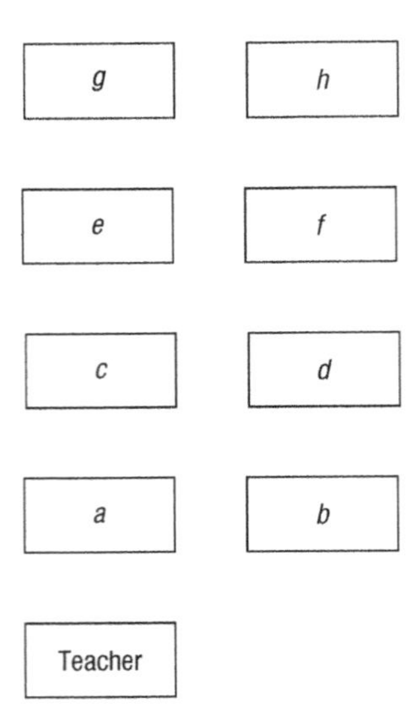

Figure 6.2 Arrangement of tables in "Take a seat"

be caught with probability 80%, at tables e and f with probability 55% and at tables g and h with probability 5%. The objective for you and Barbara is to maximize the chance of copying without being caught. The utilities for you and Barbara are given in Table 6.10. Here, the utility 45 for you means that there is a 45% chance of copying from Barbara without being caught. Similarly for the other utilities in the table. Note that if you both choose seat a or both choose seat b, then with probability 0.5 you will be seated at a and Barbara at b, and with probability 0.5 you will be seated at b and Barbara at a. So, the expected chance of successful copying would be

$$(0.5) \cdot 0\% + (0.5) \cdot 10\% = 5\%.$$

This explains the number 5 in the table. Which seats can you rationally choose under common full belief in "caution and respect of preferences"?

In order to answer this question, we apply the algorithm of iterated addition of preference restrictions.

Step 1: In the full game, your choice a is weakly dominated by b, and your choice b is weakly dominated by the randomized choice $\frac{1}{2}c + \frac{1}{2}d$. The same holds for Barbara, since the game is completely symmetric. So, we add the preference restrictions $(a, \{b\})$ and $(b, \{c, d\})$ for you and Barbara.

Step 2: Every likelihood ordering for you about Barbara's choice that respects her preference restriction $(a, \{b\})$ from step 1 must deem her choice b infinitely more likely than her choice a. Hence, every such likelihood ordering assumes a set of choices D_2 for Barbara that contains b but not a. On each of these sets of choices D_2, your choice d is weakly dominated by your choice c. So, we add your preference restriction $(d, \{c\})$.

Table 6.10. *Utilities for you and Barbara in "Take a seat"*

		Barbara							
		a	*b*	*c*	*d*	*e*	*f*	*g*	*h*
You	*a*	5,5	0,10	0,0	0,20	0,0	0,0	0,0	0,0
	b	10,0	5,5	0,20	0,0	0,0	0,0	0,0	0,0
	c	0,0	20,0	20,20	20,20	0,0	0,45	0,0	0,0
	d	20,0	0,0	20,20	20,20	0,45	0,0	0,0	0,0
	e	0,0	0,0	0,0	45,0	45,45	45,45	0,0	0,95
	f	0,0	0,0	45,0	0,0	45,45	45,45	0,95	0,0
	g	0,0	0,0	0,0	0,0	0,0	95,0	95,95	95,95
	h	0,0	0,0	0,0	0,0	95,0	0,0	95,95	95,95

Moreover, every likelihood ordering for you about Barbara's choice that respects her preference restrictions $(a,\{b\})$ and $(b,\{c,d\})$ from step 1 must deem c or d infinitely more likely than b, and must deem b infinitely more likely than a. Hence, every such likelihood ordering must assume some set of choices D_2 for Barbara that contains c or d, but not a and not b. On each of these sets of choices D_2, your choice c is weakly dominated by the randomized choice $\frac{1}{2}e + \frac{1}{2}f$. So, we add your preference restriction $(c,\{e,f\})$.

The same holds for Barbara, so we also add preference restrictions $(d,\{c\})$ and $(c,\{e,f\})$ for Barbara.

Step 3: Every likelihood ordering for you about Barbara's choice that respects her preference restriction $(d,\{c\})$ from step 2 must deem her choice c infinitely more likely than d. Hence, every such likelihood ordering assumes a set of choices D_2 for Barbara that contains her choice c but not d. On each of these sets D_2 your choice f weakly dominates your choice e. So, we add your preference restriction $(e,\{f\})$.

Moreover, every likelihood ordering for you about Barbara's choice that respects her preference restrictions $(a,\{b\}),(b,\{c,d\}),(d,\{c\})$ and $(c,\{e,f\})$ from steps 1 and 2 must deem her choice e or f infinitely more likely than her choices a,b,c and d. So, every such likelihood ordering will assume a set of choices D_2 for Barbara that contains e or f, but not any choice from $\{a,b,c,d\}$. On each of these sets D_2, your choice f is weakly dominated by your randomized choice $\frac{1}{2}g + \frac{1}{2}h$. So, we add your preference restriction $(f,\{g,h\})$.

The same holds for Barbara, so we also add preference restrictions $(e,\{f\})$ and $(f,\{g,h\})$ for Barbara.

Step 4: Every likelihood ordering for you about Barbara's choice that respects her preference restriction $(e,\{f\})$ from step 3 must deem her choice f infinitely more likely than e. Hence, every such likelihood ordering assumes a set of choices D_2 for Barbara that contains her choice f but not e. On each of these sets D_2 your choice g weakly dominates your choice h. So, we add your preference restriction $(h,\{g\})$.

Table 6.11. *An epistemic model for "Take a seat"*

Types	$T_1 = \{t_1\}, \quad T_2 = \{t_2\}$
Beliefs for you	$b_1(t_1) = ((g,t_2);(h,t_2);(f,t_2);(e,t_2);$ $(c,t_2);(d,t_2);(b,t_2);(a,t_2))$
Beliefs for Barbara	$b_2(t_2) = ((g,t_1);(h,t_1);(f,t_1);(e,t_1);$ $(c,t_1);(d,t_1);(b,t_1);(a,t_1))$

In total we have thus gathered the preference restrictions

$$(a,\{b\}),(b,\{c,d\}),(d,\{c\}),(c,\{e,f\}),(e,\{f\}),(f,\{g,h\}) \text{ and } (h,\{g\})$$

for you. This means that only choice g is optimal after iterated addition of preference restrictions. So, by our theorem above we know that under common full belief in "caution and respect of preferences" there is only one choice you can rationally make, namely g.

We finally show that you can indeed rationally choose seat g under common full belief in "caution and respect of preferences." Consider the epistemic model in Table 6.11. We only have one type for you and one type for Barbara in this model. It can be verified that both your type and Barbara's type are cautious and respect the opponent's preferences. As such, your type t_1 expresses common full belief in "caution and respect of preferences." As choice g is optimal for your type t_1, we may conclude that you can rationally choose g under common full belief in "caution and respect of preferences." Therefore, under common full belief in "caution and respect of preferences," your only rational choice is to choose seat g. □

6.7 Order independence

In the previous section we saw that the procedure of *iterated addition of preference restrictions* delivers all choices you can rationally make under common full belief in "caution and respect of preferences." However, in this procedure it may not always be easy to find, at every step, *all* preference restrictions that are generated. Especially if the game is large, it is easy to overlook some of these preference restrictions. Now, suppose that at every step of the procedure we only find *some* – but not necessarily *all* – of the preference restrictions that can be generated. Does it matter for the final result of the algorithm? The answer, as we will see, is "no." The final set of preference restrictions generated by the procedure does not depend on the order and speed used to generate the preference restrictions. Even if we overlook some preference restrictions at some steps, the final set of preference restrictions will still be the same.

To see why, let us compare, for a given game, two procedures – Procedure 1 and Procedure 2. Procedure 1 is the original algorithm, in which at every step we always add *all* preference restrictions that can be generated on the basis of the previous steps.

Procedure 2 adds, at every step, only *some* – but not necessarily *all* – of the preference restrictions that can be generated on the basis of the previous steps. We will show that both procedures eventually generate the same set of preference restrictions for both players.

Let us denote by R_i^k and $\hat{R}_i^k$ the set of preference restrictions for player i generated at round k by Procedure 1 and Procedure 2, respectively. First of all, we show that Procedure 1 will at every round have generated at least as many preference restrictions as Procedure 2 – that is, every preference restriction in $\hat{R}_i^k$ will also be in R_i^k. At round 1, Procedure 1 generates at least as many preference restrictions as Procedure 2, by definition, so every preference restriction in $\hat{R}_i^1$ is also in R_i^1. Now, take a preference restriction (c_i, A_i) generated by Procedure 2 at round 2. Then, every likelihood ordering L_i that respects all preference restrictions in $\hat{R}_j^1$ assumes some set of choices D_j on which c_i is weakly dominated by some randomized choice on A_i. As every preference restriction in $\hat{R}_j^1$ is also in R_j^1, it follows that every likelihood ordering L_i that respects all preference restrictions in R_j^1 also assumes some set of choices D_j on which c_i is weakly dominated by some randomized choice on A_i. But then, Procedure 1 will also generate preference restriction (c_i, A_i) at round 2. Hence, every preference restriction generated by Procedure 2 at round 2 will also be generated by Procedure 1 at round 2. That is, every preference restriction in $\hat{R}_i^2$ is also in R_i^2. By continuing in this way, we conclude that at every step k, every preference restriction in $\hat{R}_i^k$ is also in R_i^k. As such, every preference restriction generated by Procedure 2 will also be generated by Procedure 1.

On the other hand, it can also be shown that every preference restriction generated by Procedure 1 will also *eventually* be generated by Procedure 2. Suppose that Procedure 1 generates some preference restriction that is not generated at all by Procedure 2. Then, let k be the first step at which Procedure 1 generates a preference restriction, say (c_i, A_i), not generated by Procedure 2. By construction of Procedure 1, every likelihood ordering for player i that respects all preference restrictions generated by Procedure 1 *before* round k, must assume some set D_j on which c_i is weakly dominated by some randomized choice on A_i. By our assumption, all these preference restrictions generated by Procedure 1 *before* round k are also *eventually* generated by Procedure 2, let us say before round $m \geq k$. But then, every likelihood ordering for player i that respects all preference restrictions generated by Procedure 2 before round m, assumes a set D_j on which c_i is weakly dominated by some randomized choice on A_i. Hence, Procedure 2 must add the preference restriction (c_i, A_i) sooner or later, which is a contradiction since we assumed that Procedure 2 does not generate preference restriction (c_i, A_i) at all. We thus conclude that every preference restriction added by Procedure 1 is also finally added by Procedure 2. As such, Procedures 1 and 2 eventually generate exactly the same set of preference restrictions. We thus obtain the following general result:

Theorem 6.7.1 *(Order and speed of addition do not matter)*
The set of preference restrictions eventually generated by the procedure of iterated addition of preference restrictions, does not change if we change the order and speed by which we add preference restrictions.

So, even if at some stages of the procedure we overlook some preference restrictions, we will still end up with the same set of preference restrictions at the end, as long as we do not overlook some preference restrictions forever. In particular, the choices that are optimal after iterated addition of preference restrictions do not depend upon the order and speed of addition of such preference restrictions.

6.8 Proofs

In this section we will prove two results: Lemma 6.5.1 and Theorem 6.6.3. Lemma 6.5.1 states that if you are cautious, assume an opponent's set of choices D_j and your choice c_i is weakly dominated on D_j by some randomized choice on A_i, then you prefer some choice in A_i to c_i. Theorem 6.6.3 states that the algorithm of iterated addition of preference restrictions yields precisely those choices you can rationally make under common full belief in "caution and respect of preferences." We proceed by three parts.

In the first part we prove Lemma 6.5.1, but also show the following related result: Consider a likelihood ordering L_i on player j's choices and suppose that under *every* lexicographic belief with this particular likelihood ordering L_i, you prefer some choice in A_i to c_i. Then, L_i must assume some set of choices D_j on which c_i is weakly dominated by some randomized choice on A_i. So, in a sense, it is the converse of Lemma 6.5.1. These two results will be important for proving that the algorithm works.

In the second part we present a method of "blowing up" a lexicographic belief about the opponent's choices without changing the preferences you hold over your own choices. This method will be used in the proof of why the algorithm works.

In the final part we finally prove Theorem 6.6.3, which states that the algorithm works.

Assuming an opponent's set of choices

We first prove the following lemma.

Lemma 6.5.1 *(Preferences if you assume an opponent's set of choices)*
Suppose player i holds a cautious lexicographic belief and assumes a set of choices D_j for player j. Let A_i be some set of choices for player i. If his choice c_i is weakly dominated on D_j by some randomized choice r_i on A_i, then player i will prefer some choice in A_i to c_i.

Proof: Suppose player i holds lexicographic belief $b_i = (b_i^1; b_i^2; ...; b_i^K)$ about j's choices. Suppose also that player i assumes a set of choices D_j for player j on which his choice c_i is weakly dominated by a randomized choice r_i on A_i. So, the lexicographic belief b_i deems all choices in D_j infinitely more likely than all choices outside D_j. But then, we can find some level $k \in \{1, ..., K\}$ such that every choice in D_j has positive probability in some of the first k levels of b_i, whereas the first k levels assign probability zero to all choices outside D_j.

As r_i weakly dominates c_i on D_j, we have that $u_i^l(c_i,b_i) \leq u_i^l(r_i,b_i)$ for all $l \leq k$, and $u_i^l(c_i,b_i) < u_i^l(r_i,b_i)$ for at least some $l \leq k$. Here, $u_i^l(c_i,b_i)$ denotes the expected utility of choosing c_i under the level l belief b_i^l, and $u_i^l(r_i,b_i)$ denotes the expected utility of r_i under b_i^l. By definition,

$$u_i^l(r_i,b_i) = \sum_{a_i \in A_i} r_i(a_i) \cdot u_i^l(a_i,b_i).$$

Since $u_i^l(r_i,b_i) \geq u_i^l(c_i,b_i)$ for all $l \leq k$, we have for every $l \leq k$ that:

- either $u_i^l(a_i,b_i) = u_i^l(c_i,b_i)$ for all $a_i \in A_i$, or
- there is some $a_i \in A_i$ with $u_i^l(a_i,b_i) > u_i^l(c_i,b_i)$.

Moreover, as $u_i^l(r_i,b_i) > u_i^l(c_i,b_i)$ for some $l \leq k$, there must be an $l \leq k$ and some $a_i \in A_i$ with $u_i^l(a_i,b_i) > u_i^l(c_i,b_i)$. But then, there must be some $a_i \in A_i$ and some level $l \leq k$ such that

- $u_i^m(a_i,b_i) = u_i^m(c_i,b_i)$ for all $m < l$, and
- $u_i^l(a_i,b_i) > u_i^l(c_i,b_i)$.

Hence, player i will prefer this choice $a_i \in A_i$ to c_i. This completes the proof. $\diamond$

We now show the following related result, which can be viewed as the converse of Lemma 6.5.1.

Lemma 6.8.1 *Let L_i be a likelihood ordering on j's choices. Further, let c_i be a choice and A_i a set of choices for player i. Suppose that, under every cautious lexicographic belief with this particular likelihood ordering L_i, player i prefers some choice in A_i to c_i. Then, L_i assumes some set of choices D_j for player j on which c_i is weakly dominated by some randomized choice on A_i.*

Proof: Let $L_i = (L_i^1; L_i^2; ...; L_i^K)$ be a likelihood ordering on j's choices. Suppose that, under *every* cautious lexicographic belief with likelihood ordering L_i, player i prefers some choice in A_i to c_i. Assume, contrary to what we want to prove, that L_i does not assume any set of choices D_j on which c_i is weakly dominated by some randomized choice on A_i.

Note that L_i assumes the set of choices L_i^1. Hence, on L_i^1 choice c_i is not weakly dominated by any randomized choice on A_i. Consider now the reduced game in which player i can only choose from $\{c_i\} \cup A_i$ and player j can only choose from L_i^1. By Theorem 5.8.3 from Chapter 5 we then know that, within this reduced game, c_i is optimal for some cautious lexicographic belief b_i^1 on L_i^1 with one level only. Hence, under the cautious one-level belief b_i^1 on L_i^1, choice c_i is weakly preferred to all choices in A_i.

Note also that L_i assumes the set of choices $L_i^1 \cup L_i^2$. By the same reasoning as above, we can find some cautious one-level belief b_i^2 on $L_i^1 \cup L_i^2$ under which c_i is weakly preferred to all choices in A_i.

In fact, L_i also assumes the sets $L_i^1 \cup L_i^2 \cup L_i^3, ..., L_i^1 \cup L_i^2 \cup ... \cup L_i^K$. By applying the argument above to each of these sets, we see that

- there is a cautious one-level belief b_i^1 on L_i^1 under which c_i is weakly preferred to all choices in A_i,
- there is a cautious one-level belief b_i^2 on $L_i^1 \cup L_i^2$ under which c_i is weakly preferred to all choices in A_i,
- there is a cautious one-level belief b_i^3 on $L_i^1 \cup L_i^2 \cup L_i^3$ under which c_i is weakly preferred to all choices in A_i,

$$\vdots$$

- there is a cautious one-level belief b_i^K on $L_i^1 \cup L_i^2 \cup ... \cup L_i^K$ under which c_i is weakly preferred to all choices in A_i.

Now, we construct a cautious lexicographic belief b_i on C_j by putting the beliefs $b_i^1, b_i^2, b_i^3, ..., b_i^K$ together. That is, we define

$$b_i = (b_i^1; b_i^2; ...; b_i^K).$$

By construction, this lexicographic belief b_i has exactly the likelihood ordering L_i on j's choices.

Since c_i is weakly preferred to all choices in A_i under each of the levels $b_i^1, ..., b_i^K$, it follows that c_i is weakly preferred to all choices in A_i under the lexicographic belief b_i. So, we have constructed a cautious lexicographic belief b_i with likelihood ordering L_i under which c_i is weakly preferred to all choices in A_i. This, however, contradicts our initial assumption that player i prefers some choice in A_i to c_i under *every* cautious lexicographic belief with likelihood ordering L_i. Hence, we conclude that L_i must assume some set of choices D_j for player j on which c_i is weakly dominated by some randomized choice on A_i. This completes the proof. $\diamond$

Blowing up a lexicographic belief

Take a lexicographic belief $b_i = (b_i^1; b_i^2; ...; b_i^K)$ for player i about j's choice. Now, for some level $k \in \{1, ..., K\}$ include after belief b_i^k an additional belief $\hat{b}_i^{k+1}$ that is a copy of one of the beliefs of the previous levels $b_i^1, ..., b_i^k$. That is, $\hat{b}_i^{k+1} = b_i^l$ for some $l \in \{1, ..., k\}$. Then, the new lexicographic belief

$$\hat{b}_i = (b_i^1; ...; b_i^k; \hat{b}_i^{k+1}; b_i^{k+1}; ...; b_i^K)$$

is called a *blown-up version* of b_i. For example, take $b_i = (\frac{1}{2}a + \frac{1}{2}b; c; \frac{1}{3}d + \frac{2}{3}e)$. Then,

$$\hat{b}_i = (\tfrac{1}{2}a + \tfrac{1}{2}b; c; \tfrac{1}{2}a + \tfrac{1}{2}b; \tfrac{1}{3}d + \tfrac{2}{3}e)$$

is a blown-up version of b_i, by inserting after the secondary belief a copy of the primary belief. Also

$$\hat{b}_i = (\tfrac{1}{2}a + \tfrac{1}{2}b; c; c; \tfrac{1}{3}d + \tfrac{2}{3}e)$$

is a blown-up version of b_i, by inserting after the secondary belief a copy of the secondary belief. But

$$\hat{b}_i = (\tfrac{1}{2}a + \tfrac{1}{2}b; \tfrac{1}{3}d + \tfrac{2}{3}e; c; \tfrac{1}{3}d + \tfrac{2}{3}e)$$

is not a blown-up version of b_i, since we have inserted after the primary belief a copy of the tertiary belief, which is not allowed. If you insert a new belief after level k, then the new belief must be a copy from a *previous* level, so it cannot be a copy of a level after k. The following lemma states that blowing up a lexicographic belief in this way does not change the induced preference relation over choices. So, if player i holds a lexicographic belief about j's choice, then we can replace this lexicographic belief by a blown-up version without changing i's preferences over his own choices.

Lemma 6.8.2 *(Blowing up a lexicographic belief)*
Let b_i be a lexicographic belief about j's choices and let $\hat{b}_i$ be a blown-up version of b_i. Then, b_i and $\hat{b}_i$ induce the same preference relation on i's choices.

Proof: Let $b_i = (b_i^1; b_i^2; ...; b_i^K)$, and let $\hat{b}_i$ be some blown-up version of b_i. That is,

$$\hat{b}_i = (b_i^1; ...; b_i^k; \hat{b}_i^{k+1}; b_i^{k+1}; ...; b_i^K)$$

where $\hat{b}_i^{k+1}$ is a copy of one of the beliefs of the previous levels $b_i^1, ..., b_i^k$. So, after level k we have inserted a copy from one of the previous levels. Define

$$\hat{b}_i^l := \begin{cases} b_i^l, & \text{if } l \in \{1, ..., k\}, \\ b_i^{l-1}, & \text{if } l \in \{k+2, ..., K+1\}. \end{cases}$$

We show that b_i and $\hat{b}_i$ induce the same preference relation on i's choices.

Take some choices $c_i, c_i' \in C_i$. We show that c_i is preferred to c_i' under b_i if and only if c_i is preferred to c_i' under $\hat{b}_i$.

(a) Suppose that c_i is preferred to c_i' under b_i. Then, there is some $l \in \{1, ..., K\}$ such that $u_i^l(c_i, b_i) > u_i^l(c_i', b_i)$ and $u_i^m(c_i, b_i) = u_i^m(c_i', b_i)$ for all $m < l$.

If $l \leq k$, then $\hat{b}_i^l = b_i^l$ for all $m \leq l$. Hence, $u_i^l(c_i, \hat{b}_i) > u_i^l(c_i', \hat{b}_i)$ and $u_i^m(c_i, \hat{b}_i) = u_i^m(c_i', \hat{b}_i)$ for all $m < l$, which implies that c_i is preferred to c_i' under $\hat{b}_i$.

If $l \geq k+1$, then $\hat{b}_i^{l+1} = b_i^l$, so $u_i^{l+1}(c_i, \hat{b}_i) > u_i^{l+1}(c_i', \hat{b}_i)$. Moreover, under all levels $b_i^1, ..., b_i^{l-1}$, player i is indifferent between c_i and c_i'. Since the levels $\hat{b}_i^1, ..., \hat{b}_i^l$ are all equal to some of the levels $b_i^1, ..., b_i^{l-1}$, it follows that under all levels $\hat{b}_i^1, ..., \hat{b}_i^l$, player i is indifferent between c_i and c_i'. Hence, $u_i^m(c_i, \hat{b}_i) = u_i^m(c_i', \hat{b}_i)$ for all $m < l+1$. So, c_i is preferred to c_i' under $\hat{b}_i$.

We thus have shown that, if c_i is preferred to c_i' under b_i, then c_i is preferred to c_i' under $\hat{b}_i$.

(b) Suppose that c_i is preferred to c_i' under $\hat{b}_i$. Then, there is some $l \in \{1, ..., K+1\}$ such that $u_i^l(c_i, \hat{b}_i) > u_i^l(c_i', \hat{b}_i)$ and $u_i^m(c_i, \hat{b}_i) = u_i^m(c_i', \hat{b}_i)$ for all $m < l$.

If $l \leq k$, then $b_i^m = \hat{b}_i^m$ for all $m \leq l$. Hence, $u_i^l(c_i, b_i) > u_i^l(c_i', b_i)$ and $u_i^m(c_i, b_i) = u_i^m(c_i', b_i)$ for all $m < l$, which implies that c_i is preferred to c_i' under b_i.

If $l = k+1$, then $\hat{b}_i^l = \hat{b}_i^{k+1}$. However, $\hat{b}_i^{k+1}$ is a copy of some of the previous levels. Since $u_i^l(c_i, \hat{b}_i) > u_i^l(c_i', \hat{b}_i)$, this implies that $u_i^m(c_i, \hat{b}_i) > u_i^m(c_i', \hat{b}_i)$ for some $m < l$, which is a contradiction. So, l cannot be $k+1$.

If $l \geq k+2$, then $\hat{b}_i^l = b_i^{l-1}$. Hence, $u_i^{l-1}(c_i, b_i) > u_i^{l-1}(c_i', b_i)$. Moreover, under all levels $\hat{b}_i^1, ..., \hat{b}_i^{l-1}$, player i is indifferent between c_i and c_i'. Since the levels $b_i^1, ..., b_i^{l-2}$ are all equal to some of the levels $\hat{b}_i^1, ..., \hat{b}_i^{l-1}$, it follows that under all levels $b_i^1, ..., b_i^{l-2}$, player i is indifferent between c_i and c_i'. Hence, $u_i^m(c_i, b_i) = u_i^m(c_i', b_i)$ for all $m < l-1$. So, c_i is preferred to c_i' under b_i.

We thus have shown that, if c_i is preferred to c_i' under $\hat{b}_i$, then c_i is preferred to c_i' under b_i.

So, in general we have shown that c_i is preferred to c_i' under b_i, if and only if, c_i is preferred to c_i' under $\hat{b}_i$. Hence, b_i and $\hat{b}_i$ induce the same preference relation on i's choices. This completes the proof. $\diamond$

The algorithm works

We are now ready to prove the following theorem, which shows that the algorithm of iterated addition of preference restrictions yields exactly those choices that can rationally be made under common full belief in "caution and respect of preferences."

Theorem 6.6.3 *(The algorithm works)*
Consider a two-player static game with finitely many choices for both players.
(1) For every $k \geq 1$, the choices that can rationally be made by a cautious type that expresses up to k-fold full belief in "caution and respect of preferences" are exactly those choices that are optimal after $(k+1)$-fold addition of preference restrictions.
(2) The choices that can rationally be made by a cautious type that expresses common full belief in "caution and respect of preferences" are exactly those choices that are optimal after the full procedure of iterated addition of preference restrictions.

Proof: In order to prove this theorem, we prove the following two parts:

(a) For every type t_i that is cautious and expresses up to k-fold full belief in "caution and respect of preferences," every optimal choice c_i must be optimal after $(k+1)$-fold addition of preference restrictions.

(b) For every choice c_i that is optimal after $(k+1)$-fold addition of preference restrictions, there is a type t_i that is cautious and expresses up to k-fold full belief in "caution and respect of preferences," such that c_i is optimal for t_i. Moreover, for every choice c_i that is optimal after the *full* procedure of iterated addition of preference restrictions, there is a type t_i that is cautious and expresses *common* full belief in "caution and respect of preferences," such that c_i is optimal for t_i.

Proof of (a): We will show that for every type t_i that is cautious and expresses up to k-fold full belief in "caution and respect of preferences," every optimal choice c_i must be optimal after $(k+1)$-fold addition of preference restrictions.

For every k, let T_i^k be the set of types for player i that are cautious and express up to k-fold full belief in "caution and respect of preferences." Moreover, let R_i^k be the set of preference restrictions for player i generated by k-fold addition of preference restrictions. We prove the following lemma.

Lemma 6.8.3 *For every k and both players i, every type $t_i \in T_i^k$ respects all preference restrictions in R_j^k.*

Proof of Lemma 6.8.3: We prove the statement by induction on k.

Induction start: Take some $t_i \in T_i^1$. That is, t_i is cautious, fully believes that j is cautious and respects j's preferences. Consider a preference restriction (c_j, A_j) in R_j^1 produced in step 1 of the algorithm. Then, c_j is weakly dominated on C_i by a randomized choice in A_j. By Lemma 6.5.1, we know that j must prefer some choice in A_j to c_j, whenever j is cautious. Since t_i fully believes that j is cautious, type t_i fully believes that player j prefers some choice in A_j to c_j. As t_i respects j's preferences, he must deem some choice in A_j infinitely more likely than c_j. So, t_i respects the preference restriction (c_j, A_j). Hence, we have shown that t_i respects all preference restrictions for player j produced in step 1. That is, every $t_i \in T_i^1$ respects all preference restrictions in R_j^1.

Induction step: For some $k \geq 2$, suppose that for both players i every type $t_i \in T_i^{k-1}$ respects all preference restrictions in R_j^{k-1}. Now, take some type $t_i \in T_i^k$; that is, t_i is cautious and expresses up to k-fold full belief in "caution and respect of preferences." So, type t_i only deems possible opponent's types in T_j^{k-1} that express up to $(k-1)$-fold full belief in "caution and respect of preferences." We will show that t_i respects all preference restrictions in R_j^k.

Let (c_j, A_j) in R_j^k be a preference restriction for player j produced in step k. Then, by construction of the algorithm, every likelihood ordering L_j that respects all preference restrictions in R_i^{k-1}, assumes some set $D_i \subseteq C_i$ on which c_j is weakly dominated by some randomized choice in A_j. As t_i only deems possible opponent's types in T_j^{k-1} and – by our induction assumption – every such type in T_j^{k-1} respects all preference restrictions in R_i^{k-1}, we conclude that t_i only deems possible opponent's types that respect all preference restrictions in R_i^{k-1}. Hence, t_i only deems possible opponent's types t_j that assume some set $D_i \subseteq C_i$ on which c_j is weakly dominated by some randomized choice in A_j. By Lemma 6.5.1, every such type t_j prefers some choice in A_j to c_j. So, type t_i only deems possible opponent's types t_j that prefer some choice in A_j to c_j. As t_i respects j's preferences, we conclude that t_i must deem some choice in A_j infinitely more likely than c_j. That is, type t_i respects the preference restriction (c_j, A_j) in R_j^k. So, we see that every type $t_i \in T_i^k$ respects all preference restrictions in R_j^k. By induction, the proof of Lemma 6.8.3 is complete. ◇

Hence, we may conclude that for every k, every type $t_i \in T_i^k$ respects all preference restrictions in R_j^k. We finally show that every choice that is optimal for some type $t_i \in T_i^k$, must also be optimal after $(k+1)$-fold addition of preference restrictions. Or, equivalently, we will show that every choice that is not optimal after $(k+1)$-fold addition of preference restrictions, cannot be optimal for any type $t_i \in T_i^k$. Take some type $t_i \in T_i^k$ and some choice c_i that is not optimal after $(k+1)$-fold addition of preference restrictions. Then, c_i must be in some preference restriction (c_i,A_i) in R_i^{k+1}, generated at step $k+1$ of iterated addition of preference restrictions. So, by construction of the algorithm, every likelihood ordering L_i that respects all preference restrictions in R_j^k assumes some subset of choices $D_j \subseteq C_j$ for which c_i is weakly dominated by some randomization in A_i. As, by the lemma above, t_i respects all preference restrictions in R_j^k, we conclude that type t_i assumes some subset of choices $D_j \subseteq C_j$ for which c_i is weakly dominated by some randomization in A_i. Hence, by Lemma 6.5.1, type t_i prefers some choice in A_i to c_i, so c_i cannot be optimal for t_i. So, we see that every choice c_i that is not optimal after $(k+1)$-fold addition of preference restrictions, cannot be optimal for any type $t_i \in T_i^k$. This implies that every choice that is optimal for some type $t_i \in T_i^k$, must also be optimal after $(k+1)$-fold addition of preference restrictions. Since this holds for every k, this completes the proof of part (a).

Proof of (b): We will show that for every choice c_i that is optimal after $(k+1)$-fold addition of preference restrictions, there is a type t_i that is cautious and expresses up to k-fold full belief in "caution and respect of preferences," such that c_i is optimal for t_i. Moreover, we will show that for every choice c_i that is optimal after the *full* procedure of iterated addition of preference restrictions, there is a type t_i that is cautious and expresses *common* full belief in "caution and respect of preferences," such that c_i is optimal for t_i.

Remember from part (a) that R_i^k is the set of preference restrictions for player i generated by k-fold addition of preference restrictions.

For both players i, we define the set of types

$$T_i = \{t_i(c_i,A_i) : (c_i,A_i) \notin R_i^1\}.$$

That is, for every preference restriction (c_i,A_i) that is not generated in step 1 of the algorithm, we construct a type $t_i(c_i,A_i)$. Suppose that the algorithm of iterated addition of preference restrictions terminates after K steps – that is, $R_i^{K+1} = R_i^K$ for both players i. For every $k \in \{1,2,...,K\}$, let T_i^k be the set of types $t_i(c_i,A_i) \in T_i$ where $(c_i,A_i) \notin R_i^k$. Since $R_i^1 \subseteq R_i^2 \subseteq ... \subseteq R_i^K$, we have that $T_i^K \subseteq T_i^{K-1} \subseteq ... \subseteq T_i^1$, where $T_i^1 = T_i$. We construct the types in such a way that every type $t_i(c_i,A_i) \in T_i$ is cautious and weakly prefers choice c_i to all choices in A_i. Moreover, if $t_i \in T_i^k$ for some $k \in \{1,...,K-1\}$, then t_i expresses up to $(k-1)$-fold full belief in "caution and respect of preferences," whereas all types in T_i^K express common full belief in "caution and respect of preferences."

We proceed with the following four steps. In step 1 we will construct for every $k \geq 1$ and every $(c_i,A_i) \notin R_i^k$, some cautious lexicographic belief $b_i(c_i,A_i)$ on j's choices that respects all preference restrictions in R_j^{k-1}, and under which c_i is weakly preferred to all choices in A_i. In step 2 we will use these beliefs to define the beliefs for the types $t_i(c_i,A_i)$ in T_i. These types $t_i(c_i,A_i)$ will have the property that $t_i(c_i,A_i)$ weakly prefers c_i to all choices in A_i. In step 3 we will prove that, for every $k \geq 2$, every type $t_i \in T_i^k$ expresses up to $(k-1)$-fold full belief in "caution and respect of preferences." In step 4 we will prove that every type $t_i \in T_i^K$ expresses common full belief in "caution and respect of preferences."

These four steps will complete the proof of part (b). We now summarize the proof. Consider a choice c_i that is optimal after $(k+1)$-fold addition of preference restrictions. Then, c_i is not in any preference restriction (c_i,A_i) in R_i^{k+1}, and hence, in particular, $(c_i,C_i) \notin R_i^{k+1}$. As the associated type $t_i(c_i,C_i)$ is in T_i^{k+1}, it will be seen in step 3 that $t_i(c_i,C_i)$ expresses up to k-fold full belief in "caution and respect of preferences." Moreover, type $t_i(c_i,C_i)$ is cautious and weakly prefers c_i to all choices in C_i, and hence c_i is optimal for type $t_i(c_i,C_i)$. Summarizing, we see that choice c_i is optimal for a type $t_i(c_i,C_i)$ that is cautious and expresses up to k-fold full belief in "caution and respect of preferences."

Now, take a choice c_i that is optimal after the *full* procedure of iterated addition of preference restrictions. Then, c_i is not in any preference restriction (c_i,A_i) in R_i^K, and hence, in particular, $(c_i,C_i) \notin R_i^K$. As the associated type $t_i(c_i,C_i)$ is in T_i^K, it will follow by step 4 that $t_i(c_i,C_i)$ expresses common full belief in "caution and respect of preferences." Moreover, type $t_i(c_i,C_i)$ is cautious and weakly prefers c_i to all choices in C_i, and hence c_i is optimal for type $t_i(c_i,C_i)$. Summarizing, we see that choice c_i is optimal for a type $t_i(c_i,C_i)$ that is cautious and expresses common full belief in "caution and respect of preferences."

So, indeed, steps 1–4 will complete the proof of part (b).

Step 1: Construction of beliefs

In this step we will construct for every $k \geq 1$ and every $(c_i,A_i) \notin R_i^k$, some cautious lexicographic belief $b_i(c_i,A_i)$ on *j*'s choices that respects all preference restrictions in R_j^{k-1}, and under which c_i is weakly preferred to all choices in A_i.

Take some preference restriction $(c_i,A_i) \notin R_i^1$. We distinguish the following two cases:

(1) Suppose that $(c_i,A_i) \in R_i^K$. Then, there is some $k \in \{1,...,K-1\}$ such that $(c_i,A_i) \notin R_i^k$ but $(c_i,A_i) \in R_i^{k+1}$. By construction of the algorithm, R_i^k contains exactly those preference restrictions (c_i',A_i') such that every likelihood ordering L_i that respects all preference restrictions in R_j^{k-1}, assumes some set D_j on which c_i' is weakly dominated by some randomized choice on A_i'. Here, we use the convention that R_j^0 contains no preference restrictions at all. Since $(c_i,A_i) \notin R_i^k$, it follows that there is some likelihood ordering L_i respecting all preference restrictions in R_j^{k-1}, such that L_i does not assume any set D_j on which c_i is weakly dominated by some randomized choice on A_i. Hence, by Lemma 6.8.1, there is some cautious lexicographic belief $b_i(c_i,A_i)$ on *j*'s choices,

such that $b_i(c_i,A_i)$ has likelihood ordering L_i and c_i is weakly preferred to all choices in A_i under this belief. Recall that L_i respects all preference restrictions in R_j^{k-1}. Hence, there is some cautious lexicographic belief $b_i(c_i,A_i)$ on j's choices, such that $b_i(c_i,A_i)$ respects all preference restrictions in R_j^{k-1} and c_i is weakly preferred to all choices in A_i under this belief. So, for every $(c_i,A_i) \notin R_i^k$ with $(c_i,A_i) \in R_i^{k+1}$, we can construct a cautious lexicographic belief $b_i(c_i,A_i)$ on j's choices, such that $b_i(c_i,A_i)$ respects all preference restrictions in R_j^{k-1} and c_i is weakly preferred to all choices in A_i under this belief.
(2) Suppose that $(c_i,A_i) \notin R_i^K$. As the algorithm terminates after K steps, we know that $R_i^{K+1} = R_i^K$, and hence $(c_i,A_i) \notin R_i^{K+1}$. By a similar reasoning as in (1), we can then construct a cautious lexicographic belief $b_i(c_i,A_i)$ on j's choices, such that $b_i(c_i,A_i)$ respects all preference restrictions in R_j^K and c_i is weakly preferred to all choices in A_i under this belief.

Step 2: Construction of types
We will construct the beliefs for the types $t_i(c_i,A_i) \in T_i$ in two substeps. In step 2.1 we will assign to every type $t_i(c_i,A_i)$ a lexicographic belief $\rho_i(c_i,A_i)$ on the set $C_j \times T_j$ of the opponent's choice-type pairs such that $\rho_i(c_i,A_i)$ induces the same preference relation on i's choices as $b_i(c_i,A_i)$ does, but $\rho_i(c_i,A_i)$ is not cautious. We shall refer to the belief levels in $\rho_i(c_i,A_i)$ as the "main levels." In step 2.2 we will make a blown-up version $\sigma_i(c_i,A_i)$ of the lexicographic belief $\rho_i(c_i,A_i)$, in the sense of Lemma 6.8.2, such that $\sigma_i(c_i,A_i)$ is cautious. We will use the beliefs $\sigma_i(c_i,A_i)$ as the eventual beliefs for the types $t_i(c_i,A_i)$. Finally, in step 2.3 we will show that $\sigma_i(c_i,A_i)$ induces the same preference relation on i's choices as $\rho_i(c_i,A_i)$ does.

Step 2.1: Construction of the main levels
We define for every pair $(c_i,A_i) \notin R_i^1$ some lexicographic belief $\rho_i(c_i,A_i)$ on the set $C_j \times T_j$ of the opponent's choice-type pairs, such that $\rho_i(c_i,A_i)$ induces the same preference relation on i's choices as $b_i(c_i,A_i)$ does. We distinguish three cases.
(1) Consider a pair $(c_i,A_i) \notin R_i^1$ with $(c_i,A_i) \in R_i^2$. Consider the associated lexicographic belief $b_i(c_i,A_i) = (b_i^1;...;b_i^M)$ on j's choices from step 1. Take an arbitrary type $\hat{t}_j \in T_j$ for the opponent. The lexicographic belief $\rho_i(c_i,A_i) = (\rho_i^1;...;\rho_i^M)$ on the set $C_j \times T_j$ of the opponent's choice-type pairs is then defined as follows: For every level $m \in \{1,...,M\}$,

$$\rho_i^m(c_j,t_j) := \begin{cases} b_i^m(c_j), & \text{if } t_j = \hat{t}_j \\ 0, & \text{otherwise} \end{cases}$$

for every $(c_j,t_j) \in C_j \times T_j$. Hence, ρ_i^m induces the probability distribution b_i^m on j's choices for every level m. Consequently, $\rho_i(c_i,A_i)$ induces the same preference relation on i's choices as $b_i(c_i,A_i)$ does.
(2) Consider a pair $(c_i,A_i) \notin R_i^2$ with $(c_i,A_i) \in R_i^K$. Then, there is some $k \in \{2,...,K-1\}$ such that $(c_i,A_i) \notin R_i^k$, but $(c_i,A_i) \in R_i^{k+1}$. Consider the associated lexicographic belief $b_i(c_i,A_i) = (b_i^1;...;b_i^M)$ on j's choices from step 1. Recall from step 1 that $b_i(c_i,A_i)$ is cautious, respects all preference restrictions in R_j^{k-1} and under $b_i(c_i,A_i)$, choice c_i is

weakly preferred to all choices in A_i. For all of the opponent's choices c_j, let $A_j(c_j)$ be the set of those choices that are not deemed infinitely more likely than c_j by $b_i(c_i,A_i)$. Since $b_i(c_i,A_i)$ respects all preference restrictions in R_j^{k-1}, it must be the case that $(c_j,A_j(c_j)) \notin R_j^{k-1}$, and hence $t_j(c_j,A_j(c_j))$ is an opponent's type in T_j^{k-1}.

The lexicographic belief $\rho_i(c_i,A_i) = (\rho_i^1;...;\rho_i^M)$ on the set $C_j \times T_j$ of the opponent's choice-type pairs is then defined as follows: For every level $m \in \{1,...,M\}$,

$$\rho_i^m(c_j,t_j) := \begin{cases} b_i^m(c_j), & \text{if } t_j = t_j(c_j,A_j(c_j)) \\ 0, & \text{otherwise} \end{cases}$$

for every $(c_j,t_j) \in C_j \times T_j$. Hence, ρ_i^m induces the probability distribution b_i^m on j's choices for every level m. Consequently, $\rho_i(c_i,A_i)$ induces the same preference relation on i's choices as $b_i(c_i,A_i)$ does. Moreover, $\rho_i(c_i,A_i)$ only deems possible opponent's types $t_j(c_j,A_j(c_j))$ in T_j^{k-1}.

(3) Consider a pair $(c_i,A_i) \notin R_i^K$. Consider the associated lexicographic belief $b_i(c_i,A_i) = (b_i^1;...;b_i^M)$ on j's choices from step 1. Recall from step 1 that $b_i(c_i,A_i)$ is cautious, respects all preference restrictions in R_j^K and under $b_i(c_i,A_i)$ choice c_i is weakly preferred to all choices in A_i. For all of the opponent's choices c_j, let $A_j(c_j)$ be the set of those choices that are not deemed infinitely more likely than c_j by $b_i(c_i,A_i)$. Since $b_i(c_i,A_i)$ respects all preference restrictions in R_j^K, it must be the case that $(c_j,A_j(c_j)) \notin R_j^K$, and hence $t_j(c_j,A_j(c_j))$ is an opponent's type in T_j^K.

The lexicographic belief $\rho_i(c_i,A_i) = (\rho_i^1;...;\rho_i^M)$ on the set $C_j \times T_j$ of the opponent's choice-type pairs is then defined as follows: For every level $m \in \{1,...,M\}$,

$$\rho_i^m(c_j,t_j) := \begin{cases} b_i^m(c_j), & \text{if } t_j = t_j(c_j,A_j(c_j)) \\ 0, & \text{otherwise} \end{cases}$$

for every $(c_j,t_j) \in C_j \times T_j$. Hence, ρ_i^m induces the probability distribution b_i^m on j's choices for every level m. Consequently, $\rho_i(c_i,A_i)$ induces the same preference relation on i's choices as $b_i(c_i,A_i)$ does. Moreover, $\rho_i(c_i,A_i)$ only deems possible opponent's types $t_j(c_j,A_j(c_j))$ in T_j^K.

Step 2.2: Construction of blow-up levels
Note that the lexicographic beliefs $\rho_i(c_i,A_i)$ constructed in step 2.1 are not necessarily cautious. If $(c_i,A_i) \notin R_i^2$, then the belief $\rho_i(c_i,A_i)$ only considers possible opponent's types $t_j(c_j,A_j(c_j))$, but for each of these types $t_j(c_j,A_j(c_j))$ it only considers possible one choice, namely c_j. We will now extend the beliefs $\rho_i(c_i,A_i)$ constructed in step 2.1 to lexicographic beliefs $\sigma_i(c_i,A_i)$ on $C_j \times T_j$ which induce the same preference relation on i's choices as $\rho_i(c_i,A_i)$ does, but which *are* cautious.

If $(c_i,A_i) \notin R_i^1$, but $(c_i,A_i) \in R_i^2$, then the lexicographic belief $\rho_i(c_i,A_i)$ constructed in step 2.1 is cautious, as it only considers possible one type for the opponent, namely $\hat{t}_j$. So we can simply define $\sigma_i(c_i,A_i) = \rho_i(c_i,A_i)$.

Now, take a pair $(c_i,A_i) \notin R_i^2$, and consider the lexicographic belief $\rho_i(c_i,A_i)$ constructed in step 2.1. Then, $\rho_i(c_i,A_i)$ is not necessarily cautious as we have seen. For all of the opponent's types $t_j(c_j,A_j(c_j))$ it considers possible, there is only one choice it considers possible for this type, namely c_j. So, in order to extend $\rho_i(c_i,A_i)$ to a cautious lexicographic belief, we need to add extra belief levels that cover all pairs $(c_j', t_j(c_j,A_j(c_j)))$ with $c_j' \neq c_j$.

For every pair (c_j', c_j) with $c_j' \neq c_j$, we define a "blow-up" level $\tau_i(c_j', c_j)$ as follows: Let m be the first level such that ρ_i^m assigns positive probability to c_j'. Then, $\tau_i(c_j', c_j)$ is a copy of ρ_i^m, except for the fact that $\tau_i(c_j', c_j)$ shifts the probability that ρ_i^m assigned to the pair $(c_j', t_j(c_j', A_j(c_j')))$ completely towards the pair $(c_j', t_j(c_j, A_j(c_j)))$. In particular, $\tau_i(c_j', c_j)$ induces the same probability distribution on j's choices as ρ_i^m.

Now, fix a choice c_j for the opponent. Let l be the first level such that ρ_i^l assigns positive probability to c_j. Suppose that the lexicographic belief $b_j(c_j,A_j(c_j))$ induces the ordering $(c_j^1, ..., c_j^R)$ on j's choices, meaning that under $b_j(c_j,A_j(c_j))$ choice c_j^1 is weakly preferred to c_j^2, that c_j^2 is weakly preferred to c_j^3, and so on. Suppose further that $c_j = c_j^r$, and that all choices $c_j^1, ..., c_j^{r-1}$ are strictly preferred to c_j.

We insert blow-up levels $\tau_i(c_j^1, c_j), ..., \tau_i(c_j^{r-1}, c_j)$ between the main levels ρ_i^{l-1} and ρ_i^l in this particular order. So, $\tau_i(c_j^1, c_j)$ comes before $\tau_i(c_j^2, c_j)$, and so on. We then insert blow-up levels $\tau_i(c_j^{r+1}, c_j), ..., \tau_i(c_j^R, c_j)$ after the last main level ρ_i^M in this particular order.

If we do so for all of the opponent's choices c_j, we obtain a cautious lexicographic belief $\sigma_i(c_i,A_i)$ with main levels ρ_i^m and blow-up levels $\tau_i(c_j', c_j)$ in between. This completes the construction of the lexicographic beliefs $\sigma_i(c_i,A_i)$ for every $(c_i,A_i) \notin R_i^2$.

We use these beliefs $\sigma_i(c_i,A_i)$ as the eventual beliefs for the types $t_i(c_i,A_i)$. That is, every type $t_i(c_i,A_i) \in T_i$ holds the belief $\sigma_i(c_i,A_i)$ about the opponent's choice-type pairs. This completes the construction of the beliefs for the types.

Step 2.3: Every lexicographic belief $\sigma_i(c_i,A_i)$ induces the same preference relation on i's choices as $\rho_i(c_i,A_i)$

We now prove that every lexicographic belief $\sigma_i(c_i,A_i)$ constructed in step 2.2 induces the same preference relation on i's choices as the belief $\rho_i(c_i,A_i)$ constructed in step 2.1. If $(c_i,A_i) \in R_i^2$ this is obvious, as $\sigma_i(c_i,A_i) = \rho_i(c_i,A_i)$ in this case.

So, consider a pair $(c_i,A_i) \notin R_i^2$. By construction, the main levels in $\sigma_i(c_i,A_i)$ coincide exactly with the levels $\rho_i^1, ..., \rho_i^M$ in $\rho_i(c_i,A_i)$.

Consider a blow-up level $\tau_i(c_j', c_j)$ that is inserted between the main levels ρ_i^{l-1} and ρ_i^l. We show that $\tau_i(c_j', c_j)$ induces the same probability distribution on j's choices as some ρ_i^r with $r \leq l-1$.

By our construction above, c_j must have positive probability in ρ_i^l, and under $b_j(c_j,A_j(c_j))$ choice c_j' must be preferred to c_j. Since, by definition of $b_j(c_j,A_j(c_j))$, choice c_j is weakly preferred to every choice in $A_j(c_j)$ under $b_j(c_j,A_j(c_j))$, it must be that $c_j' \notin A_j(c_j)$. By definition, $A_j(c_j)$ contains all those choices that are not deemed infinitely more likely than c_j by $b_i(c_i,A_i)$, and hence c_j' must be deemed infinitely more likely than c_j by $b_i(c_i,A_i)$. By construction of $\rho_i(c_i,A_i)$, this implies that $\rho_i(c_i,A_i)$ deems c_j' infinitely more likely than c_j. Since c_j has positive probability in ρ_i^l, choice c_j' has positive probability for the first time in some ρ_i^r with $r \leq l-1$. But then, by construction of $\tau_i(c_j',c_j)$, the blow-up level $\tau_i(c_j',c_j)$ is a copy of ρ_i^r, except for the fact that $\tau_i(c_j',c_j)$ shifts the probability that ρ_i^r assigned to the pair $(c_j',t_j(c_j',A_j(c_j')))$ completely towards the pair $(c_j',t_j(c_j,A_j(c_j)))$. In particular, $\tau_i(c_j',c_j)$ induces the same probability distribution on j's choices as ρ_i^r. So, we have shown that every blow-up level $\tau_i(c_j',c_j)$ that is inserted between the main levels ρ_i^{l-1} and ρ_i^l induces the same probability distribution on j's choices as some main level ρ_i^r with $r \leq l-1$.

Consider next a blow-up level $\tau_i(c_j',c_j)$ that is inserted after the last main level M. Then, by construction in step 2.2, the blow-up level $\tau_i(c_j',c_j)$ induces the same probability distribution on j's choices as some main level ρ_i^r with $r \leq M$.

Now, let $\hat{\sigma}_i(c_i,A_i)$ be the lexicographic belief that $\sigma_i(c_i,A_i)$ induces on j's choices, and let $\hat{\rho}_i(c_i,A_i) = (\hat{\rho}_i^1;...;\hat{\rho}_i^M)$ be the lexicographic belief that $\rho_i(c_i,A_i)$ induces on j's choices. Let $\hat{\tau}_i(c_j',c_j)$ be the probability distribution that the blow-up level $\tau_i(c_j',c_j)$ induces on j's choices. By our insight above, we may conclude that every blow-up level $\hat{\tau}_i(c_j',c_j)$ that is inserted between main levels $\hat{\rho}_i^{l-1}$ and $\hat{\rho}_i^l$ in $\hat{\sigma}_i(c_i,A_i)$ is a copy of some $\hat{\rho}_i^r$ with $r \leq l-1$, and every blow-up level $\hat{\tau}_i(c_j',c_j)$ that is inserted after the last main level $\hat{\rho}_i^M$ in $\hat{\sigma}_i(c_i,A_i)$ is a copy of some $\hat{\rho}_i^r$ with $r \leq M$. This means, however, that $\hat{\sigma}_i(c_i,A_i)$ is obtained from $\hat{\rho}_i(c_i,A_i)$ by iteratively blowing up the lexicographic belief, in the sense of Lemma 6.8.2. Hence, by Lemma 6.8.2, the lexicographic belief $\hat{\sigma}_i(c_i,A_i)$ induces the same preference relation on i's choices as $\hat{\rho}_i(c_i,A_i)$. Consequently, $\sigma_i(c_i,A_i)$ induces the same preference relation on i's choices as $\rho_i(c_i,A_i)$, which was to be shown. As, by step 2.1, $\rho_i(c_i,A_i)$ induces the same preference relation on i's choices as $b_i(c_i,A_i)$, it follows that $\sigma_i(c_i,A_i)$ induces the same preference relation on i's choices as $b_i(c_i,A_i)$.

Step 3: Every type $t_i \in T_i^k$ expresses up to $(k-1)$-fold full belief in "caution and respect of preferences."
We prove this statement by induction on k.

Induction start: We start with $k = 2$. Take some type $t_i(c_i,A_i) \in T_i^2$, with lexicographic belief $\sigma_i(c_i,A_i)$ about the opponent's choice-type pairs. We show that $\sigma_i(c_i,A_i)$ expresses 1-fold full belief in "caution and respect of preferences"; that is, it respects j's preferences and fully believes that j is cautious.

First of all, $\sigma_i(c_i,A_i)$ only deems possible opponent's types $t_j(c_j,A_j) \in T_j$, which are all cautious. Hence, $\sigma_i(c_i,A_i)$ fully believes that j is cautious. It remains to show that $\sigma_i(c_i,A_i)$ respects j's preferences. Suppose that $\sigma_i(c_i,A_i)$ deems possible some type $t_j(c_j,A_j)$ and that $t_j(c_j,A_j)$ prefers choice c_j' to choice c_j''. We show that $\sigma_i(c_i,A_i)$ deems $(c_j',t_j(c_j,A_j))$ infinitely more likely than $(c_j'',t_j(c_j,A_j))$.

Since $t_j(c_j,A_j)$ is deemed possible by $\sigma_i(c_i,A_i)$, it must be the case that $t_j(c_j,A_j) = t_j(c_j,A_j(c_j))$. By construction, type $t_j(c_j,A_j(c_j))$ holds lexicographic belief $\sigma_j(c_j,A_j(c_j))$, which, as we saw in step 2.3, induces the same preference relation on j's choices as $b_j(c_j,A_j(c_j))$. Since, by assumption, $t_j(c_j,A_j(c_j))$ prefers c_j' to c_j'', it follows that c_j' is preferred to c_j'' under $b_j(c_j,A_j(c_j))$. So the construction of the blow-up levels in $\sigma_i(c_i,A_i)$ ensures that $\sigma_i(c_i,A_i)$ deems $(c_j',t_j(c_j,A_j))$ infinitely more likely than $(c_j'',t_j(c_j,A_j))$, which was to be shown. So, $\sigma_i(c_i,A_i)$ respects j's preferences.

Altogether, we can see that every type $t_i(c_i,A_i) \in T_i^2$ fully believes that j is cautious and respects j's preferences. That is, every type $t_i(c_i,A_i) \in T_i^2$ expresses 1-fold full belief in "caution and respect of preferences."

Induction step: Take some $k \geq 3$, and assume that for both players i, every type $t_i \in T_i^{k-1}$ expresses up to $(k-2)$-fold full belief in "caution and respect of preferences." Choose some type $t_i(c_i,A_i) \in T_i^k$. We show that $t_i(c_i,A_i)$ expresses up to $(k-1)$-fold full belief in "caution and respect of preferences." By construction in step 2, type $t_i(c_i,A_i)$ has belief $\sigma_i(c_i,A_i)$ about the opponent's choice-type pairs. Moreover, we have seen in steps 2.1 and 2.2 that type $t_i(c_i,A_i) \in T_i^k$ only deems possible opponent's types $t_j \in T_j^{k-1}$. By our induction assumption, every such type $t_j \in T_j^{k-1}$ expresses up to $(k-2)$-fold full belief in "caution and respect of preferences." This implies, however, that type $t_i(c_i,A_i)$ expresses up to $(k-1)$-fold full belief in "caution and respect of preferences." This completes the induction step.

By induction on k, we conclude that for every $k \geq 2$, for both players i, every type $t_i \in T_i^k$ expresses up to $(k-1)$-fold full belief in "caution and respect of preferences." This completes step 3.

Step 4: Every type $t_i \in T_i^K$ expresses common full belief in "caution and respect of preferences"
In order to prove this, we show the following lemma.

Lemma 6.8.4 *For all $k \geq K-1$, every type $t_i \in T_i^K$ expresses up to k-fold full belief in "caution and respect of preferences."*

Proof of Lemma 6.8.4: We prove the statement by induction on k.

Induction start: We start with $k = K-1$. Take some type $t_i \in T_i^K$. Then, by step 3 we know that t_i expresses up to $(K-1)$-fold full belief in "caution and respect of preferences."

Induction step: Take some $k \geq K$, and assume that for both players i, every type $t_i \in T_i^K$ expresses up to $(k-1)$-fold full belief in "caution and respect of preferences." Consider a type $t_i(c_i,A_i) \in T_i^K$, with lexicographic belief $\sigma_i(c_i,A_i)$. By the construction in steps 2.1 and 2.2, we know that $t_i(c_i,A_i)$ only deems possible opponent's types $t_j \in T_j^K$. By our induction assumption, every such type $t_j \in T_j^K$ expresses up to $(k-1)$-fold full belief in "caution and respect of preferences." But then, it follows that type $t_i(c_i,A_i)$ expresses up to k-fold full belief in "caution and respect of preferences." By induction on k, the proof of this lemma is complete. ◇

By step 3 and the lemma above, we conclude that every type $t_i \in T_i^K$ expresses up to k-fold full belief in "caution and respect of preferences" for every k, and hence expresses common full belief in "caution and respect of preferences." This completes the proof of step 4.

Using steps 1–4 we can now easily prove part (b), stated at the beginning of the proof of the theorem. Consider a choice c_i that is optimal after $(k+1)$-fold addition of preference restrictions. Then, c_i is not in any preference restriction (c_i, A_i) in R_i^{k+1}, and hence, in particular, $(c_i, C_i) \notin R_i^{k+1}$. As the associated type $t_i(c_i, C_i)$ is in T_i^{k+1}, it follows by step 3 that $t_i(c_i, C_i)$ expresses up to k-fold full belief in "caution and respect of preferences." Moreover, type $t_i(c_i, C_i)$ is cautious and weakly prefers c_i to all choices in C_i, and hence c_i is optimal for type $t_i(c_i, C_i)$. Summarizing, we see that choice c_i is optimal for a type $t_i(c_i, C_i)$ that is cautious and expresses up to k-fold full belief in "caution and respect of preferences."

Now, take a choice c_i that is optimal after the full procedure of iterated addition of preference restrictions. Then, c_i is not in any preference restriction (c_i, A_i) in R_i^K, and hence, in particular, $(c_i, C_i) \notin R_i^K$. As the associated type $t_i(c_i, C_i)$ is in T_i^K, it follows by step 4 that $t_i(c_i, C_i)$ expresses common full belief in "caution and respect of preferences." Moreover, type $t_i(c_i, C_i)$ is cautious and weakly prefers c_i to all choices in C_i, and hence c_i is optimal for type $t_i(c_i, C_i)$. Summarizing, we see that choice c_i is optimal for a type $t_i(c_i, C_i)$ that is cautious and expresses common full belief in "caution and respect of preferences." So, the proof of part (b) is hereby complete.

Using parts (a) and (b) we can finally prove Theorem 6.6.3. We begin with part (1) of the theorem. Take some $k \geq 1$. Then by combining parts (a) and (b) of this proof, we conclude that a choice c_i is optimal for a type t_i that is cautious and expresses up to k-fold full belief in "caution and respect of preferences," if and only if, the choice c_i is optimal after $(k+1)$-fold addition of preference restrictions. This completes the proof of part (1) of the theorem.

Let us now turn to part (2) of the theorem. Suppose that a choice c_i is optimal for a type t_i that is cautious and expresses common full belief in "caution and respect of preferences." Since t_i expresses up to k-fold full belief in "caution and respect of preferences" for every k, it follows from part (a) of the proof that c_i must be optimal after $(k+1)$-fold addition of preference restrictions, for all k. That is, c_i must be optimal after the *full* procedure of iterated addition of preference restrictions.

Next, suppose that choice c_i is optimal after the full procedure of iterated addition of preference restrictions. Then, by part (b), we know that c_i is optimal for a type t_i that is cautious and expresses common full belief in "caution and respect of preferences." So, altogether we see that a choice c_i is optimal for a type t_i that is cautious and expresses common full belief in "caution and respect of preferences," if and only if, c_i is optimal after the full procedure of iterated addition of preference restrictions. This proves part (2) of the theorem. The proof of Theorem 6.6.3 is hereby complete. ■

Practical problems

6.1 Painting a room

Recall the story from Problem 5.1 in Chapter 5.

(a) Look at the epistemic model you constructed in Problem 5.1, part (e). Which types in this model respect the opponent's preferences? Which types do not?

(b) Use the algorithm of iterated addition of preference restrictions to find those choices you can rationally make under common full belief in "caution and respect of preferences."

(c) Construct an epistemic model with lexicographic beliefs such that:

- every type in this model is cautious and expresses common full belief in "caution and respect of preferences," and
- for every choice found in (b) there is a type in this model for which this choice is optimal.

6.2 The closer the better

Recall the story from Problem 5.3 in Chapter 5.

(a) In part (f) of Problem 5.3 you constructed an epistemic model in which every type is cautious and expresses common full belief in "caution and primary belief in rationality." Which of these types respect the opponent's preferences? Which types do not?

(b) In part (e) of Problem 5.3 you computed the numbers you can rationally choose under common full belief in "caution and primary belief in rationality." Show that you can also rationally choose each of these numbers under common full belief in "caution and respect of preferences." In order to show this, construct an epistemic model such that:

- every type in this model is cautious and expresses common full belief in "caution and respect of preferences," and
- for every number found in part (e) of Problem 5.3 there is a type in this model for which this choice is optimal.

6.3 Planting a tree

You and Barbara won a beautiful tree in a lottery yesterday. Today, you have to decide where to plant it. Figure 6.3 shows the street where you both live and the location of your houses. The distance between two neighboring locations in $\{a, b, ..., g\}$ is two hundred meters. The procedure is now as follows: You both write down, on a piece of paper, one of the possible locations from $\{a, b, ..., g\}$. If you both write down the same location, then the tree will be planted exactly at that location. If you have chosen neighboring locations, then the tree will be planted exactly halfway between those locations. In all other cases the tree will not be planted, as the locations you desire are too far apart. If the tree is not planted, the utility for both of you would be zero.

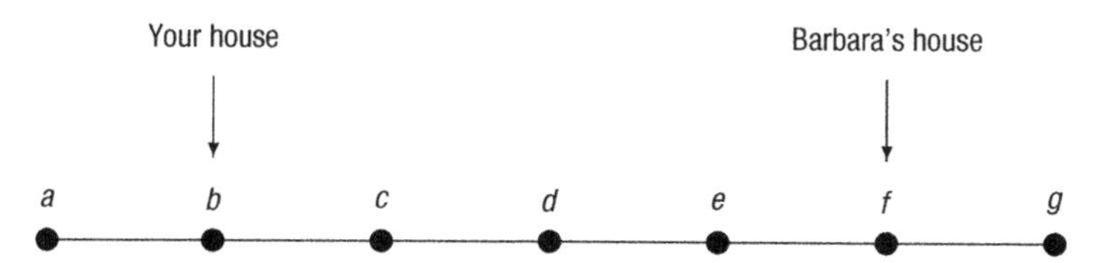

Figure 6.3 Street in "Planting a tree"

Having the tree planted would increase your utility by 6, but you only want it planted if it is close enough to your house. More precisely, every one hundred meters between the tree and your house would decrease your utility by 1. So, for instance, if the tree were planted at location f, then your utility would be $6 - 8 = -2$, which means that in this case you would prefer that the tree was not planted at all! The same holds for Barbara.

(a) Formulate this situation as a game between Barbara and you.

(b) Use the algorithm of iterated addition of preference restrictions to find those locations you can rationally choose under common full belief in "caution and respect of preferences."

(c) Construct an epistemic model with lexicographic beliefs such that:

- every type in this model is cautious and expresses common full belief in "caution and respect of preferences," and
- for every location found in (b) there is a type in this model for which this location is optimal.

6.4 A historical trip

You and Barbara are planning a trip to a country with a great history. After days of deliberation, you have reduced the possible choices to Spain, Greece and Egypt. But which country will it be? In order to decide you both write down, on a piece of paper, one or more of these countries. You can write down one, two or even all three countries if you wish. If there is only one country named by both of you, then you will go to that particular country. If there are two countries that both of you have written down, then you will toss a coin to decide which of these two countries to go to. If you both name all three countries, then you will throw a dice in order to decide which of these three countries to go to (each country will then have an equal chance). If there is no country named by both of you, then you will stay at home, being angry with each other.

The preferences for you and Barbara are shown in Table 6.12. So, a trip to Spain would give you a utility of 6 and Barbara a utility of 2. Similarly for the other countries.

(a) Formulate this situation as a game between Barbara and you.

(b) Find those choices you can rationally make under common full belief in "caution and primary belief in rationality."

Table 6.12. *Utilities for you and Barbara in "A historical trip"*

	Spain	Greece	Egypt	Home
You	6	4	2	0
Barbara	2	4	6	0

(c) Use the algorithm of iterated addition of preference restrictions to find those choices you can rationally make under common full belief in "caution and respect of preferences."

(d) Construct an epistemic model with lexicographic beliefs such that:

- every type in this model is cautious and expresses common full belief in "caution and respect of preferences," and
- for every choice found in (c) there is a type in this model for which this choice is optimal.

6.5 Lasergame

You and Barbara will play "Lasergame" in an old, mysterious castle. Figure 6.4 shows the 30 chambers in the castle. The rules of the game are as follows: You both wear a laser gun – a toy, not a real one – and the objective is to hit the other person. You may start at any one of the four chambers that begin with "1," and must walk in the direction indicated until you reach the end of the castle. Similarly, Barbara can start at any one of the four chambers that begin with "2," and must walk in the indicated direction as well. Suppose that every chamber has four open doors, one in the middle of each wall. So, from each chamber you can see into every other chamber in the same horizontal row and every other chamber in the same vertical row. Suppose that Barbara and you walk at the same speed and that you both only look ahead. If you start walking and see Barbara, then you will shoot and hit her, and your utility will be 1. If you do not see Barbara, then you will not be able to hit her, and your utility will be 0. The same holds for Barbara. In particular, if you both enter the same chamber at the same time, then you will hit each other, and the utilities for you and Barbara would be 1. The question is: From which of the four chambers beginning with a "1" would you start?

(a) Formulate this situation as a game between Barbara and you.

(b) Find those chambers where you can rationally start under common full belief in "caution and primary belief in rationality."

(c) Use the algorithm of iterated addition of preference restrictions to find those chambers where you can rationally start under common full belief in "caution and respect of preferences."

(d) Construct an epistemic model with lexicographic beliefs such that:

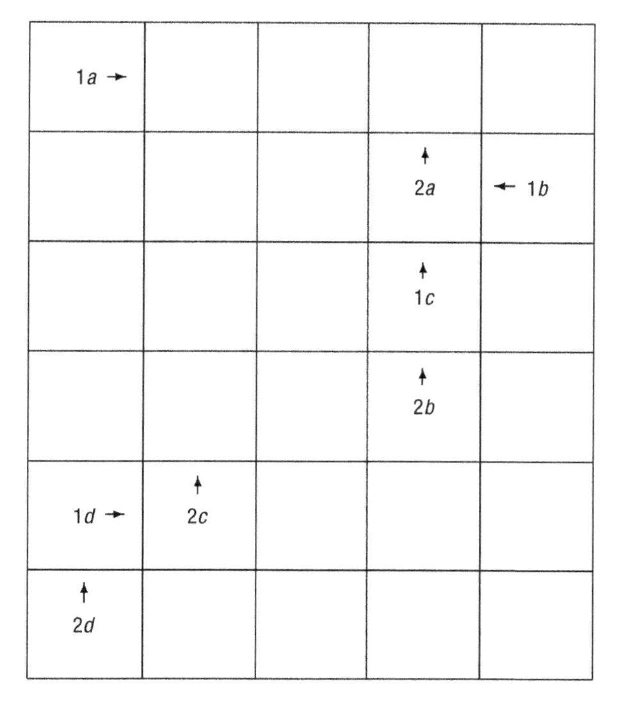

Figure 6.4 Castle in "Lasergame"

- every type in this model is cautious and expresses common full belief in "caution and respect of preferences," and
- for every chamber found in (c) there is a type in this model for which this chamber is optimal.

6.6 A first-price auction

Today, Chris will sell his impressive collection of toy cars. You and Barbara are both interested, and have agreed on the following auction with Chris: You both whisper a price between 0 and 60 euros in Chris' ear. In order to keep things simple, suppose that the price must be a multiple of 10 euros, so the set of possible prices is $\{0, 10, 20, 30, 40, 50, 60\}$. The person who names the highest price will receive the collection of toy cars and must pay the price named. We call this a *first-price auction.* If you both whisper the same price, then Chris will toss a coin to decide who gets the collection. Suppose you value the collection at 50 euros whereas Barbara values it at 40 euros. So, if you obtain the collection and pay price P then your utility will be $50 - P$. If you do not get the collection, your utility will be zero. Similarly, if Barbara receives the collection at price P, then her utility would be $40 - P$. If she does not get the collection, her utility will be zero.

(a) Formulate this situation as a game between Barbara and you.
(b) Find those prices you can rationally name under common full belief in "caution and primary belief in rationality."
(c) Use the algorithm of iterated addition of preference restrictions to find those prices you can rationally name under common full belief in "caution and respect of preferences."
(d) Construct an epistemic model with lexicographic beliefs such that:

- every type in this model is cautious and expresses common full belief in "caution and respect of preferences," and
- for every price found in (c) there is a type in this model for which this price is optimal.

6.7 Going to a party

This evening you and Barbara will be going to a party, and as usual you both have to decide which color to wear. Suppose that both of you have a choice between red, green, yellow, white and black. You have no strict preference for any color and only care about the combination of your color with Barbara's. You think that red and green only match with yellow, white and black, that yellow and white only match with red, green and black, and that black matches with every color except black itself. A good combination of colors gives you a utility of 1, whereas a bad combination yields a utility of 0. Barbara, on the other hand, only cares about her own color, as long as you do not wear the same color. More precisely, wearing red or green gives her a utility of 7, wearing yellow or white a utility of 3, and black a utility of 1, as long as you wear a different color. If you happen to wear the same color, she will not like it and her utility will be 0.
(a) Model this situation as a game between you and Barbara.
(b) Find those colors you can rationally choose under common full belief in "caution and primary belief in rationality."
(c) Use the algorithm of iterated addition of preference restrictions to find those colors you can rationally choose under common full belief in "caution and respect of preferences."
(d) Construct an epistemic model with lexicographic beliefs such that:

- every type in this model is cautious and expresses common full belief in "caution and respect of preferences," and
- for every color found in (c) there is a type in this model for which this color is optimal.

Theoretical problems

6.8 Self-proper sets of preference restrictions

Consider a game with two players and finitely many choices for both players. For both players i, let R_i be a set of preference restrictions. The pair (R_1, R_2) of sets of preference restrictions is called *self-proper* if for both players i, the following condition holds: A pair (c_i, A_i) is a preference restriction in R_i, if and only if, every likelihood ordering L_i

respecting all preference restrictions in R_j assumes a set of choices D_j for player j on which c_i is weakly dominated by some randomized choice in A_i.

Show that a choice c_i can rationally be chosen under common full belief in "caution and respect of preferences," *if and only if*, there is some self-proper pair (R_1, R_2) of sets of preference restrictions, such that c_i is not in any preference restriction (c_i, A_i) in R_i.

6.9 Proper equilibrium

Consider a game with two players and finitely many choices for both players. Let b_i be a lexicographic belief about j's choice, for both players i. The pair (b_1, b_2) of lexicographic beliefs is called a *proper equilibrium* if, for both players i,

- the lexicographic belief b_i is cautious, and
- b_i deems an opponent's choice c_j infinitely more likely than another choice c_j' whenever player j prefers c_j to c_j' under the lexicographic belief b_j.

A choice c_i is called a *proper equilibrium choice* if there is a proper equilibrium (b_1, b_2) such that c_i is optimal under the lexicographic belief b_i.

(a) Show that every proper equilibrium choice is a perfect equilibrium choice (see Problem 5.9), but not vice versa.

Consider an epistemic model with lexicographic beliefs. A type t_i *fully believes that j is of type t_j* if the lexicographic belief $b_i(t_i)$ assigns, at *each* of its levels, probability 1 to type t_j. Type t_i *fully believes that j has correct beliefs*, if the lexicographic belief $b_i(t_i)$ assigns, at *each* of its levels, only positive probability to j's types that fully believe that i is of type t_i. So, this definition is analogous to the definition of belief in the opponent's correct beliefs in Chapter 4.

Consider an epistemic model with lexicographic beliefs, and a type t_i that:

- is cautious,
- expresses common full belief in "caution and respect of preferences,"
- fully believes that j has correct beliefs,
- fully believes that j fully believes that i has correct beliefs.

(By the latter, we mean that $b_i(t_i)$, at *each* of its levels, only assigns positive probability to j's types that fully believe that i has correct beliefs.)

(b) Show that there is a single type t_j such that t_i fully believes that j is of type t_j.

(c) Show that t_i's lexicographic belief about j's choice, together with t_j's lexicographic belief about i's choice, is a proper equilibrium.

(d) Show that c_i is a proper equilibrium choice, *if and only if*, it can be chosen rationally by a type t_i that:

- is cautious,
- expresses common full belief in "caution and respect of preferences,"

- fully believes that j has correct beliefs,
- fully believes that j fully believes that i has correct beliefs.

6.10 When are all choices possible under common full belief in "caution and respect of preferences"?

Consider a game with two players and finitely many choices for both players. Show that all choices in the game can rationally be made under common full belief in "caution and respect of preferences," *if and only if*, no choice in the game is weakly dominated.

Literature

Respecting the opponent's preferences

The main idea in this chapter is that of *respecting the opponent's preferences*, which states that if you believe that choice a is better for your opponent than choice b, then you deem his choice a infinitely more likely than his choice b – even if you think that both a and b are not optimal. This idea was introduced by Myerson (1978), who used it as the key ingredient for his concept of *proper equilibrium* – a concept that we will discuss in more detail below. Myerson, however, formalized the idea without using lexicographic beliefs. More precisely, Myerson considered for both players i a probability distribution σ_i on i's choices, assigning positive probability to each of i's choices. For a given $\epsilon > 0$, this pair (σ_1, σ_2) is called an *ϵ-proper equilibrium* if, whenever choice c_i is better than c_i' under σ_j, then $\sigma_i(c_i') \leq \epsilon \cdot \sigma_i(c_i)$. If we choose $\epsilon > 0$ very small, and interpret σ_j as the belief that player i holds about j's choice, then the condition above states the following: If choice c_i is better than c_i' for player i under his belief σ_j about j's choice, then player j must deem c_i' much less likely than c_i. This, however, is exactly the idea behind respecting the opponent's preferences.

Later, Schuhmacher (1999) and Asheim (2001) incorporated the idea of respecting the opponent's preferences into the concept of *proper rationalizability*, on which we will comment more below. Schuhmacher (1999) presented an epistemic model with *standard* beliefs. He defined, for a given $\epsilon > 0$, the *ϵ-proper trembling condition*, which states that, whenever choice c_i is better than choice c_i' for type t_i, then the probability with which type t_i is expected to choose c_i' must be at most ϵ times the probability with which t_i is expected to choose c_i. So, if ϵ is very small, the ϵ-proper trembling condition says that for a given type t_i, better choices must be deemed much more likely than inferior choices. This is precisely the same as respecting the opponent's preferences. Asheim (2001), on the other hand, used an epistemic model with *lexicographic* beliefs and defined the condition of respecting the opponent's preferences in exactly the same way as we do.

Proper rationalizability

The central concept in this chapter is common full belief in "caution and respect of preferences," which states that a player is cautious and respects the opponent's preferences, fully believes that the opponent is cautious and respects his opponent's preferences, and so on. In the literature this concept is known as *proper rationalizability*. It was first proposed by Schuhmacher (1999) – within an epistemic model with *standard* beliefs – and was later characterized by Asheim (2001) within an epistemic model with *lexicographic* beliefs.

Schuhmacher defined, for every $\epsilon > 0$, the ϵ-proper trembling condition mentioned above, and then imposed common belief in the ϵ-proper trembling condition. This concept is often called ϵ-proper rationalizability. By letting ϵ tend to zero we obtain proper rationalizability, which is equivalent to common full belief in "caution and respect of preferences" in two-player games.

Asheim calls a type *properly consistent* if it is cautious and respects the opponent's preferences. He defined proper rationalizability by imposing common full belief in proper consistency – hence we essentially copied Asheim's approach in this chapter. Asheim (2001) showed, in his Proposition 3, that his definition of proper rationalizability is equivalent to Schuhmacher's definition for games with two players. A difference is that Asheim's definition is restricted to two-player games, whereas Schuhmacher also considers games with more than two players.

Proper equilibrium

As we have mentioned above, the idea of respecting the opponent's preferences goes back to Myerson (1978) who used it as the basis for his concept of *proper equilibrium.* Myerson considered for both players i a probability distribution σ_i on i's choices, which can be interpreted as player j's standard belief about i's choices. Cautious reasoning is formalized by requiring that σ_i assigns positive probability to each of i's choices. For a given $\epsilon > 0$, the pair (σ_1, σ_2) is called an ϵ-proper equilibrium if, whenever choice c_i is better than c_i' under σ_j, then $\sigma_i(c_i') \leq \epsilon \cdot \sigma_i(c_i)$. For small ϵ, this condition essentially says that both players must respect the opponent's preferences. Finally, a pair (σ_1, σ_2) is called a *proper equilibrium* if there is a sequence $(\sigma_1^k, \sigma_2^k)_{k \in \mathbb{N}}$ converging to (σ_1, σ_2), and a sequence $(\epsilon^k)_{k \in \mathbb{N}}$ of positive numbers converging to 0, such that for every k: (1) the pair (σ_1^k, σ_2^k) assigns positive probability to each of the choices, and (2) the pair (σ_1^k, σ_2^k) is an ϵ^k-proper equilibrium. Hence, a proper equilibrium is essentially obtained by taking the notion of ϵ-proper equilibrium, and letting ϵ tend to zero. Blume, Brandenburger and Dekel (1991b) provided a characterization of proper equilibrium in terms of *lexicographic beliefs* – at least for two-player games. This is the definition we used in theoretical problem 6.9.

Myerson (1978) proved that every proper equilibrium is a perfect equilibrium (Selten, 1975) – see the literature section of Chapter 5 – but not vice versa. So, proper equilibrium is more restrictive than perfect equilibrium.

Both Schuhmacher (1999) and Asheim (2001) showed that proper equilibrium leads to proper rationalizability, but not vice versa. More precisely, Schuhmacher showed in his Proposition 1 that for every ϵ-proper equilibrium (σ_1, σ_2) we can construct a pair of types (t_1, t_2) that hold these beliefs (σ_1, σ_2), and for which there is common belief in the ϵ-proper trembling condition. Asheim (2001) showed in his Proposition 2 that every choice that has positive probability in some proper equilibrium (σ_1, σ_2) is optimal for some properly rationalizable type. The converse of these statements is not true – there are choices that are allowed by proper rationalizability, but which do not have positive probability in any proper equilibrium. Hence, proper equilibrium is more restrictive than proper rationalizability.

In fact, in theoretical problem 6.9 the reader is asked to show that proper equilibrium is obtained if one takes proper rationalizability, and adds the following two requirements: (1) player i fully believes that j has correct beliefs, and (2) player i fully believes that j fully believes that i has correct beliefs. This is essentially Proposition 1 in Asheim (2001). Hence, proper equilibrium relates to proper rationalizability in exactly the same way as Nash equilibrium relates to common belief in rationality, and also in the same way as a perfect equilibrium relates to common full belief in "caution and primary belief in rationality" – see Chapter 4 and the literature section of Chapter 5.

Algorithm

The algorithm of *iterated addition of preference restrictions* was developed by Perea (2011a). In that paper, it was shown that the algorithm yields exactly those choices that can rationally be made under common full belief in "caution and respect of preferences" – see Theorem 6.6.3 in this chapter. The proof for this theorem is based on the proof in Perea (2011a). Moreover, it is shown in Perea (2011a) that the order and speed by which we add preference restrictions do not matter for the final output of the algorithm. This is Theorem 6.7.1 in this chapter.

Schuhmacher (1999) proposed a different procedure, called *iteratively proper trembling,* which for every given $\epsilon > 0$ yields the set of ϵ-properly rationalizable choices, but not necessarily the exact set of properly rationalizable choices. Schulte (2003) introduced another algorithm, called *iterated backward inference*, which for every game yields a set of choices that always includes all properly rationalizable choices. However, the output set may contain choices that are *not* properly rationalizable. So, Schulte's algorithm does not always provide exactly the set of properly rationalizable choices.

The algorithm of iterated addition of preference restrictions is different from the algorithms we saw in Chapters 3 and 5, as it does not proceed by eliminating choices from the game. Instead, it makes use of *likelihood orderings* and *preference restrictions.* Asheim and Perea (2009) show that the Dekel–Fudenberg procedure from Chapter 5 can be reformulated in terms of likelihood orderings and preference restrictions, which makes it easier to compare this procedure to iterated addition of preference restrictions. By doing so, it can easily be seen that the algorithm of iterated addition of preference restrictions is more restrictive than the Dekel–Fudenberg procedure. This must be the case, since we saw in Theorem 6.2.3 of this chapter that every choice that can rationally be made under common full belief in "caution and respect of preferences" can also rationally be made under common full belief in "caution and primary belief in rationality." Remember from Chapter 5 that the Dekel–Fudenberg procedure yields exactly those choices that can rationally be made under common full belief in "caution and primary belief in rationality."

Dividing a pizza

The example "Dividing a pizza," which we used in this chapter, is based on the bargaining procedure in Ellingsen and Miettinen (2008).

7 Assuming the opponent's rationality

7.1 Assuming the opponent's rationality

In the previous chapter we discussed the idea of *respecting the opponent's preferences*, which states that you deem an opponent's choice a infinitely more likely than some other choice b whenever you believe that the opponent prefers a to b. That is, you deem a better choice infinitely more likely than a worse choice, even if you believe that both choices are not optimal for the opponent. Various examples illustrated that this idea may put severe restrictions on your lexicographic belief – not only on your primary belief but also on your secondary and higher-level beliefs. And these restrictions – we saw – may be crucial for the eventual choice you make in a game.

The concept of *respecting the opponent's preferences* is just one reasonable condition we may impose on a lexicographic belief, but it is certainly not the only one. To illustrate this, let us return to the example "Spy game" we introduced in the previous chapter.

Example 7.1 Spy game

Recall the story from Example 6.5. For convenience, we have reproduced the utilities for you and Barbara in Table 7.1. As we have seen, common full belief in "caution and respect of preferences" leads you to choose pub c. In a nutshell, the reasoning was as follows: It is clear that Barbara prefers pub a to pub b. Hence, you must deem the event that Barbara will visit pub a infinitely more likely than the event that Barbara will visit pub b if you respect her preferences. This, in turn, implies that you will prefer pub b to pub a. So, if Barbara respects your preferences, she must deem your choice pub b infinitely more likely than your choice pub a. But then, Barbara will also prefer pub b to pub c. So, if you respect Barbara's preferences, you must deem the event that Barbara will visit pub b infinitely more likely than the event that Barbara will visit pub c. As such, your optimal choice is to visit pub c, since this is the pub you deem least likely for Barbara.

Note that for Barbara, pub b is *never* an optimal choice, whereas pub c *can* be an optimal choice for her if she deems your choice pub a sufficiently likely. So, one could argue that her choice pub c may be considered better than her choice pub b. Still, respecting Barbara's preferences does not require you to deem her choice pub c infinitely more likely than pub b. On the contrary: common full belief in "caution

Table 7.1. *Utilities for you and Barbara in "Spy game" (II)*

		Barbara		
		Pub a	Pub b	Pub c
You	Pub a	0,3	1,2	1,4
	Pub b	1,3	0,2	1,1
	Pub c	1,6	1,2	0,1

and respect of preferences" eventually concludes that you will deem her choice pub b infinitely more likely than her choice pub c, despite the fact that pub b can never be optimal for Barbara, whereas pub c can.

One could also argue differently here: Since pub c *can* be optimal for Barbara, whereas pub b *cannot*, we may consider her choice pub c better than her choice pub b. Therefore you deem her choice pub c infinitely more likely than her choice pub b.

The same argument can be made when comparing her choices pub a and pub b: For Barbara, pub a *can* be optimal for some appropriately chosen cautious belief about your choice, whereas her choice pub b *cannot*. Therefore, you also deem her choice pub a infinitely more likely than her choice pub b.

Thus, by combining these two arguments, we can see that you will deem Barbara's choices pub a and pub c both infinitely more likely than her choice pub b, since the first two are potentially optimal choices for Barbara, whereas the latter is not. We say that you *assume Barbara's rationality.*

But then, your optimal choice will be pub b since you deem pub b to be the least likely choice for Barbara. Compare this to common full belief in "caution and respect of preferences," which led you to choose pub c. So, assuming Barbara's rationality eventually leads to a different choice for you than respecting Barbara's preferences.□

In a more general setting, assuming the opponent's rationality may be described as follows: Consider two choices a and b for your opponent such that:

- choice a *can* be optimal for *some* cautious lexicographic belief he may hold about your choice, whereas
- choice b *cannot* be optimal for *any* cautious lexicographic belief.

Then, you must deem his choice a infinitely more likely than his choice b. The underlying motivation is that you deem his choice a better than his choice b, because a is a potentially optimal choice for your opponent, whereas b is not.

The question now rises: How can we formalize this idea of assuming the opponent's rationality within an epistemic model? We will see that the answer to this question is rather delicate.

Consider an epistemic model with a set of types T_i for player i and a set of types T_j for his opponent j. Take some type t_i for player i in T_i, with a cautious lexicographic belief on the set $C_j \times T_j$ of the opponent's choice-type pairs. What do we mean when we say that t_i assumes j's rationality? Intuitively, assuming j's rationality means that t_i should deem all choices for j that are optimal for *some* cautious belief infinitely more likely than j's choices that are *not* optimal for *any* cautious belief. In formal terms, this implies that type t_i should deem all of the opponent's choice-type pairs (c_j, t_j) where c_j is optimal for t_j and t_j is cautious, infinitely more likely than all choice-type pairs (c_j', t_j') where either c_j' is not optimal for t_j' or t_j' is not cautious.

But is this condition enough to formalize the idea of assuming j's rationality? Unfortunately, the answer is "no." To see this, let us return to the example "Spy game." Table 7.2 gives an epistemic model for this situation. Consider your type t_1. It only deems possible the opponent's type t_2 and the opponent's choice-type pairs (a, t_2), (b, t_2) and (c, t_2). Among these pairs only the pair (a, t_2) is such that the choice is optimal for the type. Since type t_1 deems the pair (a, t_2) infinitely more likely than the other two pairs, it satisfies our condition above. However, type t_1 does not assume Barbara's rationality: For Barbara, choice c *can* be optimal for *some* belief whereas b cannot, but nevertheless type t_1 deems Barbara's choice b infinitely more likely than c.

The reason is that your type t_1 does not include any type for Barbara for which her potentially optimal choice c is indeed optimal. In the epistemic model above, your type t_1 only considers Barbara's type t_2, for which c is not optimal. So, in order to make things work, we must require that t_1 also deems possible some other type t_2' for which Barbara's potentially optimal choice c is indeed optimal. Such a type t_1 is shown in Table 7.3. Recall that for Barbara, a and c are potentially optimal choices, whereas b is not. Type t_1 in Table 7.3 indeed deems possible a type t_2^a for Barbara for which her choice a is optimal, *and* a type t_2^c for Barbara for which her choice c is optimal. Further, type t_1 deems all "rational" choice-type pairs (c_2, t_2) for Barbara infinitely more likely than all "irrational" choice-type pairs (c_2', t_2'), where "rational" means that the choice c_2 is optimal for the type t_2, and "irrational" means that the choice c_2' is not optimal for the type t_2'. Of the six choice-type pairs that t_1 deems possible, only two are rational – (a, t_2^a) and (c, t_2^c) – and these two are deemed infinitely more likely than the other, irrational, choice-type pairs. Hence, we see that type t_1 deems Barbara's potentially optimal choices a and c infinitely more likely than her choice b, which can never be optimal. So, this type t_1 indeed formalizes the idea of "assuming Barbara's rationality."

This example makes clear that, in order to formally express the idea of assuming the opponent's rationality, we need to impose the following extra requirement:

- Consider a choice c_j for the opponent. If c_j is optimal for *some* cautious belief, then your type t_i should deem possible *some* cautious type t_j for the opponent for which c_j is indeed optimal.

Altogether, we arrive at the following formal definition for assuming the opponent's rationality:

Table 7.2. *An epistemic model for "Spy game" (II)*

Types	$T_1 = \{t_1\}, \quad T_2 = \{t_2\}$
Beliefs for you	$b_1(t_1) = ((a,t_2);(b,t_2);(c,t_2))$
Beliefs for Barbara	$b_2(t_2) = ((c,t_1);(b,t_1);(a,t_1))$

Table 7.3. *An epistemic model for "Spy game" (III)*

Types	$T_1 = \{t_1\}, \quad T_2 = \{t_2^a, t_2^c\}$
Beliefs for you	$b_1(t_1) = ((a,t_2^a);(c,t_2^c);(c,t_2^a);(b,t_2^a);(a,t_2^c);(b,t_2^c))$
Beliefs for Barbara	$b_2(t_2^a) = ((b,t_1);(c,t_1);(a,t_1))$ $b_2(t_2^c) = ((a,t_1);(b,t_1);(c,t_1))$

Definition 7.1.1 *(Assuming the opponent's rationality)*
Consider an epistemic model and a cautious type t_i in this model. Type t_i ***assumes j's rationality*** *if*
(1) for all of the opponent's choices c_j that are optimal for some cautious belief, type t_i deems possible some cautious type t_j for the opponent for which c_j is indeed optimal;
(2) type t_i deems all choice-type pairs (c_j,t_j) where t_j is cautious and c_j is optimal for t_j, infinitely more likely than all choice-type pairs (c'_j,t'_j) that do not have this property.

We will now verify that this definition indeed captures our intuition for assuming the opponent's rationality. Consider some epistemic model and a cautious type t_i within it, satisfying properties (1) and (2) above. Take two of the opponent's choices c_j and c'_j, such that c_j is optimal for some cautious belief, whereas c'_j is never optimal for any cautious belief. By property (1), type t_i must deem possible some cautious type t_j for which c_j is indeed optimal. Since t_i is cautious, it must therefore deem possible the choice-type pair (c_j,t_j), where t_j is cautious and c_j is optimal for t_j. Now, let (c'_j,t'_j) be an arbitrary choice-type pair deemed possible by t_i, where the choice is c'_j. As c'_j is not optimal for any cautious belief, either c'_j is not optimal for t'_j, or t'_j is not cautious. But then, by property (2), t_i must deem the pair (c_j,t_j) above infinitely more likely than (c'_j,t'_j). Since this is true for *every* choice-type pair (c'_j,t'_j) with choice c'_j, we may conclude that type t_i deems c_j infinitely more likely than c'_j. Summarizing, we see that if c_j is optimal for some cautious belief, whereas c'_j is never optimal for any cautious belief, then type t_i deems c_j infinitely more likely than c'_j. But this is exactly what we had in mind for assuming the opponent's rationality.

To conclude this section, let us compare the condition for assuming the opponent's rationality with the other two conditions we discussed for lexicographic beliefs: primary belief in the opponent's rationality, and respecting the opponent's preferences. First of all it is easy to see that, whenever you assume the opponent's rationality, you also primarily believe in the opponent's rationality. Consider a type t_i for player i that assumes j's rationality. Condition (1) in the definition for assuming the opponent's rationality, together with the fact that t_i is cautious, guarantees that if c_j is optimal for some cautious belief, then t_i deems possible some choice-type pair (c_j, t_j) where t_j is cautious and c_j is optimal for t_j. Moreover, condition (2) in this definition states that t_i should deem all choice-type pairs (c_j, t_j) where t_j is cautious and c_j is optimal for t_j, infinitely more likely than all choice-type pairs (c_j', t_j') that do not have this property. In particular, this implies that t_i's primary belief can only assign positive probability to pairs (c_j, t_j) where t_j is cautious and c_j is optimal for t_j. This means, however, that t_i primarily believes in j's rationality.

The converse is not true: If you primarily believe in the opponent's rationality, then you do not necessarily assume the opponent's rationality. Consider, for instance, the epistemic model for "Spy game" in Table 7.2. Your type t_1 primarily believes in Barbara's rationality. However, we have seen that type t_1 does not assume Barbara's rationality. We have thus derived the following result:

Theorem 7.1.2 *(Assuming the opponent's rationality implies primary belief in the opponent's rationality)*
Every type that assumes the opponent's rationality also primarily believes in the opponent's rationality. The converse is not true in general.

We now turn to respecting the opponent's preferences. Does assuming the opponent's rationality imply that you respect the opponent's preferences, or vice versa? We will see that neither of these is true. Consider first the epistemic model for "Spy game" in Table 7.2. You may check that your type t_1 respects Barbara's preferences. However, as we have seen, t_1 does not assume Barbara's rationality.

Consider now the alternative epistemic model for "Spy game" in Table 7.3. We have seen that your type t_1 assumes Barbara's rationality. However, t_1 does not respect Barbara's preferences. Namely, t_1 considers possible Barbara's type t_2^a, for which her choice b is better than her choice c. But type t_1 deems the pair (c, t_2^a) infinitely more likely than (b, t_2^a), and hence does not respect Barbara's preferences. So, we conclude that assuming the opponent's rationality does not imply respecting the opponent's preferences, nor vice versa.

7.2 Common assumption of rationality

So far we have discussed and formalized the idea of assuming the opponent's rationality. In intuitive terms it states that whenever an opponent's choice a is optimal for *some* cautious belief and another choice b is *never* optimal for *any* cautious belief, then you deem a infinitely more likely than b. So, in a sense you divide the opponent's choices into two categories – the *good* choices, which are optimal for some cautious belief, and the *bad* choices, which are not optimal for any cautious belief. You accordingly deem the good choices infinitely more likely than the bad choices.

We can take this argument a step further. Consider now three choices for the opponent – a, b and c – and suppose that:

- c is not optimal for any cautious belief,
- b is optimal for some cautious belief but not for any cautious belief that assumes your rationality, and
- a is optimal for some cautious belief that assumes your rationality.

As before, we would consider c a bad choice, and a and b good choices because the latter two are optimal for some cautious belief. However, choice a can be supported by a belief that assumes your rationality, whereas b cannot. For that reason, we may consider a a "better" choice than b, as it can be supported by a "better" belief. Consequently, you deem a infinitely more likely than b.

This principle may be summarized as follows: Consider two of the opponent's choices a and b such that:

- a is optimal for some cautious belief that assumes your rationality, and
- b is not optimal for any cautious belief that assumes your rationality.

Then, you deem a infinitely more likely than b. We say that you express *2-fold assumption of rationality*. In order to see the consequences of this idea, let us return to the example "Dividing a pizza."

Example 7.2 Dividing a pizza

Recall the story from Example 6.3. We have reproduced the utilities for you and Barbara in Table 7.4. Let us first see what we can say if you and Barbara assume the opponent's rationality. Note that for Barbara, claiming 0, 1 and 2 slices can never be optimal for any cautious belief, as these choices are weakly dominated by "claiming the rest." On the other hand, claiming 3 and 4 slices and claiming the rest are all optimal for some cautious belief that Barbara could hold. So, if you assume Barbara's rationality, then you must deem her "good" choices 3, 4 and "rest" infinitely more likely than her "bad" choices 0, 1 and 2. Similarly, if Barbara assumes your rationality, then she will deem your choices 3, 4 and "rest" infinitely more likely than your choices 0, 1 and 2.

We now take the argument one step further. Barbara's "good" choices 4 and "rest" are both optimal for some cautious belief that assumes your rationality, but her "good" choice 3 is never optimal for any cautious belief that assumes your rationality. We have seen that a belief for Barbara that assumes your rationality must deem your choices 3, 4 and "rest" infinitely more likely than your choices 0, 1 and 2. But then, under every such belief, her choice 3 is always worse than her choice 4. On the other hand, her choices 4 and "rest" *can* be optimal under such a belief: Her choice "rest" is optimal if she deems your choice 3 infinitely more likely than your choice "rest," your choice "rest" infinitely more likely than your choice 4, and your choice 4 infinitely more likely than your other choices. Her choice 4 is optimal if she deems your choice "rest" infinitely

Table 7.4. *Utilities for you and Barbara in "Dividing a pizza" (II)*

		Barbara					
		0	1	2	3	4	rest
You	0	0,0	0,1	0,2	0,3	0,4	0,4
	1	1,0	1,1	1,2	1,3	0,0	1,3
	2	2,0	2,1	2,2	0,0	0,0	2,2
	3	3,0	3,1	0,0	0,0	0,0	3,1
	4	4,0	0,0	0,0	0,0	0,0	4,0
	rest	4,0	3,1	2,2	1,3	0,4	2,2

more likely than your choices 3 and 4, and deems your choices 3 and 4 infinitely more likely than your other choices. So, indeed, Barbara's choices 4 and "rest" are optimal for some cautious beliefs that assume your rationality, whereas her choice 3 is not.

Therefore, if you express 2-fold assumption of rationality, you must deem her choices 4 and "rest" infinitely more likely than her choice 3. Together with our condition above, we see that you must deem Barbara's choices 4 and "rest" infinitely more likely than her choice 3, and deem her choice 3 infinitely more likely than her choices 0, 1 and 2. But then, your only optimal choice will be to claim 4 slices!

Hence, we see that if you assume Barbara's rationality and express 2-fold assumption of rationality, then you must claim 4 slices. The same applies to Barbara: If she assumes your rationality, and expresses 2-fold assumption of rationality, then she will also go for 4 slices. Eventually, you would both end up with nothing!

Compare this to the choices that are selected by common full belief in "caution and respect of preferences." In Chapter 6 we saw that under common full belief in "caution and respect of preferences," you can rationally choose 4 and "rest." Hence, assuming Barbara's rationality and expressing 2-fold assumption of rationality is more restrictive. □

Let us now see how we can formalize this idea of 2-fold assumption of rationality within an epistemic model. Consider some epistemic model M and a type t_i within it. Recall the intuitive idea: If the opponent's choice c_j can be optimal for some cautious belief that assumes your rationality, whereas some other choice c_j' cannot, then you must deem c_j infinitely more likely than c_j'. In terms of types, this can be rephrased as follows: If c_j is optimal for *some* cautious type t_j – in some epistemic model, not necessarily M – that assumes i's rationality, whereas c_j' *is not*, then you must deem c_j infinitely more likely than c_j'.

As with assuming the opponent's rationality, we need the following requirement to make things work: If c_j is optimal for some cautious type – not necessarily in M – that assumes i's rationality, then t_i must deem possible some cautious type t_j assuming i's rationality for which c_j is indeed optimal. We then obtain the following formal definition.

Definition 7.2.1 *(Two-fold assumption of rationality)*
Consider an epistemic model M and a cautious type t_i within M. Type t_i expresses ***2-fold assumption of rationality*** *if*
(1) whenever an opponent's choice c_j is optimal for some cautious type – not necessarily in M – that assumes i's rationality, type t_i deems possible some cautious type t_j assuming i's rationality for which c_j is indeed optimal;
(2) type t_i deems all choice-type pairs (c_j,t_j) where t_j is cautious, t_j assumes i's rationality and c_j is optimal for t_j, infinitely more likely than all choice-type pairs (c'_j,t'_j) that do not have this property.

Let us verify that this definition formalizes exactly what we have in mind. Consider two of the opponent's choices c_j and c'_j, such that

- c_j is optimal for some cautious belief assuming i's rationality, and
- c'_j is not optimal for any cautious belief assuming i's rationality.

Then, by property (1), type t_i must deem possible some cautious type t_j assuming i's rationality for which c_j is optimal. As t_i is cautious, it must also deem possible the choice-type combination (c_j,t_j) where c_j is optimal for t_j, and t_j is cautious and assumes i's rationality. Now, take some arbitrary choice-type combination (c'_j,t'_j) with choice c'_j. Then, either c'_j is not optimal for t'_j, or t'_j is not cautious or t'_j does not assume i's rationality. By property (2), type t_i must then deem (c_j,t_j) infinitely more likely than (c'_j,t'_j). Since this holds for every choice-type pair (c'_j,t'_j) with choice c'_j, we may conclude that t_i deems c_j infinitely more likely than c'_j. So, summarizing, if c_j is optimal for some cautious belief assuming i's rationality, and c'_j is not, then t_i deems c_j infinitely more likely than c'_j, precisely as we want.

In order to illustrate the definition of 2-fold assumption of rationality, let us return to the example "Dividing a pizza."

Example 7.3 Dividing a pizza

Consider the epistemic model in Table 7.5. We have only included some of the beliefs for the types for brevity. Suppose that the beliefs for t_1^3, t_1^4 and t_1^r are such that each of these types is cautious. Similarly for Barbara's types.

Let us first see which of the types in this model assume the opponent's rationality. Consider first your type t_1^3. Recall that Barbara's choices 0, 1 and 2 are never optimal for any cautious belief, whereas her choices 3, 4 and "rest" *are* optimal for some cautious belief. Moreover, her choice 3 is optimal for her cautious type t_2^3, her choice 4 is optimal for her cautious type t_2^4 and her choice "rest" is optimal for her cautious type t_2^r. Since your type t_1^3 considers Barbara's types t_2^3, t_2^4 and t_2^r possible, condition (1) in the definition for assuming the opponent's rationality is satisfied by t_1^3. However, t_1^3 deems the irrational choice-type pair $(1,t_2^r)$ infinitely more likely than the rational choice-type pair (r,t_2^r), and hence t_1^3 does not satisfy condition (2) in this definition. So, we conclude that t_1^3 does not assume Barbara's rationality.

Table 7.5. *An epistemic model for "Dividing a pizza" (II)*

Types	$T_1 = \{t_1^3, t_1^4, t_1^r\}$, $T_2 = \{t_2^3, t_2^4, t_2^r\}$
Beliefs for you	$b_1(t_1^3) = ((1,t_2^r);(r,t_2^r);(3,t_2^3);(4,t_2^4);...)$
	$b_1(t_1^4) = ((4,t_2^4);(r,t_2^r);(3,t_2^3);...)$
	$b_1(t_1^r) = ((3,t_2^3);(r,t_2^r);(4,t_2^4);...)$
Beliefs for Barbara	$b_2(t_2^3) = ((1,t_1^r);(r,t_1^r);(3,t_1^3);(4,t_1^4);...)$
	$b_2(t_2^4) = ((4,t_1^4);(r,t_1^r);(3,t_1^3);...)$
	$b_2(t_2^r) = ((3,t_1^3);(r,t_1^r);(4,t_1^4);...)$

We now move to your type t_1^4. By exactly the same argument as above, we can conclude that t_1^4 satisfies condition (1) in the definition for assuming the opponent's rationality. Note that within this epistemic model, the only rational choice-type pairs for Barbara are $(3,t_2^3)$, $(4,t_2^4)$ and (r,t_2^r). Since t_1^4 deems these rational choice-type pairs infinitely more likely than all irrational choice-type pairs for Barbara, we conclude that t_1^4 assumes Barbara's rationality. In fact, the same applies to your type t_1^r, and therefore also your type t_1^r assumes Barbara's rationality.

The next question we ask is: Do the types t_1^4 and t_1^r also express 2-fold assumption of rationality?

Let us start with t_1^4. We saw in Example 7.2 that for Barbara only her choices 4 and "rest" are optimal for some cautious belief that assumes your rationality. We saw that Barbara's types t_2^4 and t_2^r are indeed cautious and assume your rationality, that her choice 4 is optimal for her type t_2^4 and her choice "rest" is optimal for her type t_2^r. Since t_1^4 deems these two types t_2^4 and t_2^r possible, it satisfies condition (1) in the definition of 2-fold assumption of rationality. In the epistemic model, there are only two choice-type pairs for Barbara where the choice is optimal for the type, and the type is cautious and assumes your rationality. These are the pairs $(4,t_2^4)$ and (r,t_2^r). Since t_1^4 deems these two pairs $(4,t_2^4)$ and (r,t_2^r) infinitely more likely than the other pairs for Barbara, we conclude that t_1^4 also satisfies condition (2) in the definition of 2-fold assumption of rationality. So, we see that your type t_1^4 expresses 2-fold assumption of rationality.

What about your type t_1^r? Note that t_1^r deems the choice-type pair $(3,t_2^3)$ infinitely more likely than the pair (r,t_2^r). However, type t_2^3 does not assume your rationality, as we have seen, whereas type t_2^r does. Therefore, your type t_1^r violates condition (2) in the definition of 2-fold assumption of rationality. As such, your type t_1^r does not express 2-fold assumption of rationality.

Summarizing, we see that:

- your type t_1^3 does not assume Barbara's rationality,
- your type t_1^4 assumes Barbara's rationality and expresses 2-fold assumption of rationality, and

- your type t_1^r assumes Barbara's rationality but does not express 2-fold assumption of rationality. □

So far, we have formally defined what it means to assume your opponent's rationality and to express 2-fold assumption of rationality. Of course we can continue in this fashion and define k-fold assumption of rationality for any k. We will build this up inductively, starting with 1-fold assumption of rationality.

Definition 7.2.2 *(k-fold assumption of rationality)*
Consider an epistemic model M with lexicographic beliefs. Then, a cautious type t_i in M is said to express 1-fold assumption of rationality if t_i assumes j's rationality.
For a given $k \geq 2$, we say that a cautious type t_i in M expresses ***k-fold assumption of rationality*** *if*
(1) whenever an opponent's choice c_j is optimal for some cautious type – not necessarily in M – that expresses up to $(k-1)$-fold assumption of rationality, type t_i deems possible some cautious type t_j for the opponent expressing up to $(k-1)$-fold assumption of rationality for which c_j is indeed optimal;
(2) type t_i deems all choice-type pairs (c_j, t_j), where t_j is cautious, t_j expresses up to $(k-1)$-fold assumption of rationality and c_j is optimal for t_j, infinitely more likely than all choice-type pairs (c_j', t_j') that do not have this property.

Expressing up to $(k-1)$-fold assumption of rationality means that the type expresses 1-fold assumption of rationality, 2-fold assumption of rationality, ... , and $(k-1)$-fold assumption of rationality.

One might perhaps think that if you express k-fold assumption of rationality, then you automatically express $(k-1)$-fold assumption of rationality. This, however, is not true! Suppose that in the epistemic model of Table 7.5 we have an additional type $\hat{t}_1$ for you with belief

$$b_1(\hat{t}_1) = ((4, t_2^4); (r, t_2^r); (1, t_2^r); (3, t_2^3); ...).$$

We will show that $\hat{t}_1$ expresses 2-fold assumption of rationality but not 1-fold assumption of rationality.

Note first that $\hat{t}_1$ deems the choice-type pair $(1, t_2^r)$, where the choice is not optimal for the type, infinitely more likely than the pair $(3, t_2^3)$, where the choice is optimal for the cautious type t_2^3. Hence, $\hat{t}_1$ violates condition (2) in the definition of assuming the opponent's rationality, which means that $\hat{t}_1$ does not assume Barbara's rationality. In other words, $\hat{t}_1$ does not express 1-fold assumption of rationality.

We will see, however, that $\hat{t}_1$ *does* express 2-fold assumption of rationality. As we have already seen, the only choices for Barbara that are optimal if she is cautious and assumes your rationality, are her choices 4 and r. Barbara's types t_2^4 and t_2^r indeed assume your rationality, her choice 4 is indeed optimal for her type t_2^4 and her choice r is indeed optimal for her type t_2^r. Since $\hat{t}_1$ deems possible these types t_2^4 and t_2^r, it satisfies condition (1) in the definition of 2-fold assumption of rationality. In the epistemic model, there are only two choice-type pairs for Barbara where the choice is optimal for the type, the type is cautious and the type assumes your rationality. These are the pairs $(4,t_2^4)$ and (r,t_2^r). As your type $\hat{t}_1$ deems these choice-type pairs infinitely more likely than the other choice-type pairs, it satisfies condition (2) in the definition of 2-fold assumption of rationality. Hence, we may conclude that your type $\hat{t}_1$ does express 2-fold assumption of rationality but not 1-fold assumption of rationality.

It is now easy to define *common assumption of rationality*. This simply means that you express 1-fold assumption of rationality, 2-fold assumption of rationality, 3-fold assumption of rationality, and so on, ad infinitum.

Definition 7.2.3 *(Common assumption of rationality)*
*Consider an epistemic model M, and a cautious type t_i within M. Type t_i expresses **common assumption of rationality** if t_i expresses k-fold assumption of rationality for every k.*

To illustrate this concept, let us return again to the examples "Dividing a pizza" and "Spy game."

Example 7.4 Dividing a pizza

Consider the epistemic model in Table 7.5. Is there any type for you in this model that expresses common assumption of rationality? Clearly, the types t_1^3 and t_1^r do not, since we have seen that t_1^3 does not assume Barbara's rationality, and t_1^r does not express 2-fold assumption of rationality.

What about t_1^4? We have seen that t_1^4 assumes Barbara's rationality and expresses 2-fold assumption of rationality. So, t_1^4 expresses 1-fold and 2-fold assumption of rationality. Similarly for Barbara's type t_2^4.

We will now show that your type t_1^4 also expresses 3-fold assumption of rationality. In this game there is only one choice for Barbara, namely 4, that is optimal for a cautious belief that expresses 1-fold and 2-fold assumption of rationality. If Barbara assumes your rationality, she must deem your choices 3, 4 and "rest" infinitely more likely than your other choices, and if she expresses 2-fold assumption of rationality, then she must deem your choices 4 and "rest" infinitely more likely than your choice 3 as well. We saw this in Example 7.2. But then, her only possible optimal choice would be 4.

Note that Barbara's choice 4 is optimal for her type t_2^4, and that this type t_2^4 indeed expresses 1-fold and 2-fold assumption of rationality. Since your type t_1^4 deems possible Barbara's type t_2^4, your type t_1^4 satisfies condition (1) in the definition of 3-fold assumption of rationality.

Within this epistemic model, there is only one choice-type pair for Barbara where the choice is optimal for the type, the type is cautious and the type expresses 1-fold and 2-fold assumption of rationality. This is the pair $(4,t_2^4)$. Since your type t_1^4 deems this pair $(4,t_2^4)$ infinitely more likely than all other pairs, your type t_1^4 also satisfies condition (2) in the definition of 3-fold assumption of rationality. We may thus conclude that your type t_1^4 expresses 3-fold assumption of rationality.

By continuing in this fashion, we can actually conclude that your type t_1^4 also expresses 4-fold assumption of rationality, and further. So, your type t_1^4 actually expresses common assumption of rationality. As your choice 4 is rational for your type t_1^4, we see that you can rationally choose 4 under common assumption of rationality.

In Example 7.2 we saw that if you assume Barbara's rationality and express 2- fold assumption of rationality, then your only possible optimal choice is to claim 4 slices. Hence, we conclude that under common assumption of rationality, your only optimal choice is to claim 4 slices. □

Example 7.5 Spy game

Consider the epistemic model in Table 7.3. We have already verified that your type t_1 assumes Barbara's rationality. But does it also express *common* assumption of rationality? The answer is "no."

To see this, we first show that Barbara's type t_2^a does not assume your rationality. Barbara's type t_2^a only deems possible one type for you, namely t_1, for which only your choice b is optimal. However, your choices a and c are also optimal for *some* cautious beliefs. Hence, condition (1) for assuming the opponent's rationality requires that t_2^a should deem possible some cautious type t_1^a for you for which a is optimal, and some cautious type t_1^c for which c is optimal. But type t_2^a does not, so Barbara's type t_2^a does not assume your rationality. Actually, the same can be said about Barbara's other type t_2^c, so also her type t_2^c does not assume your rationality.

Now, let us return to your type t_1. By construction, type t_1 only deems possible Barbara's types t_2^a and t_2^c, neither of which assume your rationality. However, for Barbara's choice a there is *some* type t_2, *outside* the epistemic model in Table 7.3, such that:

- t_2 is cautious and assumes your rationality, and
- a is optimal for t_2.

Summarizing, we see that Barbara's choice a is optimal for some cautious type that assumes your rationality, but your type t_1, in the epistemic model of Table 7.3, does not deem possible any type for Barbara that assumes your rationality. But then, your type t_1 violates condition (1) in the definition of 2-fold assumption of rationality. Hence, your type t_1 does not express 2-fold assumption of rationality. In particular, t_1 does not express common assumption of rationality.

Consider the alternative epistemic model in Table 7.6. We denote the types in this new epistemic model by r, to distinguish them from the types in Table 7.3. Again, to save space, we only show the first few levels of the lexicographic beliefs. The reader may extend these beliefs so that every type is cautious.

Within this new model, Barbara's type r_2^a is cautious and assumes your rationality, and her choice a is optimal for r_2^a. To see that r_2^a assumes your rationality, note that your choices a, b and c are optimal for your cautious types r_1^a, r_1^b and r_1^c, respectively, and that Barbara's type r_2^a deems possible each of these types. Moreover, r_2^a deems your rational choice-type pairs $(a, r_1^a), (b, r_1^b)$ and (c, r_1^c) infinitely more likely than the other, irrational choice-type pairs. So, indeed, r_2^a assumes your rationality. Therefore,

Table 7.6. *An epistemic model for "Spy game" (IV)*

Types	$T_1 = \{r_1^a, r_1^b, r_1^c\}$, $T_2 = \{r_2^a, r_2^c\}$
Beliefs for you	$b_1(r_1^a) = ((c, r_2^c); (b, r_2^c); (a, r_2^a), ...)$
	$b_1(r_1^b) = ((a, r_2^a); (c, r_2^c); (b, r_2^a); ...)$
	$b_1(r_1^c) = ((a, r_2^a); (b, r_2^a); (c, r_2^c); ...)$
Beliefs for Barbara	$b_2(r_2^a) = ((b, r_1^b); (c, r_1^c); (a, r_1^a); ...)$
	$b_2(r_2^c) = ((a, r_1^a); (b, r_1^b); (c, r_1^c); ...)$

we conclude that Barbara's choice a is optimal for some cautious type that assumes your rationality, namely r_2^a.

We will now show that your type r_1^b expresses common assumption of rationality.

First, we demonstrate that r_1^b expresses 1-fold assumption of rationality, that is, assumes Barbara's rationality. Note that for Barbara only the choices a and c are optimal for some cautious belief. Your type r_1^b considers possible the cautious type r_2^a for Barbara, for which a is optimal, and the cautious type r_2^c, for which c is optimal. So, condition (1) in the definition of assuming the opponent's rationality is satisfied. Moreover, the only rational choice-type pairs for Barbara in the model are (a, r_2^a) and (c, r_2^c). Since type r_1^b deems the rational choice-type pairs (a, r_2^a) and (c, r_2^c) infinitely more likely than the other, irrational choice-type pairs, r_1^b also satisfies condition (2) in the definition of assuming the opponent's rationality. Hence, your type r_1^b assumes Barbara's rationality.

We next show that your type r_1^b also expresses 2-fold assumption of rationality. It can easily be verified that Barbara's types r_2^a and r_2^c express 1-fold assumption of rationality. As a is optimal for r_2^a and c is optimal for r_2^c, we conclude that Barbara's choices a and c are both optimal for some cautious type that expresses 1-fold assumption of rationality. Since your type r_1^b deems possible Barbara's types r_2^a and r_2^c, we see that r_1^b satisfies condition (1) in the definition of 2-fold assumption of rationality. In the epistemic model, the only choice-type pairs for Barbara where the choice is optimal for the type, the type is cautious and the type expresses 1-fold assumption of rationality, are the pairs (a, r_2^a) and (c, r_2^c). As r_1^b deems the pairs (a, r_2^a) and (c, r_2^c) infinitely more likely than the other pairs, your type r_1^b also satisfies condition (2) in the definition of 2-fold assumption of rationality. So, type r_1^b also expresses 2-fold assumption of rationality.

We now show that your type r_1^b also expresses 3-fold assumption of rationality. To see this, we first show that Barbara's type r_2^a expresses up to 2-fold assumption of rationality. We have already seen that r_2^a expresses 1-fold assumption of rationality, so it remains to show that r_2^a expresses 2-fold assumption of rationality.

As we have seen, your type r_1^b expresses 1-fold assumption of rationality. Since your choice b is optimal for r_1^b, we conclude that your choice b is optimal for some type that is cautious and expresses 1-fold assumption of rationality. Moreover, your other choices a and c cannot be optimal for such a type: If you assume Barbara's rationality, then you must deem her choices a and c infinitely more likely than b, and hence your only optimal choice will be b. So, only your choice b is optimal for some type that is cautious and expresses 1-fold assumption of rationality. Since Barbara's type r_2^a deems possible such a type, namely r_1^b, her type r_2^a satisfies condition (1) in the definition of 2-fold assumption of rationality. Within the epistemic model, only your pair (b, r_1^b) is such that the choice is optimal for the type, the type is cautious and the type expresses 1-fold assumption of rationality. Since Barbara's type r_2^a deems the pair (b, r_1^b) infinitely more likely than the other pairs, type r_2^a satisfies condition (2) in the definition of 2-fold assumption of rationality. So, indeed, Barbara's type r_2^a expresses 2-fold assumption of rationality. In total, we see that Barbara's type r_2^a expresses up to 2-fold assumption of rationality.

We can now verify that your type r_1^b indeed expresses 3-fold assumption of rationality. Above we saw that Barbara's type r_2^a is cautious and expresses up to 2-fold assumption of rationality. As her choice a is optimal for r_2^a, we see that Barbara's choice a is optimal for some cautious type that expresses up to 2-fold assumption of rationality. Moreover, her other two choices b and c cannot be optimal for a cautious type that expresses up to 2-fold assumption of rationality: If you express 1-fold assumption of rationality, then you must deem her choices a and c infinitely more likely than b, and hence your only optimal choice is b. So, if Barbara expresses 2-fold assumption of rationality, then she must deem your choice b infinitely more likely than your choices a and c, and therefore b and c cannot be optimal for Barbara in that case. So, we see that for Barbara, only choice a is optimal for a type that is cautious and expresses up to 2-fold assumption of rationality. Since your type r_1^b deems possible such a type for Barbara, namely r_2^a, your type r_1^b satisfies condition (1) in the definition of 3-fold assumption of rationality. In the epistemic model, (a, r_2^a) is the only pair for Barbara where the choice is optimal for the type, the type is cautious, and the type expresses up to 2-fold assumption of rationality. Since your type r_1^b deems this pair (a, r_2^a) infinitely more likely than the other pairs, r_1^b also satisfies condition (2) in the definition of 3-fold assumption of rationality. So, we conclude that your type r_1^b indeed expresses 3-fold assumption of rationality.

By continuing in this fashion, we can prove that type r_1^b expresses common assumption of rationality. In fact, in the epistemic model it is the only type for you that expresses common assumption of rationality. As your choice b is optimal for type r_1^b, we see that you can rationally choose b under common assumption of rationality. Earlier, we argued that under common assumption of rationality you cannot rationally choose a or c. Therefore, your only optimal choice under common assumption of rationality is to visit pub b. □

7.3 Algorithm

In the previous two sections we developed the concept of *common assumption of rationality*. We now ask: Is there an algorithm that computes all the choices you can rationally make under common assumption of rationality? As we will see, such an algorithm can indeed be found, and it will be a very simple one.

Step 1: 1-fold assumption of rationality
Let us begin with a more basic question: Which choices can you rationally make if you are cautious and express 1-fold assumption of rationality? That is, if you are cautious and assume the opponent's rationality.

We saw in Theorem 5.8.3 from Chapter 5 that any choice that is optimal for a cautious lexicographic belief must not be weakly dominated in the original game.

Moreover, if you assume the opponent's rationality, then you must deem all of the opponent's choices that are optimal for *some* cautious lexicographic belief infinitely more likely than all choices that are *not* optimal for *any* cautious lexicographic belief. From Theorem 5.8.3 in Chapter 5 we know that the choices that are optimal for *some* cautious lexicographic belief are exactly the choices that are not weakly dominated in the original game. Hence, if you assume the opponent's rationality, then you must deem all of the opponent's choices that are *not* weakly dominated infinitely more likely than all choices that *are* weakly dominated.

Let us denote by C_j^1 the set of the opponent's choices that are not weakly dominated in the original game. So, if you assume the opponent's rationality, then you must deem all choices in C_j^1 infinitely more likely than all choices outside C_j^1. Which choices can you then rationally make yourself?

Take an arbitrary cautious lexicographic belief $b_i = (b_i^1;...;b_i^K)$ for you that deems all choices in C_j^1 infinitely more likely than all choices outside C_j^1. That is, there is some level $L < K$ such that:

- the beliefs $b_i^1,...,b_i^L$ only assign positive probability to choices in C_j^1, and
- every choice in C_j^1 has a positive probability in some of the beliefs $b_i^1,..,b_i^L$.

Now, take an arbitrary choice c_i for you that is optimal under the lexicographic belief b_i. Then, in particular, c_i must be optimal under the *truncated* lexicographic belief $(b_i^1;...;b_i^L)$. Note, however, that this truncated belief is cautious on C_j^1, which means that every choice in C_j^1 has positive probability in some level of the truncated belief.

So, we see that every choice that is optimal under a lexicographic belief that assumes j's rationality, must be optimal under some cautious lexicographic belief on C_j^1. Now, consider the reduced game in which player j can only choose from C_j^1. Then, every choice that is optimal under a lexicographic belief that assumes j's rationality, must be optimal under some cautious lexicographic belief in the reduced game. However, by applying Theorem 5.8.3 to the reduced game, we know that every such choice cannot be weakly dominated in this reduced game. Hence, every choice that is optimal under a cautious lexicographic belief that assumes j's rationality cannot be weakly dominated in the reduced game where player j can only choose from C_j^1. Remember that C_j^1 are exactly the choices for player j that are not weakly dominated in the original game.

So, C_j^1 is obtained by eliminating all weakly dominated choices for player j from the original game.

We thus reach the following conclusion: If you are cautious and assume j's rationality, then every optimal choice for you:

- cannot be weakly dominated in the original game, and
- cannot be weakly dominated in the reduced game that is obtained after eliminating the weakly dominated choices from the original game.

We will call the procedure where we first eliminate all weakly dominated choices from the original game, and then eliminate all weakly dominated choices from the reduced game obtained, *2-fold elimination of weakly dominated choices*. We may thus conclude that, if you are cautious and express 1-fold assumption of rationality, then every optimal choice for you must survive 2-fold elimination of weakly dominated choices.

Step 2: Up to 2-fold assumption of rationality

We now ask the question: Which choices can you rationally make if you are cautious and express up to 2-fold assumption of rationality? That is, if you are cautious, assume the opponent's rationality and express 2-fold assumption of rationality.

Remember that 2-fold assumption of rationality means the following: If there are two choices c_j and c_j' for player j such that c_j is optimal for some cautious belief assuming i's rationality, whereas c_j' is not, then you must deem c_j infinitely more likely than c_j'. Let us denote by C_j^2 the set of those choices for player j that are optimal for some cautious lexicographic belief that assumes i's rationality. Hence, if you express 2-fold assumption of rationality, then you must deem all choices in C_j^2 infinitely more likely than all choices outside C_j^2. But then, by exactly the same argument as above, every optimal choice for you cannot be weakly dominated in the reduced game where player j can only choose from C_j^2.

From step 1, we know that C_j^2 are exactly those choices for player j that survive 2-fold elimination of weakly dominated choices. Hence, we conclude that, if you are cautious and express up to 2-fold assumption of rationality, then every optimal choice for you cannot be weakly dominated in the reduced game obtained after 2-fold elimination of weakly dominated choices. That is, every such choice must survive *3-fold elimination of weakly dominated choices.*

Of course, we can continue in the same fashion. Eventually, we reach the conclusion that, if you are cautious and express up to k-fold assumption of rationality, then every optimal choice for you must survive $(k+1)$-fold elimination of weakly dominated choices. In particular, if you are cautious and express common assumption of rationality, then every optimal choice for you must survive *iterated elimination of weakly dominated choices.* We formally define this algorithm:

Algorithm 7.3.1 *(Iterated elimination of weakly dominated choices)*

Step 1. *Within the original game, eliminate all choices that are weakly dominated.*

Step 2. *Within the reduced game obtained after step 1, eliminate all choices that are weakly dominated.*
Step 3. *Within the reduced game obtained after step 2, eliminate all choices that are weakly dominated.*

⋮

And so on.

Obviously, this algorithm stops after finitely many steps. As we have argued above, every choice that can rationally be made by a cautious type that expresses up to k-fold assumption of rationality must survive $(k+1)$-fold elimination of weakly dominated choices. Moreover, it can be shown that the converse is also true: For every choice c_i that survives $(k+1)$-fold elimination of weakly dominated choices, we can construct a cautious type t_i that expresses up to k-fold assumption of rationality, such that c_i is optimal for t_i. Hence, the choices that can rationally be made by a type that expresses up to k-fold assumption of rationality are exactly the choices that survive $(k+1)$-fold elimination of weakly dominated choices. As a consequence, the choices that can rationally be made under *common* assumption of rationality are precisely the choices that survive the *full* procedure of iterated elimination of weakly dominated choices. We thus obtain the following result.

Theorem 7.3.2 *(The algorithm works)*
Consider a two-player static game with finitely many choices for both players.
(1) For every $k \geq 1$, the choices that can rationally be made by a cautious type that expresses up to k-fold assumption of rationality are exactly those choices that survive $(k+1)$-fold elimination of weakly dominated choices.
(2) The choices that can rationally be made by a cautious type that expresses common assumption of rationality are exactly those choices that survive iterated elimination of weakly dominated choices.

The proof of this result can be found in the proofs section at the end of this chapter. We will now use this result to compare *common assumption of rationality* to the other two concepts we discussed for lexicographic beliefs, namely *common full belief in "caution and primary belief in rationality"* and *common full belief in "caution and respect of preferences."*

In Chapter 5 we saw that a choice can rationally be made under common full belief in "caution and primary belief in rationality" precisely when this choice survives the Dekel–Fudenberg procedure; that is, one round of elimination of weakly dominated choices, followed by iterated elimination of strictly dominated choices. The reader may verify that the procedure of iterated elimination of weakly dominated choices eliminates as least as many choices as the Dekel–Fudenberg procedure. This is actually theoretical problem 7.8 at the end of this chapter. So, every choice that survives iterated elimination of weakly dominated choices also survives the Dekel–Fudenberg procedure, but in general not vice versa. Together with the theorem above, this leads to the following conclusion.

Theorem 7.3.3 *(Relation between common assumption of rationality and common full belief in "caution and primary belief in rationality")*
Every choice c_i that can rationally be made under common assumption of rationality can also rationally be made under common full belief in "caution and primary belief in rationality." The converse is not true in general.

To see that the converse may not be true, consider again the example "Spy game," with the utilities as depicted in Table 7.1. We saw that under common assumption of rationality, your only optimal choice is pub b. Indeed, if we apply iterated elimination of weakly dominated choices, we first eliminate the weakly dominated choice b for Barbara, and in the reduced game that remains we eliminate the weakly dominated choices a and c for you. So, only choice b for you survives, and hence under common assumption of rationality you can only rationally choose b.

If we apply the Dekel–Fudenberg procedure to this game, then in the first round we eliminate the weakly dominated choice b for Barbara. However, in the reduced game that remains no choices are strictly dominated, so the Dekel–Fudenberg procedure stops here. So, all of your choices survive, which means that you can rationally make all of your choices under common full belief in "caution and primary belief in rationality." In particular, you can rationally choose a and c under common full belief in "caution and primary belief in rationality," but not under common assumption of rationality.

Let us now compare common assumption of rationality to common full belief in "caution and respect of preferences." In the example "Spy game," we saw that under common full belief in "caution and respect of preferences" you can only rationally choose pub c, whereas under common assumption of rationality you can only rationally choose pub b. Hence, in this example both concepts yield unique – but different – choices for you.

In the example "Dividing a pizza," common full belief in "caution and respect of preferences" allows two choices for you, namely to claim 4 slices or to claim the rest. Common assumption of rationality, on the other hand, yields a unique choice for you, which is to claim 4 slices. Hence, in this example, common assumption of rationality is more restrictive, in terms of choices, than common full belief in "caution and respect of preferences."

However, it may also be the other way around. To see this, consider again the example "Take a seat" from Chapter 6, with the utilities in Table 7.7. In Chapter 6 we saw that under common full belief in "caution and respect of preferences," your only rational choice is to take seat g. Which seats can you rationally take under common assumption of rationality? To answer this question, we apply the algorithm of iterated elimination of weakly dominated choices to the game.

Note that your choices a and b are weakly dominated by your randomized choice $\frac{1}{2}c + \frac{1}{2}d$. No other choices are weakly dominated. The same applies to Barbara. So, in the first round we eliminate the choices a and b for you and Barbara. In the reduced game that remains, your choices c and d are weakly dominated by the randomized choice $\frac{1}{2}e + \frac{1}{2}f$, and no other choices are weakly dominated. The same holds for Barbara.

Table 7.7. *Utilities for you and Barbara in "Take a seat" (II)*

		Barbara							
		a	b	c	d	e	f	g	h
You	a	5,5	0,10	0,0	0,20	0,0	0,0	0,0	0,0
	b	10,0	5,5	0,20	0,0	0,0	0,0	0,0	0,0
	c	0,0	20,0	20,20	20,20	0,0	0,45	0,0	0,0
	d	20,0	0,0	20,20	20,20	0,45	0,0	0,0	0,0
	e	0,0	0,0	0,0	45,0	45,45	45,45	0,0	0,95
	f	0,0	0,0	45,0	0,0	45,45	45,45	0,95	0,0
	g	0,0	0,0	0,0	0,0	0,0	95,0	95,95	95,95
	h	0,0	0,0	0,0	0,0	95,0	0,0	95,95	95,95

So, in round 2 we eliminate the choices c and d for you and Barbara. In the reduced game that remains, your choices e and f are weakly dominated by your randomized choice $\frac{1}{2}g + \frac{1}{2}h$, and no other choices are weakly dominated. The same holds for Barbara. Hence, in round 3 we eliminate the choices e and f for you and Barbara. In the reduced game that remains, no choice is weakly dominated, so iterated elimination of weakly dominated choices stops here. Since both of your choices g and h survive this procedure, we can conclude that under common assumption of rationality, you can rationally take the seats g or h. In contrast, under common full belief in "caution and respect of preferences" you could only rationally take seat g. So, in this example common full belief in "caution and respect of preferences" is more restrictive than common assumption of rationality.

Summarizing, we see that there is no general logical relation between common assumption of rationality on the one hand, and common full belief in "caution and respect of preferences" on the other. There are examples where they both yield unique, but different, choices, there are examples where the former concept is more restrictive than the latter, and there are other examples where the latter is more restrictive than the former.

To conclude this section, let us think about the question whether common assumption of rationality is always possible. So, for a given game, can we always construct some epistemic model and some type for you within it, that expresses common assumption of rationality? Or are there situations where common assumption of rationality is simply too demanding?

Unfortunately, we cannot deliver an easy iterative procedure that always delivers a type that expresses common assumption of rationality, as we did in Chapters 3, 5 and 6. But our Theorem 7.3.2 will be of help here. This theorem states, in particular, that every choice that survives iterated elimination of weakly dominated choices can rationally be chosen under common assumption of rationality. Now, consider a choice c_i for you that survives iterated elimination of weakly dominated choices. We know that such a choice exists, as the algorithm always stops after finitely many steps and never eliminates all choices for a given player. According to the theorem, c_i can rationally be chosen under common assumption of rationality. That is, we can construct an epistemic model M and a type t_i within M, such that t_i expresses common assumption of rationality and c_i is rational for t_i. Since this can always be done according to the theorem, we conclude that we can always construct some epistemic model and some type for you within it, that expresses common assumption of rationality. We thus obtain the following general existence result.

Theorem 7.3.4 *(Common assumption of rationality is always possible)*
Consider a game with two players and finitely many choices for both players. Then, for both players i, we can construct an epistemic model and a type t_i within it, such that t_i expresses common assumption of rationality.

In general, however, it will not be possible to construct an epistemic model in which *all* types express common assumption of rationality. To see this, consider again the example "Spy game." There, Barbara's choice c is optimal for some cautious type but not for any cautious type that expresses common assumption of rationality. We have seen that under common assumption of rationality Barbara can only rationally choose a. Now, consider some type t_1 for you that expresses common assumption of rationality. In particular, t_1 assumes Barbara's rationality. As c is optimal for Barbara for some cautious type, your type t_1 must deem possible some cautious type t_2 for Barbara for which her choice c is indeed optimal. But then, this type t_2 cannot express common assumption of rationality, as c cannot be optimal for Barbara for any cautious type that expresses common assumption of rationality. So, your type t_1 must deem possible some type t_2 for Barbara that does not express common assumption of rationality. In particular, this type t_2 is itself in the epistemic model. We thus conclude that every epistemic model for "Spy game" must contain at least one type that does not express common assumption of rationality.

In this sense, common assumption of rationality is very different from common full belief in "caution and primary belief in rationality" and common full belief in "caution and respect of preferences." For every game we can construct an epistemic model where *all* types express common full belief in "caution and respect of preferences," and therefore also express common full belief in "caution and primary belief in rationality." This is not possible for common assumption of rationality.

7.4 Order dependence

For two of the algorithms we have seen so far – iterated elimination of strictly dominated choices and iterated addition of preference restrictions – we have shown that the order of elimination or addition does not matter for the eventual output. Moreover, we have seen that for the Dekel–Fudenberg procedure, the order of elimination *after round 1* does not matter for the final result. Is this also true for the algorithm of iterated elimination of weakly dominated choices? So, is the order of elimination irrelevant as well for

Table 7.8. *Utilities for you and Barbara in "Where to read my book?" (III)*

		Barbara		
		Pub a	Pub b	Pub c
You	Pub a	0, 3	1, 2	1, 1
	Pub b	1, 3	0, 2	1, 1
	Pub c	1, 3	1, 2	0, 1

iterated elimination of weakly dominated choices? As we will see, the answer is "no"! For the algorithm in this chapter, it is absolutely crucial that at every round we always eliminate *all* weakly dominated choices, otherwise we might end up with the wrong sets of choices at the end. Here is an example that illustrates this.

Example 7.6 Where to read my book?

Recall the story from Example 5.2. We have reproduced the utilities for you and Barbara in Table 7.8. Let us first apply iterated elimination of weakly dominated choices with the *original* order of elimination – that is, at every step we always eliminate *all* weakly dominated choices. Then, at the first step we eliminate Barbara's choices b and c, after which we eliminate your choice a. Then, no further choices can be eliminated, and hence we end up with the choices b and c for you. So, under common assumption of rationality you can rationally go to pub b or pub c.

Suppose now that in the first step we only eliminate Barbara's choice c, but not choice b. Then, in step 2 we could eliminate your choices a and b, after which we would eliminate Barbara's choice b. In this case, the only choice left for you would be pub c. So, the eventual output is different than under the original order of elimination.

We could also only eliminate Barbara's choice b at step 1. Then, in step 2, we could eliminate your choices a and c, after which we would eliminate Barbara's choice c. With this particular order of elimination, the only choice surviving for you would be pub b.

Overall, we see that the set of choices that survives the algorithm of iterated elimination of weakly dominated choices crucially depends on the specific order and speed of elimination. In that sense, the algorithm is very different from the other algorithms we have seen so far. □

7.5 Proofs

In this section we will prove Theorem 7.3.2, which states that the choices that can rationally be made under common assumption of rationality are exactly the choices that survive the procedure of iterated elimination of weakly dominated choices. Before we do so, we first show an "optimality principle" for the algorithm of iterated elimination of weakly dominated choices. This optimality principle is similar to the ones we have seen for iterated elimination of strictly dominated choices and the Dekel–Fudenberg procedure.

Optimality principle

For every $k \in \{0,1,2,...\}$ let C_i^k be the set of choices for player i that survive k-fold elimination of weakly dominated choices, where $C_i^0 = C_i$. Let Γ^k be the associated reduced game in which both players i can choose from C_i^k. That is, Γ^k is the reduced game that results after k-fold elimination of weakly dominated choices.

Consider some $c_i \in C_i^k$, where $k \geq 1$. Then, of course, c_i is in each of the sets $C_i^1,...,C_i^k$. Take some arbitrary $m \in \{1,..,k\}$. Since $c_i \in C_i^m$ we know, by construction of the algorithm, that c_i is not weakly dominated in the reduced game Γ^{m-1} where i can choose from C_i^{m-1} and j can choose from C_j^{m-1}. By applying Theorem 5.8.3 to the reduced game Γ^{m-1}, we then conclude that c_i is optimal, among the choices in C_i^{m-1}, for some standard belief b_i^{m-1} that assigns positive probability to all choices in C_j^{m-1}, and probability zero to all other choices. Let us denote by $\Delta^+(C_j^{m-1})$ the set of standard beliefs that assign positive probability to all choices in C_j^{m-1}, and probability zero to all choices outside.

So, for every choice $c_i \in C_i^k$ there is some lexicographic belief

$$b_i = (b_i^{k-1}; b_i^{k-2}; ...; b_i^0)$$

such that, for every $m \in \{0,...,k-1\}$,

- the belief b_i^m is in $\Delta^+(C_j^m)$, and
- c_i is optimal for b_i^m among the choices in C_i^m.

In fact, we can say more about c_i. Not only is, for every $m \in \{0,...,k-1\}$, choice c_i optimal for b_i^m *among the choices in* C_i^m. But, c_i is even optimal for b_i^m *among all the choices in the original game.* That is, c_i is actually optimal for the lexicographic belief above *among all choices in the original game.* We refer to this result as the "optimality principle" for iterated elimination of weakly dominated choices.

Theorem 7.5.1 *(Optimality principle)*
Consider a choice $c_i \in C_i^k$ for player i that survives k-fold elimination of weakly dominated choices. Then, c_i is optimal, among the choices in the original game, for some cautious lexicographic belief

$$b_i = (b_i^{k-1}; b_i^{k-2}; ...; b_i^0)$$

where $b_i^m \in \Delta^+(C_j^m)$ for all $m \in \{0,...,k-1\}$.

Proof: Consider some choice $c_i \in C_i^k$. Then, we know from above that there is some lexicographic belief

$$b_i = (b_i^{k-1}; b_i^{k-2}; ...; b_i^0)$$

such that, for every $m \in \{0,...,k-1\}$:

- the belief b_i^m is in $\Delta^+(C_j^m)$, and
- c_i is optimal for b_i^m among the choices in C_i^m.

We will show that, for every m, the choice c_i is optimal for b_i^m *among all choices in* C_i. Consider some $m \in \{0,...,k-1\}$. Suppose, on the contrary, that c_i is not optimal for b_i^m among the choices in C_i. Then, there is some $d_i \in C_i \backslash C_i^m$ with $u_i(d_i, b_i^m) > u_i(c_i, b_i^m)$. Hence, we can find some $l < m$ and some $d_i \in C_i^l \backslash C_i^{l+1}$ such that

$$u_i(d_i, b_i^m) > u_i(c_i, b_i^m),$$

and

$$u_i(c_i, b_i^m) \geq u_i(c_i', b_i^m) \text{ for all } c_i' \in C_i^{l+1}. \qquad (*)$$

Define for every $n \geq 2$ the belief

$$b_{i,n} := (1 - \tfrac{1}{n})b_i^m + \tfrac{1}{n}b_i^l.$$

Since $b_i^m \in \Delta^+(C_j^m)$ and $b_i^l \in \Delta^+(C_j^l)$, it follows that $b_{i,n} \in \Delta^+(C_j^l)$ for all $n \geq 2$. We can find some $N \geq 2$ and some choice $d_i^* \in C_i^l$ such that, for all $n \geq N$, choice d_i^* is optimal for $b_{i,n}$ among the choices in C_i^l. Since $b_{i,n} \in \Delta^+(C_j^l)$, we know from Theorem 5.8.3 in Chapter 5 that d_i^* is not weakly dominated in the reduced game Γ^l, obtained after l-fold elimination of weakly dominated choices. Hence, $d_i^* \in C_i^{l+1}$.

As, for all $n \geq N$, choice d_i^* is optimal for $b_{i,n}$ among the choices in C_i^l, we have for all $n \geq N$ that

$$u_i(d_i^*, b_{i,n}) \geq u_i(d_i, b_{i,n}).$$

As $b_{i,n}$ converges to b_i^m as $n \to \infty$, we conclude that

$$u_i(d_i^*, b_i^m) \geq u_i(d_i, b_i^m) > u_i(c_i, b_i^m).$$

Hence, we have found some $d_i^* \in C_i^{l+1}$ with $u_i(d_i^*, b_i^m) > u_i(c_i, b_i^m)$. This, however, contradicts (*). We thus conclude that c_i must be optimal for b_i^m among the choices in C_i.

Since this holds for every $m \in \{0,...,k-1\}$, we see that choice c_i is indeed optimal, among the choices in the original game, for the lexicographic belief $b_i = (b_i^{k-1}; b_i^{k-2}...; b_i^0)$. This completes the proof. ■

The algorithm works

We will now use the optimality principle to prove that the algorithm works. That is, we will prove the following theorem.

Theorem 7.3.2 *(The algorithm works)*
Consider a two-player static game with finitely many choices for both players.
(1) For every $k \geq 1$, the choices that can rationally be made by a cautious type that expresses up to k-fold assumption of rationality are exactly those choices that survive $(k+1)$-fold elimination of weakly dominated choices.

(2) The choices that can rationally be made by a cautious type that expresses common assumption of rationality are exactly those choices that survive iterated elimination of weakly dominated choices.

Proof: As before, let C_i^k be the set of choices for player i that survive k-fold elimination of weakly dominated choices. Suppose that the procedure of iterated elimination of weakly dominated choices terminates after K rounds, and that $C_i^{K+1} = C_i^K$ for both players i.

In this proof we will construct an epistemic model M with the following properties:

- For both players i and every choice $c_i \in C_i^1$, there is a cautious type $t_i^{c_i}$ for which c_i is optimal.
- For every k, if the choice c_i is in C_i^k, then the associated type $t_i^{c_i}$ expresses up to $(k-1)$-fold assumption of rationality.
- If the choice c_i is in C_i^K, then the associated type $t_i^{c_i}$ expresses common assumption of rationality.

For every k, let T_i^k be the set of types $t_i^{c_i}$ for player i with $c_i \in C_i^k$. As $C_i^K \subseteq C_i^{K-1} \subseteq ... \subseteq C_i^1$, it follows that $T_i^K \subseteq T_i^{K-1} \subseteq ... \subseteq T_i^1$, where $T_i^1 = T_i$. In order to construct types with the properties above, we proceed by the following four steps.

In step 1 we construct, for every choice $c_i \in C_i^1$, some lexicographic belief about j's choice for which c_i is optimal. In step 2 we use these lexicographic beliefs to construct our epistemic model M. In this model, we define for every choice $c_i \in C_i^1$ some type $t_i^{c_i}$. In step 3 we show that, for every k and both players i, every type $t_i \in T_i^k$ expresses up to $(k-1)$-fold assumption of rationality. In step 4 we finally prove that, for both players i, every type $t_i \in T_i^K$ expresses common assumption of rationality.

Step 1: Construction of beliefs

For every $k \in \{1, ..., K-1\}$, let D_i^k be the set of choices that survive k-fold elimination of weakly dominated choices but not $(k+1)$-fold elimination. Hence, $D_i^k = C_i^k \backslash C_i^{k+1}$. To define the beliefs, we distinguish the following two cases:

(1) Consider first some $k \in \{1, ..., K-1\}$ and some choice c_i in D_i^k. By Theorem 7.5.1 we can find some cautious lexicographic belief

$$b_i^{c_i} := (b_i^{k-1,c_i}; b_i^{k-2,c_i}; ...; b_i^{0,c_i})$$

with $b_i^{m,c_i} \in \Delta^+(C_j^m)$ for every $m \in \{0, ..., k-1\}$, such that c_i is optimal for $b_i^{c_i}$.

(2) Consider now some choice $c_i \in C_i^K$ that survives iterated elimination of weakly dominated choices. Then, also $c_i \in C_i^{K+1}$. Hence, by Theorem 7.5.1 we can find some cautious lexicographic belief

$$b_i^{c_i} := (b_i^{K,c_i}; b_i^{K-1,c_i}; ...; b_i^{0,c_i})$$

with $b_i^{m,c_i} \in \Delta^+(C_j^m)$ for every $m \in \{0, ..., K\}$, such that c_i is optimal for $b_i^{c_i}$.

Step 2: Construction of types
We will now use these lexicographic beliefs $b_i^{c_i}$ to construct, for every choice $c_i \in C_i^1$, some type $t_i^{c_i}$. For both players i, let the set of types T_i be given by

$$T_i = \{t_i^{c_i} : c_i \in C_i^1\}.$$

For every k, let T_i^k be the set of types $t_i^{c_i}$ with $c_i \in C_i^k$. Hence, $T_i^K \subseteq T_i^{K-1} \subseteq ... \subseteq T_i^1$, where $T_i^1 = T_i$.

We will now define, for every type $t_i^{c_i}$, the corresponding lexicographic belief $b_i(t_i^{c_i})$ on $C_j \times T_j$. Take some choice $c_i \in C_i^1$. We distinguish the following two cases:
(1) Consider first the types $t_i^{c_i} \in T_i$ with $c_i \in D_i^k$ for some $k \in \{1, ..., K-1\}$. That is, $t_i^{c_i} \in T_i^k \backslash T_i^{k+1}$ for some k. From step 1 we know that c_i is optimal for some cautious lexicographic belief $b_i^{c_i} = (b_i^{k-1,c_i}; b_i^{k-2,c_i}; ...; b_i^{0,c_i})$, where $b_i^{m,c_i} \in \Delta^+(C_j^m)$ for every $m \in \{0, ..., k-1\}$. We construct the lexicographic belief

$$b_i(t_i^{c_i}) = (b_i^{k-1}(t_i^{c_i}); b_i^{k-2}(t_i^{c_i}); ...; b_i^0(t_i^{c_i}))$$

on $C_j \times T_j$ as follows.

Let us start with the beliefs $b_i^{k-1}(t_i^{c_i}), ..., b_i^1(t_i^{c_i})$. Take some $m \in \{1, ..., k-1\}$. We will define the belief $b_i^m(t_i^{c_i})$. For all of the opponent's choices $c_j \in C_j^m$, we know that c_j is optimal for the lexicographic belief $b_j^{c_j}$. But there may be some other lexicographic beliefs $b_j^{d_j}$, with $d_j \in C_j^m$, for which c_j is optimal. Let $C_j^m(c_j)$ be the set of choices $d_j \in C_j^m$ such that c_j is optimal for $b_j^{d_j}$, and let $|C_j^m(c_j)|$ denote the number of choices in $C_j^m(c_j)$.

For all of the opponent's choices $c_j \in C_j^m$, the belief $b_i^m(t_i^{c_i})$ assigns probability $b_i^{m,c_i}(c_j)/|C_j^m(c_j)|$ to every choice-type pair $(c_j, t_j^{d_j})$ with $d_j \in C_j^m(c_j)$, and probability zero to all other choice-type pairs. So, the belief $b_i^m(t_i^{c_i})$ has the same probability distribution on j's choices as b_i^{m,c_i}. Moreover, $b_i^m(t_i^{c_i})$ only assigns positive probability to choice-type pairs (c_j, t_j) where $c_j \in C_j^m$, and $t_j = t_j^{d_j}$ for some $d_j \in C_j^m$.

We define the last level, $b_i^0(t_i^{c_i})$, a little differently. Let $|T_j|$ denote the number of different types for player j in this model. For every choice $c_j \in C_j$ and type $t_j \in T_j$, belief $b_i^0(t_i^{c_i})$ assigns probability $b_i^{0,c_i}(c_j)/|T_j|$ to the choice-type pair (c_j, t_j). So, $b_i^0(t_i^{c_i})$ holds the same probability distribution on j's choices as b_i^{0,c_i}. As b_i^{0,c_i} assigns positive probability to each of j's choices in $C_j^0 = C_j$, it follows that $b_i^0(t_i^{c_i})$ assigns positive probability to *all* choice-type pairs (c_j, t_j) in $C_j \times T_j$. This guarantees that type $t_i^{c_i}$ is cautious.

This completes the construction of the lexicographic belief $b_i(t_i^c)$. Since all levels in $b_i(t_i^c)$ hold the same probability distribution on C_j as the corresponding level in $b_i^{c_i}$, and since c_i is optimal for $b_i^{c_i}$, it follows that choice c_i is optimal for type $t_i^{c_i}$.

Summarizing, we see that for every choice $c_i \in D_i^k$, with $k \in \{1, ..., K-1\}$, we can construct a cautious type $t_i^{c_i} \in T_i^k$ with lexicographic belief

$$b_i(t_i^{c_i}) = (b_i^{k-1}(t_i^{c_i}); b_i^{k-2}(t_i^{c_i}); ...; b_i^0(t_i^{c_i}))$$

such that:

- c_i is optimal for $t_i^{c_i}$,
- for every m, the belief $b_i^m(t_i^c)$ assigns positive probability to all choices in C_j^m, and probability zero to all choices outside,
- for every $m \in \{1,...,k-1\}$, the belief $b_i^m(t_i^{c_i})$ only assigns positive probability to choice-type pairs (c_j,t_j) where $c_j \in C_j^m$, choice c_j is optimal for t_j and $t_j = t_j^{d_j}$ for some $d_j \in C_j^m$.

(2) Consider next the types $t_i^{c_i} \in T_i$ with $c_i \in C_i^K$. That is, $t_i^{c_i} \in T_i^K$. From step 1, we know that c_i is optimal for some cautious lexicographic belief $b_i^{c_i} = (b_i^{K,c_i}; b_i^{K-1,c_i}; ...; b_i^{0,c_i})$ where $b_i^{m,c_i} \in \Delta^+(C_j^m)$ for every $m \in \{0,...,K\}$. We construct the lexicographic belief

$$b_i(t_i^{c_i}) = (b_i^K(t_i^{c_i}); b_i^{K-1}(t_i^{c_i}); ...; b_i^0(t_i^{c_i}))$$

on $C_j \times T_j$ in a similar way as for Case 1.

Start with the primary belief $b_i^K(t_i^{c_i})$. For every $c_j \in C_j^K$, let $C_j^K(c_j)$ be the set of choices $d_j \in C_j^K$ such that c_j is optimal for the belief $b_j^{d_j}$. For all of the opponent's choices $c_j \in C_j^K$, the primary belief $b_i^K(t_i^{c_i})$ assigns probability $b_i^{K,c_i}(c_j)/|C_j^K(c_j)|$ to every choice-type pair $(c_j, t_j^{d_j})$ with $d_j \in C_j^K(c_j)$, and probability zero to all other choice-type pairs. As before, $|C_j^K(c_j)|$ denotes the number of choices in $C_j^K(c_j)$.

The other beliefs $b_i^{K-1}(t_i^{c_i}), ..., b_i^0(t_i^{c_i})$ are defined exactly as in case 1.

In exactly the same way as for case 1, we can conclude that choice c_i is optimal for type $t_i^{c_i}$. Hence, for every choice $c_i \in C_i^K$ we can construct a cautious type $t_i^{c_i} \in T_i^K$ with lexicographic belief

$$b_i(t_i^{c_i}) = (b_i^K(t_i^{c_i}); b_i^{K-1}(t_i^{c_i}); ...; b_i^0(t_i^{c_i}))$$

such that:

- c_i is optimal for $t_i^{c_i}$,
- for every m, the belief $b_i^m(t_i^c)$ assigns positive probability to all choices in C_j^m and probability zero to all choices outside,
- for every $m \in \{1,...,K\}$, the belief $b_i^m(t_i^{c_i})$ only assigns positive probability to choice-type pairs (c_j,t_j) where $c_j \in C_j^m$, choice c_j is optimal for t_j and $t_j = t_j^{d_j}$ for some $d_j \in C_j^m$.

The construction of the epistemic model M is hereby complete.

Step 3: Every type $t_i \in T_i^k$ expresses up to $(k-1)$-fold assumption of rationality
In order to prove this step, we will show the following lemma.

Lemma 7.5.2 *For every $k \in \{1,...,K\}$, the following two statements are true:*
(a) Every choice that can rationally be chosen by a cautious type that expresses up to $(k-1)$-fold assumption of rationality must be in C_i^k.
(b) For every choice $c_i \in C_i^k$, the associated type $t_i^{c_i} \in T_i^k$ is cautious and expresses up to $(k-1)$-fold assumption of rationality.

By 0-fold assumption of rationality we simply mean no condition at all. So, every type expresses 0-fold assumption of rationality. Note that, by construction, choice c_i is optimal for the type $t_i^{c_i}$. So, by combining the two statements (a) and (b), we can show that for every k, the set C_i^k contains exactly those choices that can rationally be made by a cautious type that expresses up to $(k-1)$-fold assumption of rationality.

Proof of Lemma 7.5.2: We will prove the two statements by induction on k.

Induction start: Start with $k = 1$.
(a) Take a choice c_i that can rationally be chosen by a cautious type that expresses 0-fold assumption of rationality. Then, c_i is optimal for some cautious type. So, by Theorem 5.8.3 in Chapter 5, we know that c_i is not weakly dominated on C_j. Therefore, $c_i \in C_i^1$.
(b) Take some choice $c_i \in C_i^1$ and consider the associated type $t_i^{c_i}$. By construction, $t_i^{c_i}$ is cautious. Moreover, by definition, every type expresses 0-fold assumption of rationality, and so does $t_i^{c_i}$.

Induction step: Now consider some $k \in \{2,...,K\}$ and assume that the two statements are true for every $m \leq k-1$.
(a) Take a choice c_i that can rationally be made by a cautious type t_i that expresses up to $(k-1)$-fold assumption of rationality. We know, from the induction assumption, that the choices in C_j^{k-1} are exactly those choices for player j that can rationally be chosen by a cautious type t_j that expresses up to $(k-2)$-fold assumption of rationality. Since t_i expresses $(k-1)$-fold assumption of rationality, it follows from conditions (1) and (2) in the definition of $(k-1)$-fold assumption of rationality, that t_i must deem all choices in C_j^{k-1} infinitely more likely than all choices outside C_j^{k-1}. As c_i is optimal for t_i, it follows by a similar argument as in Section 7.3 that c_i is optimal under some truncated cautious lexicographic belief on C_j^{k-1}. Therefore, by Theorem 5.8.3, c_i is not weakly dominated on C_j^{k-1}, which means that $c_i \in C_i^k$. So, we have shown that every choice that can rationally be made by a cautious type that expresses up to $(k-1)$-fold assumption of rationality, must be in C_i^k.
(b) Now consider some choice $c_i \in C_i^k$. Then, either $c_i \in C_i^K$ or $c_i \in C_i^{m+1} \backslash C_i^{m+2}$ for some $m \in \{k-1,...,K-2\}$. Hence, by construction in step 2, the associated type $t_i^{c_i} \in T_i^k$ has a lexicographic belief

$$b_i(t_i^{c_i}) = (b_i^m(t_i^{c_i}); b_i^{m-1}(t_i^{c_i}); ...; b_i^0(t_i^{c_i}))$$

with $m \geq k-1$. We already know that $t_i^{c_i}$ is cautious. We will prove that $t_i^{c_i}$ expresses up to $(k-1)$-fold assumption of rationality. Since $t_i^{c_i} \in T_i^k$, we know that $t_i^{c_i} \in T_i^{k-1}$

and hence, by our induction assumption, $t_i^{c_i}$ expresses up to $(k-2)$-fold assumption of rationality. So, it remains to show that $t_i^{c_i}$ expresses $(k-1)$-fold assumption of rationality. Hence, we have to verify two conditions:

(1) Whenever a choice c_j is optimal for *some* cautious type t_j – not necessarily in our epistemic model M – that expresses up to a $(k-2)$-fold assumption of rationality, type $t_i^{c_i}$ should deem possible such a type t_j for which c_j is indeed optimal.

(2) Type $t_i^{c_i}$ should deem all choice-type pairs (c_j,t_j) where t_j is cautious, t_j expresses up to $(k-2)$-fold assumption of rationality and c_j is optimal for t_j, infinitely more likely than all other choice-type pairs (c_j',t_j') that do not have these properties.

We first verify condition (1). Consider any choice c_j that is optimal for some cautious type t_j that expresses up to $(k-2)$-fold assumption of rationality. By our induction assumption of part (a) of Lemma 7.5.2, it follows that c_j must be in C_j^{k-1}. By construction, the belief level $b_i^{k-1}(t_i^{c_i})$ assigns positive probability to $t_j^{c_j}$, for which c_j is optimal. Moreover, by our induction assumption of part (b) of Lemma 7.5.2 we know that $t_j^{c_j}$ expresses up to $(k-2)$-fold assumption of rationality, since $c_j \in C_j^{k-1}$.

So, whenever choice c_j is optimal for some cautious type t_j that expresses up to $(k-2)$-fold assumption of rationality, type $t_i^{c_i}$ deems possible the type $t_j^{c_j}$, which is cautious, expresses up to $(k-2)$-fold assumption of rationality and for which choice c_j is optimal. Hence, $t_i^{c_i}$ satisfies condition (1) above.

We now verify condition (2). Consider an arbitrary choice-type pair (c_j,t_j) in the epistemic model, where t_j is cautious, t_j expresses up to $(k-2)$-fold assumption of rationality and c_j is optimal for t_j. Then, by our induction assumption of part (a) of the lemma, c_j must be in C_j^{k-1}. Moreover, since t_j is in the epistemic model M, we must have that $t_j = t_j^{d_j}$ for some $d_j \in C_j$. As d_j is optimal for $t_j^{d_j}$, and by assumption $t_j^{d_j} = t_j$ is cautious and expresses up to $(k-2)$-fold assumption of rationality, we know by our induction assumption of part (a) of the lemma that $d_j \in C_j^{k-1}$. So, $t_j = t_j^{d_j}$ for some $d_j \in C_j^{k-1}$. As c_j is optimal for $t_j = t_j^{d_j}$, we conclude that $d_j \in C_j^{k-1}(c_j)$. Hence, $t_j = t_j^{d_j}$ for some $d_j \in C_j^{k-1}(c_j)$.

By construction, belief level $b_i^{k-1}(t_i^{c_i})$ assigns positive probability to the choice-type pair $(c_j,t_j^{d_j})$ whenever $c_j \in C_j^{k-1}$ and $d_j \in C_j^{k-1}(c_j)$. So, we conclude that every choice-type pair (c_j,t_j) in the epistemic model, where t_j is cautious, t_j expresses up to $(k-2)$-fold assumption of rationality and c_j is optimal for t_j, has positive probability under $b_i^{k-1}(t_i^{c_i})$.

Now consider an arbitrary choice-type pair (c_j',t_j') in the epistemic model that does not have these properties; that is, either t_j' is not cautious, t_j' does not express up to $(k-2)$-fold assumption of rationality or c_j' is not optimal for t_j'. Since every type in the model is cautious, we have that either t_j' does not express up to $(k-2)$-fold assumption of rationality or c_j' is not optimal for t_j'.

Consider first the case where c'_j is not optimal for t'_j. By construction, all belief levels in $b_i(t_i^{c_i})$ except the final one, assign positive probability only to choice-type pairs where the choice is optimal for the type. Hence, (c'_j, t'_j) has positive probability only in the final level, $b_i^0(t_i^{c_i})$.

Consider next the case where t'_j does not express up to $(k-2)$-fold assumption of rationality. Since t'_j is in the epistemic model, we must have $t'_j = t_j^{d_j}$ for some $d_j \in C_j$. By our induction assumption of part (b) of the lemma, we know that $t_j^{d_j}$ expresses up to $(k-2)$-fold assumption of rationality whenever $d_j \in C_j^{k-1}$. Since $t'_j = t_j^{d_j}$ does not express up to $(k-2)$-fold assumption of rationality, we must conclude that $d_j \notin C_j^{k-1}$. By construction, the beliefs $b_i^m(t_i^{c_i}), ..., b_i^{k-1}(t_i^{c_i})$ only assign positive probability to types $t_j^{d_j}$ where $d_j \in C_j^{k-1}$. Hence, type $t'_j = t_j^{d_j}$ has positive probability for the first time *after* belief level $b_i^{k-1}(t_i^{c_i})$.

Summarizing we conclude that, whenever (c_j, t_j) is such that t_j is cautious, t_j expresses up to $(k-2)$-fold assumption of rationality and c_j is optimal for t_j, it has positive probability under $b_i^{k-1}(t_i^{c_i})$. On the other hand, whenever (c'_j, t'_j) does not have these properties, it has positive probability for the first time *after* $b_i^{k-1}(t_i^{c_i})$ in the lexicographic belief $b_i(t_i^{c_i})$. Hence, type $t_i^{c_i}$ deems all pairs (c_j, t_j) where t_j is cautious, t_j expresses up to $(k-2)$-fold assumption of rationality and c_j is optimal for t_j, infinitely more likely than all other pairs (c'_j, t'_j) that do not have these properties. So, we may conclude that type $t_i^{c_i}$ also satisfies condition (2) above.

Overall, we see that type $t_i^{c_i}$ satisfies conditions (1) and (2) above, and hence expresses $(k-1)$-fold assumption of rationality. Since we already know, by our induction assumption, that $t_i^{c_i}$ expresses up to $(k-2)$-fold assumption of rationality, it follows that $t_i^{c_i}$ expresses up to $(k-1)$-fold assumption of rationality. So, for every choice $c_i \in C_i^k$, the associated type $t_i^{c_i} \in T_i^k$ is cautious and expresses up to $(k-1)$-fold assumption of rationality. This proves part (b) of the lemma.

By induction on k, the statements (a) and (b) hold for every $k \in \{1, ..., K\}$. This completes the proof of the lemma. ◇

Clearly, the statement in step 3 follows from this lemma. So, step 3 is complete now.

Step 4: Every type $t_i \in T_i^K$ expresses common assumption of rationality
In order to prove this, we show the following lemma.

Lemma 7.5.3 *For all choices $c_i \in C_i^K$ and all $k \geq K-1$, the associated type $t_i^{c_i} \in T_i^K$ expresses up to k-fold assumption of rationality.*

Proof of Lemma 7.5.3: We prove the statement by induction on k.

Induction start: Start with $k = K-1$. Take some $c_i \in C_i^K$. By Lemma 7.5.2 we know that type $t_i^{c_i}$ expresses up to $(K-1)$-fold assumption of rationality.

Induction step: Now consider some $k \geq K$, and assume that for both players i and for all choices $c_i \in C_i^K$, the associated type $t_i^{c_i}$ expresses up to $(k-1)$-fold assumption of rationality.

Consider an arbitrary choice $c_i \in C_i^K$ and the associated type $t_i^{c_i}$ with lexicographic belief

$$b_i(t_i^{c_i}) = (b_i^K(t_i^{c_i}); b_i^{K-1}(t_i^{c_i}); ...; b_i^0(t_i^{c_i}))$$

as constructed in step 2. We must show that $t_i^{c_i}$ expresses up to k-fold assumption of rationality. Since, by our induction assumption, $t_i^{c_i}$ expresses up to $(k-1)$-fold assumption of rationality, it is sufficient to show that $t_i^{c_i}$ expresses k-fold assumption of rationality. So, we must verify the following two conditions:

(1) Whenever a choice c_j is optimal for *some* cautious type t_j – not necessarily in our epistemic model M – that expresses up to $(k-1)$-fold assumption of rationality, type $t_i^{c_i}$ should deem possible some cautious type t_j that expresses up to $(k-1)$-fold assumption of rationality, and for which c_j is indeed optimal.

(2) Type $t_i^{c_i}$ should deem all choice-type pairs (c_j, t_j), where t_j is cautious, t_j expresses up to $(k-1)$-fold assumption of rationality and c_j is optimal for t_j, infinitely more likely than all other choice-type pairs (c_j', t_j') that do not have these properties.

We first check condition (1) for $t_i^{c_i}$. Consider an arbitrary choice c_j that is optimal for some cautious type t_j that expresses up to $(k-1)$-fold assumption of rationality. Since $k \geq K$, it follows that t_j expresses up to $(K-1)$-fold assumption of rationality. But then, by Lemma 7.5.2, part (b), we know that $c_j \in C_j^K$. By construction, the primary belief $b_i^K(t_i^{c_i})$ assigns positive probability to type $t_j^{c_j}$, for which c_j is optimal. Since $c_j \in C_j^K$, we know by our induction assumption that $t_j^{c_j}$ expresses up to $(k-1)$-fold assumption of rationality. Hence, whenever c_j is optimal for some cautious type t_j that expresses up to $(k-1)$-fold assumption of rationality, then $t_i^{c_i}$ deems possible the type $t_j^{c_j}$, which is cautious, expresses up to $(k-1)$-fold assumption of rationality and for which c_j is indeed optimal. So, $t_i^{c_i}$ satisfies condition (1).

We now turn to condition (2). Take an arbitrary pair (c_j, t_j) in the model where t_j is cautious, t_j expresses up to $(k-1)$-fold assumption of rationality and c_j is optimal for t_j. Then, first of all, c_j must be in C_j^K as we have seen. Since t_j is in the model, we must have that $t_j = t_j^{d_j}$ for some $d_j \in C_j$. Moreover, as d_j is optimal for $t_j^{d_j}$, and by assumption $t_j^{d_j} = t_j$ is cautious and expresses up to $(k-1)$-fold assumption of rationality, we must have that $d_j \in C_j^K$. Since c_j is optimal for $t_j^{d_j}$, we conclude that $d_j \in C_j^K(c_j)$. So, $t_j = t_j^{d_j}$ for some $d_j \in C_j^K(c_j)$.

By construction, the primary belief $b_i^K(t_i^{c_i})$ assigns positive probability to the pair $(c_j, t_j^{d_j})$, as $c_j \in C_j^K$ and $d_j \in C_j^K(c_j)$. We thus conclude that, whenever the pair (c_j, t_j) in the model is such that t_j is cautious, t_j expresses up to $(k-1)$-fold assumption of rationality and c_j is optimal for t_j, then the primary belief $b_i^K(t_i^{c_i})$ assigns positive probability to the pair (c_j, t_j).

Now consider some pair (c_j', t_j') that does not have these properties; that is, either c_j' is not optimal for t_j' or t_j' does not express up to $(k-1)$-fold assumption of rationality.

Consider first the case where c_j' is not optimal for t_j'. By construction, all belief levels in $b_i(t_i^{c_i})$ except the final one assign positive probability only to choice-type pairs where the choice is optimal for the type. Hence, (c_j', t_j') has positive probability only in the final level, $b_i^0(t_i^{c_i})$.

Consider next the case where t_j' does not express up to $(k-1)$-fold assumption of rationality. Since t_j' is in the epistemic model, we must have $t_j' = t_j^{d_j}$ for some $d_j \in C_j$. By our induction assumption, we know that $t_j^{d_j}$ expresses up to $(k-1)$-fold assumption of rationality whenever $d_j \in C_j^K$. Since $t_j' = t_j^{d_j}$ does not express up to $(k-1)$-fold assumption of rationality, we must conclude that $d_j \notin C_j^K$. By construction, the primary belief $b_i^K(t_i^{c_i})$ only assigns positive probability to types $t_j^{d_j}$ where $d_j \in C_j^K$. Hence, type $t_j' = t_j^{d_j}$ has positive probability for the first time *after* the primary belief $b_i^K(t_i^{c_i})$.

So, we see that if (c_j', t_j') is such that either c_j' is not optimal for t_j' or t_j' does not express up to $(k-1)$-fold assumption of rationality, then the lexicographic belief $b_i(t_i^{c_i})$ assigns positive probability to (c_j', t_j') for the first time *after* the primary belief $b_i^K(t_i^{c_i})$. On the other hand, we have seen that if (c_j, t_j) is such that t_j is cautious, t_j expresses up to $(k-1)$-fold assumption of rationality and c_j is optimal for t_j, then the primary belief $b_i^K(t_i^{c_i})$ assigns positive probability to the pair (c_j, t_j).

We may thus conclude that, if (c_j, t_j) is such that t_j is cautious, t_j expresses up to $(k-1)$-fold assumption of rationality and c_j is optimal for t_j, and (c_j', t_j') does not have these properties, then type $t_i^{c_i}$ deems (c_j, t_j) infinitely more likely than (c_j', t_j'). So, $t_i^{c_i}$ satisfies condition (2) as well, which means that type $t_i^{c_i}$ indeed expresses k-fold assumption of rationality.

Hence, we conclude that for every $c_i \in C_i^K$, the associated type $t_i^{c_i} \in T_i^K$ expresses up to k-fold assumption of rationality, which was to be shown. The statement in the lemma now follows by induction on k. ◇

The statement in step 4 now easily follows from this lemma. Consider some type $t_i \in T_i^K$. Then, $t_i = t_i^{c_i}$ for some $c_i \in C_i^K$. By Lemma 7.5.3, we know that $t_i^{c_i}$ expresses k-fold assumption of rationality for all k, and hence expresses common assumption of rationality. Hence, every type $t_i \in T_i^K$ expresses common assumption of rationality. This completes step 4.

Using Lemma 7.5.2 and Lemma 7.5.3 it is now easy to prove Theorem 7.3.2. Lemma 7.5.2 implies that the set C_i^k contains exactly those choices that can rationally be made by a cautious type that expresses up to $(k-1)$-fold assumption of rationality. Hence, a choice can rationally be made by a cautious type that expresses up to k-fold assumption of rationality precisely when it is in C_i^{k+1} – that is, survives $(k+1)$-fold elimination of weakly dominated choices. We thus have shown part (1) of Theorem 7.3.2.

Let us now turn to part (2) of the theorem. Consider some choice c_i that can rationally be made by a cautious type that expresses common assumption of rationality. Then, in particular, c_i can rationally be made by a cautious type that expresses up to $(K-1)$-fold assumption of rationality. Hence, by Lemma 7.5.2, it follows that $c_i \in C_i^K$, so c_i survives iterated elimination of weakly dominated choices. So, every choice c_i that can rationally be made by a cautious type that expresses common assumption of rationality, survives iterated elimination of weakly dominated choices.

On the other hand, consider some choice c_i that survives the iterated elimination of weakly dominated choices. Hence, $c_i \in C_i^K$. By Lemma 7.5.3 we know that the associated type $t_i^{c_i}$ expresses k-fold assumption of rationality for all k. In other words, the associated type $t_i^{c_i}$ expresses common assumption of rationality. As c_i is optimal for $t_i^{c_i}$, it follows that c_i can rationally be chosen by a cautious type that expresses common assumption of rationality. So, every choice c_i that survives iterated elimination of weakly dominated choices, can rationally be made by a cautious type that expresses common assumption of rationality.

Together with our conclusion above, we see that a choice c_i can rationally be made by a cautious type that expresses common assumption of rationality, if and only if, c_i survives iterated elimination of weakly dominated choices. This completes the proof of part (2) of the theorem. The proof of Theorem 7.3.2 is hereby complete. ■

Practical problems

7.1 The closer the better

Recall the story from Problem 5.3 in Chapter 5.

(a) In part (b) of Problem 6.2 you constructed an epistemic model in which every type is cautious and expresses common full belief in "caution and respect of preferences." Which of these types express 1-fold assumption of rationality? And which of these types express 2-fold assumption of rationality?

(b) Which numbers can you and Barbara rationally choose under 1-fold assumption of rationality?

(c) Construct an epistemic model such that, for every number c_i found in (b), there is a type $t_i^{c_i}$ such that:

- c_i is optimal for $t_i^{c_i}$, and
- $t_i^{c_i}$ expresses 1-fold assumption of rationality.

(d) Which numbers can you and Barbara rationally choose when expressing up to 2-fold assumption of rationality?

(e) Construct an epistemic model such that, for every number c_i found in (d), there is a type $t_i^{c_i}$ such that:

- c_i is optimal for $t_i^{c_i}$, and
- $t_i^{c_i}$ expresses up to 2-fold assumption of rationality.

(f) Which numbers can you and Barbara rationally choose under common assumption of rationality? Compare this to the numbers you could rationally choose under common full belief in "caution and primary belief in rationality" (Problem 5.3(e)) and the numbers you could rationally choose under common full belief in "caution and respect of preferences" (Problem 6.2(b)).
(g) Construct an epistemic model such that, for every number c_i found in (f), there is a type $t_i^{c_i}$ such that:

- c_i is optimal for $t_i^{c_i}$, and
- $t_i^{c_i}$ expresses common assumption of rationality.

Which types in your model express common assumption of rationality? Which do not?

7.2 Stealing an apple

Recall the story from Problem 5.7 in Chapter 5.
(a) For every $k \in \{0, 1, 2, ...\}$ find those chambers that you and the farmer can rationally choose if you express up to k-fold assumption of rationality. Which chambers can you rationally choose under common assumption of rationality? Compare this to the chambers you could rationally choose under common full belief in "caution and primary belief in rationality" (Problem 5.7(d)).
(b) Construct an epistemic model such that, for every choice c_i found in (a) for common assumption of rationality, there is a type $t_i^{c_i}$ such that c_i is optimal for $t_i^{c_i}$ and $t_i^{c_i}$ expresses common assumption of rationality. Which types in your model express common assumption of rationality? Which do not?
Hint for (b): For every $k \in \{0, 1, 2, ...\}$ consider those choices c_i that can rationally be made when expressing up to k-fold assumption of rationality, but not when expressing $(k+1)$-fold assumption of rationality. For each of these choices c_i, construct a type $t_i^{c_i}$ such that:

- choice c_i is optimal for type $t_i^{c_i}$, and
- type $t_i^{c_i}$ expresses up to k-fold assumption of rationality.

For this construction, start with the beliefs of the types for $k = 0$, then define the beliefs of the types for $k = 1$, and so on. Finally, consider the choices c_i that can rationally be made under *common* assumption of rationality. For each of these choices c_i, construct the belief of the type $t_i^{c_i}$ such that:

- choice c_i is optimal for type $t_i^{c_i}$, and
- type $t_i^{c_i}$ expresses common assumption of rationality.

7.3 A first-price auction

Recall the story from Problem 6.6 in Chapter 6.

(a) For every $k \in \{0, 1, 2, ...\}$ find those prices that you and Barbara can rationally name when expressing up to k-fold assumption of rationality. Which prices can you rationally name under common assumption of rationality? Compare this to the prices you could rationally name under common full belief in "caution and primary belief in rationality" (Problem 6.6(b)) and the prices you could rationally name under common full belief in "caution and respect of preferences" (Problem 6.6(c)).

(b) Construct an epistemic model such that, for every choice c_i found in (a) for common assumption of rationality, there is a type $t_i^{c_i}$ such that c_i is optimal for $t_i^{c_i}$ and $t_i^{c_i}$ expresses common assumption of rationality. Which types in your model express common assumption of rationality? Which do not? Use the hint that is given in Problem 7.2(b).

7.4 A night at the opera

You and Barbara are planning to go to the opera tomorrow evening. However, you both still need to reserve tickets. Today you told her that you would like to sit next to her. She answered that she would prefer not to, since you always talk too much during the opera. It therefore seems a good idea that you both reserve your tickets independently of each other. At this moment, you and Barbara are both sitting behind your computers, ready to reserve a ticket online. There are only nine seats available, and these nine seats happen to be in the same row, all next to each other. Suppose these seats are numbered 1 to 9. The question is: Which seat should you reserve?

If you reserve seats that are next to each other, then your utility will be 1 whereas Barbara's utility will be 0. If you reserve seats that are separated by at least one other seat, your utility will be 0, and Barbara's utility will be 1. If you both select the same seat, the computer program will take care of this and choose two seats that are next to each other.

(a) For every $k \in \{0, 1, 2, ...\}$ find those seats that you and Barbara can rationally choose when expressing up to k-fold assumption of rationality. Which seats can you and Barbara rationally choose under common assumption of rationality?

(b) Construct an epistemic model such that, for every choice c_i found in (a) for common assumption of rationality, there is a type $t_i^{c_i}$ such that c_i is optimal for $t_i^{c_i}$ and $t_i^{c_i}$ expresses common assumption of rationality. Which types in your model express common assumption of rationality? Which do not? Use the hint that is given in Problem 7.2(b).

7.5 Doing the dishes

It is 8.30 pm, you are watching TV, and your favorite movie starts right now. The movie ends at 10.30 pm. Unfortunately, there is also a huge pile of dishes waiting for you to be washed, which would take an hour. Barbara, who has been visiting a friend since 8.00 pm, said that she would take the bus home, but did not say at what time. The bus ride takes half an hour, and buses arrive at 9.00 pm, 9.30 pm, 10.00 pm, 10.30 pm and 11.00 pm in front of your house. You promised her that if she comes home and finds out that you have not finished washing the dishes, then you will do the cooking for the rest of the month. If you have not yet started the dishes when she gets in, you must do them immediately. If you have already started but not finished them, you must of course finish them. The question is: When should you start washing the dishes? To keep things simple, suppose that you only need to consider starting at 8.30, 9.00, 9.30, 10.00 and 10.30.

Obviously, you want to see as much as possible of the movie, but on the other hand you do not want to do the cooking for the rest of the month. Barbara, on the other hand, would like to stay as long as possible with her friend, but would also be very happy if you have to do the cooking for the rest of the month. More precisely, every minute that you watch your movie increases your utility by 1, but doing the cooking for the rest of the month would decrease your utility by 40. For Barbara, every minute she spends with her friend increases her utility by 1, but her utility would increase by another 140 if you have to do the cooking for the rest of the month.

(a) Formulate this situation as a game between you and Barbara.

(b) Find those times you can rationally start doing the dishes under common full belief in "caution and primary belief in rationality."

(c) For every $k \in \{0, 1, 2, ...\}$ find those times that you can rationally start doing the dishes when expressing up to k-fold assumption of rationality. Also find those times that Barbara can rationally arrive when expressing up to k-fold assumption of rationality. What times can you and Barbara rationally choose under common assumption of rationality? Compare this to you answer in (b).

(d) Construct an epistemic model such that, for every choice c_i found in (c) for common assumption of rationality, there is a type $t_i^{c_i}$ such that c_i is optimal for $t_i^{c_i}$, and $t_i^{c_i}$ expresses common assumption of rationality. Which types in your model express common assumption of rationality? Which do not? Use the hint that is given in Problem 7.2(b).

7.6 Who lets the dog out?

It is 7.00 pm in the evening and it is raining outside. The problem is: Who will let the dog out? You or Barbara? The first who says: "I will go," must go out with the dog. However, if you both wait too long then the dog may wet the carpet.

More precisely, assume that you both choose a time in your mind between 7.00 pm and 9.30 pm, which is the latest time that you plan to say: "I will go,"in case the other person has not yet done so. To make things easy, suppose that you and Barbara only need to consider the times 7.00, 7.30, 8.00, 8.30, 9.00 and 9.30. If you both simultaneously say: "I will go," then you will argue for a couple of minutes, and eventually Barbara will go out with the dog thanks to your excellent negotiation skills.

Barbara hates going out in the rain more than you do: Going out with the dog would decrease your utility by 5, but would decrease Barbara's utility by 7.

Every half an hour that goes by increases the chance that the dog will wet the carpet by 20%. If the dog does wet the carpet, this would decrease the utility for you and Barbara by 10.

The question is: At what time do you plan to say: "I will go out with the dog"?

(a) Model this situation as a game between you and Barbara.

(b) Under common assumption of rationality, at what time will you say: "I will go"? What about Barbara? Who will go out with the dog eventually?

(c) Construct an epistemic model such that for every time c_i found in (b) there is a type $t_i^{c_i}$ such that:

- c_i is optimal for $t_i^{c_i}$, and
- $t_i^{c_i}$ expresses common assumption of rationality.

How many types do we need in total in this epistemic model? Which types in your model express common assumption of rationality? Which do not? Use the hint that is given in Problem 7.2(b).

7.7 Time to clean the house

It is Saturday morning, the sun is shining and Barbara thinks it is time to clean the house. There are six rooms in total, and Barbara will tell you in a minute how many rooms you must clean. This can be any number between 1 and 6. You are thinking about a negotiating strategy. Suppose, for simplicity, that you can adopt one of the following seven strategies during the negotiation:

- Accept to clean any number of rooms.
- Accept to clean at most five rooms.
- Accept to clean at most four rooms.
- Accept to clean at most three rooms.
- Accept to clean at most two rooms.
- Accept to clean at most one room.
- Refuse to clean any room.

However, you find it hard to be tough during negotiations. So, the tougher you plan to be, the more utility you will lose because of it. More precisely, if you plan to accept at most k rooms, then your utility will be decreased by $6 - k$.

If you accept to clean at most k rooms, but Barbara asks you to clean more, then you will eventually only clean k rooms. If Barbara asks you to clean fewer than or equal to k rooms, then you will of course do precisely as Barbara says.

Suppose that cleaning a room decreases your utility by 2 and increases Barbara's utility by 2. If Barbara asks you to clean more rooms than you accept, then you will enter into a fierce discussion, which would decrease Barbara's utility by 5.

(a) Model this situation as a game between you and Barbara.

(b) What negotiating strategies can you rationally choose under common full belief in "caution and primary belief in rationality"? What about Barbara?

(c) What negotiating strategies can you rationally choose under common assumption of rationality? What about Barbara?
(d) Construct an epistemic model such that for every choice c_i found in (c) there is a type $t_i^{c_i}$ such that:

- c_i is optimal for $t_i^{c_i}$, and
- $t_i^{c_i}$ expresses common assumption of rationality.

How many types do we need in total in this epistemic model? Which types in your model express common assumption of rationality? Which do not? Use the hint that is given in Problem 7.2(b).

Theoretical problems

7.8 Dekel–Fudenberg procedure and iterated elimination of weakly dominated choices

Show that every choice that survives iterated elimination of weakly dominated choices, also survives the Dekel–Fudenberg procedure.

7.9 Common assumption of rationality and common full belief in "caution and respect of preferences"

For a given game with two-players, let C_i^k be the set of choices that survive k-fold elimination of weakly dominated choices. Now, consider a game with the following properties:

- For every k, every choice $c_i \in C_i^k \backslash C_i^{k+1}$ is weakly dominated on C_j^k by *all choices in* C_i^{k+1}.
- Iterated elimination of weakly dominated choices yields a unique choice for both players.

Show that, in such games, common assumption of rationality yields the same choices as common full belief in "caution and respect of preferences."

7.10 Self-admissible pairs of choice sets

In a given game, consider a choice c_i and a randomized choice r_i for player i. Say that c_i and r_i are *equivalent* for player i if they always induce the same expected utility for player i, that is, $u_i(c_i,c_j) = u_i(r_i,c_j)$ for every $c_j \in C_j$.
(a) Suppose that the choice c_i for player i is equivalent to the randomized choice r_i. Let b_i be a lexicographic belief for player i under which c_i is optimal. Show that every choice with a positive probability under r_i is also optimal under the lexicographic belief b_i.

Let $D_i \subseteq C_i$ and $D_j \subseteq C_j$ be subsets of choices for players i and j. The pair (D_i,D_j) of choice sets is called *self-admissible* if

- every $c_i \in D_i$ is not weakly dominated on C_j,
- every $c_i \in D_i$ is not weakly dominated on D_j,
- whenever $c_i \in D_i$, and c_i for player i is equivalent to a randomized choice r_i, then every choice with a positive probability under r_i must also be in D_i,
- likewise for player j.

(b) Let D_i and D_j be the sets of choices that survive iterated elimination of weakly dominated choices. Show that (D_i, D_j) is self-admissible. Show that in the example "Spy game" this is not the only self-admissible pair of choice sets.

Consider an epistemic model with lexicographic beliefs, and a cautious type t_i within it. Say that t_i *weakly assumes the opponent's rationality* if t_i deems all choice-type pairs (c_j, t_j) where t_j is cautious and c_j is optimal for t_j, infinitely more likely than all choice-type pairs (c'_j, t'_j) that do not have these properties. So, we do not require condition (1) in the definition of assuming the opponent's rationality. In the obvious way, we can then a define k-fold weak assumption of rationality for all k and common weak assumption of rationality.
(c) Let c_i be a choice that can rationally be made under common weak assumption of rationality. Show that there is some self-admissible pair (D_i, D_j) of choice sets with $c_i \in D_i$.
(d) Let (D_i, D_j) be some self-admissible pair of choice sets and $c_i \in D_i$. Show that c_i can rationally be made under common weak assumption of rationality.
Hints for (d): It is sufficient to construct lexicographic beliefs with two levels only. Moreover, you may use the following for self-admissible pairs (D_i, D_j): For every $c_i \in D_i$ there is a standard belief b_i on C_j, assigning positive probability to all choices in C_j, such that c_i is optimal under b_i, and all other choices that are optimal under b_i are in D_i. Also, you can construct the epistemic model in such a way that all types weakly assume the opponent's rationality.

So, by parts (c) and (d) we conclude that you can rationally make a choice c_i under common weak assumption of rationality, if and only if, c_i is in some self-admissible pair of choice sets. In the definition of self-admissible pairs of choice sets, we do need the third condition, otherwise part (d) above would not be true. To see this, consider the game in Table 7.9. The pair of choice sets $(\{a\}, \{e, f\})$ clearly satisfies the first two conditions in the definition of self-admissible sets. The next question explores the third condition.
(e) Show that $(\{a\}, \{e, f\})$ does not satisfy the third condition in the definition of self-admissible pairs of choice sets. Explain why you cannot rationally choose a under common weak assumption of rationality.
(f) Show that, in general, every choice that can rationally be made under common assumption of rationality, can also rationally be made under common weak assumption of rationality.
(g) Find a game where there is some choice c_i that can rationally be made under common weak assumption of rationality but not under common assumption of rationality.

Table 7.9. *On self-admissible pairs of choice sets*

		Barbara	
		e	f
You	a	1,4	1,4
	b	−1,3	−1,0
	c	2,0	0,3
	d	0,0	2,3

Literature

Assuming the opponent's rationality

The main idea in this chapter is that of *assuming the opponent's rationality*,which states that if an opponent's choice a is optimal for some cautious belief, whereas another choice b is not, then you deem a infinitely more likely than b. The idea of assuming the opponent's rationality was developed by Brandenburger, Friedenberg and Keisler (2008). The way they formalized this notion is different, however, from our approach in this chapter. In order to explain these differences, let us first recall how we formally defined the idea of assuming the opponent's rationality.

We took a cautious type t_i within an epistemic model M. We then said that t_i assumes the opponent's rationality if (1) for all of the opponent's choices c_j that are optimal for some cautious belief, type t_i deems possible some cautious type t_j for which c_j is indeed optimal, and (2) type t_i deems all choice-type pairs (c_j, t_j) where t_j is cautious and c_j is optimal for t_j, infinitely more likely than all choice-type pairs (c_j', t_j') that do not have these properties.

So, this definition consists of two parts. Part (1) states that type t_i must deem possible sufficiently many types for the opponent, whereas part (2) says that t_i must deem all "rational" choice-type pairs for the opponent infinitely more likely than all other choice-type pairs. We call condition (1) a "richness" condition, as it requires the set of types in the epistemic model M to be "rich enough."

Brandenburger, Friedenberg and Keisler (2008), for their definition of assuming the opponent's rationality, essentially adopted condition (2) above, but imposed a richness condition that is stronger than our condition (1). They required the type t_i to deem possible *all of the* opponent's choice-type pairs (c_j, t_j) in M, and, in turn, required the epistemic model M to contain *all* types t_i' that deem possible all of the opponent's choice-type pairs in M. Moreover, they imposed that the epistemic model M also contains at least one "different" type for each player, which does not deem possible all of the opponent's choice-type pairs in M. Epistemic models M with these properties are called *complete*.

Brandenburger, Friedenberg and Keisler then considered a type t_i within an epistemic model M, and say that t_i assumes the opponent's rationality if (1′) the epistemic model

M is complete, and type t_i deems possible all of the opponent's choice-type pairs in M, and (2′) type t_i deems all choice-type pairs (c_j, t_j), where t_j deems possible all of the opponent's choice-type pairs and c_j is optimal for t_j, infinitely more likely than all choice-type pairs (c_j', t_j') that do not have this property.

It can be shown that condition (1′) is stronger than our condition (1). Consider a type t_i that satisfies condition (1′) in Brandenburger, Friedenberg and Keisler. Consider now an opponent's choice c_j that is optimal for some cautious type t_j – possibly outside the epistemic model M. By condition (1′) the epistemic model M is complete, and hence includes a cautious type t_j for which c_j is indeed optimal. Moreover, condition (1′) guarantees that t_i deems possible all of the opponent's choice-type pairs in M, and therefore, in particular, deems possible the cautious type t_j in M for which c_j is optimal. As such, we may conclude that t_i satisfies condition (1). So, indeed, the richness condition (1′) in Brandenburger, Friedenberg and Keisler is more restrictive than the richness condition (1) used in this chapter. The conditions (2) and (2′) are essentially equivalent, as they both state that "rational" choice-type pairs for the opponent should all be deemed infinitely more likely than the opponent's "irrational" choice-type pairs.

So, the key difference between the model in Brandenburger, Friedenberg and Keisler and our model is that the former requires the epistemic model M to be complete, whereas we do not insist on M being complete. In fact, for our definition of assuming the opponent's rationality it is sufficient to include only epistemic models with *finitely* many types. The definition in Brandenburger, Friedenberg and Keisler, in contrast, requires *infinitely* many – in fact, *uncountably* many – types in M, as every complete epistemic model necessarily contains uncountably many types.

More precisely, what we call assumption of rationality in this chapter, is in Brandenburger, Friedenberg and Keisler called assumption of rationality within a *complete type structure*. They also define, in the first part of their paper, assumption of rationality within an *arbitrary* epistemic model M, by means of condition (2′), but without the richness condition (1′). We will come back to this weaker definition later, when we discuss self-admissible pairs of choice sets.

Common assumption of rationality

By iterating the definition of assumption of rationality within a complete type structure, Brandenburger, Friedenberg and Keisler define, for every k, the concept of kth-order assumption of rationality within a complete type structure. This is done in essentially the same way as in this chapter for k-fold assumption of rationality. In Theorem 9.1 they show that the choices that can rationally be made by a type that expresses up to kth-order assumption of rationality within a complete type structure, are precisely the choices that survive $(k+1)$-fold elimination of weakly dominated choices. That is, in terms of choices selected, kth-order assumption of rationality within a complete type structure in Brandenburger, Friedenberg and Keisler is equivalent to k-fold assumption of rationality in this chapter.

However, an important difference with our model is the fact that *common* assumption of rationality within a complete and continuous type structure is typically *impossible*.

Here, by "continuous" we mean that the functions that map types to lexicographic beliefs on the opponent's choice-type pairs are continuous. In Theorem 10.1, Brandenburger, Friedenberg and Keisler show that for most games there is no complete continuous epistemic model M that contains a type that expresses common assumption of rationality within a complete type structure. In other words, for most games every complete continuous epistemic model M does not contain a type that expresses kth-order assumption of rationality within a complete type structure for every k.

The reason for this non-existence is that, typically, within a complete continuous epistemic model M, the set of types T_i^k that express up to kth-order assumption of rationality within a complete type structure becomes strictly smaller whenever k increases, and this process never stops. So, there is no number K such that $T_i^{K+1} = T_i^K$. For every k, let us denote by $(C_i \times T_i^k)^{\text{opt}}$ the set of choice-type pairs (c_i, t_i) where c_i is optimal for t_i, and $t_i \in T_i^k$. Now, take an arbitrary type t_i within M. If t_i were to express common assumption of rationality within a complete type structure, then t_i must deem, for every k, all choice-type pairs in $(C_j \times T_j^k)^{\text{opt}}$ infinitely more likely than all choice-type pairs outside $(C_j \times T_j^k)^{\text{opt}}$. That is, all choice-type pairs in $(C_j \times T_j^2)^{\text{opt}} \backslash (C_j \times T_j^3)^{\text{opt}}$ must be deemed infinitely more likely than all choice-type pairs in $(C_j \times T_j^1)^{\text{opt}} \backslash (C_j \times T_j^2)^{\text{opt}}$, all choice-type pairs in $(C_j \times T_j^3)^{\text{opt}} \backslash (C_j \times T_j^4)^{\text{opt}}$ must be deemed infinitely more likely than all choice-type pairs in $(C_j \times T_j^2)^{\text{opt}} \backslash (C_j \times T_j^3)^{\text{opt}}$, and so on, ad infinitum. However, as the sets $C_j \times T_j^k$ become strictly smaller with every k, it follows that the sets $(C_j \times T_j^k)^{\text{opt}} \backslash (C_j \times T_j^{k+1})^{\text{opt}}$ are non-empty for every k, and hence it would take infinitely many levels in the lexicographic belief to meet all these conditions. But the lexicographic belief of type t_i only has finitely many levels, and can therefore never satisfy each of these conditions. That is, no type t_i within the complete continuous epistemic model can meet all the conditions imposed by common assumption of rationality within a complete type structure. Keisler and Lee (2011) demonstrated, however, that one can always construct a complete but *discontinuous* epistemic model M within which common assumption of rationality within a complete type structure is possible. That is, to make common assumption of rationality within a complete type structure possible, one has to move to complete epistemic models where the functions that map types to beliefs are *not* continuous.

In Theorem 7.3.4 we have shown, however, that common assumption of rationality is always possible within our framework. The reason we do not have a non-existence problem is that for our definition of common assumption of rationality, it is sufficient to work with epistemic models that contain *finitely* many types. In such epistemic models there will always be a number K such that, for both players i, the set T_i^{K+1} of types expressing up to $(K+1)$-fold assumption of rationality is the same as the set T_i^K of types expressing up to K-fold assumption of rationality. So, there will always be a number K such that every additional level of iterated assumption of rationality does not further restrict the set of types within the epistemic model M. This is not true for common assumption of rationality within a complete type structure in Brandenburger,

Friedenberg and Keisler. And it is precisely this phenomenon that causes the non-existence problem in their framework.

Algorithm

In Theorem 7.3.2 we saw that the algorithm of *iterated elimination of weakly dominated choices* delivers exactly those choices that can rationally be made under common assumption of rationality. In fact, the procedure of iterated elimination of weakly dominated choices has a long tradition in game theory, and was discussed in early books by Luce and Raiffa (1957) and Farquharson (1969). The latter book used the procedure of iterated elimination of weakly dominated choices to analyze voting schemes, just as Moulin (1979) did some years later. Ever since, the procedure has often been applied to narrow down the choices for the players in a game, although an epistemic foundation for the procedure only appeared in 2008, with the paper of Brandenburger, Friedenberg and Keisler.

Lexicographic rationalizability

Suppose we apply the algorithm of iterated elimination of weakly dominated choices to a given game. If the procedure terminates after K steps, then for every $k \in \{0, ..., K-1\}$ we may denote by D_i^k the set of choices for player i that survive step k but not step $k+1$ of the algorithm. Moreover, we may denote by D_i^K the set of choices that survive the complete algorithm. We know by Theorem 7.3.2 that for every $k \in \{1, ..., K-1\}$, the set D_i^k contains precisely those choices that can rationally be made by a cautious type that expresses up to $(k-1)$-fold assumption of rationality, but not by a cautious type that expresses up to k-fold assumption of rationality. Moreover, D_i^K contains exactly those choices that can rationally be made by a cautious type that expresses common assumption of rationality. But then, every type that expresses common assumption of rationality must deem all of the opponent's choices in D_j^K infinitely more likely than all choices in D_j^{K-1}, must deem all choices in D_j^{K-1} infinitely more likely than all choices in D_j^{K-2}, and so on. So, choices that are ruled out later in the procedure must be deemed infinitely more likely than choices that have been ruled out earlier in the procedure!

This property is the basis for the concept of *lexicographic rationalizability*, as proposed by Stahl (1995). More precisely, Stahl defined for every $k \in \{0, 1, 2, ...\}$ a set of choices $C_i^k \subseteq C_i$ using the following inductive procedure: We set $C_i^0 := C_i$ for both players i. For every $k \geq 1$, let B_i^k be the set of cautious lexicographic beliefs b_i on C_j that deem all choices in C_j^{k-1} infinitely more likely than all choices in $C_j^{k-2} \backslash C_j^{k-1}$, deem all choices in $C_j^{k-2} \backslash C_j^{k-1}$ infinitely more likely than all choices in $C_j^{k-3} \backslash C_j^{k-2}$, and so on. That is, B_i^k contains the cautious lexicographic beliefs that: (1) deem choices that are not currently ruled out infinitely more likely than choices that have already been ruled out, and (2) deem choices that are ruled out later in the procedure infinitely more likely than choices that are ruled out earlier in the procedure. Moreover, let C_i^k be the set of choices that are optimal for some lexicographic belief b_i in B_i^k. This procedure thus uniquely defines a sequence of choice sets $(C_i^0, C_i^1, C_i^2, ...)$ where $C_i^{k+1} \subseteq C_i^k$ for all k. A choice c_i is called *lexicographically rationalizable* if $c_i \in C_i^k$ for all k.

Stahl (1995) showed that the lexicographically rationalizable choices are precisely those choices that survive the procedure of iterated elimination of weakly dominated choices. Moreover, for every k the choice set C_i^k contains exactly those choices that survive k-fold elimination of weakly dominated choices. Hence, in terms of choices selected, lexicographic rationalizability is equivalent to common assumption of rationality.

Self-admissible pairs of choice sets

In theoretical problem 7.10 we defined *self-admissible pairs of choice sets.* This concept was developed by Brandenburger, Friedenberg and Keisler (2008), who call them *self-admissible sets.* Recall from Problem 7.10 that a pair of choice sets (D_1, D_2) is called a self-admissible pair of choice sets if for both players i, (1) every $c_i \in D_i$ is not weakly dominated on C_j, (2) every $c_i \in D_i$ is not weakly dominated on D_j, and (3) whenever $c_i \in D_i$, and c_i for player i is equivalent to a randomized choice r_i, then every choice with a positive probability under r_i must also be in D_i.

In Problem 7.10 we also introduced *weak* assumption of rationality. A type t_i *weakly* assumes the opponent's rationality if t_i deems all choice-type pairs (c_j, t_j) where t_j is cautious and c_j is optimal for t_j, infinitely more likely than all choice-type pairs (c_j', t_j') that do not have these properties. So, we do not require the "richness" condition (1) in the definition of assuming the opponent's rationality. This definition corresponds to what Brandenburger, Friedenberg and Keisler (2008) call *assumption of rationality* – but without the specification of a *complete* type structure.

In Problem 7.10 the reader is asked to show that a choice can rationally be made under common *weak* assumption of rationality, if and only if, the choice is in a self-admissible pair of choice sets. This result is due to Brandenburger, Friedenberg and Keisler (2008) who proved it in Theorem 8.1. That is, if we do not insist on the richness condition in the definition of (common) assumption of rationality, then common assumption of rationality is equivalent, in terms of choices selected, to self-admissible pairs of choice sets. The sets of choices that survive iterated elimination of weakly dominated choices just constitute one particular self-admissible pair of choice sets, but there may be many more self-admissible pairs of choice sets in a game.

The example of Table 7.9 in Problem 7.10 is also taken from Brandenburger, Friedenberg and Keisler (2008). Finally, self-admissible pairs of choice sets were also investigated in Brandenburger and Friedenberg (2010), who applied the concept to some particular games of interest.

Part III

Conditional beliefs in dynamic games

8 Belief in the opponents' future rationality

8.1 Belief revision

In the first two parts of this book we considered so-called *static games*. These are situations where every player only makes one choice, and at the time of his choice a player does not have any information about the choices of his opponents. It may be, for instance, that all players decide simultaneously and independently. Another possibility is that the players choose at different points in time, but that they have not learnt anything about past choices of the opponents when it is their time to choose. In such situations there is no need to change your belief during a game, as you do not receive any information about your opponents up to the moment when you must make your choice. We can therefore model a player's belief in a static game by a *single*, *static* belief hierarchy, which does not change over time.

In this part of the book we will study *dynamic games*, which are situations where a player may have to make several choices over a period of time, and where he may learn about the opponents' previous choices during the game. This will cover a much broader spectrum of situations than static games, but the analysis will also be a bit harder here. The main reason is that a player in a dynamic game may have to *revise* his beliefs about the opponents during the game, as he may observe choices by the opponents that he did not expect before the game started. So it will no longer be sufficient to model a player's beliefs by a single, static belief hierarchy. Instead, we have to take into account that his beliefs may change during the game. The following example will illustrate this.

Example 8.1 Painting Chris' house

Chris is planning to paint his house tomorrow and he would like you or Barbara to help him with this job. You and Barbara are both interested, but Chris only needs one person to help him. Chris therefore proposes the following procedure: This evening you must both come to his house, and whisper a price in his ear, which can be 200, 300, 400 or 500 euros. Chris will then give the job to the person who whispered the lowest price, and this person will receive exactly this price for the job. If you both choose the same price, Chris will toss a coin to decide which friend to choose.

Just when you and Barbara are about to leave for Chris' house, Barbara receives a phone call from a colleague. He asks her whether she could repair his car tomorrow, for a price of 350 euros. So now Barbara has to decide whether to accept the colleague's offer or not. If she agrees to repair his car, she cannot help Chris tomorrow, which means that you will paint his house for a price of 500 euros.

The question is: What price will you whisper in Chris' ear if Barbara rejects the colleague's offer? In order to answer this question, we must first formally represent this situation by a game. Note that this is no longer a static game, since Barbara must first decide whether or not to accept the colleague's offer, and if she rejects it, you and Barbara must subsequently choose a price for Chris' job. So, Barbara must possibly make two decisions instead of only one. Moreover, you will learn whether Barbara has accepted or rejected the colleague's offer before going to Chris' house. So, you will learn about Barbara's first decision during the game, which means that this situation must be modeled as a dynamic game and not as a static game.

A possible way to represent this situation is shown in Figure 8.1. This picture should be read as follows: Initially, Barbara must decide between "reject" – rejecting the colleague's offer, and "accept" – accepting the colleague's offer. If she accepts the

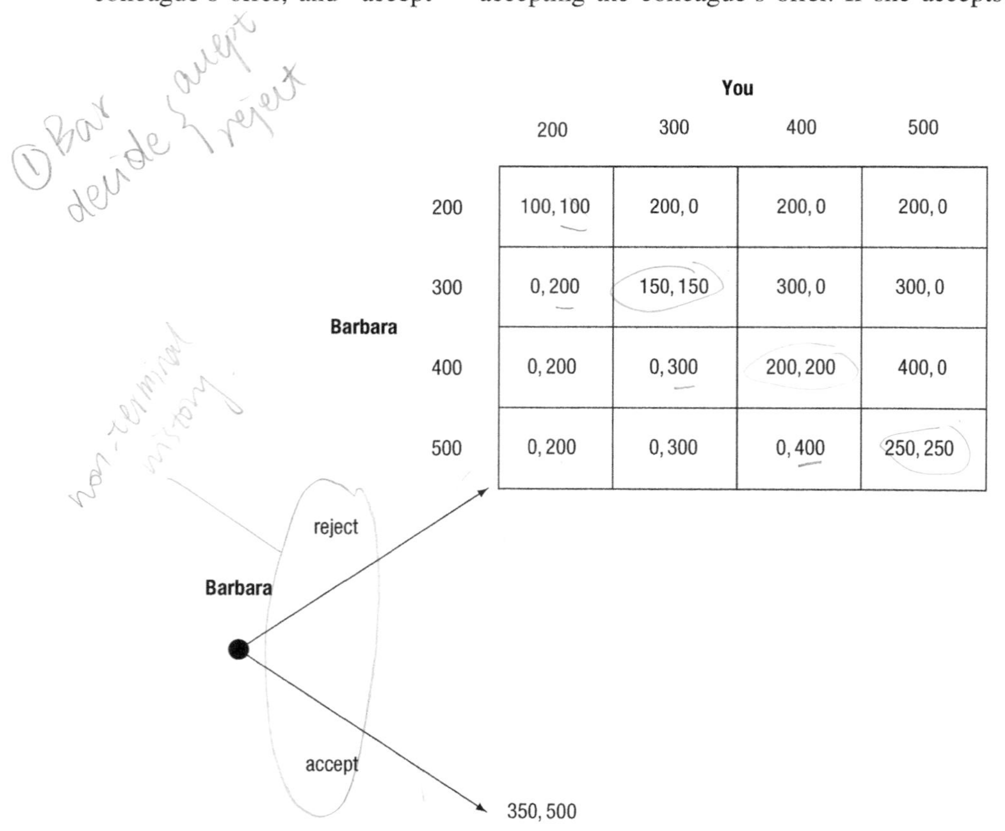

Figure 8.1 Painting Chris' house

colleague's offer, Barbara will repair the colleague's car and receive 350 euros, and you will paint Chris' house and receive 500 euros. So, if Barbara chooses "accept" then the game ends, and the utilities are 350 for Barbara and 500 for you. If Barbara rejects the colleague's offer, then you will know this, and Barbara and you must both whisper a price in Chris' ear. The resulting utilities for Barbara and you are shown in the table that follows "reject." Remember that Chris will toss a coin if you both choose the same price. So, if you both choose 300, then you will be picked with probability 0.5, and your expected utility will be

$$(0.5) \cdot 300 + (0.5) \cdot 0 = 150,$$

and the same for Barbara. Similarly, you can calculate the other expected utilities if you both choose the same price. In this figure, the first number always represents Barbara's utility and the second number corresponds to your utility.

Suppose now that it is time for you to whisper a price into Chris' ear. What price would you choose? First, it is clear that choosing a price of 500 can never be optimal for you. Your choice 500 is strictly dominated by the randomized choice in which you choose 200 and 400 with probability 0.5. But if Barbara indeed believes that you will not choose 500, then she believes that after rejecting the colleague's offer she will not get more than 300. So, Barbara would then be better off by accepting the colleague's offer at the beginning. It is therefore natural for you to believe, at the beginning of the game, that Barbara will accept the colleague's offer.

But when it is your time to choose a price, then you *know* that Barbara has in fact *rejected* the colleague's offer, which should certainly come as a surprise. Hence, your initial belief about Barbara, namely that she would accept the colleague's offer, has been contradicted if you are in the situation where you must choose your price. You must therefore *revise* your initial belief about Barbara when it is your time to choose. But how?

We cannot give a unique answer to this question, since there are at least two plausible scenarios for revising your belief, which we will call scenario 1 and scenario 2.

Scenario 1: Suppose that, after seeing Barbara reject the colleague's offer, you believe that Barbara has made a *mistake* by rejecting that offer. However, suppose you still believe that:

- Barbara will choose rationally in the *remainder of the game*,
- Barbara believes that you will choose rationally,
- Barbara believes that you believe that she will choose rationally in the *remainder of the game*,

and so on.

Then, after seeing her reject the colleague's offer, you believe that Barbara will not choose a price of 500, since that price can never be optimal for her. So, if you choose rationally, then you will never choose a price of 400 or 500, since neither can be optimal if you believe that Barbara will not choose 500. After discounting Barbara's choice 500, your choices 400 and 500 are both strictly dominated by the randomized choice in which you choose 200 and 300 with probability 0.5. But if Barbara believes that you will not choose 400 or 500, then Barbara will choose 200 if she has rejected the colleague's offer.

So, under this scenario, after observing that Barbara has rejected the colleague's offer, you believe that Barbara will choose price 200. Therefore, you would choose a price of 200 as well.

Scenario 2: Suppose that, after observing that Barbara has rejected the colleague's offer, you believe that it was a *rational choice* by Barbara to reject that offer. That is, you believe that Barbara has rejected the colleague's offer because she expects to obtain more than 350 euros by rejecting that offer. This, however, is only possible if you believe that Barbara will choose a price of 400 after rejecting the offer, since that is her only chance of receiving more than 350 euros. But in that case, your optimal choice would be to choose a price of 300, and not 200. □

So in this example we have seen two plausible belief revision scenarios, which lead to opposite choices for you: Under the first scenario you would choose a price of 200, whereas under the second scenario you would choose a price of 300, after observing Barbara reject the colleague's offer. The crucial difference between the two scenarios is how you understand Barbara's rejection of the colleague's offer. In the first scenario you believe that Barbara has made a mistake by rejecting that offer, but you maintain your belief that Barbara will choose rationally then, and that Barbara believes that you will choose rationally, and so on. In the second scenario, you do not believe that Barbara has made a mistake, but instead you believe that it was her intention to turn down her colleague's offer.

It is difficult to say which of these two scenarios is more reasonable. In fact, we think both scenarios make intuitive sense. That is the reason why we will consider both belief revision scenarios in this part of the book. Scenario 1 will be formalized in this chapter, and will eventually lead to the concept of *common belief in future rationality.* Scenario 2 will be studied in the next chapter, and will finally result in the concept of *common strong belief in rationality.*

Before we can formally present these two concepts, we must first define precisely what we mean by a dynamic game, how we model beliefs in dynamic games and how we can incorporate these beliefs into an epistemic model. This will be our task for the next few sections.

8.2 Dynamic games

In this section we shall formally define dynamic games. We saw a dynamic game in Example 8.1, and we will use this as a reference for our discussion. Consider Figure 8.1. There are two points in time where players must reach a decision: At the beginning of the game, where Barbara must choose between "reject" and "accept," and after her choice "reject," where Barbara and you must both name a price. These two points in time are called *non-terminal histories.*

The beginning of the game, which is the first non-terminal history, will be denoted by $\emptyset$, as it is not preceded by any choice by the players. In other words, it is preceded by the *empty* set of choices, and hence the symbol $\emptyset$. The second non-terminal history, where both players must choose a price, will be denoted by *reject*, as it is preceded by Barbara's choice "reject."

In the game, there are 17 possible different ways in which it could end: One after Barbara chooses to accept her colleague's offer, and 16 after Barbara chooses to reject her colleague's offer – one for every combination of prices. These situations where the game ends are called *terminal histories*. Similarly to non-terminal histories, we shall denote every terminal history by its sequence of preceding choices. So, $(reject, (300, 400))$ is the terminal history that results if Barbara rejects her colleague's offer, Barbara chooses price 300 and you choose price 400. This terminal history leads to utility 300 for Barbara and utility 0 for you. On the other hand, *accept* is the terminal history that results if Barbara chooses to accept the colleague's offer.

So, we see that this game has 2 non-terminal histories and 17 terminal histories, and each history can be identified with the sequence of preceding choices. At every non-terminal history, one or more players make a choice, and every terminal history gives a utility to each of the players.

Suppose now that Barbara must make a plan for what she will do in this game. What would such a plan look like? Intuitively, such a choice plan should specify what Barbara will do in every possible situation in the game. A possible plan for Barbara would be *accept*, which states that Barbara would accept the colleague's offer at the beginning. Another possibility is to start with *reject*, which represents a plan where Barbara would reject her colleague's offer at the beginning. This plan is not complete, however, since after rejecting her colleague's offer Barbara must also decide what price to whisper in Chris' ear. A possible complete plan would be, for instance, $(reject, 300)$ which states that Barbara will first reject her colleague's offer and will then choose price 300.

A complete choice plan is called a *strategy*. In this particular game, Barbara has five possible strategies:

$$accept, \ (reject, 200), \ (reject, 300), \ (reject, 400) \text{ and } (reject, 500).$$

Your only decision in this game is to choose a price after observing Barbara reject her colleague's offer. So, you have four possible strategies in this game:

$$200, 300, 400 \text{ and } 500.$$

Here, the strategy 200 means that you would choose price 200 if Barbara rejects her colleague's offer.

The game in Figure 8.1 has a relatively simple structure. We can also model situations that are more complex. Consider, for instance, the game in Figure 8.2. This is an abstract game, without any underlying story, to illustrate some of the ideas. The players in this game are player 1 and player 2. At the start of the game, player 1 chooses between a and b, and player 2 chooses between c and d. If player 1 chooses b, the game ends and the utilities are as given in the row that corresponds to b. If player 1 chooses a and

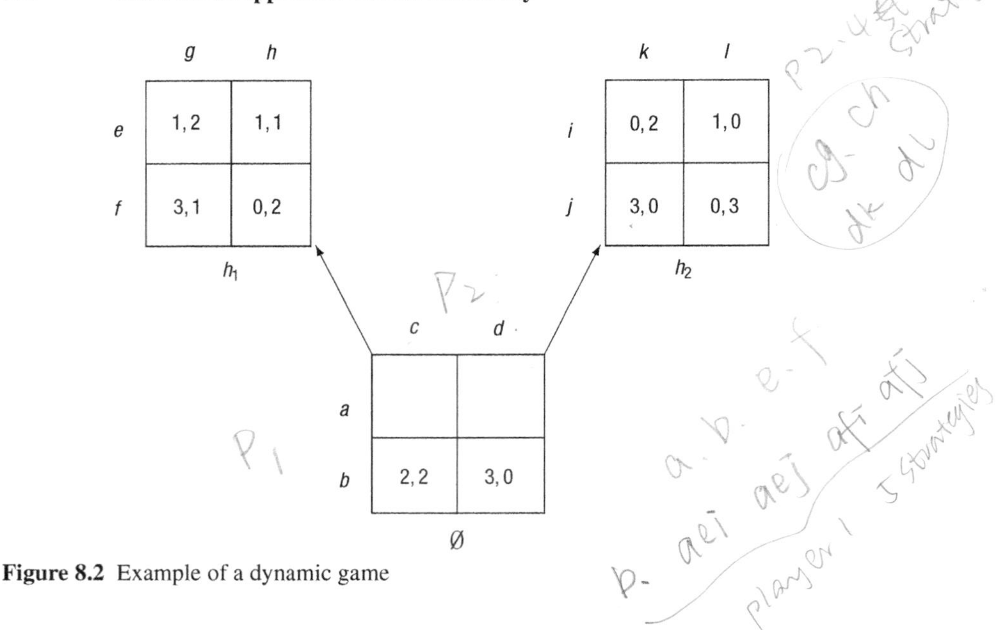

Figure 8.2 Example of a dynamic game

player 2 chooses c, the game moves to the table in the upper left corner. There, player 1 chooses between e and f, player 2 chooses between g and h, and the game ends. The resulting utilities are as depicted in that table. If at the beginning player 1 chooses a and player 2 chooses d, the game moves to the table in the upper right corner. There, player 1 chooses between i and j, player 2 chooses between k and l, and the game ends with the utilities as depicted in that table.

Recall that the non-terminal histories are the situations in the game where one or more players must make a choice. In this game, players must make a choice at the beginning of the game, at the upper left table, and at the upper right table. The upper left table is preceded by the choices (a,c), whereas the upper right table is preceded by (a,d). So, the three non-terminal histories in this game are Ø (beginning), (a,c) and (a,d). Note that there are ten terminal histories in this game. For instance, $((a,c),(f,g))$ is the terminal history that results if at the beginning, the players choose a and c, and at the upper left table they choose f and g. The utilities for that terminal history are 3 and 1.

What are the possible strategies for player 1 in this game? For player 1, a possible strategy is b. Namely, if he chooses b then he knows that the game will end immediately, so b is a complete choice plan. Suppose that player 1 starts with choice a. Then, he knows that the game will continue after this choice, but he does not know whether the game will move to the upper left table or the upper right table. This depends on player 2's choice, about which he is uncertain. So, in order for the choice plan to be complete player 1 must specify a choice for the upper left table, and a choice for the upper right table. A possible strategy for player 1 would be (a,e,j), which states that player 1 starts by choosing a, that he would choose e if player 2 chooses c at the beginning, and that he would choose j if player 2 chooses d at the beginning. As an exercise, try to find all the other strategies for player 1 in this game.

What can we say about the strategies for player 2? Suppose that player 2 starts by choosing c. Player 2 does not know whether the game will stop after this choice, or whether the game will continue. This depends on player 1's choice, about which player 2 is uncertain. But player 2 knows that, if the game continues, then the game will move to the upper left table, since player 2 has choosen c. So, (c,h) would be a complete choice plan, and hence a strategy, for player 2. Note that there is no need here for player 2 to specify what he would do at the upper right table, since this table cannot be reached if he chooses c. As an exercise, find all the other strategies for player 2 in this game.

In the game of Figure 8.2 it is assumed that, at the upper left table, player 1 knows that player 2 has chosen c at the beginning, and that player 2 knows that player 1 has chosen a at the beginning. Similarly, at the upper right table player 1 knows that player 2 has chosen d, and player 2 knows that player 1 has chosen a. That is, at the second stage of the game player knows exactly what choice has already been made by his opponent. This need not always be the case, however! It may also be that a player, at a certain stage, does not exactly know what choices have been made by his opponent so far. Consider, for instance, the game in Figure 8.3. This figure should be read as follows: If player 1 chooses a initially, and the game moves to the second stage, then player 1 does not know whether player 2 has chosen c or d at the beginning. That is, player 1 does not know whether the game has moved to non-terminal history (a,c) (the upper left table) or to non-terminal history (a,d) (the upper right table). This is modeled by the set h_1, which contains the histories (a,c) and (a,d). The set h_1 is called an *information*

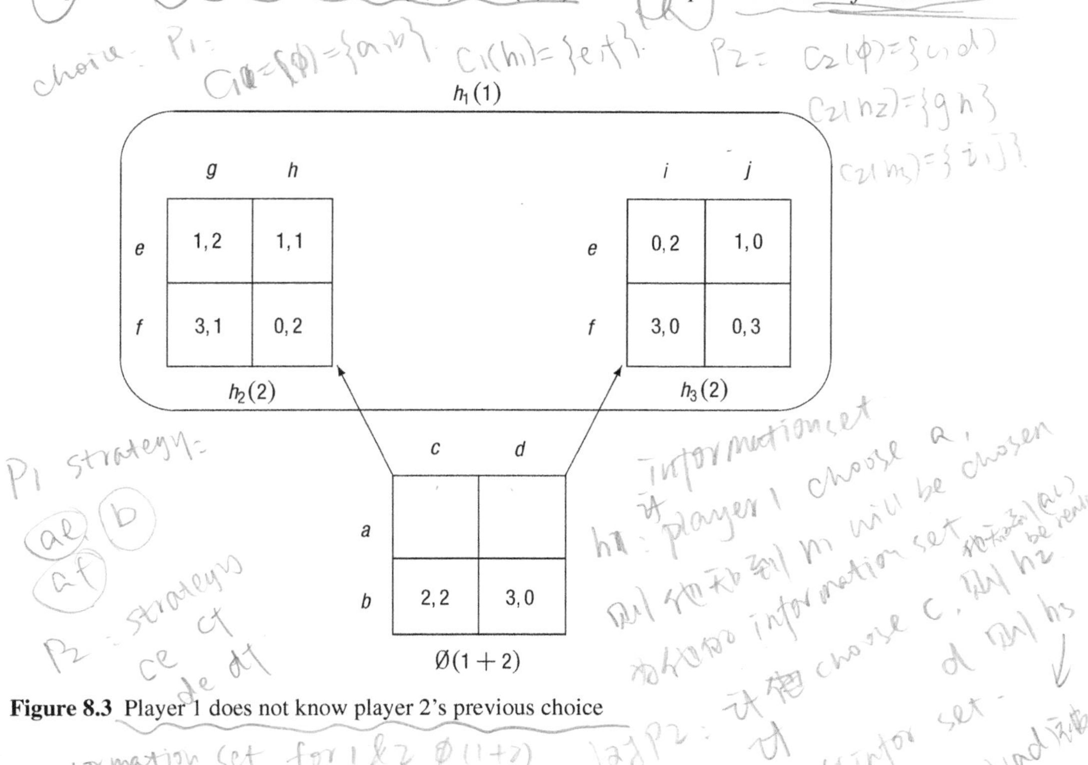

Figure 8.3 Player 1 does not know player 2's previous choice

set for player 1. The interpretation is that player 1, at h_1, does not know which of the histories in h_1 have been realized. In this case, player 1 does not know whether history (a,c) or history (a,d) has been realized.

Player 2, on the other hand, knows at the second stage whether the game is at history (a,c) or at history (a,d). This is because player 2 knows that player 1 has chosen a, and knows whether he has chosen c or d himself. Therefore, the histories (a,c) and (a,d) correspond to two different information sets, h_2 and h_3, for player 2.

So the information sets in this game represent the information that the players have about the choices made by the opponent in the past. Player 1's information set h_1, containing the histories (a,c) and (a,d), indicates that player 1 is uncertain whether history (a,c) or (a,d) has been realized. He knows, however, that one of these two histories must have occurred. Player 2's information set h_2, containing only the history (a,c), indicates that player 2 knows at that stage that history (a,c) has been realized. Similarly, player 2's information set h_3, containing only the history (a,d), indicates that player 2 knows that history (a,d) has been realized.

The numbers in brackets for the information sets indicate the player, or players, to which this information set belongs. So, the information set at the beginning of the game, indicated by $\emptyset$, belongs to players 1 and 2. Information set h_1 belongs only to player 1, as it represents the information that player 1 has (and not the information that player 2 has) at that stage of the game. The information sets h_2 and h_3, in turn, belong only to player 2, as they represent only the information that player 2 has (and not the information that player 1 has) at that stage of the game.

Obviously, a player can base his decision only upon the opponent's choices he has *observed* so far and not on choices he has not observed. In this particular game, for instance, player 1 does not know at h_1 whether player 2 has chosen c or d. At h_1, therefore, player 1 cannot base his decision on whether player 2 has chosen c or d, since he does not know. Consequently, player 1's choices at the histories (a,c) and (a,d) in h_1 must be the same, since he does not know which of these two histories has occurred. Note that player 1's available choices at histories (a,c) and (a,d) are the same, namely e and f. Player 2, on the other hand, knows whether history (a,c) or (a,d) has been realized, and therefore player 2 can base his decision on whether (a,c) or (a,d) has occurred. For that reason, player 2's available choices at history (a,c), which are g and h, are different from his available choices at (a,d), which are i and j.

In fact, the information sets for a player represent precisely the situations where this player must make a choice. In this game, player 1 has two information sets, $\emptyset$ (beginning) and h_1, and these are precisely the points in the game where player 1 must make a choice. Moreover, the player can only base his choice on the information set that is reached, and not on the precise history of events in the game (if the information set contains at least two histories).

Now, what would be the strategies for player 1 in this game? Obviously, b is a strategy, as by choosing b player 1 will be sure of ending the game, so there is no need to specify other choices. Now, suppose that player 1 chooses a at the beginning, so he knows that the game will move to his information set h_1. Remember that player 1, at h_1, does not know whether player 2 has chosen c or d, so player 1 cannot say that he will choose e after c and will choose f after d. So, (a,e,f) is not a strategy here, as player 1 cannot base his decision at the second stage on whether player 2 has chosen c or d. In fact, the only two strategies for player 1 that start with a are (a,e) and (a,f). The strategy (a,e), for instance, states that player 1 will choose a at the beginning and e at information set h_1, without knowing whether player 2 has chosen c or d. The possible strategies for player 2 are still the same as for Figure 8.2, as his information about past choices has not changed.

Could we change the information structure in this game by assigning information set h_1 to player 2, and information sets h_2 and h_3 to player 1? If that were the case, then player 2 would not know at h_1 whether he, himself, has chosen c or d, and that seems very strange! We should at least require that a player always remembers the choices he has made himself. After all, a player can always write down every choice he made on a notepad. Moreover, a player can always write down the information about the opponent's past choices he has at each stage, thereby making sure that he will never forget this information.

Consider, for instance, the game in Figure 8.4. At information set h_1, player 1 knows that player 2 has chosen c at the beginning, whereas at information set h_2 he knows that player 2 has chosen d at the beginning. Suppose now that at h_1, player 1 chooses e and player 2 chooses h, or that at h_2 player 1 chooses i and player 2 chooses k. Then, the game moves to the third stage, were player 1 must choose between m and n. There, player 1 has an information set h_3, at which he no longer knows whether player 2 initially chose c or d! At h_3 player 1 does not know whether the game has reached the upper left table or the upper right table. Since the upper left table can only be reached if player 2 chooses c at the beginning, and the upper right table can only be reached if player 2 chooses d at the beginning, player 1 at h_3 has not remembered whether player 2 has initially chosen c or d. So, this is an example of a situation where a player has forgotten information he previously had.

In general, a player who always remembers his own past choices and always remembers the information he previously had about his opponent's past choices, is said to have *perfect recall.* Throughout this part of the book, we will always assume that all players in the game have perfect recall.

After these examples, we are now ready to define in general what we mean by a dynamic game. Formally, a *dynamic game* has the following ingredients:

- First, there are *non-terminal histories,* which represent the situations where one or more players must make a choice. Every non-terminal history x consists of a sequence of choices, namely those choices that have been made by the players in the past and that lead to x. The beginning of the game is also a non-terminal history and is denoted by $\emptyset$. The set of non-terminal histories in the game is denoted by X. In the game of Figure 8.2, the set of non-terminal histories is $X = \{\emptyset, (a,c), (a,d)\}$.
- *Terminal histories* represent the situations where the game ends. Every terminal history z consists of the sequence of choices that eventually leads to z. The set of

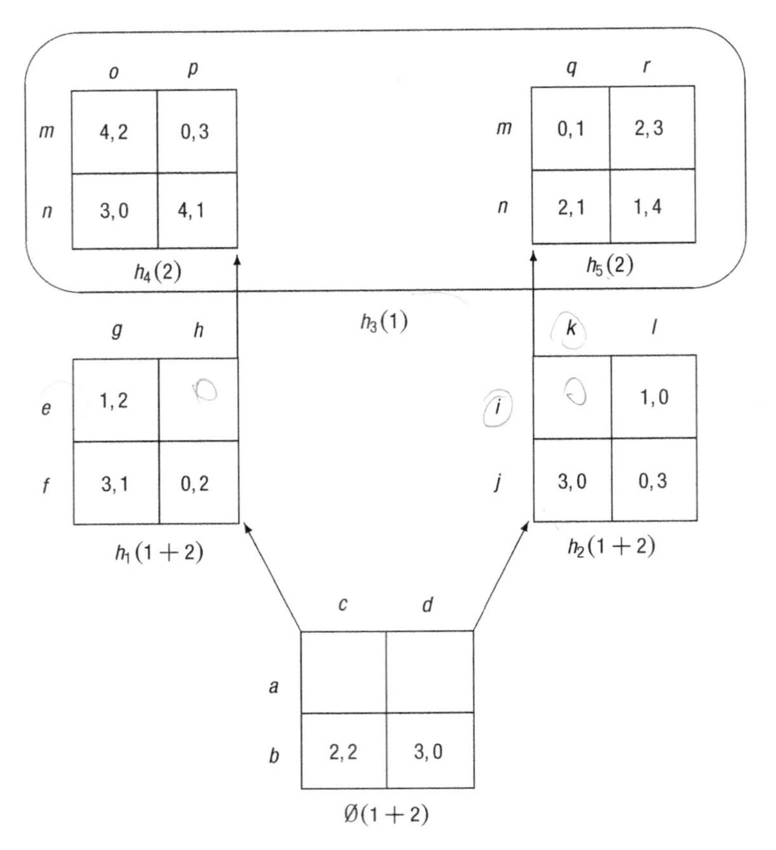

Figure 8.4 Player 1 forgets information he previously had

terminal histories is denoted by Z. In the game of Figure 8.2, a terminal history is, for instance, $z = ((a,c),(f,h))$.

- The set of *players* is denoted by I. For every non-terminal history $x \in X$, we denote by $I(x)$ the set of players who must make a choice at x. We also call $I(x)$ the set of *active* players at x. For a given player i, we denote by X_i the set of non-terminal histories where i is active. In the game of Figure 8.1, the non-terminal histories are $\emptyset$ and *reject*, and the corresponding sets of active players are $I(\emptyset) = \{\text{Barbara}\}$ and $I(reject) = \{\text{Barbara, You}\}$. Initially, only Barbara can make a choice. So, $X_{\text{Barbara}} = \{\emptyset, reject\}$ and $X_{\text{You}} = \{reject\}$.
- For every player i there is a collection of *information sets,* that represent the information that player i has about the opponents' past choices at the various stages of the game. Every information set h for player i consists of a set of non-terminal histories $\{x_1, ..., x_k\} \subseteq X_i$. If h has at least two histories, then at h player i knows that some history from $\{x_1, ..., x_k\}$ has been realized, without knowing precisely which one. If h has only one history, that is, $h = \{x\}$, then player i knows at h that history x has

been realized. The collection of information sets for player i is denoted by H_i. The collection H_i must be such that (1) no history $x \in X_i$ can be in two different information sets in H_i, and (2) every history $x \in X_i$ must be in at least one information set in H_i. In the game of Figure 8.3, the collection of information sets for player 1 is $H_1 = \{\emptyset, h_1\}$ and for player 2 it is $H_2 = \{\emptyset, h_2, h_3\}$. The information sets have the following sets of non-terminal histories:

$$\emptyset = \{\emptyset\},\ h_1 = \{(a,c),(a,d)\},$$
$$h_2 = \{(a,c)\} \text{ and } h_3 = \{(a,d)\}.$$

Here, $\emptyset = \{\emptyset\}$ means that the beginning of the game, $\emptyset$, is an information set for both players, which only contains the non-terminal history $\emptyset$.

- For every player i and every information set $h \in H_i$ for player i, there is a set of *available choices* $C_i(h)$. This means that player i can make a choice from $C_i(h)$ if the game reaches information set h. In the game of Figure 8.3, the sets of available choices are

$$C_1(\emptyset) = \{a,b\},\ C_1(h_1) = \{e,f\},$$
$$C_2(\emptyset) = \{c,d\},\ C_2(h_2) = \{g,h\} \text{ and } C_2(h_3) = \{i,j\}.$$

- At every terminal history $z \in Z$, every player i receives a *utility* $u_i(z)$. That is, if the game ends at z, then player i will derive a utility $u_i(z)$ from this particular outcome.
- We assume that every player i, at every information set $h \in H_i$, *remembers his own past choices*. Formally, if x and y are two different histories in h, then x and y must contain the same past choices for player i. In the game of Figure 8.3, h_1 cannot be an information set for player 2 since h_1 contains two histories (a,c) and (a,d) with different past choices for player 2.
- We assume that every player i, at every information set $h \in H_i$, *remembers the information he previously had* about the opponents' past choices. Formally, if x and y are two different histories in h, then x and y must pass through exactly the same past information sets for player i. In the game of Figure 8.4, player 1 knows at the second stage of the game whether player 2 has chosen c or d, but he does not remember this information at his information set h_3. Consider the histories $((a,c),(e,h))$ and $((a,d),(i,k))$ in h_3. The first history $((a,c),(e,h))$ passes through player 1's past information set h_1, where he knew that player 2 had chosen c, whereas the second history $((a,d),(i,k))$ passes through a different past information set h_2 for player 1, where he knew that player 2 had chosen d. So, the two histories $((a,c),(e,h))$ and $((a,d),(i,k))$ in h_3 pass through different sequences of player 1's information sets, which means that at h_3 player 1 has forgotten information he previously had at h_1 or h_2.
- We say that the game has *perfect recall* if every player always remembers his own past choices and always remembers the information he previously had about his opponents' past choices.

Note that a static game, which we considered in the first two parts of this book, is a special case of a dynamic game. In a static game, the only non-terminal history is $\emptyset$ and all players are active at $\emptyset$. In such a game, no player observes any past choices by his opponents, and hence every player only has one information set, which is $\emptyset$.

As we have seen in the examples, a *strategy* for player i is a complete plan of his choices throughout the game. Formally, this means that player i selects a choice for each of his information sets $h \in H_i$, unless h cannot be reached due to a choice selected at some earlier information set. Consider, for instance, the game in Figure 8.3. If player 1 starts by choosing b, then his information set h_1 can no longer be reached, so player 1 does not have to select a choice for h_1 in this case. For that reason, b is a strategy for player 1. Similarly, if player 2 starts by choosing c at $\emptyset$, then his information set h_3 can no longer be reached, so he does not have to specify a choice for it. So, (c, g) is a complete choice plan, and hence a strategy, for player 2. So we obtain the following formal definition of a strategy.

Definition 8.2.1 *(Strategy)*
*A **strategy** for player i is a function s_i that assigns to each of his information sets $h \in H_i$ some available choice $s_i(h) \in C_i(h)$, unless h cannot be reached due to some choice $s_i(h')$ at an earlier information set $h' \in H_i$. In the latter case, no choice needs to be specified at h.*

We denote the set of strategies for player i by S_i. This completes the formal description of a dynamic game. The reader will have noticed that defining a dynamic game formally requires much work, especially when compared to defining a static game formally. However, the graphical representation of a dynamic game is usually easy to understand, so the reader should not be discouraged by the lengthy definition of a dynamic game.

8.3 Conditional beliefs

In Part I on static games, we modeled the belief of a player about his opponents' choices by a *single* probability distribution over the opponents' choices. This probability distribution describes the belief he has at the beginning of the game. In a static game a player cannot observe any of the opponents' choices before he must make a choice himself, so there is no need for him to change his beliefs, and hence a single probability distribution is sufficient. In that setting, we called a choice *rational* for a player if it is optimal for the (single) belief he holds about the opponents' choices.

In a dynamic game things are more complicated. Suppose that a player must make a choice at one of his information sets. To see whether this choice is optimal for this player or not, we must consider the belief this player holds *at the moment he must make this choice.* And this belief may very well be different from the one he held at the beginning of the game! To see this, consider again the example "Painting Chris' house," with the associated dynamic game shown in Figure 8.1. Before the game starts, it seems natural for you to believe that Barbara will accept her colleague's offer. Namely, it cannot be optimal for you to choose a price of 500, and if Barbara indeed believes you will not choose a price of 500, then it will be optimal for her to accept her colleague's offer at the beginning. Now, suppose you initially believe that Barbara will accept her colleague's offer, and you are planning to choose a price of 300 if she does not accept it. Can we say, on the basis of this information, whether planning this price of 300 is optimal for you? The answer is "no." At the moment you actually have to choose a price, it is clear that Barbara has rejected her colleague's offer, and hence your initial belief about Barbara has been contradicted. At that moment in time, you have to form a *new* belief about what Barbara will do, and your plan to choose a price of 300 must be evaluated according to this new belief, and not with respect to your initial belief.

Suppose that when you learn that Barbara has rejected her colleague's offer, you believe that, with probability 0.5, she will choose price 300, and that with probability 0.5 she will choose price 400. We call this a *conditional belief* for you about Barbara, since you will only hold this belief conditional on the event that Barbara rejects her colleague's offer. Initially you hold a different belief, namely that Barbara will accept the colleague's offer.

Given this conditional belief, your plan to choose a price of 300 is indeed optimal: If Barbara does indeed reject her colleague's offer, and you hold the conditional belief above, then choosing a price of 300 would give you an expected utility of 225, whereas choosing any other price would give you a lower expected utility.

So, in this example there are two moments in time at which you hold a belief about Barbara's strategy choice: At the beginning of the game, and when you learn that Barbara has rejected her colleague's offer. For evaluating your strategy choice, however, only your second belief should be used, as it is precisely this belief you would hold at the moment you must choose a price.

The example above thus shows that in a dynamic game, we must be careful to use the right beliefs in order to evaluate a choice by a player. Another difficulty is that in a dynamic game, a player typically chooses more than once, so we do not just evaluate the optimality of *one* choice, but usually of *several* choices. To illustrate this, consider the game from Figure 8.2. In this game there are three information sets, namely $\emptyset$ (beginning), h_1 and h_2.

Suppose that player 2 is planning to choose the strategy (c, g), and that he believes at $\emptyset$ that player 1 will choose b, and thereby terminate the game. Is this enough to conclude that player 2's strategy (c, g) is optimal for him? The answer should be "no": From player 2's initial belief – that player 1 will choose b – we can only conclude that choosing c is optimal for him at $\emptyset$. However, player 2 cannot be sure that player 1 will indeed choose b – there is always the possibility that he would choose a and thereby move the game to h_1. Hence, in order to judge whether strategy (c, g) is optimal for player 2, we must also evaluate the optimality of his choice g at h_1, even though player 2 initially believes that h_1 will not be reached.

But in order to decide whether g is optimal for player 2 at h_1, his initial belief that player 1 will choose b can no longer be used. At the moment when player 2 must decide between g and h, he knows that player 1 has chosen a, and not b, and his initial belief would thus have been contradicted. So, if the game reaches h_1, player 2 must form a new, conditional belief about player 1's strategy choice, and the optimality of g must be evaluated according to player 2's conditional belief at h_1. Suppose that at h_1, player 2's conditional belief is that player 1 will choose e with probability 0.75 and f with probability 0.25. Then, under this conditional belief it would be optimal to choose g at h_1.

Summarizing, we conclude that strategy (c,g) is optimal for player 2 if he initially believes that player 1 will choose b, and at h_1 holds the conditional belief that player 1 will choose e and f with probabilities 0.75 and 0.25, respectively.

So we see that, in order to evaluate whether a given strategy for a player is optimal in a dynamic game, we need to know his conditional belief for *every* information set where this strategy prescribes a choice. We can also turn the problem around: For a given player i, we can first state his conditional belief for each of his information sets, and then ask which strategies would be optimal for i under these conditional beliefs. Consider again the game from Figure 8.2. Suppose that player 2:

- initially believes that player 1 will choose strategy (a,f,j) with probability 0.5, and strategy b with probability 0.5, and
- holds at h_1 and h_2 the conditional belief that player 1 will choose strategy (a,f,j) with probability 1.

Note that at h_1 and h_2, player 2 can no longer assign a positive probability to player 1 choosing b, as it is clear at these information sets that player 1 has chosen a. Hence, player 2 must revise his initial belief at h_1 and h_2, and form a new, conditional belief.

The question is: What strategy is optimal for player 2 under these conditional beliefs? Let us start at the information sets h_1 and h_2. At h_1, player 2 believes that player 1 will choose strategy (a,f,j). So, effectively, player 2 believes at h_1 that player 1 will choose f at h_1, and hence player 2's optimal choice at h_1 is h. At h_2, player 2 also believes that player 1 will choose strategy (a,f,j), so he believes that player 1 will choose j at h_2. Hence, player 2's optimal choice at h_2 is l.

Now, we turn to the beginning of the game, $\emptyset$. There, player 2 believes that player 1 will choose strategies (a,f,j) and b with probability 0.5. In order to judge the optimality of player 2's behavior at $\emptyset$, we can no longer look at 2's choice between c and d in isolation, as his expected utility will also depend on his future choices at h_1 and h_2. What we can do is to consider each of player 2's *strategies*, and see which strategy would give him the highest expected utility at $\emptyset$.

Suppose player 2 chooses strategy (c,g). What would be player 2's expected utility under his initial belief that player 1 will choose (a,f,j) and b with probability 0.5? Player 2 believes that, with probability 0.5, player 1 will choose b, and hence the game would end after the choices (b,c), yielding player 2 a utility of 2. Moreover, player 2 believes that, with probability 0.5, player 1 will choose a at $\emptyset$ and f at h_1. Since player 2 is planning to choose (c,g), this would lead to the choice combination $((a,c),(f,g))$, which would give him a utility of 1. So, overall, player 2 expects utility

2 with probability 0.5 and utility 1 with probability 0.5, and hence player 2's expected utility would be 1.5.

In the same way, the reader can check that under player 2's initial belief, the strategy (c,h) gives him an expected utility of 2, strategy (d,k) gives him an expected utility of 0 and (d,l) gives him 1.5. Hence, at $\emptyset$ only strategy (c,h) is optimal for player 2, given his initial belief at $\emptyset$ that player 1 will choose (a,f,j) and b with probability 0.5.

Summarizing, we see that strategy (c,h) is optimal for player 2 at $\emptyset$, given his initial belief, and that choice h is optimal for player 2 at h_1, given his conditional belief. So, we may conclude that strategy (c,h) is indeed optimal for player 2, given his beliefs at the various information sets.

The examples above should have given the reader some intuition for what we mean by *conditional beliefs* in a dynamic game. Let us now try to be more formal about this. Consider some player i in a dynamic game and an information set h for this player. If the game reaches h, then player i knows that his opponents can only have chosen a combination of strategies that *leads to* h, and he may discard those strategy combinations that could never lead to h. Consider, for instance, the game in Figure 8.2, and suppose that the game reaches h_1, which is the information set following the choices (a,c). Then, player 1 knows that player 2 could only have chosen strategy (c,g) or (c,h), since these are the only strategies for player 2 that lead to h_1. Similarly, player 2 knows at h_1 that player 1 could only have chosen a strategy from $(a,e,i),(a,e,j),(a,f,i)$ and (a,f,j), as these are the only strategies for player 1 that lead to h_1. Formally, the opponents' strategy combinations that lead to an information set can be defined as follows.

Definition 8.3.1 *(Opponents' strategy combinations that lead to an information set) Consider an information set h for player i and a strategy combination* $(s_1,...,s_{i-1},s_{i+1},...,s_n)$ *for i's opponents. This* ***strategy combination leads to*** *h, if there is some strategy for player i that together with this strategy combination would lead to some history in h.*

To illustrate this definition, consider the game in Figure 8.5 which has three players. In this figure we have left out the utilities for the players because these are not important here. The figure should be read as follows. At the beginning, player 1 chooses between a and b, and player 2 chooses between c and d. If the players choose a and c, the game moves to information set h_1, and if the players choose a and d, the game moves to information set h_2. If player 1 chooses b at the beginning, the game stops. At h_1, player 1 will choose between e and f, and player 2 will choose between g and h. If the players choose e and h at h_1, the game moves to information set h_3. Otherwise, the game stops after h_1. Similarly, at h_2 player 1 will choose between i and j, and player 2 will choose between k and l. If the players choose i and k at h_2, the game moves to h_3. Otherwise, the game stops after h_2. At information set h_3, player 3 chooses between m and n. Notice that information set h_3 contains two histories, $((a,c),(e,h))$ and $((a,d),(i,k))$. So, at h_3 player 3 knows that one of these two histories has occurred but does not know which of them.

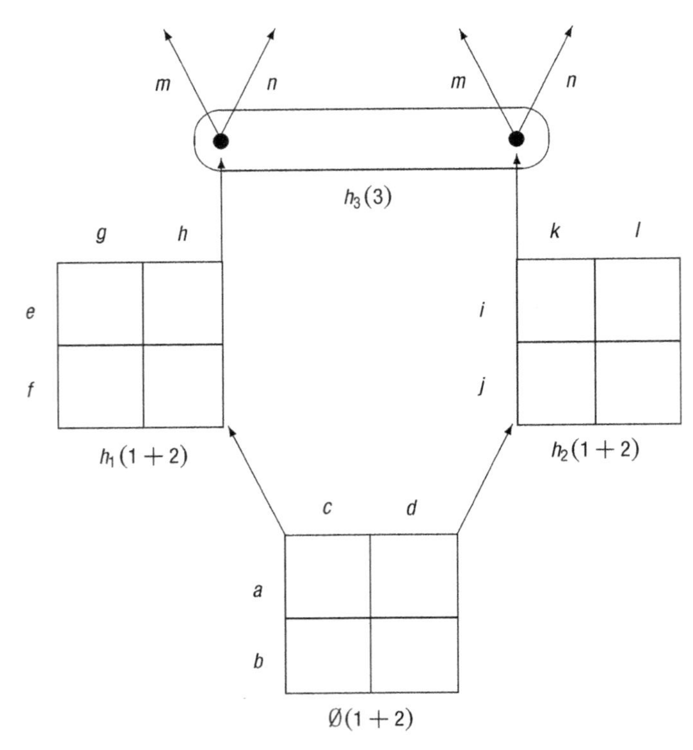

Figure 8.5 Strategy combinations that lead to an information set

What are the strategy combinations for players 1 and 2 that lead to h_3? Two such strategy combinations are:

$$((a,e,j),(c,h)) \text{ and } ((a,f,i),(d,k)).$$

The first strategy combination leads to history $((a,c),(e,h))$ in h_3, whereas the other strategy combination leads to the other history $((a,d),(i,k))$ in h_3. As an exercise, try to find all the other strategy combinations for players 1 and 2 that lead to h_3.

In a similar way, we can define *individual* strategies, not strategy combinations, that lead to a given information set.

Definition 8.3.2 (*Strategies that lead to an information set*)
Consider an information set h for player i and a strategy s_i for the same player i. Strategy s_i ***leads to*** *h, if there is some strategy combination for the opponents that together with s_i would lead to some history in h.*

In general, if a game reaches an information set h for player i, then this player will form a belief at h about the possible strategy combinations that could have been chosen by his opponents. Of course, such a belief should only assign positive probability to the opponents' strategy combinations that lead to h. We call such a belief a *conditional belief* for player i at h about the opponents' strategies.

Definition 8.3.3 *(Conditional belief)*
Consider an information set h for player i. A ***conditional belief*** *for player i at h about the opponents' strategies is a probability distribution $b_i(h)$ over the set of the opponents' strategy combinations, assigning positive probability only to strategy combinations that lead to h.*

In the game of Figure 8.5, for instance, a possible conditional belief for player 3 at h_3 would be

$$b_3(h_3) = (0.7)((a,e,j),(c,h)) + (0.3)((a,f,i),(d,k)),$$

which indicates that at h_3, player 3 believes that with probability 0.7 his opponents will choose strategies (a,e,j) and (c,h), and that with probability 0.3 his opponents will choose strategies (a,f,i) and (d,k). On the other hand,

$$b_3(h_3) = (0.7)((a,e,j),(c,g)) + (0.3)((a,f,i),(d,k))$$

would not be a valid conditional belief, since it assigns positive probability to the strategy combination $((a,e,j),(c,g))$ which does not lead to h_3.

Our next step will be to define formally what it means for a strategy to be *optimal* at an information set. Consider a player i who holds a conditional belief $b_i(h)$ at h about the opponents' strategies, and take some strategy s_i for player i that leads to h. We say that strategy s_i is *optimal at h* if it maximizes player i's expected utility at h, given the conditional belief $b_i(h)$ he holds at h about the opponents' strategies.

Definition 8.3.4 *(Optimal strategy)*
Consider an information set h for player i, at which he holds a conditional belief $b_i(h)$ about the opponents' strategies. Consider a strategy s_i for player i that leads to h. Strategy s_i is ***optimal*** *for player i at h, given his conditional belief $b_i(h)$, if*

$$u_i(s_i, b_i(h)) \geq u_i(s_i', b_i(h))$$

for every other strategy s_i' that leads to h.

Here, $u_i(s_i, b_i(h))$ denotes the expected utility that player i obtains by choosing strategy s_i under the conditional belief $b_i(h)$ about the opponents' strategy choices. The reader may wonder why there is a restriction to strategies for player i that lead to h. The reason is that information set h can only be reached if player i himself chooses a strategy that leads to h. So, if information set h is reached by the game, then it is clear that player i is following a strategy that leads to h. Therefore, these are the only strategies we need to consider for player i if h is reached.

As an illustration, consider the game from Figure 8.2. Suppose that at h_1 player 1 holds the conditional belief

$$b_1(h_1) = (0.4)(c,g) + (0.6)(c,h)$$

about player 2's strategy choice. For player 1, the only strategies that lead to h_1 are (a,e,i), (a,e,j), (a,f,i) and (a,f,j), so these are the only strategies we need to consider for player 1 at h_1. The expected utility that each of these strategies generates at h_1 under the conditional belief $b_1(h_1)$ are

$$u_1((a,e,i),b_1(h_1)) = (0.4)\cdot 1 + (0.6)\cdot 1 = 1,$$
$$u_1((a,e,j),b_1(h_1)) = (0.4)\cdot 1 + (0.6)\cdot 1 = 1,$$
$$u_1((a,f,i),b_1(h_1)) = (0.4)\cdot 3 + (0.6)\cdot 0 = 1.2,$$
$$u_1((a,f,j),b_1(h_1)) = (0.4)\cdot 3 + (0.6)\cdot 0 = 1.2.$$

So, we conclude that strategies (a,f,i) and (a,f,j) are both optimal for player 1 at h_1 under his conditional belief $b_1(h_1)$. It is not surprising that (a,f,i) and (a,f,j) give the same expected utility at h_1, since both strategies prescribe the same choice at h_1. The choice prescribed at h_2 is no longer relevant if h_1 has been reached. Similarly for (a,e,i) and (a,e,j).

However, if player 1 holds the same belief $(0.4)(c,g) + (0.6)(c,h)$ at the beginning of the game, then (a,f,i) and (a,f,j) would not be optimal for player 1! Under this belief player 1's strategy b would give an expected utility of 2, which is higher than any of the expected utilities above. So, if player 1 holds the conditional belief

$$b_1(\emptyset) = (0.4)(c,g) + (0.6)(c,h)$$

at the beginning, then b would be optimal for player 1 at the beginning, and not (a,f,i) or (a,f,j).

Summarizing, we see that if player 1 holds the same conditional belief $(0.4)(c,g) + (0.6)(c,h)$ at $\emptyset$ and at h_1, then strategy b would be optimal for player 1 at the beginning, whereas strategies (a,f,i) and (a,f,j) would be optimal at h_1. The reason for this difference is that at $\emptyset$, all strategy choices are still open for player 1, and a strategy can only be optimal if it is at least as good as *any* other available strategy for player 1, including strategy b. At h_1, however, it is evident that player 1 did not choose b at the beginning, which may have been a mistake since b was the only optimal strategy at the beginning under the beliefs above. At that stage, we say that the strategies (a,f,i) and (a,f,j) are optimal at h_1, since they provide the best way to continue for player 1 at h_1 after mistakenly having chosen a at the beginning.

So, in general, if we say that a strategy s_i for player i is optimal at h given his conditional belief $b_i(h)$, then what we mean is that s_i provides the best way for player i to continue from information set h. It may well be that some earlier choices in s_i, which were made before information set h was reached, were in fact suboptimal given the earlier conditional beliefs.

In a dynamic game, a player typically not only holds a conditional belief at one information set, but he holds a conditional belief at *each of his information sets.* Such a combination of conditional beliefs is called a *conditional belief vector.*

Definition 8.3.5 *(Conditional belief vector)*
A ***conditional belief vector*** $b_i = (b_i(h))_{h \in H_i}$ *for player i about his opponents' strategies, prescribes at every information set* $h \in H_i$ *some conditional belief* $b_i(h)$ *about the opponents' strategies.*

Recall that H_i denotes the collection of all information sets for player i in the game. So the important difference from a static game is that in a dynamic game, a player will not typically hold a particular belief throughout a game, but will in general hold different beliefs at different times, gathered in a conditional belief vector. In many cases, a player will be forced to revise his belief about the opponents' strategies during the course of the game, as he may observe some choices that he did not expect at the beginning.

Given his conditional belief vector, a player would ideally like to choose a strategy that is optimal *for each of his information sets* that can be reached by this strategy. So, whatever information set is reached, the strategy should always describe the best way to continue from that information set. The question is whether such a strategy can always be found.

The answer is "not always." To see this, consider again the game from Figure 8.2. Suppose that player 2 holds the conditional belief vector b_2 with

$$b_2(\emptyset) = (0.5) \cdot (a,f,j) + (0.5) \cdot b,$$

$$b_2(h_1) = (a,e,i), \text{ and } b_2(h_2) = (a,e,i).$$

At $\emptyset$, the expected utilities that player 2 obtains from his strategies are:

$$u_2((c,g), b_2(\emptyset)) = (0.5) \cdot 1 + (0.5) \cdot 2 = 1.5,$$

$$u_2((c,h), b_2(\emptyset)) = (0.5) \cdot 2 + (0.5) \cdot 2 = 2,$$

$$u_2((d,k), b_2(\emptyset)) = (0.5) \cdot 0 + (0.5) \cdot 0 = 0,$$

$$u_2((d,l), b_2(\emptyset)) = (0.5) \cdot 3 + (0.5) \cdot 0 = 1.5.$$

So, at $\emptyset$ only strategy (c,h) is optimal. Clearly, at h_1 only strategy (c,g) is optimal under the conditional belief $b_2(h_1) = (a,e,i)$, whereas at h_2 only strategy (d,k) is optimal under the conditional belief $b_2(h_2) = (a,e,i)$. This means, however, that there is no strategy for player 2 that is optimal at all of his information sets! At the beginning, only (c,h) is optimal, but choice h is no longer optimal if h_1 is reached, given the new conditional belief there.

The reason for this problem is that the conditional belief vector above is rather strange. At the beginning, player 2 assigns probability 0.5 to player 1's strategies (a,f,j) and b. If information set h_1 is reached, player 2 knows that player 1 must have choosen a. But then, it seems natural to assign probability 1 to (a,f,j) at h_1, since that is the only strategy starting with choice a to which player 2 assigned a positive probability initially. The same problem occurs at h_2.

Formally speaking, the conditional belief vector above does not satisfy *Bayesian updating*. We will give a precise definition of Bayesian updating at the end of this chapter, and we will show that the problem above can never occur if the beliefs of a player satisfy Bayesian updating.

8.4 Epistemic model

In the previous section we saw that in a dynamic game, a player holds for each of his information sets a conditional belief about his opponents' strategies. As for static games, we want to investigate which of these conditional belief vectors may be considered *reasonable.* Consider the example "Painting Chris' house" with the dynamic game shown in Figure 8.1. Suppose that after seeing Barbara reject her colleague's offer, you hold the conditional belief that Barbara will choose a price of 400. Is this a reasonable belief to hold?

This crucially depends on what you believe, for that information set, about what Barbara believes you will do. To facilitate our discussion, let us denote by $\emptyset$ the beginning of the game, and by h_1 the information set that follows after Barbara chooses to reject the colleague's offer. If you believe, at h_1, that Barbara believes, at $\emptyset$, that you will choose price 300 at h_1, then it is not reasonable to believe, at h_1, that Barbara will choose a price of 400, as she would be better off by choosing a price 200 instead. On the other hand, if you believe, at h_1, that Barbara believes, at $\emptyset$, that you will choose a price of 500, then it makes sense to believe, at h_1, that Barbara will reject her colleague's offer, and subsequently choose a price of 400, since that would be the best thing for her to do.

Hence, whether a particular conditional belief about your opponent's strategy choice is reasonable or not depends crucially on the belief you have, for that information set, about the *opponent's* conditional beliefs. And whether your beliefs about your opponent's beliefs are reasonable or not, depends, in turn, on what you believe your opponent believes about his opponents' beliefs, and so on. That is, in order to develop a meaningful theory of reasonable beliefs and strategy choices, we must model not just the players' conditional beliefs about their opponents' strategy choices, but also the players' conditional beliefs about the opponents' conditional beliefs about their opponents' strategy choices, and so on. We must thus find a convenient way to model the players' *belief hierarchies* in a dynamic game.

Remember how we modeled belief hierarchies in a static game: A belief hierarchy for player i must specify a belief about:

- the opponents' choices,
- the opponents' beliefs about their opponents' choices,
- the opponents' beliefs about their opponents' beliefs about their opponents' choices,

and so on. Hence, a belief hierarchy for player i in a static game specifies a belief about:

- the opponents' choices, and
- the opponents' *belief hierarchies.*

This insight was the key idea that eventually led to our construction of an epistemic model for static games. If we call every belief hierarchy a *type*, then every type for player i can be linked with a probabilistic belief about his opponents' choices *and types*. Such a list of types, together with their beliefs about the opponents' choices and types, gives us an epistemic model. From this epistemic model, we can then derive for every type the complete belief hierarchy that corresponds to it. That is how we modeled belief hierarchies for static games.

For dynamic games, we can apply a similar reasoning. A conditional belief hierarchy for player i must specify, for each of his information sets, a conditional belief about:

- the opponents' strategy choices,
- the conditional beliefs every opponent has, for each of his information sets, about his opponents' strategy choices,
- the conditional beliefs every opponent has, for each of his information sets, about the conditional beliefs each of his opponents has, for each of his information sets, about his opponents' strategy choices,

and so on. So, a *conditional belief hierarchy* for player i specifies, for each of his information sets, a conditional belief about:

- his opponents' strategy choices, and
- his opponents' *conditional belief hierarchies.*

As with static games, let us call every belief hierarchy a *type.* Then, a *type* for player i in a dynamic game specifies, for each of his information sets, a conditional belief about his opponents' strategy choices and *types*. This leads to the following definition of an epistemic model for dynamic games.

Definition 8.4.1 *(Epistemic model for dynamic games)*
An ***epistemic model*** *for a dynamic game specifies for every player i a set T_i of possible types. Moreover, every type t_i for player i specifies for every information set $h \in H_i$ a probability distribution $b_i(t_i, h)$ over the set*

$$(S_1 \times T_1) \times ... \times (S_{i-1} \times T_{i-1}) \times (S_{i+1} \times T_{i+1}) \times ... \times (S_n \times T_n)$$

of the opponents' strategy-type combinations. This probability distribution $b_i(t_i, h)$ must only assign positive probability to the opponents' strategy combinations that lead to h. The probability distribution $b_i(t_i, h)$ represents the conditional belief that type t_i has at h about the opponents' strategies and types.

Let us illustrate this definition by means of some examples. Consider first our example "Painting Chris' house" with the game in Figure 8.1. There, Barbara has two information sets, $\emptyset$ and h_1, and you have one information set, h_1. Remember that $\emptyset$ is the beginning of the game, and h_1 is the information set following Barbara's choice "reject." Consider the epistemic model in Table 8.1. Barbara is player 1 and you are player 2. We consider two different types for Barbara and two for you. Let us try to derive from this epistemic model the belief hierarchy for your type t_2^{200}. First, your type t_2^{200} believes at h_1 that Barbara will choose strategy (*reject*, 200) and that Barbara is of type t_1^a. So, in particular, your type t_2^{200} believes, at h_1, that Barbara will choose price 200.

Table 8.1. *An epistemic model for "Painting Chris' house"*

Types	$T_1 = \{t_1^a, t_1^{400}\}$, $T_2 = \{t_2^{200}, t_2^{300}\}$
Beliefs for Barbara	$b_1(t_1^a, \emptyset) = (300, t_2^{300})$ $b_1(t_1^a, h_1) = (300, t_2^{300})$ $b_1(t_1^{400}, \emptyset) = (500, t_2^{200})$ $b_1(t_1^{400}, h_1) = (500, t_2^{200})$
Beliefs for you	$b_2(t_2^{200}, h_1) = ((reject, 200), t_1^a)$ $b_2(t_2^{300}, h_1) = ((reject, 400), t_1^{400})$

Barbara's type t_1^a believes, at $\emptyset$ and at h_1, that you will choose price 300 and that you are of type t_2^{300}. So, your type t_2^{200} believes, at h_1, that Barbara believes, at $\emptyset$ and h_1, that you would choose price 300 at h_1.

Your type t_2^{300} believes, at h_1, that Barbara will choose strategy $(reject, 400)$ and that Barbara is of type t_1^{400}. So, your type t_2^{200} believes, at h_1, that Barbara believes, at $\emptyset$ and h_1, that you believe, at h_1, that Barbara chooses price 400. And so on.

In this way we can construct the complete belief hierarchy for your type t_2^{200}, and for the other three types in the epistemic model as well.

As a further illustration, consider the game in Figure 8.2 and the epistemic model in Table 8.2. What can we say about the belief hierarchy of type t_1 in this game? At $\emptyset$, type t_1 believes that player 2 will choose (c, h), whereas at h_2 it believes that player 2 will choose strategy (d, k). So, type t_1 will revise his belief about player 2's strategy choice if the game moves from $\emptyset$ to h_2.

Also, at $\emptyset$ type t_1 believes that player 2 has type t_2, whereas at h_2 it believes that player 2 has the other type, $\hat{t}_2$. From the table, we can read that type t_2 believes at h_2 that player 1 will choose (a, f, i), whereas type $\hat{t}_2$ believes at h_2 that player 1 will choose (a, e, j). Hence, type t_1 believes at $\emptyset$ that player 2 will believe at h_2 that player 1 will choose (a, f, i), whereas it believes at h_2 that player 2 believes at h_2 that player 1 will choose (a, e, j). So, type t_1 will also revise his belief about player 2's belief if the game moves from $\emptyset$ to h_2.

From the table we can also deduce that type t_1 believes at h_2 that player 2 is of type $\hat{t}_2$, type $\hat{t}_2$ believes at h_1 that player 1 is of type $\hat{t}_1$, and type $\hat{t}_1$ believes at $\emptyset$ that, with probability 0.3, player 2 will choose (c, g), and with probability 0.7 player 2 will choose (d, l). As such, type t_1 believes at h_2 that player 2 believes at h_1 that player 1 believes at $\emptyset$ that player 2 will choose (c, g) and (d, l) with probabilities 0.3 and 0.7. So, from this epistemic model we can derive for type t_1 and also for the other three types all the conditional beliefs we want.

Table 8.2. *An epistemic model for the game in Figure 8.2*

Types	$T_1 = \{t_1, \hat{t}_1\}, T_2 = \{t_2, \hat{t}_2\}$
Beliefs for player 1	$b_1(t_1, \emptyset) = ((c,h), t_2)$
	$b_1(t_1, h_1) = ((c,h), t_2)$
	$b_1(t_1, h_2) = ((d,k), \hat{t}_2)$
	$b_1(\hat{t}_1, \emptyset) = (0.3) \cdot ((c,g), t_2) + (0.7) \cdot ((d,l), \hat{t}_2)$
	$b_1(\hat{t}_1, h_1) = ((c,g), t_2)$
	$b_1(\hat{t}_1, h_2) = ((d,l), \hat{t}_2)$
Beliefs for player 2	$b_2(t_2, \emptyset) = (b, t_1)$
	$b_2(t_2, h_1) = ((a,f,i), t_1)$
	$b_2(t_2, h_2) = ((a,f,i), t_1)$
	$b_2(\hat{t}_2, \emptyset) = ((a,e,j), \hat{t}_1)$
	$b_2(\hat{t}_2, h_1) = ((a,e,j), \hat{t}_1)$
	$b_2(\hat{t}_2, h_2) = ((a,e,j), \hat{t}_1)$

8.5 Belief in the opponents' future rationality

The previous section has shown how we can model the players' belief hierarchies in a dynamic game by means of an epistemic model. The idea is to identify a belief hierarchy for player i with a type, which holds, for each of i's information sets, a conditional belief about the opponents' strategy choices and types. With this technology, we can now start to impose restrictions on the players' belief hierarchies and thereby develop some concept of plausible reasoning and choice for dynamic games.

For static games with standard beliefs, the main idea was to require that a player believes that his opponents all choose rationally. This was the content of Chapter 2. How can we translate this idea to the context of dynamic games? A first attempt could be to require that a player, at each of his information sets, believes that his opponents will choose optimal strategies. The problem, however, is that in many dynamic games this is not possible!

Consider, for instance, the example "Painting Chris' house" at the beginning of this chapter. Suppose that both you and Barbara can only choose prices 200, 300 and 400, but not 500, if Barbara rejects her colleague's offer. This leads to the game shown in Figure 8.6. At your information set h_1 you can no longer believe that Barbara will choose an optimal strategy, since in this game Barbara can never obtain more than 300 by rejecting the colleague's offer. That is, at h_1 you *must* believe that Barbara has chosen irrationally. But then, what conditions could we reasonably impose on your conditional belief at h_1 about Barbara's chosen price?

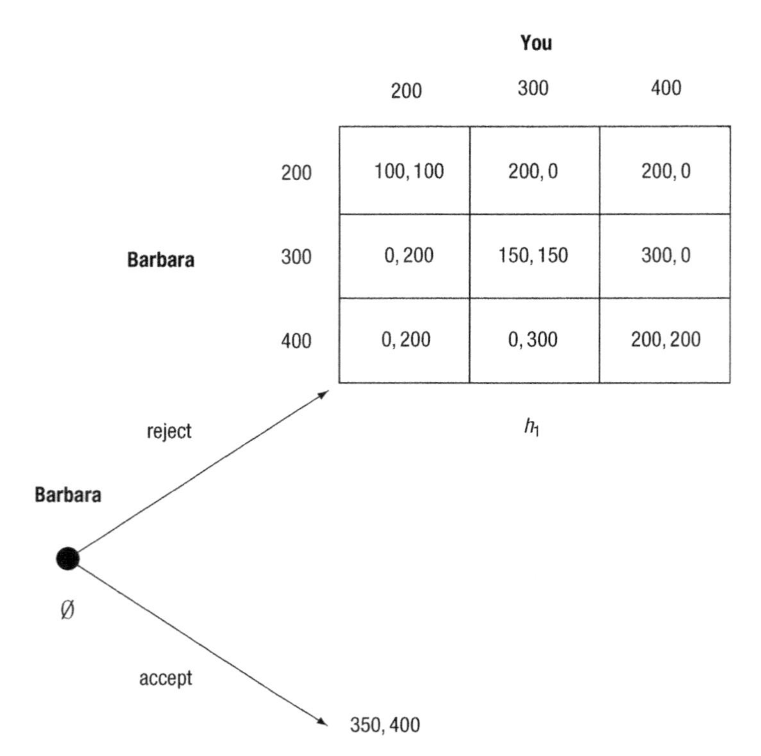

Figure 8.6 Painting Chris' house with restricted price sets

One possibility is to require you to believe at h_1 that Barbara will continue by choosing an optimal price at h_1, even if you believe that rejecting the offer at ∅ was a mistake by Barbara. This means that, after seeing Barbara reject the offer, you believe that Barbara will certainly not choose a price of 400, since this price can never be optimal for her at h_1. We say that you *believe in Barbara's future rationality*: Even if you believe that Barbara has chosen irrationally in the past (by rejecting the offer), you still believe that she will choose rationally now and in the future. If you indeed believe in Barbara's future rationality at h_1, and thus believe at h_1 that Barbara will choose either 200 or 300 as a price, then your unique optimal strategy is to choose a price of 200.

Suppose now that, initially, you also receive an offer from a (different) colleague to repair her car, for a price of 250 euros. So, the new situation is that at the first stage of the game you and Barbara must simultaneously decide whether or not to accept the colleagues' offers. If you both reject the offers, then at the next stage you both must name a price, as before. If only one of you accepts, the person that rejected the colleague's offer will paint Chris' house for a price of 500 euros. The new situation leads to the game shown in Figure 8.7 and the associated epistemic model in Table 8.3. Your type t_2 believes at ∅ that Barbara will reject the colleague's offer at ∅. Moreover, your type

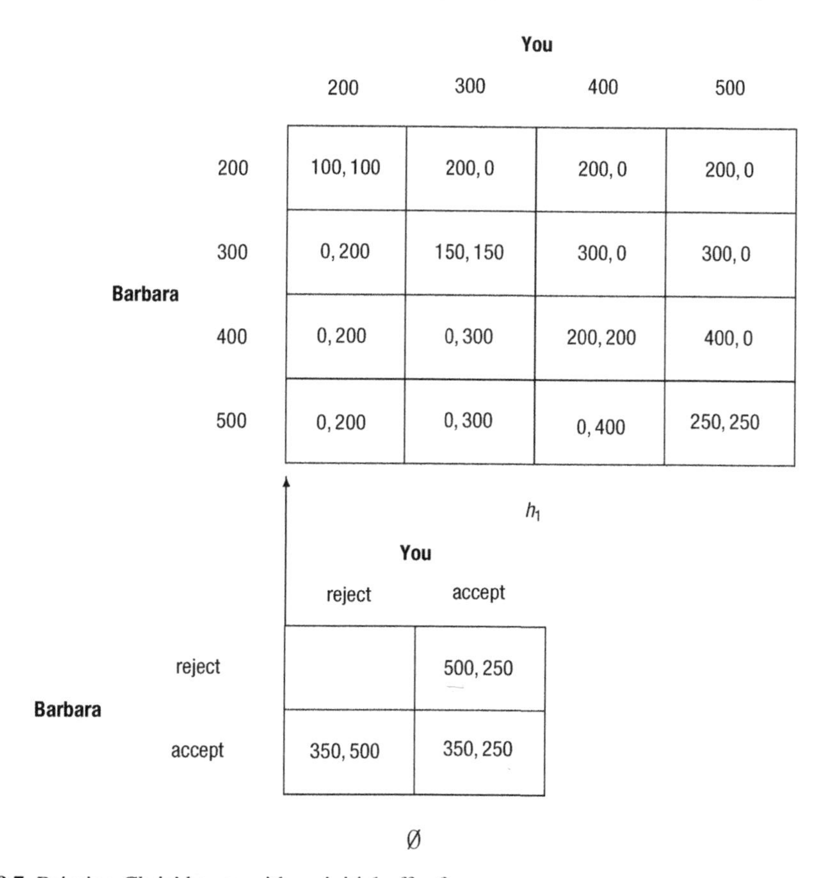

Figure 8.7 Painting Chris' house with an initial offer for you

t_2 believes at $\emptyset$ that Barbara believes at $\emptyset$ that you will accept your colleague's offer. If Barbara indeed believes so at $\emptyset$, then it is optimal for her to reject her colleague's offer. So, your type t_2 believes at $\emptyset$ that Barbara chooses rationally at $\emptyset$. At the same time, your type t_2 believes at $\emptyset$ that Barbara will choose a price of 400 at h_1, and your type t_2 believes at $\emptyset$ that Barbara will believe at h_1 that you will choose a price of 300 at h_1. This means, however, that your type t_2 believes at $\emptyset$ that Barbara will choose irrationally at the future information set h_1. So, t_2 does not believe in Barbara's future rationality.

Consider now your other type t_2'. By the same reasoning as above, your type t_2' believes at $\emptyset$ that Barbara will choose rationally at $\emptyset$. Moreover, your type t_2' believes at $\emptyset$ that Barbara will choose price 300 at h_1, and believes at $\emptyset$ that Barbara would believe at h_1 that you will choose a price of 400 at h_1. If Barbara believes at h_1 that you would choose a price of 400, then it would indeed be optimal for her to choose a price of 300 there. Hence, your type t_2' believes at $\emptyset$ that Barbara chooses rationally at the future information set h_1. Finally, your type t_2' believes at h_1 that Barbara will choose price

Table 8.3. *An epistemic model for the game in Figure 8.7*

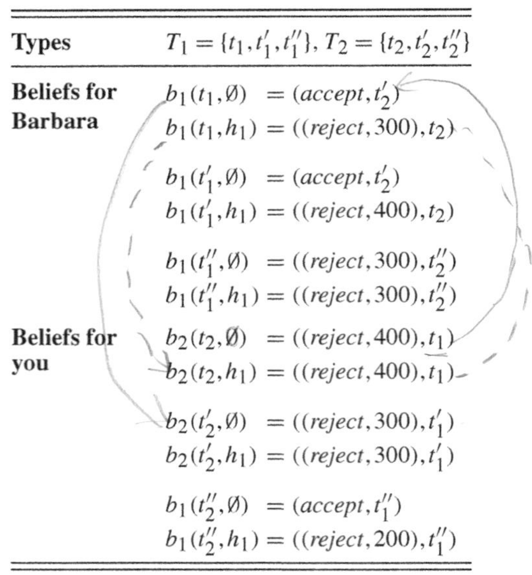

Types	$T_1 = \{t_1, t_1', t_1''\}$, $T_2 = \{t_2, t_2', t_2''\}$
Beliefs for Barbara	$b_1(t_1, \emptyset) = (accept, t_2')$
	$b_1(t_1, h_1) = ((reject, 300), t_2)$
	$b_1(t_1', \emptyset) = (accept, t_2')$
	$b_1(t_1', h_1) = ((reject, 400), t_2)$
	$b_1(t_1'', \emptyset) = ((reject, 300), t_2'')$
	$b_1(t_1'', h_1) = ((reject, 300), t_2'')$
Beliefs for you	$b_2(t_2, \emptyset) = ((reject, 400), t_1)$
	$b_2(t_2, h_1) = ((reject, 400), t_1)$
	$b_2(t_2', \emptyset) = ((reject, 300), t_1')$
	$b_2(t_2', h_1) = ((reject, 300), t_1')$
	$b_1(t_2'', \emptyset) = (accept, t_1'')$
	$b_1(t_2'', h_1) = ((reject, 200), t_1'')$

300 at h_1, and believes at h_1 that Barbara believes at h_1 that you will choose a price of 400 at h_1. Therefore, your type t_2' believes at h_1 that Barbara chooses rationally at h_1.

So, we see that your type t_2' believes at $\emptyset$ that Barbara chooses rationally at $\emptyset$ and at h_1, and believes at h_1 that Barbara chooses rationally at h_1. That is, for each of your information sets, your type t_2' believes that Barbara chooses rationally now and in the future. We can thus conclude that t_2' believes in Barbara's future rationality.

Finally, consider your type t_2''. This type believes at $\emptyset$ that Barbara will accept her colleague's offer at $\emptyset$, and believes at $\emptyset$ that Barbara believes at $\emptyset$ that you will reject your colleague's offer and choose a price of 300. If Barbara indeed believes so at $\emptyset$, then it is optimal for her to accept her colleague's offer. So, your type t_2'' believes at $\emptyset$ that Barbara chooses rationally at $\emptyset$. Does t_2'' also believe at $\emptyset$ that Barbara chooses rationally at h_1? This question is in fact irrelevant here: Since your type t_2'' believes at $\emptyset$ that Barbara will accept her colleague's offer, your type t_2'' believes at $\emptyset$ that Barbara will choose in such a way that h_1 can never be reached. So, your type t_2'' does not reason at $\emptyset$ about the rationality of Barbara at h_1. The reader may verify that the type t_2'' believes at h_1 that Barbara chooses rationally at h_1.

Summarizing, we see that t_2'' believes at $\emptyset$ that Barbara rationally chooses "accept" at $\emptyset$, which is the only information set for Barbara that can be reached if she chooses "accept." Moreover, t_2'' believes at h_1 that Barbara will choose rationally at h_1. So, your type t_2'' believes in Barbara's future rationality.

However, your type t_2'' believes at h_1 that Barbara has chosen *irrationally in the past*, at $\emptyset$. Namely, your type t_2'' believes at h_1 that Barbara has rejected the colleague's offer at $\emptyset$, and believes at h_1 that Barbara believes at $\emptyset$ that you will reject your colleague's offer and will choose a price of 300. However, if Barbara indeed believes so at $\emptyset$, then she should have accepted her colleague's offer at $\emptyset$. Hence, your type t_2'' always believes that Barbara will choose rationally now and in the future, but does *not* believe at h_1 that Barbara has chosen rationally in the past.

So, in the game of Figure 8.7, a type t_i for player i is said to believe in the opponent's future rationality if

- at $\emptyset$ it believes that the opponent chooses rationally now (at $\emptyset$) and in the future (at h_1), and
- at h_1 it believes that the opponent chooses rationally now (at h_1).

This can actually be stated more generally: In an arbitrary dynamic game, a type t_i is said to believe in the opponents' future rationality if at every information set h for player i it believes that every opponent chooses rationally now and in the future.

To see how we can formalize "now and in the future," let us reconsider the game from Figure 8.3. In that game, what would it mean precisely for player 1's type t_1 to believe in player 2's future rationality? At $\emptyset$, type t_1 must believe that player 2 will choose rationally now (at $\emptyset$) and in the future (at h_2 and h_3), and at h_1 it must believe that player 2 will choose rationally now (at h_2 and h_3).

Information sets h_1 for player 1 and h_2 for player 2 are called *simultaneous*, as the history (a,c) is present in both h_1 and h_2. So, we cannot say that h_1 comes before h_2, or that h_2 comes before h_1 – both information sets simply occur at the same time. Formally, simultaneous information sets can be defined as follows.

Definition 8.5.1 *(Simultaneous information sets)*
Two information sets h and h' are ***simultaneous*** *if there is a history that is present in both h and h'.*

So, if we say that a type t_i for player i believes at information set h that his opponent j will choose rationally *now*, then we mean that t_i believes at h that player j will choose rationally at every information set h' that is *simultaneous* with h. The following definition states formally what it means to say that an information set *follows* another information set.

Definition 8.5.2 *(Information set following another information set)*
Information set h' ***follows*** *another information set h if there is a history x in h, and a history x' in h', such that history x' follows history x.*

Consider, for instance, the game in Figure 8.5. Information set h_3 for player 3 follows information set h_1 for players 1 and 2. History (a,c) is in h_1, history $((a,c),(e,h))$ is in h_3, and obviously history $((a,c),(e,h))$ follows history (a,c).

Throughout this book we will assume that there is an *unambiguous* chronological ordering between any two information sets. More precisely we will assume that if an information set h' follows another information set h, then it cannot be the case that h

follows h' or that h is simultaneous with h'. That is, for any two information sets h and h' there cannot be histories x,y in h, and histories x',y' in h', such that x' follows x and y follows y', nor can it be the case that x' follows x, and y is present in both h and h'. Moreover, if an information set h is simultaneous with h', then it cannot be the case that h follows h' or vice versa.

The following definition will make our terminology a bit shorter.

Definition 8.5.3 *(Information set weakly following another information set)*
Information set h' ***weakly follows*** *another information set h if h' either follows h or is simultaneous with h.*

In general, whenever we say that type t_i for player i believes at h that opponent j will choose rationally *in the future*, then we mean that t_i believes at h that j will choose rationally at every information set h' for j that *weakly follows h*.

Moreover, we say that t_i believes in opponent j's future rationality if at every information set h for player i, type t_i believes that j will choose rationally at every information set h' for player j that weakly follows h. So, t_i always believes that j will choose rationally now and in the future.

We still must formally define what we mean by the phrase that "t_i believes at information set h that opponent j will choose rationally at information set h'." Remember how we formalized for static games the requirement that t_i believes in opponent j's rationality: It means that type t_i only assigns positive probability to choice-type pairs (c_j,t_j) for player j where c_j is optimal for t_j. In a similar fashion we can now define, for dynamic games, what it means to say that t_i believes at h that opponent j will choose rationally at h'.

Definition 8.5.4 *(Belief in the opponent's rationality)*
Consider a type t_i for player i, an information set h for player i and an information set h' for player j. Type t_i ***believes at*** *h* ***that opponent*** *j* ***will choose rationally at*** *h' if his conditional belief $b_i(t_i,h)$ at h only assigns positive probability to strategy-type pairs (s_j,t_j) for player j where strategy s_j is optimal for type t_j at information set h' (if s_j leads to h').*

When we say that s_j is optimal for type t_j at h', we mean that s_j is optimal for player j at h' given the conditional belief that type t_j holds at h' about the opponents' strategy choices. If strategy s_j does not lead to h' then we say, by convention, that s_j is optimal for t_j at h', although s_j does not prescribe a choice at h'. With the above definitions, we can finally define precisely what we mean by belief in the opponents' future rationality.

Definition 8.5.5 *(Belief in the opponents' future rationality)*
Consider a type t_i for player i, an information set h for player i and an opponent j. Type t_i believes at h in opponent j's future rationality if t_i believes at h that j will choose rationally at every information set h' for player j that weakly follows h. Type t_i ***believes***

in the opponents' future rationality *if* t_i *believes, at every information set* h *for player* i*, in every opponent's future rationality.*

This definition will be the main building block for this chapter. In the following section we will use this definition to formally present the concept of common belief in future rationality.

8.6 Common belief in future rationality

So far we have formally presented the idea that you always believe that your opponents will choose rationally now and in the future. We have called this condition *belief in the opponents' future rationality.* Sometimes, this condition leads to a unique decision. Consider, for instance, the game shown in Figure 8.6, which is based on the example "Painting Chris' house," but in which you and Barbara can only choose prices 200, 300 and 400. If you believe in Barbara's future rationality, then you must believe at h_1 that Barbara will choose rationally at h_1. That is, at h_1 you must believe that Barbara will either choose price 200 or 300, since a price of 400 can never be optimal for her at h_1. But then, your unique optimal choice would be to choose a price of 200. So, in the game of Figure 8.6, if you believe in Barbara's future rationality, then you will definitely choose a price of 200.

In most examples, however, believing in the opponents' future rationality is only the beginning of the story. Let us return to the original example "Painting Chris' house," with the game as shown in Figure 8.1. If you believe in Barbara's future rationality, then you will believe at h_1 that Barbara will choose a rational price at h_1. Hence, at h_1 you will believe that Barbara will choose a price of 200, 300 or 400 at h_1, since a price of 500 can never be optimal for her, whereas the prices 200, 300 and 400 *are* optimal at h_1 for Barbara under *some* belief about your price. If you believe so at h_1, then you will either choose a price of 200 or 300 at h_1, since a price of 400 or 500 can never be optimal for you if you believe that Barbara will not choose a price of 500. So, if you believe in Barbara's future rationality, then you will either choose 200 or 300 at h_1.

However, if we consider it reasonable to believe in Barbara's future rationality, then it would also be reasonable for you to believe, at h_1, that Barbara believes in *your* future rationality. Since a price of 500 can never be optimal for you at h_1, this would mean that you believe, at h_1, that Barbara believes that you will certainly not choose a price of 500. Hence, only prices 200 and 300 are optimal for Barbara. So, we see that if you believe in Barbara's future rationality, and at the same time believe that Barbara believes in your future rationality, then you will believe, at h_1, that Barbara will either choose a price of 200 or 300 at h_1. Therefore, you would certainly choose a price of 200 at h_1. Hence, if you believe in Barbara's future rationality, and believe that Barbara believes in your future rationality, there is only one possible price you could rationally choose, namely 200.

Of course, we can iterate the argument above: If it is reasonable to believe in your opponents' future rationality, then it would also be reasonable for you to believe, at every information set, that:

- your opponents always believe in their opponents' future rationality,
- your opponents always believe that their opponents always believe in their opponents' future rationality,

and so on. This leads us to *common belief in future rationality*. As usual, we will define this concept inductively, by first defining 1-fold belief in future rationality, then 2-fold belief in future rationality, and so on.

Definition 8.6.1 *(Common belief in future rationality)*
(1) Type t_i expresses ***1-fold belief in future rationality*** *if t_i believes in the opponents' future rationality.*
(2) Type t_i expresses ***2-fold belief in future rationality*** *if t_i only assigns, at every information set $h \in H_i$, positive probability to opponents' types that express 1-fold belief in future rationality.*
(3) Type t_i expresses ***3-fold belief in future rationality*** *if t_i only assigns, at every information set $h \in H_i$, positive probability to opponents' types that express 2-fold belief in future rationality.*
And so on.
Type t_i expresses ***common belief in future rationality*** *if t_i expresses k-fold belief in future rationality for every k.*

This will be the central concept for this chapter. We say that you can rationally choose a strategy under common belief in future rationality, if there is a type expressing common belief in future rationality for which this strategy is optimal at every information set.

Definition 8.6.2 *(Rational strategy under common belief in future rationality)*
Player i can ***rationally choose some strategy s_i under common belief in future rationality****, if there is some epistemic model and some type t_i for player i in this epistemic model, such that t_i expresses common belief in future rationality and strategy s_i is optimal for type t_i at every information set $h \in H_i$ that s_i leads to.*

To illustrate the formal definition of common belief in future rationality, let us reconsider the example "Painting Chris' house."

Example 8.2 Painting Chris' house

Consider the game in Figure 8.1 and the epistemic model in Table 8.1. What can we say about your two types t_2^{200} and t_2^{300}? Do they express common belief in future rationality or not?

Consider first your type t_2^{200}. This type believes at h_1 that Barbara is of type t_1^a, and that Barbara will choose a price of 200. Barbara's type t_1^a, in turn, believes at h_1 that you will choose a price of 300, and hence price 200 is optimal for Barbara's type t_1^a at h_1. So, your type t_2^{200} believes at h_1 that Barbara will choose rationally at h_1, and hence your type t_2^{200} believes in Barbara's future rationality.

Barbara's type t_1^a believes, at $\emptyset$ and h_1, that you are of type t_2^{300} and that you will choose price 300. Since price 300 is optimal at h_1 for your type t_2^{300}, Barbara's type t_1^a

Table 8.4. *An alternative epistemic model for "Painting Chris' house"*

Types	$T_1 = \{t_1\}, T_2 = \{t_2\}$
Beliefs for Barbara	$b_1(t_1, \emptyset) = (200, t_2)$ $b_1(t_1, h_1) = (200, t_2)$
Beliefs for you	$b_2(t_2, h_1) = ((reject, 200), t_1)$

believes in your future rationality. As your type t_2^{200} believes at h_1 that Barbara is of type t_1^a, we conclude that your type t_2^{200} believes that Barbara believes in your future rationality.

Note that your type t_2^{300} believes at h_1 that Barbara is of type t_1^{400} and that Barbara will choose price 400. As price 400 is optimal at h_1 for Barbara's type t_1^{400}, your type t_2^{300} believes in Barbara's future rationality. From the epistemic model, we can deduce that your type t_2^{200} believes at h_1 that Barbara believes that you are of type t_2^{300}. Hence, your type t_2^{200} believes that Barbara believes that you believe in Barbara's future rationality.

Consider now Barbara's type t_1^{400}, which believes at $\emptyset$ and h_1 that you are of type t_2^{200} and that you will choose price 500. Obviously, price 500 is not optimal at h_1 for your type t_2^{200}. So, Barbara's type t_1^{400} does not believe in your future rationality. As your type t_2^{200} believes at h_1 that Barbara believes that you believe at h_1 that Barbara is of type t_1^{400}, we conclude that your type t_2^{200} believes at h_1 that Barbara believes that you believe at h_1 that Barbara does not believe in your future rationality.

Summarizing, we see that your type t_2^{200} believes in Barbara's future rationality, believes that Barbara believes in your future rationality, believes that Barbara believes that you believe in Barbara's future rationality, but does not believe that Barbara believes that you believe that Barbara believes in your future rationality. In particular, your type t_2^{200} does not express common belief in future rationality.

In a similar fashion, you can verify that your other type t_2^{300} believes in Barbara's future rationality, but does not believe that Barbara believes in your future rationality. So t_2^{300} also does not express common belief in future rationality. In the same way, we can conclude that both of Barbara's types t_1^a and t_1^{400} do not express common belief in future rationality either. Hence, no type in the epistemic model of Table 8.1 expresses common belief in future rationality.

Consider, as an alternative, the new epistemic model in Table 8.4. In this epistemic model, we only consider one type for you and Barbara. The reader may verify that your type t_2 believes in Barbara's future rationality, and that Barbara's type t_1 believes in your future rationality. Since your type t_2 believes at h_1 that Barbara is of type t_1, and Barbara's type t_1 believes, at $\emptyset$ and h_1, that you are of type t_2, it follows that your type t_2 expresses common belief in future rationality. As price 200 is optimal for your type t_2, we conclude that you can rationally choose price 200 under common belief in future rationality.

We have already seen that there is only one price that could possibly be optimal for you under common belief in future rationality, namely price 200. Therefore, price 200 is the only price you can rationally choose under common belief in future rationality.

Let us now return to the game in Figure 8.7, which is based on the original example "Painting Chris' house," but in which you also receive an offer from a colleague before going to Chris' house. Consider the epistemic model in Table 8.3. Which of your types t_2, t_2' and t_2'' express common belief in future rationality? We have seen that your type t_2 does not believe in Barbara's future rationality, and hence does not express common belief in future rationality. Based on this, we can also conclude that your type t_2' does not express common belief in future rationality. Namely, your type t_2' believes at $\emptyset$ that Barbara is of type t_1', which believes at h_1 that you are of type t_2, which, we have seen, does not believe in Barbara's future rationality. So, your type t_2' believes at $\emptyset$ that Barbara believes at h_1 that you do not believe in Barbara's future rationality. Therefore, your type t_2' also does not express common belief in future rationality.

What about your type t_2''? Your type t_2'' believes throughout the game that Barbara is of type t_1''. Note, however, that Barbara's type t_1'' does not believe in your future rationality. Her type t_1'' believes at $\emptyset$ that you are of type t_2'' and that you will choose price 300 at h_1. At h_1, however, your type t_2'' believes that Barbara will choose price 200, and hence price 300 is not optimal for your type t_2'' at h_1. So, Barbara's type t_1'' believes at $\emptyset$ that you will not choose rationally at h_1. We thus see that your type t_2'' believes throughout the game that Barbara believes at $\emptyset$ that you will not choose rationally at h_1. Hence, your type t_2'' does not express common belief in future rationality either.

Our conclusion is thus that none of your types in the epistemic model of Table 8.3 expresses common belief in future rationality. Let us have a look at an alternative epistemic model for this game, shown in Table 8.5. As you may verify, all types in this epistemic model believe in the opponent's future rationality. As such, each of your types t_2^a and t_2^{200} expresses common belief in future rationality. Since your strategy *accept* is optimal for your type t_2^a, and your strategy (*reject*, 200) is optimal, at every information set, for your type t_2^{200}, it follows that you can rationally choose *accept* and (*reject*, 200) under common belief in future rationality.

Are there any other strategies you can rationally choose under common belief in future rationality? We will see that there are not. If you believe in Barbara's future rationality then you must believe at h_1 that Barbara will not choose price 500 at h_1, and hence you would not choose prices 400 and 500 at h_1. Hence, if Barbara believes in your future rationality, and believes that you believe in her future rationality, then Barbara would believe at h_1 that you will not choose prices 400 and 500, and hence she would only choose price 200 at h_1. So, if you express common belief in future rationality, then you will believe at h_1 that Barbara will choose price 200, and hence your unique optimal choice at h_1 would be to choose price 200 as well. Hence, under common belief in future rationality your strategies (*reject*, 300), (*reject*, 400) and (*reject*, 500) cannot be optimal at h_1. Summarizing, we may thus conclude that *accept* and (*reject*, 200) are the only strategies that you can rationally choose under common belief in future rationality. □

Table 8.5. *An alternative epistemic model for the game in Figure 8.7*

Types	$T_1 = \{t_1^a, t_1^{200}\}$, $T_2 = \{t_2^a, t_2^{200}\}$
Beliefs for Barbara	$b_1(t_1^a, \emptyset) = ((reject, 200), t_2^{200})$ $b_1(t_1^a, h_1) = ((reject, 200), t_2^{200})$ $b_1(t_1^{200}, \emptyset) = (accept, t_2^a)$ $b_1(t_1^{200}, h_1) = ((reject, 200), t_2^a)$
Beliefs for you	$b_2(t_2^a, \emptyset) = ((reject, 200), t_1^{200})$ $b_2(t_2^a, h_1) = ((reject, 200), t_1^{200})$ $b_2(t_2^{200}, \emptyset) = (accept, t_1^a)$ $b_2(t_2^{200}, h_1) = ((reject, 200), t_1^a)$

8.7 Existence

In the last few sections we developed the idea of common belief in future rationality, which states that you always believe that your opponents will choose rationally now and in the future, that you always believe that your opponents always believe that their opponents will choose rationally now and in the future, and so on. An important question is whether common belief in future rationality is always possible. Can we always construct, for every dynamic game and every player in that game, a type that expresses common belief in future rationality? Or are there games where common belief in future rationality leads to logical contradictions? We will demonstrate in this section that common belief in future rationality is always possible in every dynamic game.

To get an idea for how we can show this, let us return to the game shown in Figure 8.7. In that game, there are two information sets for you and Barbara, $\emptyset$ and h_1. Start with an arbitrary choice for you at h_1, say 500, and an arbitrary choice for you at $\emptyset$, say *accept*, to give the choice-combination (*accept*, 500) for you.

Suppose now that Barbara believes at h_1 that you will choose 500, and that she believes at $\emptyset$ that you will *accept*. Then, Barbara would choose 400 at h_1 and choose *reject* at $\emptyset$, yielding the choice-combination (*reject*, 400) for Barbara.

Suppose you believe at h_1 that Barbara will choose 400, and believe at $\emptyset$ that Barbara will choose (*reject*, 400). Then, you would choose 300 at h_1, and choose *reject* at $\emptyset$, leading to the choice-combination (*reject*, 300) for you.

Suppose Barbara believes at h_1 that you will choose 300, and believes at $\emptyset$ that you will choose (*reject*, 300). Then, Barbara would choose 200 at h_1, and *accept* at $\emptyset$, yielding the choice-combination (*accept*, 200) for Barbara.

If you believe at h_1 that Barbara will choose 200, and believe at $\emptyset$ that Barbara will choose *accept*, then you would choose 200 at h_1 and *reject* at $\emptyset$. We obtain the choice-combination (*reject*, 200) for you.

If Barbara believes at h_1 that you will choose 200, and believes at $\emptyset$ that you will *reject*, then Barbara would choose 200 at h_1 and *accept* at $\emptyset$. We obtain the choice-combination $(accept, 200)$ for Barbara.

However, we have seen the choice-combination $(accept, 200)$ for Barbara before. So, we now have the following cycle of choice-combinations:

$$(accept_1, 200_1) \longrightarrow (reject_2, 200_2) \longrightarrow (accept_1, 200_1),$$

where the subindex 1 means that these choices belong to player 1 (Barbara) and the subindex 2 means that these choices belong to player 2 (you).

Our next step will be to transform this cycle of choice-combinations into conditional belief hierarchies for players 1 and 2 that express common belief in future rationality. The idea is that every choice-combination for player i in this cycle can be interpreted as a conditional belief vector for opponent j about player i's strategy choice. For instance, the choice-combination $(accept_1, 200_1)$ for player 1 can be viewed as the conditional belief vector for player 2 in which he believes at $\emptyset$ that player 1 will choose $accept_1$ and in which he believes at h_1 that player 1 will choose $(reject_1, 200_1)$. Now, for every conditional belief vector in the cycle we can construct a type with exactly that belief vector. More precisely, we define a type t_1 for player 1 and a type t_2 for player 2, such that

$$b_1(t_1, \emptyset) = ((reject_2, 200_2), t_2)$$
$$b_1(t_1, h_1) = ((reject_2, 200_2), t_2)$$

$$b_2(t_2, \emptyset) = (accept_1, t_1),$$
$$b_2(t_2, h_1) = ((reject_1, 200_1), t_1).$$

By construction of our cycle above, choice 200_2 is optimal at h_1 for type t_2, and strategy $(reject_2, 200_2)$ is optimal at $\emptyset$ for type t_2. Since type t_1 believes at h_1 that player 2 is of type t_2 and will choose 200_2, and believes at $\emptyset$ that player 2 is of type t_2 and will choose $(reject_2, 200_2)$, we conclude that type t_1 believes in player 2's future rationality.

Similarly, choice 200_1 is optimal at h_1 for type t_1 and strategy $accept_1$ is optimal at $\emptyset$ for type t_1. Since type t_2 believes at h_1 that player 1 is of type t_1 and will choose 200_1, and believes at $\emptyset$ that player 1 is of type t_1 and will choose $accept_1$, we conclude that type t_2 believes in player 1's future rationality.

Hence, both types t_1 and t_2 constructed from the cycle above believe in the opponent's future rationality. As type t_1 always believes that player 2 is of type t_2, and type t_2 always believes that player 1 is of type t_1, we can conclude that both t_1 and t_2 express common belief in future rationality. In this way, we have constructed an epistemic model in which every type expresses common belief in future rationality.

The procedure above can be generalized to any dynamic game. Consider a dynamic game with n players. For every player i and every information set $h \in H_i$ for player i, specify an arbitrary choice $c_i^1(h) \in C_i(h)$ at h. By putting all these choices, for all players and all information sets, together, we obtain a choice-combination c^1, which specifies for every player i, and every information set $h \in H_i$, the choice $c_i^1(h)$.

Consider some player i, and assume that player i believes, for every information set $h \in H_i$, that every opponent j will choose $c_j^1(h')$ for every information set $h' \in H_j$ that weakly follows h. So, for every information set $h \in H_i$, player i believes that his opponents will choose according to c^1 in the present and in the future.

Take an ultimate information set h for player i, that is, an information set $h \in H_i$ that is not followed by any other information set for player i. At h, select a choice $c_i^2(h)$ for player i that can be optimal at h, given that player i believes at h that his opponents will choose according to c^1 now and in the future, and given some *arbitrary* collection of opponents' past choices that leads to h. Then, consider a penultimate information set h for player i, that is, an information set $h \in H_i$ that is only followed by ultimate information sets for player i. At h, select a choice $c_i^2(h)$ for player i that can be optimal at h, given that player i believes at h that his opponents will choose according to c^1 now and in the future, *given his own optimal choices $c_i^2(h')$ at ultimate information sets h' for player i that follow h*, and given some arbitrary collection of the opponents' past choices that leads to h. Then, consider a pen-penultimate information set h for player i, that is, an information set $h \in H_i$ that is only followed by ultimate and penultimate information sets for player i. At h, select a choice $c_i^2(h)$ for player i that can be optimal at h, given that player i believes at h that his opponents will choose according to c^1 now and in the future, *given his own optimal choices $c_i^2(h')$ at ultimate and penultimate information sets h' for player i that follow h*, and given some arbitrary collection of the opponents' past choices that leads to h.

By continuing in this fashion, we will define for every information set $h \in H_i$ some choice $c_i^2(h)$, such that $c_i^2(h)$ can be optimal for player i at h, given that player i believes at h that his opponents will choose according to c^1 now and in the future, *given his own optimal choices $c_i^2(h')$ at information sets h' for player i that follow h*, and given some arbitrary collection of the opponents' past choices that leads to h. If we do so for every player i, then we obtain a new choice-combination c^2, which specifies for every player i and every information set $h \in H_i$ some choice $c_i^2(h)$. Moreover, the new choice-combination $c_i^2(h)$ is such that, for every player i and every information set $h \in H_i$, choice $c_i^2(h)$ can be optimal for player i at h, given that player i believes at h that his opponents will choose according to c^1 now and in the future, given his own optimal choices $c_i^2(h')$ at information sets h' for player i that follow h, and given some arbitrary collection of opponents' past choices that leads to h. We say that the new choice-combination c^2 is *locally optimal* under the old choice-combination c^1.

We can construct in the same way a choice-combination c^3 that is locally optimal under c^2, a choice-combination c^4 that is locally optimal under c^3, and so on. So, we obtain an infinite chain of choice-combinations $(c^1, c^2, c^3, ...)$, where c^{k+1} is locally optimal under c^k for every k. As there are only finitely many choices in the game, we will eventually find a choice-combination in the chain that occurs for the second time. From that moment on, however, the chain will repeat itself over and over again. That is, there must be some choice-combination c^m in the chain after which we end up in a cycle of choice-combinations:

$$c^m \longrightarrow c^{m+1} \longrightarrow c^{m+2} \longrightarrow ... \longrightarrow c^{m+K} = c^m.$$

In a similar way as our example above, we can construct *types* from this cycle of choice-combinations. For every choice-combination c^{m+k} in this cycle and every player i, we can construct a type t_i^{m+k} which:

- believes at every $h \in H_i$ that his opponents will choose according to c^{m+k} now and in the future,
- believes for every $h \in H_i$ that every opponent j is of type t_j^{m+k-1}.

Every type t_i^{m+k} can be constructed so as to believe in the opponents' future rationality. At information set $h \in H_i$ type t_i^{m+k} believes that every opponent j will choose $c_j^{m+k}(h')$ for every $h' \in H_j$ that weakly follows h. By construction, c^{m+k} is locally optimal under c^{m+k-1}, and hence $c_j^{m+k}(h')$ can be optimal for player j at h' if player j believes at h' that his opponents will choose according to c^{m+k-1} now and in the future, given some arbitrary collection of opponents' past choices that leads to h'. Note that t_i^{m+k} believes at h that player j is of type t_j^{m+k-1}, who indeed believes at h' that his opponents will choose according to c^{m+k-1} now and in the future. This means, however, that type t_i^{m+k} can be defined such that it believes at h in player j's future rationality. Since this applies to every information set $h \in H_i$ and every opponent j, we conclude that type t_i^{m+k} can be defined such that it believes in the opponents' future rationality.

Hence, the epistemic model we construct from the cycle of choice-combinations will be such that every type t_i^{m+k} in this model believes in the opponents' future rationality. As a type t_i^{m+k} in this model always believes that his opponents' types are t_j^{m+k-1}, which also belong to this model, we can conclude that every type t_i^{m+k} in this model expresses common belief in future rationality.

Moreover, the types t_i^{m+k} so constructed have some very special properties. First, a type t_i^{m+k} never changes his belief about the opponents' types, as t_i^{m+k} always believes that every opponent j is of type t_j^{m+k-1}. Intuitively, this means that type t_i^{m+k} never changes his beliefs about the opponents' belief hierarchies. Second, every type t_i^{m+k} in this model always assigns probability 1 to one specific strategy-type combination for each opponent.

As this construction is always possible for every dynamic game, we may conclude that for every dynamic game we can always construct an epistemic model where all types express common belief in future rationality, and where all types have the properties described. In particular, common belief in future rationality is always possible for every dynamic game. We thus obtain the following result.

Theorem 8.7.1 *(Common belief in future rationality is always possible)*
For every dynamic game we can construct an epistemic model in which every type expresses common belief in future rationality.

Moreover, we can construct the epistemic model in such a way that every type never changes his belief about the opponents' types during the game, and where every type, at every information set, always assigns probability 1 to one specific strategy-type combination for each opponent.

Be careful, however. If a strategy s_i can rationally be chosen by player i under common belief in future rationality, then it may often be necessary to support strategy s_i by a type t_i that, at some information sets, assigns positive probability to *various different* strategy-type combinations for the opponents. The second part of the theorem only states that we can always find *some* epistemic model with these special properties. But in general not every strategy that can rationally be chosen under common belief in future rationality can necessarily be supported by such an epistemic model with these special properties.

8.8 Algorithm

In this section we ask: Is there an algorithm that will help us to compute those strategies that can rationally be chosen under common belief in future rationality? To see what such an algorithm would look like, let us return to the example "Painting Chris' house."

Example 8.3 Painting Chris' house

Consider the scenario from Figure 8.7, in which you and Barbara both receive an initial offer from a colleague. At the beginning of the game, $\emptyset$, all strategies are still possible for you and Barbara. So, for $\emptyset$ we can build the matrix $\Gamma^0(\emptyset)$ shown in Table 8.6. We call $\Gamma^0(\emptyset)$ the *full decision problem* at $\emptyset$. In the rows we list all strategies for Barbara and in the columns we list all strategies for you. For every pair of strategies, the matrix shows the utilities for Barbara and you. In this matrix, the letter r stands for "reject."

For information set h_1, it is clear that Barbara has chosen a strategy from $(r,200),(r,300),(r,400)$ and $(r,500)$, and that you have chosen a strategy from $(r,200),(r,300),(r,400)$ and $(r,500)$ as well. For h_1 we can thus build the matrix $\Gamma^0(h_1)$ also shown in Table 8.6, which we call the full decision problem at h_1. The strategies in the rows are the strategies for Barbara that are still possible at h_1, and the strategies in the columns are the strategies for you that are still possible at h_1.

If you believe in Barbara's future rationality, then in particular you believe at $\emptyset$ that Barbara will choose rationally at h_1. Note that Barbara's strategy $(r,500)$ is strictly dominated in the full decision problem $\Gamma^0(h_1)$ by the randomized strategy $(0.5)\cdot(r,200)+(0.5)\cdot(r,400)$, and hence her strategy $(r,500)$ cannot be optimal at h_1. So, if you believe at $\emptyset$ that Barbara will choose rationally at h_1, then you believe at $\emptyset$ that Barbara will not choose strategy $(r,500)$. This can be represented by eliminating Barbara's strategy $(r,500)$ from the decision problem $\Gamma^0(\emptyset)$. Summarizing, we see that we must eliminate Barbara's strategy $(r,500)$ from the decision problem $\Gamma^0(\emptyset)$, since this strategy is strictly dominated at some *future* decision problem $\Gamma^0(h_1)$. The interpretation of this operation is that you believe at $\emptyset$ that Barbara will not choose $(r,500)$, as this strategy can never be optimal for her at h_1.

Table 8.6. *Full decision problems in the game of Figure 8.7*

$\Gamma^0(\emptyset)$	$(r,200)$	$(r,300)$	$(r,400)$	$(r,500)$	*accept*
$(r,200)$	100, 100	200, 0	200, 0	200, 0	500, 250
$(r,300)$	0, 200	150, 150	300, 0	300, 0	500, 250
$(r,400)$	0, 200	0, 300	200, 200	400, 0	500, 250
$(r,500)$	0, 200	0, 300	0, 400	250, 250	500, 250
accept	350, 500	350, 500	350, 500	350, 500	350, 250

$\Gamma^0(h_1)$	$(r,200)$	$(r,300)$	$(r,400)$	$(r,500)$
$(r,200)$	100, 100	200, 0	200, 0	200, 0
$(r,300)$	0, 200	150, 150	300, 0	300, 0
$(r,400)$	0, 200	0, 300	200, 200	400, 0
$(r,500)$	0, 200	0, 300	0, 400	250, 250

If you believe in Barbara's future rationality, then you will also believe at h_1 that Barbara will choose rationally at h_1. As Barbara's strategy $(r,500)$ cannot be optimal at h_1, we must also eliminate her strategy $(r,500)$ from the decision problem $\Gamma^0(h_1)$ at h_1. The meaning of this is that you believe at h_1 that Barbara will not choose strategy $(r,500)$.

The same reasoning can be applied to Barbara's beliefs about you. If she believes in your future rationality, then she will believe at $\emptyset$ and h_1 that you will choose rationally at h_1. So, Barbara believes at $\emptyset$ and h_1 that you will not choose strategy $(r,500)$, and hence we can eliminate your strategy $(r,500)$ from the decision problem $\Gamma^0(\emptyset)$ at $\emptyset$ and from the decision problem $\Gamma^0(h_1)$ at h_1.

So, in step 1 of our procedure we will eliminate the strategies $(r,500)$ for Barbara and you from the decision problems $\Gamma^0(\emptyset)$ and $\Gamma^0(h_1)$, as both strategies are strictly dominated in $\Gamma^0(h_1)$. This leads to the *reduced* decision problems $\Gamma^1(\emptyset)$ and $\Gamma^1(h_1)$ shown in Table 8.7. The meaning of these reduced decision problems $\Gamma^1(\emptyset)$ and $\Gamma^1(h_1)$ is the following: If a player believes in his opponent's future rationality, then this player believes at $\emptyset$ that the opponent will only choose strategies from $\Gamma^1(\emptyset)$, and believes at h_1 that the opponent will only choose strategies from $\Gamma^1(h_1)$.

Suppose now that you not only believe in Barbara's future rationality, but that you also believe at $\emptyset$ and h_1 that Barbara believes in your future rationality. Then, you believe at $\emptyset$ and h_1 that Barbara believes at h_1 that you will only choose strategies from $\Gamma^1(h_1)$. But then, if you believe in Barbara's future rationality, you believe at $\emptyset$ and h_1 that Barbara will choose a strategy that is optimal for her at h_1, given that she believes at h_1 that you will choose strategies from $\Gamma^1(h_1)$ only. However, these optimal strategies for Barbara are exactly the strategies that are not strictly dominated for Barbara in $\Gamma^1(h_1)$.

Table 8.7. *Decision problems after step 1 in the game of Figure 8.7*

$\Gamma^1(\emptyset)$	$(r,200)$	$(r,300)$	$(r,400)$	*accept*
$(r,200)$	100,100	200,0	200,0	500,250
$(r,300)$	0,200	150,150	300,0	500,250
$(r,400)$	0,200	0,300	200,200	500,250
accept	350,500	350,500	350,500	350,250

$\Gamma^1(h_1)$	$(r,200)$	$(r,300)$	$(r,400)$
$(r,200)$	100,100	200,0	200,0
$(r,300)$	0,200	150,150	300,0
$(r,400)$	0,200	0,300	200,200

So, if you believe in Barbara's future rationality, and believe that Barbara believes in your future rationality, we must eliminate from $\Gamma^1(\emptyset)$ and $\Gamma^1(h_1)$ those strategies for Barbara that are strictly dominated within $\Gamma^1(h_1)$. The reason is that these strategies can never be optimal for Barbara at h_1, if she believes in your future rationality. Therefore, you must believe at $\emptyset$ and h_1 that Barbara will not choose such strategies. Now, the only strategy for Barbara that is strictly dominated in $\Gamma^1(h_1)$ is $(r,400)$. So, we eliminate Barbara's strategy $(r,400)$ from $\Gamma^1(\emptyset)$ and $\Gamma^1(h_1)$ as it is strictly dominated in $\Gamma^1(h_1)$.

We can apply a similar reasoning to Barbara's beliefs about you: If she believes in your future rationality, and believes that you believe in her future rationality, we can eliminate your strategy $(r,400)$ from $\Gamma^1(\emptyset)$ and $\Gamma^1(h_1)$, since it is strictly dominated in $\Gamma^1(h_1)$.

In total, we will eliminate in step 2 of our procedure the strategies $(r,400)$ for you and Barbara from the decision problems $\Gamma^1(\emptyset)$ and $\Gamma^1(h_1)$, as these are strictly dominated in $\Gamma^1(h_1)$. This leads to the reduced decision problems $\Gamma^2(\emptyset)$ and $\Gamma^2(h_1)$ shown in Table 8.8. The interpretation is that if a player believes in his opponent's future rationality, and believes that his opponent believes in his future rationality, then this player will believe at $\emptyset$ that the opponent will only choose strategies from $\Gamma^2(\emptyset)$, and will believe at h_1 that the opponent will only choose strategies from $\Gamma^2(h_1)$.

Suppose now that you believe in Barbara's future rationality, that you believe that Barbara believes in your future rationality, and that you believe that Barbara believes that you believe in her future rationality. Then, you believe at $\emptyset$ and h_1 that Barbara believes at h_1 that you will only choose strategies from $\Gamma^2(h_1)$, that is $(r,200)$ or $(r,300)$. Hence, $(r,300)$ can no longer be optimal for her at h_1, since it is strictly dominated by $(r,200)$ at $\Gamma^2(h_1)$. So, if you believe in Barbara's future rationality, believe that Barbara believes in your future rationality, and believe that Barbara believes that you believe

Table 8.8. *Decision problems after step 2 in the game of Figure 8.7*

$\Gamma^2(\emptyset)$	$(r,200)$	$(r,300)$	*accept*
$(r,200)$	100, 100	200, 0	500, 250
$(r,300)$	0, 200	150, 150	500, 250
accept	350, 500	350, 500	350, 250

$\Gamma^2(h_1)$	$(r,200)$	$(r,300)$
$(r,200)$	100, 100	200, 0
$(r,300)$	0, 200	150, 150

in her future rationality, then you believe, at $\emptyset$ and h_1, that Barbara will not choose $(r,300)$. So, we must eliminate Barbara's strategy $(r,300)$ from the decision problems $\Gamma^2(\emptyset)$ and $\Gamma^2(h_1)$, as it is strictly dominated at $\Gamma^2(h_1)$.

If we apply a similar reasoning for Barbara's beliefs about you, then we must also eliminate your strategy $(r,300)$ from $\Gamma^2(\emptyset)$ and $\Gamma^2(h_1)$, as it is strictly dominated for you at $\Gamma^2(h_1)$.

So, in total we will eliminate in step 3 of our procedure the strategies $(r,300)$ for you and Barbara from the decision problems $\Gamma^2(\emptyset)$ and $\Gamma^2(h_1)$, since these strategies are strictly dominated in $\Gamma^2(h_1)$. This leads to the new decision problems $\Gamma^3(\emptyset)$ and $\Gamma^3(h_1)$ shown in Table 8.9. Since in the decision problems $\Gamma^3(\emptyset)$ and $\Gamma^3(h_1)$ no strategy is strictly dominated, the procedure would stop here.

The strategies that survive for you at $\emptyset$ at the end of this procedure are $(r,200)$ and *accept*. We have seen before that these are exactly the strategies that you can rationally choose under common belief in future rationality. So, the algorithm just described provides, for this specific example, exactly those strategies for you that can rationally be chosen under common belief in future rationality. □

We will now generalize the algorithm above to any dynamic game, and explain – at least partially – why it yields precisely those strategies that can rationally be chosen under common belief in future rationality. In order to define the algorithm in a precise way, we must first formalize the notion of a *full decision problem* and a *reduced decision problem* at an information set – ingredients that play a central role in the algorithm. Take an arbitrary dynamic game with n players, consider some player i and some information set h for player i. The *full decision problem* for player i at h consists of his strategies leading to h, denoted by the set $S_i(h)$, and the set of opponents' strategy combinations leading to h, denoted by $S_{-i}(h)$. The interpretation is that at information set h, player i knows that his opponents must have chosen some strategy combination from $S_{-i}(h)$,

Table 8.9. *Decision problems after step 3 in the game of Figure 8.7*

	$\Gamma^3(\emptyset)$	
	$(r, 200)$	*accept*
$(r, 200)$	$100, 100$	$500, 250$
accept	$350, 500$	$350, 250$

	$\Gamma^3(h_1)$
	$(r, 200)$
$(r, 200)$	$100, 100$

and that he is implementing some strategy in $S_i(h)$. A *reduced decision problem* for player i at h is obtained by eliminating some of the strategies of player i from $S_i(h)$, or eliminating some of the opponents' strategy combinations from $S_{-i}(h)$, or both.

Definition 8.8.1 *(Full and reduced decision problem at an information set)*
Consider an information set $h \in H_i$ for player i. Let $S_i(h)$ be the set of player i's strategies that lead to h, and let $S_{-i}(h)$ be the set of opponents' strategy combinations that lead to h. The pair $\Gamma^0(h) = (S_i(h), S_{-i}(h))$ is called the ***full decision problem*** *for player i at h. A* ***reduced decision problem*** *for player i at h is a pair $\Gamma(h) = (D_i(h), D_{-i}(h))$, where $D_i(h)$ is a subset of the set $S_i(h)$ of player i's strategies leading to h, and $D_{-i}(h)$ is a subset of the set $S_{-i}(h)$ of the opponents' strategy combinations leading to h.*

Let us now try to formulate the algorithm for an arbitrary dynamic game. As in previous chapters we will do so in steps. In the first step we will characterize those strategies you can rationally choose under 1-fold belief in future rationality. Subsequently, in step 2, we will identify those strategies you can rationally choose if you express 1-fold and 2-fold belief in future rationality, and so on. This will eventually give us the complete algorithm.

Step 1: 1-fold belief in future rationality

We first wish to characterize those strategies you can rationally choose under 1-fold belief in future rationality. Consider a type t_i for player i that expresses 1-fold belief in future rationality – that is, which believes in the opponents' future rationality. Take some information set h for player i, and some other information set h' for opponent j that weakly follows h. Let $\Gamma^0(h)$ be the full decision problem at h, and let $\Gamma^0(h')$ be the full decision problem at h'. Since t_i believes in the opponents' future rationality, he believes, in particular, at h that player j will choose rationally at h'. Now, the strategies for player j that can be optimal at h' are exactly those strategies for player j in $\Gamma^0(h')$ that are not strictly dominated in $\Gamma^0(h')$. So, if player i believes in his opponents' future rationality, then he believes at h that player j will not choose a strategy that is strictly dominated in $\Gamma^0(h')$. We can therefore eliminate from $\Gamma^0(h)$ those strategies for player j that are strictly dominated in $\Gamma^0(h')$.

This argument can be generalized. If type t_i believes in his opponents' future rationality, then at every information set $h \in H_i$ he believes that every opponent j will not choose a strategy that is strictly dominated in some decision problem $\Gamma^0(h')$ for player j that weakly follows h. By a decision problem for player j we mean a decision problem at an information set h' at which player j is active. So, if t_i believes in his opponents' future rationality, then for every information set $h \in H_i$ we can delete from $\Gamma^0(h)$ those strategies for player j that are strictly dominated in some decision problem $\Gamma^0(h')$ for player j that weakly follows h. This leads to a *reduced* decision problem $\Gamma^1(h)$ at h. The interpretation is that a type t_i who believes in his opponents' future rationality, will believe at h that his opponents will choose strategies from $\Gamma^1(h)$.

But then, every optimal strategy s_i for type t_i must be optimal, at every information set $h \in H_i$ that s_i leads to, within the reduced decision problem $\Gamma^1(h)$. Consequently, every optimal strategy s_i for type t_i, at every information set $h \in H_i$ that s_i leads to, cannot be strictly dominated within the reduced decision problem $\Gamma^1(h)$.

Now, let $\Gamma^2(\emptyset)$ be the reduced decision problem at the beginning of the game, which is obtained from $\Gamma^1(\emptyset)$ by deleting all strategies s_i that are strictly dominated, at some $h \in H_i$ that s_i leads to, within the reduced decision problem $\Gamma^1(h)$. So, we see that every optimal strategy s_i for a type t_i that believes in the opponents' future rationality, must be in $\Gamma^2(\emptyset)$. In other words, every strategy that can rationally be chosen under 1-fold belief in future rationality must be in $\Gamma^2(\emptyset)$.

Step 2: Up to 2-fold belief in future rationality

We now concentrate on the strategies you can rationally choose if you express up to 2-fold belief in future rationality – that is, express 1-fold and 2-fold belief in future rationality. Consider a type t_i that expresses up to 2-fold belief in future rationality. So, at every information set $h \in H_i$, the type t_i only assigns positive probability to opponents' strategy-type pairs (s_j, t_j) where s_j is optimal for t_j at every $h' \in H_j$ that weakly follows h, and where type t_j expresses 1-fold belief in future rationality. Now, consider such an information set $h \in H_i$, an information set $h' \in H_j$ weakly following h, and a type t_j that expresses 1-fold belief in future rationality. We have seen in step 1 that every such type t_j assigns at h' only positive probability to opponents' strategies in $\Gamma^1(h')$. So, every strategy s_j that is optimal for t_j at h' cannot be strictly dominated in $\Gamma^1(h')$. Hence, we conclude that type t_i assigns at h only positive probability to opponents' strategies s_j that are not strictly dominated in any reduced decision problem $\Gamma^1(h')$ that weakly follows h, and at which j is active.

Now, let $\Gamma^2(h)$ be the reduced decision problem at h which is obtained from $\Gamma^1(h)$ by deleting those strategies s_j that are strictly dominated within some reduced decision problem $\Gamma^1(h')$ weakly following h at which j is active. Then, we conclude that t_i assigns at h only positive probability to opponents' strategies in $\Gamma^2(h)$. Hence, every type t_i that expresses up to 2-fold belief in future rationality, assigns at every $h \in H_i$ only positive probability to opponents' strategies in $\Gamma^2(h)$.

But then, every optimal strategy s_i for such a type t_i must be optimal, at every information set $h \in H_i$ that s_i leads to, for some belief within the reduced decision problem $\Gamma^2(h)$. Consequently, every optimal strategy s_i for type t_i, at every information set $h \in H_i$ that s_i leads to, cannot be strictly dominated within the reduced decision problem $\Gamma^2(h)$.

Now, let $\Gamma^3(\emptyset)$ be the reduced decision problem at the beginning of the game, which is obtained from $\Gamma^2(\emptyset)$ by deleting all strategies s_i that are strictly dominated, at some $h \in H_i$ that s_i leads to, within the reduced decision problem $\Gamma^2(h)$. So, we see that every optimal strategy s_i for a type t_i that expresses up to 2-fold belief in future rationality, must be in $\Gamma^3(\emptyset)$.

By continuing in this fashion we would inductively generate reduced decision problems $\Gamma^k(h)$ for every k, and every information set h, and conclude that every strategy that can rationally be chosen by a type that expresses up to k-fold belief in future rationality, must be in $\Gamma^{k+1}(\emptyset)$. This leads to the following algorithm, which we call the *backward dominance procedure.*

Algorithm 8.8.2 *(Backward dominance procedure)*
Step 1: *For every full decision problem $\Gamma^0(h)$, eliminate for every player i those strategies that are strictly dominated at some full decision problem $\Gamma^0(h')$ that weakly follows $\Gamma^0(h)$ and at which player i is active. This leads to reduced decision problems $\Gamma^1(h)$ for every information set h.*
Step 2: *For every reduced decision problem $\Gamma^1(h)$, eliminate for every player i those strategies that are strictly dominated at some reduced decision problem $\Gamma^1(h')$ that weakly follows $\Gamma^1(h)$ and at which player i is active. This leads to new reduced decision problems $\Gamma^2(h)$ for every information set.*
And so on. Continue until no more strategies can be eliminated in this way.

Whenever we say that we eliminate the strategy s_i for player i from the full decision problem $\Gamma^0(h)$, we precisely mean the following: If player i is active at h, and the full decision problem $\Gamma^0(h)$ is given by $(S_i(h), S_{-i}(h))$, then we simply eliminate strategy s_i from $S_i(h)$. If player j is active at h but not i, and the full decision problem $\Gamma^0(h)$ is given by $(S_j(h), S_{-j}(h))$, then we eliminate from $S_{-j}(h)$ every strategy combination that contains strategy s_i for player i. Similarly for eliminating a strategy from a *reduced* decision problem.

It is easily seen that the backward dominance procedure will always stop after finitely many steps. The reason is that we only have finitely many strategies in every decision problem, and at every round we eliminate a strategy from at least one decision problem. So, there must be a round in which no further strategy can be eliminated.

Another way to state the algorithm is as follows: If at step k we find a strategy s_i for player i that is strictly dominated at some decision problem $\Gamma^{k-1}(h)$ at which player i is active, then we eliminate s_i from decision problem $\Gamma^{k-1}(h)$ and from every decision problem $\Gamma^{k-1}(h')$ that comes either before h, or is simultaneous with h. So, we eliminate strategy s_i *backwards,* starting from the decision problem $\Gamma^{k-1}(h)$. This explains the term *backward* in the backward dominance procedure.

We have seen above that every strategy that can rationally be chosen by a type that expresses up to k-fold belief in future rationality, must be in $\Gamma^{k+1}(\emptyset)$. It can be shown that the converse is also true: Every strategy in $\Gamma^{k+1}(\emptyset)$ is optimal for a type that expresses up to k-fold belief in future rationality. So, the strategies that can rationally be chosen by a type that expresses up to k-fold belief in future rationality are exactly the strategies in $\Gamma^{k+1}(\emptyset)$, surviving the first $k+1$ steps of the backward dominance procedure. Moreover, the strategies that can rationally be chosen under *common* belief in future rationality are precisely the strategies that survive the *full* backward dominance procedure. So we obtain the following general result.

Theorem 8.8.3 *(The algorithm works)*
(1) For every $k \geq 1$, the strategies that can rationally be chosen by a type that expresses up to k-fold belief in future rationality are exactly the strategies in $\Gamma^{k+1}(\emptyset)$, surviving the first $k+1$ steps of the backward dominance procedure.
(2) The strategies that can rationally be chosen by a type that expresses common belief in future rationality are exactly the strategies that survive the full backward dominance procedure – that is, the strategies that are in $\Gamma^k(\emptyset)$ for every k.

The formal proof of this theorem can be found in the proofs section at the end of this chapter. As an illustration of our algorithm and our result, let us return to the original version of the example "Painting Chris' house."

Example 8.4 Painting Chris' house
Consider the original game from Figure 8.1, in which only Barbara receives an offer from a colleague before going to Chris' house. Let us run the backward dominance procedure and see which strategies survive.

In this game there are two information sets: $\emptyset$, at which only Barbara is active, and h_1 (the information set after Barbara's choice "reject"), at which you and Barbara are both active. So, the full decision problems at the beginning of the procedure are given by Table 8.10.

Step 1: At Barbara's full decision problem $\Gamma^0(\emptyset)$, her strategies $(r, 200)$, $(r, 300)$ and $(r, 500)$ are all strictly dominated by her strategy *accept*. So, we eliminate these strategies for Barbara from $\Gamma^0(\emptyset)$, but not from $\Gamma^0(h_1)$ as h_1 comes after $\emptyset$. For the full decision problem $\Gamma^0(h_1)$, where both you and Barbara are active, $(r, 500)$ and 500 are the only strategies for you and Barbara that are strictly dominated. Hence, we eliminate your strategy 500 from $\Gamma^0(\emptyset)$ and $\Gamma^0(h_1)$, and we eliminate Barbara's strategy $(r, 500)$ from $\Gamma^0(h_1)$. We have already eliminated her strategy $(r, 500)$ from $\Gamma^0(\emptyset)$. Hence we obtain the new decision problems $\Gamma^1(\emptyset)$ and $\Gamma^1(h_1)$ shown in Table 8.11.

Step 2: For Barbara's decision problem $\Gamma^1(\emptyset)$, her strategy $(r, 400)$ is strictly dominated by *accept*, so we eliminate her strategy $(r, 400)$ from $\Gamma^1(\emptyset)$, but not (yet) from $\Gamma^1(h_1)$. For the decision problem $\Gamma^1(h_1)$, the strategies $(r, 400)$ and 400 are strictly dominated for you and Barbara. So, we eliminate your strategy 400 from $\Gamma^1(\emptyset)$ and $\Gamma^1(h_1)$, and

Table 8.10. *Full decision problems in the game of Figure 8.1*

	$\Gamma^0(\emptyset)$: Barbara active			
	200	300	400	500
$(r,200)$	100, 100	200, 0	200, 0	200, 0
$(r,300)$	0, 200	150, 150	300, 0	300, 0
$(r,400)$	0, 200	0, 300	200, 200	400, 0
$(r,500)$	0, 200	0, 300	0, 400	250, 250
accept	350, 500	350, 500	350, 500	350, 500

	$\Gamma^0(h_1)$: Barbara and you active			
	200	300	400	500
$(r,200)$	100, 100	200, 0	200, 0	200, 0
$(r,300)$	0, 200	150, 150	300, 0	300, 0
$(r,400)$	0, 200	0, 300	200, 200	400, 0
$(r,500)$	0, 200	0, 300	0, 400	250, 250

Table 8.11. *Decision problems after step 1 in the game of Figure 8.1*

	$\Gamma^1(\emptyset)$: Barbara active		
	200	300	400
$(r,400)$	0, 200	0, 300	200, 200
accept	350, 500	350, 500	350, 500

	$\Gamma^1(h_1)$: Barbara and you active		
	200	300	400
$(r,200)$	100, 100	200, 0	200, 0
$(r,300)$	0, 200	150, 150	300, 0
$(r,400)$	0, 200	0, 300	200, 200

we eliminate Barbara's strategy $(r,400)$ from $\Gamma^1(h_1)$. This leads to the new decision problems in Table 8.12.

Step 3: In the decision problem $\Gamma^2(h_1)$, the strategies $(r,300)$ and 300 are strictly dominated for you and Barbara. So, we eliminate the strategy 300 for you from $\Gamma^2(\emptyset)$ and $\Gamma^2(h_1)$, and we eliminate Barbara's strategy $(r,300)$ from $\Gamma^2(h_1)$.

This finally leads to the decision problems in Table 8.13, in which no further strategies can be eliminated. So, the backward dominance procedure stops after step 3. The only strategy that is left for you at $\emptyset$ is 200, and hence 200 is the only strategy for you that survives the backward dominance procedure. According to our theorem

Table 8.12. *Decision problems after step 2 in the game of Figure 8.1*

	$\Gamma^2(\emptyset)$: Barbara active	
	200	300
accept	350, 500	350, 500
	$\Gamma^2(h_1)$: Barbara and you active	
	200	300
$(r, 200)$	100, 100	200, 0
$(r, 300)$	0, 200	150, 150

Table 8.13. *Decision problems after step 3 in the game of Figure 8.1*

	$\Gamma^3(\emptyset)$: Barbara active
	200
accept	350, 500
	$\Gamma^3(h_1)$: Barbara and you active
	200
$(r, 200)$	100, 100

above, 200 must be the only strategy that can you can rationally choose under common belief in future rationality. This is indeed true, as we know from our analysis in Section 8.6. □

8.9 Order independence

The backward dominance procedure tells us that, at every round k, we must eliminate from *every* decision problem $\Gamma^{k-1}(h)$, and for *every* player i, *every* strategy that is strictly dominated for some decision problem $\Gamma^{k-1}(h')$ that weakly follows h and at which i is active. Suppose now that we do not eliminate, at some round, and for some decision problem $\Gamma^{k-1}(h)$, *all* strategies for player i that we can eliminate in this way, but only *some*. Would it matter for the eventual result of the procedure? We will see in this section that it will not. That is, the precise order and speed with which we eliminate strategies in the backward dominance procedure is irrelevant for the output of this

Table 8.14. *Changing the order of elimination in the game of Figure 8.1*

	$\Gamma^0(h_1)$			
	200	300	400	500
$(r,200)$	100, 100	200, 0	200, 0	200, 0
$(r,300)$	0, 200	150, 150	300, 0	300, 0
$(r,400)$	0, 200	0, 300	200, 200	400, 0
$(r,500)$	0, 200	0, 300	0, 400	250, 250

	$\hat{\Gamma}^1(h_1)$		
	200	300	400
$(r,200)$	100, 100	200, 0	200, 0
$(r,300)$	0, 200	150, 150	300, 0
$(r,400)$	0, 200	0, 300	200, 200

	$\hat{\Gamma}^2(h_1)$	
	200	300
$(r,200)$	100, 100	200, 0
$(r,300)$	0, 200	150, 150

	$\hat{\Gamma}^3(h_1)$
	200
$(r,200)$	100, 100

procedure. To see why, let us first reconsider the example "Painting Chris' house," which we know so well by now.

Example 8.5 Painting Chris' house

Consider the original game from Figure 8.1. Suppose that we change the order of elimination in the backward dominance procedure as follows: We first apply the entire procedure to the decision problem at h_1, and only after finishing do we apply the procedure to the decision problem at $\emptyset$. So we start with the full decision problem at h_1, which is depicted in Table 8.14. In the first few rounds of the procedure, we only eliminate strategies at h_1, and not at $\emptyset$.

Step 1: We eliminate from $\Gamma^0(h_1)$ those strategies that are strictly dominated at $\Gamma^0(h_1)$. That is, we eliminate strategy $(r,500)$ for Barbara and 500 for you from $\Gamma^0(h_1)$. This gives the new decision problem $\hat{\Gamma}^1(h_1)$ shown in Table 8.14.

Step 2: Next, we eliminate from $\hat{\Gamma}^1(h_1)$ those strategies that are strictly dominated at $\hat{\Gamma}^1(h_1)$. That is, we eliminate strategy $(r,400)$ for Barbara and 400 for you from $\hat{\Gamma}^1(h_1)$. This gives the new decision problem $\hat{\Gamma}^2(h_1)$ shown in Table 8.14.

Table 8.15. *Changing the order of elimination in the game of Figure 8.1 (II)*

	$\Gamma^0(\emptyset)$			
	200	300	400	500
$(r, 200)$	100, 100	200, 0	200, 0	200, 0
$(r, 300)$	0, 200	150, 150	300, 0	300, 0
$(r, 400)$	0, 200	0, 300	200, 200	400, 0
$(r, 500)$	0, 200	0, 300	0, 400	250, 250
accept	350, 500	350, 500	350, 500	350, 500

	$\hat{\Gamma}^4(\emptyset)$
	200
$(r, 200)$	100, 100
accept	350, 500

	$\hat{\Gamma}^5(\emptyset)$
	200
accept	350, 500

Step 3: Then, we eliminate from $\hat{\Gamma}^2(h_1)$ those strategies that are strictly dominated at $\hat{\Gamma}^2(h_1)$. That is, we eliminate strategy $(r, 300)$ for Barbara and 300 for you from $\hat{\Gamma}^2(h_1)$. This gives the new decision problem $\hat{\Gamma}^3(h_1)$ shown in Table 8.14.

After this step, we can eliminate no more strategies at h_1. So now, we consider $\emptyset$ and start eliminating there. We start from the full decision problem $\Gamma^0(\emptyset)$ at $\emptyset$, which can be found in Table 8.15.

Step 4: From $\Gamma^0(\emptyset)$, we can eliminate the strategies $(r, 300), (r, 400)$ and $(r, 500)$ for Barbara and 300, 400 and 500 for you, because each of these strategies was strictly dominated for some of the previous decision problems at h_1. This leads to the new decision problem $\hat{\Gamma}^4(\emptyset)$ shown in Table 8.15.

Step 5: Finally, we eliminate from $\hat{\Gamma}^4(\emptyset)$ strategy $(r, 200)$ for Barbara, as it is strictly dominated for Barbara at $\hat{\Gamma}^4(\emptyset)$. This gives the decision problem $\hat{\Gamma}^5(\emptyset)$ shown in Table 8.15.

After step 5, no further strategies can be eliminated at $\emptyset$ and h_1, so the procedure stops here. Note that the strategies that survive this procedure, namely *accept* for Barbara, and 200 for you, are exactly the same strategies that survive the original backward dominance procedure, with the original order of elimination. So, changing the order of elimination in this particular way has no effect on the output of the procedure in this example. □

We will now explain why, in general, the order of elimination is of no importance to the backward dominance procedure. Take an arbitrary dynamic game, and let us compare two elimination procedures for this game:

- **Procedure 1:** This is the original backward dominance procedure in which we eliminate, at every round k, from *every* decision problem $\Gamma^{k-1}(h)$, and for *every* player i, *every* strategy that is strictly dominated at some decision problem $\Gamma^{k-1}(h')$ that weakly follows h and at which i is active.
- **Procedure 2:** This is a procedure in which we eliminate, at every round k, from *some* decision problems $\hat{\Gamma}^{k-1}(h)$, and for *some* players i, *some* strategies that are strictly dominated at some decision problem $\hat{\Gamma}^{k-1}(h')$ that weakly follows h and at which i is active.

So, in Procedure 1 we always eliminate all strategies we can, whereas in Procedure 2 we only eliminate *some* strategies – but not necessarily all of them. Suppose that Procedure 1 yields decision problems $\Gamma^k(h)$ at every round k, whereas Procedure 2 leads to decision problems $\hat{\Gamma}^k(h)$ at every round k.

We will first show that every strategy that survives Procedure 1 must also survive Procedure 2. A strategy survives Procedure 1 if it is in $\Gamma^k(\emptyset)$ for all k, and similarly for Procedure 2. In order to prove this, we show that, for every round k, and for every information set h, every strategy that survives round k of Procedure 1 at h, must also survive round k of Procedure 2 at h.

To show this, let us first compare round 1 of both procedures. Consider some information set h, and let strategy s_i for player i be in $\Gamma^1(h)$. That is, s_i survives round 1 of Procedure 1 at h. This is only possible if s_i is not strictly dominated in any decision problem $\Gamma^0(h')$ that weakly follows h, and where i is active. But then, s_i will certainly not be eliminated by Procedure 2 at h in round 1. So, s_i will survive round 1 of Procedure 2 at h as well, and hence s_i will be in $\hat{\Gamma}^1(h)$. So, we see that, for every information set h, every strategy in $\Gamma^1(h)$ will also be in $\hat{\Gamma}^1(h)$.

Let us now compare round 2 of both procedures. Consider some information set h, and let strategy s_i for player i be in $\Gamma^2(h)$. So, s_i has not been eliminated from $\Gamma^1(h)$ in round 2 of Procedure 1. This means that s_i is not strictly dominated in any decision problem $\Gamma^1(h')$ that weakly follows h, and where i is active. Above we have seen that for every such h', all strategies in $\Gamma^1(h')$ are also in $\hat{\Gamma}^1(h')$. But then, if s_i is not strictly dominated in the smaller decision problem $\Gamma^1(h')$, it will certainly not be strictly dominated in the larger decision problem $\hat{\Gamma}^1(h')$. Hence, we may conclude that s_i is also not strictly dominated in any decision problem $\hat{\Gamma}^1(h')$ that weakly follows h, and where i is active. This means, in turn, that s_i cannot be eliminated at h in round 2 of Procedure 2, and hence s_i will be in $\hat{\Gamma}^2(h)$. So, we see that, for every information set h, every strategy in $\Gamma^2(h)$ will also be in $\hat{\Gamma}^2(h)$.

By repeating this argument we finally conclude that, in every round k, and for every information set h, every strategy in $\Gamma^k(h)$ will also be in $\hat{\Gamma}^k(h)$. That is, in every round

k, and for every information set h, every strategy that survives round k of Procedure 1 at h, must also survive round k of Procedure 2 at h.

A direct consequence of this insight is that every strategy that is in $\Gamma^k(\emptyset)$ for all k, must also be in $\hat{\Gamma}^k(\emptyset)$ for all k. In other words, every strategy that survives Procedure 1 must also survive Procedure 2.

We now prove that the converse is also true: Every strategy that survives Procedure 2 must also survive Procedure 1. In order to prove this, we will show that, for every round k, and for every information set h, every strategy that is eliminated in round k of Procedure 1 at h, will also *eventually* be eliminated in Procedure 2 at h – maybe already at round k, but maybe at some later round.

Suppose that this were not true. So, suppose that there is some round k, some information set h, and some strategy s_i for player i, such that s_i is eliminated in round k of Procedure 1 at h, but is never eliminated by Procedure 2 at h. Then, let k^* be the *first* round where this happens. So, let k^* be the *first* round where there is some information set h and some strategy s_i, such that s_i is eliminated at round k^* of Procedure 1 at h, but is never eliminated by Procedure 2 at h.

Since s_i is eliminated at round k^* of Procedure 1 at h, there must be some decision problem $\Gamma^{k^*-1}(h')$ weakly following h where i is active, such that s_i is strictly dominated in $\Gamma^{k^*-1}(h')$. By our choice of k^*, we know that all strategies that are eliminated up to round $k^* - 1$ of Procedure 1 at h', are also eventually eliminated by Procedure 2 at h'. This means that there is some round m, such that all strategies that are eliminated up to round $k^* - 1$ of Procedure 1 at h', are eliminated up to round m of Procedure 2 at h'. So, all strategies that are in $\hat{\Gamma}^m(h')$ are also in $\Gamma^{k^*-1}(h')$. As strategy s_i is strictly dominated in $\Gamma^{k^*-1}(h')$, it must also be strictly dominated in $\hat{\Gamma}^m(h')$, because the set of strategies in $\hat{\Gamma}^m(h')$ is a subset of the set of strategies in $\Gamma^{k^*-1}(h')$. This means, however, that strategy s_i must be eliminated sooner or later by Procedure 2 at h. This contradicts our assumption that s_i is never eliminated by Procedure 2 at h.

So, we conclude that every strategy that is eliminated by Procedure 1 for some information set h at some round k, must also eventually be eliminated by Procedure 2 at h. Hence, every strategy that survives Procedure 2 for some information set h must also survive Procedure 1 for this information set. In particular, every strategy that is in $\hat{\Gamma}^k(\emptyset)$ for all k, must also be in $\Gamma^k(\emptyset)$ for all k. This means, however, that every strategy that survives Procedure 2 must also survive Procedure 1.

As we have already seen above that the converse is also true, we conclude that the set of strategies that survive Procedure 1 is exactly the same as the set of strategies that survive Procedure 2. So, the order and speed with which we eliminate strategies has no influence on the eventual output of the backward dominance procedure. We thus have shown the following result.

Theorem 8.9.1 *(Order and speed of elimination does not matter)*
The set of strategies that survive the backward dominance procedure does not change if we change the order and speed of elimination.

This result has some important practical implications. In some dynamic games, it may be easier to use a different order and speed of elimination than prescribed by the original backward dominance procedure. The theorem above ensures that choosing this "easier" order of elimination does not change the eventual output of the procedure. The following section will provide a couple of examples that illustrate this.

8.10 Backwards order of elimination

We learned in the previous section that for the backward dominance procedure it does not matter in what order we eliminate strategies. For many dynamic games we may exploit this fact since the original order of elimination prescribed by the backward dominance procedure is often not the most practical one. According to the original order of elimination, we must scan at every round through *all* the decision problems in the dynamic game, and see whether there are some strategies that are strictly dominated. In many games, it is more practical to first analyze only the *final* decision problems in the game, and then to work *backwards* towards the beginning of the game. We call this the *backwards order of elimination.* According to our theorem above, choosing this more practical order of elimination will not affect the outcome. Here is a first illustration.

Example 8.6 Two friends and a treasure

You, Barbara and Chris are on holiday in the Caribbean. This morning, you and Barbara found 60 pieces of gold on the beach. As you both claim to have seen the treasure first, you enter into a fierce fight about how to divide it between the two of you. Then, Chris jumps in and proposes the following procedure: You and Barbara both name the number of pieces you want to have. This number must be either 10, 20, 30 or 40, to make things easy. If the sum of the numbers does not exceed 60, then you will both get the number you named, and the rest of the pieces (if there are any) will go to Chris. If the sum of the numbers exceeds 60, then it is impossible to give both of you the numbers you named. In that case, you must both name a new number, different from the number you chose before. The rules in this second round are the same as before: If the sum of the new numbers does not exceed 60, then you both get your number. If the sum of the new number still exceeds 60, then you must name a number again, different from the numbers you named at the previous two attempts. This dynamic game is represented graphically by Figure 8.8. Suppose, for instance, that you choose 30 and Barbara chooses 40 at the beginning of the game. Then, it is impossible to give you the numbers you named, and the game would move to information set h_1, where you can no longer choose 30 and Barbara can no longer choose 40. If at h_1, you name 40 and Barbara names 30, then the game would move to information set h_4, where you and Barbara can no longer choose the amounts 30 and 40. There, the game will end, as the sum of the numbers can no longer exceed 60.

We want to find those strategies you can rationally choose under common belief in future rationality, and to that purpose we will use the backward dominance procedure.

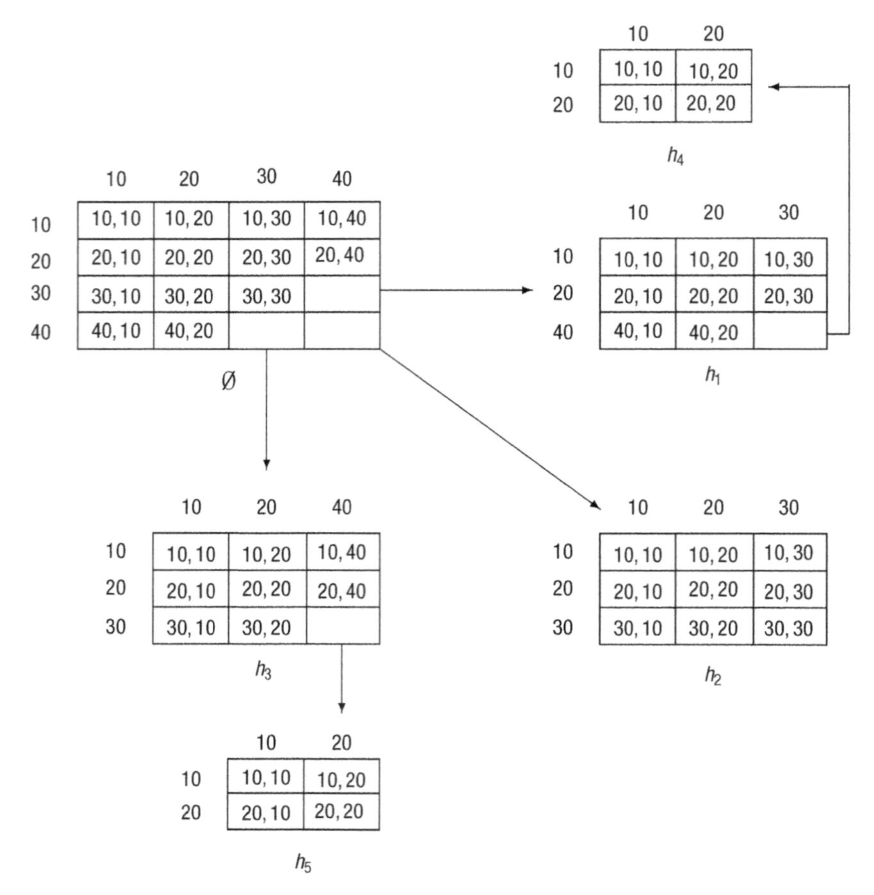

Figure 8.8 Graphical representation of "Two friends and a treasure"

In this game, however, using the *original* order of elimination as stipulated by the backward dominance procedure is very inconvenient. If you wish to apply the original order of elimination, then you must in particular find in the full decision problem $\Gamma^0(\emptyset)$ at the beginning of the game those strategies that are strictly dominated at $\Gamma^0(\emptyset)$ or at some other full decision problem $\Gamma^0(h)$. But constructing the full decision problem $\Gamma^0(\emptyset)$ at the beginning of the game is an enormous task, since both players have 18 strategies in $\Gamma^0(\emptyset)$ (check this)!

A more practical order of elimination is to start at the *ultimate* information sets in the game, and then work backwards towards the beginning of the game. Here, we call an information set "ultimate" if it is not followed by another information set in the game. In this example, the ultimate information sets are h_2, h_4 and h_5.

Let us start at h_4. The strategies for you that lead to h_4 are

$$(30_0, 40_1, 10_4) \text{ and } (30_0, 40_1, 20_4).$$

Table 8.16. *Full decision problem at h_4 in "Two friends and a treasure"*

	$(40_0, 30_1, 10_2, 10_4)$	$(40_0, 30_1, 20_2, 10_4)$	$(40_0, 30_1, 30_2, 10_4)$	$(40_0, 30_1, 10_2, 20_4)$	$(40_0, 30_1, 20_2, 20_4)$	$(40_0, 30_1, 30_2, 20_4)$
$(30_0, 40_1, 10_4)$	10,10	10,10	10,10	10,20	10,20	10,20
$(30_0, 40_1, 20_4)$	20,10	20,10	20,10	20,20	20,20	20,20

Here, $(30_0, 40_1, 10_4)$ denotes the strategy where you choose 30 at Ø, 40 at h_1 and 10 at h_4. Similarly for $(30_0, 40_1, 20_4)$. The strategies for Barbara that lead to h_4 are

$$(40_0, 30_1, 10_2, 10_4), (40_0, 30_1, 20_2, 10_4), (40_0, 30_1, 30_2, 10_4),$$
$$(40_0, 30_1, 10_2, 20_4), (40_0, 30_1, 20_2, 20_4), (40_0, 30_1, 30_2, 20_4).$$

So, the full decision problem $\Gamma^0(h_4)$ at h_4 is given by Table 8.16. To simplify the notation, let us denote your strategies in $\Gamma^0(h_4)$ by 10_4 and 20_4, which contains sufficient information. A strategy for you that reaches h_4 must choose 30 at Ø and 40 at h_1, so these are common features of *all* strategies for you in $\Gamma^0(h_4)$, and can therefore be left out when analyzing $\Gamma^0(h_4)$. The only important information about your strategy choice in $\Gamma^0(h_4)$ is the choice you make at h_4.

Within the decision problem $\Gamma^0(h_4)$, Barbara's strategies

$$(40_0, 30_1, 10_2, 10_4), (40_0, 30_1, 20_2, 10_4) \text{ and } (40_0, 30_1, 30_2, 10_4)$$

are equivalent, since they always yield the same outcome against any strategy for you in $\Gamma^0(h_4)$. This is no surprise, as these three strategies prescribe the same choice at h_4, which is the only choice that counts for Barbara within the decision problem at h_4. So, we can group these three strategies together, and call this group of strategies 10_4. Similarly, we can group the other three strategies for Barbara in $\Gamma^0(h_4)$ together since they are equivalent within $\Gamma^0(h_4)$, and call this group 20_4.

These simplifications lead to the "simplified" full decision problem $\Gamma^0(h_4)$ shown in Table 8.17. At $\Gamma^0(h_4)$ your strategy 10_4 and Barbara's group of strategies 10_4 are strictly dominated. Hence, we can eliminate these strategies from h_4, leaving only the strategy 20_4 for you and the group of strategies 20_4 for Barbara at h_4.

We can analyze the full decision problem at h_5 in a similar way. Within $\Gamma^0(h_5)$, we can eliminate your group of strategies 10_5 and Barbara's strategy 10_5. This leaves only your group of strategies 20_5 and Barbara's strategy 20_5, at h_5.

The full decision problem at h_2 is depicted in Table 8.18. Here, 10_2 represents the group of three strategies for you that choose 40 at Ø, choose 10 at h_2, and choose any number from 10, 20, 30 at h_3. Similarly for the other groups of strategies in this table.

Table 8.17. *Simplified full decision problem at h_4 in "Two friends and a treasure"*

	10_4	20_4
10_4	10, 10	10, 20
20_4	20, 10	20, 20

Table 8.18. *Full decision problem at h_2 in "Two friends and a treasure"*

	10_2	20_2	30_2
10_2	10, 10	10, 20	10, 30
20_2	20, 10	20, 20	20, 30
30_2	30, 10	30, 20	30, 30

Table 8.19. *Full decision problem at h_1 in "Two friends and a treasure"*

	10_1	20_1	$(30_1, 10_4)$	$(30_1, 20_4)$
10_1	10, 10	10, 20	10, 30	10, 30
20_1	20, 10	20, 20	20, 30	20, 30
$(40_1, 10_4)$	40, 10	40, 20	10, 10	10, 20
$(40_1, 20_4)$	40, 10	40, 20	20, 10	20, 20

Within $\Gamma^0(h_2)$, the groups of strategies 10_2 and 20_2 are strictly dominated for you and Barbara, and hence we can eliminate these strategies from $\Gamma^0(h_2)$. This leaves only the group of strategies 30_2 for you and Barbara at h_2.

Now that we have finished the elimination procedures for the ultimate information sets h_2, h_4 and h_5, we move to the penultimate information sets. These are the information sets that are followed by at least one other information set, but where all information sets that follow are ultimate information sets. In this game, the penultimate information sets are h_1 and h_3.

Let us first look at the full decision problem at h_1, which is depicted in Table 8.19. Here, $(40_1, 10_4)$ is an abbreviation for your strategy $(30_0, 40_1, 10_4)$, and similarly for your other strategies. For Barbara, $(30_1, 10_4)$ denotes the group of three strategies that choose 40 at $\varnothing$, choose 30 at h_1, choose 10 at h_4, and make some choice from 10, 20, 30

Table 8.20. *Reduced decision problem at h_1 in "Two friends and a treasure" after first round of elimination*

	10_1	20_1	$(30_1, 20_4)$
10_1	10, 10	10, 20	10, 30
20_1	20, 10	20, 20	20, 30
$(40_1, 20_4)$	40, 10	40, 20	20, 20

Table 8.21. *Final decision problem at h_1 in "Two friends and a treasure"*

	20_1	$(30_1, 20_4)$
20_1	20, 20	20, 30
$(40_1, 20_4)$	40, 20	20, 20

at h_2. We have grouped these three strategies together as they are equivalent within the decision problem at h_1. Also the other three columns for Barbara represent groups of three strategies that are equivalent at h_1.

From the decision problem at h_1, we can first eliminate the strategies that we have eliminated already for the future information set h_4. If we eliminate a strategy at a given information set, then according to the backward dominance procedure we can also eliminate this strategy at all information sets that come before it. At h_4, we had eliminated the strategy 10_4 for you, which corresponds to your strategy $(40_1, 10_4)$ in Table 8.19. So, we eliminate your strategy $(40_1, 10_4)$ from the decision problem at h_1. At h_4, we also eliminated Barbara's group of strategies 10_4, which corresponds to Barbara's group of strategies $(30_1, 10_4)$ in Table 8.19. So, we also eliminate Barbara's group of strategies $(30_1, 10_4)$ from the decision problem at h_1. This leads to the reduced decision problem shown in Table 8.20.

In the next step, we can eliminate at h_1 the strategy 10_1 for you and the group of strategies 10_1 for Barbara, as these are strictly dominated within this reduced decision problem at h_1. This leads to the final decision problem at h_1 in Table 8.21, from which no further strategies can be eliminated.

In exactly the same way, the analysis at information set h_3 leads to the final decision problem in Table 8.22.

Up to this stage we have applied the elimination procedure to the ultimate information sets h_2, h_4 and h_5, and to the penultimate information sets h_1 and h_3. We finally move to the beginning of the game $\emptyset$. From the full decision problem at $\emptyset$ we can eliminate the strategies we have previously eliminated at h_1, h_2 and h_3.

Table 8.22. *Final decision problem at h_3 in "Two friends and a treasure"*

	20_3	$(40_3, 20_5)$
20_3	20, 20	20, 40
$(30_3, 20_5)$	30, 20	20, 20

Table 8.23. *Reduced decision problem at ∅ in "Two friends and a treasure" after first round of elimination*

	10_0	20_0	$(30_0, 20_3)$	$(30_0, 40_3, 20_5)$	$(40_0, 20_1, 30_2)$	$(40_0, 30_1, 30_2, 20_4)$
10_0	10, 10	10, 20	10, 30	10, 30	10, 40	10, 40
20_0	20, 10	20, 20	20, 30	20, 30	20, 40	20, 40
$(30_0, 20_1)$	30, 10	30, 20	30, 30	30, 30	20, 20	20, 30
$(30_0, 40_1, 20_4)$	30, 10	30, 20	30, 30	30, 30	40, 20	20, 20
$(40_0, 30_2, 20_3)$	40, 10	40, 20	20, 20	20, 40	30, 30	30, 30
$(40_0, 30_2, 30_3, 20_5)$	40, 10	40, 20	30, 20	20, 20	30, 30	30, 30

Remember that at h_1, only the strategies 20_1 and $(40_1, 20_4)$ survived for you, and only the groups of strategies 20_1 and $(30_1, 20_4)$ survived for Barbara. At h_3, only the groups of strategies 20_3 and $(30_3, 20_5)$ survived for you, and only the strategies 20_3 and $(40_3, 20_5)$ survived for Barbara. At h_2, only the group of strategies 30_2 survived for you and Barbara. Hence, we can already eliminate all other strategies that reach h_1, h_2 or h_3 from the decision problem at ∅. This leads to the reduced decision problem at ∅ in Table 8.23.

Within this decision problem, your strategy 10_0 is clearly strictly dominated, and your strategy 20_0 is strictly dominated by the randomized strategy

$$(0.5) \cdot (30_0, 20_1) + (0.5) \cdot (40_0, 30_2, 20_3).$$

Similarly for Barbara. So, we eliminate the strategies 10_0 and 20_0 for you and Barbara at ∅, which leads to the final decision problem at ∅ in Table 8.24, from which no further strategies can be eliminated.

Hence, there are four strategies that you can rationally choose under common belief in future rationality, namely the four that have survived in Table 8.24. In particular, under common belief in future rationality you will either name 30 or 40 at the beginning. □

Table 8.24. *Final decision problem at ∅ in "Two friends and a treasure"*

	$(30_0, 20_3)$	$(30_0, 40_3, 20_5)$	$(40_0, 20_1, 30_2)$	$(40_0, 30_1, 30_2, 20_4)$
$(30_0, 20_1)$	30, 30	30, 30	20, 20	20, 30
$(30_0, 40_1, 20_4)$	30, 30	30, 30	40, 20	20, 20
$(40_0, 30_2, 20_3)$	20, 20	20, 40	30, 30	30, 30
$(40_0, 30_2, 30_3, 20_5)$	30, 20	20, 20	30, 30	30, 30

The order of elimination we have used in the example above is called the *backwards order of elimination.* The idea is that you start at the ultimate information sets in the game, and then work backwards towards the beginning of the game. We will show in this section that the backwards order of elimination yields the same outcome as the original backward dominance procedure if the game is with *observed past choices.*

Definition 8.10.1 *(Game with observed past choices)*
A dynamic game is ***with observed past choices*** *if at every information set, the active players know precisely the choices made by the opponents in the past.*

On a technical level, a dynamic game is with observed past choices precisely when every information set in the game consists of a single non-terminal history. Namely, if an information set h for player i consists of two different non-terminal histories x and y, then player i does not know at h whether the opponents have chosen according to x or according to y in the past, and hence player i will not have observed the opponents' past choices completely. As a consequence, in a game with observed past choices there is for every information set h no other information set that is simultaneous with it. Note that the example "Two friends and a treasure" above corresponds to a dynamic game with observed past choices.

We will now define the backwards order of elimination more formally. Assume we consider a dynamic game with observed past choices. If we use the *backwards order of elimination*, we start by considering an *ultimate* information set h. In the first round at h, you delete from the full decision problem $\Gamma^0(h)$, and for every player i who is active at h, those strategies that are strictly dominated at $\Gamma^0(h)$. This gives $\Gamma^1(h)$. At the second round, you delete from the reduced decision problem $\Gamma^1(h)$, and for every player i who is active at h, those strategies that are strictly dominated at $\Gamma^1(h)$, and so on, until no further strategies can be eliminated in this way. That is, we perform iterated elimination of strictly dominated strategies within $\Gamma^0(h)$ for the *active* players at h *only*. We do this for every ultimate information set h in the game.

We then turn to a *penultimate* information set h. That is, h is followed by at least one information set h', but every information set h' following h is an ultimate information set. In the first round, we delete from the full decision problem $\Gamma^0(h)$ those strategies that were previously eliminated at some ultimate information set h' that follows h. This gives $\Gamma^1(h)$. In the second round, we delete from the reduced decision problem $\Gamma^1(h)$, and for every player i who is active at h, those strategies that are strictly dominated at $\Gamma^1(h)$. This gives $\Gamma^2(h)$. In the third round, we eliminate from the decision problem $\Gamma^2(h)$, and for every player i who is active at h, those strategies that are strictly dominated at $\Gamma^2(h)$. And so on, until no further strategies can be eliminated in this way. That is, we perform iterated elimination of strictly dominated strategies within $\Gamma^1(h)$ for the active players at h only. We do this for every penultimate information set h.

We then turn to a *pen-penultimate* information set h. That is, h is followed by at least one penultimate information set, and all information sets that follow h are either ultimate or penultimate information sets. In the first round, we delete from the full decision problem $\Gamma^0(h)$ those strategies that were previously eliminated at some penultimate or ultimate information set h' that follows h. This gives $\Gamma^1(h)$. Subsequently, we perform iterated elimination of strictly dominated strategies within $\Gamma^1(h)$ for the active players at h only. We do this for every pen-penultimate information set h.

By continuing in this fashion, we will eventually arrive at the beginning of the game $\emptyset$. In the first round at $\emptyset$, we delete from the full decision problem $\Gamma^0(\emptyset)$ those strategies that were previously eliminated at some information set h that follows $\emptyset$. This gives $\Gamma^1(\emptyset)$. We then perform iterated elimination of strictly dominated strategies within $\Gamma^1(\emptyset)$ for the active players at $\emptyset$ only. This concludes the backwards order of elimination.

So, the backwards order of elimination can formally be described as follows.

Definition 8.10.2 *(Backwards order of elimination)*
Consider a dynamic game with observed past choices. The backward dominance procedure with the backwards order of elimination proceeds as follows.
We start at the ***ultimate*** *information sets. For every ultimate information set h we perform iterated elimination of strictly dominated strategies within $\Gamma^0(h)$ for the active players at h.*
We then move to the ***penultimate*** *information sets. For every penultimate information set h, we first remove from $\Gamma^0(h)$ the strategies that have already been removed at the ultimate information sets following h, and then perform iterated elimination of strictly dominated strategies for the active players at h.*
And so on, until we reach the ***beginning*** *of the game $\emptyset$. From $\Gamma^0(\emptyset)$ we first remove the strategies that have already been removed at the previous rounds, and then perform iterated elimination of strictly dominated strategies for the active players at $\emptyset$.*

The strategies that survive at ∅ are the strategies that survive the backward dominance procedure with the backwards order of elimination.

We will now explain why, in a dynamic game with observed past choices, the backwards order of elimination yields exactly the same output as the original order of elimination in the backward dominance procedure.

Consider first an ultimate information set h in the game. We will show that the backwards order of elimination deletes exactly the same strategies from $\Gamma^0(h)$ as the original order of elimination. Since the game is with observed past choices, and the information set h is ultimate, the information set h is not followed by any other information set, nor is h simultaneous with any other information set. That is, h is not weakly followed by any other information set. Hence, the original order of elimination will start by eliminating from $\Gamma^0(h)$ those strategies s_i where player i is active at h, and where s_i is strictly dominated within $\Gamma^0(h)$. This gives the reduced decision problem $\Gamma^1(h)$. Then, the original order of elimination will eliminate from $\Gamma^1(h)$ those strategies s_i where player i is active at h, and where s_i is strictly dominated within $\Gamma^1(h)$. And so on. Hence, the original order of elimination will perform iterated elimination of strictly dominated strategies within $\Gamma^0(h)$ for the active players at h – precisely as the backwards order of elimination does. Afterwards, the original order of elimination will not eliminate any more strategies at an ultimate information set h, as h is not weakly followed by any other information set. Hence, we conclude that at every ultimate information set, the original order of elimination deletes exactly the same strategies as the backwards order of elimination – no more and no less.

We now turn to a penultimate information set h. That is, h is followed by at least one other information set h', and every such information set h' is ultimate. As the game is with observed past choices, h is not simultaneous with any other information set. That is, every other information set h' that weakly follows h must be an ultimate information set that (strictly) follows h. What strategies does the original order of elimination remove at h? We can distinguish two types of elimination here.

First, we eliminate at h those strategies s_i that are strictly dominated at an ultimate information set h' following h, within the current reduced decision problem at h', for some active player i at h'. But, as we have seen above, these are precisely the strategies that have been eliminated at h' by performing iterated elimination of strictly dominated strategies within $\Gamma^0(h')$ for the active players at h'. Hence, these are exactly the strategies that were eliminated at h' with the backwards order of elimination. So, the original order of elimination removes from h all strategies that will be removed for ultimate information sets h' following h under the backwards order of elimination.

The second type of elimination under the original order of elimination is that we remove at h all strategies s_i where i is active at h, and strategy s_i is strictly dominated within the current decision problem at h. This, however, amounts to performing iterated elimination of strictly dominated strategies within the decision problem at h for the active players at h.

Overall, we see that the original order of elimination proceeds as follows at h: We eliminate at h all strategies that will be removed for ultimate information sets h' following h under the backwards order of elimination, and moreover we perform iterated elimination of strictly dominated strategies at h for the active players at h. But this is precisely what the backwards order of elimination does at h – not necessarily in the same order, but this does not matter for the eventual output as we saw in Theorem 8.9.1. So, we conclude that for every penultimate information set, the original order of elimination and the backwards order of elimination eventually remove exactly the same sets of strategies.

By continuing in this fashion, we see that for every information set h in the game, the original order of elimination will eventually delete exactly the same sets of strategies as the backwards order of elimination. The two procedures may do so in a different order, but this does not matter for the eventual output. In particular, the two orders of elimination will eventually delete the same strategies at the beginning of the game. Hence, both orders of elimination yield the same output if the game is with observed past choices. These findings are summarized by the following theorem.

Theorem 8.10.3 *(Backwards order of elimination)*
Consider a dynamic game with observed past choices. Then, for every information set h, the sets of strategies that survive the backward dominance procedure at h with the original order of elimination, are the same as the sets of strategies that survive the backward dominance procedure at h with the backwards order of elimination.

Summarizing, we see that for dynamic games with observed past choices, the backwards order of elimination is a "valid" procedure, which yields exactly the same strategies at the end as the original backward dominance procedure. As such, we may use the backwards order of elimination for these types of games to find the strategies that can rationally be chosen under common belief in future rationality.

Corollary 8.10.4 *(Backwards order of elimination and common belief in future rationality)*
Consider a dynamic game with observed past choices. Then, the strategies that can rationally be chosen under common belief in future rationality are precisely the strategies that survive the backward dominance procedure with the backwards order of elimination.

The big practical advantage of the backwards order of elimination is that we can analyze the information sets in the game on a one-by-one basis, and in a clear predetermined order. Once we have finished the elimination process for a given information set, the analysis of this information set is closed, and we will never return to it in the remainder of the procedure. On the other hand, in the original backward dominance procedure we must at every round analyze each of the information sets again, and this often makes the analysis more involved than by using the backwards order of elimination.

To conclude this section, we will discuss another example which shows how useful this backwards order of elimination can be, especially when the game we want to analyze is very large.

Example 8.7 Bargaining with commitment

The beginning of the story is the same as in "Two friends and a treasure." However, Chris now proposes another procedure to decide how you and Barbara will divide the treasure of the 60 pieces of gold you found on the beach. At the beginning, you and Barbara simultaneously name a number between 0 and 60, which must be a multiple of 10. This is the number of pieces you *commit to*. That is, you both commit not to accept any outcome that gives you less than this number. You and Barbara must pay 10% of your commitments to Chris at the end of the day. If the sum of these commitments exceeds 60, then it is impossible to satisfy both your claims, and you will both receive 15 gold pieces. If the sum of these numbers is 60 or less, then you make a proposal to Barbara on how to divide the 60 pieces, and Barbara can either accept or reject this proposal. Again, this proposal can only use numbers that are multiples of 10. The number you propose for yourself must be at least the commitment you named at the beginning of the procedure. Similarly, Barbara will only accept an offer that would give her at least the commitment she named at the beginning. If Barbara rejects the offer, the negotiation breaks down and you both receive 15 gold pieces. If she accepts the offer, you both receive the numbers you agreed.

So this game has three stages. In the first stage, you and Barbara simultaneously name a commitment. If these two commitments together exceed 60, then the game ends at the first stage, and you will both receive 15 pieces of gold minus the number you must pay to Chris. If the sum of the two numbers is 60 or less, the game goes to stage 2. There, you make a proposal to Barbara for how to divide the treasure. At stage 3, finally, Barbara decides whether to accept or to reject this proposal.

The first stage of the game is depicted in Table 8.25. The blank cells in the matrix are the situations where the game would move to stage 2, because the sum of the numbers named does not exceed 60. In the other cells, we show the utilities for you and Barbara, since the game would stop there at stage 1. For instance, if you name the commitment 30 and Barbara names the commitment 50 at stage 1, then you both must pay 10% of these numbers to Chris, but only receive 15 pieces of gold in return because it is impossible to satisfy both your claims. This results in utilities of 12 and 10, respectively.

The question we wish to answer is which commitment numbers you and Barbara can rationally choose under common belief in future rationality, and which division is likely to occur. Note that this is a game with *observed past choices*. So, to answer this question we can use the backward dominance procedure with the *backwards order of elimination*.

We start at the ultimate information sets in the game. These are exactly the information sets at stage 3, where you have made a proposal to Barbara, and Barbara must decide whether to accept or reject that proposal. Clearly, Barbara will only accept the proposal if it gives her more than 15, since that is what she can get by rejecting your offer.

Table 8.25. *Stage 1 of "Bargaining with commitment"*

	0	10	20	30	40	50	60
0							
10							14,9
20						13,10	13,9
30					12,11	12,10	12,9
40				11,12	11,11	11,10	11,9
50			10,13	10,12	10,11	10,10	10,9
60		9,14	9,13	9,12	9,11	9,10	9,9

Moreover, the rules of the procedure force her to reject any proposal that gives her less than the commitment she named at stage 1. Hence, if Barbara chooses rationally at stage 3, then she will accept a proposal precisely when this proposal gives her at least 20, and at least the commitment she named at stage 1.

We then move to the penultimate information sets in the game. These are exactly the information sets in stage 2, where you must make a proposal to Barbara. We have seen above that, if you believe that Barbara chooses rationally at stage 3, then you believe that Barbara will accept a proposal only when it gives her at least 20 and at least the commitment she named at stage 1. We distinguish three classes of information sets.

Case 1. Suppose that Barbara has named the commitment 50 or 60 at stage 1. In that case, you believe that Barbara will only accept your offer if you give her at least 50. But that would leave at most 10 for yourself, which is less than the 15 you can get if Barbara rejects your offer. So in this case it is better to make Barbara reject your offer, for instance by offering her 0. The utilities you expect here are 15 for you and 15 for Barbara, minus the 10% of the commitments in stage 1 that must be paid to Chris.

Case 2. Suppose that Barbara has named the commitment 0 or 10 at stage 1. In that case, you believe that Barbara will accept your offer only when you offer her at least 20. So, if your own commitment is at most 40, then the best you can do here is to offer exactly 20 to Barbara and 40 to yourself, which you believe will be accepted by her. The utilities you expect here are 40 for yourself and 20 for Barbara, minus the 10% of the commitments in stage 1 that must be paid to Chris. However, if your commitment at stage 1 is $x \geq 50$, then you can propose at most 10 to Barbara, which she will surely reject. So, in that case the utilities you expect are 15 for you and 15 for Barbara, minus the 10% of the commitments in stage 1 that must be paid to Chris.

Case 3. Suppose that Barbara has named the commitment $y \in \{20, 30, 40\}$ at stage 1. In that case, you believe that Barbara will accept your offer only if you offer her at least the commitment y she named at stage 1. So, the best you can do is to offer Barbara precisely the commitment y she named at stage 1, and to propose $60 - y$ for yourself. Since the sum of the commitments named at stage 1 does not exceed 60, we know that your share $60 - y$ is at least the commitment you named at stage 1. Moreover, since y

is at most 40, your share $60 - y$ is more than 15, which is what you would obtain if Barbara rejected your offer. Hence, proposing y to Barbara and $60 - y$ for yourself is indeed the optimal thing to do here. The utilities you expect are $60 - y$ for yourself and y for Barbara, minus the 10% of the commitments in stage 1 that must be paid to Chris.

We finally turn to the beginning of the game, where you and Barbara must name your commitments. Let us call these commitments x (for you) and y (for Barbara). By our analysis above, we know that for every combination of commitments (x, y) there is a *unique* pair of utilities that you expect after (x, y) if you express common belief in future rationality. These utilities are as follows:

- If $x + y > 60$, then it is impossible to satisfy both your claims. In that case, you will both receive 15 pieces of gold, minus 10% of the commitment you named, which must be paid to Chris. The utilities for you and Barbara are given by Table 8.25.
- If $x + y \leq 60$ and $y \geq 50$, then we know from case 1 that you will make Barbara reject your offer. So, you both receive 15 pieces of gold, minus 10% of the commitment you named at stage 1. Hence, the utilities for you and Barbara are $15 - x/10$ and $15 - y/10$, respectively.
- If $x + y \leq 60$, $y \leq 10$ and $x \leq 40$, then we know from case 2 that you will propose 40 for yourself and 20 for Barbara, and you expect Barbara to accept. The utilities you expect for yourself and Barbara are $40 - x/10$ and $20 - y/10$, respectively.
- If $x + y \leq 60$, $y \leq 10$ and $x \geq 50$, then we know from case 2 that Barbara will reject your offer. Hence, the utilities for you and Barbara are $15 - x/10$ and $15 - y/10$, respectively.
- If $x + y \leq 60$ and $20 \leq y \leq 40$, then we know from case 3 that you will propose y to Barbara and $60 - y$ for yourself, and you expect Barbara to accept. The utilities you expect for yourself and Barbara are $60 - y - x/10$ and $y - y/10$, respectively.

So, the reduced decision problem at $\emptyset$ is given by Table 8.26. Here, the cells with a * are situations where it is impossible to satisfy both your claims, and where the game ends at stage 1. The six cells with ** are situations where Barbara rejects your offer at stage 3. So, in cells with a * or **, you and Barbara both receive 15 pieces of gold, minus 10% of the commitment you must pay to Chris. In the reduced decision problem, 30 for you represents the strategy where you choose the commitment 30 at stage 1, and then choose optimally at stage 2, as described above. Similarly for your other strategies in Table 8.26. For Barbara, we represent by 30 the strategy where she chooses the commitment 30 in stage 1, and then chooses optimally at stage 3, as described above.

Within the decision problem, your strategy 0 strictly dominates all your other strategies. So, we may eliminate all these other strategies for you from this decision problem. But then, within the decision problem that remains, the unique optimal strategy for Barbara is 40. Hence, the only strategies that survive the backward dominance procedure are 0 for you and 40 for Barbara.

So, under common belief in future rationality at stage 1 you will commit to 0 and you expect Barbara to commit to 40. At stage 2 you will propose 40 to Barbara and

Table 8.26. *Reduced decision problem at ∅ in "Bargaining with commitment"*

	0	10	20	30	40	50	60
0	40,20	40,19	40,18	30,27	20,36	15,10**	15,9**
10	39,20	39,19	39,18	29,27	19,36	14,10**	14,9*
20	38,20	38,19	38,18	28,27	18,36	13,10*	13,9*
30	37,20	37,19	37,18	27,27	12,11*	12,10*	12,9*
40	36,20	36,19	36,18	11,12*	11,11*	11,10*	11,9*
50	10,15**	10,14**	10,13*	10,12*	10,11*	10,10*	10,9*
60	9,15**	9,14*	9,13*	9,12*	9,11*	9,10*	9,9*

20 for yourself, which you expect Barbara to accept. That is, under common belief in future rationality you expect to receive only 20 pieces of gold, whereas you expect Barbara to receive much more, namely 40 pieces (minus the 4 pieces she must pay to Chris). Hence, this bargaining procedure gives a clear advantage to Barbara. This is quite surprising, as in this procedure you have the privilege of making a proposal to Barbara, but apparently this privilege does not work in your favor. □

8.11 Backward induction

In the previous section we have seen that in games with observed past choices there is an easy way to use the backward dominance procedure, namely by applying the backwards order of elimination. Suppose now that the game is not just with observed past choices, but also that for every information set there is exactly *one* player who makes a choice. Such games are said to have *perfect information.*

Definition 8.11.1 *(Game with perfect information)*
A dynamic game has ***perfect information*** *if at every information set there is only one active player and this player always knows exactly what choices have been made by his opponents in the past.*

On a technical level, this means that every information set h has only one non-terminal history, and there is only one player who is active at h. We will see that for games with perfect information, applying the backward dominance procedure is especially easy. To illustrate this, let us consider the following example.

Example 8.8 The shrinking treasure

The beginning of the story is the same as in "Two friends and a treasure." However, Chris now proposes yet another bargaining procedure to divide the treasure between Barbara and you. At the beginning, Chris takes 10 gold pieces from the treasure, which he may give back to you and Barbara later. At round 1, you must make a proposal to Barbara on how to divide the remaining 50 pieces of gold. In this proposal, the number you suggest for yourself and the number you offer to Barbara must both be multiples of 10. Barbara can then either accept or reject your offer. If she accepts, then you will both receive the numbers as specified in the proposal, plus a bonus of 5 pieces each from Chris for reaching an agreement so early. If Barbara rejects your proposal, the procedure goes to round 2.

At the beginning of round 2, Chris takes away another 10 pieces from the treasure as a punishment for not having reached an agreement at round 1. It is then Barbara's turn to make a proposal on how to divide the remaining 40 pieces, again in multiples of 10 pieces. You may accept or reject her proposal. If you accept her proposal, you both receive the numbers specified by this proposal, plus a bonus of 4 pieces (instead of 5) for having reached an agreement in round 2. If you reject her proposal, the procedure moves to round 3.

At the beginning of round 3, Chris takes away another 10 pieces from the treasure, and you must make a proposal on how to divide the remaining 30 pieces. And so on.

This procedure can last for at most five rounds, since at round 5 there will only be 10 pieces left to divide, which Chris would take if you do not reach an agreement. Hence, if no agreement is reached at round 5, then you and Barbara will both receive nothing. If you reach an agreement at round 3, then you will both receive a bonus of 3 pieces. Similarly, reaching an agreement at round 4 would give a bonus of 2 pieces each, and reaching an agreement at the final round would yield a bonus of 1 piece.

The question is: What strategy, or strategies, can you rationally choose under common belief in future rationality, and how would you and Barbara divide the treasure?

Note that this is a game with *perfect information*, since at every stage of the bargaining procedure only one player makes a choice, and this player always knows exactly what the opponent has done previously. In particular, this is a game with observed past choices, and hence we may apply the backward dominance procedure with the backwards order of elimination to find those strategies you can rationally choose under common belief in future rationality.

Let us start at the end of the game, which is round 5. There, you must make a proposal to Barbara on how to divide the remaining 10 pieces of gold, and Barbara must decide whether to accept or to reject that proposal. Since the proposed amounts must be multiples of 10, you can only propose the divisions $(10, 0)$ and $(0, 10)$, where $(10, 0)$ means 10 for you and 0 for Barbara, and $(0, 10)$ means 0 for you and 10 for Barbara. If Barbara accepts, you will both receive a bonus of 1 piece. If not, you will both end up with nothing. The procedure at round 5 is shown graphically by the first game on the left in Figure 8.9. Here, a stands for "accept," and r stands for "reject." In fact, at round 5 there are many copies of this game, since for every history of proposals at rounds 1 to 4, there will be a new subgame starting at round 5, which is a copy of the first game in Figure 8.9.

The ultimate information sets in the whole game are the histories at round 5, where Barbara must decide whether to accept or reject your proposal. At those ultimate information sets, the backward dominance procedure eliminates the choices for Barbara that are strictly dominated. Clearly, her choice r is strictly dominated by a at each of these

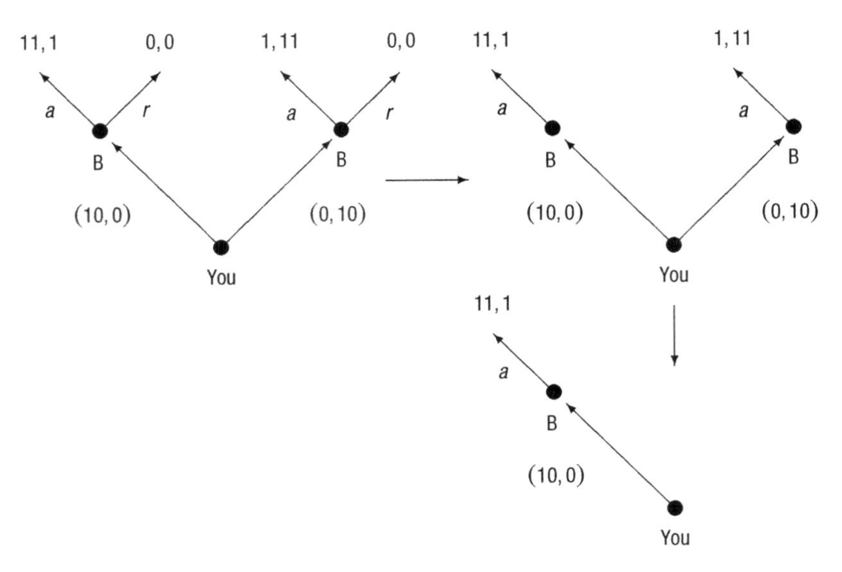

Figure 8.9 Round 5 of "The shrinking treasure"

histories at round 5, so we eliminate Barbara's choice *r* from every ultimate information set in round 5. This yields the second, reduced decision problem in Figure 8.9, shown at top right.

We then move to the penultimate information sets in the whole game, which are the histories at round 5 where you must decide between proposing $(10, 0)$ or $(0, 10)$. Within the reduced decision problem for that history – that is, the second game in Figure 8.9 – your proposal $(0, 10)$ is clearly strictly dominated by $(10, 0)$, and hence we may eliminate your proposal $(0, 10)$ from this reduced decision problem. This finally yields the third game in Figure 8.9, shown at bottom right, which is the fully reduced decision problem at round 5. So, the backward dominance procedure uniquely selects the outcome $(11, 1)$ in round 5.

We then move to the pen-penultimate information sets in the whole game, which are the final histories at round 4 where you must decide whether to accept or reject Barbara's proposal. Note that rejecting her proposal at these information sets would move the game to round 5. At these information sets, we start by removing the future choices at round 5 we have previously eliminated. As we have seen, this uniquely yields the outcome $(11, 1)$ whenever the game would reach round 5. This then gives the initial reduced decision problem at round 4, depicted by the first game in Figure 8.10. Remember that in round 4 there are 20 pieces to divide, and hence Barbara can make three different proposals: $(0, 20)$, $(10, 10)$ and $(20, 0)$. Here, $(0, 20)$ means 0 for you and 20 for Barbara, and so on. Also recall that you will both receive a bonus of 2 pieces if there is an agreement in round 4.

Within the initial reduced decision problem at round 4, let us first analyze the final histories in round 4, where you must decide whether to accept or reject Barbara's

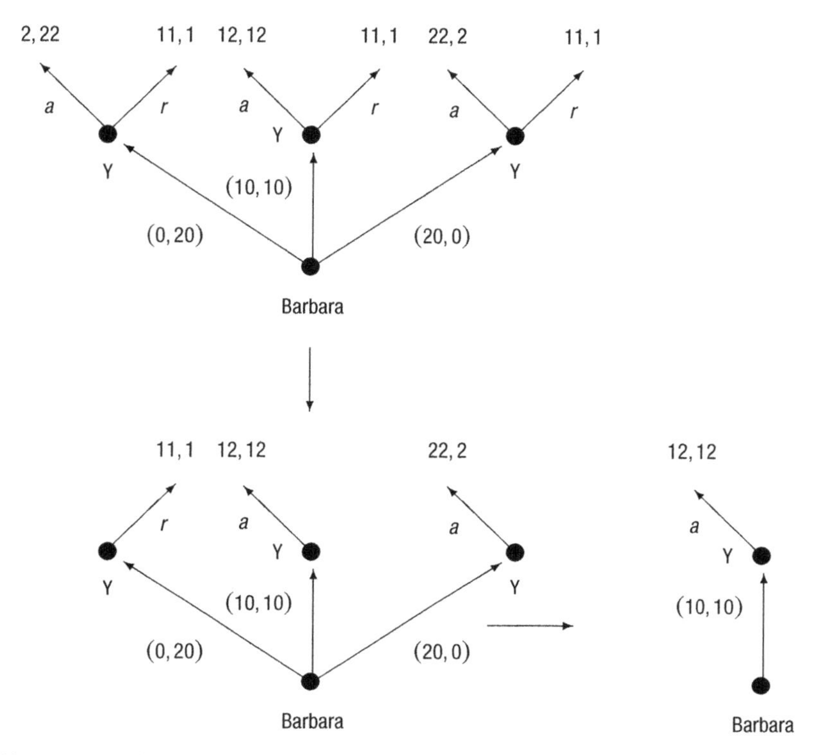

Figure 8.10 Round 4 of "The shrinking treasure"

proposal. According to the backward dominance procedure, we must eliminate your strictly dominated strategies. As you can see from Figure 8.10, accepting proposal (0, 20) is strictly dominated, and rejecting the proposals (10, 10) and (20, 0) are strictly dominated. If we eliminate these choices for you we obtain the second reduced decision problem in Figure 8.10.

We now move to the histories at the beginning of round 4, where Barbara must choose a proposal. Within the second reduced decision problem in Figure 8.10, her proposals (0, 20) and (20, 0) are both strictly dominated by her proposal (10, 10). So, we eliminate her proposals (0, 20) and (20, 0) from the game, resulting in the third game in Figure 8.10. This is the fully reduced decision problem at round 4, uniquely yielding the outcome (12, 12).

We can analyze round 3 in similar fashion. At round 3, you must make a proposal to Barbara on how to divide the remaining 30 pieces. We have seen that Barbara, by rejecting, would receive 12 pieces at round 4. As such, you believe that she will only accept proposals giving her at least 10 pieces, because she would also receive a bonus of 3 pieces. So, for you any proposal except (20, 10) is strictly dominated, since 10 is the minimum amount that you expect Barbara to accept, and of course you do not want to give her more than that. If you propose (20, 10), this would result in the outcome

(23, 13), as you would both receive a bonus of 3 pieces. Hence, the backward dominance procedure at round 3 would uniquely yield the outcome (23, 13) if round 3 is reached.

At round 2, Barbara must make a proposal on how to divide the remaining 40 pieces. If you reject her offer, then round 3 would be reached in which, as we have seen, the outcome would be (23, 13). So, you will only accept a proposal if it gives you at least 20 pieces, because you will receive a bonus of 4 pieces. Hence, for Barbara any proposal except (20, 20) is strictly dominated. If she offers (20, 20) then you would accept, yielding the outcome (24, 24).

Let us finally analyze round 1 more formally, as we did in rounds 5 and 4. You must make a proposal to Barbara on how to divide the 50 coins. If Barbara rejects your offer, the game would move to round 2 where the backward dominance procedure uniquely yields the outcome (24, 24), as we have seen. The initial reduced decision problem at round 1 is shown in Figure 8.11. Here, we have written your final amount *above* Barbara's final amount, instead of *next* to it, to save some space.

Within this initial reduced decision problem, let us first analyze the histories at the end of round 1, where Barbara must decide whether to accept or to reject your proposal. It can be seen from Figure 8.11 that after your proposals (50, 0) and (40, 10), Barbara's choice a is strictly dominated by r, and after each of your other proposals, her choice r is strictly dominated by her choice a. If we eliminate these choices, we obtain the reduced decision problem shown at the top of Figure 8.12. Within that reduced decision problem, all proposals except (30, 20) are strictly dominated for you. After eliminating these proposals, we are left with the final reduced decision problem at ∅, which is the lower game in Figure 8.12. There, you will propose the division (30, 20) to Barbara, and Barbara will accept, resulting in 35 for you and 25 for Barbara. Hence, applying the backward dominance procedure to this game yields a unique outcome, in which you receive 35 pieces of gold and Barbara 25. So, apparently the privilege of making the first proposal has worked to your advantage here. □

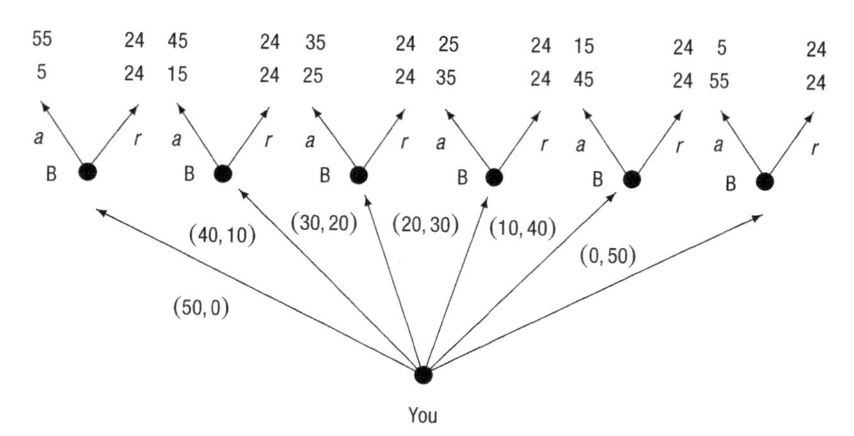

Figure 8.11 Round 1 of "The shrinking treasure"

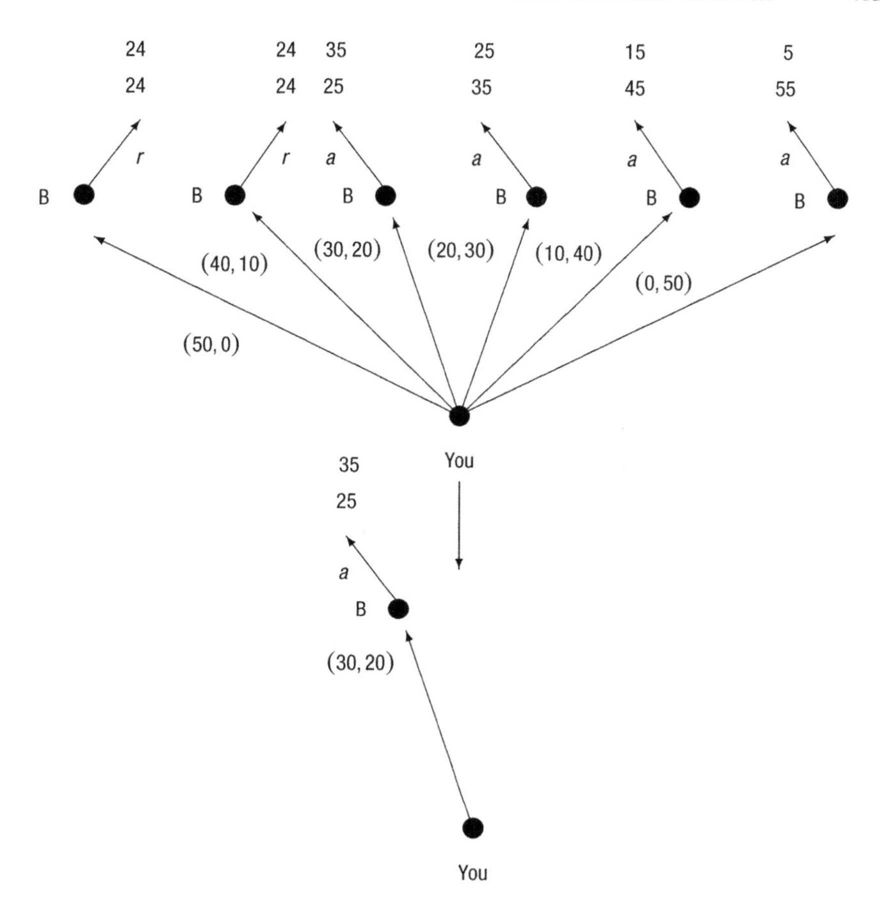

Figure 8.12 Reduced decision problem for round 1 of "The shrinking treasure"

You will have noticed that applying the backward dominance procedure to the game above was especially easy. The reason was that at every information set we only had to care about the strategy choices of *one* player, namely the single player who must choose at that information set, and see which strategy choices are strictly dominated. This applies to *every* game with perfect information: If we use the backward dominance procedure, then at every information set we need only consider the strategy choices of the *single* player who is active, and investigate which strategy choices can be eliminated. Hence, applying the backward dominance procedure to games with perfect information is a relatively easy task.

Let us go back, for a moment, to our example "The shrinking treasure" and see how the backward dominance procedure worked. Since we could use the backwards order of elimination, we started at the ultimate information sets, which were the information sets in round 5 where Barbara must decide whether or not to accept your final proposal. There, we eliminated Barbara's choice "reject" from each of these information sets, since it was always better for her to accept any of your proposals in round 5. In other words, for every ultimate information set we kept the unique optimal choice "accept" for Barbara. This is illustrated by the second game in Figure 8.9.

Then, we moved to the penultimate information sets, which were your information sets at the beginning of round 5, where you have to make a proposal to Barbara. Given that we selected Barbara's optimal choice "accept" at the end of round 5, your proposal $(0, 10)$ was strictly dominated by your other proposal $(10,0)$, and hence we could eliminate your proposal $(0, 10)$. This means that at the beginning of round 5 we kept your proposal $(10,0)$, which is the only optimal proposal for you if Barbara chooses optimally at the end of round 5 – that is, she always accepts. Since this proposal $(10,0)$ leads to the outcome $(11, 1)$, we thus have the unique outcome $(11, 1)$ at round 5.

We then turned to the pen-penultimate information sets in the game, which were the information sets in round 4 where you must decide whether or not to accept Barbara's proposal. Given the optimal strategy choices from round 5, leading to 11 pieces for you, the optimal strategy choices for you at the end of round 4 are to reject the proposal $(0,20)$, and to accept the proposals $(10, 10)$ and $(20,0)$. Accepting the first proposal would eventually give you fewer than 11 pieces, whereas accepting the other two proposals would eventually give you more than 11 pieces.

Afterwards, we moved to the information sets just before these, which are the information sets at round 4 where Barbara must make a proposal. Given the optimal strategy choices, where you reject the proposal $(0,20)$, and accept the proposals $(10, 10)$ and $(20,0)$, the optimal proposal for Barbara is to offer $(10, 10)$, finally leading to the outcome $(12, 12)$. And so on.

So, what the backward dominance procedure does in this example is:

- It starts at the ultimate information sets and selects the player's optimal strategies.
- It then turns to the penultimate information sets and selects the player's optimal strategies *given* the optimal strategies previously selected at the ultimate information sets.
- It then moves to the pen-penultimate information sets and selects the player's optimal strategies *given* the optimal strategies previously selected at the penultimate and ultimate information sets.

And so on, until we reach the beginning of the game. This procedure is called the *backward induction* procedure. So, in the example "The shrinking treasure" above, applying the backward dominance procedure with the backwards order of elimination is in fact equivalent to applying the backward induction procedure. In general, the backward induction procedure for dynamic games with perfect information can be described as follows.

Algorithm 8.11.2 *(Backward induction procedure)*
Consider a dynamic game with perfect information. At the beginning, we select at every ***ultimate*** *information set, all strategies for the active player that are optimal at*

this information set. These are called the backward induction strategies at the ultimate information sets.

We then select, at every ***penultimate*** *information set, all strategies for the active player that are optimal for* ***some*** *configuration of the opponents' backward induction strategies at the ultimate information sets. These are called the backward induction strategies at the penultimate information sets.*

We then select, at every ***pen-penultimate*** *information set, all strategies for the active player that are optimal for* ***some*** *configuration of the opponents' backward induction strategies at the penultimate and ultimate information sets. These are called the backward induction strategies at the pen-penultimate information sets.*

And so on, until we reach the beginning of the game.

A strategy s_i for player i is called a *backward induction strategy* if it is a backward induction strategy at $\emptyset$ – the beginning of the game. Note that in the example "The shrinking treasure," you and Barbara both have a unique backward induction strategy. Moreover, we saw that in this example, applying the backward dominance procedure with the backwards order of elimination is equivalent to using the backward induction procedure. Consequently, the strategies that are selected by the backward dominance procedure in this example are exactly the backward induction strategies. So, by our Theorem 8.8.3, the backward induction strategies in this example are precisely the strategies that can rationally be chosen under common belief in future rationality. In fact, this is true for *every* dynamic game with perfect information, including games where there are several backward induction strategies for a player.

Theorem 8.11.3 *(Common belief in future rationality leads to backward induction) Consider a dynamic game with perfect information. Then, the strategies that can rationally be chosen under common belief in future rationality are exactly the backward induction strategies.*

Hence, in a dynamic game with perfect information, applying the relatively simple backward induction procedure is sufficient for finding the strategies that can rationally be chosen under common belief in future rationality. We will now show why this is true in general.

Consider some arbitrary dynamic game with perfect information. We know, from Theorem 8.8.3, that the strategies that can rationally be chosen under common belief in future rationality are given by the backward dominance procedure. So, we must show that applying the backward dominance procedure leads to exactly the same strategies as using the backward induction procedure. As every game with perfect information is with observed past choices, we may use the backwards order of elimination.

Let us start at some ultimate information set h, where player i must make a choice. That is, after i's choice at h the game ends. The backward dominance procedure eliminates at h all strategies for player i that are strictly dominated within the full decision problem at h. These are precisely those strategies for player i that are not a backward induction strategy for player i at h. So, removing these strategies amounts to selecting the backward induction strategies for player i at h. But this is exactly what the backward induction procedure does at h. Hence, for every ultimate information set, the backward dominance procedure does exactly the same as the backward induction procedure.

Next, consider some penultimate information set h, where some player i must make a choice. The backward dominance procedure at h starts by removing from the full decision problem $\Gamma^0(h)$ at h those strategies that were strictly dominated for some ultimate information set h' following h. We have seen that removing these strategies is the same as selecting the backward induction strategies for the ultimate information sets. So, the backward dominance procedure at h starts by selecting the backward induction strategies at the ultimate information sets. This gives the reduced decision problem $\Gamma^1(h)$ at h. In the backward dominance procedure, we then remove from the reduced decision problem $\Gamma^1(h)$ those strategies for the active player – player i – that are strictly dominated within $\Gamma^1(h)$. We know from Chapter 2 that a strategy is strictly dominated within $\Gamma^1(h)$ precisely when it is not optimal for any probabilistic belief about the opponents' strategies in $\Gamma^1(h)$. So, the backward dominance procedure keeps in $\Gamma^1(h)$ those strategies for player i that are optimal for *some* probabilistic belief about the opponents' strategies in $\Gamma^1(h)$. Now, dynamic games with *perfect information* have the special property that a strategy that is optimal for some probabilistic belief about the opponents' strategies, is also optimal for some belief that assigns *probability 1* to one particular strategy combination. This result is summarized in the following lemma.

Lemma 8.11.4 *(Probability 1 beliefs are sufficient for games with perfect information) Consider a dynamic game with perfect information and an information set h in this game where player i is active. Suppose a strategy s_i is optimal at h for some probabilistic belief about the opponents' strategies. Then, s_i is optimal at h for some belief that assigns probability 1 to one particular strategy combination.*

The proof can be found in the proofs section at the end of this chapter. Hence, for identifying optimal strategies in a dynamic game with perfect information, it suffices to consider probability 1 beliefs only. Since we have seen above that the backward dominance procedure keeps in $\Gamma^1(h)$ precisely those strategies for player i that are optimal for some probabilistic belief in $\Gamma^1(h)$, we keep, by the lemma above, exactly those strategies for player i in $\Gamma^1(h)$ that are optimal for some belief that assigns probability 1 to one particular strategy combination in $\Gamma^1(h)$. In other words, we keep in $\Gamma^1(h)$ precisely those strategies for player i that are optimal for *some* strategy combination in $\Gamma^1(h)$. We have seen above that $\Gamma^1(h)$ contains exactly the backward induction strategies at the ultimate information sets. Hence, we keep at $\Gamma^1(h)$ precisely those strategies for player i that are optimal for *some* configuration of the opponents' backward induction strategies at the ultimate information sets. But this is exactly what the backward induction procedure does at the penultimate information set h. So, at every penultimate information set, the backward dominance procedure does exactly the same as the backward induction procedure.

By continuing this argument we will eventually reach the conclusion that for all information sets, the backward dominance procedure and the backward induction procedure do the same. Therefore, they both select, for every player, the same set of strategies, namely the backward induction strategies. This would then prove Theorem 8.11.3.

8.12 Games with unobserved past choices

In the previous two sections we analyzed games with observed past choices. For such games there is an easy way to apply the backward dominance procedure by using the very convenient *backwards order of elimination.* That is, we start with the decision problems at the ultimate information sets in the game, and analyze these *in isolation* without looking at the other information sets in the game. We then move to the decision problems for the penultimate information sets, and analyze these *in isolation* without looking at the other information sets in the game, and so on.

Now, why could we use the backwards order of elimination in these games with observed past choices? Recall that in such games, all players always know precisely what the other players have done so far in the game. This means that, whenever an information set h is reached, then all players who must choose at h or after h, know that h has been reached. Moreover, all players know that all players know that h has been reached, and so on. That is, at h there is *common knowledge* among the active players at, and after, h that the game has reached h. For that reason, we can analyze the decision problem at h in *isolation*, without having to look at other information sets, when applying the backward dominance procedure.

Things may become more complicated, however, if the players have *not* observed some of the past choices in the game. As an illustration, consider the following example.

Example 8.9 Bargaining with unobserved past choices

The story is almost the same as in "Two friends and a treasure" – see Example 8.6. The only difference is that at stages 2 and 3 of the bargaining procedure, you and Barbara do not know the amount(s) that your opponent has chosen in the past. Except for this difference, the bargaining procedure is exactly the same as in Example 8.6. This dynamic game is represented graphically in Figure 8.13. Here, for an information set the number in brackets indicates the player to which it belongs. So, $h_1(1)$ means that information set h_1 only belongs to you (player 1) but not to Barbara (player 2). Similarly for $h_2(2)$, $h_3(2)$ and $h_4(1)$. The information sets $\emptyset, h_5$ and h_6 have no number in brackets, which means that these information sets belong to both players.

To understand the information sets in this game, let us first look at your information set h_1. Since you chose 30 at $\emptyset$, you know that Barbara must have chosen 40 at $\emptyset$, otherwise the game would have ended at $\emptyset$. So, even though you have not *observed* Barbara's choice at $\emptyset$, you can conclude at h_1, by logical reasoning, that Barbara must have chosen 40 at $\emptyset$. That is, you know that the game has reached the matrix at h_1. However, at h_1 Barbara does *not* know whether the game is at h_1 or at the matrix below it. Barbara only knows that she has chosen 40 at $\emptyset$, and that the game has not stopped at $\emptyset$. Therefore, she knows that you have chosen either 30 or 40 at $\emptyset$, without knowing precisely which. In other words, Barbara knows that the game has reached h_1 or the

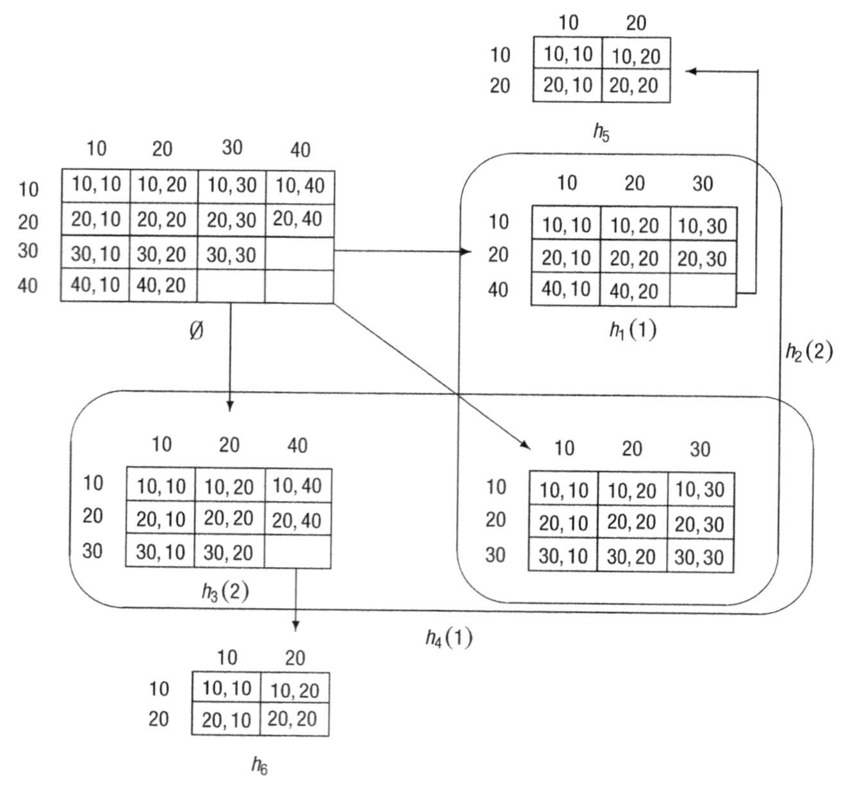

Figure 8.13 Graphical representation of "Bargaining with unobserved past choices"

matrix below it, without knowing which one has been reached. So, the matrix at h_1 and the matrix below it form an information set for Barbara, which we call h_2. In exactly the same way, we can explain the information sets h_3 and h_4.

Note that at h_5, both players know exactly what choices have been made by the opponent in the past, even though they have not *observed* these choices. At h_5 you know that you have chosen 30 at Ø and 40 at h_1. So, you know that Barbara must have chosen 40 at Ø and 30 at h_1, otherwise the game would have ended before stage 3. Similarly, at h_5 Barbara knows that she has chosen 40 at Ø and 30 at h_2. So, she knows that you must have chosen 30 at Ø and 40 at h_2, since otherwise the game would have ended before stage 3. So, at h_5 there is common knowledge that the game has reached h_5, and the same applies to h_6.

The question is: What strategies can you rationally choose under common belief in future rationality? To answer this question we apply the backward dominance procedure.

From which information sets do we start? Preferably, we would like to start at ultimate information sets to make the analysis easier. Consider the ultimate information set h_5. Since at h_5 there is common knowledge that the game has indeed reached h_5, we may start the backward dominance procedure at h_5, and analyze the decision problem there in isolation. As for the example "Two friends and a treasure," we would eliminate at the full decision problem $\Gamma^0(h_5)$ the strategy $(30_0, 40_1, 10_5)$ for you and the strategy $(40_0, 30_2, 10_5)$ for Barbara, as these are strictly dominated in $\Gamma^0(h_5)$. This would thus leave at h_5 the strategy $(30_0, 40_1, 20_5)$ for you and the strategy $(40_0, 30_2, 20_5)$ for Barbara. Here, strategy $(40_0, 30_2, 20_5)$ for Barbara means that she chooses 40 at $\emptyset$, chooses 30 at her information set h_2 and chooses 20 at h_5. Similarly for your strategy $(30_0, 40_1, 20_5)$.

In exactly the same way we can analyze the ultimate information set h_6, leaving the strategies $(40_0, 30_4, 20_6)$ for you and $(30_0, 40_3, 20_6)$ for Barbara.

But how should we continue? The problem is that the matrix in the lower right corner (with choices 10, 20 and 30 for both of you), although being an instance where the game ends, can no longer be analyzed in isolation within the backward dominance procedure! The reason is that you do not know at that matrix whether the game has reached it or the matrix to the left of it (this is reflected by information set h_4 for you). And Barbara does not know whether the game has reached this matrix or the matrix above it (this is reflected by information set h_2 for Barbara). This means that we must analyze the information sets h_1, h_2, h_3 and h_4 simultaneously, and not independently of each other!

In Table 8.27 we represent the decision problems at h_1, h_2, h_3 and h_4. Note that in each of these decision problems we have eliminated the strategies for you and Barbara that choose 10 at the future information sets h_5 or h_6, since these strategies were eliminated for h_5 and h_6.

In the next step, we look for your strategies that are strictly dominated within the decision problems at h_1 and h_4, at which you are active. First, your strategy $(30_0, 10_1)$ is strictly dominated within your decision problem at h_1. Hence, we eliminate your strategy $(30_0, 10_1)$ from the decision problem at h_1 but also from the decision problem at h_2, since h_1 and h_2 are simultaneous! Also, your strategy $(40_0, 10_4)$ is strictly dominated for your decision problem at h_4. Hence, we eliminate $(40_0, 10_4)$ from the decision problem at h_4, but also from the decision problems at h_3 and h_2, since h_4 and h_3 are simultaneous and also h_4 and h_2 are simultaneous!

Similarly, we then look for Barbara's strategies that are strictly dominated within the decision problems at h_2 and h_3, where she is active. Barbara's strategy $(40_0, 10_2)$ is strictly dominated in her decision problem at h_2. Hence, we eliminate her strategy $(40_0, 10_2)$ from the decision problem at h_2, but also from the decision problems at h_1 and h_4, since h_2 is simultaneous with h_1 and h_4. Moreover, Barbara's strategy $(30_0, 10_3)$ is strictly dominated within her decision problem at h_3, and hence we eliminate this strategy from h_3 and h_4, as these two information sets are simultaneous.

After these eliminations, no further strategies can be eliminated from h_1, h_2, h_3 and h_4. Remember that at every information set h, we only look for strictly dominated strategies for the player that is *active* at h, not for the other player! The resulting reduced decision problems at h_1, h_2, h_3 and h_4 can be found in Table 8.28.

Table 8.27. *Decision problems at h_1, h_2, h_3 and h_4 in "Bargaining with unobserved past choices"*

$\Gamma(h_1)$: Only you active

	$(40_0, 10_2)$	$(40_0, 20_2)$	$(40_0, 30_2, 20_5)$
$(30_0, 10_1)$	10, 10	10, 20	10, 30
$(30_0, 20_1)$	20, 10	20, 20	20, 30
$(30_0, 40_1, 20_5)$	40, 10	40, 20	20, 20

$\Gamma(h_2)$: Only Barbara active

	$(40_0, 10_2)$	$(40_0, 20_2)$	$(40_0, 30_2, 20_5)$
$(30_0, 10_1)$	10, 10	10, 20	10, 30
$(30_0, 20_1)$	20, 10	20, 20	20, 30
$(30_0, 40_1, 20_5)$	40, 10	40, 20	20, 20
$(40_0, 10_4)$	10, 10	10, 20	10, 30
$(40_0, 20_4)$	20, 10	20, 20	20, 30
$(40_0, 30_4, 20_6)$	30, 10	30, 20	30, 30

$\Gamma(h_3)$: Only Barbara active

	$(30_0, 10_3)$	$(30_0, 20_3)$	$(30_0, 40_3, 20_6)$
$(40_0, 10_4)$	10, 10	10, 20	10, 40
$(40_0, 20_4)$	20, 10	20, 20	20, 40
$(40_0, 30_4, 20_6)$	30, 10	30, 20	20, 20

$\Gamma(h_4)$: Only you active

	$(30_0, 10_3)$	$(30_0, 20_3)$	$(30_0, 40_3, 20_6)$	$(40_0, 10_2)$	$(40_0, 20_2)$	$(40_0, 30_2, 20_5)$
$(40_0, 10_4)$	10, 10	10, 20	10, 40	10, 10	10, 20	10, 30
$(40_0, 20_4)$	20, 10	20, 20	20, 40	20, 10	20, 20	20, 30
$(40_0, 30_4, 20_6)$	30, 10	30, 20	20, 20	30, 10	30, 20	30, 30

We finally move to the beginning of the game, $\emptyset$. Within the decision problem at $\emptyset$, we can eliminate the strategies we eliminated at $h_1, ..., h_6$. This yields the decision problem $\Gamma(\emptyset)$ shown in Table 8.29. Within this decision problem, your strategy 10_0 is clearly strictly dominated, and your strategy 20_0 is strictly dominated by your randomized choice

$$(0.5) \cdot (30_0, 20_1) + (0.5) \cdot (40_0, 30_4, 20_6).$$

Table 8.28. *Reduced decision problems at h_1, h_2, h_3 and h_4 in "Bargaining with unobserved past choices"*

$\Gamma(h_1)$: Only you active

	$(40_0, 20_2)$	$(40_0, 30_2, 20_5)$
$(30_0, 20_1)$	20, 20	20, 30
$(30_0, 40_1, 20_5)$	40, 20	20, 20

$\Gamma(h_2)$: Only Barbara active

	$(40_0, 20_2)$	$(40_0, 30_2, 20_5)$
$(30_0, 20_1)$	20, 20	20, 30
$(30_0, 40_1, 20_5)$	40, 20	20, 20
$(40_0, 20_4)$	20, 20	20, 30
$(40_0, 30_4, 20_6)$	30, 20	30, 30

$\Gamma(h_3)$: Only Barbara active

	$(30_0, 20_3)$	$(30_0, 40_3, 20_6)$
$(40_0, 20_4)$	20, 20	20, 40
$(40_0, 30_4, 20_6)$	30, 20	20, 20

$\Gamma(h_4)$: Only you active

	$(30_0, 20_3)$	$(30_0, 40_3, 20_6)$	$(40_0, 20_2)$	$(40_0, 30_2, 20_5)$
$(40_0, 20_4)$	20, 20	20, 40	20, 20	20, 30
$(40_0, 30_4, 20_6)$	30, 20	20, 20	30, 20	30, 30

Hence, we may eliminate the strategies 10_0 and 20_0 for you, and similarly for Barbara. So, we obtain the reduced decision problem for $\emptyset$ shown in Table 8.30. Within this reduced decision problem, your strategy $(40_0, 20_4)$ is strictly dominated by the randomized choice

$$(0.5) \cdot (30_0, 20_1) + (0.5) \cdot (40_0, 30_4, 20_6),$$

and hence we may eliminate your strategy $(40_0, 20_4)$ at $\emptyset$. Similarly, we may eliminate Barbara's strategy $(40_0, 20_2)$ at $\emptyset$. This yields the final decision problem at $\emptyset$ in Table 8.31, from which no further strategies can be eliminated.

Hence, under common belief in future rationality you can rationally choose one of the strategies $(30_0, 20_1)$, $(30_0, 40_1, 20_5)$ and $(40_0, 30_4, 20_6)$.

□

Table 8.29. *Decision problem at ∅ in "Bargaining with unobserved past choices"*

	10_0	20_0	$(30_0, 20_3)$	$(30_0, 40_3, 20_6)$	$(40_0, 20_2)$	$(40_0, 30_2, 20_5)$
10_0	10, 10	10, 20	10, 30	10, 30	10, 40	10, 40
20_0	20, 10	20, 20	20, 30	20, 30	20, 40	20, 40
$(30_0, 20_1)$	30, 10	30, 20	30, 30	30, 30	20, 20	20, 30
$(30_0, 40_1, 20_5)$	30, 10	30, 20	30, 30	30, 30	40, 20	20, 20
$(40_0, 20_4)$	40, 10	40, 20	20, 20	20, 40	20, 20	20, 30
$(40_0, 30_4, 20_6)$	40, 10	40, 20	30, 20	20, 20	30, 20	30, 30

Table 8.30. *Reduced decision problem at ∅ in "Bargaining with unobserved past choices"*

	$(30_0, 20_3)$	$(30_0, 40_3, 20_6)$	$(40_0, 20_2)$	$(40_0, 30_2, 20_5)$
$(30_0, 20_1)$	30, 30	30, 30	20, 20	20, 30
$(30_0, 40_1, 20_5)$	30, 30	30, 30	40, 20	20, 20
$(40_0, 20_4)$	20, 20	20, 40	20, 20	20, 30
$(40_0, 30_4, 20_6)$	30, 20	20, 20	30, 20	30, 30

Table 8.31. *Final decision problem at ∅ in "Bargaining with unobserved past choices"*

	$(30_0, 20_3)$	$(30_0, 40_3, 20_6)$	$(40_0, 30_2, 20_5)$
$(30_0, 20_1)$	30, 30	30, 30	20, 30
$(30_0, 40_1, 20_5)$	30, 30	30, 30	20, 20
$(40_0, 30_4, 20_6)$	30, 20	20, 20	30, 30

8.13 Bayesian updating

In a conditional belief vector, a player holds a belief about the opponents' strategy choices – and possibly beliefs – at *every* information set where he is active. In principle, there need not be any logical connection between the beliefs held at the various information sets. Consider, for instance, the game in Figure 8.2. Suppose that player 2 holds the conditional belief vector b_2 with

$$b_2(\emptyset) = (0.5) \cdot (a,f,j) + (0.5) \cdot b,$$
$$b_2(h_1) = (a,e,i), \text{ and } b_2(h_2) = (a,e,i).$$

Then, as we saw at the end of Section 8.3, there is no strategy for player 2 that is optimal at *each* of his information sets $\emptyset, h_1$ and h_2. The reason is that the beliefs that player 2 holds at h_1 and h_2 are not consistent with the belief he held at $\emptyset$. At $\emptyset$, player 2 assigns probability 0.5 to the event that player 1 will choose b, and probability 0.5 to the event that player 1 will choose strategy (a,f,j). Now, at h_1 it is evident that player 1 has chosen a and not b, so among the strategies that player 2 deemed possible at $\emptyset$ only (a,f,j) is still possible. But then, it would be natural for player 2 to assign probability 1 to player 1's strategy (a,f,j) once h_1 is reached, simply by *updating* the initial belief he held at $\emptyset$. But this is not what player 2 does at h_1! If h_1 is reached, player 2 now assigns probability 1 to the "completely new" strategy (a,e,i) for player 1, although the strategy (a,f,j) he deemed possible at $\emptyset$ is still possible at h_1.

So, there is something strange about the belief that player 2 holds at h_1, compared to the belief he held at $\emptyset$, because he refuses to simply update his initial belief once h_1 is reached. In general, when player i holds a belief at information set $h \in H_i$ about the opponents' strategies – and also possibly the opponents' types – and the game moves from h to another information set $h' \in H_i$, then it makes sense for player i to base his belief at h' upon the belief he held at h by updating it, whenever this is possible. This condition is known as *Bayesian updating* and may be formalized as follows.

Definition 8.13.1 *(Bayesian updating)*
A type t_i satisfies ***Bayesian updating*** *if for every two information sets $h,h' \in H_i$ where h' follows h, and $b_i(t_i,h)(S_{-i}(h') \times T_{-i}) > 0$, it holds that*

$$b_i(t_i,h')(s_{-i},t_{-i}) = \frac{b_i(t_i,h)(s_{-i},t_{-i})}{b_i(t_i,h)(S_{-i}(h') \times T_{-i})}$$

for every strategy-type combination $(s_{-i},t_{-i}) \in S_{-i}(h') \times T_{-i}$ of his opponents.

Here, $S_{-i}(h')$ denotes the set of the opponents' strategy combinations that lead to h' and T_{-i} denotes the set of the opponents' type combinations. Hence, $S_{-i}(h') \times T_{-i}$ is the set of the opponents' strategy-type combinations where the strategy combination leads to h'. By $b_i(t_i,h)(S_{-i}(h') \times T_{-i})$ we denote the total probability that the conditional belief $b_i(t_i,h)$ assigns to all of the opponents' strategy-type combinations in $S_{-i}(h') \times T_{-i}$.

So, Bayesian updating says that, whenever type t_i assigns at h some positive probability to an opponents' strategy combination that leads to h', then t_i must base his conditional belief at h' upon his previous belief at h. More precisely, when the game moves from h to h', type t_i must maintain the relative likelihood he assigns to the opponents' strategy-type combinations in $S_{-i}(h') \times T_{-i}$. If, however, type t_i assigns at h probability 0 to the set of opponents' strategy combinations leading to h' – that is, the type deems it impossible at h that h' will be reached – then the belief at h' can no longer be based upon the belief held at h, and hence there is no restriction imposed by Bayesian updating.

It can be shown that for every type which satisfies Bayesian updating, there will always be a strategy that is optimal at each of his information sets. Hence, the problem of not finding an optimal strategy can only occur if the type does not satisfy Bayesian updating.

Lemma 8.13.2 *(Optimal strategies exist for types that satisfy Bayesian updating) Consider a type t_i for player i that satisfies Bayesian updating. Then, there is a strategy s_i that is optimal for t_i at every information set $h \in H_i$ that s_i leads to.*

The proof for this result can be found in the proofs section at the end of this chapter. Note that for common belief in future rationality, which we investigated in this chapter, we did not require the players to satisfy Bayesian updating. Would it make a difference to the eventual strategy choices selected, if we additionally insist on Bayesian updating? The answer, as we will see, is "yes." There are strategies that a player can rationally choose under common belief in future rationality, which he cannot rationally choose if we impose (common belief in) Bayesian updating. Here is an example that illustrates this.

Consider the game in Figure 8.14 between player 1 and player 2. Initially players 1 and 2 simultaneously choose between a and b, and between c and d, respectively. If player 1 chooses b, the game stops. If he chooses a, then the game moves either to the upper left table or the upper right table, depending on whether player 2 chooses c or d. The information set h_1, which belongs to player 1, indicates that player 1 does not know whether player 2 has chosen c or d. Player 2, on the other hand, knows whether he has chosen c or d, which is reflected by his information sets h_2 and h_3.

It can be shown that under common belief in future rationality – but *without* insisting on Bayesian updating – player 2 can rationally choose strategy (c,h). Consider the epistemic model in Table 8.32. The reader may verify that both types t_1^b and t_2^{ch} believe in the opponent's future rationality. As a consequence, type t_2^{ch} expresses common belief in future rationality. Since strategy (c,h) is optimal for type t_2^{ch} at $\emptyset$ and h_2, it follows that player 2 can rationally choose strategy (c,h) under common belief in future rationality.

Note, however, that player 1's type t_1^b in this model does not satisfy Bayesian updating. Initially, type t_1^b believes that player 2 will choose strategy (c,h), but at h_1 he suddenly believes that player 2 will choose a completely different strategy (d,i), whereas strategy (c,h) is still possible for player 2 when h_1 is reached. Since player 2's type t_2^{ch} believes, throughout the game, that player 1 is of type t_1^b, type t_2^{ch} does not believe that player 1 satisfies Bayesian updating.

In fact, it can be shown that under common belief in future rationality *and* common belief in Bayesian updating, player 2 cannot rationally choose strategy (c,h). Initially player 1 can only assign positive probability to player 2's strategies (c,g) and (c,h), as player 2's strategies (d,i) and (d,j) can never be optimal for player 2 at $\emptyset$. But then, if

Table 8.32. *An epistemic model for the game in Figure 8.14*

Types	$T_1 = \{t_1^b\},\ T_2 = \{t_2^{ch}\}$
Beliefs for player 1	$b_1(t_1^b, \emptyset) = ((c,h), t_2^{ch})$ $b_1(t_1^b, h_1) = ((d,i), t_2^{ch})$
Beliefs for player 2	$b_2(t_2^{ch}, \emptyset) = (b, t_1^b)$ $b_2(t_2^{ch}, h_2) = ((a,f), t_1^b)$ $b_2(t_2^{ch}, h_3) = ((a,f), t_1^b)$

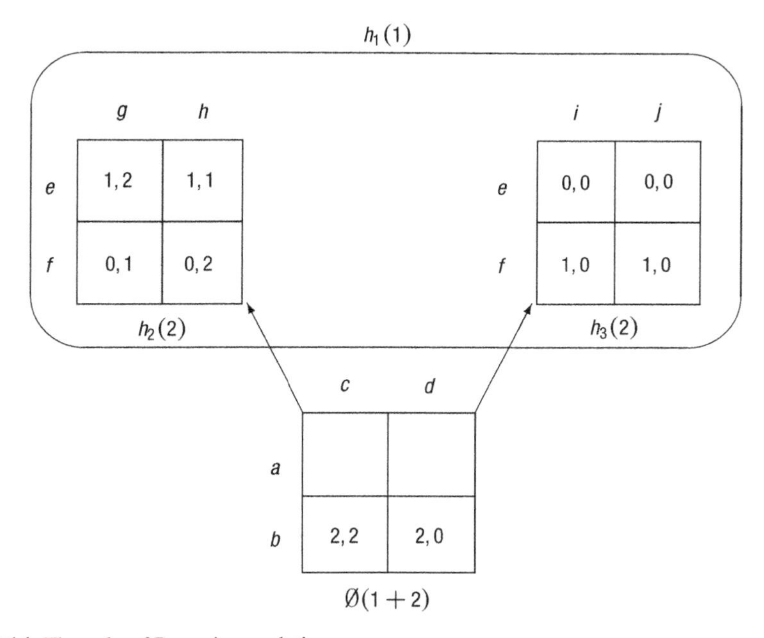

Figure 8.14 The role of Bayesian updating

player 1 satisfies Bayesian updating, he must at h_1 *still* assign positive probability only to strategies (c,g) and (c,h) for player 2. Hence, the only strategy that can be optimal for player 1 at h_1 is (a,e). So, if player 2 expresses common belief in future rationality, and believes that player 1 satisfies Bayesian updating, then player 2 must believe at h_2 that player 1 will choose strategy (a,e), and hence the only strategy that can be optimal for player 2 at h_2 is (c,g). This means, in particular, that player 2 cannot rationally choose strategy (c,h) if he expresses common belief in future rationality and believes that player 1 satisfies Bayesian updating.

Summarizing, we see that for common belief in future rationality it is important whether we impose (common belief in) Bayesian updating or not. Throughout this chapter we have not imposed Bayesian updating, because we wanted to keep the concept as basic as possible. Moreover, by not imposing Bayesian updating, we obtain a concept that is completely *forward looking* in the sense that we only require players to reason about the opponents' present and future choices and beliefs, and not about events that happened in the past. If we additionally impose Bayesian updating, then the players must also critically look at the beliefs they held in the past, as they must base their current belief upon them, whenever possible. So, Bayesian updating requires the players to be looking at the past also.

8.14 Proofs

In this section we will prove three results: First Lemma 8.13.2, which shows that for every type that satisfies Bayesian updating we can always find a strategy that is optimal at every information set. We also prove Theorem 8.8.3, stating that in every dynamic game the backward dominance procedure always yields exactly those strategies that can rationally be chosen under common belief in future rationality. Finally, we will prove Lemma 8.11.4, stating that in a dynamic game with perfect information, every strategy that is optimal for some probabilistic belief is also optimal for a belief that assigns probability 1 to one specific strategy combination for the opponents.

In order to prove these results we have divided this section into four parts. In the first part we prove Lemma 8.13.2, and use it to show an important property of the backward dominance procedure in Lemma 8.14.5 concerning the existence of optimal strategies within this procedure. In the second part we prove an optimality principle for the backward dominance procedure, similar to the optimality principles we saw in earlier chapters for other procedures. In the third part we use our results from the first two parts to prove Theorem 8.8.3, showing that the backward dominance procedure works for common belief in future rationality. In the fourth and final part, we prove Lemma 8.11.4 for games with perfect information.

Bayesian updating and optimal strategies

We start this first part by showing that under *Bayesian updating* we can always find a strategy that is optimal at each information set. More precisely, if a conditional belief vector about the opponents' strategy choices satisfies Bayesian updating, then there will always be a strategy s_i that is optimal under this conditional belief vector at every information set $h \in H_i$ that s_i leads to. A conditional belief vector $(b_i(h))_{h \in H_i}$ about the opponents' strategy choices satisfies *Bayesian updating* if for every two information sets $h, h' \in H_i$ where h' follows h and $b_i(h)(S_{-i}(h')) > 0$, it holds that

$$b_i(h')(s_{-i}) = \frac{b_i(h)(s_{-i})}{b_i(h)(S_{-i}(h'))}$$

for all of the opponents' strategy combinations $s_{-i} \in S_{-i}(h')$. Here, $S_{-i}(h')$ denotes the set of opponents' strategy combinations that lead to h', and $b_i(h)(S_{-i}(h'))$ is the total probability that $b_i(h)$ assigns to strategy combinations in $S_{-i}(h')$. Remember the definition of Bayesian updating in Section 8.13. The difference with the definition above is that the version of Bayesian updating in Section 8.13 applies to *types* – which hold conditional beliefs about the opponents' strategy choices *and types* – whereas the definition above only applies to conditional beliefs about the opponents' strategy choices. It is easily seen, however, that a type that satisfies Bayesian updating – in the sense of Section 8.13 – always induces a conditional belief vector about the opponents' strategy choices that satisfies Bayesian updating in the above sense. We saw in Section 8.3 that for types that do not satisfy Bayesian updating there may not always be a strategy that is optimal for each of the information sets. The following lemma shows that this problem cannot occur if the conditional belief vector about the opponents' strategy choices satisfies Bayesian updating.

Lemma 8.14.1 *(Optimal strategies exist for conditional belief vectors that satisfy Bayesian updating)*
Consider a conditional belief vector $(b_i(h))_{h \in H_i}$ for player i about the opponents' strategy choices that satisfies Bayesian updating. Then, there is a strategy s_i that is optimal under this conditional belief vector at every information set $h \in H_i$ that s_i leads to.

Proof: Consider a conditional belief vector $(b_i(h))_{h \in H_i}$ about the opponents' strategy choices that satisfies Bayesian updating. We will construct a strategy s_i^* that, at every information set $h \in H_i$ that s_i^* leads to, is optimal for the belief $b_i(h)$.

We start with information sets $h \in H_i$ that are not followed by any other information set for player i. Let us denote this collection of information sets by H_i^1. For every $h \in H_i^1$, we specify a choice $c_i^*(h) \in C_i(h)$ that is optimal at h, given the conditional belief $b_i(h)$ at h. More precisely, we select a choice $c_i^*(h)$ with

$$u_i(c_i^*(h), b_i(h)) \geq u_i(c_i, b_i(h)) \text{ for all } c_i \in C_i(h).$$

We next consider information sets $h \in H_i$ that are only followed by information sets in H_i^1. That is, we look at information sets $h \in H_i$ that are followed by at least one other information set for player i, all of which are in H_i^1. Let us denote this collection of information sets by H_i^2. So, for every $h \in H_i^2$ we have already specified a choice $c_i^*(h')$ for every $h' \in H_i^1$ that follows h. For every $h \in H_i^2$ we specify a choice $c_i^*(h)$ that is optimal at h, given the conditional belief at h, and given the future choices $c_i^*(h')$ for information sets $h' \in H_i^1$ that follow h. Let us denote by $H_i(h)$ the collection of information sets for player i that follow h. So, for every $h \in H_i^2$ we specify a choice $c_i^*(h)$ with

$$u_i((c_i^*(h), (c_i^*(h'))_{h' \in H_i(h)}), b_i(h)) \geq u_i((c_i, (c_i^*(h'))_{h' \in H_i(h)}), b_i(h))$$

for all $c_i \in C_i(h)$.

Now look at information sets for player i that are followed by at least one information set $h' \in H_i^2$, and where all information sets following h are either in H_i^1 or H_i^2. Let us denote this collection of information sets by H_i^3. So, for every $h \in H_i^3$ we have already specified a choice $c_i^*(h')$ for every $h' \in H_i$ that follows h. For $h \in H_i^3$ we then specify a choice $c_i^*(h)$ that is optimal at h, given the conditional belief at h, and given the future choices $c_i^*(h')$ for information sets $h' \in H_i$ that follow h. That is, we specify a choice $c_i^*(h)$ with

$$u_i((c_i^*(h),(c_i^*(h'))_{h' \in H_i(h)}),b_i(h)) \geq u_i((c_i,(c_i^*(h'))_{h' \in H_i(h)}),b_i(h))$$

for all $c_i \in C_i(h)$.

By proceeding in this way, we will finally specify for *every* information set $h \in H_i$ a choice $c_i^*(h)$ that is optimal at h, given the conditional belief at h, and given the future choices $c_i^*(h')$ for information sets $h' \in H_i$ that follow h. That is, for every information set $h \in H_i$ we have that

$$u_i((c_i^*(h),(c_i^*(h'))_{h' \in H_i(h)}),b_i(h)) \geq u_i((c_i,(c_i^*(h'))_{h' \in H_i(h)}),b_i(h))$$

for all $c_i \in C_i(h)$. We say that the collection $(c_i^*(h))_{h \in H_i}$ of choices is *locally optimal* for every $h \in H_i$, given the conditional belief vector $(b_i(h))_{h \in H_i}$.

Now, let s_i^* be the strategy that specifies the locally optimal choice $c_i^*(h)$ for every $h \in H_i$ that s_i^* leads to. We will show that the strategy s_i^* is optimal for every $h \in H_i$ that s_i^* leads to, given the conditional belief $b_i(h)$ at h.

To prove this result, we categorize the information sets for player i into groups $H_i^1, H_i^2, \ldots$ where, as before, H_i^1 contains all information sets for player i that are not followed by any other information set for player i, the collection H_i^2 contains all information sets for player i that are only followed by information sets in H_i^1, and so on. We prove, by induction on k, that for every k, and every $h \in H_i^k$ that s_i^* leads to, the strategy s_i^* is optimal at h, given the conditional belief $b_i(h)$ at h.

We start with information sets $h \in H_i^1$. Consider an arbitrary alternative strategy $s_i \in S_i(h)$ that leads to h. Since $s_i^*(h) = c_i^*(h)$, and the collection $(c_i^*(h))_{h \in H_i}$ of choices is *locally optimal* at h given $b_i(h)$, we have that

$$u_i(s_i^*,b_i(h)) = u_i(c_i^*(h),b_i(h)) \geq u_i(s_i(h),b_i(h)) = u_i(s_i,b_i(h)).$$

As this holds for every alternative strategy $s_i \in S_i(h)$, we conclude that s_i^* is optimal at h, given the conditional belief $b_i(h)$.

Suppose now that $h \in H_i^k$, and assume that s_i^* is optimal for every $h' \in H_i^1, \ldots, H_i^{k-1}$ given the conditional belief $b_i(h')$. We show that s_i^* is optimal at h given the conditional belief $b_i(h)$.

Consider an arbitrary alternative strategy s_i that leads to h. We show that

$$u_i(s_i,b_i(h)) \leq u_i(s_i^*,b_i(h)).$$

Suppose that s_i prescribes choice c_i at h. Let $H_i(h,c_i)$ be the collection of information sets h' for player i that follow choice c_i at h, and for which there is no player i information set between h and h'. Let $S_{-i}^0(h,c_i)$ be the set of opponents' strategy combinations $s_{-i} \in S_{-i}(h)$ for which choice c_i at h, in combination with s_{-i}, does not lead to any

$h' \in H_i(h, c_i)$. Then, all of the opponents' strategy combinations $s_{-i} \in S_{-i}(h)$ are either in $S^0_{-i}(h, c_i)$, or in $S_{-i}(h')$ for exactly one $h' \in H_i(h, c_i)$. Hence,

$$\begin{aligned} u_i(s_i, b_i(h)) &= \sum_{h' \in H_i(h,c_i)} \sum_{s_{-i} \in S_{-i}(h')} b_i(h)(s_{-i}) \cdot u_i(s_i, s_{-i}) \\ &+ \sum_{s_{-i} \in S^0_{-i}(h,c_i)} b_i(h)(s_{-i}) \cdot u_i(s_i, s_{-i}) \\ &= \sum_{h' \in H_i(h,c_i)} b_i(h)(S_{-i}(h')) \sum_{s_{-i} \in S_{-i}(h')} b_i(h')(s_{-i}) \cdot u_i(s_i, s_{-i}) \\ &+ \sum_{s_{-i} \in S^0_{-i}(h,c_i)} b_i(h)(s_{-i}) \cdot u_i(s_i, s_{-i}) \\ &= \sum_{h' \in H_i(h,c_i)} b_i(h)(S_{-i}(h')) \cdot u_i(s_i, b_i(h')) \\ &+ \sum_{s_{-i} \in S^0_{-i}(h,c_i)} b_i(h)(s_{-i}) \cdot u_i(s_i, s_{-i}). \end{aligned}$$

Here, the second equality follows from the assumption that the conditional belief vector $(b_i(h))_{h \in H_i}$ satisfies Bayesian updating, which implies that

$$b_i(h)(s_{-i}) = b_i(h)(S_{-i}(h')) \cdot b_i(h')(s_{-i})$$

for every $h' \in H_i(h, c_i)$ and every $s_{-i} \in S_{-i}(h')$.

Since every $h' \in H_i(h, c_i)$ follows h, h' must be in $H_i^1, ..., H_i^{k-1}$, and hence we know by our induction assumption that

$$u_i(s_i, b_i(h')) \leq u_i(s_i^*, b_i(h'))$$

for every $h' \in H_i(h, c_i)$. Moreover, for all of the opponents' strategy combinations $s_{-i} \in S^0_{-i}(h, c_i)$, the choice c_i in combination with s_{-i} does not reach any player i information set after h. Hence, the utility $u_i(s_i, s_{-i})$ for such $s_{-i} \in S^0_{-i}(h, c_i)$ depends only on the choice c_i at h, and not on the other choices prescribed by s_i. By combining the two insights above, we obtain that

$$\begin{aligned} u_i(s_i, b_i(h)) &\leq \sum_{h' \in H_i(h,c_i)} b_i(h)(S_{-i}(h')) \cdot u_i(s_i^*, b_i(h')) \\ &+ \sum_{s_{-i} \in S^0_{-i}(h,c_i)} b_i(h)(s_{-i}) \cdot u_i(c_i, s_{-i}) \\ &= u_i((c_i, (c_i^*(h'))_{h' \in H_i(h)}), b_i(h)) \\ &\leq u_i((c_i^*(h), (c_i^*(h'))_{h' \in H_i(h)}), b_i(h)) \\ &= u_i(s_i^*, b_i(h)). \end{aligned}$$

Here, the second inequality follows from the assumption that the combination $(c_i^*(h))_{h\in H_i}$ of choices is locally optimal at h, given the conditional belief $b_i(h)$. Both equalities follow from the fact that s_i^* prescribes the locally optimal choice $c_i^*(h')$ for every $h' \in H_i$ that s_i^* leads to.

So, we have shown that

$$u_i(s_i, b_i(h)) \leq u_i(s_i^*, b_i(h))$$

for every alternative strategy s_i that leads to h, which means that s_i^* is optimal at h given the conditional belief $b_i(h)$. By induction on k, it follows that s_i^* is optimal at *every* information set $h \in H_i$ that s_i^* leads to, given the conditional belief $b_i(h)$. This completes the proof. ◇

With the help of the lemma above, it is now easy to prove Lemma 8.13.2 in Section 8.13, which states that for every type that satisfies Bayesian updating there is always a strategy that is optimal at every information set.

Lemma 8.13.2 *(Optimal strategies exist for types that satisfy Bayesian updating)*
Consider a type t_i for player i that satisfies Bayesian updating. Then, there is a strategy s_i that is optimal for t_i at every information set $h \in H_i$ that s_i leads to.

Proof: Consider a type t_i for player i that satisfies Bayesian updating. Suppose that t_i holds the conditional belief vector $(b_i(h))_{h\in H_i}$ about the opponents' strategy combinations. Then, this conditional belief vector $(b_i(h))_{h\in H_i}$ will satisfy Bayesian updating. Hence, by Lemma 8.14.1 there is a strategy s_i that is optimal under the conditional belief vector $(b_i(h))_{h\in H_i}$ at every information set $h \in H_i$ that s_i leads to. But then, strategy s_i is optimal for type t_i at every information set $h \in H_i$ that s_i leads to. This completes the proof. ◇

Lemma 8.14.1 above thus shows that, if a player's conditional belief vector satisfies Bayesian updating, then he has at least one strategy that is optimal for him at each of his information sets, given his conditional beliefs there. We now ask a slightly different question: Suppose that for every information set $h \in H_i$, player i is required to only assign positive probability to a *subset* $D_{-i}(h) \subseteq S_{-i}(h)$ of the opponents' strategy combinations that lead to h. Under what conditions on the sets $D_{-i}(h)$ does player i have a strategy that is optimal at each of his information sets, given his conditional beliefs there?

In order to answer this question, we need some further terminology. For a given player i and information set $h \in H_i$, let $D_{-i}(h) \subseteq S_{-i}(h)$ be a subset of the opponents' strategy combinations that lead to h. Let $(D_{-i}(h))_{h\in H_i}$ be the collection of all these subsets $D_{-i}(h)$ with $h \in H_i$. Consider a conditional belief vector $(b_i(h))_{h\in H_i}$ for player i. We say that $(b_i(h))_{h\in H_i}$ is a *conditional belief vector* on $(D_{-i}(h))_{h\in H_i}$ if for every $h \in H_i$, the conditional belief $b_i(h)$ only assigns positive probability to opponents' strategy combinations in $D_{-i}(h)$.

Consider some information set $h^* \in H_i$ and some conditional belief $b_i(h^*) \in \Delta(D_{-i}(h^*))$. The question is: Can we extend $b_i(h^*)$ to a conditional belief vector $(b_i(h))_{h\in H_i}$ on $(D_{-i}(h))_{h\in H_i}$, such that there exists a strategy s_i leading to h^* that is optimal, at every $h \in H_i$ weakly following h^*, for the belief $b_i(h)$? We provide a sufficient condition on $(D_{-i}(h))_{h\in H_i}$ under which this is indeed possible.

Definition 8.14.2 *(Forward inclusion property)*
The collection $(D_{-i}(h))_{h\in H_i}$ of opponents' strategy subsets $D_{-i}(h) \subseteq S_{-i}(h)$ satisfies the ***forward inclusion property*** *if for every $h, h' \in H_i$ where h' follows h, $D_{-i}(h) \cap S_{-i}(h') \subseteq D_{-i}(h')$.*

That is, if s_{-i} is an opponents' strategy combination in $D_{-i}(h)$ that leads to the future information set h', then s_{-i} is also a member of $D_{-i}(h')$.

Lemma 8.14.3 *(Bayesian updating and forward inclusion)*
For a given player i, consider a collection $(D_{-i}(h))_{h\in H_i}$ of opponents' strategy subsets satisfying the forward inclusion property. At a given information set $h^* \in H_i$ consider some conditional belief $b_i(h^*) \in \Delta(D_{-i}(h^*))$. Then, $b_i(h^*)$ can be extended to a conditional belief vector $(b_i(h))_{h\in H_i}$ on $(D_{-i}(h))_{h\in H_i}$ that satisfies Bayesian updating at every $h \in H_i$ weakly following h^*.

Proof: Consider some information set $h^* \in H_i$ and some conditional belief $b_i(h^*) \in \Delta(D_{-i}(h^*))$. We will extend $b_i(h^*)$ to some conditional belief vector $(b_i(h))_{h\in H_i}$ on $(D_{-i}(h))_{h\in H_i}$, satisfying Bayesian updating at information sets weakly following h^*.

Let $H_i(h^*)$ be the collection of information sets for player i that follow h^*. Let $H_i^+(h^*)$ be those information sets $h \in H_i(h^*)$ with $b_i(h^*)(S_{-i}(h)) > 0$. For every $h \in H_i^+(h^*)$ we define the conditional belief $b_i(h) \in \Delta(D_{-i}(h))$ by

$$b_i(h)(s_{-i}) := \frac{b_i(h^*)(s_{-i})}{b_i(h^*)(S_{-i}(h))}$$

for all of the opponents' strategy combinations $s_{-i} \in S_{-i}(h)$ leading to h. So, $b_i(h)$ is obtained from $b_i(h^*)$ by Bayesian updating. To see that $b_i(h) \in \Delta(D_{-i}(h))$, note that $b_i(h)$ only assigns positive probability to $s_{-i} \in S_{-i}(h)$ that have positive probability under $b_i(h^*)$. Since, by construction, $b_i(h^*) \in \Delta(D_{-i}(h^*))$, it follows that $b_i(h)$ only assigns positive probability to $s_{-i} \in D_{-i}(h^*) \cap S_{-i}(h)$. However, by the forward inclusion property, $D_{-i}(h^*) \cap S_{-i}(h) \subseteq D_{-i}(h)$, and hence $b_i(h) \in \Delta(D_{-i}(h))$.

Now, consider an information set $h \in H_i(h^*) \backslash H_i^+(h^*)$ that is not preceded by any $h' \in H_i(h^*) \backslash H_i^+(h^*)$. That is, $b_i(h^*)(S_{-i}(h)) = 0$, but $b_i(h^*)(S_{-i}(h')) > 0$ for every $h' \in H_i$ between h^* and h. For every such h, choose some arbitrary conditional belief $b_i(h) \in \Delta(D_{-i}(h))$.

Let $H_i^+(h)$ be those information sets $h' \in H_i$ weakly following h with $b_i(h)(S_{-i}(h')) > 0$. For every $h' \in H_i^+(h)$, let the conditional belief $b_i(h')$ be obtained from $b_i(h)$ by Bayesian updating. By the same argument as above, it can be shown that $b_i(h') \in \Delta(D_{-i}(h'))$ for every $h' \in H_i^+(h)$.

By continuing in this fashion, we will finally define for every $h \in H_i$ following h^* some conditional belief $b_i(h) \in \Delta(D_{-i}(h))$, such that these conditional beliefs, together with $b_i(h^*)$, satisfy Bayesian updating. For every information set $h \in H_i$ not weakly following

h^*, define $b_i(h) \in \Delta(D_{-i}(h))$ arbitrarily. So, $(b_i(h))_{h \in H_i}$ is a conditional belief vector on $(D_{-i}(h))_{h \in H_i}$ that extends $b_i(h^*)$, and satisfies Bayesian updating at information sets weakly following h^*. $\diamond$

The lemma above implies in particular that, whenever the collection $(D_{-i}(h))_{h \in H_i}$ satisfies the forward inclusion property, then it allows for a conditional belief vector $(b_i(h))_{h \in H_i}$ that satisfies Bayesian updating at all information sets. Hence, by Lemma 8.14.1 we conclude that, if the collection $(D_{-i}(h))_{h \in H_i}$ satisfies the forward inclusion property, then we can always find a strategy that is optimal, at all information sets, for some conditional belief vector on $(D_{-i}(h))_{h \in H_i}$.

Our next result shows that the sets of strategies surviving a particular round of the backward dominance procedure satisfy the forward inclusion property. This result thus guarantees that we can apply Lemma 8.14.3 to every round of the backward dominance procedure – which will be important for proving Theorem 8.8.3 later. In this lemma, whenever we say that a strategy s_i survives step k of the backward dominance procedure at h, we mean that s_i is in the reduced decision problem $\Gamma^k(h)$ obtained at h after step k of the backward dominance procedure.

Lemma 8.14.4 *(Backward dominance procedure satisfies the forward inclusion property)*
For every player i and information set $h \in H_i$, let $S^k_{-i}(h)$ be the set of opponents' strategy combinations that survive step k of the backward dominance procedure at h. Then, the collection $(S^k_{-i}(h))_{h \in H_i}$ of strategy subsets satisfies the forward inclusion property.

Proof: For $k = 0$ the statement is trivial since $S^0_{-i}(h) = S_{-i}(h)$ for all h. So, take some $k \geq 1$. Suppose that $h, h' \in H_i$ and that h' follows h. Take some opponent's strategy s_j in $S^k_{-i}(h) \cap S_{-i}(h')$, that is, $s_j \in S^k_j(h) \cap S_j(h')$. Here, $S^k_j(h)$ denotes the set of strategies for player j that survive step k of the backward dominance procedure at h. Then, since $s_j \in S^k_j(h)$, we have that s_j is not strictly dominated in any decision problem $\Gamma^{k-1}(h'')$ where $h'' \in H_j$ and h'' weakly follows h. As h' follows h, it holds in particular that s_j is not strictly dominated in any decision problem $\Gamma^{k-1}(h'')$ where $h'' \in H_j$ and h'' weakly follows h'. Together with the fact that $s_j \in S_j(h')$, this implies that $s_j \in S^k_j(h')$. So, $S^k_{-i}(h) \cap S_{-i}(h') \subseteq S^k_{-i}(h')$, and hence the forward inclusion property holds. $\diamond$

By combining all of the results above, we can prove the following important property of the backward dominance procedure.

Lemma 8.14.5 *(Backward dominance procedure and optimal strategies)*
For every player i and information set $h \in H_i$, let $S^k_{-i}(h)$ be the set of opponents' strategy combinations that survive step k of the backward dominance procedure at h. For a given information set $h^* \in H_i$ consider some conditional belief $b_i(h^*) \in \Delta(S^k_{-i}(h^*))$. Then, we can extend $b_i(h^*)$ to a conditional belief vector $(b_i(h))_{h \in H_i}$ on $(S^k_{-i}(h))_{h \in H_i}$, and find a strategy s_i leading to h^*, such that s_i is optimal, at every $h \in H_i$ weakly following h^* that s_i leads to, under the belief $b_i(h)$.

Proof: Consider an information set $h^* \in H_i$ and a conditional belief $b_i(h^*) \in \Delta(S^k_{-i}(h^*))$. By Lemma 8.14.4 we know that the collection $(S^k_{-i}(h))_{h \in H_i}$ of strategy subsets satisfies the forward inclusion property. Hence, by Lemma 8.14.3, the belief $b_i(h^*)$ can be extended to a conditional belief vector $(b_i(h))_{h \in H_i}$ on $(S^k_{-i}(h))_{h \in H_i}$ that satisfies Bayesian updating at every $h \in H_i$ weakly following h^*. But then, by Lemma 8.14.1, we can find a strategy s_i leading to h^* that is optimal, at every $h \in H_i$ weakly following h^* that s_i leads to, under the conditional belief $b_i(h)$. ◇

Optimality principle

In the second part of this proofs section we show an important optimality property of the backward dominance procedure. Recall that in the backward dominance procedure, $\Gamma^k(h)$ denotes the reduced decision problem at h produced at the end of step k. For every player i, let us denote by $S^k_i(h)$ the set of strategies for player i in $\Gamma^k(h)$. By construction of the algorithm, $S^k_i(h)$ contains exactly those strategies in $S^{k-1}_i(h)$ that, for every $h' \in H_i$ weakly following h, are not strictly dominated in $\Gamma^{k-1}(h')$. We know from Chapter 2 that s_i is not strictly dominated in $\Gamma^{k-1}(h')$, if and only if, there is some belief $b_i(h') \in \Delta(S^{k-1}_{-i}(h'))$ such that s_i is optimal for $b_i(h')$ *among all strategies* in $S^{k-1}_i(h')$. That is,

$$u_i(s_i, b_i(h')) \geq u_i(s_i', b_i(h')) \text{ for all } s_i' \in S^{k-1}_i(h').$$

However, we can show a little more about s_i. Not only is s_i optimal for the belief $b_i(h')$ among the strategies in $S^{k-1}_i(h')$, it is even optimal among *all strategies that lead to* h'. That is, for every $h' \in H_i$ weakly following h that s_i leads to, we have that

$$u_i(s_i, b_i(h')) \geq u_i(s_i', b_i(h')) \text{ for all } s_i' \in S_i(h').$$

We call this the *optimality principle* for the backward dominance procedure, and it will play a crucial role in proving that the backward dominance procedure selects precisely those strategies that can rationally be chosen under common belief in future rationality.

Lemma 8.14.6 *(Optimality principle)*
Consider a strategy $s_i \in S^k_i(h)$ that survives step k of the backward dominance procedure at h. Then, for every $h' \in H_i$ that s_i leads to, and that weakly follows h, there is some belief $b_i(h') \in \Delta(S^{k-1}_{-i}(h'))$ such that s_i is optimal for $b_i(h')$ among the strategies in $S_i(h')$.

Proof: Consider some information set h, some player i, some strategy $s_i \in S^k_i(h)$ and some $h' \in H_i$ that s_i leads to, and that weakly follows h. Then we know from our argument above that there is some belief $b_i(h') \in \Delta(S^{k-1}_{-i}(h'))$ such that

$$u_i(s_i, b_i(h')) \geq u_i(s_i', b_i(h')) \text{ for all } s_i' \in S^{k-1}_i(h'). \tag{8.1}$$

We will prove that, in fact,

$$u_i(s_i, b_i(h')) \geq u_i(s_i', b_i(h')) \text{ for all } s_i' \in S_i(h').$$

Suppose, on the contrary, that there is some $s_i' \in S_i(h')$ such that

$$u_i(s_i, b_i(h')) < u_i(s_i', b_i(h')). \tag{8.2}$$

We will show that in this case there is some $s_i^* \in S_i^{k-1}(h')$ with $u_i(s_i', b_i(h')) \leq u_i(s_i^*, b_i(h'))$, which together with (8.2) contradicts (8.1).

Since $b_i(h') \in \Delta(S_{-i}^{k-1}(h'))$, we know by Lemma 8.14.5 that we can extend $b_i(h')$ to a conditional belief vector $(b_i(h''))_{h'' \in H_i}$ on $(S_{-i}^{k-1}(h''))_{h'' \in H_i}$, and find a strategy $s_i^* \in S_i(h')$ that is optimal, at every $h'' \in H_i$ weakly following h' that s_i^* leads to, under the belief $b_i(h'')$. As each of these conditional beliefs $b_i(h'')$ is in $\Delta(S_{-i}^{k-1}(h''))$, it follows that $s_i^* \in S_i^k(h')$, and hence in particular $s_i^* \in S_i^{k-1}(h')$. Moreover, s_i^* is optimal at h' for the belief $b_i(h')$. And hence, we have by (8.2) that

$$u_i(s_i, b_i(h')) < u_i(s_i', b_i(h')) \leq u_i(s_i^*, b_i(h')) \text{ for some } s_i^* \in S_i^{k-1}(h').$$

This, however, contradicts (8.1). So, (8.2) must be incorrect, and hence s_i is optimal at h' for the belief $b_i(h')$ among the strategies in $S_i(h')$. $\diamond$

The algorithm works

With the help of the first two parts we can now prove the following theorem, which shows that the backward dominance procedure "works" for the concept of common belief in future rationality.

Theorem 8.8.3 *(The algorithm works)*
(1) For every $k \geq 1$, the strategies that can rationally be chosen by a type that expresses up to k-fold belief in future rationality are exactly the strategies in $\Gamma^{k+1}(\emptyset)$, surviving the first $k+1$ steps of the backward dominance procedure.
(2) The strategies that can rationally be chosen by a type that expresses common belief in future rationality are exactly the strategies that survive the full backward dominance procedure – that is, the strategies that are in $\Gamma^k(\emptyset)$ for every k.

Proof: For every k and every player i, let us denote by S_i^k the set of strategies for player i that survive step k of the backward dominance procedure. That is, S_i^k contains the strategies for player i in the reduced decision problem $\Gamma^k(\emptyset)$ obtained at $\emptyset$ after step k of the algorithm. Moreover, we denote by BR_i^k the set of strategies for player i that are optimal for some type that expresses up to k-fold belief in future rationality. We will show that $BR_i^k = S_i^{k+1}$ for every $k \geq 1$. To prove this, we proceed by two parts. In part (a) we will show that $BR_i^k \subseteq S_i^{k+1}$ for all $k \geq 1$. In part (b) we will prove that $S_i^{k+1} \subseteq BR_i^k$ for all $k \geq 1$.

(a) Show that $BR_i^k \subseteq S_i^{k+1}$

To prove this, we will use the following notation. For every k, every player i and every information set $h \in H_i$, let $S_{-i}^k(h)$ be the set of the opponents' strategy combinations that survive step k of the backward dominance procedure at h. That is, $S_{-i}^k(h)$ contains the opponents' strategy combinations in the reduced decision problem $\Gamma^k(h)$ obtained at h after step k of the algorithm. For every opponent $j \neq i$, let $S_j^k(h)$ be the set of strategies for player j in $S^k_{-i}(h)$. By $B^k_{-i}(h)$ we denote the set of opponents' strategy combinations $(s_j)_{j\neq i} \in S_{-i}(h)$, such that there is some type t_i expressing up to k-fold belief in future rationality that at h assigns positive probability to s_{-i}. In other words, $B^k_{-i}(h)$ contains those strategy combinations that may have positive probability at h when expressing up to k-fold belief in future rationality. We prove the following lemma.

Lemma 8.14.7 *For every $k \geq 1$, every player i and every information set $h \in H_i$, we have that $B^k_{-i}(h) \subseteq S^k_{-i}(h)$.*

Proof of Lemma 8.14.7: We will prove the statement by induction on k.

Induction start: We start with $k = 1$. Take some player i, some information set $h \in H_i$ and some strategy combination $(s_j)_{j\neq i} \in B^1_{-i}(h)$. Then, there is some type t_i expressing 1-fold belief in future rationality, such that $b_i(t_i, h)$ assigns positive probability to $(s_j)_{j\neq i}$. Consider some opponent j. As t_i assigns at h positive probability to strategy s_j, and t_i believes in j's future rationality, we can find, for every $h' \in H_j$ weakly following h that s_j leads to, some conditional belief $b_j(h')$ for which s_j is optimal at h'. Hence, by Theorem 2.5.3 in Chapter 2 we know that for every $h' \in H_j$ weakly following h that s_j leads to, strategy s_j is not strictly dominated within $\Gamma^0(h')$. So, by construction of the backward dominance procedure we have that s_j is in the reduced decision problem $\Gamma^1(h)$ – that is, $s_j \in S^1_j(h)$. As this holds for all of the opponents' strategies s_j in $(s_j)_{j\neq i}$, we conclude that $(s_j)_{j\neq i} \in S^1_{-i}(h)$. So, we have shown that all of the opponents' strategy combinations $(s_j)_{j\neq i} \in B^1_{-i}(h)$ are in $S^1_{-i}(h)$, and hence $B^1_{-i}(h) \subseteq S^1_{-i}(h)$.

Induction step: Take some $k \geq 2$, and assume that for every player i and every information set $h \in H_i$, we have that $B^{k-1}_{-i}(h) \subseteq S^{k-1}_{-i}(h)$. Take some player i and some information set $h \in H_i$. We will show that $B^k_{-i}(h) \subseteq S^k_{-i}(h)$.

Take some strategy combination $(s_j)_{j\neq i}$ in $B^k_{-i}(h)$. Then, there is some type t_i expressing up to k-fold belief in future rationality, such that $b_i(t_i, h)$ assigns positive probability to $(s_j)_{j\neq i}$. Consider some opponent j. As t_i assigns at h positive probability to s_j, and t_i expresses up to k-fold belief in future rationality, there must be some type t_j expressing up to $(k-1)$-fold belief in future rationality, such that s_j is optimal for t_j at every $h' \in H_j$ weakly following h that s_j leads to. By our induction assumption, every such type t_j assigns at every $h' \in H_j$ only positive probability to the opponents' strategy combinations in $S^{k-1}_{-j}(h')$. So, strategy s_j must be optimal, for every $h' \in H_j$ weakly following h that s_j leads to, for some conditional belief $b_j(h')$ on $S^{k-1}_{-j}(h')$. By Theorem 2.5.3 in Chapter 2 we may thus conclude that at every $h' \in H_j$ weakly following h that s_j leads to, strategy s_j is not strictly dominated within $\Gamma^{k-1}(h')$. Hence, by construction of the backward dominance procedure we have that $s_j \in S^k_j(h)$. Since this holds for all the opponents' strategies s_j in $(s_j)_{j\neq i}$, it follows that $(s_j)_{j\neq i} \in S^k_{-i}(h)$. So, we have shown that all of the opponents' strategy combinations $(s_j)_{j\neq i} \in B^k_{-i}(h)$ are in $S^k_{-i}(h)$, and hence $B^k_{-i}(h) \subseteq S^k_{-i}(h)$. By induction on k, the proof of this lemma is complete. $\diamond$

With Lemma 8.14.7 it is now easy to show part (a) of the proof. We must show that $BR_i^k \subseteq S_i^{k+1}$ for all k and all players i. Take some strategy $s_i \in BR_i^k$. Then, there is some type t_i expressing up to k-fold belief in future rationality, such that s_i is optimal for t_i at every $h \in H_i$ that s_i leads to. By definition, type t_i assigns at every information set $h \in H_i$ only positive probability to opponents' strategy combinations in $B_{-i}^k(h)$. So, by Lemma 8.14.7, type t_i assigns at every information set $h \in H_i$ only positive probability to opponents' strategy combinations in $S_{-i}^k(h)$. Hence, strategy s_i is optimal, at every $h \in H_i$ that s_i leads to, for some conditional belief on $S_{-i}^k(h)$. By Theorem 2.5.3 in Chapter 2 we thus know that, at every $h \in H_i$ that s_i leads to, strategy s_i is not strictly dominated within $\Gamma^k(h)$. But then, by construction of the backward dominance procedure, we have that $s_i \in S_i^{k+1}(\emptyset)$ – that is, $s_i \in S_i^{k+1}$. Hence, we have shown that every $s_i \in BR_i^k$ is in S_i^{k+1}. So, $BR_i^k \subseteq S_i^{k+1}$. This completes part (a) of the proof.

(b) Show that $S_i^{k+1} \subseteq BR_i^k$

We must show that every strategy that survives the first $k+1$ steps of the backward dominance procedure can rationally be chosen by a type that expresses up to k-fold belief in future rationality. For every k, every player i and every information set h, let $S_i^k(h)$ be the set of strategies that survive step k of the backward dominance procedure at h. So, $S_i^k(h)$ contains exactly the strategies for player i in the reduced decision problem $\Gamma^k(h)$ obtained at h after step k. Moreover, $S_i^k(\emptyset) = S_i^k$. Suppose that the backward dominance procedure terminates after K steps – that is, $S_i^{K+1}(h) = S_i^K(h)$ for all players i and all information sets h. This part of the proof will construct an epistemic model M with the following properties:

- For every information set h, every player i and every strategy $s_i \in S_i^1(h)$, there is a type $t_i^{s_i,h}$ such that s_i is optimal for $t_i^{s_i,h}$ at every $h' \in H_i$ weakly following h that s_i leads to.
- For every $k \geq 2$, if $s_i \in S_i^k(h)$, then the associated type $t_i^{s_i,h}$ expresses up to $(k-1)$-fold belief in future rationality.
- If $s_i \in S_i^K(h)$, then the associated type $t_i^{s_i,h}$ expresses common belief in future rationality.

For this construction we proceed by the following steps. In step 1 we will construct, for every information set h and every strategy $s_i \in S_i^1(h)$, some conditional belief vector $b_i^{s_i,h}$ about the opponents' strategy choices, such that s_i is optimal for $b_i^{s_i,h}$ at every $h' \in H_i$ weakly following h that s_i leads to. In step 2 we will use these beliefs to construct our epistemic model M, where we define for every information set h and every strategy $s_i \in S_i^1(h)$ some type $t_i^{s_i,h}$ such that s_i is optimal for $t_i^{s_i,h}$ at every $h' \in H_i$ weakly following h that s_i leads to. In step 3 we will show that, for every $k \geq 2$ and every $s_i \in S_i^k(h)$, the associated type $t_i^{s_i,h}$ expresses up to $(k-1)$-fold belief in future rationality. Finally, in step 4 we will prove that for every $s_i \in S_i^K(h)$, the associated type $t_i^{s_i,h}$ expresses common belief in future rationality.

Step 1: Construction of beliefs
We will construct, for every information set h and every strategy $s_i \in S_i^1(h)$, some conditional belief vector $b_i^{s_i,h}$ about the opponents' strategy choices, such that s_i is optimal for $b_i^{s_i,h}$ at every $h' \in H_i$ weakly following h that s_i leads to. Consider an information set h. For every $k \in \{1,...,K-1\}$, let $D_i^k(h)$ be the set of strategies for player i that are in $S_i^k(h)$ but not in $S_i^{k+1}(h)$. To define the conditional belief vector $b_i^{s_i,h}$, we distinguish the following two cases:
(1) First consider some $k \in \{1,...,K-1\}$ and some strategy $s_i \in D_i^k(h)$. Then, by Lemma 8.14.6, for every $h' \in H_i$ that s_i leads to, and that weakly follows h, there is some conditional belief $b_i^{s_i,h}(h') \in \Delta(S_{-i}^{k-1}(h'))$ such that s_i is optimal for $b_i^{s_i,h}(h')$ among the strategies in $S_i(h')$. For every other information set $h' \in H_i$, define the conditional belief $b_i^{s_i,h}(h')$ such that $b_i^{s_i,h}(h') \in \Delta(S_{-i}^{k-1}(h'))$.
(2) Consider next some strategy $s_i \in S_i^K(h)$, surviving the full backward dominance procedure at h. Then, also $s_i \in S_i^{K+1}(h)$. So, by Lemma 8.14.6, for every $h' \in H_i$ that s_i leads to, and that weakly follows h, there is some conditional belief $b_i^{s_i,h}(h') \in \Delta(S_{-i}^K(h'))$ such that s_i is optimal for $b_i^{s_i,h}(h')$ among the strategies in $S_i(h')$. For every other information set $h' \in H_i$, define the conditional belief $b_i^{s_i,h}(h')$ such that $b_i^{s_i,h}(h') \in \Delta(S_{-i}^K(h'))$.

This completes the construction of the conditional belief vectors $b_i^{s_i,h}$ in step 1.

Step 2: Construction of types
We will now use the conditional belief vectors $b_i^{s_i,h}$ from step 1 to construct, for every information set h and every strategy $s_i \in S_i^1(h)$, some type $t_i^{s_i,h}$ such that s_i is optimal for $t_i^{s_i,h}$ at every $h' \in H_i$ weakly following h that s_i leads to. For every player i let the set of types T_i be given by

$$T_i = \{t_i^{s_i,h} : h \in H \text{ and } s_i \in S_i^1(h)\},$$

where H denotes the collection of all information sets in the game. For every information set h and every $k \in \{1,...,K\}$, let $T_i^k(h)$ be the set of types $t_i^{s_i,h}$ with $s_i \in S_i^k(h)$. As $S_i^K(h) \subseteq S_i^{K-1}(h) \subseteq ... \subseteq S_i^1(h)$, we have that $T_i^K(h) \subseteq T_i^{K-1}(h) \subseteq ... \subseteq T_i^1(h)$ for every information set h.

Consider an information set h. To define the beliefs of every type in $T_i^1(h)$, we distinguish the following three cases:
(1) First consider a type $t_i^{s_i,h}$ with $s_i \in D_i^1(h)$. That is, $t_i^{s_i,h} \in T_i^1(h)\backslash T_i^2(h)$. We define the conditional belief vector $b_i(t_i^{s_i,h})$ in the following way: Take some arbitrary type $\hat{t}_j$ for every opponent j and an information set $h' \in H_i$. For every opponents' strategy profile $(s_j)_{j\neq i}$, let $b_i^{s_i,h}(h')((s_j)_{j\neq i})$ be the probability that $b_i^{s_i,h}(h')$ assigns to $(s_j)_{j\neq i}$. Here, $b_i^{s_i,h}(h')$ is the conditional belief about the opponents' strategies constructed in step 1. Let $b_i(t_i^{s_i,h}, h')$ be the conditional belief at h' about the opponents' strategy-type

pairs given by

$$b_i(t_i^{s_i,h},h')((s_j,t_j)_{j\neq i}) := \begin{cases} b_i^{s_i,h}(h')((s_j)_{j\neq i}), & \text{if } t_j = \hat{t}_j \text{ for every } j \neq i \\ 0, & \text{otherwise.} \end{cases}$$

So, for every $h' \in H_i$, type $t_i^{s_i,h}$ holds the same belief about the opponents' strategy choices as $b_i^{s_i,h}$. Remember that, by construction, strategy s_i is optimal for $b_i^{s_i,h}$ at every $h' \in H_i$ weakly following h that s_i leads to. Therefore, strategy s_i is optimal for type $t_i^{s_i,h}$ at every $h' \in H_i$ weakly following h that s_i leads to.
(2) Consider next the types $t_i^{s_i,h}$ with $s_i \in D_i^k(h)$ for some $k \in \{2,...,K-1\}$. Hence, $t_i^{s_i,h} \in T_i^k(h)\backslash T_i^{k+1}(h)$ for some $k \in \{2,...,K-1\}$. We define the conditional belief vector $b_i(t_i^{s_i,h})$ as follows: For every information set $h' \in H_i$, let $b_i(t_i^{s_i,h},h')$ be the conditional belief at h' about the opponents' strategy-type pairs given by

$$b_i(t_i^{s_i,h},h')((s_j,t_j)_{j\neq i}) := \begin{cases} b_i^{s_i,h}(h')((s_j)_{j\neq i}), & \text{if } t_j = t_j^{s_j,h'} \text{ for every } j \neq i \\ 0, & \text{otherwise.} \end{cases}$$

Remember that, by construction in step 1, strategy s_i is optimal for $b_i^{s_i,h}$ at every $h' \in H_i$ weakly following h that s_i leads to. Therefore, strategy s_i is optimal for type $t_i^{s_i,h}$ at every $h' \in H_i$ weakly following h that s_i leads to.

Moreover, we know from step 1 that for every $h' \in H_i$, the belief $b_i^{s_i,h}(h')$ only assigns positive probability to opponent's strategies in $S_j^{k-1}(h')$. Therefore, the type $t_i^{s_i,h}$ assigns, at every $h' \in H_i$, only positive probability to opponents' types $t_j^{s_j,h'}$ where $s_j \in S_j^{k-1}(h')$. That is, the type $t_i^{s_i,h}$ assigns, at every $h' \in H_i$, only positive probability to opponents' types in $T_j^{k-1}(h')$.
(3) Consider finally the types $t_i^{s_i,h}$ with $s_i \in S_i^K(h)$. That is, $t_i^{s_i,h} \in T_i^K(h)$. We define the conditional belief vector $b_i(t_i^{s_i,h})$ as follows: For every information set $h' \in H_i$, let $b_i(t_i^{s_i,h},h')$ be the conditional belief at h' about the opponents' strategy-type pairs given by

$$b_i(t_i^{s_i,h},h')((s_j,t_j)_{j\neq i}) := \begin{cases} b_i^{s_i,h}(h')((s_j)_{j\neq i}), & \text{if } t_j = t_j^{s_j,h'} \text{ for every } j \neq i \\ 0, & \text{otherwise.} \end{cases}$$

So, for every $h' \in H_i$, type $t_i^{s_i,h}$ holds the same belief about the opponents' strategy choices as $b_i^{s_i,h}$. Remember that, by construction in step 1, strategy s_i is optimal for $b_i^{s_i,h}$ at every $h' \in H_i$ weakly following h that s_i leads to. Therefore, strategy s_i is optimal for type $t_i^{s_i,h}$ at every $h' \in H_i$ weakly following h that s_i leads to.

Moreover, we know from step 1 that for every $h' \in H_i$, the belief $b_i^{s_i,h}(h')$ only assigns positive probability to opponents' strategies in $S_j^K(h')$. Therefore, the type $t_i^{s_i,h}$ assigns, at every $h' \in H_i$, only positive probability to opponents' types $t_j^{s_j,h'}$ where $s_j \in S_j^K(h')$.

That is, the type $t_i^{s_i,h}$ assigns, at every $h' \in H_i$, only positive probability to opponents' types in $T_j^K(h')$.

This completes the construction of the beliefs for the types.

Step 3: Every type $t_i \in T_i^k(h)$ expresses up to $(k-1)$-fold belief in future rationality
We will prove this statement by induction on k.

Induction start: We start with $k = 2$. Take some type $t_i \in T_i^2(h)$. That is, $t_i = t_i^{s_i,h}$ for some $s_i \in S_i^2(h)$. By construction in step 2, the type $t_i^{s_i,h}$ assigns at every information set $h' \in H_i$ only positive probability to opponents' strategy-type pairs $(s_j, t_j^{s_j,h'})$ where $s_j \in S_j^1(h')$ and $t_j^{s_j,h'} \in T_j^1(h')$. Moreover, we have also seen in step 2 that for every such strategy-type pair $(s_j, t_j^{s_j,h'})$, the strategy s_j is optimal for the type $t_j^{s_j,h'}$ at every $h'' \in H_j$ weakly following h' that s_j leads to. So, the type $t_i^{s_i,h}$ assigns at every information set $h' \in H_i$ only positive probability to opponents' strategy-type pairs $(s_j, t_j^{s_j,h'})$ where s_j is optimal for the type $t_j^{s_j,h'}$ at every $h'' \in H_j$ weakly following h' that s_j leads to. This means, however, that $t_i^{s_i,h}$ believes in the opponents' future rationality. That is, $t_i^{s_i,h}$ expresses 1-fold belief in future rationality, as was to be shown.

Induction step: Take some $k \geq 3$, and assume that for every player i and every information set h, every type $t_i \in T_i^{k-1}(h)$ expresses up to $(k-2)$-fold belief in future rationality. Now, take some player i, some information set h and some type $t_i \in T_i^k(h)$. That is, $t_i = t_i^{s_i,h}$ for some $s_i \in S_i^k(h)$. By construction in step 2, the type $t_i^{s_i,h}$ assigns at every information set $h' \in H_i$ only positive probability to opponents' strategy-type pairs $(s_j, t_j^{s_j,h'})$ where $s_j \in S_j^{k-1}(h')$ and $t_j^{s_j,h'} \in T_j^{k-1}(h')$. We have also seen in step 2 that for every such strategy-type pair $(s_j, t_j^{s_j,h'})$, the strategy s_j is optimal for the type $t_j^{s_j,h'}$ at every $h'' \in H_j$ weakly following h' that s_j leads to. Moreover, by our induction assumption, every such opponent's type $t_j^{s_j,h'} \in T_j^{k-1}(h')$ expresses up to $(k-2)$-fold belief in future rationality. We thus conclude that type $t_i^{s_i,h}$ assigns at every information set $h' \in H_i$ only positive probability to opponents' strategy-type pairs $(s_j, t_j^{s_j,h'})$ where: (1) s_j is optimal for the type $t_j^{s_j,h'}$ at every $h'' \in H_j$ weakly following h' that s_j leads to, and (2) the type $t_j^{s_j,h'} \in T_j^{k-1}(h')$ expresses up to $(k-2)$-fold belief in future rationality. This means, however, that type $t_i^{s_i,h}$ expresses up to $(k-1)$-fold belief in future rationality, which was to be shown. So we see that every type $t_i^{s_i,h} \in T_i^k(h)$ expresses up to $(k-1)$-fold belief in future rationality. By induction on k, the statement in Step 3 follows.

Step 4: Every type $t_i \in T_i^K(h)$ expresses common belief in future rationality
To prove this statement, we will show the following lemma.

Lemma 8.14.8 *For every $k \geq K-1$, every type $t_i \in T_i^K(h)$ expresses up to k-fold belief in future rationality.*

Proof of Lemma 8.14.8: We will prove the statement by induction on k.

Induction start: We begin with $k = K - 1$. We know from step 3 that every type $t_i \in T_i^K(h)$ expresses up to a $(K-1)$-fold belief in future rationality.

Induction step: Take some $k \geq K$, and assume that for every player i and every information set h, every type $t_i \in T_i^K(h)$ expresses up to $(k-1)$-fold belief in future rationality. Now, take some player i, some information set h and some type $t_i \in T_i^K(h)$. That is, $t_i = t_i^{s_i,h}$ for some $s_i \in S_i^K(h)$. By construction in step 2, the type $t_i^{s_i,h}$ assigns at every information set $h' \in H_i$ only positive probability to opponents' strategy-type pairs $(s_j, t_j^{s_j,h'})$ where $s_j \in S_j^K(h')$ and $t_j^{s_j,h'} \in T_j^K(h')$. We have seen in step 2 that for every such strategy-type pair $(s_j, t_j^{s_j,h'})$, the strategy s_j is optimal for the type $t_j^{s_j,h'}$ at every $h'' \in H_j$ weakly following h' that s_j leads to. Moreover, by our induction assumption, every such type $t_j^{s_j,h'} \in T_j^K(h')$ expresses up to $(k-1)$-fold belief in future rationality. We thus conclude that type $t_i^{s_i,h}$ assigns at every information set $h' \in H_i$ only positive probability to opponents' strategy-type pairs $(s_j, t_j^{s_j,h'})$ where: (1) s_j is optimal for the type $t_j^{s_j,h'}$ at every $h'' \in H_j$ weakly following h' that s_j leads to, and (2) the type $t_j^{s_j,h'} \in T_j^K(h')$ expresses up to $(k-1)$-fold belief in future rationality. This means, however, that type $t_i^{s_i,h}$ expresses up to k-fold belief in future rationality, which was to be shown. So we see that every type $t_i^{s_i,h} \in T_i^K(h)$ expresses up to k-fold belief in future rationality. By induction on k, the statement in the lemma follows. ◇

Obviously, the statement in step 4 follows directly from the lemma above.

With steps 1–4, it is now easy to show part (b) in this proof, namely that $S_i^{k+1} \subseteq BR_i^k$ for every player i and every k. Take some strategy $s_i \in S_i^{k+1}$. That is, $s_i \in S_i^{k+1}(\emptyset)$. Then, the associated type $t_i^{s_i,\emptyset}$ is in $T_i^{k+1}(\emptyset)$, and hence we know by step 3 that this type $t_i^{s_i,\emptyset}$ expresses up to k-fold belief in future rationality. Moreover, by our construction in steps 1 and 2, the strategy s_i is optimal for the type $t_i^{s_i,\emptyset}$ at every $h' \in H_i$ weakly following $\emptyset$ that s_i leads to. Hence, strategy s_i is rational for the type $t_i^{s_i,\emptyset}$. Since the type $t_i^{s_i,\emptyset}$ expresses up to k-fold belief in future rationality, we may conclude that $s_i \in BR_i^k$. So, we see that every strategy $s_i \in S_i^{k+1}$ is also in BR_i^k. That is, $S_i^{k+1} \subseteq BR_i^k$, as was to be shown. This completes the proof of part (b).

Now that we have shown parts (a) and (b), it is easy to prove the statements in Theorem 8.8.3. Let us start with part (1) of the theorem. By parts (a) and (b) in this proof, we know that $BR_i^k = S_i^{k+1}$ for all k. That is, the strategies that can rationally be chosen by a type that expresses up to k-fold belief in future rationality are exactly the strategies that survive the first $k+1$ steps of the backward dominance procedure. This completes the proof of part (1) in the theorem.

Let us now move to part (2) of the theorem. Consider first a strategy s_i that can rationally be chosen by a type that expresses common belief in future rationality. Then, $s_i \in BR_i^k$ for all k, and hence – by part (a) of the proof – $s_i \in S_i^{k+1}$ for all k. So, s_i survives the full backward dominance procedure. Hence, we see that every strategy s_i that can rationally be chosen by a type that expresses common belief in future rationality, survives the full backward dominance procedure.

Next, take some strategy s_i that survives the full backward dominance procedure. Then, $s_i \in S_i^K(\emptyset)$. By step 4 we know that the associated type $t_i^{s_i,\emptyset}$ expresses common belief in future rationality. Moreover, by construction in steps 1 and 2, the strategy s_i is optimal for the type $t_i^{s_i,\emptyset}$ at every $h' \in H_i$ weakly following $\emptyset$ that s_i leads to. Hence, strategy s_i is rational for the type $t_i^{s_i,\emptyset}$. Since the type $t_i^{s_i,\emptyset}$ expresses common belief in future rationality, we conclude that strategy s_i is rational for a type that expresses common belief in future rationality. So, we see that every strategy s_i that survives the full backward dominance procedure, is rational for a type that expresses common belief in future rationality.

Altogether we see that a strategy is rational for a type that expresses common belief in future rationality, if and only if, the strategy survives the full backward dominance procedure. This completes the proof of part (2) of the theorem. ■

Games with perfect information

For games with perfect information we will prove Lemma 8.11.4, which shows that for such games it is sufficient to only consider probability 1 beliefs about the opponents' strategy combinations. In order to prove that result, we will first show the following lemma, which holds for arbitrary dynamic games – not only for those with perfect information.

Lemma 8.14.9 *(Bayesian updating and optimality at future information sets)*
Consider an arbitrary dynamic game, not necessarily with perfect information. Consider a player i and two information sets h, h' for player i where h' follows h. Take a strategy s_i for player i that leads to h'. Suppose that s_i is optimal for player i at h given the conditional belief $b_i(h)$, and suppose that $b_i(h)(S_{-i}(h')) > 0$. Let $b_i(h')$ be the conditional belief at h' obtained from $b_i(h)$ by Bayesian updating. Then, strategy s_i is optimal for player i at h' given the conditional belief $b_i(h')$.

Proof: Suppose that s_i is optimal for player i at h given the conditional belief $b_i(h)$, and assume that $b_i(h)(S_{-i}(h')) > 0$. Then,

$$\begin{aligned} u_i(s_i, b_i(h)) &= \sum_{s_{-i} \in S_{-i}(h)} b_i(h)(s_{-i}) \cdot u_i(s_i, s_{-i}) \\ &= \sum_{s_{-i} \in S_{-i}(h')} b_i(h)(s_{-i}) \cdot u_i(s_i, s_{-i}) + \\ &\quad + \sum_{s_{-i} \in S_{-i}(h) \setminus S_{-i}(h')} b_i(h)(s_{-i}) \cdot u_i(s_i, s_{-i}) \end{aligned}$$

$$= b_i(h)(S_{-i}(h')) \sum_{s_{-i} \in S_{-i}(h')} b_i(h')(s_{-i}) \cdot u_i(s_i, s_{-i}) +$$
$$+ \sum_{s_{-i} \in S_{-i}(h) \backslash S_{-i}(h')} b_i(h)(s_{-i}) \cdot u_i(s_i, s_{-i})$$
$$= b_i(h)(S_{-i}(h')) \cdot u_i(s_i, b_i(h')) +$$
$$+ \sum_{s_{-i} \in S_{-i}(h) \backslash S_{-i}(h')} b_i(h)(s_{-i}) \cdot u_i(s_i, s_{-i}).$$

Here, the third equality follows from the assumption that $b_i(h')$ is obtained from $b_i(h)$ by Bayesian updating, and hence

$$b_i(h)(s_{-i}) = b_i(h)(S_{-i}(h')) \cdot b_i(h')(s_{-i})$$

for every $s_{-i} \in S_{-i}(h')$.

Assume now, contrary to what we want to prove, that s_i is not optimal at h' given the belief $b_i(h')$. That is, there is some strategy $s_i' \in S_i(h')$ for which

$$u_i(s_i, b_i(h')) < u_i(s_i', b_i(h')).$$

Let s_i'' be the strategy that coincides with s_i' at all player i information sets weakly following h', and that coincides with s_i at all other information sets for player i. Then, $s_i'' \in S_i(h)$ and

$$u_i(s_i'', b_i(h)) = b_i(h)(S_{-i}(h')) \cdot u_i(s_i'', b_i(h')) +$$
$$+ \sum_{s_{-i} \in S_{-i}(h) \backslash S_{-i}(h')} b_i(h)(s_{-i}) \cdot u_i(s_i'', s_{-i})$$
$$= b_i(h)(S_{-i}(h')) \cdot u_i(s_i', b_i(h')) +$$
$$+ \sum_{s_{-i} \in S_{-i}(h) \backslash S_{-i}(h')} b_i(h)(s_{-i}) \cdot u_i(s_i, s_{-i})$$
$$> b_i(h)(S_{-i}(h')) \cdot u_i(s_i, b_i(h')) +$$
$$+ \sum_{s_{-i} \in S_{-i}(h) \backslash S_{-i}(h')} b_i(h)(s_{-i}) \cdot u_i(s_i, s_{-i})$$
$$= u_i(s_i, b_i(h)).$$

For the inequality, we have been using the assumption that $b_i(h)(S_{-i}(h')) > 0$. This, however, would mean that $u_i(s_i'', b_i(h)) > u_i(s_i, b_i(h))$, which contradicts our assumption that s_i is optimal at h for the belief $b_i(h)$. Hence, s_i must be optimal at h' given the belief $b_i(h')$. This completes the proof. ◇

For the proof of Lemma 8.11.4 we also need to introduce the notion of *minimax choices* and *minimax utility* for a player in a game with perfect information. Consider a dynamic game with perfect information and a player i in this game. For every information set h, controlled by some player j, we define a *minimax choice* $c_j^*(h)$ and the *minimax utility* $u_i^*(h)$ for player i at h. We do so recursively, starting at the ultimate information sets in the game.

For every ultimate information set h, we distinguish two cases: If player i is active at h, then the minimax choice $c_i^*(h)$ is a choice at h that *maximizes* i's utility. If player $j \neq i$ is active at h, then the minimax choice $c_j^*(h)$ is a choice at h that *minimizes* i's utility. The minimax utility $u_i^*(h)$ at h is the utility for player i induced by the minimax choice at h.

For every penultimate information set h we again distinguish two cases: If player i is active at h, then the minimax choice $c_i^*(h)$ is a choice at h that *maximizes* i's utility, given that at every ultimate information set h' following h a minimax choice is played. If player $j \neq i$ is active at h, then the minimax choice $c_j^*(h)$ is a choice at h that *minimizes* i's utility, given that at every ultimate information set h' following h a minimax choice is played. The minimax utility $u_i^*(h)$ is the utility for player i in the subgame starting at h if at h, and all subsequent information sets, only minimax choices are played.

We continue in this fashion until we reach the beginning of the game. In this way we define for every information set h in the game a minimax choice $c_j^*(h)$ and the minimax utility $u_i^*(h)$ for player i. Note that there may be several minimax choices at a given information set h, but all of them will eventually induce the same minimax utility $u_i^*(h)$. So, the minimax utility $u_i^*(h)$ does not depend upon the specific minimax choices selected at the various information sets in the game.

Hence, for every information set h, the minimax utility $u_i^*(h)$ is the lowest expected utility that player i can achieve in the subgame following h, if player i chooses optimally in this subgame given his belief about the opponents' strategy combinations.

We are now ready to prove the following lemma.

Lemma 8.11.4 *(Probability 1 beliefs are sufficient for games with perfect information) Consider a dynamic game with perfect information and an information set h in this game where player i is active. Suppose a strategy s_i is optimal at h for some probabilistic belief about the opponents' strategies. Then, s_i is optimal at h for some belief that assigns probability 1 to one particular strategy combination.*

Proof: Suppose that strategy s_i is optimal at information set h for some probabilistic belief $b_i(h) \in \Delta(S_{-i}(h))$ about the opponents' strategy combinations. Let $Z(s_i, b_i(h))$ be the set of terminal histories that have positive probability according to $(s_i, b_i(h))$, and let z^* be the terminal history in $Z(s_i, b_i(h))$ that gives the highest utility for player i.

We now construct the opponents' strategy combination s_{-i}^* as follows: Let $H(z^*)$ be the collection of information sets on the path to z^*. At every information set $h' \in H(z^*)$ not controlled by player i, let s_{-i}^* prescribe the choice that leads to z^*. At every information set $h' \in H(z^*)$ controlled by player i, and every choice c_i at h' which differs from $s_i(h')$, let s_{-i}^* prescribe the opponents' minimax choices in the subgame that follows h' and c_i.

We show that s_i is optimal at h if player i believes at h that, with probability 1, his opponents will choose s^*_{-i}. By construction,

$$u_i(s_i, s^*_{-i}) = u_i(z^*),$$

as both s_i and s^*_{-i} prescribe the choices that lead to z^*. Consider now some alternative strategy s'_i that leads to h. We will show that $u_i(s_i, s^*_{-i}) \geq u_i(s'_i, s^*_{-i})$. We distinguish two cases:

Suppose first that (s'_i, s^*_{-i}) reaches z^*. Then, we have that

$$u_i(s'_i, s^*_{-i}) = u_i(z^*) = u_i(s_i, s^*_{-i}).$$

Assume next that (s'_i, s^*_{-i}) does not reach z^*. Then, there must be some information set h' for player i between h and z^*, reached by both (s_i, s^*_{-i}) and (s'_i, s^*_{-i}), such that $s'_i(h') \neq s_i(h')$. That is, s'_i prescribes a different choice at h' than s_i does. Let h'' be the information set that immediately follows choice $s'_i(h')$ at h'. By construction, s^*_{-i} prescribes the opponents' minimax choices in the subgame that follows h'', and hence

$$u_i(s'_i, s^*_{-i}) \leq u^*_i(h''),$$

where $u^*_i(h'')$ is player i's minimax utility at h''. Since player i is active at h', the minimax utility $u^*_i(h')$ at h' is the maximum of all minimax utilities $u^*_i(\tilde{h})$, by ranging over all information sets $\tilde{h}$ that immediately follow h'. Since h'' immediately follows h', we have that

$$u^*_i(h'') \leq u^*_i(h').$$

Together with the inequality above, we conclude that

$$u_i(s'_i, s^*_{-i}) \leq u^*_i(h').$$

We will now show that $u_i(s_i, s^*_{-i}) \geq u^*_i(h')$. Remember that the terminal history z^* follows h', and that z^* has positive probability under $(s_i, b_i(h))$. But then, the conditional belief $b_i(h)$ must assign positive probability to some strategy combination s_{-i} that leads to h'. That is, $b_i(h)(S_{-i}(h')) > 0$. Since s_i is optimal at h for the conditional belief $b_i(h)$, we conclude from Lemma 8.14.9 that s_i is also optimal at h' for the conditional belief $b_i(h')$, obtained from $b_i(h)$ by Bayesian updating. Remember that $u^*_i(h')$ is the lowest expected utility player i can achieve at h' by choosing optimally, given his belief, and hence

$$u_i(s_i, b_i(h')) \geq u^*_i(h').$$

Also remember that $Z(s_i, b_i(h))$ was the set of terminal histories with positive probability under $(s_i, b_i(h))$, and that z^* was the terminal history in $Z(s_i, b_i(h))$ with the highest utility for player i. Denote by $Z(s_i, b_i(h'))$ the set of terminal histories with positive probability under $(s_i, b_i(h'))$. As h' follows h, and $b_i(h')$ is obtained from h by Bayesian updating, it follows that $Z(s_i, b_i(h')) \subseteq Z(s_i, b_i(h))$. Since z^* follows h', and z^* is in $Z(s_i, b_i(h))$, we conclude that z^* must also be in $Z(s_i, b_i(h'))$. Hence, z^* is the terminal history in $Z(s_i, b_i(h'))$ with the highest utility for player i. This means, however, that

$$u_i(z^*) \geq u_i(s_i, b_i(h')).$$

As $u_i(s_i, s_{-i}^*) = u_i(z^*)$, it follows that

$$u_i(s_i, s_{-i}^*) \geq u_i(s_i, b_i(h')) \geq u_i^*(h').$$

Since we know, from above, that $u_i(s_i', s_{-i}^*) \leq u_i^*(h')$, we conclude that

$$u_i(s_i, s_{-i}^*) \geq u_i(s_i', s_{-i}^*).$$

This holds for every alternative strategy s_i' that leads to h. Hence, s_i is optimal at h under the belief that his opponents, with probability 1, choose s_{-i}^*. This completes the proof. ◇

Practical problems

8.1 Two parties in a row

After ten consecutive attempts Chris finally passed his driving test. In order to celebrate this memorable event, he has organized two parties in a row – one on Friday and one on Saturday. You and Barbara are both invited to the first party on Friday. The problem, as usual, is to decide which color to wear for that evening. You can both choose from *blue*, *green*, *red* or *yellow*. The utilities that you and Barbara derive from wearing these colors are given by Table 8.33. As before, you both feel unhappy when you wear the same color as your friend. In that case, the utilities for you and Barbara would only be 1.

In order to decide which people to invite to the second party, Chris will apply a very strange selection criterion: Only those people dressed in yellow at the first party will be invited to the party on Saturday. Moreover, you will only go to the party on Saturday if Barbara is invited as well, and similarly for Barbara. If you are both invited to the party on Saturday, you face the same decision problem as on Friday, namely which color to wear for that evening.

Suppose that the total utility for you and Barbara is simply the sum of the utilities you derive from the parties you go to.

(a) Model this situation as a dynamic game between you and Barbara. How many strategies do you have in this game?

(b) Is this a game with observed past choices? Is this a game with perfect information?

(c) Which strategies can you rationally choose under common belief in future rationality? Which colors can you rationally wear on Friday under common belief in future rationality? And on Saturday, provided you go to the party on that day? What about Barbara?

(d) Construct an epistemic model such that, for every strategy found in (c), there is some type that expresses common belief in future rationality, and for which that strategy is rational.

Table 8.33. *The utilities for you and Barbara in "Two parties in a row"*

	blue	green	red	yellow	same color as friend
you	6	4	3	2	1
Barbara	6	2	4	3	1

8.2 Selling ice cream

It is Monday morning. The weather forecast for today is excellent and you and Barbara are both thinking of selling ice cream on the beach. Figure 8.15 is a map of the beach, where $a, b, ..., g$ are the possible locations where you and Barbara can sell ice cream. The map gives the expected number of people between every two locations on the beach. So, for today we expect 200 people between locations b and c, 300 people between c and d, and so on. Assume that everybody on the beach wants to buy exactly one ice cream, and because of the hot weather will simply go to the nearest seller.

You and Barbara both have only a one-day license for selling ice cream. Barbara will definitely sell ice cream today, since she will go on holiday tomorrow. You are still thinking about whether to sell ice cream today, facing the competition of Barbara, or go tomorrow and be the only seller. However, the weather forecast for tomorrow is not as good, and you expect all the numbers in Figure 8.15 to drop by 40%.

If you decide to sell today, then both you and Barbara must simultaneously decide where to sell ice cream. If you choose the same location as Barbara, you will get into a fight with her and nobody will dare to buy any ice cream from you or Barbara.

(a) Model this situation as a dynamic game between you and Barbara.

(b) Find those strategies you can rationally choose under common belief in future rationality. Do the same for Barbara.

(c) Construct an epistemic model such that, for every strategy found in (b), there is some type that expresses common belief in future rationality, and for which that strategy is rational.

(d) How would the answer to (b) change if the weather forecast for tomorrow is more optimistic, such that you would expect the numbers in Figure 8.15 to drop by only 30%?

Barbara has called you and told you that she has cancelled her holiday trip. So, she could also decide to sell ice cream tomorrow. Suppose that you tell Barbara whether you will sell ice cream today or tomorrow, and then Barbara will decide whether to go today or tomorrow. Assume that tomorrow, the number of people on the beach will drop by 40%.

(e) Model this new situation as a dynamic game between you and Barbara.

(f) Find those strategies you can rationally choose under common belief in future rationality. Do the same for Barbara.

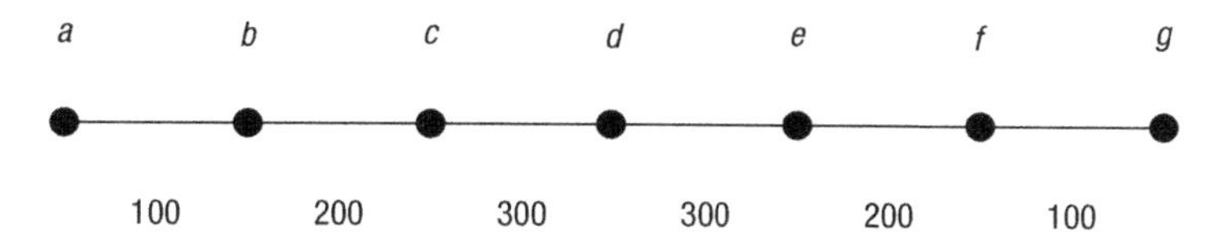

Figure 8.15 Selling ice cream

8.3 A nightmare with sharks

Last night you had a terrible nightmare. In your dream, you were one of five pirates who survived a terrible fight with a Spanish ship. During this battle you managed to capture the Spanish treasure, 1000 pieces of silver. However, the other four pirates, Barbara, Chris, Diana and Edgar, also want a share of this treasure. The pirates eventually settle on the following procedure: In round 1 you must make a proposal to the other pirates on how to divide the treasure between the five of you. Afterwards, Barbara, Chris, Diana and Edgar will sequentially vote either in favor or against your proposal. That is, Barbara votes first, then Chris votes, then Diana and finally Edgar. Every time a pirate votes, he or she knows how the others have voted so far.

Your proposal will only be accepted if at least 50% of the other pirates vote in favor of it. If it is accepted, the pirates who voted for it will receive an additional bonus of 4 pieces of silver each, whereas the other pirates will just receive their share, and the dream will end. However, if your proposal is not accepted, you will be thrown overboard and will fight for the rest of the dream against hungry sharks! In that case, the procedure will go to round 2, in which the remaining 4 pirates try to reach an agreement.

In round 2, it is Barbara's turn to make a proposal on how to divide the 1000 pieces between the four remaining pirates. Chris, Diana and Edgar will sequentially vote either in favor or against the proposal. Barbara's proposal will only be accepted if at least 50% of these pirates vote in favor. If it is accepted, the pirates who voted for it will receive an additional bonus of 3 pieces, whereas the other pirates will just receive their share, and the dream will end. If Barbara's proposal is not accepted, she will be thrown overboard and the procedure will continue with the three remaining pirates, and so on.

In round 3, the bonus for voting in favor of the proposal is 2 pieces (if the proposal is accepted), whereas in round 4 it is 1 piece. By round 5, when you, Barbara, Chris and Diana have all been thrown overboard, only Edgar remains, and he will of course keep the entire treasure for himself and live as a rich man ever after.

(a) Is this a game with observed past choices? Is this a game with perfect information?
(b) Suppose we want to find those strategies you can rationally choose under common belief in future rationality. Which procedure should we use?
(c) What proposal(s) can you rationally make in round 1 under common belief in future rationality? How do you expect the other pirates to vote? What proposals do you expect the other pirates to make if their respective rounds are reached? Do you expect to be thrown overboard? Does this nightmare have a happy ending for you?

8.4 The love letter

During a beautiful holiday on Mallorca you met Charlie, to whom you have been writing love letters ever since. This morning Barbara told you that she has found out, and you can see that she is extremely jealous. She is even threatening to burn your letter if you write to Charlie again! Now, you are sitting at your desk and must decide whether or not to write a letter to Charlie. If you write a letter then Barbara will certainly find out, and she can either decide to search for the letter and burn it or let it go. Charlie, both upon receiving a letter and upon observing that no letter has arrived, must choose between writing to you or not. In case no letter has arrived, Charlie will not know whether this is because you did not send one or because Barbara has burned it.

Assume that the utilities for you, Barbara and Charlie are as follows: Writing a letter requires a considerable effort from you and decreases your utility by 2. On the other hand, receiving a letter from Charlie would increase your utility by 5. However, if Barbara were to burn your letter, then this would put you into a state of shock and would decrease your utility by 5.

For Barbara, searching for the letter and burning it is a difficult task and decreases her utility by 2. However, if Charlie does not write to you, then Barbara would have succeeded in her mission and her utility would increase by 10.

Charlie only wants to write to you if you have written a letter to her. More precisely, Charlie's utility decreases by 5 if Charlie writes a letter when you have not, or if Charlie does not write a letter when you have written one.

(a) Model this situation as a dynamic game among you, Barbara and Charlie. Is this a game with observed past choices? Is this a game with perfect information?

(b) Which strategy, or strategies, can you, Barbara and Charlie rationally choose under common belief in future rationality?

(c) Construct an epistemic model such that, for every strategy found in (b), there is some type that expresses common belief in future rationality, and for which that strategy is rational.

(d) Suppose now that you not only express common belief in future rationality, but that you also believe that Barbara holds the same belief about Charlie's strategy choice as you do. Which strategy, or strategies, can you then rationally choose?

8.5 The street musicians

The weather is lovely and Barbara wants to perform as a street musician today. She plays the violin, but she would love to perform with a trumpet player: both you and Chris play the trumpet. So, the question is: Who will play with Barbara, and how will the money collected be divided between the two musicians?

Suppose that the amount expected to be collected is 10 euros. In order to decide who will play with Barbara, and how to divide these 10 euros, you propose the following procedure:

In round 1, you and Barbara will simultaneously name an odd number between 1 and 9. The person who names the highest number, x, will propose to the other person how to divide the 10 euros. If you and Barbara name the same number x at the beginning, you will toss a coin to decide who makes the proposal. This proposal must only use even numbers, where 0 also counts as an even number. The other person may accept or reject this proposal. If the other person accepts the proposal, you and Barbara will play together, and divide the money according to this proposal. However, if the other person rejects the proposal, then the proposer must pay the amount x named at the beginning to the other person, and the procedure will move to round 2. In that case, you will not play.

In round 2, Barbara will negotiate with Chris about how to divide the money. The rules are the same as above. If they reach an agreement, Barbara and Chris will perform together, and divide the money according to the accepted proposal. If the proposal is rejected, nobody will play, and hence nobody will earn anything. However, in that case the proposer must still pay the amount x named at the beginning to the other person.

Everyone's objective is to maximize the expected amount of money to be earned as a street musician. However, Barbara has a slight preference for performing with you. More precisely, if she were to perform with Chris then her utility would be decreased by 0.5.

(a) Is this a game with observed past choices? Is this a game with perfect information?
(b) Find the strategies that you, Barbara and Chris can rationally choose under common belief in future rationality. Whom do you expect to perform as street musicians? How do you expect the money to be divided? How do you explain this outcome?

8.6 The three mountains

You and Barbara are on holiday in the French Alps. Since you both like cycling in the mountains, you have organized a three-day cycling competition between the two of you. On the first day the stage ends on top of the *Col de la Madeleine.* The second stage leads to the top of the *Alpe d'Huez,* whereas on the last day you must climb to the *Col du Galibier.* The winner at the *Col de la Madeleine* gets 4 points, the winner at the *Alpe d'Huez* gets 6 points, whereas the winner at the *Col du Galibier* gets 8 points. Moreover, if the winner beats the loser by more than 20 minutes, then the winner will get 1 bonus point whereas the loser will get 1 penalty point. The objective for you and Barbara is to collect as many points as possible.

These three climbs are extremely tough, and the difficulty is how you must decide to divide your energy over these three stages. Suppose that your total energy is two units, which you must divide over these three days. So, for the first climb you can use either zero energy units, meaning that you take it easy, or one unit, meaning that you go fast but without going to the limit, or the maximum of two units, meaning that you completely exhaust yourself on that day. The same applies to Barbara. At the beginning of each day, you and Barbara simultaneously decide how many of your remaining energy units to consume on that day. The winner of a stage will be the person who uses the most energy on that particular day. Moreover, the winner will beat the other person by more than 20 minutes if the winner expends two energy units and the other person zero. If you and Barbara use the same amount of energy during a stage, then there is an equal probability of winning for both of you (but not by more than 20 minutes).

(a) Which strategies can you rationally choose under common belief in future rationality? How much energy can you rationally consume during the first stage under common belief in future rationality?

(b) From the strategies found in (a), which strategy would you really choose and why?

8.7 The Walkin' Fridge

You, Barbara and Chris have a wonderful idea for a new invention: A mobile refrigerator – also called the "Walkin' Fridge." Instead of you going to the refrigerator for a beer, the refrigerator comes to you! In order to turn this project into a success, it is necessary that each person dedicates enough time to it. You have agreed on the following procedure: First *you* will start working on the project, then *Barbara* and finally *Chris*. Each member must decide, at the beginning of his or her term, how many days to dedicate to the project. For simplicity, suppose you can dedicate at most 100 days to the project, and Barbara and Chris at most 200 days each. Assume moreover that Barbara will know how many days you have worked on the idea, and that Chris will know how many days you and Barbara have dedicated to the project.

Suppose that for each of you individually, the expected income from the project will be equal to the product of the numbers of days spent on the project by the three members. That is, if you work for a days, Barbara works for b days, and Chris works for c days, then each member will receive an expected income of $a \cdot b \cdot c$ from the project. Assume that the cost of working for x days is $x^3/3$ for each person. The final utility for every person is the expected income from the project minus the cost of working.

(a) How many days will you rationally dedicate to the Walkin' Fridge project under common belief in future rationality? How many days do you expect Barbara and Chris to dedicate to the project under common belief in future rationality? (In order to make computations easier, assume that the numbers of days you can choose can be any real number between 0 and 100, and that the numbers of days that Barbara and Chris can choose can be any real number between 0 and 200. So, for instance, you can choose to work for $\sqrt{37}$ days if you wish.)

(b) Suppose now that the cost of working for x days is not $x^3/3$, but x^3 for every person. So, the cost of working is three times higher than before. How many days will you now rationally dedicate to the Walkin' Fridge project under common belief in future rationality? How many days do you expect Barbara and Chris to dedicate to the project under common belief in future rationality?

(c) Suppose that the cost of working for x days is $\alpha \cdot x^3$ for each person, where α is some real number greater than zero. Determine, for all possible values of α, the number of days you will rationally dedicate to the Walkin' Fridge project under common belief in future rationality.

Theoretical problems

8.8 Strategy subsets closed under belief in future rationality

Consider an arbitrary dynamic game. For every information set h and every player i, let $D_i(h) \subseteq S_i(h)$ be some subset of strategies that lead to h. The collection $(D_i(h))_{h \in H, i \in I}$ of strategy subsets is called *closed under belief in future rationality* if for every $s_i \in D_i(h)$, and every $h' \in H_i$ weakly following h that s_i leads to, there is some belief $b_i(h') \in \Delta(D_{-i}(h'))$ for which s_i is optimal.

Show that a strategy s_i can rationally be chosen under common belief in future rationality, if and only if, there is a collection $(D_i(h))_{h \in H, i \in I}$ of strategy subsets that is closed under belief in future rationality, and in which $s_i \in D_i(\emptyset)$.

8.9 Repeated games

Consider some static game G. Consider the dynamic game in which: (1) the same game G is played at periods $1, 2, ..., M$, (2) the players know, at the beginning of every period, which choices have been made by their opponents at previous periods, and (3) the final utility for a player is the sum of the utilities achieved at periods 1 to M. This dynamic game is denoted by $\Gamma(G, M)$, and is called a *repeated game* with observed past choices.

(a) Suppose that within the static game G, there is only one choice c_i^* for every player i that survives iterated elimination of strictly dominated choices. Show that within the repeated game $\Gamma(G, M)$, there is only one strategy for each player that he can rationally choose under common belief in future rationality. What does this strategy look like?

(b) Show that every choice c_i that survives iterated elimination of strictly dominated choices within the static game G, can rationally be chosen by player i under common belief in future rationality at every period of the repeated game $\Gamma(G, M)$.

(c) Give an example of a static game G with the following property: For some player i there is a choice c_i which: (1) is strictly dominated in the static game G, but (2) can rationally be chosen by player i under common belief in future rationality at period 1 of the twice-repeated game $\Gamma(G, 2)$.

8.10 Strictly and weakly dominated strategies

(a) Consider an arbitrary dynamic game and some information set $h \in H_i$ in that game. Suppose that strategy s_i is *strictly* dominated within the full decision problem $\Gamma^0(h)$ at h. Show that strategy s_i is then *weakly* dominated within the full decision problem $\Gamma^0(\emptyset)$ at the beginning of the game.

(b) Consider an arbitrary dynamic game. Suppose that no strategy is weakly dominated within the full decision problem $\Gamma^0(\emptyset)$ at the beginning of the game. Show that every strategy can rationally be chosen under common belief in future rationality.

Literature

Dynamic games

In this chapter we introduced a new class of games – dynamic games – in which players may have to choose at several points in time, and may receive full or partial information about the choices made by their opponents in the past. In the literature such games are also often called *extensive form games* as opposed to *normal form games*, which is another name for static games. Von Neumann (1928) and von Neumann and Morgenstern (1944) provided a formal and general mathematical model for dynamic games. However, the model for dynamic games that nowadays is most often used is due to Kuhn (1950, 1953), who described a dynamic game by means of a *tree* with *vertices* and *edges*. The *vertices* represent the non-terminal and terminal histories in the game, whereas the edges between the vertices represent the choices that can be made, moving the game from one vertex to another. The model we use in this chapter can be seen as an extension of Kuhn's model, as we also allow for *simultaneous* choices by various players at a given non-terminal history (or vertex). In that sense, our model closely corresponds to the notion of a *multi-stage* game, as used in Fudenberg and Tirole (1991a), which also explicitly allows for simultaneous choices at the various stages of the game. In this chapter we considered dynamic games with *observed past choices* as a special class. In the literature, such games are also known as dynamic games with *almost perfect information* – see for instance Asheim (2002).

Strategy

The definition of a strategy we use in this book is different from the usual definition found in the literature. Traditionally, a strategy for player i in a dynamic game is defined as a mapping that prescribes at *every* information set of player i some available choice – including those information sets for player i that cannot be reached if the player implements the strategy! In the game of Figure 8.1, for instance, the pair (*accept*, 200) would be a strategy for Barbara in the *traditional* sense, stating that Barbara will accept the colleague's offer at the beginning of the game, but would choose a price of 200 if she were to reject the colleague's offer. But what is the meaning of 200 in this strategy? In our opinion, 200 has no meaning at all, since Barbara will never have to choose a price if she correctly implements this strategy – which is to accept the colleague's offer. In fact, we think that the choice 200 would only cause confusion on the reader's part, as he may start to wonder why we specify this choice of 200 at all, when it is clear that this choice can never occur if Barbara uses this strategy.

For that reason we have decided to adopt a different definition of a strategy for player i, which only prescribes a choice for player i at information sets that are *possible* if player i implements the strategy. In a sense we have removed the redundant information from a strategy in the traditional sense, and only kept those choices that are really needed to make a plan in a dynamic game. We think this corresponds more closely to the intuitive idea behind planning your choices in a dynamic game, which is what a strategy is supposed to do. Our definition of a strategy corresponds to the notion of a *plan of action* in Rubinstein (1991), who also offers a critical discussion of the traditional definition of a strategy in dynamic games. It should be mentioned, however, that for the formal analysis in this book it makes no difference whether one uses the traditional definition of a strategy or the definition we use.

Conditional beliefs

A major difference with static games is that in a dynamic game, a player will in general not just have a single belief, but at each of his information sets he will hold a separate conditional belief about the opponents' strategy choices. In particular, a player may have to *revise* his beliefs about the opponents' strategy choices if a new information set is reached. In the years before epistemic game theory, such conditional beliefs about the opponents' strategy choices were typically modeled by *behavioral strategies* – due to Kuhn (1953) – and *systems of beliefs* in the sense of Kreps and Wilson (1982).

Formally, a *behavioral strategy* σ_i for player i, as defined by Kuhn (1953), assigns to each of i's information sets $h \in H_i$ a probability distribution $\sigma_i(h)$ over i's available choices at h. Within the setting of this book, the probability distribution $\sigma_i(h)$ can naturally be interpreted as the conditional belief that i's opponents hold, at information sets *before* h, about i's *future* choice at h. In that sense, the behavioral strategy σ_i can be interpreted as the conditional beliefs that i's opponents hold, at their respective information sets, about i's *future* choices.

A *system of beliefs* for player i, as defined by Kreps and Wilson (1982), is a mapping μ_i that assigns to every information set $h \in H_i$ some probability distribution $\mu_i(h)$ on the set of non-terminal histories in h. Note that a non-terminal history is characterized by the sequence of past choices that leads to it. In that sense, $\mu_i(h)$ describes the conditional belief that player i holds at h about the opponents' *past* choices. Hence, a system of beliefs *à la* Kreps and Wilson (1982) describes the conditional beliefs that players have about the opponents' past choices. From now on we will refer to such systems of beliefs as *Kreps–Wilson beliefs*, to distinguish them from the conditional beliefs we use in this book.

Hence, if we put together the behavioral strategies σ_i and the Kreps–Wilson beliefs μ_i for all players i, then we describe for every player i and every information set $h \in H_i$, the conditional belief for player i about the opponents' *past* choices – given by $\mu_i(h)$ – and the conditional belief for player i about the opponents' *future* choices – given by the combination $(\sigma_j)_{j \neq i}$ of opponents' behavioral strategies. That is, the combination $((\sigma_i)_{i \in I}, (\mu_i)_{i \in I})$ fully describes the players' conditional beliefs about the opponents' past and future choices at all stages in the game. Such a combination $((\sigma_i)_{i \in I}, (\mu_i)_{i \in I})$ of behavioral strategies and Kreps–Wilson beliefs is called an *assessment* in Kreps and Wilson (1982).

In the definition of a conditional belief used in this book, we make no distinction between the conditional belief about the opponents' *past* choices and the conditional belief about the opponents' *future* choices. We just speak about the conditional belief that player i has at $h \in H_i$ about the opponents' *strategies*, which involves both past and future choices. This way of defining conditional beliefs was used by Pearce (1984), who called them *conjectures* instead of conditional beliefs. Nowadays it is common – especially within epistemic game theory – to define conditional beliefs as in this book, that is, by describing at every information set the probabilistic belief about the opponents' *full* strategies, without making a distinction between beliefs about past choices and beliefs about future choices.

Epistemic model

For the purposes of this book, we do not just describe the players' conditional beliefs about the opponents' strategy choices, but also the conditional beliefs about the opponents' conditional beliefs about their opponents' strategy choices, and so on. One way to do so – which is the road we have taken in this book – is to develop an epistemic model in which there is a set of types for every player, and every type holds, at each of his information sets, a conditional belief about the opponents' strategy choices *and types*. From such an epistemic model we can then derive, for every type, the complete conditional belief hierarchy – that is, the conditional beliefs about the opponents' strategies, the conditional beliefs about the opponents' conditional beliefs about their opponents' strategies, and so on. This epistemic model is based upon the models by Ben-Porath (1997) and Battigalli and Siniscalchi (1999). Ben-Porath (1997) proposed an epistemic model for dynamic games with *perfect information*, which is very similar to ours. The main difference is that Ben-Porath assumed within his model that all types satisfy *Bayesian updating* – a condition we do not impose within our model. Although Ben-Porath's model is restricted to games with perfect information, it can easily be extended to general dynamic games as well. Battigalli and Siniscalchi (1999) developed an epistemic model for general dynamic games which is very similar to our model and that of Ben-Porath. However, Battigalli and Siniscalchi do more, as they construct a *complete* – and in fact *terminal* – epistemic model for dynamic games. That is, they develop for a given dynamic game an epistemic model that contains *all* conditional belief hierarchies one can possibly think of. The construction of such a terminal epistemic model is analogous to the construction of terminal epistemic models for standard beliefs in static games, as we described in the literature section of Chapter 3. But the construction for dynamic games is more involved, as one has to take into account that the players hold a belief at each of their information sets.

The problem with "common belief in rationality" in dynamic games

For static games we explored the idea of *common belief in rationality*, which states that you believe that every opponent chooses rationally, believe that every opponent believes that each of his opponents chooses rationally, and so on. A crucial question is how we can extend this idea to dynamic games. As a first attempt, one could look for a concept which states that a player, at each of his information sets: (1) believes that his opponents have chosen rationally in the past and will choose rationally in the future, (2) believes that every opponent, at each of his information sets, believes that his opponents have chosen rationally in the past and will choose rationally in the future, and so on. Let us call this concept "common belief in past and future rationality." The problem with this concept is that in many dynamic games, it may not be possible to satisfy all of its requirements. That is, in many dynamic games this concept is simply too restrictive, as it will lead to logical contradictions at some of the information sets. Consider, for instance, the game in Figure 8.1 from the example "Painting Chris' house." Let h_1 be the information set where you must choose a price. Then, common belief in past and future rationality is simply not possible at h_1. If you believe at h_1 that Barbara believes that you will choose rationally at h_1, then you believe at h_1 that Barbara believes that you will not choose 500 at h_1. But then, you believe that it can never be optimal for Barbara to reject her colleague's offer. Hence, if information set h_1 is reached, and it has thus become clear that Barbara *did* reject her colleague's offer, then you can no longer believe that Barbara has chosen rationally in the past, and that Barbara believes that you will choose rationally at h_1. So, at h_1 you can no longer express common belief in past and future rationality.

This problem with extending the idea of "common belief in rationality" to dynamic games was investigated by Reny (1992a, 1993) for the special class of dynamic games with perfect information. Reny (1992a) focused on a particular game with perfect information, called "Take it or leave it," and demonstrated that in this game common belief in past and future rationality is not possible if player 1 leaves the dollar at the beginning of the game. What would player 2 believe about player 1's future choices and player 1's beliefs, once player 1 has decided to leave the dollar at the beginning? As Reny rightly points out, there is no unique obvious answer to that question. In this chapter we assume that in this case, player 2 would still believe that player 1 will choose rationally at all *future* stages, and that player 2 would still believe that player 1 will believe that player 2 will choose rationally at all *future* stages, and so on. But this is certainly not the only plausible way in which player 2 could react to player 1 leaving the dollar at the beginning. After all, we do not say in this chapter that people *must* reason in accordance with common belief in future rationality; we only investigate what would happen if people *were* to reason in accordance with common belief in future rationality. Reny's main message in his paper is that if common belief in past and future rationality is not possible at all stages of a game, then there is no obvious way for the players to reason within it. In Reny (1993) the discussion is extended to all two-player dynamic games with perfect information. More precisely, Reny (1993) showed that for a given two-player game with perfect information in which no two terminal histories yield the same utility for a player, common belief in past and future rationality is possible throughout the game, if and only if, the unique combination of backward induction strategies reaches *all* information sets in the game. Since this constitutes a very small class of games, Reny thus showed that there are only very few games with perfect information in which common belief in past and future rationality is possible at all stages. Other papers that address the difficulties in designing a sound concept of rationality for dynamic games include Binmore (1987), Bicchieri (1989) and Basu (1990).

Throughout this book we assume that the players' utility functions are completely transparent to everyone, and that players in a dynamic game will never revise their belief about the opponents' utility functions. So, even if a player observes a surprising choice by an opponent, it will never be a reason for him to change his mind about the opponent's utility function. There are models that allow players in a dynamic game to change their beliefs about the opponents' utility functions as the game unfolds. See, for instance, Perea (2006, 2007b). In such a setting, common belief in past and future rationality will always be possible, as a player can always interpret observed past choices by an opponent as *optimal* choices by changing his belief about the opponent's utility function if necessary.

Belief in the opponents' future rationality

The main idea in this chapter is that of *believing in the opponents' future rationality*, which states that at each of your information sets you believe that your opponents will choose rationally now and in the future. At the same time, we do not impose any restriction on the beliefs you hold about your opponents' *past* choices. This idea has been floating around in the game theory literature for a long time, and has implicitly been assumed in most of the more traditional concepts for dynamic games, such as *subgame perfect equilibrium* (Selten, 1965) and *sequential equilibrium* (Kreps and Wilson, 1982). We will comment more on these concepts below. The idea of believing in the opponents' future rationality has also been incorporated – either explicitly or implicitly – in the various epistemic foundations for the *backward induction procedure* that have been proposed in the literature. We will also discuss these foundations in more detail below. Perea (2011b) explicitly formulated the idea of belief in the opponents' future rationality for *general* dynamic games, and incorporated it into the concept of *common belief in future rationality* which is the basis for this chapter. Common belief in future rationality is very closely related to *sequential rationalizability* as developed by Dekel, Fudenberg and Levine (1999, 2002) and Asheim and Perea (2005). As we will see below, sequential rationalizability is more restrictive than common belief in future rationality, but they essentially reveal the same idea.

Algorithm

In this chapter we developed an algorithm – the *backward dominance procedure* – which always yields precisely those strategies that can rationally be chosen under common belief in future rationality. Both the backward dominance procedure and Theorem 8.8.3 – stating that the algorithm works for common belief in future rationality – are taken from Perea (2011b). The backward dominance procedure is closely related to Penta's (2009) *backwards rationalizability* procedure. The essential difference between the two procedures is that the backwards rationalizability procedure imposes Bayesian updating whereas the backward dominance procedure does not.

More precisely, the backwards rationalizability procedure inductively generates, at every step k, some set B_i^k of conditional belief vectors for player i, and for every information set $h \in H$ some set $S_i^k(h)$ of strategies for every player i. At the beginning, let B_i^0 be the set of all conditional belief vectors for player i that satisfy Bayesian updating, and let $S_i^0(h)$ be the set of all strategies for player i that lead to h. At every further step k, let B_i^k be the set of conditional belief vectors in B_i^{k-1} that assign, at every $h \in H_i$, only positive probability to opponents' strategies in $S_j^{k-1}(h)$, and let $S_i^k(h)$ be the set of strategies in $S_i^{k-1}(h)$ that are optimal, at every $h' \in H_i$ weakly following h, for some conditional belief vector in B_i^k.

A strategy s_i is said to survive the backwards rationalizability procedure if $s_i \in S_i^k(\emptyset)$ for every k. If we omit the condition that every conditional belief vector in B_i^0 must satisfy Bayesian updating, then the procedure above is equivalent to the backward dominance procedure. In that case, the set of strategies surviving the procedure would be identical to the set of strategies surviving the backward dominance procedure. However, we have seen in Section 8.13 of this chapter that the requirement of (common belief in) Bayesian updating changes the strategy choices that can rationally be chosen under common belief in future rationality. Consequently, adding the condition of (common belief in) Bayesian updating to the backward dominance procedure – which would lead to the backwards rationalizability procedure – has important consequences for the strategies that survive the procedure. For instance, in the game of Figure 8.14 in Section 8.13, we saw that player 2's strategy (c,h) can rationally be chosen under common belief in future rationality, but without insisting on Bayesian updating. That is, strategy (c,h) survives the backward dominance procedure. But it may be checked that (c,h) does *not* survive the backwards rationalizability procedure.

Chen and Micali (2011) presented an algorithm, resulting in the *backward-robust solution* to a dynamic game, that is equivalent to the backward dominance procedure with the backwards order of elimination, as discussed in Section 8.10 of this chapter.

Subgame perfect equilibrium

As we mentioned above, the idea of believing in the opponents' future rationality was implicitly present in the concept of *subgame perfect equilibrium*, as developed in Selten (1965). This concept is especially appropriate for dynamic games with observed past choices, so we will define the concept only for such games. Consider a dynamic game with observed past choices, and take for every player i some behavioral strategy σ_i, assigning to every information set $h \in H_i$ some probability distribution $\sigma_i(h)$ on i's available choices at h. For every information set h in the game, let σ_i^h be the restriction of σ_i to information sets $h' \in H_i$ that weakly follow h. The combination $(\sigma_i)_{i \in I}$ of behavioral strategies is a *subgame perfect equilibrium* if for every information set h in the game, the combination $(\sigma_i^h)_{i \in I}$ of restricted behavioral strategies is a Nash equilibrium in the subgame that starts at h. That is, at every information set h in the game and for every player i, the restricted behavioral strategy σ_i^h only assigns positive probability to choices for player i that are optimal in the subgame starting at h, given the opponents' restricted behavioral strategies $(\sigma_j^h)_{j \neq i}$ in that subgame.

Remember from above that the probability distribution $\sigma_i(h)$ on i's choices at h can be interpreted as the beliefs that i's opponents hold, at information sets weakly before h, about i's *future* choice at h. In that light, the restricted behavioral strategy

σ_i^h in the definition of a subgame perfect equilibrium can be interpreted as the beliefs that i's opponents hold at h about i's future choices in the subgame that follows h. Moreover, the opponents' restricted behavioral strategies $(\sigma_j^h)_{j \neq i}$ in that subgame can be seen as i's beliefs about the opponents' future choices in that subgame. Hence, if $(\sigma_i)_{i \in I}$ is a subgame perfect equilibrium then, for every information set h, player i's opponents believe that player i will only make choices in the future which are optimal, in the subgame starting at h, given the beliefs that player i holds about the opponents' future choices. That is, in a subgame perfect equilibrium i's opponents will always believe in i's future rationality. As this holds for every player i, we conclude that in a subgame perfect equilibrium every player always believes in his opponents' future rationality. More than this, in a subgame perfect equilibrium every player always expresses *common belief in future rationality*. So, subgame perfect equilibrium is more restrictive than common belief in future rationality within the class of dynamic games with observed past choices. In fact, there are dynamic games with observed past choices where some strategy can rationally be chosen under common belief in future rationality, but which is not optimal in any subgame perfect equilibrium. Hence, subgame perfect equilibrium can rule out strategy choices that are allowed by common belief in future rationality.

In fact, for dynamic games with observed past choices, the difference between subgame perfect equilibrium and common belief in future rationality is analogous to the difference between Nash equilibrium and common belief in rationality as discussed in Chapter 4. For instance, in subgame perfect equilibrium a player is required to believe that his opponents have correct beliefs about his own beliefs, whereas this is not required under common belief in future rationality. Moreover, in games with more than two players, subgame perfect equilibrium requires that player i's belief about j's future choices is independent of i's belief about k's future choices – something that is not assumed by common belief in future rationality. As for Nash equilibrium, we believe that subgame perfect equilibrium is often too restrictive, as it may rule out strategy choices which seem perfectly reasonable if one takes belief in the opponents' future rationality as a starting point. Especially the requirement that a player must believe that his opponents hold correct beliefs about his own beliefs seems far too restrictive, and is certainly not a basic condition to impose.

Sequential equilibrium

Kreps and Wilson (1982) extended the main idea behind subgame perfect equilibrium to general dynamic games – not necessarily with observed past choices – which finally resulted in the concept of *sequential equilibrium*. Within a general dynamic game, consider for every player i a behavioral strategy σ_i – assigning to every information set $h \in H_i$ some probability distribution $\sigma_i(h)$ on i's choices at h – and consider, in addition, a *system of beliefs* μ_i which assigns to every information set $h \in H_i$ some probability distribution $\mu_i(h)$ on the non-terminal histories within h. The interpretation is that $\mu_i(h)$ describes player i's conditional belief at h about the opponents' past choices. As before, $\sigma_i(h)$ may be interpreted as the beliefs that i's opponents have, at information sets before h, about i's *future* choice at h. The combination $((\sigma_i)_{i\in I}, (\mu_i)_{i\in I})$ of behavioral strategies and systems of beliefs is called an *assessment.*

The assessment $((\sigma_i)_{i\in I}, (\mu_i)_{i\in I})$ is *sequentially rational* if for every player i, the behavioral strategy σ_i assigns, at every $h \in H_i$, only positive probability to i's choices that are optimal, given the belief $\mu_i(h)$ that i holds at h about the opponents' past choices, and the belief $(\sigma_j)_{j\neq i}$ that i holds at h about the opponents' future choices. Now, sequential rationality is just a way of expressing belief in the opponents' future rationality. Namely, $\sigma_i(h)$ represents the belief that i's opponents have, at information sets before h, about i's future choice at h. So, if $((\sigma_i)_{i\in I}, (\mu_i)_{i\in I})$ is sequentially rational, then for every player i and every information set $h \in H_i$, the opponents of player i believe prior to information set h that i will choose optimally at h, given the belief that i holds at h about the opponents' past and future choices. That is, i's opponents believe before information set h that player i will choose rationally at information set h. Since this holds for every player i and every information set $h \in H_i$, sequential rationality implies that players always believe in the opponents' future rationality.

The assessment $((\sigma_i)_{i\in I}, (\mu_i)_{i\in I})$ is *consistent* if there is a sequence of assessments $((\sigma_i^n)_{i\in I}, (\mu_i^n)_{i\in I})_{n\in\mathbb{N}}$ converging to it, such that for every $n \in \mathbb{N}$ the following holds: (1) the behavioral strategies σ_i^n assign positive probability to *all* choices in the game, and (2) the belief systems $(\mu_i^n)_{i\in I}$ are derived from $(\sigma_i^n)_{i\in I}$ by Bayesian updating. Condition (2) implies that the conditional beliefs of the players about the opponents' strategy choices satisfy Bayesian updating. So, consistency in the sense above implies Bayesian updating, but is actually stronger than this.

Finally, an assessment $((\sigma_i)_{i\in I}, (\mu_i)_{i\in I})$ is a *sequential equilibrium* if it is both sequentially rational and consistent. It can be shown that for dynamic games with observed past choices, sequential equilibrium is equivalent to subgame perfect equilibrium. More precisely, for such games a behavioral strategy profile $(\sigma_i)_{i\in I}$ is a subgame perfect equilibrium, if and only if, it can be extended to a sequential equilibrium by adding an appropriate system of beliefs. However, sequential equilibrium can be more restrictive than subgame perfect equilibrium if the game contains unobserved past choices.

The most important ingredient for sequential equilibrium is the sequential rationality requirement, which – we have seen – corresponds to imposing belief in the opponents' future rationality throughout the game. In fact, the sequential rationality requirement makes sure that the players always express *common belief in future rationality.* So, sequential equilibrium implies common belief in future rationality. The converse, however, is not true. There are games where some strategy can rationally be chosen under common belief in future rationality, but which is not optimal if the conditional beliefs of the players constitute sequential equilibrium. In fact, the difference between a sequential equilibrium and common belief in future rationality is essentially analogous to the difference between subgame perfect equilibrium and common belief in future rationality. The same critique that we had concerning subgame perfect equilibrium also applies to sequential equilibrium: the concept may in some games rule out strategies that seem perfectly reasonable if one uses belief in the opponents' future rationality as the basis.

Some authors have provided weaker forms of the sequential equilibrium concept by relaxing the somewhat complicated consistency requirement above. See, for instance, Fudenberg and Tirole (1991b) and Bonanno (2010).

Sequential rationalizability
The concept of *sequential rationalizability* was proposed independently by Dekel, Fudenberg and Levine (1999, 2002) and Asheim and Perea (2005). Although they formulated the concept in very different ways, both formulations finally lead to the same strategy choices for each player. We will now use the formulation by Dekel, Fudenberg and Levine to describe the concept. For every player i, consider a *strategy-belief triple* $(\sigma_i, \mu_i, \sigma^i_{-i})$, where σ_i is a behavioral strategy for player i, μ_i is a system of beliefs for player i in the sense of Kreps and Wilson (1982) and σ^i_{-i} is a combination of behavioral strategies for i's opponents. As before, the interpretation is that μ_i describes i's conditional beliefs about the opponents' past choices, whereas σ^i_{-i} represents i's conditional beliefs about the opponents' future choices. Now take for every player i some set V_i of strategy-belief triples. The combination $(V_i)_{i\in I}$ of sets of strategy-belief triples is called *sequentially rationalizable* if for every player i and every strategy-belief triple $(\sigma_i, \mu_i, \sigma^i_{-i})$ in V_i, we have: (1) (μ_i, σ^i_{-i}) is *consistent* in the sense of Kreps and Wilson (1982) above, (2) for every information set $h \in H_i$, the behavioral strategy σ_i only assigns positive probability to choices that are optimal at h, given the conditional belief $\mu_i(h)$ about the opponents' past choices, and the belief σ^i_{-i} about the opponents' future choices, and (3) the belief σ^i_{-i} about the opponents' future choices only assigns positive probability to opponents' choices c_j that have positive probability in some behavioral strategy σ_j that forms part of a strategy-belief triple $(\sigma_j, \mu_j, \sigma^j_{-j})$ in V_j.

By combining the conditions (2) and (3) we see that for every strategy-belief triple $(\sigma_i, \mu_i, \sigma^i_{-i})$ in V_i, the belief σ^i_{-i} about the opponents' future choices only assigns positive probability to opponents' choices c_j that are optimal, for their respective information set $h \in H_j$, given the conditional belief at h about the opponents' past and future choices. Hence, one could say that in every strategy-belief triple $(\sigma_i, \mu_i, \sigma^i_{-i})$ in V_i, the belief σ^i_{-i} about the opponents' future choices "expresses belief in the opponents' future rationality." As every such belief σ^i_{-i} only refers to opponents' choices c_j that "are part" of some strategy-belief triple $(\sigma_j, \mu_j, \sigma^j_{-j})$ in V_j, it can in fact be shown that in every strategy-belief triple $(\sigma_i, \mu_i, \sigma^i_{-i})$ in V_i, the belief σ^i_{-i} expresses common belief in future rationality.

A strategy s_i is *sequentially rationalizable* if there is some sequentially rationalizable combination $(V_i)_{i\in I}$ of sets of strategy-belief triples, and some strategy-belief triple $(\sigma_i, \mu_i, \sigma^i_{-i})$ in V_i, such that every choice in s_i has positive probability in σ_i. In view of the insights above, we can conclude that every sequentially rationalizable strategy can rationally be chosen under common belief in future rationality. The converse is not true, however. In the game of Figure 8.14 in Section 8.13, for instance, the strategy (c, h) can rationally be chosen by player 2 under common belief in future rationality, but (c, h) is not sequentially rationalizable. The reason is that sequential rationalizability assumes (common belief in) Bayesian updating – because of condition (1) above – and we have seen in Section 8.13 that under common belief in future rationality and common belief in Bayesian updating, player 2 cannot rationally choose (c,h). Hence, in terms of strategies selected, sequential rationalizability is more restrictive than common belief in future rationality. The essential difference between the two concepts is that sequential rationalizability assumes (common belief in) Bayesian updating and assumes that the players hold independent beliefs about the opponents' future choices if there are more than two players, whereas common belief in future rationality does not impose any of these conditions.

Backward induction

The class of dynamic games with perfect information – in which the players choose sequentially and a player always knows all choices made by the opponents in the past – is an important special class of games with many applications. In Section 8.11 we saw that for such games, common belief in future rationality selects precisely those strategies that survive the *backward induction procedure.* Many people – including myself – have claimed that Zermelo (1913) was the first one to introduce the backward induction procedure, to prove his famous theorem about chess. But I recently discovered – through the paper of Schwalbe and Walker (2001) – that this is not true! Zermelo (1913) modeled the game of chess as a *potentially infinite* game that could go on forever so long as nobody achieves a winning position. Therefore, certain parts of the game do not have ultimate information sets needed to start the backward induction procedure, and as such the backward induction procedure cannot be applied to his version of the game of chess. Instead, Zermelo used a purely set-theoretic proof to show his theorem about chess, which states that: (1) either white has a strategy that guarantees a win, or (2) black has a strategy that guarantees a win, or (3) both white and black have strategies that guarantee at least a draw. Schwalbe and Walker (2001) give an English translation of Zermelo's original paper, and discuss his paper in depth.

To the best of my knowledge, von Neumann and Morgenstern were the first to use the backward induction procedure, in the third edition of their book *Theory of Games and Economic Behavior* from 1953. In their Chapter 15, they used backward induction to prove that every two-person zero-sum dynamic game with perfection information is strictly determined.

For a long time, backward induction has been used by game theorists and other scientists to analyze dynamic games with perfect information, without knowing precisely which patterns of reasoning lead to backward induction. In particular, when a player observes an unexpected move by an opponent, how should this player reason about the opponent's future moves and beliefs so as to choose in accordance with backward induction? Or, more generally, which conditions must we impose upon a player's conditional belief hierarchy such that this player will choose in accordance with backward induction? By answering this question, one would be providing an *epistemic foundation* for backward induction.

Starting with Aumann (1995), different authors have provided different epistemic foundations for backward induction. See, for instance, Samet (1996), Balkenborg and Winter (1997), Stalnaker (1998), Asheim (2002), Quesada (2002, 2003), Clausing (2003, 2004), Asheim and Perea (2005), Feinberg (2005), Perea (2008), Baltag, Smets and Zvesper (2009) and Bach and Heilmann (2011). Perea (2007c) provided an overview of most of these epistemic foundations and compared the conditions that these various foundations impose on the conditional belief hierarchy of a player. Most of these foundations assume – either explicitly or implicitly – that the conditional belief hierarchy of a player expresses common belief in future rationality or some variant thereof. Almost all of the foundations above focus on so-called *generic* games with perfect information, in which different terminal histories for a given player induce different utilities. In such games, the backward induction procedure always selects a unique choice for every information set, and hence selects a unique backward induction strategy for each player. In contrast, our definition of the backward induction procedure in Section 8.11 applies to *all* games with perfect information, also non-generic ones.

Backward induction paradoxes

In some particular dynamic games, the backward induction procedure – and hence common belief in future rationality – selects strategies that seem somewhat counter-intuitive. Famous examples of this kind are Rosenthal's (1981) centipede game, the finitely repeated prisoner's dilemma as discussed in Kreps *et al.* (1982), Selten's (1978) chain-store paradox, and Reny's (1992a) "Take it or leave it" game. See Binmore (1996) and Aumann (1996, 1998), amongst others, for further discussions of the centipede game. It should be noted that the finitely repeated prisoner's dilemma is not a game with perfect information, but still this game can be analyzed by a backward induction-like procedure.

A common feature of the examples above is that the underlying game has a highly repetitive structure. Now suppose that you are a player in such a game and observe that your opponent has consistently made non-backward induction choices in the past. Would you then believe that this same opponent will make backward induction choices in the remainder of the game? Probably not. But the backward induction procedure assumes that even in such situations you would still believe that your opponent will make backward induction choices in the future, even though this seems somewhat counter-intuitive. This is what generates the "backward induction paradoxes" in these examples.

The same holds for common belief in future rationality in general dynamic games. Suppose you observe that your opponent has consistently not acted in accordance with common belief in future rationality. Would you still believe that this opponent will choose in accordance with common belief in future rationality in the remainder of the game? Not necessarily, although common belief in future rationality suggests that even under such circumstances one should not drop one's belief that this opponent will choose in accordance with common belief in future rationality in the game that lies ahead. This shows again that – at least in our view – there is no such thing as an ultimate concept in game theory. There will always be problems with every concept.

Common initial belief in rationality

Ben-Porath (1997) proposed a concept for dynamic games with perfect information that is weaker than common belief in future rationality. It states that a player believes, at the beginning of the game: (1) that every opponent will choose rationally at every information set, (2) that every opponent believes, at the beginning of the game, that each of his opponents will choose rationally at every information set, and so on. Ben-Porath calls this concept *common certainty of rationality at the beginning of the game,* but we will refer to it as *common initial belief in rationality.* This concept can easily be extended to *general* dynamic games. See theoretical problem 9.9 in Chapter 9 for a more restrictive notion, which we call *common belief in "initial belief in rationality,"* and which applies to all dynamic games.

Implicit in the above concept is the assumption that players do not just have conditional beliefs at each of their information sets, but also hold initial beliefs at the beginning of the game. If, in this chapter, we allowed players to also hold initial beliefs at the beginning of the game, then it could be verified that common belief in future rationality is more restrictive than common initial belief in rationality.

Ben-Porath (1997) showed that in *generic* games with perfect information, the strategies that can rationally be chosen under common initial belief in rationality are precisely those strategies that survive the Dekel–Fudenberg procedure when applied to $\Gamma^0(\emptyset)$. Remember that $\Gamma^0(\emptyset)$ represents the full decision problem at the beginning of the game, and that the Dekel–Fudenberg procedure starts by eliminating all weakly dominated strategies from $\Gamma^0(\emptyset)$, and then proceeds by iteratively eliminating strictly dominated strategies. For generic games with perfect information this procedure is less restrictive than the backward dominance procedure. That is, in such games, every strategy that survives the backward dominance procedure also survives the Dekel–Fudenberg procedure at $\Gamma^0(\emptyset)$, but the converse is not true in general. As a consequence, in every generic game with perfect information, every strategy that can rationally be chosen under common belief in future rationality can also rationally be chosen under common initial belief in rationality. In theoretical problem 9.9 in Chapter 9 the reader is asked to show that this is true in general – for every dynamic game.

Reny (1992b) proposed *weak sequential rationality*, which is similar in spirit to common initial belief in rationality, but phrased in terms of behavioral strategies and beliefs in the sense of Kreps and Wilson (1982). Consider an assessment $((\sigma_i)_{i\in I}, (\mu_i)_{i\in I})$, consisting of a combination $(\sigma_i)_{i\in I}$ of behavioral strategies and a system of beliefs $(\mu_i)_{i\in I}$ *à la* Kreps and Wilson (1982). The assessment is *weakly sequentially rational* if for every player i, the behavioral strategy σ_i assigns, at every $h \in H_i$ *that can be reached under* σ_i, only positive probability to i's choices that are optimal at h, given the belief $\mu_i(h)$ that i holds at h about the opponents' past choices, and the belief $(\sigma_j)_{j\neq i}$ that i holds at h about the opponents' future choices. So, the difference from sequential rationality as defined by Kreps and Wilson (1982) is that the optimality condition for

σ_i is now only imposed at those information sets $h \in H_i$ that can actually be reached by σ_i, whereas sequential rationality also imposes such an optimality condition for information sets $h \in H_i$ that are avoided by σ_i. Remember that σ_i can be interpreted as the belief that i's opponents hold about i's future choices. Now, let $H_i(\sigma_i)$ be the collection of information sets $h \in H_i$ that can actually be reached by σ_i. Then, the restriction of σ_i to information sets in $H_i(\sigma_i)$ can be viewed as the *initial* belief that i's opponents hold about i's strategy choice. The weak sequential rationality condition thus states that the initial belief about i's strategy choice must only assign positive probability to i's choices that are optimal – that is, there must be initial belief in i's rationality. More than this, the weak sequential rationality condition actually implies that there is *common* initial belief in rationality. In that sense, the weak sequential rationality condition is very similar to common initial belief in rationality.

Probability 1 beliefs in games with perfect information

In Lemma 8.11.4 we showed that in games with perfect information, every strategy that is optimal at a given information set for some probabilistic belief about the opponents' strategy choices, is also optimal for a belief that assigns probability 1 to one particular strategy combination of the opponents. This lemma was important for showing that in games with perfect information, the backward induction procedure selects precisely those strategies that can rationally be chosen under common belief in future rationality. Lemma 8.11.4 is based on Lemma 1.2.1 in Ben-Porath (1997) and the associated proof is largely based on Ben-Porath's proof.

One-deviation property

In the proofs section of this chapter we proved Lemma 8.14.1, which states that for every conditional belief vector that satisfies Bayesian updating there will always be a strategy that is optimal at every information set. The proof starts by recursively constructing, for every information set h for player i, some choice $c_i^*(h)$, such that $c_i^*(h)$ is optimal at h, given i's conditional belief at h about the opponents' strategy choices, and given his optimal choices $c_i^*(h')$ in the future. We say that $c_i^*(h)$ is *locally* optimal at h. The second step in the proof showed that the strategy s_i^* so constructed is not only *locally* optimal at every information set, but is in fact (globally) optimal at every information set $h \in H_i$ that s_i^* leads to. The latter result – which holds whenever the conditional belief vector satisfies Bayesian updating – is known as the *one-shot deviation principle* or *one-deviation property*. That is, if the conditional belief vector for player i about the opponents' strategy choices satisfies Bayesian updating, then every strategy for player i that is locally optimal at every information set, is also (globally) optimal at every information set. Here, the difference between local and global optimality is that under local optimality we see whether a given *choice* for player i is optimal at an information set, fixing his choices at future information sets, whereas under global optimality we verify whether player i's *complete strategy* – including his future choices – is optimal at an information set. So, local optimality is always implied by global optimality, but the converse is not always true. The one-deviation property states that under Bayesian updating the converse is also true – that is, under Bayesian updating, local optimality also implies global optimality. See Hendon, Jacobsen and Sloth (1996) and Perea (2002) for a discussion and proof of the one-deviation property. The proof of Lemma 8.14.1 is based on the proofs in Hendon, Jacobsen and Sloth (1996) and Perea (2002).

Examples and exercises

Example 8.7, "Bargaining with commitment," is based on a bargaining procedure discussed in Miettinen and Perea (2010). Example 8.8, "The shrinking treasure," basically resembles the alternating offers bargaining procedure as proposed by Ståhl (1972) and Rubinstein (1982). The practical problem 8.3, "A nightmare with sharks," is also known as the pirates example. Practical problem 8.5, "The street musicians," is based upon a bargaining procedure proposed in Navarro and Perea (2011). Finally, theoretical problem 8.8, "Strategy subsets closed under belief in future rationality," is based on Perea (2011b).

9 Strong belief in the opponents' rationality

9.1 Strong belief in the opponents' rationality

In the previous chapter we introduced the idea of *common belief in future rationality* for dynamic games, which means that you always believe that your opponents choose rationally now and in the future, you always believe that your opponents always believe that their opponents choose rationally now and in the future, and so on. This concept thus puts no restrictions on what you think about the opponents' choices that were made in the past – in fact you are free to conclude anything you want from what your opponents have done so far, as long as you still believe that these opponents will choose rationally now and in the future, that you still believe that these opponents also reason in this way about *their* opponents, and so on. So, common belief in future rationality makes a very sharp distinction between reasoning about the past and reasoning about the future: Anything goes for reasoning about the past, but very severe conditions are imposed on how you reason about the future.

In many dynamic situations, this is not the only plausible way to reason about your opponents. It often makes intuitive sense to also think critically about what your opponent has done so far, and to use his past behavior to draw conclusions about how he may act now and in the future. To illustrate this, let us go back to the example "Painting Chris' house."

Example 9.1 Painting Chris' house

Recall the story from Example 8.1 in the previous chapter. For convenience, we have reproduced the graphical representation of the game in Figure 9.1. Let us first repeat how common belief in future rationality applies to this game. Suppose you observe that Barbara has rejected her colleague's offer. If you believe that Barbara will choose rationally afterwards, then you believe that she will certainly not choose a price of 500, as it is strictly dominated for her by the randomized choice in which she chooses the prices 200 and 400 with probability 0.5. However, if you believe that Barbara will not choose a price of 500, then 400 and 500 can no longer be optimal prices for you. Hence, if Barbara believes that you choose rationally, and that you believe in Barbara's future rationality, then Barbara believes that you will not choose prices 400 and 500. But then,

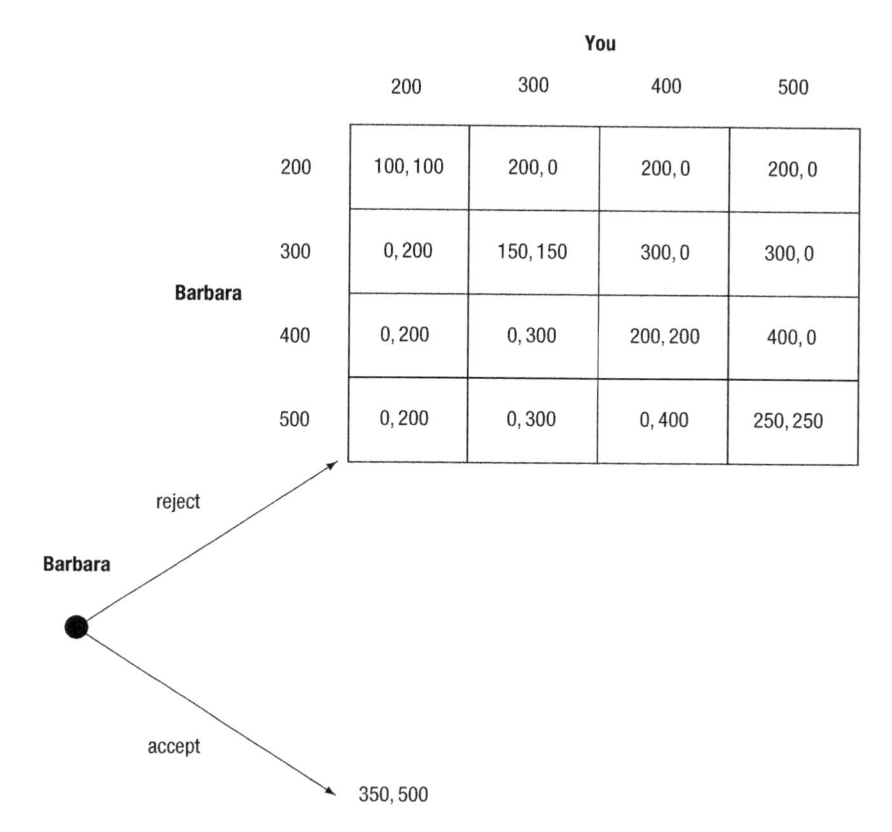

Figure 9.1 Painting Chris' house (II)

only a price of 200 can be optimal for Barbara after rejecting her colleague's offer. That is, if you believe in Barbara's future rationality, believe that Barbara believes in your rationality, and believe that Barbara believes that you believe in her future rationality, then you believe, upon observing that Barbara has rejected her colleague's offer, that Barbara will choose a price of 200. Therefore, you will choose a price of 200 as well. As such, your only rational strategy under common belief in future rationality is to choose a price of 200.

Consequently, under common belief in future rationality you are led to conclude that Barbara has made an *irrational* choice when you see that she has rejected her colleague's offer! Under common belief in future rationality, namely, you believe that Barbara believes that you will choose a price of 200. But then, you believe that Barbara would do best by accepting her colleague's offer, so if you observe that she has rejected the offer you conclude that she must have made a mistake.

You could also reason differently here. On observing Barbara reject her colleague's offer, you could alternatively ask whether this decision could be part of a *rational* strategy for Barbara, that is, a strategy that is also optimal at the beginning of the game for some belief about your price choice. If so, then you believe, on observing her reject the offer, that she is indeed using such a rational strategy that includes rejecting the colleague's offer as a rational choice. We say that you *strongly believe in Barbara's rationality.*

Let us investigate the consequences of *strong belief in Barbara's rationality* for you. For Barbara, rejecting the colleague's offer can only be part of a rational strategy if she were to choose a price of 400 afterwards. Only then will she have a chance of receiving more than the 350 euros she would get by accepting her colleague's offer. So, if you strongly believe in Barbara's rationality, then you must conclude, after observing Barbara reject the offer, that she will continue by choosing a price of 400, as this is the only way for her to turn it into a rational strategy. But then, you will choose a price of 300, and not 200 as you would under common belief in future rationality! So, under strong belief in Barbara's rationality you would choose differently than under common belief in future rationality.

Again, it is difficult to say which of these two lines of reasoning is more appealing – they are simply different, and in our opinion both make a lot of intuitive sense. So, we give both patterns of reasoning next to each other in this book, and it is up to the reader to decide which line of reasoning is more attractive. □

In the example above we described the idea of *strongly believing in Barbara's rationality*, which means that *if* it is possible for you to believe that rejecting her colleague's offer is part of a rational strategy for Barbara, then on observing that she did reject the offer, you *must* believe that Barbara has a rational strategy that includes rejecting the offer as a rational choice. We can generalize this idea to arbitrary dynamic games in the following way: Suppose that player i finds himself at an information set $h \in H_i$, and that he reasons about the opponents' strategy choices. We say that player i *strongly believes in the opponents' rationality* at information set h if, *whenever there is* a combination of opponents' strategies that leads to h and that is optimal at all of opponents' information sets, player i *must* at h *only assign positive probability* to these optimal strategies. That is, if it is possible for player i to believe at h that each of his opponents chooses a rational strategy, he *must* believe at h that every opponent chooses a rational strategy.

Note that it may not always be possible at a player to believe, at a given information set, that his opponents are using rational strategies. Consider, for instance, the example "Painting Chris' house," and suppose that Barbara could earn 450 euros (instead of 350) by accepting her colleague's offer. Then, on observing Barbara reject her colleague's offer, it is no longer possible for you to believe that Barbara is using a rational strategy. In that case, strong belief in Barbara's rationality would not impose any restrictions on what you believe after observing Barbara reject the offer. In general, if it is not possible for player i to believe at information set $h \in H_i$ that each of his opponents chooses a rational strategy, then strong belief in the opponents' rationality does not restrict player i's belief at information set h at all.

Table 9.1. *An epistemic model for "Painting Chris' house" (II)*

Types	$T_1 = \{t_1\}, T_2 = \{t_2\}$
Beliefs for Barbara	$b_1(t_1, \emptyset) = (200, t_2)$ $b_1(t_1, h_1) = (200, t_2)$
Beliefs for you	$b_2(t_2, h_1) = ((reject, 200), t_1)$

Let us now see how we can formalize the idea of strong belief in the opponents' rationality, which we have described somewhat informally above. Let us model the players' conditional belief hierarchies formally by means of an epistemic model M, in which T_i is the set of types for player i, and $b_i(t_i, h)$ specifies t_i's conditional belief at information set $h \in H_i$ about the opponents' strategy-type combinations. How can we formalize within the epistemic model M the requirement that, *if* it is possible for player i to believe at h that every opponent chooses a rational strategy, then player i *must* at h only assign positive probability to opponents' rational strategies? As a first attempt, we could try the following condition: Consider a type t_i within the epistemic model M and an information set $h \in H_i$. We say that type t_i strongly believes in the opponents' rationality at h if:

- *whenever* there is an opponents' strategy-type combination in M where: (a) the opponents' strategy combination leads to h, and (b) for every opponent j, the strategy is optimal for the type at every $h' \in H_j$ that the strategy leads to, then
- type t_i *must* at h only assign positive probability to strategy-type combinations in M that satisfy conditions (a) and (b).

However, this condition is not enough to correctly model strong belief in the opponents' rationality. Let us return to the example "Painting Chris' house" with the game in Figure 9.1 and consider the epistemic model M in Table 9.1. Here, h_1 is the information set after Barbara rejects her colleague's offer. Barbara is player 1, whereas you are player 2. Within the epistemic model M, there is no type for Barbara for which rejecting the offer is part of an optimal strategy. The only type t_1 for Barbara in M believes that you will choose a price of 200, and hence rejecting the offer is not optimal for t_1. So, the condition above does not restrict your belief at information h_1 within the epistemic model M.

However, it is clear that there *is* a type for Barbara, outside M, for which rejecting the offer is part of an optimal strategy – namely any type for Barbara that assigns a large enough probability to you choosing a price of 500. So, strong belief in the opponents' rationality means in this case that you *must* believe at h_1 that Barbara is of such a type for which rejecting the offer is part of an optimal strategy. The problem, however, is that such a type for Barbara is not included in this epistemic model, which prevents us from correctly implementing the idea of strong belief in the opponents' rationality.

In order to solve this problem we must require that the epistemic model contains "sufficiently many types" for the players. That is, we must add the following condition to the one above:

- If at information set $h \in H_i$ we can find a combination of opponents' types, possibly outside M, for which there is a combination of optimal strategies leading to h, then M must contain at least one such combination of types.

Together with the condition above, we arrive at the following formal definition of *strong belief in the opponents' rationality.*

Definition 9.1.1 *(Strong belief in the opponents' rationality)*
Consider an epistemic model M, a type t_i for player i within M, and an information set $h \in H_i$. Type t_i ***strongly believes in the opponents' rationality*** *at h if, whenever we can find a combination of opponents' types, possibly outside M, for which there is a combination of optimal strategies leading to h, then:*
(1) the epistemic model M must contain at least one such combination of types, and
(2) type t_i must at h only assign positive probability to opponents' strategy-type combinations where the strategy combination leads to h and the strategies are optimal for the types.
Finally, type t_i is said to strongly believe in the opponents' rationality if he does so at every information set $h \in H_i$.

Here, whenever we say that a strategy s_j is optimal for a type t_j, we mean that s_j is optimal for t_j at every information set $h \in H_j$ that s_j leads to. Note that there is a remarkable similarity between this definition and the one we gave for *assuming the opponent's rationality* in Chapter 7. In both definitions we require the epistemic model to contain sufficiently many types – condition (1) in both definitions – otherwise the definitions "do not work" as we have seen.

With this formal definition it can now easily be verified that in the epistemic model of Table 9.1, your type t_2 does not strongly believe in Barbara's rationality. We have seen that for Barbara we can find a belief hierarchy for which there is an optimal strategy leading to h_1, namely any belief hierarchy that assigns a large enough probability to you choosing a price of 500. However, the epistemic model M does not contain any type for Barbara for which there is an optimal strategy leading to h_1, and hence condition (1) in the definition of strong belief in the opponents' rationality is violated.

Consider now another epistemic model for "Painting Chris' house" shown in Table 9.2. We will verify that within this new epistemic model M, your type t_2 strongly believes in Barbara's rationality. Note that for Barbara's type t_1^r there is an optimal strategy leading to h_1, namely the strategy (*reject*,400). Since this type t_1^r is contained in M, condition (1) in the definition of strong belief in Barbara's rationality is satisfied. Your type t_2 assigns at h_1 probability 1 to Barbara's strategy-type pair ((*reject*,400), t_1^r),

Table 9.2. *Another epistemic model for "Painting Chris' house"*

Types	$T_1 = \{t_1^a, t_1^r\}$, $T_2 = \{t_2\}$
Beliefs for Barbara	$b_1(t_1^a, \emptyset) = (300, t_2)$ $b_1(t_1^a, h_1) = (300, t_2)$ $b_1(t_1^r, \emptyset) = (500, t_2)$ $b_1(t_1^r, h_1) = (500, t_2)$
Beliefs for you	$b_2(t_2, h_1) = ((reject, 400), t_1^r)$

where the strategy (*reject*,400) leads to h_1, and is optimal for type t_1^r. Hence, your type t_2 also satisfies condition (2) in the definition, and therefore we may conclude that your type t_2 strongly believes in Barbara's rationality.

9.2 Common strong belief in rationality

We have introduced the idea of strongly believing in the opponents' rationality, which means that, whenever possible, you believe that your opponents are implementing rational strategies. We can carry this argument one step further, however. Consider a dynamic game with two players, say i and j. Suppose that player i is at information set $h \in H_i$, and he is considering two strategies, s_j and s_j', for opponent j that both lead to h. Assume that both strategies are optimal for some types of player j. However, s_j is also optimal for a type that strongly believes in i's rationality, whereas s_j' is only optimal for types that do not strongly believe in i's rationality. Then, intuitively, s_j is a "more plausible" strategy for player j than s_j', as it can be supported by a "more plausible" belief hierarchy. Consequently, player i must at information set h assign probability 0 to the "less plausible" strategy s_j' as there is a "more plausible" strategy for player j leading to h. We say that player i expresses *2-fold strong belief in rationality*. Before attempting to formalize this requirement, let us first apply it to the following example.

Example 9.2 Watching TV with Barbara

It is Wednesday evening, and Barbara and you must decide which TV program to watch this evening. The only interesting programs on TV tonight are *Blackadder* and *Dallas*. The problem is that you prefer *Blackadder* whereas Barbara wants to see *Dallas*. More precisely, watching *Blackadder* will give you a utility of 6 and Barbara a utility of 3, whereas for *Dallas* it is the other way around. As a possible resolution to this problem, you both simultaneously write down a program on a piece of paper and compare them. If you both write down the same program, you will watch it together. If you write down different programs, then the TV will remain switched off and you will play a game of cards together. In that case, your utilities will only be 2. However, before writing down a program on a piece of paper, you have the option of starting a fight with Barbara

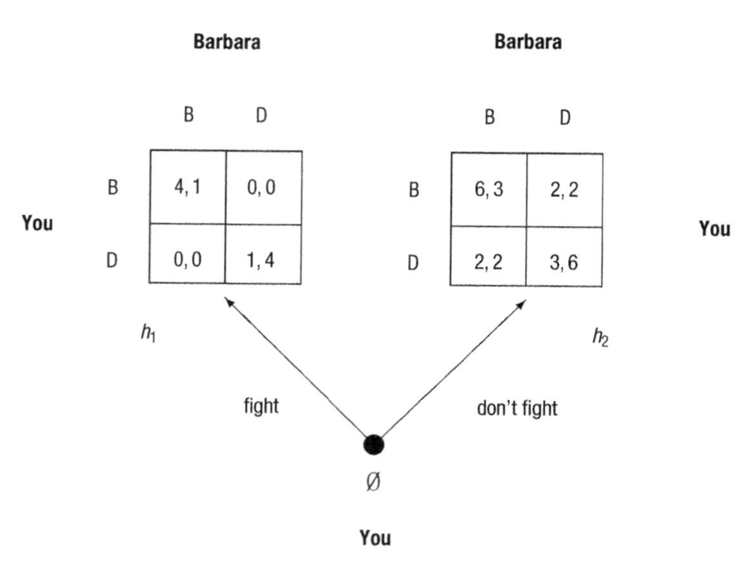

Figure 9.2 Watching TV with Barbara

in order to convince her to watch your favorite program – something that you have done repeatedly in the past, so Barbara is well aware of this option. If you start a fight, then this would reduce both your utility and Barbara's utility by 2. This situation can be modeled by the dynamic game in Figure 9.2. Here, B stands for *Blackadder* and D stands for *Dallas*.

Suppose that the game reaches h_1, that is, you chose to start a fight with Barbara. Of the two strategies for you that reach h_1, (*fight*, B) and (*fight*, D), only strategy (*fight*, B) can be optimal at the beginning of the game, since (*fight*, D) is always worse than not starting a fight at the beginning of the game. So, if Barbara strongly believes in your rationality, then at h_1 she must believe that you have chosen the strategy (*fight*, B). Hence, if Barbara strongly believes in your rationality, then she can only rationally choose B at h_1. Consequently, under strong belief in your rationality, Barbara can only rationally choose the strategies (B,B) and (B,D), where (B,D) means that she chooses B at h_1 and D at h_2.

Now, what do you believe about Barbara's strategy choice in this game? Notice first that each of Barbara's strategies is optimal for some conditional belief hierarchy. That is, if you are only required to strongly believe in Barbara's rationality, then you are free to hold any belief about Barbara's strategy choice. However, we have just seen that Barbara's strategies (B,B) and (B,D) can both be supported by types for Barbara that strongly believe in your rationality, whereas her strategies (D,B) and (D,D) cannot. So, in a sense, her strategies (B,B) and (B,D) are "more plausible" than (D,B) and (D,D). But then, if you express 2-fold strong belief in rationality, you can only assign positive probability to Barbara's strategies (B,B) and (B,D). That is, under 2-fold strong belief in rationality you believe that Barbara will write down *Blackadder* after starting a fight with her. As a consequence, you expect a utility of 4 by starting a fight with her, and choosing *Blackadder* afterwards. Hence, under 2-fold strong belief in rationality it is no longer optimal to choose strategy $(don't, D)$, as it would yield you at most a utility of 3. Summarizing, we see that under 2-fold strong belief in rationality you can only rationally choose the strategies $(fight, B)$ and $(don't, B)$. □

Let us now try to formalize the notion of 2-fold strong belief in rationality within an epistemic model. Consider an arbitrary dynamic game, and suppose player i finds himself at information set $h \in H_i$. The idea is that player i, if possible, only assigns positive probability to those strategy combinations for the opponents that lead to h, and where every strategy can be supported by a type that strongly believes in his opponents' rationality. In order to formalize this idea correctly, we must again make sure that the epistemic model contains sufficiently many types. More precisely, if at information set h we can indeed find some combination of opponents' strategies and types, possibly outside the model, where: (a) the strategy combination leads to h, (b) every strategy is optimal for the corresponding type, and (c) every type strongly believes in the opponents' rationality, then the epistemic model must contain such a combination of opponents' types.

Definition 9.2.1 *(2-fold strong belief in rationality)*
Consider an epistemic model M, a type t_i for player i within M, and an information set $h \in H_i$. Type t_i expresses ***2-fold strong belief in rationality*** *at h if, whenever we can find a combination of opponents' types, possibly outside M, that strongly believe in their opponents' rationality, and for which there is a combination of optimal strategies leading to h, then:*
(1) the epistemic model M must contain at least one such combination of types, and
(2) type t_i must at h only assign positive probability to opponents' strategy-type combinations where the strategy combination leads to h, the types strongly believe in their opponents' rationality and the strategies are optimal for the types.
Finally, type t_i is said to express 2-fold strong belief in rationality if he does so at every information set $h \in H_i$.

To illustrate this definition, let us return to the example "Watching TV with Barbara," and consider the epistemic model in Table 9.3. In this model we have not explicitly specified the information sets for your conditional beliefs. This is because we assume that your types hold the same conditional belief at $\emptyset, h_1$ and h_2. Indeed, there is no reason for you to change your initial belief when the game reaches h_1 or h_2, because Barbara does not get to make a choice before h_1 and h_2.

We will show that your types t_1^{fB} and t_1^{dB} both express 2-fold strong belief in rationality. To do so, we first verify that Barbara's types t_2^{BB} and t_2^{BD} strongly believe in your rationality. Barbara's types t_2^{BB} and t_2^{BD} assign, at h_1 and h_2, probability 1 to a strategy-type pair for you where the strategy is optimal for the type. Therefore, we

Table 9.3. *An epistemic model for "Watching TV with Barbara"*

Types	$T_1 = \{t_1^{fB}, t_1^{dB}, t_1^{dD}\}$, $T_2 = \{t_2^{BB}, t_2^{BD}, t_2^{DD}\}$
Beliefs for you	$b_1(t_1^{fB}) = ((B,D), t_2^{BD})$ $b_1(t_1^{dB}) = ((B,B), t_2^{BB})$ $b_1(t_1^{dD}) = ((D,D), t_2^{DD})$
Beliefs for Barbara	$b_2(t_2^{BB}, h_1) = ((fight, B), t_1^{fB})$ $b_2(t_2^{BB}, h_2) = ((don't, B), t_1^{dB})$ $b_2(t_2^{BD}, h_1) = ((fight, B), t_1^{fB})$ $b_2(t_2^{BD}, h_2) = ((don't, D), t_1^{dD})$ $b_2(t_2^{DD}, h_1) = ((fight, D), t_1^{fB})$ $b_2(t_2^{DD}, h_2) = ((don't, D), t_1^{dD})$

can immediately conclude that Barbara's types t_2^{BB} and t_2^{BD} strongly believe in your rationality.

Your type t_1^{fB} assigns probability 1 to Barbara's strategy-type pair $((B,D), t_2^{BD})$. Since, as we have seen, Barbara's type t_2^{BD} strongly believes in your rationality, and strategy (B,D) is optimal (at h_1 and h_2) for type t_2^{BD}, it follows that your type t_1^{fB} expresses 2-fold strong belief in rationality. In a similar fashion, you may verify that your type t_1^{dB} also expresses 2-fold strong belief in rationality.

However, your type t_1^{dD} does not express 2-fold strong belief in rationality. To show this, we first verify that Barbara's type t_2^{DD} does not strongly believe in your rationality. At h_1, there is a strategy-type pair for you where the strategy leads to h_1 and the strategy is optimal for the type, namely the strategy-type pair $((fight, B), t_1^{fB})$. Hence, for Barbara's type t_2^{DD} to strongly believe in your rationality, it must at h_1 assign positive probability only to strategy-type pairs for you where the strategy is optimal for the type. But this is not what t_2^{DD} does. At h_1, type t_2^{DD} assigns probability 1 to your strategy-type pair $((fight, D), t_1^{fB})$, where $(fight, D)$ is not optimal for t_1^{fB} at the beginning of the game. Hence, Barbara's type t_2^{DD} does not strongly believe in your rationality.

Your type t_1^{dD} always assigns probability 1 to Barbara's type t_2^{DD} which, as we have seen, does not strongly believe in your rationality. This means, however, that your type t_1^{dD} does not express 2-fold strong belief in rationality.

Hence, the only types for you in the epistemic model that express 2-fold strong belief in rationality are t_1^{fB} and t_1^{dB}. The optimal strategies for these two types are $(fight, B)$ and $(don't, B)$ which, as we have seen above, are the only strategies that you can rationally choose under 2-fold strong belief in rationality.

At first sight, one might be tempted to conclude that if a type expresses 2-fold strong belief in rationality, then it also automatically strongly believes in the opponent's rationality. This, however, is wrong! There are types that express 2-fold strong belief in rationality, but which do not strongly believe in the opponents' rationality! As an illustration, let us go back to the example "Painting Chris' house" with the epistemic model in Table 9.1. We have already seen above that your type t_2 does not strongly believe in Barbara's rationality. We will show, however, that this same type t_2 of yours expresses 2-fold strong belief in rationality.

Notice first that there is no type for Barbara that strongly believes in your rationality, and for which rejecting her colleague's offer is optimal. If Barbara strongly believes in your rationality, then she believes that you will not choose a price of 500, and hence she would never reject her colleague's offer. So, in the definition of 2-fold strong belief in rationality, there is no type for Barbara that strongly believes in your rationality, and for which there is an optimal strategy leading to information set h_1. But then, 2-fold strong belief in rationality does not impose any restrictions on the beliefs you may hold at information set h_1, which means in particular that your type t_2 expresses 2-fold strong belief in rationality. That is, your type t_2 expresses 2-fold strong belief in rationality, but does not strongly believe in Barbara's rationality.

Now that we have established a formal definition for 2-fold strong belief in rationality, we can easily extend this idea to formally define *k-fold strong belief in rationality* for every $k \geq 3$. The way we build this definition is very similar to how we defined *k-fold assumption of rationality* for lexicographic beliefs in Chapter 7. We will first define 1-fold strong belief in rationality – which will simply be strong belief in the opponents' rationality – and we will then inductively define k-fold strong belief in rationality for every $k \geq 2$. For defining k-fold strong belief in rationality we will always use the definitions already given for 1-fold through to $(k-1)$-fold strong belief in rationality – as we did for k-fold assumption of rationality.

Definition 9.2.2 *(k-fold strong belief in rationality)*
Consider an epistemic model M and a type t_i for player i within M. Type t_i is said to express 1-fold strong belief in rationality if t_i strongly believes in the opponents' rationality.
*Now, consider a $k \geq 2$ and an information set $h \in H_i$. We say that type t_i expresses k-**fold strong belief in rationality** at h if, whenever we can find a combination of opponents' types, possibly outside M, that express up to $(k-1)$-fold strong belief in rationality, and for which there is a combination of optimal strategies leading to h, then:*
(1) the epistemic model M must contain at least one such combination of types, and
(2) type t_i must at h only assign positive probability to opponents' strategy-type combinations where the strategy combination leads to h, the types express up to $(k-1)$-fold strong belief in rationality and the strategies are optimal for the types.

Finally, type t_i is said to express k-fold strong belief in rationality if he does so at every information set $h \in H_i$.

Here, whenever we say that a type expresses "up to $(k-1)$-fold strong belief in rationality," we mean that it expresses 1-fold strong belief in rationality, 2-fold strong belief in rationality, ..., to $(k-1)$-fold strong belief in rationality. Note that if we choose $k = 2$, then we obtain the definition of 2-fold strong belief in rationality as already given.

We have seen above that a type that expresses 2-fold strong belief in rationality need not express 1-fold strong belief in rationality. Hence, in general, a type that expresses k-fold strong belief in rationality need not express $(k-1)$-fold strong belief in rationality – a phenomenon we have already seen for k-fold assumption of rationality in Chapter 7.

With the definition of k-fold strong belief in rationality, it is now easy to define *common* strong belief in rationality, which simply means that the type expresses k-fold strong belief in rationality for every k.

Definition 9.2.3 *(Common strong belief in rationality)*
Consider an epistemic model M and a type t_i for player i within M. Type t_i is said to express **common strong belief in rationality** *if t_i expresses k-fold strong belief in rationality for every k.*

Player i can rationally choose strategy s_i under common strong belief in rationality if there is some epistemic model M and some type t_i within M such that
(1) type t_i expresses common strong belief in rationality, and
(2) strategy s_i is optimal for type t_i at every information set $h \in H_i$ that s_i leads to.

Let us illustrate this definition by means of the example "Watching TV with Barbara."

Example 9.3 Watching TV with Barbara

Consider the graphical representation in Figure 9.2 and the epistemic model in Table 9.3. We have seen above that your types t_1^{fB} and t_1^{dB} express 2-fold strong belief in rationality but that your type t_1^{dD} does not. Your type t_1^{dD} does express 1-fold strong belief in rationality, however. Moreover, we have seen that Barbara's types t_2^{BB} and t_2^{BD} express 1-fold strong belief in rationality, but that her type t_2^{DD} does not. We will now show that your types t_1^{fB} and t_1^{dB} also express 1-fold strong belief in rationality, and that Barbara's types t_2^{BB} and t_2^{BD} also express 2-fold strong belief in rationality.

Your types t_1^{fB} and t_1^{dB} both assign, at each information set, probability 1 to a strategy-type pair for Barbara where the strategy is optimal for the type. This is enough to conclude that these two types express 1-fold strong belief in rationality.

Let us now turn to Barbara's type t_2^{BB}. At information set h_1 (if you decide to fight) there is a type for you that expresses 1-fold strong belief in rationality, and an optimal strategy for that type leading to h_1, namely your type t_1^{fB} with optimal strategy (*fight*, B). So, in order for t_2^{BB} to express 2-fold strong belief in rationality at h_1, it must at h_1 assign positive probability only to strategy-type pairs for you where the type expresses 1-fold strong belief in rationality and the strategy is optimal for the type. But that is what t_2^{BB} does at h_1. Type t_2^{BB} assigns at h_1 probability 1 to the strategy-type pair $((fight, B), t_1^{fB})$, where t_1^{fB} expresses 1-fold strong belief in rationality and $(fight, B)$ is optimal for t_1^{fB}.

At information set h_2 (if you decide not to fight) there is also a type for you that expresses 1-fold strong belief in rationality, and an optimal strategy for that type leading to h_2, namely your type t_1^{dB} with optimal strategy $(don't, B)$, or your type t_1^{dD} with optimal strategy $(don't, D)$. So, in order for t_2^{BB} to express 2-fold strong belief in rationality at h_2, it must at h_2 assign positive probability only to strategy-type pairs for you where the type expresses 1-fold strong belief in rationality and the strategy is optimal for the type. But that is what t_2^{BB} does at h_2. Namely, type t_2^{BB} assigns at h_2 probability 1 to the strategy-type pair $((don't, B), t_1^{dB})$, where t_1^{dB} expresses 1-fold strong belief in rationality and $(don't, B)$ is optimal for t_1^{dB}. Hence, we may conclude that Barbara's type t_2^{BB} expresses 2-fold strong belief in rationality at h_1 and h_2, and therefore expresses 2-fold strong belief in rationality overall.

In the same way, it can be verified that Barbara's type t_2^{BD} also expresses 2-fold strong belief in rationality.

So, overall we see that your types t_1^{fB} and t_1^{dB} and Barbara's types t_2^{BB} and t_2^{BD} express up to 2-fold strong belief in rationality, whereas the other types t_1^{dD} and t_2^{DD} do not.

We will now show that your types t_1^{fB} and t_1^{dB} also express 3-fold strong belief in rationality.

Let us first turn to your type t_1^{fB}. At every information set, your type t_1^{fB} assigns probability 1 to Barbara's strategy-type pair $((B, D), t_2^{BD})$, where t_2^{BD} expresses up to 2-fold strong belief in rationality and (B, D) is optimal for t_2^{BD}. This is enough to conclude that your type t_1^{fB} expresses 3-fold strong belief in rationality. In a similar way, it can be checked that your type t_1^{dB} also expresses 3-fold strong belief in rationality.

Next, we show that Barbara's type t_2^{BB} also expresses 3-fold strong belief in rationality. By definition, Barbara's type t_2^{BB} assigns at h_1 probability 1 to your strategy-type pair $((fight, B), t_1^{fB})$, and assigns at h_2 probability 1 to your strategy-type pair $((don't, B), t_1^{dB})$. Since t_1^{fB} and t_1^{dB} express up to 2-fold strong belief in rationality, strategy $(fight, B)$ is optimal for t_1^{fB} and strategy $(don't, B)$ is optimal for t_1^{dB}, we may immediately conclude that Barbara's type t_2^{BB} expresses 3-fold strong belief in rationality.

We will show, however, that Barbara's type t_2^{BD} does not express 3-fold strong belief in rationality. At h_2 there is a strategy-type pair for you where the strategy leads to h_2, the type expresses up to 2-fold strong belief in rationality and the strategy is optimal for the type, namely the strategy-type pair $((don't, B), t_1^{dB})$. So, for t_2^{BD} to express 3-fold strong belief in rationality, it must assign at h_2 only positive probability to strategy-type pairs for you where the strategy is optimal for the type and the type expresses up to 2-fold strong belief in rationality. This, however, is not what t_2^{BD} does. At h_2, type t_2^{BD}

assigns probability 1 to your strategy-type pair $((don't, D), t_1^{dD})$, but we have seen that t_1^{dD} does not express up to 2-fold strong belief in rationality. Hence, Barbara's type t_2^{BD} does not express 3-fold strong belief in rationality.

Summarizing, we see that your types t_1^{fB} and t_1^{dB} and Barbara's type t_2^{BB} express up to 3-fold strong belief in rationality, whereas the other types do not.

We will now prove that your type t_1^{dB} and Barbara's type t_2^{BB} also express 4-fold strong belief in rationality, but that your type t_1^{fB} does not.

For your type t_1^{dB} this is easily seen, as t_1^{dB} always assigns probability 1 to Barbara's strategy-type pair $((B,B), t_2^{BB})$, where t_2^{BB} expresses up to 3-fold strong belief in rationality and strategy (B,B) is optimal for type t_2^{BB}. This is enough to conclude that your type t_1^{dB} expresses 4-fold strong belief in rationality.

Consider now your type t_1^{fB}. Clearly, at each of your information sets there is a strategy-type pair for Barbara where the strategy leads to that information set, the strategy is optimal for the type and the type expresses up to 3-fold strong belief in rationality, namely the strategy-type pair $((B,B), t_2^{BB})$. So, in order for t_1^{fB} to express 4-fold strong belief in rationality, it must at every information set only assign positive probability to strategy-type pairs for Barbara where the strategy is optimal for the type and the type expresses up to 3-fold strong belief in rationality. But this is not what t_1^{fB} does. Type t_1^{fB} assigns probability 1 to Barbara's type t_2^{BD} which, as we have seen, does not express up to 3-fold strong belief in rationality. Hence, your type t_1^{fB} does not express 4-fold strong belief in rationality.

We will now show that Barbara's type t_2^{BB} does express 4-fold strong belief in rationality. Note that type t_2^{BB} assigns at h_1 probability 1 to your strategy-type pair $((fight, B), t_1^{fB})$ and assigns at h_2 probability 1 to your strategy-type pair $((don't, B), t_1^{dB})$. We know, from before, that $(fight, B)$ is optimal for t_1^{fB}, that $(don't, B)$ is optimal for t_1^{dB}, and that the types t_1^{fB} and t_1^{dB} both express up to 3-fold strong belief in rationality. This is enough to conclude that Barbara's type t_2^{BB} expresses 4-fold strong belief in rationality.

Summarizing, we see that your type t_1^{dB} and Barbara's type t_2^{BB} express up to 4-fold strong belief in rationality, and that the other types do not.

We will finally show that your type t_1^{dB} and Barbara's type t_2^{BB} also express 5-fold strong belief in rationality and further.

Let us start with your type t_1^{dB} and see why it expresses 5-fold strong belief in rationality. This is easily seen, in fact, as t_1^{dB} always assigns probability 1 to Barbara's strategy-type pair $((B,B), t_2^{BB})$, where (B,B) is optimal for t_2^{BB} and, as we have seen, type t_2^{BB} expresses up to 4-fold strong belief in rationality.

To verify the same for Barbara's type t_2^{BB} is more difficult, however. Consider the information set h_1, after you started a fight with Barbara. We will show that there is *no* type for you that expresses up to 4-fold strong belief in rationality and that has an optimal strategy leading to h_1.

We have seen that only your strategies (*fight*, B), (*don't*, B) and (*don't*, D) are optimal at the beginning for some belief about Barbara's strategy choices. Hence, under 1-fold strong belief in rationality Barbara must conclude at h_1 that you chose (*fight*, B). So, the only optimal strategies for Barbara under 1-fold strong belief in rationality are (B,B) and (B,D).

Consequently, under 2-fold strong belief in rationality you must believe, throughout the game, that Barbara chose either (B,B) or (B,D). As such, the only optimal strategies for you when expressing up to 2-fold strong belief in rationality are (*fight*, B) and (*don't*, B).

But then, under 3-fold strong belief in rationality Barbara must conclude at h_1 that you chose (*fight*, B) and must conclude at h_2 that you chose (*don't*, B). Hence, the only optimal strategy for Barbara if she expresses up to 3-fold strong belief in rationality is (B,B).

So, under 4-fold strong belief in rationality you must believe that Barbara chose (B,B). This means that there is only one strategy you can rationally choose when expressing up to 4-fold strong belief in rationality, namely (*don't*, B). Hence, indeed, there is no type for you that expresses up to 4-fold strong belief in rationality and that has an optimal strategy leading to h_1.

So, 5-fold strong belief in rationality will not impose any restrictions on Barbara's beliefs at h_1. In particular, Barbara's type t_2^{BB} expresses 5-fold strong belief in rationality at h_1. At information set h_2, Barbara's type t_2^{BB} assigns probability 1 to your strategy-type pair ((*don't*, B), t_1^{dB}). Since type t_1^{dB} expresses up to 4-fold strong belief in rationality, as we have seen, and strategy (*don't*, B) is optimal for type t_1^{dB}, it follows that Barbara's type t_2^{BB} expresses 5-fold strong belief in rationality at h_2. Hence, we may conclude that t_2^{BB} expresses 5-fold strong belief in rationality overall.

Summarizing, we see that your type t_1^{dB} and Barbara's type t_2^{BB} express up to 5-fold strong belief in rationality.

In a similar fashion, it can be verified that both types also express 6-fold strong belief in rationality and further. That is, both your type t_1^{dB} and Barbara's type t_2^{BB} express common strong belief in rationality. Since your strategy (*don't*, B) is optimal for your type t_1^{dB}, we see that you can rationally choose strategy (*don't*, B) under common strong belief in rationality.

We can actually show a little more: Under common strong belief in rationality, your *only* optimal strategy is (*don't*, B)! We argued that if you express up to 4-fold strong belief in rationality, then your only optimal strategy is (*don't*, B), which will certainly be the only optimal strategy under *common* strong belief in rationality. Moreover, we have also seen that if you express up to 4-fold strong belief in rationality, then you believe that Barbara will choose (B,B), that is, that Barbara will always write down *Blackadder*. But then, under common strong belief in rationality, your only possible belief is that Barbara will always write down *Blackadder*, which means that you expect to obtain the maximum utility of 6 by not starting a fight and writing down *Blackadder*. Hence, we see that under common strong belief in rationality you expect to end up with the best possible scenario – that you will watch your favorite program *Blackadder* without having to start a fight with Barbara! □

Let us finally go back to the example "Painting Chris' house" and apply common strong belief in rationality.

Example 9.4 Painting Chris' house

Reconsider the dynamic game in Figure 9.1 and the epistemic model in Table 9.2. We have already seen that your type t_2 expresses 1-fold strong belief in rationality. We will show that your type t_2 and Barbara's type t_1^a both express common strong belief in rationality, but Barbara's type t_1^r does not.

In fact, Barbara's type t_1^r does not even express 1-fold strong belief in rationality. Her type t_1^r assigns at the beginning of the game probability 1 to your strategy-type pair $(500, t_2)$. But strategy 500 is not optimal for your type t_2, and hence Barbara's type t_1^r does not strongly believe in your rationality, that is, it does not express 1-fold strong belief in rationality.

We next show that Barbara's type t_1^a expresses 1-fold strong belief in rationality. This is easily seen, as t_1^a always assigns probability 1 to your strategy-type pair $(300, t_2)$, and strategy 300 is optimal for your type t_2. So, both your type t_2 and Barbara's type t_1^a express 1-fold strong belief in rationality.

Next, we will prove that your type t_2 and Barbara's type t_1^a also express 2-fold strong belief in rationality. Let us start with your type t_2. Is there a strategy-type pair for Barbara where the type expresses 1-fold strong belief in rationality, the strategy is optimal for the type and the strategy leads to h_1? The answer is "no." If Barbara expresses 1-fold strong belief in rationality – that is, she strongly believes in your rationality – then she believes that you will not choose a price of 500. Hence, it cannot be optimal for Barbara to reject her colleague's offer. So, there is no type for Barbara that expresses 1-fold strong belief in rationality and that has an optimal strategy leading to h_1. Thus 2-fold strong belief in rationality will not impose any restrictions on your beliefs at h_1. In particular, your type t_2 expresses 2-fold strong belief in rationality.

It is easily verified that Barbara's type t_1^a expresses 2-fold strong belief in rationality: Her type t_1^a always assigns probability 1 to your strategy-type pair $(300, t_2)$, where t_2 expresses 1-fold strong belief in rationality and strategy 300 is optimal for t_2. So, we conclude that both your type t_2 and Barbara's type t_1^a express up to 2-fold strong belief in rationality.

In a similar fashion it can be verified that the types t_2 for you and t_1^a for Barbara also express 3-fold strong belief in rationality and further. That is, t_2 and t_1^a both express common strong belief in rationality. As strategy 300 is optimal for your type t_2, we conclude that you can rationally choose a price of 300 under common strong belief in rationality.

In fact, 300 is the *only* price you can rationally choose under common strong belief in rationality. If you strongly believe in Barbara's rationality, then you must believe, after observing Barbara reject her colleague's offer, that Barbara will then choose a price of 400, as this is the only price that gives her more than 350 – the amount she could have obtained by accepting her colleague's offer. So, the only optimal price for you would be to choose 300.

If Barbara expresses common strong belief in rationality, then she must believe that you will choose a price of 300 if she rejects her colleague's offer. Therefore, under common strong belief in rationality there is only one optimal strategy for Barbara: to *accept* the colleague's offer at the beginning of the game. □

9.3 Algorithm

We have introduced and formalized the idea of common strong belief in rationality. We will now investigate whether there is an easy algorithm that generates precisely those strategies you can rationally choose under common strong belief in rationality. It turns out that we can indeed find such an algorithm, and in fact it will be very similar to the backward dominance procedure for common belief in future rationality. On the way to this algorithm we will start by characterizing those strategies you can rationally choose under 1-fold strong belief in rationality, then we will characterize the strategies you can rationally choose when expressing up to 2-fold strong belief in rationality, and so on.

Step 1: 1-fold strong belief in rationality

We will start by asking the following question: Which strategies can you rationally choose under 1-fold strong belief in rationality, that is, if you strongly believe in the opponents' rationality? Remember that 1-fold strong belief in rationality means the following: If player i finds himself at information set $h \in H_i$, and if there is some strategy combination for the opponent that leads to h and where every strategy is optimal, then player i should at h only assign positive probability to such optimal strategy combinations. If there is no such optimal strategy combination leading to h, then no conditions can be imposed on player i's beliefs at h.

By an "optimal strategy" we mean a strategy s_j that is optimal, at every information set $h' \in H_j$ it leads to, for *some* conditional belief. We can rephrase the definition of 1-fold strong belief in rationality as follows: Player i should at information set h assign probability 0 to all of the opponent's strategies that are not optimal, *unless* there is no optimal strategy combination that leads to h. In the latter case, no conditions are imposed on player i's beliefs at h.

From Chapter 8 we know that an opponent's strategy s_j is optimal at an information set $h' \in H_j$ for *some* conditional belief, when s_j is not strictly dominated within the *full decision problem* $\Gamma^0(h')$. Remember that the full decision problem $\Gamma^0(h') = (S_j(h'), S_{-j}(h'))$ contains for player j only the set $S_j(h')$ of strategies leading to h', and contains for the opponents only the set $S_{-j}(h')$ of strategy combinations leading to h'. This means that an opponent's strategy s_j is not optimal when it is strictly dominated within some full decision problem $\Gamma^0(h')$ at which j is active. Hence, 1-fold strong belief in rationality can alternatively be stated as follows: Player i should at information set h assign probability 0 to every strategy for opponent j that is strictly dominated in some full decision problem $\Gamma^0(h')$ at which j is active, *unless* this would rule out all possible beliefs for player i at h. In the latter case, no conditions are imposed on player i's beliefs at h.

Now, the requirement that player i should at h assign probability 0 to these strategies for opponent j can be mimicked by "removing" these strategies from the full decision problem $\Gamma^0(h)$ at h. Hence, 1-fold strong belief in rationality can be mimicked by the following elimination step: For the full decision problem $\Gamma^0(h)$, eliminate for every opponent j those strategies that are strictly dominated within some full decision problem $\Gamma^0(h')$ at which j is active, *unless* this would remove all strategy combinations leading to h. In the latter case, we do not remove any strategies from h. Let $\Gamma^1(h)$ be the reduced decision problem at h that remains after removing strategies from $\Gamma^0(h)$ in this way.

So far we have shown that, if player i expresses 1-fold strong belief in rationality, then at every information set $h \in H_i$ he should assign positive probability only to opponents' strategy combinations in $\Gamma^1(h)$. In that case, the strategies he can rationally choose are precisely those strategies that are optimal, at every information set $h \in H_i$ they lead to, for some conditional belief in $\Gamma^1(h)$. But, as we know from Chapter 8, these are precisely the strategies that are not strictly dominated within any decision problem $\Gamma^1(h)$ at which player i is active.

We thus see that for every player i, the strategies he can rationally choose under 1-fold strong belief in rationality are precisely the strategies that are not strictly dominated in any reduced decision problem $\Gamma^1(h)$ at which he is active. In turn, these are precisely the strategies that remain after removing, from the reduced decision problem $\Gamma^1(\emptyset)$ at the beginning of the game, those strategies for player i that *are* strictly dominated in some reduced decision problem $\Gamma^1(h)$ at which i is active.

Summarizing, the strategies that can rationally be chosen under 1-fold strong belief in rationality are obtained by the following two-step elimination procedure:

First, eliminate from every full decision problem $\Gamma^0(h)$ all strategies s_j that are strictly dominated in some full decision problem $\Gamma^0(h')$ at which player j is active, *unless* this would remove all strategy combinations leading to h. In the latter case, do not remove any strategies from $\Gamma^0(h)$. This leads to the new decision problems $\Gamma^1(h)$ at every information set h.

Second, eliminate from the decision problem $\Gamma^1(\emptyset)$ at the beginning of the game, and for every player i, those strategies that are strictly dominated in some reduced decision problem $\Gamma^1(h)$ at which i is active. This leads to the new decision problem $\Gamma^2(\emptyset)$ at the beginning of the game.

The strategies that can rationally be chosen under 1-fold strong belief in rationality are exactly the strategies in $\Gamma^2(\emptyset)$.

Step 2: Up to 2-fold strong belief in rationality

We now go one step further, and we will characterize those strategies you can rationally choose when expressing up to 2-fold strong belief in rationality. Consider some player i and an information set $h \in H_i$ at which he is active. Then, for player i to express 2-fold strong belief in rationality at h means the following: If there is a strategy combination for the opponents leading to h that can rationally be chosen under 1-fold strong belief in rationality, then player i must at h only assign positive probability to such strategy combinations.

We have seen in step 1 that the players' strategies that can rationally be chosen under 1-fold strong belief in rationality are precisely the strategies in $\Gamma^2(\emptyset)$. So, 2-fold strong belief in rationality at h then means the following for player i: If there is a strategy combination for the opponents in $\Gamma^2(\emptyset)$ leading to h, then player i must at h only assign positive probability to opponents' strategy combinations in $\Gamma^2(\emptyset)$. Or, equivalently, we remove from the decision problem $\Gamma^1(h)$ those strategies that are not in $\Gamma^2(\emptyset)$, *unless* this would remove all strategy combinations leading to h.

By construction, a strategy s_j for player j is not in $\Gamma^2(\emptyset)$ if it is strictly dominated in some reduced decision problem $\Gamma^1(h')$ at which player j is active. So, we remove from $\Gamma^1(h)$ those strategies s_j that are strictly dominated in some reduced decision problem $\Gamma^1(h')$ at which player j is active, *unless* this would remove all strategy combinations leading to h. In the latter case, we do not remove any more strategies from $\Gamma^1(h)$. By doing so for every h, we obtain the newly reduced decision problems $\Gamma^2(h)$ for every information set h. Hence, a player who expresses up to 2-fold strong belief in rationality assigns at each of his information sets h only positive probability to opponents' strategy combinations in $\Gamma^2(h)$.

The strategies that player i can rationally choose when expressing up to 2-fold strong belief in rationality will be precisely those strategies that are optimal, at every information set $h \in H_i$ they lead to, for some conditional belief in $\Gamma^2(h)$. These, in turn, are exactly the strategies s_i that are not strictly dominated within any reduced decision problem $\Gamma^2(h)$ they lead to.

So, the strategies that can rationally be chosen when expressing up to 2-fold strong belief in rationality are obtained by the following procedure:

First, eliminate from every reduced decision problem $\Gamma^1(h)$ all strategies s_j that are strictly dominated in some reduced decision problem $\Gamma^1(h')$ at which player j is active, *unless* this would remove all strategy combinations leading to h. In the latter case, do not remove any more strategies from $\Gamma^1(h)$. This leads to the newly reduced decision problems $\Gamma^2(h)$ for every information set h.

Second, eliminate from the decision problem $\Gamma^2(\emptyset)$ at the beginning of the game, and for every player i, those strategies that are strictly dominated in some reduced decision problem $\Gamma^2(h)$ at which i is active. This leads to the new decision problem $\Gamma^3(\emptyset)$ at the beginning of the game.

The strategies that can rationally be chosen when expressing up to 2-fold strong belief in rationality are exactly the strategies in $\Gamma^3(\emptyset)$.

By repeating this argument, we can construct for every $k \geq 3$ and every information set h a reduced decision problem $\Gamma^k(h)$, and conclude that $\Gamma^k(\emptyset)$ contains exactly those strategies that can rationally be chosen by players who express up to $(k-1)$-fold strong belief in rationality. So, by iterating this reduction process until no further strategies can be removed, we obtain an algorithm that yields precisely those strategies that can rationally be chosen under common strong belief in rationality. This algorithm is called the *iterated conditional dominance procedure,* and is formally presented below.

Algorithm 9.3.1 *(Iterated conditional dominance procedure)*
Step 1: *From every full decision problem $\Gamma^0(h)$, eliminate for every player i those strategies that are strictly dominated in some full decision problem $\Gamma^0(h')$ at which player i is active, unless this would remove all strategy combinations that lead to h. In the latter case, we do not remove any strategies from $\Gamma^0(h)$. This leads to reduced decision problems $\Gamma^1(h)$ at every information set h.*
Step 2: *From every reduced decision problem $\Gamma^1(h)$, eliminate for every player i those strategies that are strictly dominated in some reduced decision problem $\Gamma^1(h')$ at which player i is active, unless this would remove all strategy combinations that lead to h. In the latter case, we do not remove any strategies from $\Gamma^1(h)$. This leads to the new reduced decision problems $\Gamma^2(h)$ at every information set.*
And so on. Continue until no more strategies can be eliminated in this way.

As in Chapter 8, eliminating a strategy s_i from a full decision problem $\Gamma^0(h)$ formally means the following: If player i is active at h, and the full decision problem $\Gamma^0(h)$ is given by $(S_i(h), S_{-i}(h))$, then we simply eliminate strategy s_i from $S_i(h)$. If player j is active at h but not i, and the full decision problem $\Gamma^0(h)$ is given by $(S_j(h), S_{-j}(h))$, then we eliminate from $S_{-j}(h)$ every strategy combination that contains strategy s_i for player i. Similarly for eliminating a strategy from a *reduced* decision problem.

Note that this algorithm is very similar to the backward dominance procedure discussed in the previous chapter, as it also proceeds by successively eliminating strategies from decision problems at information sets. However, the criterion to eliminate a strategy from a decision problem is different: In the backward dominance procedure, we eliminate a strategy s_i in a decision problem at h if s_i is strictly dominated within a decision problem *weakly following* h at which player i is active. In the iterated conditional dominance procedure, we would also eliminate s_i if it is strictly dominated within a decision problem that comes *before* h, or that comes *neither before nor after* h – as long as this would not remove all strategy combinations leading to h.

As for the backward dominance procedure, it can easily be seen that the iterated conditional dominance procedure will always stop within finitely many steps: there are only finitely many strategies for each player. Since the algorithm proceeds by successively eliminating strategies from decision problems at information sets, there must be a step in the procedure in which no further strategies can be eliminated, and this is where the algorithm stops.

We say that a strategy *survives* the iterated conditional dominance procedure if it is never eliminated from the decision problem at the beginning of the game.

Definition 9.3.2 *(Strategy surviving the iterated conditional dominance procedure)*
For every information set h and every k, let $\Gamma^k(h)$ be the reduced decision problem produced in step k of the iterated conditional dominance procedure. Strategy s_i for

player i ***survives the iterated conditional dominance procedure*** *if s_i is in the decision problem $\Gamma^k(\emptyset)$ for every k.*

It is clear that for every player i there will be at least one strategy that survives the iterated conditional dominance procedure, as we can never eliminate *all* strategies for a player at $\emptyset$. Our main theorem below states that the strategies that can rationally be chosen under common strong belief in rationality are exactly the strategies that survive the iterated conditional dominance procedure. That is, the algorithm gives us precisely what we want: the strategies that are optimal under common strong belief in rationality.

Theorem 9.3.3 *(The algorithm works)*
(1) For every $k \geq 1$, the strategies that can rationally be chosen by a type that expresses up to k-fold strong belief in rationality are precisely the strategies in $\Gamma^{k+1}(\emptyset)$, that is, those strategies that survive step $k+1$ of the iterated conditional dominance procedure at $\emptyset$.
(2) The strategies that can rationally be chosen by a type that expresses common strong belief in rationality are exactly the strategies that survive the full iterated conditional dominance procedure at $\emptyset$.

The proof can be found in the proofs section at the end of this chapter. This theorem shows, in particular, that common strong belief in rationality is always possible. That is, in every dynamic game we can find for every player a type that expresses common strong belief in rationality. We have seen that the iterated conditional dominance procedure gives for every player at least one strategy. In view of the theorem, this strategy can rationally be chosen by a type that expresses common strong belief in rationality. So, in particular we can build an epistemic model in which some type of this player expresses common strong belief in rationality. In other words, common strong belief in rationality is always possible in every game. Unfortunately, we cannot provide an easy method to prove the existence, as we did for common belief in future rationality.

Corollary 9.3.4 *(Common strong belief in rationality is always possible)*
We can always build an epistemic model which contains, for every player i, some type t_i that expresses common strong belief in rationality.

However, for most games we cannot construct an epistemic model in which *all* types express common strong belief in rationality. Consider, for instance, the example "Painting Chris' house." Suppose we build an epistemic model in which there is a type t_2 for you that expresses common strong belief in rationality. In particular, this type t_2 must strongly believe in Barbara's rationality. So, at information set h_1 – after Barbara has decided to reject her colleague's offer – type t_2 must believe that Barbara is of a type t_1 for which it is optimal to indeed reject the offer. But, as we have seen, there is no type for Barbara that expresses common strong belief in rationality and for which rejecting her colleague's offer is optimal. That is, your type t_2 must at information set h_1 believe that Barbara is of a type t_1 that does not express common strong belief in rationality. In particular, this epistemic model must contain at least one type for Barbara that does not express common strong belief in rationality. Summarizing, we see that every epistemic model for "Painting Chris' house" contains at least one type that does *not* express common strong belief in rationality.

Compare this to common belief in future rationality. In Theorem 8.7.1 of Chapter 8, we saw that we can always construct an epistemic model in which *all* types express common belief in future rationality. This is not possible if we use common strong belief in rationality!

Let us now apply the iterated conditional dominance procedure to the two examples we have discussed so far in this chapter, and verify that it indeed yields precisely those strategies that can rationally be chosen under common strong belief in rationality.

Example 9.5 Painting Chris' house

Consider the graphical representation of this game in Figure 9.1. Let h_1 be the information set that follows Barbara's choice "reject." The two full decision problems $\Gamma^0(\emptyset)$ and $\Gamma^0(h_1)$ are shown in Table 9.4.

Step 1: In the decision problem $\Gamma^0(\emptyset)$, the strategies $(r,200)$, $(r,300)$ and $(r,500)$ for Barbara are all strictly dominated by the strategy *accept*. We therefore eliminate these strategies from the decision problem $\Gamma^0(\emptyset)$ and also from the *future* decision problem $\Gamma^0(h_1)$, since by doing so we do not remove all of Barbara's strategies leading to h_1. Note, namely, that strategy $(r,400)$ for Barbara leads to h_1, but is not eliminated at h_1. This is crucially different from how the backward dominance procedure works in this example. According to the backward dominance procedure, we cannot eliminate the strategies $(r,200)$ and $(r,300)$ from $\Gamma^0(h_1)$, since these strategies are only strictly dominated in the decision problem $\Gamma^0(\emptyset)$ at the beginning, but not at $\Gamma^0(h_1)$.

For you, strategy 500 is strictly dominated by the randomized strategy $(0.5)\cdot 200 + (0.5)\cdot 400$ in $\Gamma^0(h_1)$. We therefore eliminate your strategy 500 from the decision problems at $\emptyset$ and h_1. This leads to the reduced decision problems $\Gamma^1(\emptyset)$ and $\Gamma^1(h_1)$ in Table 9.5.

Step 2: Within the reduced decision problem $\Gamma^1(\emptyset)$ at the beginning, Barbara's strategy $(r,400)$ is strictly dominated by *accept*. We therefore eliminate her strategy $(r,400)$ from $\Gamma^1(\emptyset)$. However, we cannot eliminate this strategy $(r,400)$ from the future decision problem $\Gamma^1(h_1)$, since by doing so we would remove all of Barbara's strategies leading to h_1!

For you, strategies 200 and 400 are both strictly dominated by 300 in the reduced decision problem $\Gamma^1(h_1)$ at h_1. We therefore eliminate your strategies 200 and 400 from $\Gamma^1(h_1)$ and also from $\Gamma^1(\emptyset)$. This leads to the final decision problems in Table 9.6, from which no further strategies can be eliminated. So, the only strategies that survive the iterated conditional dominance procedure are *accept* for Barbara, and 300 for you, which, as we have seen, are precisely the strategies that Barbara and you can rationally choose under common strong belief in rationality. □

Table 9.4. *Full decision problems in "Painting Chris' house"*

	$\Gamma^0(\emptyset)$: Barbara active			
	200	300	400	500
$(r,200)$	100, 100	200, 0	200, 0	200, 0
$(r,300)$	0, 200	150, 150	300, 0	300, 0
$(r,400)$	0, 200	0, 300	200, 200	400, 0
$(r,500)$	0, 200	0, 300	0, 400	250, 250
accept	350, 500	350, 500	350, 500	350, 500

	$\Gamma^0(h_1)$: Barbara and you active			
	200	300	400	500
$(r,200)$	100, 100	200, 0	200, 0	200, 0
$(r,300)$	0, 200	150, 150	300, 0	300, 0
$(r,400)$	0, 200	0, 300	200, 200	400, 0
$(r,500)$	0, 200	0, 300	0, 400	250, 250

Table 9.5. *Reduced decision problems after step 1 in "Painting Chris' house"*

	$\Gamma^1(\emptyset)$: Barbara active		
	200	300	400
$(r,400)$	0, 200	0, 300	200, 200
accept	350, 500	350, 500	350, 500

	$\Gamma^1(h_1)$: Barbara and you active		
	200	300	400
$(r,400)$	0, 200	0, 300	200, 200

Example 9.6 Watching TV with Barbara

Consider the graphical representation of this game in Figure 9.2. The three full decision problems in this game are $\Gamma^0(\emptyset), \Gamma^0(h_1)$ and $\Gamma^0(h_2)$, as depicted in Table 9.7.

Step 1: In the decision problem $\Gamma^0(\emptyset)$ at the beginning, only your strategy (*fight*, D) is strictly dominated. We therefore eliminate (*fight*, D) from $\Gamma^0(\emptyset)$ and also from the future decision problem $\Gamma^0(h_1)$. By doing so we have not removed all your strategies leading to h_1, as (*fight*, B) still remains. No other strategies can be eliminated in round 1. This leads to the reduced decision problems in Table 9.8.

Table 9.6. *Final decision problems after step 2 in "Painting Chris' house"*

	$\Gamma^2(\emptyset)$: Barbara active
	300
accept	350, 500
	$\Gamma^2(h_1)$: Barbara and you active
	300
$(r, 400)$	0, 300

Table 9.7. *Full decision problems in "Watching TV with Barbara"*

	$\Gamma^0(\emptyset)$: You active			
	(B,B)	(B,D)	(D,B)	(D,D)
$(fight, B)$	4, 1	4, 1	0, 0	0, 0
$(fight, D)$	0, 0	0, 0	1, 4	1, 4
$(don't, B)$	6, 3	2, 2	6, 3	2, 2
$(don't, D)$	2, 2	3, 6	2, 2	3, 6
	$\Gamma^0(h_1)$: You and Barbara active			
	(B,B)	(B,D)	(D,B)	(D,D)
$(fight, B)$	4, 1	4, 1	0, 0	0, 0
$(fight, D)$	0, 0	0, 0	1, 4	1, 4
	$\Gamma^0(h_2)$: You and Barbara active			
	(B,B)	(B,D)	(D,B)	(D,D)
$(don't, B)$	6, 3	2, 2	6, 3	2, 2
$(don't, D)$	2, 2	3, 6	2, 2	3, 6

Step 2: In decision problem $\Gamma^1(h_1)$, Barbara's strategies (D, B) and (D, D) are strictly dominated. We therefore eliminate these two strategies from $\Gamma^1(h_1)$, and also from $\Gamma^1(\emptyset)$ and $\Gamma^1(h_2)$. No other strategies can be eliminated at step 2. This leads to the new decision problems in Table 9.9.

Step 3: In decision problem $\Gamma^2(\emptyset)$ at the beginning of the game, your strategy $(don't, D)$ is strictly dominated. We therefore eliminate this strategy from $\Gamma^2(\emptyset)$ and also from the future decision problem $\Gamma^2(h_2)$, as by doing so we have not removed all your strategies

Table 9.8. *Reduced decision problems after step 1 in "Watching TV with Barbara"*

	$\Gamma^1(\emptyset)$: You active			
	(B,B)	(B,D)	(D,B)	(D,D)
$(fight,B)$	4,1	4,1	0,0	0,0
$(don't,B)$	6,3	2,2	6,3	2,2
$(don't,D)$	2,2	3,6	2,2	3,6
	$\Gamma^1(h_1)$: You and Barbara active			
	(B,B)	(B,D)	(D,B)	(D,D)
$(fight,B)$	4,1	4,1	0,0	0,0
	$\Gamma^1(h_2)$: You and Barbara active			
	(B,B)	(B,D)	(D,B)	(D,D)
$(don't,B)$	6,3	2,2	6,3	2,2
$(don't,D)$	2,2	3,6	2,2	3,6

Table 9.9. *Reduced decision problems after step 2 in "Watching TV with Barbara"*

	$\Gamma^2(\emptyset)$: You active	
	(B,B)	(B,D)
$(fight,B)$	4,1	4,1
$(don't,B)$	6,3	2,2
$(don't,D)$	2,2	3,6
	$\Gamma^2(h_1)$: You and Barbara active	
	(B,B)	(B,D)
$(fight,B)$	4,1	4,1
	$\Gamma^2(h_2)$: You and Barbara active	
	(B,B)	(B,D)
$(don't,B)$	6,3	2,2
$(don't,D)$	2,2	3,6

Table 9.10. *Reduced decision problems after step 3 in "Watching TV with Barbara"*

	$\Gamma^3(\emptyset)$: You active	
	(B,B)	(B,D)
$(fight,B)$	4,1	4,1
$(don't,B)$	6,3	2,2
	$\Gamma^3(h_1)$: You and Barbara active	
	(B,B)	(B,D)
$(fight,B)$	4,1	4,1
	$\Gamma^3(h_2)$: You and Barbara active	
	(B,B)	(B,D)
$(don't,B)$	6,3	2,2

Table 9.11. *Reduced decision problems after step 4 in "Watching TV with Barbara"*

	$\Gamma^4(\emptyset)$: You active
	(B,B)
$(fight,B)$	4,1
$(don't,B)$	6,3
	$\Gamma^4(h_1)$: You and Barbara active
	(B,B)
$(fight,B)$	4,1
	$\Gamma^4(h_2)$: You and Barbara active
	(B,B)
$(don't,B)$	6,3

leading to h_2. No other strategies can be eliminated in step 3. This leads to the new decision problems in Table 9.10.

Step 4: In decision problem $\Gamma^3(h_2)$, Barbara's strategy (B,D) is strictly dominated by (B,B). We thus remove Barbara's strategy (B,D) from the decision problems $\Gamma^3(\emptyset), \Gamma^3(h_1)$ and $\Gamma^3(h_2)$. No other strategies can be eliminated in step 4. This leads to the new decision problems in Table 9.11.

Table 9.12. *Final decision problems after step 5 in "Watching TV with Barbara"*

	$\Gamma^5(\emptyset)$: You active
	(B,B)
$(don't,B)$	6,3
	$\Gamma^5(h_1)$: You and Barbara active
	(B,B)
$(fight,B)$	4,1
	$\Gamma^5(h_2)$: You and Barbara active
	(B,B)
$(don't,B)$	6,3

Step 5: In decision problem $\Gamma^4(\emptyset)$, your strategy $(fight,B)$ is strictly dominated by $(don't,B)$. We thus eliminate your strategy $(fight,B)$ from $\Gamma^4(\emptyset)$. However, we cannot eliminate this strategy from $\Gamma^4(h_1)$, since by doing so we would remove at h_1 all your strategies leading to h_1. No further strategies can be eliminated at step 5. This leads to the final decision problems in Table 9.12, from which no strategies can be removed. So, the only strategies that survive the iterated conditional dominance procedure are the strategies $(don't,B)$ for you, and (B,B) for Barbara. We have seen that these are also the only strategies that can rationally be chosen under common strong belief in rationality. □

9.4 Comparison with backward dominance procedure

In this section we will compare in more detail the iterated conditional dominance procedure with the backward dominance procedure from Chapter 8. The backward dominance procedure is characterized by eliminating strategies backwards – that is, if a strategy s_i is strictly dominated in a decision problem at information set h where i is active, then we eliminate s_i at h, but also at all decision problems *before* h.

The iterated conditional dominance procedure works differently: If strategy s_i is strictly dominated in a decision problem at information set h where i is active, then we eliminate s_i at h and all decision problems before h, and also from all decision problems *after* h – as long as we do not remove all strategy combinations leading to that information set. In other words, the iterated conditional dominance procedure eliminates *backwards and forwards.*

Additionally, if s_i is strictly dominated at h, then it is also eliminated at information sets that do not come before, nor after, h. Consider, for instance, the example "Watching

TV with Barbara." In step 2 of the iterated conditional dominance procedure, Barbara's strategies (D,B) and (D,D) are strictly dominated at h_1, and we also eliminate these strategies at h_2, which is an information set that does not come before, nor after, h_1. We say that h_1 and h_2 are "parallel" information sets. We can thus say that the iterated conditional dominance procedure works by eliminating strategies backwards, forwards and in parallel.

At first sight, this could suggest that in the iterated conditional dominance procedure we always eliminate more strategies at every step than in the backward dominance procedure, and that therefore the iterated conditional dominance procedure is always more restrictive. This is not true, however! Note that in the iterated conditional dominance procedure we always have the "unless" at every elimination step. From the decision problem $\Gamma^k(h)$ at h we eliminate the strategies as specified in the algorithm *unless* this would remove all strategy combinations leading to h. In the latter case, we do not remove any strategies from $\Gamma^k(h)$. At the same time, it is possible that in the latter case we would still eliminate strategies at h in the backward dominance procedure. So, it is not true that the iterated conditional dominance procedure always eliminates more than the backward dominance procedure.

In fact, there is no logical relation, in terms of strategy choices, between the outputs of the iterated conditional dominance procedure and the backward dominance procedure – sometimes the former is more restrictive, sometimes the latter, and it may also happen that both procedures yield completely different sets of strategies for a given player. Consider, for instance, the example "Painting Chris' house" with the graphical representation in Figure 9.1. We have seen that the iterated conditional dominance procedure uniquely selects the strategy 300 for you, whereas the backward dominance procedure uniquely selects the strategy 200 for you. Hence, in this example the two procedures yield completely different results.

Now, suppose that in this example Barbara would receive 500 instead of 350 by accepting her colleague's offer. Then, it can be shown that the iterated conditional dominance procedure would select the strategies 200, 300 and 400 for you, whereas the backward dominance procedure would still uniquely select your strategy 200. So, in this modified example, the backward dominance procedure would be more restrictive in terms of strategies for you.

Next, consider the example "Watching TV with Barbara." As we have seen, the iterated conditional dominance procedure uniquely filters the strategy $(don't,B)$ for you. The reader may verify that the backward dominance procedure only eliminates the strategy $(fight,D)$ for you at $\Gamma^0(\emptyset)$, after which the procedure terminates. Hence, the backward dominance procedure selects the strategies $(fight,B)$, $(don't,B)$ and $(don't,D)$ for you. That is, in this example the iterated conditional dominance procedure is more restrictive in terms of possible strategy choices for you.

Overall, we may thus conclude that, in terms of strategy choices, there is no general logical relation between the iterated conditional dominance procedure and the backward dominance procedure.

What can we say about dynamic games with *perfect information* – that is, games in which the players choose one at a time, and always observe the choices made by their opponents in the past? We saw in the previous chapter that in such games, the backward dominance procedure is equivalent to the backward induction procedure. As a consequence, in a game with perfect information the strategies that can rationally be chosen under common belief in future rationality are precisely the backward induction strategies.

Is this also true for common strong belief in rationality and the associated iterated conditional dominance procedure? The answer is "no"! To show this, we will now give an example of a dynamic game with perfect information, in which the strategies that can rationally be chosen under common strong belief in rationality are different from the backward induction strategies. Even stronger, for one of the players there is a unique strategy that can rationally be chosen under common strong belief in rationality, and this strategy is different from the unique backward induction strategy in the game.

Example 9.7 The heat of the fight

It is Wednesday evening, and you and Barbara are facing the weekly problem of deciding which program to watch on TV: *Blackadder* or *Dallas*. Remember that you prefer *Blackadder* whereas Barbara prefers *Dallas.* Suppose, as before, that watching *Blackadder* would give you a utility of 6 and Barbara a utility of 3, whereas for watching *Dallas* it is the other way around. At the beginning of the evening, Barbara can choose either to be *nice* to you and let you choose your favorite program, or to *argue* with you about the program. If she starts to *argue* then either you can be *nice* to her and let her choose her favorite program, in order to avoid any escalation of the conflict, or you can decide to *shout* at her in response. If you decide to *shout* at her then Barbara has the option of being *nice* to you and letting you choose your favorite program, to prevent the situation from getting out of hand, or she can start throwing *dishes* on the floor as a sign of her anger. If she throws *dishes* on the floor then, because the situation has gone completely out of hand, you can only choose between two extremes: either you apologize to her and say that you are *sorry* about this whole fight, and addition let her choose her favorite program, or you walk *out* the door and watch *Blackadder* at Chris' house.

Assume that the utility for both you and Barbara decreases by 5 every time the conflict escalates. However, if you apologize to Barbara then this would increase Barbara's utility by 15. Moreover, if you decide to walk out the door and visit Chris, then this would increase your utility by 15 since Chris will serve some nice beer and potato chips to make you feel better. This situation can be represented graphically as in Figure 9.3. Here, the first utility corresponds to Barbara (player 1) and the second utility to you (player 2). Note that if you walk out the door at the end, then Barbara will watch her favorite program *Dallas* but the conflict would have escalated three times. So, her utility in that case is $6 - 15 = -9$.

Let us first analyze which strategy, or strategies, you can rationally choose under common belief in future rationality. Since this is a dynamic game with perfect information, we can find these strategies with the backward induction procedure. So we start at the

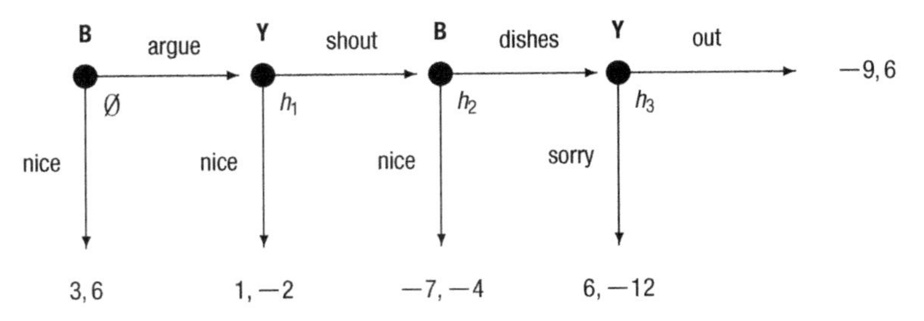

Figure 9.3 The heat of the fight

end of the game, where you must choose between saying *sorry* and walking *out* the door. Clearly, the optimal choice for you is to walk *out.*

We then move to the penultimate information set, where Barbara must choose between being *nice* to you and throwing *dishes* on the floor. The optimal choice for Barbara, given that we have selected the unique backward induction choice *out* for you at the final information set, is to be *nice* to you.

Next, we analyze the information set where you must choose between being *nice* to Barbara or to *shout* at her. Given that we have selected the backward induction choice *nice* for Barbara at the information set that follows, the optimal choice for you here is to be *nice* to Barbara.

Finally, we investigate the beginning of the game. As we have selected the backward induction choice *nice* for you at the information set that follows, the backward induction choice for Barbara at the beginning of the evening is to be *nice* to you, and let you choose your favorite program.

So, we see that the unique backward induction strategy for you in this game is to be *nice* to Barbara in case she starts arguing about the program. Hence, the only strategy that you can rationally choose under common belief in future rationality is to be *nice* to Barbara.

We will now explore which strategy, or strategies, you can rationally choose under common strong belief in rationality. To answer this question we will use the iterated conditional dominance procedure. Let us denote the four information sets in this game by $\emptyset$ (the beginning), h_1, h_2 and h_3. The full decision problems for these four information sets are given in Table 9.13. The strategies for Barbara are in the rows, and the strategies for you in the columns.

Step 1: Note that in $\Gamma^0(\emptyset)$, Barbara's strategy (*argue*, *nice*) is strictly dominated by her strategy *nice*. So, we eliminate her strategy (*argue*, *nice*) from the decision problems at $\emptyset, h_1$ and h_2. Moreover, in $\Gamma^0(h_1)$ your strategy (*shout*, *sorry*) is strictly dominated by your strategy *nice*, and in $\Gamma^0(h_3)$ the same strategy (*shout*, *sorry*) is strictly dominated by your strategy (*shout*, *out*). Hence, we eliminate your strategy (*shout*, *sorry*) from the decision problems at $\emptyset, h_1, h_2$ and h_3. This leads to the reduced decision problems shown in Table 9.14.

Table 9.13. *Full decision problems in "The heat of the fight"*

	$\Gamma^0(\emptyset)$: Barbara active		
	nice	(*shout*, *sorry*)	(*shout*, *out*)
nice	3, 6	3, 6	3, 6
(*argue*, *nice*)	1, −2	−7, −4	−7, −4
(*argue*, *dishes*)	1, −2	6, −12	−9, 6
	$\Gamma^0(h_1)$: You active		
	nice	(*shout*, *sorry*)	(*shout*, *out*)
(*argue*, *nice*)	1, −2	−7, −4	−7, −4
(*argue*, *dishes*)	1, −2	6, −12	−9, 6
	$\Gamma^0(h_2)$: Barbara active		
	(*shout*, *sorry*)	(*shout*, *out*)	
(*argue*, *nice*)	−7, −4	−7, −4	
(*argue*, *dishes*)	6, −12	−9, 6	
	$\Gamma^0(h_3)$: You active		
	(*shout*, *sorry*)	(*shout*, *out*)	
(*argue*, *dishes*)	6, −12	−9, 6	

Step 2: In the reduced decision problem in $\Gamma^1(\emptyset)$ at the beginning, Barbara's strategy (*argue*, *dishes*) is strictly dominated by her strategy *nice*. So, we eliminate her strategy (*argue*, *dishes*) from $\Gamma^1(\emptyset)$ but *not* from the decision problems at h_1, h_2 and h_3, since otherwise we would delete all of Barbara's strategies from the reduced decision problems at h_1, h_2 and h_3. Moreover, in $\Gamma^1(h_1)$ your strategy *nice* is strictly dominated by your strategy (*shout*, *out*), and hence we eliminate your strategy *nice* from $\Gamma^1(\emptyset)$ and $\Gamma^1(h_1)$. This leads to the reduced decision problems shown in Table 9.15, from which no further strategies can be eliminated.

So, the only strategies that survive the iterated conditional dominance procedure are the strategy *nice* for Barbara and the strategy (*shout*, *out*) for you. As a consequence, (*shout*, *out*) is the only strategy you can rationally choose under common strong belief in rationality. This is different from the unique strategy you can rationally choose under common belief in future rationality, which was the strategy *nice* as we have seen.

There is an easy intuitive argument for why common belief in future rationality and common strong belief in rationality lead to different strategy choices for you in this example. Suppose that at h_1 you observe that Barbara is *arguing* with you about the TV program. Under common belief in future rationality, you believe that Barbara will choose rationally at h_2, and you believe that Barbara believes that you will chooses

Table 9.14. *Reduced decision problems after step 1 of the iterated conditional dominance procedure in "The heat of the fight"*

	$\Gamma^1(\emptyset)$: Barbara active	
	nice	*(shout, out)*
nice	3, 6	3, 6
(argue, dishes)	1, −2	−9, 6

	$\Gamma^1(h_1)$: You active	
	nice	*(shout, out)*
(argue, dishes)	1, −2	−9, 6

	$\Gamma^1(h_2)$: Barbara active
	(shout, out)
(argue, dishes)	−9, 6

	$\Gamma^1(h_3)$: You active
	(shout, out)
(argue, dishes)	−9, 6

rationally at h_3. Your only optimal choice at h_3 is to walk *out*, so you believe at h_1 that Barbara believes at h_2 that you will choose *out* at h_3. If Barbara indeed believes so, her optimal choice at h_2 would be to be *nice* to you. Since you believe at h_1 that Barbara chooses rationally at h_2, you believe at h_1 that Barbara will indeed be *nice* to you at h_2, and hence you will be *nice* to her at h_1. So, under common belief in future rationality, your optimal strategy is to be *nice* to Barbara in case she starts to *argue*.

Under common strong belief in rationality your reasoning will be crucially different at h_1. Suppose that at h_1 you observe that Barbara is *arguing* about the program, and assume you strongly believe in Barbara's rationality at h_1. Then you ask whether her choice to *argue* could be part of an optimal strategy for Barbara. This is indeed possible, but only if she will go on to throw *dishes* on the floor, hoping that you will apologize. This is the only way for her to get a utility higher than 3 – the utility she gets by being *nice* to you at the beginning. Hence, if you strongly believe in Barbara's rationality, then you must believe at h_1 that Barbara is implementing the strategy (*argue*, *dishes*), and therefore your only optimal strategy choice at h_1 is to choose (*shout*, *out*). So, under common strong belief in rationality, the optimal thing for you to do is to *shout* if she starts arguing with you, and walk *out* the door if she throws dishes on the floor. □

The example above thus shows that even in dynamic games with perfect information, common strong belief in rationality and common belief in future rationality – or

Table 9.15. *Final decision problems for the iterated conditional dominance procedure in "The heat of the fight"*

	$\Gamma^2(\emptyset)$: Barbara active
	(*shout*, *out*)
nice	3,6
	$\Gamma^2(h_1)$: You active
	(*shout*, *out*)
(*argue*, *dishes*)	−9,6
	$\Gamma^2(h_2)$: Barbara active
	(*shout*, *out*)
(*argue*, *dishes*)	−9,6
	$\Gamma^2(h_3)$: You active
	(*shout*, *out*)
(*argue*, *dishes*)	−9,6

equivalently the iterated conditional dominance procedure and the backward induction procedure – may yield completely different strategy choices for you. Hence, in terms of strategy choices there is no general logical relation between the two concepts, even with a restriction to dynamic games with perfect information.

Things are different, however, if we focus on the outcomes – that is, the terminal histories – you deem possible before the game starts. In each of the examples we discussed above, every outcome you deem possible before the game starts under common strong belief in rationality is also initially deemed possible under common belief in future rationality! In the example "Painting Chris' house," for instance, there is only one outcome you initially deem possible under common strong belief in rationality and under common belief in future rationality: namely that Barbara will accept her colleague's offer at the beginning. The same holds if, in this example, Barbara will receive 500 euros instead of 350 euros by accepting her colleague's offer. Note, however, that both concepts differ in the restrictions on your possible choices if you surprisingly observe that Barbara has rejected her colleague's offer – something you initially believe will not happen. In the example "Watching TV with Barbara" there is only one outcome you initially deem possible under common strong belief in rationality, namely that you will not start a fight with Barbara, and that you will both watch your favorite program *Blackadder* together. This is also one of the outcomes you initially deem possible under common belief in future rationality. In the example "The heat of the fight," the only outcome you initially deem possible under common strong belief in rationality and under common belief in future rationality, is that Barbara will be *nice* to you at the beginning of the game. We have seen, however, that both concepts differ in the restrictions on your behavior if you surprisingly observe that Barbara has started to *argue* with you – something you initially believe will not happen.

In fact, we can show that this is always true. That is, in every dynamic game, any outcome you initially deem possible under common strong belief in rationality is also initially deemed possible under common belief in future rationality. Before we state this result formally, we will first define precisely what we mean by "outcomes that are initially deemed possible under common strong belief in rationality and under common belief in future rationality."

Definition 9.4.1 *(Outcomes that are initially deemed possible)*
*You **initially** deem an outcome z **possible** under common strong belief in rationality (common belief in future rationality) if there is a strategy combination leading to z where every strategy can rationally be chosen under common strong belief in rationality (common belief in future rationality).*

Here, by an outcome z we formally mean a terminal history z in the dynamic game – that is, a situation where the game ends. Note that under common strong belief in rationality you believe, before the start of the game, that every opponent will choose a strategy that can rationally be chosen under common strong belief in rationality, and the same applies to common belief in future rationality. So, the definition above matches exactly our intuition. We can now formally state the result we announced above.

Theorem 9.4.2 *(Outcomes under common strong belief in rationality and common belief in future rationality)*
Every outcome you initially deem possible under common strong belief in rationality is also initially deemed possible under common belief in future rationality.

The proof for this result can be found in the proofs section at the end of this chapter. Note that the converse of this theorem is not true: In the example "Watching TV with Barbara," there is only one outcome you initially deem possible under common strong belief in rationality, namely that you will not start a fight with Barbara and that you will both watch *Blackadder.* However, under common belief in future rationality initially you also deem possible other outcomes – for instance that you will start a fight with Barbara, that you will choose *Blackadder* and that Barbara will choose *Dallas.*

The theorem above has important consequences for games with perfect information. We saw in Chapter 8 that in games with perfect information, the strategies that can rationally be chosen under common belief in future rationality are exactly the backward induction strategies. Now, call an outcome in such a game a *backward induction outcome* if there is a combination of backward induction strategies that leads to this outcome. So, in a game with perfect information, the outcomes you initially deem possible under common belief in future rationality are precisely the backward induction outcomes. By the theorem above, we may thus conclude that in games with perfect information, every outcome you initially deem possible under common strong belief in rationality must be a backward induction outcome.

Corollary 9.4.3 *(Common strong belief in rationality leads to backward induction outcomes)*
In a game with perfect information, every outcome you initially deem possible under common strong belief in rationality must be a backward induction outcome.

However, under common strong belief in rationality you may no longer reason in accordance with backward induction if you surprisingly observe that your opponent has not chosen in accordance with backward induction – something you initially believe will not happen. Consider, for instance, the example "The heat of the fight," where you initially believe that Barbara will make the backward induction choice to be *nice* to you at the beginning. However, if you observe at h_1 that Barbara has not chosen the backward induction strategy *nice*, then under common strong belief in rationality you believe that she will subsequently make the non-backward induction choice to throw *dishes* on the floor. Consequently, you would choose $(shout, out)$ rather than your backward induction strategy *nice*.

9.5 Order dependence

In this section we will discuss the issue of changing the order and speed of elimination in the iterated conditional dominance procedure. A very important property of the backward dominance procedure, as we saw in Chapter 8, is that its outcome does not depend on the order and speed of elimination. Even if we do not eliminate, at a given round and at a given information set, all strategies we can, then the final outcome of the procedure will not be affected – at the end, the set of strategies that survive will be exactly the same as when we would always eliminate everything we can at every round and every information set.

This does not hold for the iterated conditional dominance procedure, however! For this procedure to give the "correct" result, it is absolutely crucial that at every round and every information set, we always eliminate *all* strategies that we can. Otherwise, the set of surviving strategies at the end could be different. As an illustration, consider the example "Painting Chris' house."

Example 9.8 Painting Chris' house

Consider the graphical representation in Figure 9.1, and the full decision problems $\Gamma^0(\emptyset)$ and $\Gamma^0(h_1)$ in Table 9.4. We have seen that for Barbara, the strategies $(r, 200)$, $(r, 300)$ and $(r, 500)$ are all strictly dominated by *accept* in the decision problem $\Gamma^0(\emptyset)$. Therefore, according to the original iterated conditional dominance procedure, we must eliminate each of these strategies from $\Gamma^0(\emptyset)$ and $\Gamma^0(h_1)$. Suppose now that from the decision problem $\Gamma^0(h_1)$ at h_1 we eliminate *only* the strategies $(r, 300)$ and

Table 9.16. *Changing the order of elimination in "Painting Chris' house": Step 1*

	$\hat{\Gamma}^1(\emptyset)$: Barbara active		
	200	300	400
$(r,400)$	$0,200$	$0,300$	$200,200$
accept	$350,500$	$350,500$	$350,500$
	$\hat{\Gamma}^1(h_1)$: Barbara and you active		
	200	300	400
$(r,200)$	$100,100$	$200,0$	$200,0$
$(r,400)$	$0,200$	$0,300$	$200,200$

Table 9.17. *Changing the order of elimination in "Painting Chris' house": Step 2*

	$\hat{\Gamma}^2(\emptyset)$: Barbara active	
	200	300
accept	$350,500$	$350,500$
	$\hat{\Gamma}^2(h_1)$: Barbara and you active	
	200	300
$(r,200)$	$100,100$	$200,0$

$(r,500)$ for Barbara, but not the strategy $(r,200)$, in step 1. Assume that we still remove your strategy 500 from $\Gamma^0(\emptyset)$ and $\Gamma^0(h_1)$, as it is strictly dominated in $\Gamma^0(h_1)$. This gives the reduced decision problems $\hat{\Gamma}^1(\emptyset)$ and $\hat{\Gamma}^1(h_1)$ shown in Table 9.16.

In the decision problem $\hat{\Gamma}^1(\emptyset)$, Barbara's strategy $(r,400)$ is strictly dominated, and hence we can eliminate it from $\hat{\Gamma}^1(\emptyset)$ and $\hat{\Gamma}^1(h_1)$. Note that by doing so we will not eliminate all of Barbara's strategies leading to h_1, as $\hat{\Gamma}^1(h_1)$ still contains another strategy $(r,200)$ for Barbara. Moreover, at $\hat{\Gamma}^1(h_1)$, your strategy 400 is strictly dominated by the randomized strategy $(0.5)\cdot 200+(0.5)\cdot 300$, and therefore we can eliminate your strategy 400 from $\hat{\Gamma}^1(\emptyset)$ and $\hat{\Gamma}^1(h_1)$. This leads to the new decision problems $\hat{\Gamma}^2(\emptyset)$ and $\hat{\Gamma}^2(h_1)$ in Table 9.17.

In the decision problem $\hat{\Gamma}^2(h_1)$, your strategy 300 is strictly dominated, and hence we can eliminate it from $\hat{\Gamma}^2(\emptyset)$ and $\hat{\Gamma}^2(h_1)$. This leads to the final decision problems shown in Table 9.18.

Table 9.18. *Changing the order of elimination in "Painting Chris' house": Step 3*

	$\hat{\Gamma}^3(\emptyset)$: Barbara active
	200
accept	350, 500

	$\hat{\Gamma}^3(h_1)$: Barbara and you active
	200
$(r, 200)$	100, 100

So, by changing the order of elimination in this way, the unique strategy for you that would survive is 200. This, however, is different from the unique strategy for you that survives the *original* iterated conditional dominance procedure – namely 300. Hence, changing the order and speed of elimination may drastically change the outcome of the iterated conditional dominance procedure! □

This example thus shows that, if we use the iterated conditional dominance procedure, there is no other way than to eliminate, at every round and every information set, all strategies we can. If we do not, we run the danger of ending up with a different set of strategies. In particular, we cannot look for a "most convenient order of elimination" – as we often did for the backward dominance procedure – when working with the iterated conditional dominance procedure. For instance, we cannot use the "backwards order of elimination," which turned out to be very useful for the backward dominance procedure in many games. This makes the iterated conditional dominance procedure somewhat harder to work with than the backward dominance procedure, especially when a dynamic game is large.

9.6 Rationality orderings

In a sense, common strong belief in rationality groups the strategies of each player into classes, ranging from "most rational" to "least rational," with typically some classes in between. First, we have the strategies that cannot be optimal under any conditional belief vector – the "least rational strategies" or 0-rational strategies. Then, we have the strategies that are optimal for some conditional belief vector, but which cannot rationally be chosen under 1-fold strong belief in rationality. We call these strategies the 1-rational strategies, as they are "more rational" than the 0-rational strategies – they can rationally be chosen under *some* conditional belief vector – but they are less rational than those which can rationally be chosen under 1-fold strong belief in rationality. We then have the 2-rational strategies, which are the ones that can rationally be chosen under 1-fold strong belief in rationality, but not when expressing up to 2-fold strong belief in rationality, and so on.

This process must stop eventually, as for every game there is a number K such that common strong belief in rationality is equivalent to expressing up to K-fold strong belief in rationality. So, in this way we obtain for every player i an ordering $(D_i^0, D_i^1, ..., D_i^K)$ of his strategies into subclasses, where D_i^0 is the set of 0-rational strategies, D_i^1 is the set of 1-rational strategies, and so on, until we reach D_i^K, which is the class of K-rational, or "most rational" strategies. For every $k \in \{1, ..., K-1\}$, the class D_i^k is the set of k-rational strategies, and it contains those strategies that can rationally be chosen by a player who expresses up to $(k-1)$-fold strong belief in rationality, but not by a player who expresses up to k-fold strong belief in rationality. The class D_i^K – containing the "most rational" strategies – consists of those strategies that can rationally be chosen under common strong belief in rationality. Finally, the class D_i^0 – the "least rational" strategies – contains those strategies that are not optimal for any conditional belief vector. Such an ordering of strategies into subclasses is called a *rationality ordering.*

The rationality ordering $(D_i^0, D_i^1, ..., D_i^K)$ induced by common strong belief in rationality has a clear intuitive meaning: For every $k \in \{1, ..., K-1\}$, the k-rational strategies are "more rational" than the $(k-1)$-rational strategies, but "less rational" than the $(k+1)$-rational strategies. So, it orders player i's strategies from "most rational" to "least rational," with usually some levels in between. Of course, the terms "more rational," "less rational," "most rational" and "least rational" are all subjective, as it reflects the particular viewpoint taken by the concept of common strong belief in rationality.

As an illustration, let us return to the example "Watching TV with Barbara."As we know from Theorem 9.3.3, there is an easy way to find those strategies that player i can rationally choose by expressing up to k-fold strong belief in rationality. These are precisely the strategies in $\Gamma^{k+1}(\emptyset)$ – the reduced decision problem at the beginning of the game after the first $k+1$ rounds of the iterated conditional dominance procedure. In Example 9.6 we computed these decision problems $\Gamma^k(\emptyset)$, and for convenience we have reproduced the results in Table 9.19.

With the help of these decision problems $\Gamma^k(\emptyset)$ we can now easily derive the rationality ordering induced on your strategies. The set D_1^0 – containing your "least rational" or 0-rational strategies – are those strategies that are not optimal for any conditional belief vector. These are precisely your strategies which are in $\Gamma^0(\emptyset)$ but not in $\Gamma^1(\emptyset)$, which in this case is only your strategy (fight, D). So, $D_1^0 = \{(\mathit{fight}, D)\}$.

The set D_1^1 of 1-rational strategies contains those strategies for you that can rationally be chosen under some conditional belief vector, but not if you express 1-fold strong belief in rationality. These are the strategies for you that are in $\Gamma^1(\emptyset)$ but not in $\Gamma^2(\emptyset)$, and there are none. So, the set D_1^1 of 1-rational strategies is empty.

The set D_1^2 of 2-rational strategies contains those strategies for you that can rationally be chosen when expressing 1-fold strong belief in rationality, but not when

Table 9.19. *The reduced decision problems $\Gamma^k(\emptyset)$ for the example "Watching TV with Barbara"*

	Your strategies	Barbara's strategies
$\Gamma^0(\emptyset)$	$(fight,B),(fight,D),$ $(don't,D),(don't,B)$	$(B,B),(B,D),(D,B),(D,D)$
$\Gamma^1(\emptyset)$	$(fight,B),(don't,D),(don't,B)$	$(B,B),(B,D),(D,B),(D,D)$
$\Gamma^2(\emptyset)$	$(fight,B),(don't,D),(don't,B)$	$(B,B),(B,D)$
$\Gamma^3(\emptyset)$	$(fight,B),(don't,B)$	$(B,B),(B,D)$
$\Gamma^4(\emptyset)$	$(fight,B),(don't,B)$	(B,B)
$\Gamma^5(\emptyset)$	$(don't,B)$	(B,B)

expressing up to 2-fold strong belief in rationality. These are precisely the strategies for you that are in $\Gamma^2(\emptyset)$ but not in $\Gamma^3(\emptyset)$, that is, your strategy $(don't,D)$. So, $D_1^2 = \{(don't,D)\}$. By continuing in this fashion, you will find that D_1^3 is empty and that $D_1^4 = \{(fight,B)\}$.

Let us finally focus on D_1^5 – the set of "most rational" or 5-rational strategies for you. This set contains those strategies that you can rationally choose under common strong belief in rationality. These are precisely the strategies for you in $\Gamma^5(\emptyset)$ – that is, $(don't,B)$ – as the iterated conditional dominance procedure stops after step 5. Hence, $D_1^5 = \{(don't,B)\}$.

So, the rationality ordering on your strategies induced by common strong belief in rationality is

$$(\{(fight,D)\},\emptyset,\{(don't,D)\},\emptyset,\{(fight,B)\},\{(don't,B)\}).$$

In a similar way, the reader may verify that the rationality ordering on Barbara's strategies is given by

$$(\emptyset,\{(D,B),(D,D)\},\emptyset,\{(B,D)\},\emptyset,\{(B,B)\}).$$

That is, the set of 0-rational strategies for Barbara is empty, the set of 1-rational strategies is $\{(D,B),(D,D)\}$, the set of 2-rational strategies is empty, the set of 3-rational strategies is $\{(B,D)\}$, the set of 4-rational strategies is empty, and the set of 5-rational, or "most rational," strategies is $\{(B,B)\}$.

In general, a rationality ordering for player i in an arbitrary dynamic game is an ordering of his strategies into subclasses, ranging from the "least rational" to the "most rational strategies." Here is a formal definition.

Definition 9.6.1 *(Rationality ordering)*
*A **rationality ordering** for player i in a dynamic game is a sequence $R_i = (D_i^0, D_i^1, ..., D_i^K)$ where $D_i^0, ..., D_i^K$ are disjoint sets of strategies for player i whose union is equal to the full set of strategies S_i. Some of these sets may be empty, but not D_i^K. For every k, all strategies in D_i^k are called k-**rational** strategies. The interpretation is that D_i^0 contains the "least rational" strategies, D_i^K the "most rational strategies," and all strategies in D_i^{k+1} are "more rational" than all strategies in D_i^k.*

Here, by "disjoint" we mean that the sets $D_i^0, ..., D_i^K$ have no overlap. That is, every strategy for player i is in exactly one of the sets from $D_i^0, ..., D_i^K$.

Above we have seen how the concept of common strong belief in rationality naturally induces a rationality ordering for every player. This is just one example of a rationality ordering, however, as a rationality ordering may in fact be *any* ordering of strategies according to the definition above. But not all rationality orderings make intuitive sense! If one takes a rationality ordering $R_i = (D_i^0, D_i^1, ..., D_i^K)$, then there must be a good reason why D_i^0 contains the "least rational" strategies, why D_i^K contains the "most rational strategies," and why strategies in D_i^{k+1} are deemed "more rational" than strategies in D_i^k.

Throughout this book, we have taken the perspective that "more rational" strategies are the ones that are supported by "more rational" beliefs. Now, in order to classify the beliefs that player i has about his opponents' strategy combinations, we must look at the rationality orderings of opponents' strategies. Suppose that for every opponent j we have a rationality ordering $R_j = (D_j^0, D_j^1, ..., D_j^K)$ of his strategies. Then, a "most rational" belief for player i assigns at every information set $h \in H_i$ only positive probability to the "most rational" strategies of the opponents, if possible. That is, if at information set $h \in H_i$ there is a combination of opponents' strategies leading to h consisting only of "most rational" strategies, then a "most rational" belief should at h only assign positive probability to opponents' strategy combinations consisting of "most rational" strategies. Moreover, if there is no combination of opponents' strategies leading to h consisting only of "most rational" strategies, but there is a combination of opponents' strategies leading to h consisting of at-least-$(K-1)$-rational strategies, then a "most rational" belief for player i should at h only assign positive probability to opponents' strategy combinations consisting of at-least-$(K-1)$-rational strategies. Here, by " at-least-$(K-1)$-rational" we mean $(K-1)$-rational or K-rational. And if at h there is no combination of opponents' strategies leading to h consisting of at-least-$(K-1)$-rational strategies, but there is a combination of opponents' strategies leading to h consisting of at-least-$(K-2)$-rational strategies, then a "most rational" belief for player i should at h only assign positive probability to opponents' strategy combinations consisting of at-least-$(K-2)$-rational strategies, and so on. That is, a "most rational" belief for player i looks at every information set h for the highest degree k such that there is a combination of at-least-k-rational strategies leading to h, and assigns at h only positive probability to at-least-k-rational strategies for the opponents. In other words, at every information set $h \in H_i$ player i allocates the highest possible rationality level to his opponents, according to their respective rationality orderings. We say that player i *strongly believes in the opponents' rationality orderings.*

Definition 9.6.2 *(Strong belief in the opponents' rationality orderings)*
Consider a player i in a dynamic game. For every opponent j, let $R_j = (D_j^0, D_j^1, ..., D_j^K)$ be a rationality ordering, with the same K for every opponent. At information set $h \in H_i$, let k be the highest degree such that there is a combination of at-least-k-rational strategies leading to h. Then, a conditional belief vector for player i ***strongly believes in the opponents' rationality orderings*** *at h if it assigns at h only positive probability to opponents' combinations of at-least-k-rational strategies.*
A conditional belief vector for player i strongly believes in the opponents' rationality orderings if it does so at every information set $h \in H_i$.

Here, by "at-least-k-rational" we mean a strategy that is k-rational or higher. So, our requirement will be that a "most rational" belief for player i should strongly believe in the opponents' rationality orderings.

But what about "less rational" beliefs, which would support "less rational" strategies? We can weaken the requirement of strong belief in the opponents' rationality orderings in the following way: Choose some level $m < K$, and require that player i only strongly believes in the opponents' rationality orderings *up to level m*. More precisely, for every information set $h \in H_i$ we first check whether there is a combination of at-least-m-rational strategies for the opponents that leads to h. If so, then we require player i at h to only assign positive probability to at-least-m-rational strategies for the opponents, but *not necessarily to K*-rational strategies. So, even if there is a combination of K-rational strategies leading to h, we only require player i to restrict at h to at-least-m-rational strategies for the opponents. If there is no combination of at-least-m-rational strategies leading to h, we check whether there is a combination of at-least-$(m-1)$-rational strategies for the opponents leading to h. If so, then we require player i at h to only assign positive probability to at-least-$(m-1)$-rational strategies for the opponents. And so on. This leads to the following formal definition.

Definition 9.6.3 *(Strong belief in the opponents' rationality orderings up to level m)*
Consider a player i in a dynamic game. For every opponent j, let $R_j = (D_j^0, D_j^1, ..., D_j^K)$ be a rationality ordering, with the same K for every opponent. Fix a level $m \leq K$. At information set $h \in H_i$, let k be the highest degree **less than or equal to** *m such that there is a combination of at-least-k-rational strategies for the opponents leading to h. Then, a conditional belief vector for player i* ***strongly believes in the opponents' rationality orderings up to level*** *m at h if it assigns at h only positive probability to opponents' combinations of at-least-k-rational strategies.*
A conditional belief vector for player i strongly believes in the opponents' rationality orderings up to level m if it does so at every information set $h \in H_i$.

So, one could say that under this weaker requirement, player i views all of opponents' strategies that are m-rational or higher as "equally rational." For instance, if $m = K - 2$, then player i does not make a distinction between the opponents' strategies that are $(K-2)$-rational, $(K-1)$-rational or K-rational – they are all viewed as being "equally rational" by player i. Note that strong belief in the opponents' rationality orderings, as formulated in the previous definition, is the same as strong belief in the opponents' rationality orderings up to level K – the highest possible level. At the other extreme, by choosing $m = 0$, we impose no conditions at all on the conditional belief vectors for player i.

With the above definition, we can now naturally classify the players' beliefs from "most rational" to "least rational." For each player, consider a rationality ordering on strategies with maximum level K. Then, the "most rational" belief vectors for player i would be those that strongly believe in the opponents' rationality orderings up to level K. The "least rational" belief vectors would be those that strongly believe in the opponents' rationality orderings up to level 0, but not up to level 1. And for every level k in between, the k-rational belief vectors are those that strongly believe in the opponents' rationality orderings up to level k, but not up to level $k+1$.

This, in turn, induces a natural classification of player i's *strategies* from "most rational" to "least rational"! Namely, player i's "most rational" strategies would intuitively be those that are optimal under a "most rational " belief vector – that is, a belief vector that strongly believes in the opponents' rationality orderings up to level K. Moreover, for every k between 1 and $K-1$, the k-rational strategies for player i are intuitively those that are optimal under a conditional belief vector that strongly believes in the opponents' rationality orderings up to level $k-1$, but not up to level k.

Summarizing, we see that if we start with a rationality ordering $R_i = (D_i^0, D_i^1, ..., D_i^K)$ for every player i, with the same maximum level K for all players, then this induces for every player i a classification of his conditional belief vectors, from 0-rational ("least rational") to K-rational ("most rational"), which in turn induces for every player i a classification of his strategies from 0-rational to K-rational. Intuitively, the classification of strategies for player i from 0-rational to K-rational must be the same as the classification in the rationality ordering R_i we started with! Combinations of rationality orderings with this property are called *self-confirming*, as the *initial* classifications of strategies given by the original rationality orderings are *confirmed* by the *induced* classifications of strategies as described above. So, in a sense, the self-confirming combinations of rationality orderings are those where the classifications of strategies from "most rational" to "least rational" are well founded.

Definition 9.6.4 *(Self-confirming combinations of rationality orderings)*
*Consider for every player i a rationality ordering $R_i = (D_i^0, D_i^1, ..., D_i^K)$ on his strategies, with the same maximum level K for all players. Then, this combination of rationality orderings is **self-confirming** if for every player i:*

- *the set D_i^K contains exactly those strategies that are optimal under a conditional belief vector that strongly believes in the opponents' rationality orderings up to level K,*

- *for every k between 1 and K − 1, the set D_i^k contains exactly those strategies that are optimal under a conditional belief vector that strongly believes in the opponents' rationality orderings up to level k − 1, but not under a conditional belief vector that strongly believes in the opponents' rationality orderings up to level k, and*
- *the set D_i^0 contains those strategies that are not optimal under any conditional belief vector.*

The reader may verify that in the example "Watching TV with Barbara," the rationality orderings induced by common strong belief in rationality are indeed self-confirming. In fact, we can show that this is true for every dynamic game! That is, in every dynamic game, the combination of rationality orderings induced by common strong belief in rationality is always self-confirming. We can even show that this is the *only* self-confirming combination of rationality orderings in a game! Here is the reason why.

Consider an arbitrary dynamic game, and for each player i a rationality ordering $R_i = (D_i^0, D_i^1, ..., D_i^K)$ on his strategies, with the same K for all players. Suppose that this combination of rationality orderings is self-confirming. Then, by definition, the sets D_i^0 of 0-rational strategies should contain precisely those strategies that are not optimal under *any* conditional belief vector.

Turn now to the sets D_i^1. Since the combination of rationality orderings is self-confirming, D_i^1 contains precisely those strategies that are optimal under a conditional belief vector that strongly believes in the opponents' rationality orderings up to level 0, but which are not optimal under any belief vector that strongly believes in the opponents' rationality orderings up to level 1. Now, the strategies that are optimal under a conditional belief vector that strongly believes in the opponents' rationality orderings up to level 0 are simply the strategies that are optimal under *some* conditional belief vector, without any further restrictions.

But which strategies are optimal under a belief vector that strongly believes in the opponents' rationality orderings up to level 1? Strong belief in the opponents' rationality orderings up to level 1 means for player i that at every information set $h \in H_i$ he must check whether there is a combination of at-least-1-rational strategies for the opponents leading to h. If so, then he must assign at h positive probability only to at-least-1-rational strategies for the opponents. Since for every opponent j, the set D_j^0 of 0-rational strategies contains precisely the strategies that are not optimal for any conditional belief vector, it follows that for every opponent j the at-least-1-rational strategies must be precisely those strategies that are optimal for some belief vector. That is strong belief in the opponents' rationality orderings up to level 1 means for player i that at every information set $h \in H_i$ he must check whether there is a strategy combination for the opponents leading to h where every strategy is optimal for some belief vector. If so, then he must assign at h positive probability only to strategies for the opponents that are optimal for some belief vector. This, however, is exactly the condition of 1-fold strong belief in rationality! Thus, the strategies that are optimal under a belief vector that strongly believes in the opponents' rationality orderings up to level 1, are precisely the strategies that can rationally be chosen under 1-fold strong belief in rationality.

Remember that D_i^1 contains precisely those strategies that are optimal under a conditional belief vector that strongly believes in the opponents' rationality orderings up to level 0, but that are not optimal under any belief vector that strongly believes in the opponents' rationality orderings up to level 1. But then, the set D_i^1 of 1-rational strategies contains precisely those strategies that can rationally be chosen under some conditional belief vector, but not under 1-fold strong belief in rationality. So, these are exactly the 1-rational strategies induced by common strong belief in rationality!

Next, consider the sets D_i^2. As the combination of rationality orderings is self-confirming, D_i^2 contains precisely those strategies that are optimal under a conditional belief vector that strongly believes in the opponents' rationality orderings up to level 1, but that are not optimal under any belief vector that strongly believes in the opponents' rationality orderings up to level 2. We have seen above that the strategies that are optimal under a conditional belief vector that strongly believes in the opponents' rationality orderings up to level 1 are precisely those strategies that can rationally be chosen under 1-fold strong belief in rationality.

But which strategies can rationally be chosen under strong belief in the opponents' rationality orderings up to level 2? For player i, strong belief in the opponents' rationality orderings up to level 2 means that at every information set $h \in H_i$ he must check whether there is a strategy combination leading to h consisting of at-least-2-rational strategies. If so, he must at h only assign positive probability to opponents' strategies that are at-least-2-rational. We have seen above that the set D_j^0 of 0-rational strategies contains those strategies that are not optimal for any beliefs and the set D_j^1 of 1-rational strategies contains those strategies that are optimal under some conditional belief vector, but not under 1-fold strong belief in rationality. But then, the at-least-2-rational strategies for opponent j are precisely those strategies that are optimal under 1-fold strong belief in rationality. Hence, for player i strong belief in the opponents' rationality orderings up to level 2 means that at every information set $h \in H_i$ he must check whether there is a strategy combination for the opponents leading to h where every strategy is optimal under 1-fold strong belief in rationality. If so, he must at h only assign positive probability to opponents' strategies that are optimal under 1-fold strong belief in rationality. But this is precisely the condition of 2-fold strong belief in rationality! Moreover, strong belief in the opponents' rationality orderings up to level 2 also means that at information set $h \in H_i$, player i must check whether there is a combination for the opponents of at-least-1-rational strategies leading to h. If so, then he must assign at h positive probability only to the at-least-1-rational strategies for the opponents. This, as we have seen, is equivalent to 1-fold strong belief in rationality. That is, strong belief in the opponents' rationality orderings up to level 2 is equivalent to expressing up to 2-fold strong belief in rationality. So, the strategies that can rationally be chosen under strong belief in the opponents' rationality orderings up to level 2 are precisely those strategies that can rationally be chosen when expressing up to 2-fold strong belief in rationality.

Remember that D_i^2 contains precisely those strategies that are optimal under a conditional belief vector that strongly believes in the opponents' rationality orderings up to level 1, but that are not optimal under any belief vector that strongly believes in the opponents' rationality orderings up to level 2. Hence, D_i^2 contains exactly those strategies that can rationally be chosen under 1-fold strong belief in rationality, but not when expressing up to 2-fold strong belief in rationality. So, these are exactly the 2-rational strategies induced by common strong belief in rationality.

By repeatedly using the arguments above, we conclude that for every k between 1 and $K-1$, the sets D_i^k of k-rational strategies contain precisely those strategies that can rationally be chosen by expressing up to $(k-1)$-fold strong belief in rationality, but not when expressing up to k-fold strong belief in rationality. That is, these sets D_i^k would be precisely the sets of k-rational strategies induced by common strong belief in rationality.

Let us finally consider the sets D_i^K of "most rational" strategies. As the combination of rationality orderings is self-confirming, D_i^K contains exactly those strategies that are optimal under strong belief in the opponents' rationality orderings up to level K. In particular, all strategies in D_i^K are optimal under strong belief in the opponents' rationality orderings up to level $K-1$. By using a similar argument as above, it can be shown that all these strategies are optimal when expressing up to $(K-1)$-fold strong belief in rationality.

Moreover, all strategies in D_i^K are also optimal under strong belief in the opponents' rationality orderings up to level K. Strong belief in the opponents' rationality orderings up to level K means in particular that player i, at every information set $h \in H_i$, must check whether there is a strategy combination for the opponents leading to h consisting of K-rational strategies – that is, strategies in D_j^K. If so, he must at h only assign positive probability to opponents' strategies s_j that are in D_j^K. We have seen that all of opponents' strategies in D_j^K are optimal when expressing up to $(K-1)$-fold strong belief in rationality. Hence, strong belief in the opponents' rationality orderings up to level K means that player i, at every information set $h \in H_i$, checks whether there is a strategy combination for the opponents leading to h in which every strategy is optimal when expressing up to $(K-1)$-fold strong belief in rationality. If so, he must at h only assign positive probability to opponents' strategies that are optimal when expressing up to $(K-1)$-fold strong belief in rationality. This, however, is exactly K-fold strong belief in rationality. That is, strong belief in the opponents' rationality orderings up to level K implies expressing up to K-fold strong belief in rationality. We have seen that all strategies in D_i^K are optimal under strong belief in the opponents' rationality orderings up to level K. It follows that all strategies in D_i^K can rationally be chosen when expressing up to K-fold strong belief in rationality.

So, not only can all strategies in D_i^K rationally be chosen when expressing up to $(K-1)$-fold strong belief in rationality as we have seen, but they can also rationally be chosen when expressing up to K-fold strong belief in rationality. By repeating this argument, we can show that all strategies in D_i^K can also rationally be chosen when expressing up to $(K+1)$-fold strong belief in rationality, when expressing up to $(K+2)$-fold strong belief in rationality, and so on. This, however, means that all strategies in D_i^K can rationally be chosen under common strong belief in rationality!

Summarizing, we see that:

- the sets D_i^0 contain exactly those strategies that are not optimal under any belief vector,
- for every k between 1 and $K-1$, the sets D_i^k contain precisely those strategies that can rationally be chosen when expressing up to $(k-1)$-fold strong belief in rationality, but not when expressing up to k-fold strong belief in rationality, and
- the sets D_i^K contain precisely those strategies that can rationally be chosen under common strong belief in rationality.

This means, however, that the rationality orderings $R_i = (D_i^0, D_i^1, ..., D_i^K)$ we started with, must be exactly the rationality orderings induced by common strong belief in rationality. So, we have shown that there is only one combination of rationality orderings that is self-confirming, namely the combination of rationality orderings induced by common strong belief in rationality. We have thus established the following general result.

Theorem 9.6.5 *(Characterization of self-confirming combinations of rationality orderings)*
Consider an arbitrary dynamic game, and let K be the number of steps after which the iterated conditional dominance procedure stops. Then, there is only one self-confirming combination of rationality orderings, that in which each rationality ordering R_i is given by $R_i = (D_i^0, D_i^1, ..., D_i^K)$, where: (1) the sets D_i^0 contain exactly those strategies that are not optimal under any belief vector, (2) for every k between 1 and $K-1$, the sets D_i^k contain precisely those strategies that can rationally be chosen when expressing up to $(k-1)$-fold strong belief in rationality, but not when expressing up to k-fold strong belief in rationality, and (3) the sets D_i^K contain precisely those strategies that can rationally be chosen under common strong belief in rationality.

We know that for every k, the strategies that can rationally be chosen when expressing up to k-fold strong belief in rationality are exactly the strategies in $\Gamma^{k+1}(\emptyset)$ given by the iterated conditional dominance procedure. So, the unique self-confirming combination of rationality orderings will always be such that: (1) for every k between 0 and $K-1$, the sets D_i^k contain precisely those strategies that are in $\Gamma^k(\emptyset)$ but not in $\Gamma^{k+1}(\emptyset)$, and (2) the sets D_i^K contain exactly the strategies in $\Gamma^K(\emptyset)$, where K is the number of steps after which the algorithm stops. That is, the iterated conditional dominance procedure provides a fast and easy method to generate the unique self-confirming combination of rationality orderings in a dynamic game.

On a conceptual level, the theorem shows that common strong belief in rationality can be characterized by the use of rationality orderings on strategies. More precisely, a player who expresses common strong belief in rationality has in mind a rationality ordering at each of his opponents, and uses these rationality orderings to form his beliefs at each of his information sets. So, at every information set h he checks for the highest degree of rationality for his opponents – as given by these rationality orderings – that makes reaching h possible, and uses this highest possible degree of rationality to form his beliefs at h. This is called the *best rationalization principle.* In particular, a player who expresses common strong belief in rationality always uses the *same* rationality orderings on his opponents' strategies throughout the game.

Let us compare this to the concept of common belief in future rationality which we discussed in the previous chapter. Is there a way to characterize this concept in terms of rationality orderings and the best rationalization principle? We will see that this is not possible.

To see this, let us consider the example "Watching TV with Barbara," which we know so well by now. Under common belief in future rationality, you can rationally choose the strategies $(fight, B)$, $(don't, B)$ and $(don't, D)$, whereas Barbara can rationally choose any of her strategies under this concept. Suppose now that Barbara expresses common belief in future rationality, and that her beliefs are induced by a rationality ordering on your strategies. Since each of your strategies $(fight, B), (don't, B)$ and $(don't, D)$ are optimal under common belief in future rationality, the "most rational" strategies for you in this ordering must be $\{(fight, B), (don't, B), (don't, D)\}$ whereas the "least rational" strategy for you must be $(fight, D)$. At information set h_1 – that is, after you have chosen "fight" – Barbara can see that there is only one "most rational" strategy for you leading to h_1, which is $(fight, B)$. So, if her beliefs are induced by this rationality ordering, she must believe at h_1 that you have chosen $(fight, B)$. So, Barbara can no longer rationally choose (D, B) or (D, D), which is a contradiction since the latter two strategies can rationally be chosen under common belief in future rationality. This shows that common belief in future rationality cannot be characterized by rationality orderings on strategies.

The intuitive reason for this is that under common belief in future rationality, Barbara's ordering of your strategies may actually *change* once the game is under way. Initially, that is, before the game starts, Barbara deems your strategies $(fight, B), (don't, B)$ and $(don't, D)$ indeed "more rational" than your strategy $(fight, D)$. But at information set h_1, after you have started a fight with her, she does not necessarily deem your strategy $(fight, B)$ "more rational" than your strategy $(fight, D)$, as both strategies can still be optimal for you *at information set* h_1 under common belief in future rationality. Remember that under common belief in future rationality, players do not need to believe that their opponents have chosen rationally in the past, only that they choose rationally now and in the future. Hence, we may conclude that rationality orderings are typical for common strong belief in rationality, and cannot be applied to common belief in future rationality.

9.7 Bayesian updating

In this chapter we presented, and explored, *common strong belief in rationality.* As in Chapter 8, we can ask what would happen if we require, in addition, that players satisfy *Bayesian updating* when revising their beliefs. That is, suppose we restrict common strong belief in rationality by additionally imposing common belief in Bayesian updating. What can we say about the strategies that can rationally be chosen under this more restrictive concept?

We saw in Chapter 8 that for *common belief in future rationality*, it crucially matters whether we impose (common belief in) Bayesian updating or not. There are games in which a strategy can rationally be chosen under common belief in future rationality, but where the same strategy cannot be rationally chosen if we additionally impose common belief in Bayesian updating. See the example of Figure 8.14 in that chapter.

For common strong belief in rationality, the story is different. It can be shown that the sets of strategies that can rationally be chosen under common strong belief in rationality are the same if we additionally impose common belief in Bayesian updating. That is, additionally requiring common belief in Bayesian updating has no consequences for the eventual strategy choices selected. We thus obtain the following result.

Theorem 9.7.1 *(Bayesian updating is irrelevant for common strong belief in rationality)*
Every strategy that can rationally be chosen under common strong belief in rationality, can also rationally be chosen under common strong belief in rationality and common belief in Bayesian updating.

We say that a type t_i expresses *common belief in Bayesian updating* if t_i assigns, at every $h \in H_i$, only positive probability to opponents' types that satisfy Bayesian updating, if t_i assigns, at every $h \in H_i$, only positive probability to opponents' types t_j that, at every $h' \in H_j$, only assign positive probability to opponents' types that satisfy Bayesian updating, and so on. Moreover, we say that a strategy s_i can rationally be chosen under common strong belief in rationality and common belief in Bayesian updating if s_i is optimal for some type t_i that expresses common strong belief in rationality, satisfies Bayesian updating and expresses common belief in Bayesian updating.

The idea for proving Theorem 9.7.1 is the following. In Theorem 9.3.3 we showed that the strategies that can rationally be chosen under common strong belief in rationality are exactly the strategies that survive the iterated conditional dominance procedure at $\emptyset$. In the proof of Theorem 9.3.3 we will construct an epistemic model M in which, for each of the strategies s_i that survive the iterated conditional dominance procedure at $\emptyset$, there is some type $t_i^{s_i}$ that expresses common strong belief in rationality, and for which s_i is optimal. In this epistemic model, we could as well have constructed these types $t_i^{s_i}$ in such a way that they satisfy Bayesian updating and express common belief in Bayesian updating. We leave it to the interested reader to adapt the proof in this way. We conclude that every strategy s_i that survives the iterated conditional dominance procedure at $\emptyset$, is optimal for some type $t_i^{s_i}$ that expresses common strong belief in rationality, satisfies Bayesian updating, and expresses common belief in Bayesian updating.

That is, every strategy which survives the iterated conditional dominance procedure at $\emptyset$, can rationally be chosen under common strong belief in rationality and common belief in Bayesian updating. By Theorem 9.7.1 it then follows that the strategies that can rationally be chosen under common strong belief in rationality are the same as the strategies that can rationally be chosen under common strong belief in rationality and common belief in Bayesian updating. In both cases we obtain exactly those strategies that survive the iterated conditional dominance procedure at $\emptyset$.

9.8 Proofs

In this section we will first prove Theorem 9.3.3, which states that the iterated conditional dominance procedure delivers exactly those strategies that can rationally be chosen under common strong belief in rationality. We will proceed in three steps: We will first derive some important properties of the iterated conditional dominance procedure. In the second subsection we will use these properties to prove an *optimality principle* for the iterated conditional dominance procedure. As you will see, proving this optimality principle will be far from easy. In the third subsection we will prove Theorem 9.3.3, relying heavily on the optimality principle we derived in the second subsection.

In the fourth and final subsection we will prove Theorem 9.4.2, which states that every outcome you initially deem possible under common strong belief in rationality, is also initially deemed possible under common belief in future rationality.

Properties of the iterated conditional dominance procedure

In this first subsection we will prove some important properties of the iterated conditional dominance procedure, which we will use in the following subsection to show the optimality principle. Recall that $\Gamma^k(\emptyset)$ is the decision problem that remains after step k of the iterated conditional dominance procedure at $\emptyset$. For every player i and every information set h, let $\Gamma^k(h) = (S_i^k(h), S_{-i}^k(h))$ be the decision problem that remains after step k of the iterated conditional dominance procedure at h. We will first show how $\Gamma^k(h)$ relates to the decision problems $\Gamma^m(\emptyset)$ for $m \leq k$.

Lemma 9.8.1 *(Structure of the decision problems $\Gamma^k(h)$)*
For some player i and some information set $h \in H_i$, let $\Gamma^k(h) = (S_i^k(h), S_{-i}^k(h))$ be the decision problem that remains after step k of the iterated conditional dominance procedure at h. Let m be the highest number less than or equal to k such that some strategy combination in $\Gamma^m(\emptyset)$ leads to h. Then, $S_i^k(h) = S_i^m(\emptyset) \cap S_i(h)$ and $S_{-i}^k(h) = S_{-i}^m(\emptyset) \cap S_{-i}(h)$.

Remember that $S_i(h)$ contains exactly those strategies for player i that lead to h, and $S_{-i}(h)$ contains exactly those strategy combinations for the opponents that lead to h.

Proof: We prove the statement by induction on k.

Induction start: Start with $k = 0$. By definition, $S_i^0(h) = S_i(h)$ and $S_{-i}^0(h) = S_{-i}(h)$. Moreover, $S_i^0(\emptyset) = S_i$ and $S_{-i}^0(\emptyset) = S_{-i}$. So, the statement trivially follows for $k = m = 0$.

Induction step: Take some $k \geq 1$, and assume that the statement is true for every $k' \leq k - 1$. Take some information set $h \in H_i$, and let m be the highest number less than or equal to k such that some strategy combination in $\Gamma^m(\emptyset)$ leads to h. There are two cases:
Case 1. If $m < k$, then there is no strategy combination in $\Gamma^k(\emptyset)$ that leads to h. So, every strategy combination that leads to h contains at least one strategy that is not in $\Gamma^k(\emptyset)$. By construction, $\Gamma^k(\emptyset)$ contains precisely those strategies s_j that are not strictly dominated in any decision problem $\Gamma^{k-1}(h')$ of step $k - 1$ at which j is active. Hence, every strategy combination that leads to h contains at least one strategy s_j that is strictly dominated in some decision problem $\Gamma^{k-1}(h')$ of step $k - 1$ at which j is active. By definition of the iterated conditional dominance procedure, we cannot eliminate any strategies from $\Gamma^{k-1}(h)$ in step k. That is, $\Gamma^k(h)$ is the same as $\Gamma^{k-1}(h)$. However, m is clearly the highest number less than or equal to $k - 1$ such that some strategy combination in $\Gamma^m(\emptyset)$ leads to h. So, we know by our induction assumption that $\Gamma^{k-1}(h)$ contains precisely those strategy combinations in $\Gamma^m(\emptyset)$ that lead to h. In other words, $S_i^{k-1}(h) = S_i^m(\emptyset) \cap S_i(h)$, and $S_{-i}^{k-1}(h) = S_{-i}^m(\emptyset) \cap S_{-i}(h)$. As $\Gamma^{k-1}(h) = \Gamma^k(h)$, the same holds for $\Gamma^k(h)$. Hence, $S_i^k(h) = S_i^m(\emptyset) \cap S_i(h)$ and $S_{-i}^k(h) = S_{-i}^m(\emptyset) \cap S_{-i}(h)$.
Case 2. Suppose that $m = k$. So, there is a strategy combination in $\Gamma^k(\emptyset)$ that leads to h. We will show that $\Gamma^k(h)$ contains precisely those strategy combinations in $\Gamma^k(\emptyset)$ that lead to h.

By definition of the iterated conditional dominance procedure, $\Gamma^k(\emptyset)$ contains precisely those strategies s_j that are not strictly dominated in any decision problem $\Gamma^{k-1}(h')$ of step $k - 1$ at which j is active. By our assumption, at least one such strategy combination leads to h. Hence, there is a combination of strategies s_j leading to h (including i's strategy) where every strategy s_j is not strictly dominated in any decision problem $\Gamma^{k-1}(h')$ of step $k - 1$ at which j is active. By definition of the iterated conditional dominance procedure, $\Gamma^k(h)$ then contains precisely those strategy combinations leading to h where every strategy s_j is not strictly dominated in any decision problem $\Gamma^{k-1}(h')$ of step $k - 1$ at which j is active. So, $\Gamma^k(h)$ contains precisely those strategy combinations in $\Gamma^k(\emptyset)$ that lead to h. That is, $S_i^k(h) = S_i^k(\emptyset) \cap S_i(h)$ and $S_{-i}^k(h) = S_{-i}^k(\emptyset) \cap S_{-i}(h)$, which was to be shown.

We may thus conclude that, in general, $S_i^k(h) = S_i^m(\emptyset) \cap S_i(h)$, and $S_{-i}^k(h) = S_{-i}^m(\emptyset) \cap S_{-i}(h)$. By induction on k, the statement in the lemma follows. ◇

In the proofs section of Chapter 8 we introduced the so-called forward inclusion property. We will repeat its definition here for convenience. For a given player i, consider a collection $(D_{-i}(h))_{h \in H_i}$ of strategy subsets, specifying at every information set $h \in H_i$ some subset $D_{-i}(h) \subseteq S_{-i}(h)$ of opponents' strategy combinations leading to h. The collection $(D_{-i}(h))_{h \in H_i}$ of strategy subsets is said to satisfy the *forward inclusion property* if for every $h, h' \in H_i$ where h' follows h, $D_{-i}(h) \cap S_{-i}(h') \subseteq D_{-i}(h')$. We will now show that the collections of strategy subsets selected by the iterated conditional dominance procedure satisfy the forward inclusion property.

Lemma 9.8.2 *(Iterated conditional dominance procedure satisfies the forward inclusion property)*
For every player i and information set $h \in H_i$, let $\Gamma^k(h) = (S_i^k(h), S_{-i}^k(h))$ be the decision problem that remains after step k of the iterated conditional dominance procedure at h. Then, the collection $(S_{-i}^k(h))_{h \in H_i}$ of strategy subsets satisfies the forward inclusion property.

Proof: Consider some information sets $h, h' \in H_i$ where h' follows h. We must show that $S_{-i}^k(h) \cap S_{-i}(h') \subseteq S_{-i}^k(h')$. Let m be the highest number less than or equal to k such that there is a strategy combination in $\Gamma^m(\emptyset)$ leading to h. Then, by Lemma 9.8.1 we know that $S_{-i}^k(h) = S_{-i}^m(\emptyset) \cap S_{-i}(h)$. Similarly, let m' be the highest number less than or equal to k such that there is a strategy combination in $\Gamma^{m'}(\emptyset)$ leading to h'. Since h' follows h we have that $m' \leq m$, and by Lemma 9.8.1 we know that $S_{-i}^k(h') = S_{-i}^{m'}(\emptyset) \cap S_{-i}(h')$.

Since $S_{-i}^k(h) = S_{-i}^m(\emptyset) \cap S_{-i}(h)$ and $S_{-i}(h') \subseteq S_{-i}(h)$, we have that $S_{-i}^k(h) \cap S_{-i}(h') = S_{-i}^m(\emptyset) \cap S_{-i}(h')$. Moreover, $S_{-i}^m(\emptyset) \subseteq S_{-i}^{m'}(\emptyset)$ as $m \geq m'$. Hence, we conclude that $S_{-i}^k(h) \cap S_{-i}(h') \subseteq S_{-i}^{m'}(\emptyset) \cap S_{-i}(h') = S_{-i}^k(h')$, which was to be shown. We have thus proved that $(S_{-i}^k(h))_{h \in H_i}$ satisfies the forward inclusion property. $\diamond$

Optimality principle
Recall that for every player i and information set $h \in H_i$, we denote by $\Gamma^k(h) = (S_i^k(h), S_{-i}^k(h))$ the reduced decision problem for player i at h after applying step k of the iterated conditional dominance procedure. Moreover, we denote by $S_i^k(\emptyset)$ the set of strategies for player i that survive step k of the iterated conditional dominance procedure at $\emptyset$. By definition of the iterated conditional dominance procedure, $S_i^k(\emptyset)$ contains precisely those strategies s_i for player i that are not strictly dominated within any decision problem $\Gamma^{k-1}(h)$ for player i that s_i leads to. For every such information set h, the decision problem $\Gamma^{k-1}(h)$ is given by $(S_i^{k-1}(h), S_{-i}^{k-1}(h))$. By Theorem 2.5.3 in Chapter 2 we know that a strategy s_i is not strictly dominated in $\Gamma^{k-1}(h)$ precisely when it is optimal, among the strategies in $S_i^{k-1}(h)$, for some belief $b_i(h) \in \Delta(S_{-i}^{k-1}(h))$. That is,

$$u_i(s_i, b_i(h)) \geq u_i(s_i', b_i(h)) \text{ for all } s_i' \in S_i^{k-1}(h).$$

Hence, $S_i^k(\emptyset)$ contains precisely those strategies s_i for player i that, at every $h \in H_i$ that s_i leads to, are optimal for some belief $b_i(h) \in \Delta(S_{-i}^{k-1}(h))$ *among the strategies in $S_i^{k-1}(h)$*.

We can show, however, that these strategies s_i are not only optimal at h among the strategies in $S_i^{k-1}(h)$ only, but in fact *among the strategies in $S_i(h)$*. That is,

$$u_i(s_i, b_i(h)) \geq u_i(s_i', b_i(h)) \text{ for all } s_i' \in S_i(h).$$

We will refer to this as the *optimality principle* for the iterated conditional dominance procedure. As in Chapter 8, it will play an important role in proving that the iterated conditional dominance procedure yields exactly those strategies that can rationally be chosen under common strong belief in rationality.

Lemma 9.8.3 *(Optimality principle)*
For some $k \geq 1$, consider a strategy $s_i \in S_i^k(\emptyset)$ that survives step k of the iterated conditional dominance procedure at $\emptyset$. Then, at every $h \in H_i$ that s_i leads to, there is some belief $b_i(h) \in \Delta(S_{-i}^{k-1}(h))$ such that s_i is optimal for $b_i(h)$ among the strategies in $S_i(h)$.

Proof: Consider a strategy $s_i^* \in S_i^k(\emptyset)$ and an information set $h^* \in H_i$ that s_i^* leads to. We will show that there is some belief $b_i(h^*) \in \Delta(S_{-i}^{k-1}(h^*))$ such that s_i^* is optimal for $b_i(h^*)$ among the strategies in $S_i(h^*)$. We will prove this statement by induction on the number of player i information sets that precede h^*.

Induction start: Consider an information set $h^* \in H_i$ that s_i^* leads to, which is not preceded by any other player i information set. As $s_i^* \in S_i^k(\emptyset)$, we know by the argument above that there is some belief $b_i(h^*) \in \Delta(S_{-i}^{k-1}(h^*))$ such that s_i^* is optimal for $b_i(h^*)$ among the strategies in $S_i^{k-1}(h^*)$. That is,

$$u_i(s_i^*, b_i(h^*)) \geq u_i(s_i', b_i(h^*)) \text{ for all } s_i' \in S_i^{k-1}(h^*). \tag{9.1}$$

We will prove that, in fact,

$$u_i(s_i^*, b_i(h^*)) \geq u_i(s_i', b_i(h^*)) \text{ for all } s_i' \in S_i(h^*).$$

Suppose, on the contrary, that there is some $s_i' \in S_i(h^*)$ such that

$$u_i(s_i^*, b_i(h^*)) < u_i(s_i', b_i(h^*)). \tag{9.2}$$

We will show that in this case there is some $s_i \in S_i^{k-1}(h^*)$ with $u_i(s_i', b_i(h^*)) \leq u_i(s_i, b_i(h^*))$, which together with (9.2) would contradict (9.1).

Remember from Lemma 9.8.2 that the collection of strategy subsets $(S_{-i}^{k-1}(h))_{h \in H_i}$ satisfies the forward inclusion property. Since $b_i(h^*) \in \Delta(S_{-i}^{k-1}(h^*))$, we know from Lemma 8.14.3 in the proofs section of Chapter 8 that we can choose at every information set $h \in H_i$ following h^* some belief $b_i(h) \in \Delta(S_{-i}^{k-1}(h))$ such that these beliefs, together with $b_i(h^*)$, satisfy Bayesian updating at all player i information sets following h^*.

Let H_i^{first} be the collection of player i information sets that are not preceded by any other player i information set. For every $h \in H_i^{\text{first}}$ other than h^*, choose some arbitrary belief $b_i(h) \in \Delta(S_{-i}^{k-1}(h))$. Then, by the same argument as above, we can choose for every information set $h' \in H_i$ following h some belief $b_i(h') \in \Delta(S_{-i}^{k-1}(h'))$ such that these beliefs, together with $b_i(h)$, satisfy Bayesian updating at all player i information sets following h.

So, in this way we can extend the belief $b_i(h^*)$ to a conditional belief vector $(b_i(h))_{h \in H_i}$ on $(S_{-i}^{k-1}(h))_{h \in H_i}$, which satisfies Bayesian updating everywhere. But then, by Lemma 8.13.2 of Chapter 8, we know that there is a strategy s_i which, at every $h \in H_i$ that s_i leads to, is optimal for $b_i(h)$ among the strategies in $S_i(h)$. Hence, at every $h \in H_i$ that s_i leads to, strategy s_i is not strictly dominated on $S_{-i}^{k-1}(h)$ among the strategies in $S_i(h)$. So, $s_i \in S_i^k(\emptyset)$, and in particular $s_i \in S_i^{k-1}(\emptyset)$. Moreover, as there is no player i information set preceding h^*, it must be true that $s_i \in S_i(h^*)$. Hence, $s_i \in S_i^{k-1}(\emptyset) \cap S_i(h^*)$. By Lemma 9.8.1 we know that $S_i^{k-1}(\emptyset) \cap S_i(h^*) \subseteq S_i^{k-1}(h^*)$, and therefore we conclude that $s_i \in S_i^{k-1}(h^*)$.

By construction, strategy s_i is optimal at h^*, among the strategies in $S_i(h^*)$, for the belief $b_i(h^*)$. But then, $u_i(s_i', b_i(h^*)) \leq u_i(s_i, b_i(h^*))$. Together with (9.2) this would imply that

$$u_i(s_i^*, b_i(h^*)) < u_i(s_i', b_i(h^*)) \leq u_i(s_i, b_i(h^*)) \text{ for some } s_i \in S_i^{k-1}(h^*).$$

This, however, contradicts (9.1). So, strategy s_i^* must be optimal for the belief $b_i(h^*)$ among the strategies in $S_i(h^*)$, which is what we wanted to show.

Induction step: Consider some information set $h^* \in H_i$ that s_i^* leads to, and assume that at every $h \in H_i$ *preceding* h^* there is some belief $b_i(h) \in \Delta(S_{-i}^{k-1}(h))$ such that s_i^* is optimal for $b_i(h)$ among the strategies in $S_i(h)$. We will prove that there is a belief $b_i(h^*) \in \Delta(S_{-i}^{k-1}(h^*))$ such that s_i^* is optimal for $b_i(h^*)$ among the strategies in $S_i(h^*)$. To do so, we distinguish two cases:

Case 1. Suppose that there is a player i information set h preceding h^* with $b_i(h)(S_{-i}(h^*)) > 0$. By our induction assumption, we know that s_i^* is optimal at h for the belief $b_i(h) \in \Delta(S_{-i}^{k-1}(h))$ among the strategies in $S_i(h)$. Let $b_i(h^*)$ be the belief at h^* obtained from $b_i(h)$ by Bayesian updating. But then, by Lemma 8.14.9 from the proofs section in Chapter 8, it follows that s_i^* is also optimal at h^* for the belief $b_i(h^*)$ among the strategies in $S_i(h^*)$.

We will now show that $b_i(h^*) \in \Delta(S_{-i}^{k-1}(h^*))$. As $b_i(h^*)$ is obtained from $b_i(h)$ by Bayesian updating, we know that

$$b_i(h^*)(s_{-i}) = \frac{b_i(h)(s_{-i})}{b_i(h)(S_{-i}(h^*))} \tag{9.3}$$

for all of opponents' strategy combinations $s_{-i} \in S_{-i}(h^*)$. By assumption, $b_i(h) \in \Delta(S_{-i}^{k-1}(h))$, and hence $b_i(h)$ only assigns positive probability to opponents' strategy combinations in $S_{-i}^{k-1}(h)$. But then, it follows by (9.3) that $b_i(h^*)$ only assigns positive probability to opponents' strategy combinations in $S_{-i}^{k-1}(h) \cap S_{-i}(h^*)$. From Lemma 9.8.2 we know that the collection of strategy subsets $(S_{-i}^{k-1}(h'))_{h' \in H_i}$ satisfies the forward inclusion property, and hence $S_{-i}^{k-1}(h) \cap S_{-i}(h^*) \subseteq S_{-i}^{k-1}(h^*)$. So, we conclude that $b_i(h^*)$ only assigns positive probability to opponents' strategy combinations in $S_{-i}^{k-1}(h^*)$, that is, $b_i(h^*) \in \Delta(S_{-i}^{k-1}(h^*))$.

Hence, we have constructed a belief $b_i(h^*) \in \Delta(S_{-i}^{k-1}(h^*))$ such that s_i^* is optimal for $b_i(h^*)$ among the strategies in $S_i(h^*)$. This completes case 1.

Case 2. Suppose that $b_i(h)(S_{-i}(h^*)) = 0$ for every information set h for player i preceding h^*.

Since $s_i^* \in S_i^k(\emptyset)$ we know from above that at h^* there is some belief $b_i(h^*) \in \Delta(S_{-i}^{k-1}(h^*))$ such that s_i^* is optimal for $b_i(h^*)$ among the strategies in $S_i^{k-1}(h^*)$. That is,

$$u_i(s_i^*, b_i(h^*)) \geq u_i(s_i', b_i(h^*)) \text{ for all } s_i' \in S_i^{k-1}(h^*). \tag{9.4}$$

We will prove that, in fact,

$$u_i(s_i^*, b_i(h^*)) \geq u_i(s_i', b_i(h^*)) \text{ for all } s_i' \in S_i(h^*).$$

Suppose, on the contrary, that there is some $s_i' \in S_i(h^*)$ such that

$$u_i(s_i^*, b_i(h^*)) < u_i(s_i', b_i(h^*)). \tag{9.5}$$

We will show that in this case there is some $s_i \in S_i^{k-1}(h^*)$ with $u_i(s_i', b_i(h^*)) \leq u_i(s_i, b_i(h^*))$, which together with (9.5) would contradict (9.4).

Denote by $H_i^{\text{pre}}(h^*)$ the collection of player i information sets that precede h^*. Let H_i^* be the collection of player i information sets $h \notin H_i^{\text{pre}}(h^*)$ such that: (1) h follows some information set $h' \in H_i^{\text{pre}}(h^*)$, and (2) $b_i(h')(S_{-i}(h)) > 0$ for this information set $h' \in H_i^{\text{pre}}(h^*)$. Remember that, by assumption in case 2, $b_i(h')(S_{-i}(h^*)) = 0$ for every $h' \in H_i^{\text{pre}}(h^*)$, and hence $h^* \notin H_i^*$. Take some information set $h \in H_i^*$, preceded by some $h' \in H_i^{\text{pre}}(h^*)$ with $b_i(h')(S_{-i}(h)) > 0$. Define the belief $b_i(h)$ at h to be the belief obtained from $b_i(h')$ by Bayesian updating. Since $b_i(h') \in \Delta(S_{-i}^{k-1}(h'))$, it can be shown in the same way as above that $b_i(h) \in \Delta(S_{-i}^{k-1}(h))$. Since $h' \in H_i^{\text{pre}}(h^*)$, we know by our induction assumption that s_i^* is optimal at h', among the strategies in $S_i(h')$, for the belief $b_i(h')$. But then, it follows from Lemma 8.14.9 from the proofs section in Chapter 8 that s_i^* is also optimal at h for $b_i(h)$ among the strategies in $S_i(h)$, if s_i^* leads to h. So, we see that for every $h \in H_i^*$ there is a belief $b_i(h) \in \Delta(S_{-i}^{k-1}(h))$ such that s_i^* is optimal at h for $b_i(h)$ among the strategies in $S_i(h)$, if s_i^* leads to h.

Denote by H_i^0 the collection of player i information sets that are not in $H_i^{\text{pre}}(h^*)$ and not in H_i^*. So, H_i^0 contains those player i information sets h that do not precede h^*, and that are either not preceded by any information set $h' \in H_i^{\text{pre}}(h^*)$ or for which $b_i(h')(S_{-i}(h)) = 0$ for every $h' \in H_i^{\text{pre}}(h^*)$ that precedes h. Note that, by assumption for case 2, $b_i(h')(S_{-i}(h^*)) = 0$ for every $h' \in H_i^{\text{pre}}(h^*)$, and hence $h^* \in H_i^0$. By construction, if $h \in H_i^0$ and $h' \in H_i$ follows h, then also h' is in H_i^0. By $H_i^{0\text{first}}$ we denote the collection of those information sets $h \in H_i^0$ that are not preceded by any other information set in H_i^0. As $h^* \in H_i^0$ and every player i information set preceding h^* is in $H_i^{\text{pre}}(h^*)$, we conclude that $h^* \in H_i^{0\text{first}}$.

For information set $h^* \in H_i^{0\text{first}}$, we have defined the belief $b_i(h^*) \in \Delta(S_{-i}^{k-1}(h^*))$ as above, satisfying (9.4). For every other information set $h \in H_i^{0\text{first}}$ we choose some arbitrary belief $b_i(h) \in \Delta(S_{-i}^{k-1}(h))$. Now, take some arbitrary information set $h \in H_i^{0\text{first}}$.

As the collection of strategy subsets $(S_{-i}^{k-1}(h'))_{h' \in H_i}$ satisfies the forward inclusion property, we know from Lemma 8.14.3 in the proofs section of Chapter 8 that we can find beliefs $b_i(h') \in \Delta(S_{-i}^{k-1}(h'))$ for all $h' \in H_i$ following h such that these beliefs, together with $b_i(h)$, satisfy Bayesian updating for all $h' \in H_i$ following h. But then, by Lemma 8.13.2 from Chapter 8, we can find a strategy $s_i^h \in S_i(h)$ such that, for every $h' \in H_i$ weakly following h that s_i^h leads to, s_i^h is optimal at h' for $b_i(h')$ among the strategies in $S_i(h')$. So, at every $h \in H_i^{0\text{first}}$ there is a strategy $s_i^h \in S_i(h)$ and beliefs $b_i(h') \in \Delta(S_{-i}^{k-1}(h'))$ at every $h' \in H_i$ weakly following h, such that, at every $h' \in H_i$ weakly following h that s_i^h leads to, s_i^h is optimal at h' for $b_i(h')$ among the strategies in $S_i(h')$.

We will now construct a strategy s_i as follows. For every $h \in H_i^{\text{pre}}(h^*)$, and every $h \in H_i^*$ that s_i^* leads to, let s_i select the same choice as s_i^*. For every $h \in H_i^0$, weakly preceded by some $h' \in H_i^{0\text{first}}$, let s_i select the same choice as $s_i^{h'}$, if $s_i^{h'}$ leads to h. So, in short, strategy s_i coincides with s_i^* at information sets in $H_i^{\text{pre}}(h^*)$ and H_i^*, and, for every $h \in H_i^{0\text{first}}$, coincides with s_i^h at information sets that weakly follow h.

We will now show that, at every $h \in H_i$ that s_i leads to, the strategy s_i is optimal for the belief $b_i(h) \in \Delta(S_{-i}^{k-1}(h))$ among the strategies in $S_i(h)$. To do so, we distinguish three cases:

Case a. Suppose that $h \in H_i^{\text{pre}}(h^*)$. By construction, every information set $h' \in H_i$ following h with $b_i(h)(S_{-i}(h')) > 0$ is in H_i^*. That is, in order to verify the optimality of a strategy s_i'' for the belief $b_i(h)$ at h we only need the choices that s_i'' prescribes at h, and at information sets in H_i^*. By construction, the strategy s_i coincides with s_i^* at h, and at information sets in H_i^*. Moreover, by the induction assumption we know that strategy s_i^* is optimal at h for the belief $b_i(h)$ among the strategies in $S_i(h)$. Therefore, we conclude that strategy s_i is also optimal at h for the belief $b_i(h)$ among the strategies in $S_i(h)$.

Case b. Suppose that $h \in H_i^*$. We will show that every information set $h' \in H_i$ following h with $b_i(h)(S_{-i}(h')) > 0$ is also in H_i^*. Take some information set $h' \in H_i$ following h with $b_i(h)(S_{-i}(h')) > 0$. As $h \in H_i^*$, we know that h is preceded by some $h'' \in H_i^{\text{pre}}(h^*)$ with $b_i(h'')(S_{-i}(h)) > 0$, and that $b_i(h)$ is obtained from $b_i(h'')$ by Bayesian updating. As $b_i(h)(S_{-i}(h')) > 0$, and $b_i(h)$ is obtained from $b_i(h'')$ by Bayesian updating, it follows that also $b_i(h'')(S_{-i}(h')) > 0$. This, however, means that $h' \in H_i^*$, since $h'' \in H_i^{\text{pre}}(h^*)$. So, indeed, every information set $h' \in H_i$ following h with $b_i(h)(S_{-i}(h')) > 0$ is in H_i^*.

Hence, in order to verify the optimality of a strategy s_i'' for the belief $b_i(h)$ at h we only need the choices that s_i'' prescribes at information sets in H_i^*. By construction, s_i coincides with s_i^* at information sets in H_i^*. Moreover, we know by construction that s_i^* is optimal at h for the belief $b_i(h)$ among the strategies in $S_i(h)$. We thus conclude that strategy s_i is also optimal at h for the belief $b_i(h)$ among the strategies in $S_i(h)$.

Case c. Suppose that $h \in H_i^0$. Let h be weakly preceded by some $h' \in H_i^{0\text{first}}$. By construction, strategy s_i coincides with $s_i^{h'}$ at h and all player i information sets following h. Moreover, we know that strategy $s_i^{h'}$ is optimal at h for $b_i(h)$ among the strategies in $S_i(h)$. Hence, we conclude that strategy s_i is also optimal at h for $b_i(h)$ among the strategies in $S_i(h)$.

In total, we see that, at every $h \in H_i$ that s_i leads to, the strategy s_i is optimal for the belief $b_i(h) \in \Delta(S_{-i}^{k-1}(h))$ among the strategies in $S_i(h)$. Hence, strategy s_i is not strictly dominated in any decision problem $\Gamma^{k-1}(h) = (S_i^{k-1}(h), S_{-i}^{k-1}(h))$ for player i that s_i leads to. This means, however, that $s_i \in S_i^k(\emptyset)$, and hence in particular $s_i \in S_i^{k-1}(\emptyset)$. Moreover, it is easy to see that strategy s_i leads to h^*. By construction, strategy s_i coincides with s_i^* at all player i information sets that precede h^*. Since s_i^* leads to h^*, we may conclude that s_i leads to h^* as well. Hence, we see that $s_i \in S_i^{k-1}(\emptyset) \cap S_i(h^*)$. By Lemma 9.8.1 we know that $S_i^{k-1}(\emptyset) \cap S_i(h^*) \subseteq S_i^{k-1}(h^*)$, and hence $s_i \in S_i^{k-1}(h^*)$.

So, we have constructed a strategy $s_i \in S_i^{k-1}(h^*)$ that is optimal, at every $h \in H_i$ that s_i leads to, for the belief $b_i(h)$ among the strategies in $S_i(h)$. In particular, strategy s_i is optimal at h^* for the belief $b_i(h^*)$ among the strategies in $S_i(h^*)$. But then, by (9.5) it follows that

$$u_i(s_i^*, b_i(h^*)) < u_i(s_i', b_i(h^*)) \leq u_i(s_i, b_i(h^*)) \text{ for some } s_i \in S_i^{k-1}(h^*),$$

which contradicts (9.4). Hence, strategy s_i^* must be optimal at h^* for the belief $b_i(h^*)$ among the strategies in $S_i(h^*)$. So, we have constructed a belief $b_i(h^*) \in \Delta(S_{-i}^{k-1}(h^*))$ such that s_i^* is optimal for $b_i(h^*)$ among the strategies in $S_i(h^*)$. This completes case 2.

By induction, this statement will then hold for every information set $h^* \in H_i$ that s_i^* leads to. So, at every $h^* \in H_i$ that s_i^* leads to, we can construct some belief $b_i(h^*) \in \Delta(S_{-i}^{k-1}(h^*))$ such that s_i^* is optimal for $b_i(h^*)$ among the strategies in $S_i(h^*)$. This completes the proof of the optimality principle. ◇

The algorithm works

We will now use the optimality principle to prove the main theorem of this chapter.

Theorem 9.3.3 *(The algorithm works)*

(1) For every $k \geq 1$, the strategies that can rationally be chosen by a type that expresses up to k-fold strong belief in rationality are precisely the strategies in $\Gamma^{k+1}(\emptyset)$, that is, those strategies that survive step $k+1$ of the iterated conditional dominance procedure at $\emptyset$.

(2) The strategies that can rationally be chosen by a type that expresses common strong belief in rationality are exactly the strategies that survive the full iterated conditional dominance procedure at $\emptyset$.

Proof: For every $k \geq 1$, every player i and every information set $h \in H_i$, let $\Gamma^k(h) = (S_i^k(h), S_{-i}^k(h))$ be the decision problem that remains after step k of the iterated conditional dominance procedure at h. Suppose that the iterated conditional dominance procedure terminates after K rounds, that is, $\Gamma^{K+1}(h) = \Gamma^K(h)$ for every information set h.

In this proof we will construct an epistemic model M with the following properties:

- For all players i and every strategy s_i in $S_i^1(\emptyset)$, there is a type $t_i^{s_i}$ for which s_i is optimal (at all information sets $h \in H_i$ that s_i leads to).
- For every $k \geq 1$, if the strategy s_i is in $S_i^k(\emptyset)$, then the associated type $t_i^{s_i}$ expresses up to $(k-1)$-fold strong belief in rationality.
- If the strategy s_i is in $S_i^K(\emptyset)$, then the associated type $t_i^{s_i}$ expresses common strong belief in rationality.

In order to achieve this, we will carry out four steps. In step 1 we will construct, for every strategy s_i in $S_i^1(\emptyset)$, some conditional belief vector about the opponents' strategy choices for which s_i is optimal. In step 2 we will use these conditional belief vectors to construct our epistemic model M. In this model, we will define for every strategy s_i in $S_i^1(\emptyset)$ some type $t_i^{s_i}$ for which s_i is optimal. In step 3 we will show that, for every $k \geq 1$ and every s_i in $S_i^k(\emptyset)$, the associated type $t_i^{s_i}$ expresses up to $(k-1)$-fold strong belief in rationality. In step 4 we will finally prove that, for every strategy s_i in $S_i^K(\emptyset)$, the associated type $t_i^{s_i}$ expresses common strong belief in rationality.

Step 1: Construction of beliefs

In this step we will construct for every strategy $s_i \in S_i^1(\emptyset)$ some conditional belief vector $b_i^{s_i}$ about the opponents' strategy choices, such that s_i is optimal for $b_i^{s_i}$ at every information set $h \in H_i$ that s_i leads to. For every $k \in \{0, ..., K-1\}$, let D_i^k be the set of strategies for player i that are in $S_i^k(\emptyset)$ but not in $S_i^{k+1}(\emptyset)$. To define the conditional belief vectors $b_i^{s_i}$ we distinguish the following two cases:

(1) Consider first some $k \in \{1, ..., K-1\}$ and some strategy s_i for player i in D_i^k. By Lemma 9.8.3, we can find at every $h \in H_i$ that s_i leads to, some conditional belief $b_i^{s_i}(h) \in \Delta(S_{-i}^{k-1}(h))$ such that s_i is optimal for $b_i^{s_i}(h)$ (among the strategies in $S_i(h)$).

For every $h \in H_i$ that s_i does not lead to, we will construct the conditional belief $b_i^{s_i}(h) \in \Delta(S_{-i}(h))$ as follows: Consider the largest m less than or equal to $k-1$ such that some strategy combination for the opponents in $S_{-i}^m(\emptyset)$ leads to h. Then, choose some arbitrary belief $b_i^{s_i}(h) \in \Delta(S_{-i}^m(\emptyset) \cap S_{-i}(h))$.

In this way, we can construct a complete conditional belief vector $b_i^{s_i} = (b_i^{s_i}(h))_{h \in H_i}$ such that s_i is optimal for $b_i^{s_i}(h)$ at all $h \in H_i$ that s_i leads to. Or, in short, s_i is optimal for $b_i^{s_i}$.

(2) Consider next some strategy $s_i \in S_i^K(\emptyset)$ that survives the full iterated conditional dominance procedure. Then, s_i is also in $S_i^{K+1}(\emptyset)$. Hence, by Lemma 9.8.3 we can find at every $h \in H_i$ that s_i leads to, some conditional belief $b_i^{s_i}(h) \in \Delta(S_{-i}^K(h))$ such that s_i is optimal for $b_i^{s_i}(h)$.

For every $h \in H_i$ that s_i does not lead to, we will construct the conditional belief $b_i^{s_i}(h) \in \Delta(S_{-i}(h))$ as follows: Consider the largest m less than or equal to K such that some strategy combination for the opponents in $S_{-i}^m(\emptyset)$ leads to h. Then, choose some arbitrary belief $b_i^{s_i}(h) \in \Delta(S_{-i}^m(\emptyset) \cap S_{-i}(h))$.

In this way, we can construct a complete conditional belief vector $b_i^{s_i} = (b_i^{s_i}(h))_{h \in H_i}$ such that s_i is optimal for $b_i^{s_i}(h)$ at all $h \in H_i$ that s_i leads to.

Step 2: Construction of types

We will now use these conditional belief vectors $b_i^{s_i}$ to construct, for every strategy $s_i \in S_i$, some type $t_i^{s_i}$. For all players i, let the set of types T_i be given by

$$T_i = \{t_i^{s_i} : s_i \in S_i\}.$$

We will define, for every type $t_i^{s_i}$ and every information set $h \in H_i$, the corresponding conditional belief $b_i(t_i^{s_i}, h)$ on $S_{-i}(h) \times T_{-i}$.

Before we do so, let us denote by T_i^k the set of types $t_i^{s_i}$ with $s_i \in S_i^k(\emptyset)$, for every $k \in \{0, ..., K\}$. Here, $S_i^k(\emptyset)$ contains those strategies that survive step k of the iterated conditional dominance procedure at $\emptyset$. By definition, we set $S_i^0(\emptyset) = S_i$. As $S_i^K(\emptyset) \subseteq S_i^{K-1}(\emptyset) \subseteq ... \subseteq S_i^0(\emptyset)$, it follows that $T_i^K \subseteq T_i^{K-1} \subseteq ... \subseteq T_i^0$, where $T_i^0 = T_i$.

In order to define the conditional beliefs $b_i(t_i^{s_i}, h)$, we distinguish the following three cases:

(1) Take first a type $t_i^{s_i}$ with $s_i \in D_i^0$. That is, $t_i^{s_i} \in T_i^0 \backslash T_i^1$. We define the conditional beliefs of $t_i^{s_i}$ in an arbitrary way.

(2) Consider next some type $t_i^{s_i}$ with $s_i \in D_i^k$ for some $k \in \{1, ..., K-1\}$. That is, $t_i^{s_i} \in T_i^k \backslash T_i^{k+1}$ for some $k \in \{1, ..., K-1\}$. In Step 1 we constructed a conditional belief vector $b_i^{s_i}$ for which s_i is optimal. For every information set $h \in H_i$, let $b_i(t_i^{s_i}, h)$ be the conditional belief about the opponents' strategy-type combinations given by

$$b_i(t_i^{s_i}, h)((s_j, t_j)_{j \neq i}) := \begin{cases} b_i^{s_i}(h)((s_j)_{j \neq i}), & \text{if } t_j = t_j^{s_j} \text{ for every } j \neq i \\ 0, & \text{otherwise.} \end{cases}$$

So, for every $h \in H_i$, type $t_i^{s_i}$ holds the same belief about the opponents' strategy choices as $b_i^{s_i}(h)$. Since strategy s_i is optimal for the conditional belief vector $b_i^{s_i}$, it follows that strategy s_i is optimal for type $t_i^{s_i}$.

Remember that $b_i^{s_i}(h)$ assigns, at every $h \in H_i$ that s_i leads to, only positive probability to opponents' strategy combinations in $S_{-i}^{k-1}(h)$. Hence, at every information set $h \in H_i$ that s_i leads to, type $t_i^{s_i}$ assigns only positive probability to opponents' strategy-type pairs (s_j, t_j) where s_j is in $S_{-i}^{k-1}(h)$ and $t_j = t_j^{s_j}$. By Lemma 9.8.1 we know that for every such information set $h \in H_i$ that s_i leads to, $S_{-i}^{k-1}(h) = S_{-i}^m(\emptyset) \cap S_{-i}(h)$, where m is the highest number less than or equal to $k-1$ such that $S_{-i}^m(\emptyset) \cap S_{-i}(h)$ is non-empty. As a consequence, at every information set $h \in H_i$ that s_i leads to, type $t_i^{s_i}$ assigns only positive probability to opponents' types $t_j \in T_j^m$, where m is the highest number less than or equal to $k-1$ such that $S_{-i}^m(\emptyset) \cap S_{-i}(h)$ is non-empty.

(3) Consider finally some type $t_i^{s_i}$ with $s_i \in S_i^K(\emptyset)$. That is, $t_i^{s_i} \in T_i^K$. In step 1 we constructed a conditional belief vector $b_i^{s_i}$ for which s_i is optimal. For every information set $h \in H_i$, let $b_i(t_i^{s_i}, h)$ be the conditional belief about the opponents' strategy-type

combinations given by

$$b_i(t_i^{s_i},h)((s_j,t_j)_{j\neq i}) := \begin{cases} b_i^{s_i}(h)((s_j)_{j\neq i}), & \text{if } t_j = t_j^{s_j} \text{ for every } j \neq i \\ 0, & \text{otherwise.} \end{cases}$$

Hence, for every $h \in H_i$, type $t_i^{s_i}$ holds the same belief about the opponents' strategy choices as $b_i^{s_i}(h)$. Since strategy s_i is optimal for the conditional belief vector $b_i^{s_i}$, it follows that strategy s_i is optimal for type $t_i^{s_i}$.

Remember that $b_i^{s_i}(h)$ assigns, at every $h \in H_i$ that s_i leads to, only positive probability to opponents' strategy combinations in $S_{-i}^K(h)$. Hence, at every information set $h \in H_i$ that s_i leads to, type $t_i^{s_i}$ assigns only positive probability to opponents' strategy-type pairs (s_j,t_j) where s_j is in $S_{-i}^K(h)$ and $t_j = t_j^{s_j}$. By Lemma 9.8.1 we know that for every such information set $h \in H_i$ that s_i leads to, $S_{-i}^K(h) = S_{-i}^m(\emptyset) \cap S_{-i}(h)$, where m is the highest number less than or equal to K such that $S_{-i}^m(\emptyset) \cap S_{-i}(h)$ is non-empty. As a consequence, at every information set $h \in H_i$ that s_i leads to, type $t_i^{s_i}$ assigns only positive probability to opponents' types $t_j \in T_j^m$, where m is the highest number less than or equal to K such that $S_{-i}^m(\emptyset) \cap S_{-i}(h)$ is non-empty.

The construction of the epistemic model M is hereby complete.

Step 3: Every type $t_i \in T_i^k$ with $k \geq 1$ expresses up to $(k-1)$-fold strong belief in rationality
In order to prove this step, we will show the following lemma.

Lemma 9.8.4 *For every $k \geq 1$, the following two statements are true:*
(a) Every strategy that can rationally be chosen when expressing up to $(k-1)$-fold strong belief in rationality must be in $\Gamma^k(\emptyset)$.
(b) For every strategy s_i in $\Gamma^k(\emptyset)$, the associated type $t_i^{s_i}$ in the epistemic model M expresses up to $(k-1)$-fold strong belief in rationality.

Note that, by construction, strategy s_i is optimal for the type $t_i^{s_i}$ if s_i is in $\Gamma^1(\emptyset)$. So, by combining the two statements (a) and (b), we can show that for every $k \geq 1$, the decision problem $\Gamma^k(\emptyset)$ contains exactly those strategies that can rationally be chosen when expressing up to $(k-1)$-fold strong belief in rationality.

Proof of Lemma 9.8.4: We will prove the two statements by induction on k.

Induction start: Start with $k = 1$.
(a) Take a strategy s_i that can rationally be chosen under 0-fold strong belief in rationality. Then, s_i is optimal for some conditional belief vector. So, s_i is not strictly dominated in any full decision problem $\Gamma^0(h)$ at which i is active, and hence s_i is in $\Gamma^1(\emptyset)$.
(b) Take some strategy s_i in $\Gamma^1(\emptyset)$ and consider the associated type $t_i^{s_i}$. By definition, every type expresses 0-fold strong belief in rationality and so does $t_i^{s_i}$.

Induction step. Take now some $k \in \{2,...,K\}$, and assume that the two statements (a) and (b) are true for every $m \leq k-1$.

(a) Consider a strategy s_i that can rationally be chosen when expressing up to $(k-1)$-fold strong belief in rationality. So, s_i is optimal for some type t_i that expresses up to $(k-1)$-fold strong belief in rationality. Hence, type t_i satisfies the following property at every information set $h \in H_i$: Let m be the highest number, less than or equal to $k-2$, such that there is an opponents' strategy combination leading to h that can rationally be chosen when expressing up to m-fold strong belief in rationality. Then, $b_i(t_i, h)$ only assigns positive probability to such strategy combinations.

We know, from the induction assumption, that for every $m \leq k-2$, the strategies that can rationally be chosen when expressing up to m-fold strong belief in rationality are exactly the strategies in $\Gamma^{m+1}(\emptyset)$. So, type t_i satisfies the following property for every information set $h \in H_i$: If m is the highest number, less than or equal to $k-1$, such that there is an opponents' strategy combination in $\Gamma^m(\emptyset)$ leading to h, then $b_i(t_i, h)$ only assigns positive probability to opponents' strategy combinations in $\Gamma^m(\emptyset)$ that lead to h.

Now, consider an information set $h \in H_i$ that strategy s_i leads to. Since strategy s_i can rationally be chosen when expressing up to $(k-1)$-fold strong belief in rationality, we have in particular that s_i can rationally be chosen when expressing up to $(k-2)$-fold strong belief in rationality. So, by the induction assumption on (a) we know that s_i is in $\Gamma^{k-1}(\emptyset)$. As s_i leads to h, the m we choose above is also the highest m less than or equal to $k-1$ such that there is a strategy combination for the players (including player i's strategy) in $\Gamma^m(\emptyset)$ leading to h. So, type t_i satisfies the following property at every information set $h \in H_i$ that strategy s_i leads to: If m is the highest number, less than or equal to $k-1$, such that there is a players' strategy combination in $\Gamma^m(\emptyset)$ leading to h, then $b_i(t_i, h)$ only assigns positive probability to opponents' strategy combinations in $\Gamma^m(\emptyset)$ that lead to h. However, by Lemma 9.8.1, the opponents' strategy combinations in $\Gamma^m(\emptyset)$ that lead to h are exactly the opponents' strategy combinations in $\Gamma^{k-1}(h)$.

Hence, we conclude that type t_i assigns, at every information set $h \in H_i$ that s_i leads to, only positive probability to opponents' strategy combinations in $\Gamma^{k-1}(h)$. Since strategy s_i is optimal for type t_i at every $h \in H_i$ that s_i leads to, we may conclude that s_i is not strictly dominated within any decision problem $\Gamma^{k-1}(h)$ at which i is active. But then, by definition of the iterated conditional dominance procedure, we conclude that s_i is in $\Gamma^k(\emptyset)$, which proves part (a) of the lemma.

(b) Consider now some strategy s_i in $\Gamma^k(\emptyset)$ and the associated type $t_i^{s_i}$. We will prove that $t_i^{s_i}$ expresses up to $(k-1)$-fold strong belief in rationality. By the induction assumption, $t_i^{s_i}$ expresses up to $(k-2)$-fold strong belief in rationality. So, it remains to show that $t_i^{s_i}$ expresses $(k-1)$-fold strong belief in rationality. Consider an information set $h \in H_i$ and suppose that there is a combination of opponents' types, possibly outside M, that express up to $(k-2)$-fold strong belief in rationality, and for which there is a combination of optimal strategies leading to h. Then, we have to verify two conditions:

(1) The epistemic model M must contain at least one such combination of opponents' types.

(2) Type $t_i^{s_i}$ must at h only assign positive probability to opponents' strategy-type combinations where the strategy combinations lead to h, the types express up to $(k-2)$-fold strong belief in rationality, and the strategies are optimal for the types.

We will first verify condition (1). So, we assume that for every opponent j there is some type t_j (possibly outside M) expressing up to $(k-2)$-fold strong belief in rationality, and some strategy s_j which is optimal for that type, such that this combination of opponents' strategies leads to h. By the induction assumption on (a), it follows that s_j must be in $\Gamma^{k-1}(\emptyset)$. But then, by the induction assumption on (b), it follows that the associated type $t_j^{s_j}$ in M expresses up to $(k-2)$-fold strong belief in rationality. Moreover, we know that s_j is optimal for $t_j^{s_j}$. Hence, M contains a combination of opponents' types that express up to $(k-2)$-fold strong belief in rationality, and for which there is a combination of optimal strategies leading to h, namely the combination of types $t_j^{s_j}$ for every opponent j constructed above. Hence, condition (1) is satisfied.

We will now verify condition (2). Again, assume that at h there is a combination of opponents' types that express up to $(k-2)$-fold strong belief in rationality, and for which there is a combination of optimal strategies leading to h. By the induction assumption on (a), we know that these strategies must be in $\Gamma^{k-1}(\emptyset)$. Hence, there is a combination of opponents' strategies in $\Gamma^{k-1}(\emptyset)$ that leads to h.

We will show that type $t_i^{s_i}$ assigns at h only positive probability to opponents' strategy-type combinations where the strategy combinations lead to h, the types express up to $(k-2)$-fold strong belief in rationality and the strategies are optimal for the types. We distinguish two cases.

Case 1. Suppose that strategy s_i leads to h.

Since, by assumption, s_i is in $\Gamma^k(\emptyset)$, then $s_i \in D_i^{k'}$ for some $k' \geq k$. But then, by construction of the types, $t_i^{s_i}$ assigns at h only positive probability to opponents' strategy combinations in $\Gamma^{k'-1}(h)$, since s_i leads to h. As $\Gamma^{k'-1}(h)$ is contained in $\Gamma^{k-1}(h)$, we conclude that type $t_i^{s_i}$ assigns at h only positive probability to opponents' strategy combinations in $\Gamma^{k-1}(h)$.

We have seen that there is a combination of opponents' strategies in $\Gamma^{k-1}(\emptyset)$ that leads to h. Moreover, as s_i is in $\Gamma^k(\emptyset)$ and s_i leads to h, we know that there is a player i strategy in $\Gamma^{k-1}(\emptyset)$ that leads to h, namely s_i. So, we can conclude that there is a strategy combination in $\Gamma^{k-1}(\emptyset)$ that leads to h. Hence, we know by Lemma 9.8.1 that $\Gamma^{k-1}(h)$ contains precisely those strategy combinations in $\Gamma^{k-1}(\emptyset)$ that lead to h. So, type $t_i^{s_i}$ assigns at h only positive probability to opponents' strategies in $\Gamma^{k-1}(\emptyset)$.

Moreover, by construction of the types, $t_i^{s_i}$ assigns at h only positive probability to opponents' strategy-type pairs (s_j, t_j) with $t_j = t_j^{s_j}$. Together with the above, we may thus conclude that type $t_i^{s_i}$ assigns at h only positive probability to opponents' strategy-type pairs (s_j, t_j) where s_j is in $\Gamma^{k-1}(\emptyset)$ and $t_j = t_j^{s_j}$. By the induction assumption on (b), we know that every such type $t_j^{s_j}$ expresses up to $(k-2)$-fold strong belief in rationality. We also know that for every s_j in $\Gamma^{k-1}(\emptyset)$, the strategy s_j is optimal for the type $t_j^{s_j}$

since $s_j \in D_j^m$ for some $m \geq 1$. Putting all these facts together, it follows that type $t_i^{s_i}$ assigns at h only positive probability to opponents' strategy-type pairs $(s_j, t_j^{s_j})$ where s_j is optimal for $t_j^{s_j}$ and type $t_j^{s_j}$ expresses up to $(k-2)$-fold strong belief in rationality. This is what we had to show.

Case 2. Suppose that strategy s_i does not lead to h.

Since, by assumption, s_i is in $\Gamma^k(\emptyset)$, then $s_i \in D_i^{k'}$ for some $k' \geq k$. As s_i does not lead to h we have, by construction in step 1, that the conditional belief $b_i^{s_i}$ assigns at h only positive probability to opponents' strategy combinations in $S_{-i}^m(\emptyset)$, where m is the largest number less than or equal to $k'-1$ such that some strategy combination for the opponents in $S_{-i}^m(\emptyset)$ leads to h. Remember the assumption above that there is a combination of opponents' strategies in $\Gamma^{k-1}(\emptyset)$ that leads to h. In other words, there is a strategy combination in $S_{-i}^{k-1}(\emptyset)$ leading to h, which means that $m \geq k-1$. Since $S_{-i}^m(\emptyset) \subseteq S_{-i}^{k-1}(\emptyset)$, we conclude that the conditional belief $b_i^{s_i}$ assigns at h only positive probability to opponents' strategy combinations in $S_{-i}^{k-1}(\emptyset)$. By construction, the type $t_i^{s_i}$ holds at h the same belief about the opponents' strategies as $b_i^{s_i}$, and hence type $t_i^{s_i}$ also assigns at h only positive probability to opponents' strategy combinations in $S_{-i}^{k-1}(\emptyset)$.

Moreover, the type $t_i^{s_i}$ assigns at h only positive probability to opponents' strategy-type pairs (s_j, t_j) with $t_j = t_j^{s_j}$. Together with the above, we may thus conclude that type $t_i^{s_i}$ assigns at h only positive probability to opponents' strategy-type pairs (s_j, t_j) where s_j is in $S_j^{k-1}(\emptyset)$ and $t_j = t_j^{s_j}$. By the induction assumption on (b), we know that every such type $t_j^{s_j}$ expresses up to $(k-2)$-fold strong belief in rationality. We also know that for every s_j in $S_j^{k-1}(\emptyset)$, the strategy s_j is optimal for the type $t_j^{s_j}$ since $s_j \in D_j^m$ for some $m \geq 1$. Putting all these facts together, it follows that type $t_i^{s_i}$ assigns at h only positive probability to opponents' strategy-type pairs $(s_j, t_j^{s_j})$ where s_j is optimal for $t_j^{s_j}$ and type $t_j^{s_j}$ expresses up to $(k-2)$-fold strong belief in rationality. This is what we had to show.

Overall, we see that type $t_i^{s_i}$ satisfies conditions (1) and (2) above, and hence expresses $(k-1)$-fold strong belief rationality. Since we already knew, by the induction assumption, that $t_i^{s_i}$ expresses up to $(k-2)$-fold strong belief in rationality, it follows that $t_i^{s_i}$ expresses up to $(k-1)$-fold strong belief in rationality. So, for every strategy s_i in $\Gamma^k(\emptyset)$, the associated type $t_i^{s_i}$ expresses up to $(k-1)$-fold strong belief in rationality. This proves part (b) of the lemma.

By induction on k, the statements (a) and (b) hold for every $k \geq 1$. This completes the proof of Lemma 9.8.4. ◇

With Lemma 9.8.4 we can now easily show the statement in step 3. Consider some type $t_i \in T_i^k$. Then, $t_i = t_i^{s_i}$ for some s_i in $\Gamma^k(\emptyset)$. By Lemma 9.8.4, part (b), it follows that t_i expresses up to $(k-1)$-fold strong belief in rationality. So, step 3 is complete.

Step 4: Every type $t_i \in T_i^K$ expresses common strong belief in rationality

In order to prove this, we will show the following lemma.

Lemma 9.8.5 *For every $k \geq K-1$, every type $t_i \in T_i^K$ expresses up to k-fold strong belief in rationality.*

Proof of Lemma 9.8.5: We will prove the statement by induction on k.

Induction start: Start with $k = K-1$. Consider some $t_i \in T_i^K$. Then, $t_i = t_i^{s_i}$ for some s_i in $\Gamma^K(\emptyset)$. By Lemma 9.8.4 we know that type $t_i^{s_i}$ expresses up to $(K-1)$-fold strong belief in rationality.

Induction step: Consider now some $k \geq K$, and assume that for all players i, every type $t_i \in T_i^K$ expresses up to $(k-1)$-fold strong belief in rationality.

Consider an arbitrary type $t_i \in T_i^K$. That is, $t_i = t_i^{s_i}$ for some s_i in $\Gamma^K(\emptyset)$. We must show that $t_i^{s_i}$ expresses up to k-fold strong belief in rationality. Since, by the induction assumption, $t_i^{s_i}$ expresses up to $(k-1)$-fold strong belief in rationality, it is sufficient to show that $t_i^{s_i}$ expresses k-fold strong belief in rationality. Consider an information set $h \in H_i$ and suppose that there is a combination of opponents' types, possibly outside M, that express up to $(k-1)$-fold strong belief in rationality, and for which there is a combination of optimal strategies leading to h. Then, we have to verify two conditions:

(1) The epistemic model M must contain at least one such combination of opponents' types.

(2) Type $t_i^{s_i}$ must at h only assign positive probability to opponents' strategy-type combinations where the strategy combinations lead to h, the types express up to $(k-1)$-fold strong belief in rationality and the strategies are optimal for the types.

We will first verify condition (1). So, assume that for every opponent j there is some type t_j (possibly outside M) expressing up to $(k-1)$-fold strong belief in rationality, and some strategy s_j which is optimal for that type, such that this combination of opponents' strategies leads to h. By Lemma 9.8.4 we know that every such strategy s_j must be in $\Gamma^K(\emptyset)$, as $k \geq K$. But then, by the induction assumption, it follows that the associated type $t_j^{s_j}$ in M expresses up to $(k-1)$-fold strong belief in rationality. Moreover, we know that s_j is optimal for $t_j^{s_j}$. Hence, M contains a combination of opponents' types that express up to $(k-1)$-fold strong belief in rationality, and for which there is a combination of optimal strategies leading to h, namely the combination of types $t_j^{s_j}$ for every opponent j. Hence, condition (1) is satisfied.

We will now verify condition (2). Assume that at h there is a combination of opponents' types that express up to $(k-1)$-fold strong belief in rationality, and for which there is a combination of optimal strategies leading to h. By Lemma 9.8.4, we know that such strategies must be in $\Gamma^K(\emptyset)$. Hence, there is a combination of opponents' strategies in $\Gamma^K(\emptyset)$ that lead to h.

We will show that type $t_i^{s_i}$ assigns at h only positive probability to opponents' strategy-type combinations where the strategy combinations lead to h, the types express up to $(k-1)$-fold strong belief in rationality, and the strategies are optimal for the types. Again, we distinguish two cases:

Case 1. Suppose that strategy s_i leads to h.

Since, by assumption, s_i is in $\Gamma^K(\emptyset)$, we have that $s_i \in S_i^K(\emptyset)$. As s_i leads to h we have, by construction, that $t_i^{s_i}$ assigns at h only positive probability to opponents' strategy combinations in $\Gamma^K(h)$. We have seen that there is a combination of opponents' strategies in $\Gamma^K(\emptyset)$ that leads to h. Moreover, as s_i is in $\Gamma^K(\emptyset)$ and s_i leads to h, we know that there is a player i strategy in $\Gamma^K(\emptyset)$ that leads to h, namely s_i. So, we can conclude that there is a strategy combination in $\Gamma^K(\emptyset)$ that leads to h. Hence, we know by Lemma 9.8.1 that $\Gamma^K(h)$ contains precisely those strategy combinations for the opponents in $\Gamma^K(\emptyset)$ that lead to h. So, type $t_i^{s_i}$ assigns at h only positive probability to opponents' strategies s_j in $\Gamma^K(\emptyset)$.

Moreover, by construction of the types, $t_i^{s_i}$ assigns at h only positive probability to opponents' strategy-type pairs (s_j, t_j) with $t_j = t_j^{s_j}$. Together with the above, we may thus conclude that type $t_i^{s_i}$ assigns at h only positive probability to opponents' strategy-type pairs (s_j, t_j) where s_j is in $\Gamma^K(\emptyset)$ and $t_j = t_j^{s_j}$. By the induction assumption, we know that every such type $t_j^{s_j}$ expresses up to $(k-1)$-fold strong belief in rationality. We also know that for every s_j in $\Gamma^K(\emptyset)$, the strategy s_j is optimal for the type $t_j^{s_j}$. Putting all these facts together, it follows that type $t_i^{s_i}$ assigns at h only positive probability to opponents' strategy-type pairs $(s_j, t_j^{s_j})$ where s_j is optimal for $t_j^{s_j}$ and type $t_j^{s_j}$ expresses up to $(k-1)$-fold strong belief in rationality. This is what was to be shown.

Case 2. Suppose that strategy s_i does not lead to h.

Since, by assumption, s_i is in $\Gamma^K(\emptyset)$, we have that $s_i \in S_i^K(\emptyset)$. As s_i does not lead to h we have, by construction in step 1, that the conditional belief $b_i^{s_i}$ assigns at h only positive probability to opponents' strategy combinations in $S_{-i}^m(\emptyset)$, where m is the largest number less than or equal to K such that some strategy combination in $S_{-i}^m(\emptyset)$ leads to h. Remember our assumption above that there is a combination of opponents' strategies in $\Gamma^K(\emptyset)$ that leads to h. In other words, there is a strategy combination for the opponents in $S_{-i}^K(\emptyset)$ leading to h, which means that $m = K$. Hence, we conclude that the conditional belief $b_i^{s_i}$ assigns at h only positive probability to opponents' strategy combinations in $S_{-i}^K(\emptyset)$. By construction, the type $t_i^{s_i}$ holds at h the same belief about the opponents' strategies as $b_i^{s_i}$, and hence type $t_i^{s_i}$ also assigns at h only positive probability to opponents' strategy combinations in $S_{-i}^K(\emptyset)$.

Moreover, the type $t_i^{s_i}$ assigns at h only positive probability to opponents' strategy-type pairs (s_j, t_j) with $t_j = t_j^{s_j}$. Together with the above, we may thus conclude that type $t_i^{s_i}$ assigns at h only positive probability to opponents' strategy-type pairs (s_j, t_j) where s_j is in $S_j^K(\emptyset)$ and $t_j = t_j^{s_j}$. By the induction assumption, we know that every such type $t_j^{s_j}$ expresses up to $(k-1)$-fold strong belief in rationality. We also know that for every

s_j in $S_j^K(\emptyset)$, the strategy s_j is optimal for the type $t_j^{s_j}$. Putting all these facts together, it follows that type $t_i^{s_i}$ assigns at h only positive probability to opponents' strategy-type pairs $(s_j, t_j^{s_j})$ where s_j is optimal for $t_j^{s_j}$ and type $t_j^{s_j}$ expresses up to $(k-1)$-fold strong belief in rationality. This is what we had to show.

Overall, we see that type $t_i^{s_i}$ satisfies conditions (1) and (2) above, and hence expresses k-fold strong belief in rationality. Since we already know, by the induction assumption, that $t_i^{s_i}$ expresses up to $(k-1)$-fold strong belief in rationality, it follows that $t_i^{s_i}$ expresses up to k-fold strong belief in rationality. So, for every strategy s_i in $\Gamma^K(\emptyset)$, the associated type $t_i^{s_i}$ expresses up to k-fold strong belief in rationality. By induction on k, the statement in the lemma follows. ◇

Obviously, the statement in step 4 follows from Lemma 9.8.5, and hence step 4 is hereby complete.

With Lemma 9.8.4 and Lemma 9.8.5 it is now easy to prove Theorem 9.3.3. Consider some strategy s_i that can rationally be chosen when expressing up to k-fold strong belief in rationality, for some $k \geq 0$. Then, by Lemma 9.8.4, part (a), it follows that s_i is in $\Gamma^{k+1}(\emptyset)$, so s_i survives $k+1$ rounds of the iterated conditional dominance procedure at $\emptyset$. On the other hand, consider some strategy in $\Gamma^{k+1}(\emptyset)$. Then, by Lemma 9.8.4, part (b), we know that the associated type $t_i^{s_i}$ expresses up to k-fold strong belief in rationality. As s_i is optimal for $t_i^{s_i}$, we conclude that s_i can rationally be chosen when expressing up to k-fold strong belief in rationality. So, we see that $\Gamma^{k+1}(\emptyset)$ contains precisely those strategies that can rationally be chosen when expressing up to k-fold strong belief in rationality. This proves part (1) of Theorem 9.3.3.

Now, take some strategy s_i that can rationally be chosen under common strong belief in rationality. Then, in particular, s_i can rationally be chosen when expressing up to $(K-1)$-fold strong belief in rationality. Hence, by Lemma 9.8.4, part (a), it follows that s_i is in $\Gamma^K(\emptyset)$, so s_i survives the iterated conditional dominance procedure at $\emptyset$. So, every strategy s_i that can rationally be chosen under common strong belief in rationality survives the iterated conditional dominance procedure at $\emptyset$.

On the other hand, consider some strategy s_i that survives the iterated conditional dominance procedure at $\emptyset$. Hence, s_i is in $\Gamma^K(\emptyset)$. By Lemma 9.8.5 we know that the associated type $t_i^{s_i}$ expresses k-fold strong belief in rationality for all k. In other words, the associated type $t_i^{s_i}$ expresses common strong belief in rationality. As s_i is optimal for $t_i^{s_i}$, it follows that s_i can rationally be chosen under common strong belief in rationality. So, every strategy s_i that survives the iterated conditional dominance procedure at $\emptyset$ can rationally be chosen under common strong belief in rationality.

Together with our conclusion above, we see that a strategy s_i can rationally be chosen under common strong belief in rationality, if and only if, s_i survives the iterated conditional dominance procedure at $\emptyset$. This proves part (2) of Theorem 9.3.3. The proof of Theorem 9.3.3 is hereby complete. ■

Outcomes under common strong belief in rationality and common belief in future rationality

We will finally prove the following theorem.

Theorem 9.4.2 *(Outcomes under common strong belief in rationality and common belief in future rationality)*
Every outcome you initially deem possible under common strong belief in rationality, is also initially deemed possible under common belief in future rationality.

Proof: We will use the following terminology and notation in the proof. For every information set h, let $\Gamma^{\text{icd}}(h)$ be the reduced game that remains at h after applying the iterated conditional dominance procedure, and let $\Gamma^{\text{bd}}(h)$ be the reduced game that remains at h after applying the backward dominance procedure. We say that an information set h is *reachable* under common strong belief in rationality if there is a strategy combination in $\Gamma^{\text{icd}}(\emptyset)$ that leads to h. This makes sense as we know from Theorem 9.3.3 that $\Gamma^{\text{icd}}(\emptyset)$ contains precisely those strategies that can rationally be chosen under common strong belief in rationality. Similarly, an information set h is called *reachable* under common belief in future rationality if there is a strategy combination in $\Gamma^{\text{bd}}(\emptyset)$ that leads to h. Denote by H_i^{icd} the collection of information sets for player i that are reachable under common strong belief in rationality, and let H_i^{bd} be the collection of information sets for player i that are reachable under common belief in future rationality. Finally, an outcome z is said to be reachable under common strong belief in rationality if there is a strategy combination in $\Gamma^{\text{icd}}(\emptyset)$ leading to z. Similarly, an outcome z is called reachable under common belief in future rationality if there is a strategy combination in $\Gamma^{\text{bd}}(\emptyset)$ leading to z. Hence, we must prove that every outcome z that is reachable under common strong belief in rationality is also reachable under common belief in future rationality.

We will proceed by the following steps. In step 1 we will transform every strategy s_i in $\Gamma^{\text{icd}}(\emptyset)$ into some new strategy $\sigma_i(s_i)$, and show that it is optimal for some particular belief vector $b_i(\sigma_i(s_i))$. In step 2, we will use the transformed strategies and belief vectors to construct, for every s_i in $\Gamma^{\text{icd}}(\emptyset)$, some type $t_i^{\sigma_i(s_i)}$. In step 3 we will prove that, for every s_i in $\Gamma^{\text{icd}}(\emptyset)$, strategy $\sigma_i(s_i)$ is optimal for type $t_i^{\sigma_i(s_i)}$. In step 4 we will prove that every type $t_i^{\sigma_i(s_i)}$ so constructed expresses common belief in future rationality. Hence, every strategy $\sigma_i(s_i)$ induced by some strategy s_i in $\Gamma^{\text{icd}}(\emptyset)$ can rationally be chosen under common belief in future rationality, and is therefore in $\Gamma^{\text{bd}}(\emptyset)$. In step 5 we will finally show that, whenever a strategy combination $(s_1, ..., s_n)$ in $\Gamma^{\text{icd}}(\emptyset)$ leads to an outcome z, then the induced strategy combination $(\sigma_1(s_1), ..., \sigma_n(s_n))$ in $\Gamma^{\text{bd}}(\emptyset)$ leads to outcome z as well. That is, every outcome z that is reachable under common strong belief in rationality is also reachable under common belief in future rationality.

Step 1: Transformation of strategies

Consider a strategy s_i in $\Gamma^{\text{icd}}(\emptyset)$. Remember that H_i^{icd} contains those information sets $h \in H_i$ that are reachable under common strong belief in rationality. That is, H_i^{icd} contains those information sets $h \in H_i$ for which there is some strategy combination in $\Gamma^{icd}(\emptyset)$ leading to h. Then, we know from the optimality principle in Lemma 9.8.3 that, for every $h \in H_i^{icd}$ that s_i leads to, there is some belief $b_i(\sigma_i(s_i),h) \in \Delta(S_{-i}^{icd}(h))$ such that s_i is optimal at h for $b_i(\sigma_i(s_i),h)$. Here, $S_{-i}^{icd}(h)$ denotes the set of strategy combinations for the opponents in $\Gamma^{icd}(h)$. Now, take an information set $h \in H_i^{icd}$ that s_i leads to. Then, by definition, there is a strategy combination in $\Gamma^{icd}(\emptyset)$ that leads to h. Hence, by Lemma 9.8.1 we know that $S_{-i}^{icd}(h) = S_{-i}^{icd}(\emptyset) \cap S_{-i}(h)$. So, for every $h \in H_i^{icd}$ that s_i leads to, there is a belief $b_i(\sigma_i(s_i),h) \in \Delta(S_{-i}^{icd}(\emptyset) \cap S_{-i}(h))$, such that s_i is optimal at h for $b_i(\sigma_i(s_i),h)$.

Consider now an information set $h \in H_i$ that s_i leads to, such that: (1) $h \notin H_i^{icd}$ and (2) every $h' \in H_i$ preceding h, if there is any, is in H_i^{icd}. So, h is the first information set for player i that is not reachable under common strong belief in rationality. Then, it is clear that every information set that follows h is also not reachable under common strong belief in rationality. Now, define at h an arbitrary belief $b_i(\sigma_i(s_i),h) \in \Delta(S_{-i}^{bd}(h))$, where $S_{-i}^{bd}(h)$ is the set of opponents' strategy combinations in $\Gamma^{bd}(h)$. By Lemma 8.14.5 in the proofs section of Chapter 8, we can find for every $h' \in H_i$ following h a belief $b_i(\sigma_i(s_i),h') \in \Delta(S_{-i}^{bd}(h'))$, and a strategy $\tilde{s}_i(h)$ leading to h, such that $\tilde{s}_i(h)$ is optimal at every $h' \in H_i$ weakly following h for the belief $b_i(\sigma_i(s_i),h')$.

Now, let $\sigma_i(s_i)$ be the strategy that: (1) at every $h \in H_i^{icd}$ that $\sigma_i(s_i)$ leads to, selects the same choice as s_i, and (2) at every $h \notin H_i^{icd}$ that $\sigma_i(s_i)$ leads to, selects the same choice as $\tilde{s}_i(h')$, where h' is the first information set for player i weakly preceding h that is not in H_i^{icd}.

Above, we have constructed for every $h \in H_i$ that $\sigma_i(s_i)$ leads to, some belief $b_i(\sigma_i(s_i),h) \in \Delta(S_{-i}(h))$. We will show that, in fact, $\sigma_i(s_i)$ is optimal at each of these information sets h for the belief $b_i(\sigma_i(s_i),h)$. We distinguish two cases:

Case 1. Suppose that $h \in H_i^{icd}$. Then we have, by construction, that $\sigma_i(s_i)$ coincides with s_i at h, and at every $h' \in H_i^{icd}$ that follows h. Moreover, $b_i(\sigma_i(s_i),h) \in \Delta(S_{-i}^{icd}(\emptyset) \cap S_{-i}(h))$ as we have seen. But then, every information set $h' \in H_i$ that can be reached with a positive probability under $\sigma_i(s_i)$ and $b_i(\sigma_i(s_i),h)$ is in H_i^{icd}. So, to evaluate the optimality of $\sigma_i(s_i)$ at h under the belief $b_i(\sigma_i(s_i),h)$, we only need the choices that $\sigma_i(s_i)$ prescribes at information sets in H_i^{icd}. But these choices are the same as the choices prescribed by s_i. Since we know that s_i is optimal at h for the belief $b_i(\sigma_i(s_i),h)$, we conclude that $\sigma_i(s_i)$ is also optimal at h for the belief $b_i(\sigma_i(s_i),h)$.

Case 2. Suppose that $h \notin H_i^{icd}$. Let h' be the first information set for player i weakly preceding h that is not in H_i^{icd}. Then, by construction, $\sigma_i(s_i)$ coincides with $\tilde{s}_i(h')$ at h, and at all player i information sets that follow h. Since we know that $\tilde{s}_i(h')$ is optimal at h for the belief $b_i(\sigma_i(s_i),h)$, it follows that also $\sigma_i(s_i)$ is optimal at h for the belief $b_i(\sigma_i(s_i),h)$.

Finally, define for every $h \in H_i$ that $\sigma_i(s_i)$ does not lead to, some arbitrary belief $b_i(\sigma_i(s_i),h) \in \Delta(S_{-i}^{bd}(h))$. Then, we obtain a complete conditional belief vector $b_i(\sigma_i(s_i))$ for player i.

Summarizing, we have transformed the strategy $s_i \in S_i^{icd}(\emptyset)$ into a new strategy $\sigma_i(s_i)$, and we have defined a conditional belief vector $b_i(\sigma_i(s_i))$, such that:

- the new strategy $\sigma_i(s_i)$ coincides with s_i at all information sets in H_i^{icd} that $\sigma_i(s_i)$ leads to,
- at every $h \in H_i^{\text{icd}}$ that $\sigma_i(s_i)$ leads to, we have that $b_i(\sigma_i(s_i),h) \in \Delta(S_{-i}^{\text{icd}}(\emptyset) \cap S_{-i}(h))$,
- at every other information set $h \in H_i$, we have that $b_i(\sigma_i(s_i),h) \in \Delta(S_{-i}^{\text{bd}}(h))$,
- the new strategy $\sigma_i(s_i)$ is optimal for the belief vector $b_i(\sigma_i(s_i))$.

Step 2: Construction of types

For every player i we will construct a set of types T_i which consists of: (1) types $t_i^{\sigma_i(s_i)}$ for every strategy $s_i \in S_i^{\text{icd}}(\emptyset)$, and (2) types $\tau_i^{s_i}$ for every strategy $s_i \in S_i$.

(1) We will start by defining the beliefs for the types $t_i^{\sigma_i(s_i)}$ for every strategy $s_i \in S_i^{\text{icd}}(\emptyset)$. Consider some strategy $s_i \in S_i^{\text{icd}}(\emptyset)$, and let $\sigma_i(s_i)$ be the transformed strategy constructed in step 1.

From above, we know that at every $h \in H_i^{\text{icd}}$ that $\sigma_i(s_i)$ leads to, strategy $\sigma_i(s_i)$ is optimal for some belief $b_i(\sigma_i(s_i),h) \in \Delta(S_{-i}^{\text{icd}}(\emptyset) \cap S_{-i}(h))$. So, the belief $b_i(\sigma_i(s_i),h)$ only assigns positive probability to opponents' strategies s_j' in $S_j^{\text{icd}}(\emptyset)$. For every $h \in H_i^{\text{icd}}$ that $\sigma_i(s_i)$ leads to, define the conditional belief $b_i(t_i^{\sigma_i(s_i)},h)$ about the opponents' strategy-type pairs by

$$b_i(t_i^{\sigma_i(s_i)},h)((s_j',t_j)_{j\neq i}) := \begin{cases} b_i(\sigma_i(s_i),h)((s_j)_{j\neq i}), & \text{if } s_j' = \sigma_j(s_j) \text{ and } t_j = t_j^{\sigma_j(s_j)} \text{ for every } j \neq i \\ 0, & \text{otherwise.} \end{cases}$$

Hence, for every $h \in H_i^{\text{icd}}$ that $\sigma_i(s_i)$ leads to, type $t_i^{\sigma_i(s_i)}$ only assigns positive probability to strategy-type pairs $(\sigma_j(s_j), t_j^{\sigma_j(s_j)})$ with $s_j \in S_j^{\text{icd}}(\emptyset)$.

For every other information set $h \in H_i$, we know that strategy $\sigma_i(s_i)$ is optimal for some belief $b_i(\sigma_i(s_i),h) \in \Delta(S_{-i}^{\text{bd}}(h))$, if $\sigma_i(s_i)$ leads to h. At each of these other information sets $h \in H_i$, define the conditional belief $b_i(t_i^{\sigma_i(s_i)},h)$ about the opponents' strategy-type pairs by

$$b_i(t_i^{\sigma_i(s_i)},h)((s_j,t_j)_{j\neq i}) := \begin{cases} b_i(\sigma_i(s_i),h)((s_j)_{j\neq i}), & \text{if } t_j = \tau_j^{s_j} \text{ for every } j \neq i \\ 0, & \text{otherwise.} \end{cases}$$

Hence, at each of these other information sets $h \in H_i$, type $t_i^{\sigma_i(s_i)}$ holds the same belief about the opponents' strategy choices as $b_i(\sigma_i(s_i),h)$, and only assigns positive probability to strategy-type pairs $(s_j, \tau_j^{s_j})$ with $s_j \in S_j^{\text{bd}}(h)$.

(2) We will now define the beliefs for the types $\tau_i^{s_i}$ for every strategy $s_i \in S_i$. For every strategy s_i, let $H_i^{\text{bd}}(s_i)$ be the (possibly empty) collection of information sets $h \in H_i$ for which s_i is in $\Gamma^{\text{bd}}(h)$. That is, $H_i^{\text{bd}}(s_i)$ contains all those information sets for player i at which s_i survives the backward dominance procedure. By Lemma 8.14.6 in the proofs section of Chapter 8, we can find for every $s_i \in S_i$ some conditional belief

vector $\beta_i(s_i) = (\beta_i(s_i,h))_{h\in H_i}$ such that: (1) $\beta_i(s_i,h) \in \Delta(S^{\text{bd}}_{-i}(h))$ for every $h \in H_i$, and (2) s_i is optimal at every $h \in H^{\text{bd}}_i(s_i)$ for the belief $\beta_i(s_i,h)$.

Now, consider $s_i \in S_i$ and an information set $h \in H_i$. For type $\tau_i^{s_i}$, let $b_i(\tau_i^{s_i},h)$ be the conditional belief about the opponents' strategy-type pairs given by

$$b_i(\tau_i^{s_i},h)((s_j,t_j)_{j\neq i}) := \begin{cases} \beta_i(s_i,h)((s_j)_{j\neq i}), & \text{if } t_j = \tau_j^{s_j} \text{ for every } j \neq i \\ 0, & \text{otherwise.} \end{cases}$$

Hence, at every information set $h \in H_i$, type $\tau_i^{s_i}$ holds the same belief about the opponents' strategy choices as $\beta_i(s_i,h)$, and only assigns positive probability to strategy-type pairs $(s_j, \tau_j^{s_j})$ with $s_j \in S_j^{\text{bd}}(h)$.

This completes the description of the epistemic model.

Step 3: Strategy $\sigma_i(s_i)$ is optimal for type $t_i^{\sigma_i(s_i)}$

Consider some strategy $s_i \in S_i^{\text{icd}}(\emptyset)$. We will show that the induced strategy $\sigma_i(s_i)$ is optimal at type $t_i^{\sigma_i(s_i)}$ at every $h \in H_i$ that $\sigma_i(s_i)$ leads to.

Consider first an information set $h \in H_i^{\text{icd}}$. Then, by construction, the belief $b_i(t_i^{\sigma_i(s_i)},h)$ only assigns positive probability to opponents' strategy combinations $(\sigma_j(s_j))_{j\neq i}$ with $s_j \in S_j^{\text{icd}}(\emptyset)$ for all j. In step 1 we saw that these strategies $\sigma_j(s_j)$ coincide with s_j at information sets in H_j^{icd}, and that strategy $\sigma_i(s_i)$ coincides with s_i at information sets in H_i^{icd}. So, strategy $\sigma_i(s_i)$, in combination with the belief $b_i(t_i^{\sigma_i(s_i)},h)$, can only lead with positive probability to information sets in H^{icd}.

By construction, the probability that the belief $b_i(t_i^{\sigma_i(s_i)},h)$ assigns to an opponents' strategy combination $(\sigma_j(s_j))_{j\neq i}$ is the same as the probability that the belief $b_i(\sigma_i(s_i),h)$ assigns to $(s_j)_{j\neq i}$. As $\sigma_j(s_j)$ coincides with s_j at H_j^{icd}, and since we know from step 1 that strategy $\sigma_i(s_i)$ is optimal for the belief $b_i(\sigma_i(s_i),h)$ at h, it follows that strategy $\sigma_i(s_i)$ is also optimal for the belief $b_i(t_i^{\sigma_i(s_i)},h)$ at h.

Consider now an information set $h \in H_i$ that is not in H_i^{icd}. Then, by construction, $b_i(t_i^{\sigma_i(s_i)},h)$ holds the same belief about the opponents' strategy choices as $b_i(\sigma_i(s_i),h)$. From step 1 we know that strategy $\sigma_i(s_i)$ is optimal for $b_i(\sigma_i(s_i),h)$ at h. Hence, strategy $\sigma_i(s_i)$ is then also optimal for $b_i(t_i^{\sigma_i(s_i)},h)$ at h.

Summarizing, we may conclude that strategy $\sigma_i(s_i)$ is optimal for the belief $b_i(t_i^{\sigma_i(s_i)},h)$ at every $h \in H_i$ that $\sigma_i(s_i)$ leads to. That is, $\sigma_i(s_i)$ is optimal for type $t_i^{\sigma_i(s_i)}$.

Step 4: All types express common belief in future rationality

We will next show that every type constructed above expresses common belief in future rationality. To do so, it is sufficient to prove that every type in the epistemic model believes in the opponents' future rationality. We distinguish two cases:

Case 1. Consider first a type $t_i^{\sigma_i(s_i)}$ for some $s_i \in S_i^{\text{icd}}(\emptyset)$. We will show that $t_i^{\sigma_i(s_i)}$ believes in the opponents' future rationality.

Take first some information set $h \in H_i^{\text{icd}}$ that $\sigma_i(s_i)$ leads to. We have seen that at h, type $t_i^{\sigma_i(s_i)}$ only assigns positive probability to strategy-type pairs $(\sigma_j(s_j), t_j^{\sigma_j(s_j)})$ with $s_j \in S_j^{\text{icd}}(\emptyset)$. From step 3 we know that strategy $\sigma_j(s_j)$ is optimal for $t_j^{\sigma_j(s_j)}$ at all $h' \in H_j$ that $\sigma_j(s_j)$ leads to. In particular, $\sigma_j(s_j)$ is optimal for $t_j^{\sigma_j(s_j)}$ at all $h' \in H_j$ weakly following h that $\sigma_j(s_j)$ leads to. Hence, at information set h, type $t_i^{\sigma_i(s_i)}$ only assigns positive probability to strategy-type pairs $(\sigma_j(s_j), t_j^{\sigma_j(s_j)})$ where $\sigma_j(s_j)$ is optimal for $t_j^{\sigma_j(s_j)}$ at all $h' \in H_j$ weakly following h that $\sigma_j(s_j)$ leads to. That is, type $t_i^{\sigma_i(s_i)}$ believes at h in the opponents' future rationality.

Now consider some other information set $h \in H_i$. By construction, the belief $b_i(t_i^{\sigma_i(s_i)}, h)$ only assigns positive probability to opponents' strategy-type pairs $(s_j, \tau_j^{s_j})$ with $s_j \in S_j^{\text{bd}}(h)$.

Consider some $s_j \in S_j^{\text{bd}}(h)$. Then, by construction of the backward dominance procedure, we have that $s_j \in S_j^{\text{bd}}(h')$ for every h' weakly following h that s_j leads to. In other words, if $s_j \in S_j^{\text{bd}}(h)$, then every $h' \in H_j$ weakly following h that s_j leads to is in $H_j^{\text{bd}}(s_j)$. Remember that $H_j^{\text{bd}}(s_j)$ is the collection of information sets $h' \in H_j$ with $s_j \in S_j^{\text{bd}}(h')$. By construction, at every $h' \in H_j^{\text{bd}}(s_j)$, type $\tau_j^{s_j}$ holds the same belief about the opponents' strategy choices as $\beta_j(s_j, h')$. Moreover, at every $h' \in H_j^{\text{bd}}(s_j)$, strategy s_j is optimal under the belief $\beta_j(s_j, h')$. So, at every $h' \in H_j^{\text{bd}}(s_j)$, strategy s_j is optimal for type $\tau_j^{s_j}$. Since we have seen that every $h' \in H_j$ weakly following h that s_j leads to, is in $H_j^{\text{bd}}(s_j)$, it follows that s_j is optimal for type $\tau_j^{s_j}$ at every $h' \in H_j$ weakly following h that s_j leads to.

So, we have shown for every $s_j \in S_j^{\text{bd}}(h)$ that s_j is optimal for type $\tau_j^{s_j}$ at every $h' \in H_j$ weakly following h that s_j leads to. Since $b_i(t_i^{\sigma_i(s_i)}, h)$ only assigns positive probability to opponents' strategy-type pairs $(s_j, \tau_j^{s_j})$ where $s_j \in S_j^{\text{bd}}(h)$, we may conclude that type $t_i^{\sigma_i(s_i)}$ believes at h in the opponents' future rationality.

Hence, type $t_i^{\sigma_i(s_i)}$ believes in the opponents' future rationality.

Case 2. Consider now a type $\tau_i^{s_i}$ for some $s_i \in S_i$. We will show that type $\tau_i^{s_i}$ also believes in the opponents' future rationality.

Consider an arbitrary information set $h \in H_i$. Then, by construction, $b_i(\tau_i^{s_i}, h)$ only assigns positive probability to opponents' strategy-type pairs $(s_j, \tau_j^{s_j})$ with $s_j \in S_j^{\text{bd}}(h)$. From above, we know that every strategy $s_j \in S_j^{\text{bd}}(h)$ is optimal for type $\tau_j^{s_j}$ at every $h' \in H_j$ weakly following h that s_j leads to. This means, however, that type $\tau_i^{s_i}$ assigns at h only positive probability to opponents' strategy-type pairs $(s_j, \tau_j^{s_j})$ where s_j is optimal for type $\tau_j^{s_j}$ at every $h' \in H_j$ weakly following h that s_j leads to. That is, type $\tau_i^{s_i}$ believes at h in the opponents' future rationality.

Summarizing, we thus see that every type in the epistemic model believes in the opponents' future rationality. As a consequence, all types in the epistemic model express common belief in future rationality.

Step 5: $(\sigma_1(s_1),...,\sigma_n(s_n))$ **leads to the same outcome as** $(s_1,...,s_n)$
Finally, consider a strategy combination $(s_1,...,s_n)$ in $\Gamma^{\text{icd}}(\emptyset)$ that leads to the outcome z. We will show that the induced strategy combination $(\sigma_1(s_1),...,\sigma_n(s_n))$ leads to the same outcome z.

By construction, the strategy combination $(s_1,...,s_n)$ only leads to information sets in H^{icd}. By step 1, we know that at every information set $h \in H_i^{\text{icd}}$, the strategy $\sigma_i(s_i)$ prescribes the same choice as s_i. So, the strategy combination $(\sigma_1(s_1),...,\sigma_n(s_n))$ must lead to the same outcome z. This completes step 5.

By using steps 3, 4 and 5, it is now easy to prove Theorem 9.4.2. Consider an outcome z you initially deem possible under common strong belief in rationality. Then, there is a strategy combination $(s_1,...,s_n)$ in $\Gamma^{\text{icd}}(\emptyset)$ that leads to the outcome z. By step 5, we know that the induced strategy combination $(\sigma_1(s_1),...,\sigma_n(s_n))$ also leads to z. By step 3 we know that every strategy $\sigma_i(s_i)$ is optimal for the type $t_i^{\sigma_i(s_i)}$ constructed in step 2. Moreover, by step 4, the type $t_i^{\sigma_i(s_i)}$ expresses common belief in future rationality. Hence, every strategy $\sigma_i(s_i)$ can rationally be chosen under common belief in future rationality. So, the induced strategy combination $(\sigma_1(s_1),...,\sigma_n(s_n))$ is in $\Gamma^{\text{bd}}(\emptyset)$. As $(\sigma_1(s_1),...,\sigma_n(s_n))$ leads to z, it follows that under common belief in future rationality, you initially deem possible the outcome z. This completes the proof of Theorem 9.4.2. ■

Practical problems

9.1 Two parties in a row

Recall the story from Problem 8.1 in Chapter 8.

(a) What strategies can you rationally choose under common strong belief in rationality? Compare this result to the strategies you can rationally choose under common belief in future rationality, found in Problem 8.1, part (c).
(b) Under common strong belief in rationality, what colors can you rationally wear at the second party, provided you go to that party? Compare this result to the colors you can rationally wear at the second party under common belief in future rationality, found in Problem 8.1, part (c). How do you explain this difference intuitively?
(c) Find the unique self-confirming pair of rationality orderings.
(d) Construct an epistemic model such that, for every strategy found in (a), there is a type that expresses common strong belief in rationality, and for which this strategy is optimal.

9.2 Selling ice cream

Recall the story from Problem 8.2 in Chapter 8.

(a) Which strategies can you and Barbara rationally choose under common strong belief in rationality? Compare this result to the strategies you and Barbara can rationally choose under common belief in future rationality, found in Problem 8.2, part (b).

(b) Suppose now that the weather forecast for tomorrow is more optimistic, such that you would expect the numbers in Figure 8.15 only to drop by 30% tomorrow. Which strategies can you and Barbara rationally choose under common strong belief in rationality? Again, compare this result to the strategies you and Barbara can rationally choose under common belief in future rationality, found in Problem 8.2, part (d).

(c) For the scenario in (b), under common strong belief in rationality, what belief can Barbara hold *initially* about your strategy choice? And what beliefs can Barbara hold about your strategy choice if she observes you selling ice cream today?

(d) For the scenario in (b), find the unique self-confirming pair of rationality orderings.

(e) For the scenario in (b), construct an epistemic model such that, for every strategy found in (b), there is a type that expresses common strong belief in rationality and for which this strategy is optimal. Which types in your model express common strong belief in rationality? Which do not?

Hint for (e): Build an epistemic model with the following properties: For every $k \in \{0, 1, 2, ...\}$ and every strategy s_i that can rationally be chosen when expressing up to k-fold strong belief in rationality, construct a type $t_i^{s_i}$ such that:

- strategy s_i is optimal for type $t_i^{s_i}$, and
- type $t_i^{s_i}$ expresses up to k-fold strong belief in rationality.

For the construction of these types, use the self-confirming pair of rationality orderings found in (d).

9.3 Watching TV with Barbara

Recall the story from Example 9.2 in this chapter. Before you both write down your choice of program on a piece of paper, you have the option of starting a fight with Barbara. Suppose now that if you decide *not* to start a fight with her, Barbara will have the option of starting a fight with *you* about the program to be watched. If Barbara does start a fight with you then, similarly as before, this would reduce both your utility and Barbara's utility by 2.

(a) Model this situation as a dynamic game between you and Barbara.

(b) Find the strategies you can rationally choose under common strong belief in rationality. What about Barbara?

(c) Describe verbally the reasoning that leads to these strategy choices.

(d) What outcome do you initially expect under common strong belief in rationality? Compare this result to the outcome you initially expected in the original situation of Example 9.2. Who has an advantage here under common strong belief in rationality, and why?

(e) Find the unique self-confirming pair of rationality orderings.

(f) Based on these rationality orderings, specify for each of your information sets what the possible beliefs are that you can hold about Barbara's strategy choice under common strong belief in rationality. Do the same for Barbara.

9.4 Never let a lady wait

It is Saturday afternoon, and Barbara and you want to have dinner this evening at 8.00 pm. In the village where you live there are only two restaurants – an Italian restaurant and a Chinese restaurant. The problem is that you prefer the Italian restaurant whereas Barbara prefers the Chinese restaurant. More precisely, having dinner in the Italian restaurant gives you a utility of 10 and eating in the Chinese restaurant yields you a utility of 7, but for Barbara it is the other way around. Even after a long discussion this morning you could not reach an agreement about which restaurant to go to. For this reason you will both go to one of the restaurants this evening, without knowing which restaurant the other person is going to, and hope that you are lucky enough to find your friend there.

You have a reputation for arriving late, and Barbara knows that you have often used it as a strategic weapon. This evening you will strategically choose between arriving on time, or arriving one hour late. That is, at 7.45 pm you must decide between walking to one of the restaurants and arriving at 8.00 pm precisely, or waiting until 8.45 pm before leaving for one of the restaurants. In the latter case you would not know, however, in which restaurant Barbara is. Barbara, on the other hand, is always on time, and so she will be this evening. So, at 7.45 pm she must decide which restaurant to go to, and she will be there exactly at 8.00 pm. If you are both in the same restaurant at 8.00 pm then you will have dinner together, and the utilities will be as described above.

The other possibility is that Barbara, at 8.45 pm, will still be waiting next to an empty chair in her restaurant of choice. In that case, Barbara will not know whether you are at the other restaurant or waiting until 8.45 pm before leaving your house. She then has two options: to stay in the restaurant and hope that you will arrive at 9.00 pm, or to go to the other restaurant and hope that you will be there at 9.00 pm. On the other hand, you will never change restaurants if you left your house at 7.45 pm and are still alone at 8.45 pm – and Barbara knows this! If you are both in the same restaurant at 9.00 pm then you will have dinner together at 9.00 pm, but the utilities for both of you will be decreased by 2 because of the one-hour delay. Assume, moreover, that walking to a second restaurant would decrease Barbara's utility by 1 because it is raining outside. If you are at different restaurants at 9.00 pm, then you will both be disappointed, go home and have a sandwich, yielding both of you a utility of 0 in total.

(a) Model this situation as a dynamic game between you and Barbara. Be careful how to model the information sets!

(b) Find the strategies that you and Barbara can rationally choose under common strong belief in rationality. Do you let Barbara wait?

(c) Describe verbally the reasoning that leads to these strategy choices. Who has an advantage under common strong belief in rationality, and why?

(d) Find the unique self-confirming pair of rationality orderings.
(e) Based on these rationality orderings, specify for each of your information sets what the possible beliefs are that you can hold about Barbara's strategy choice under common strong belief in rationality. Do the same for Barbara.

9.5 Dinner for three

It is Saturday afternoon and you, Barbara and Chris would like to have dinner this evening. As in Problem 9.4, you have a strong preference for the Italian restaurant, whereas Barbara would rather go to the Chinese restaurant. Recently, Chris has discovered a new Greek restaurant in town which he likes very much, but you and Barbara are not very enthusiastic about it. Chris, on the other hand, is allergic to soy sauce and he therefore would like to avoid Chinese food at all costs. Table 9.20 specifies the utilities that you, Barbara and Chris would derive from eating in each of the three restaurants. During the whole afternoon you tried to reach an agreement about where to have dinner, but without success. Barbara therefore told you that she will be waiting in her favorite restaurant at 8.00 pm and similarly Chris announced that he will be waiting in his favorite restaurant around that time. You told Barbara and Chris that you still have to make up your mind, and that at 8.00 pm you will decide which of the three restaurants to go to. At 8.15 pm Barbara will know whether you have chosen her favorite restaurant or not. In either case she may decide to stay or to switch to one of the other two restaurants, which would take her around 15 minutes. Similarly for Chris. You, however, will stay where you are. If at 8.30 pm there are at least two friends in the same restaurant, then these friends will have dinner together. Anybody who is alone in a restaurant at 8.30 pm will go home, eat a sandwich and go to bed early, yielding him or her a utility of 0.

(a) Model this situation as a dynamic game between you, Barbara and Chris. How many strategies does each of you have?
(b) Find the strategies that you, Barbara and Chris can rationally choose under common strong belief in rationality. Do you expect to have dinner with at least one of your friends? If so, with whom, and in which restaurant?
Hint for (b): The full decision problems in this game are very large, and therefore explicitly writing them down is not a very good idea. Instead, use the graphical representation of the game you constructed in (a), and find a way to eliminate strategies directly from the graphical representation. This will save you a lot of writing.
(c) Describe verbally the reasoning that leads to these strategy choices. Who has an advantage under common strong belief in rationality, and why?
(d) Find the unique self-confirming combination of rationality orderings.
(e) Based on these rationality orderings, specify for each of Barbara's information sets what the possible beliefs are that she can hold about your strategy choice and Chris' strategy choice under common strong belief in rationality. Do the same for Chris.

Table 9.20. *Utilities for you, Barbara and Chris in "Dinner for three"*

	Italian	Chinese	Greek
You	3	2	1
Barbara	2	3	1
Chris	2	−1	3

9.6 Read my mind

You and Barbara are participating in a TV show called "Read my mind." The rules are very simple: On a table there are six different objects, numbered from 1 to 6. You must take one of the six objects, but Barbara cannot see which one since she is blindfolded. However, the host tells Barbara whether the number you have chosen is odd or even and Barbara must try to guess the number of the object. If you choose object number k and Barbara guesses correctly, then both you and Barbara get $1000k^2$ euros. If Barbara is wrong, you both get nothing.

(a) Model this situation as a game between you and Barbara.

(b) Find the strategies that you and Barbara can rationally choose under common strong belief in rationality. What outcomes do you deem possible under common strong belief in rationality?

Hint for (b): As in Problem 9.5, the full decision problems in this game are fairly large, and therefore explicitly writing them down is not a very good idea. Instead, use the graphical representation of the game you constructed in (a), and find a way to eliminate strategies directly from the graphical representation. This will save you a lot of writing.

(c) Describe verbally the reasoning that leads to the strategy choices in (b).

(d) Find the unique self-confirming pair of rationality orderings.

(e) Construct an epistemic model such that, for every strategy found in (b), there is a type that expresses common strong belief in rationality, and for which this strategy is optimal.

Hint for (e): Build an epistemic model with the following properties: For every $k \in \{0, 1, 2, ...\}$ and every strategy s_i that can rationally be chosen when expressing up to k-fold strong belief in rationality, construct a type $t_i^{s_i}$ such that:

- strategy s_i is optimal for type $t_i^{s_i}$, and
- type $t_i^{s_i}$ expresses up to k-fold strong belief in rationality.

For the construction of these types, use the self-confirming pair of rationality orderings found in (d).

(f) Find the strategies that you and Barbara can rationally choose under common belief in future rationality. What outcomes do you deem possible under common belief in future rationality? Compare this result to your answers in (b).

9.7 Time to say goodbye

After many years of friendship and numerous adventures, it is now time to say goodbye to your friends Barbara and Chris. They have decided to leave for another country, and this afternoon they will both catch flights from the local airport. A map of the airport is shown in Figure 9.4. The airport has eight gates. Barbara's flight is at 3.30 pm at gate 5, whereas Chris must catch his flight at 3.50 pm at gate 2, and everybody knows this. You have promised Barbara and Chris that you will be waiting at one of the eight gates at 3.00 pm to say goodbye, but you forgot to tell them at which gate. Barbara and Chris will be waiting at 3.00 pm at their respective gates of departure. It is now 2.00 pm, and you must decide the gate at which you will be waiting.

Due to the shape of the airport, Barbara will only be able to see you standing at your gate at 3.00 pm if you are waiting at gate 4, 5 or 6. Similarly, Chris will only be able to see you standing if you have chosen gate 1, 2 or 3. In either case, Barbara and Chris have the option of leaving for another gate at 3.00 pm in order to meet you, or staying where they are. However, both must make sure that they will be back in time to catch their flight! Suppose that it takes 10 minutes to walk from one gate to the next.

The utilities are as follows: If you meet only one of your friends at a gate, then your utility is the number of minutes you spend with that friend before he or she has to leave. If you meet both of your friends, then your utility is the sum of the numbers of minutes you spend with Barbara and Chris, plus a bonus of 40 because you are all together. If you do not meet any of your friends, you will be very disappointed and your utility will be zero.

Barbara and Chris had a final fight yesterday evening, and are therefore only interested in seeing you. More precisely, the utility for Barbara and Chris is the number of minutes spent with you before leaving.

(a) Which gates can Barbara and Chris walk to at 3.00 pm?

(b) Model this situation as a dynamic game between you, Barbara and Chris.

(c) Find the strategies that you can rationally choose under common strong belief in rationality. How many friends do you expect to meet?

Hint for (c): As in Problems 9.5 and 9.6, the full decision problems in this game are fairly large, and therefore explicitly writing them down is not a very good idea. Instead,

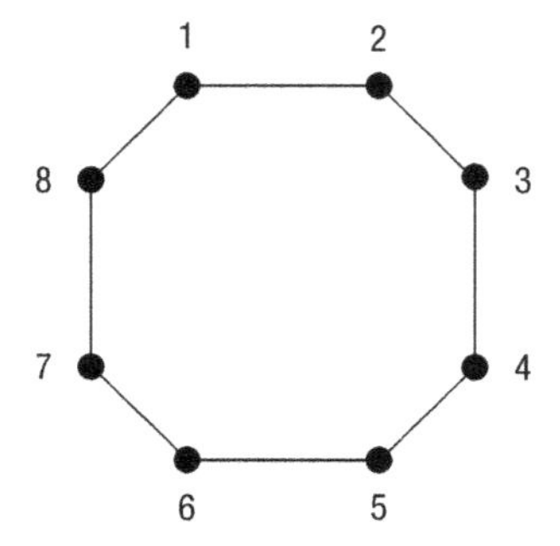

Figure 9.4 A map of the airport in "Time to say goodbye"

use the graphical representation of the game you constructed in (b), and find a way to eliminate strategies directly from the graphical representation. This will save you a lot of writing.
(d) Describe verbally the reasoning that leads to the strategy choices in (c).

Theoretical problems

9.8 Correct beliefs and common strong belief in rationality

Consider a dynamic game with two players, say i and j. A type t_i for player i is said to *believe that j has correct beliefs* if at every information set $h \in H_i$, type t_i only assigns positive probability to types t_j for player j that, at every $h' \in H_j$, assign probability 1 to i's type t_i. That is, t_i always believes that j is correct about i's type, and hence about i's beliefs.
(a) Show that, in general, it is impossible to construct a type t_i for player i that: (1) expresses common strong belief in rationality, and (2) believes that j has correct beliefs.
Hint for (a): It is sufficient to take an example from Chapter 9 and show that in that particular example there is no type t_i for player i that satisfies the two conditions above.
(b) For some of the concepts in this book, we were able to transform a concept into an associated "equilibrium concept" by additionally requiring that a player believes that his opponent is correct about his beliefs, and that a player believes that his opponent believes that he is correct about the opponent's beliefs (in the case of two players). In this way, common belief in rationality could be transformed into Nash equilibrium (Chapter 4), common full belief in "caution and primary belief in rationality" could be transformed into perfect equilibrium (Problem 5.9), and common full belief in "caution and respect of preferences" could be transformed into proper equilibrium (Problem 6.9). Can we transform common strong belief in rationality into an associated equilibrium concept? Explain your answer.

9.9 Initial belief in the opponents' rationality

Suppose that we extend the epistemic model for a dynamic game in the following way: Not only does a type t_i hold a conditional belief about the opponents' strategy-type combinations at each of his information sets $h \in H_i$, but *also* at the beginning of the game $\emptyset$. That is, before the game starts every type t_i holds an *initial* belief $b_i(t_i,\emptyset) \in \Delta(S_{-i} \times T_{-i})$ about the opponents' strategy choices and beliefs. Here, S_{-i} denotes the set of opponents' strategy combinations and T_{-i} the set of opponents' type combinations.

In this extended epistemic model, we say that a strategy s_i is *optimal* for type t_i if: (1) s_i is optimal for t_i's *initial* belief about the opponents' strategy choices, and (2) at every information set $h \in H_i$ that s_i leads to, strategy s_i is optimal for t_i's conditional belief at h about the opponents' strategy choices. Moreover, type t_i is said to *initially believe in his opponents' rationality* if t_i's initial belief $b_i(t_i,\emptyset)$ only assigns positive probability to opponents' strategy-type pairs (s_j,t_j) where the strategy s_j is optimal for the type t_j. We say that type t_i expresses *common belief in "initial belief in rationality"* if

- t_i initially believes in his opponents' rationality,
- t_i assigns, at $\emptyset$ and at every $h \in H_i$, only positive probability to opponents' types that initially believe in their opponents' rationality,
- t_i assigns, at $\emptyset$ and at every $h \in H_i$, only positive probability to opponents' types t_j, that assign, at $\emptyset$ and at every $h' \in H_j$, only positive probability to opponents' types that initially believe in their opponents' rationality,

and so on.

Finally, a strategy s_i can rationally be chosen under common belief in "initial belief in rationality" if there is an extended epistemic model, and a type t_i within it, such that strategy s_i is optimal for t_i and type t_i expresses common belief in "initial belief in rationality."

(a) Construct an algorithm, similar to those in Chapters 8 and 9, that selects for every player precisely those strategies he can rationally choose under common belief in "initial belief in rationality."

(b) Show that every strategy that can rationally be chosen under common belief in future rationality, can also rationally be chosen under common belief in "initial belief in rationality."

(c) Give an example of a dynamic game in which there is an outcome z such that: (1) you can initially deem outcome z possible under common belief in "initial belief in rationality," but (2) you never initially deem outcome z possible under common belief in future rationality.

(d) Show that every strategy that can rationally be chosen under common strong belief in rationality, can also rationally be chosen under common belief in "initial belief in rationality."

(e) Give an example of a dynamic game in which there is an outcome z such that: (1) you can initially deem outcome z possible under common belief in "initial belief in rationality," but (2) you never initially deem outcome z possible under common strong belief in rationality.

9.10 A property of the iterated conditional dominance procedure

We say that an information set h is initially deemed possible under common strong belief in rationality, if for every player i there is a strategy s_i that can rationally be chosen under common strong belief in rationality, such that this combination of strategies leads to h.

For every player i and every information set $h \in H_i$, let $\Gamma^{\text{icd}}(h) = (S_i^{\text{icd}}(h), S_{-i}^{\text{icd}}(h))$ be the reduced decision problem at h that remains after applying the iterated conditional dominance procedure. Let $S_i^{\text{icd}}(\emptyset)$ be the set of strategies for player i that survive the iterated conditional dominance procedure at $\emptyset$. Finally, let $S_i(h)$ be the set of strategies for player i that lead to h, and let $S_{-i}(h)$ be the set of strategy combinations for the opponents that lead to h.

(a) Show that an information set $h \in H_i$ is initially deemed possible under common strong belief in rationality, if and only if, $S_i^{\text{icd}}(h) = S_i^{\text{icd}}(\emptyset) \cap S_i(h)$ and $S_{-i}^{\text{icd}}(h) = S_{-i}^{\text{icd}}(\emptyset) \cap S_{-i}(h)$.
(b) Does the property in (a) also hold for common belief in future rationality and the associated backward dominance procedure? If so, then provide a proof. If not, then provide a counterexample.

Literature

Common strong belief in rationality
The main idea we explored in this chapter is that of *strong belief in the opponents' rationality,* which states that, whenever it is possible for you to believe at an information set that your opponents' have chosen rationally, then you *must* believe at that information set that your opponents have chosen rationally. This idea was first formalized by Battigalli and Siniscalchi (2002). By iterating this condition, Battigalli and Siniscalchi developed a concept which we have called *common strong belief in rationality.* Compared to our approach in this chapter, there are two major differences with the construction by Battigalli and Siniscalchi. First, Battigalli and Siniscalchi use an epistemic model for dynamic games in which it is assumed that the players' conditional beliefs satisfy *Bayesian updating,* whereas we do not. More precisely, Battigalli and Siniscalchi use *conditional probability systems* (see also the literature section of Chapter 5) as the basic ingredient of their epistemic model, which assumes that the conditional beliefs satisfy Bayesian updating whenever possible. More than this, the epistemic model they use actually assumes that the players express *common belief in Bayesian updating.* Secondly, Battigalli and Siniscalchi use a *complete* and *terminal* epistemic model (see the literature section of Chapter 8) for dynamic games, which contains all possible belief hierarchies one can think of – provided they express common belief in Bayesian updating – whereas we do not require the epistemic model to be complete. In fact, for our purposes in this book it is sufficient to work with *finite* epistemic models, containing only finitely many belief hierarchies – or types – for each player. What we do need, however, is that the epistemic model contains *sufficiently many* types. Remember that, in order to properly define *strong belief in the opponents' rationality,* we need the following "richness" condition on the epistemic model M: If at a given information set h for player i there is a combination of types for the opponents – possibly outside M – for which there is a combination of optimal strategies leading to h, then the epistemic model M must contain at least one such combination of types for which there is a combination of optimal strategies leading to h. Battigalli and Siniscalchi guarantee this "richness" condition by insisting on an epistemic model M that contains *all* possible belief hierarchies – or types – which satisfy common belief in Bayesian updating, that is, by using epistemic models M that are terminal. Similarly, the richness conditions which we need to define k-fold strong belief in rationality, for $k \geq 2$, are automatically satisfied by Battigalli and Siniscalchi's model, as they assume the epistemic model to be complete and terminal.

Extensive form rationalizability

Common strong belief in rationality was implicitly present in the concept of *extensive form rationalizability*, as proposed by Pearce (1984). In fact, Pearce (1984) presented extensive form rationalizability using an iterated elimination procedure, which recursively eliminates strategies and conditional belief vectors from the game. Later, Battigalli (1997) developed a simpler elimination procedure, which he proved to be equivalent to Pearce's original procedure.

The procedures by Pearce and Battigalli can be summarized as follows: For every player i, let S_i^0 be the set of all strategies, and let B_i^0 be the set of all conditional belief vectors b_i about the opponents' strategies satisfying Bayesian updating. At every step $k \geq 1$, let B_i^k be the set of conditional belief vectors $b_i = (b_i(h))_{h \in H_i}$ from B_i^{k-1} satisfying the following condition: If at information set $h \in H_i$ there is some strategy combination for the opponents in S_{-i}^{k-1} leading to h, then $b_i(h)$ must assign positive probability only to opponents' strategy combinations in S_{-i}^{k-1} leading to h. Let S_i^k be the set of strategies s_i for player i that are optimal, at every information set $h \in H_i$ that s_i leads to, for some conditional belief vector b_i in B_i^k. A strategy s_i is then called *extensive form rationalizable* if s_i is in S_i^k for every k.

Note that the inductive step, which selects from B_i^{k-1} those conditional belief vectors that belong to B_i^k, is very similar in spirit to the condition of k-fold strong belief in rationality. Indeed, Battigalli and Siniscalchi (2002) showed in their Proposition 6 that the strategies that can rationally be chosen under common strong belief in rationality, are exactly the extensive form rationalizable strategies in the sense of Pearce (1984) and Battigalli (1997). That is, common strong belief in rationality can be viewed as an epistemic foundation for the procedure of extensive form rationalizability.

Algorithm

In this chapter we presented an algorithm – the *iterated conditional dominance procedure* – which yields precisely those strategies that can rationally be chosen under common strong belief in rationality. The algorithm is due to Shimoji and Watson (1998), who show that their algorithm yields exactly the extensive form rationalizable strategies in a game. If we combine this result with Battigalli and Siniscalchi's theorem, showing that the extensive form rationalizable strategies are exactly the strategies that can rationally be chosen under common strong belief in rationality, we obtain Theorem 9.3.3 in this chapter, which shows that the iterated conditional dominance procedure generates precisely those strategies that can rationally be chosen under common strong belief in rationality.

Order (in) dependence

We have seen that for the iterated conditional dominance procedure, the order of elimination crucially matters for the strategies that survive the algorithm. That is, if at some step of the algorithm we either do not eliminate all strategies we can at a given information set, or do not scan through all information sets, then we may end up with different sets of strategies for the players at the end. This is in contrast with the backward dominance procedure, where the order of elimination is not relevant for the strategies that survive the algorithm. Chen and Micali (2011) and Robles (2006) showed, however, that the order of elimination in the iterated conditional dominance procedure does not affect the *outcomes* that can be reached. More precisely, if we change the order of elimination in the iterated conditional dominance procedure, then we may change the sets of strategies that survive for the various players in the game, but the *outcomes* that can be reached by the surviving strategy combinations will remain the same!

Comparison with common belief in future rationality

In Section 9.4 of this chapter we compared common strong belief in rationality with common belief in future rationality. And in particular we compared the two associated algorithms with each other – the iterated conditional dominance procedure and the backward dominance procedure. These comparisons are largely based on Perea (2010). Among other things, we showed that in terms of *outcomes* common strong belief in rationality is more restrictive than common belief in future rationality. More precisely, we proved in Theorem 9.4.2 that every outcome that is initially deemed possible under common strong belief in rationality, is also initially deemed possible under common belief in future rationality.

This theorem can also be proved by making use of the result by Chen and Micali (2011) and Robles (2006) above, which states that the *outcomes* that can be reached under the strategies that survive the iterated conditional dominance procedure, are independent of the order of elimination. It can be shown that the backward dominance procedure corresponds to the first few steps of the iterated conditional dominance procedure – by eliminating strategies in a very particular, different order – but *without completing* the elimination procedure. Since we know, from Chen and Micali (2011) and Robles (2006), that the order of elimination in the iterated conditional dominance procedure does not change the outcomes that can be reached, it follows that the outcomes that can be reached using the iterated conditional dominance procedure will be a subset of the outcomes that can be reached using the backward dominance procedure. But then, the outcomes that are initially deemed possible under common strong belief in rationality will be a subset of the outcomes that are initially deemed possible under common belief in future rationality. The result in Theorem 9.4.2 follows from this argument.

Common strong belief in rationality and backward induction

In Corollary 9.4.3 we showed that in every game with perfect information, every outcome that is initially deemed possible under common strong belief in rationality must be a backward induction outcome in that game. We showed this result by the following steps: First, we use the result mentioned above, stating that every outcome that is initially deemed possible under common strong belief in rationality, is also initially deemed possible under common belief in future rationality. Moreover, we use the insight that in games with perfect information, the backward dominance procedure – and hence common belief in future rationality – is equivalent to the backward induction procedure.

By combining these ideas we conclude that in games with perfect information, every outcome that is initially deemed possible under common strong belief in rationality, must be an outcome that is reachable under the backward induction procedure – that is, it must be a backward induction outcome.

This result was first proved by Battigalli (1997) – for *generic* games with perfect information – in his Theorem 4. A game with perfect information is called *generic* if for a given player, all terminal histories yield different utilities. In fact, Battigalli (1997) showed that in every generic game with perfect information, there is only one outcome that can be reached under extensive form rationalizability, namely the unique backward induction outcome in that game. The proof that Battigalli delivers is very different from ours, however. He uses the properties of fully stable sets of Kohlberg and Mertens (1986) to show the result, whereas we use a more direct and constructive way to prove the statement.

Gradwohl and Heifetz (2011) investigated *non-generic* games with perfect information – where different terminal histories may induce the same utility for a player – and compared common strong belief in rationality (both in the usual sense and for the agent form), common belief in future rationality and subgame perfect equilibrium for such games. The *agent form* is the game that is obtained by assigning a different player to every information set. They found that for these games, in terms of outcomes reached, subgame perfect equilibrium is more restrictive than common strong belief in rationality, which is more restrictive than common belief in future rationality, which in turn is more restrictive than common strong belief in rationality for the agent form.

Rationality orderings

The concept of *rationality orderings*, as discussed in Section 9.6, was introduced and analyzed by Battigalli (1996). The term *self-confirming combination of rationality orderings* that we use, corresponds to what Battigalli calls *correlated sequential rationality orderings.* The main idea behind rationality orderings is that a player, at each of his information sets, always looks for the "most rational" strategies for the opponents that could have led to this information set. Battigalli (1996) calls this the *best rationalization principle.* The theorem in Section 9.6, which states that there is unique self-confirming combination of rationality orderings – the one induced by common strong belief in rationality – is based on Battigalli's Theorem 2.2.

Bayesian updating

In Theorem 9.7.1 we stated that for common strong belief in rationality, it is not relevant whether we additionally impose (common belief in) Bayesian updating or not: the eventual strategy choices selected are the same. This property was first shown by Shimoji and Watson (1998) in their Theorem 2. More precisely, they show that for the strategy choices selected by the iterated conditional dominance procedure, it is inessential whether we impose (common belief in) Bayesian updating or not. As the iterated conditional dominance procedure delivers exactly the strategies that can rationally be chosen under common strong belief in rationality, the result in Theorem 9.7.1 follows.

Forward induction

Common strong belief in rationality – or equivalently, the concept of extensive form rationalizability – is considered to be a *forward-induction* concept. In the literature there is no unique definition of *forward induction*, but intuitively it describes a way of reasoning in which a player, when confronted with an unexpected choice by an opponent, tries to find the "most plausible" explanation for this surprising move by his opponent.

Indeed, common strong belief in rationality is perfectly in line with such a way of reasoning. According to this concept, a player asks at each of his information sets whether there are optimal strategies for his opponents that could have led to this information set. If so, he will ascribe positive probability only to such optimal strategies at that information set. He then asks whether there are optimal strategies for his opponent, leading to his information set, which can rationally be chosen by opponents that strongly believe in their opponents' rationality. If so, he will ascribe positive probability only to such strategies that are optimal for opponents that strongly believe in their opponents' rationality. And so on. So, the player looks for the most plausible strategies for his opponents that could have led to his information set, and assigns positive probability only to such strategies. It therefore matches precisely our intuition of forward induction reasoning.

In the literature, most forward-induction concepts presented so far are *equilibrium concepts*. More precisely, most of these concepts take the concept of *sequential equilibrium* as defined by Kreps and Wilson (1982) – see the literature section of Chapter 8 for a discussion – and impose additional restrictions on the beliefs about the opponents' *past* choices. Examples include: *forward induction equilibrium* (Cho, 1987), *justifiable sequential equilibrium* (McLennan, 1985) and *stable sets of beliefs* (Hillas, 1994) for general dynamic games, and the *intuitive criterion* (Cho and Kreps, 1987) and its various refinements for the special class of signaling games. See Chapter 5 in Perea (2001) for an overview of most of these forward-induction refinements of sequential equilibrium.

What is a bit problematic about these forward-induction refinements of sequential equilibrium is that they are encapsulated within a *backward-induction* concept – namely *sequential equilibrium*. In the literature section of Chapter 8, we saw that sequential equilibrium implicitly assumes *common belief in future rationality*, which is a typical *backward-induction* type of reasoning, as it requires players to reason critically only about the choices that opponents will make now and in the future. So, in a sense, the forward-induction refinements of sequential equilibrium are a mix of backward-induction and forward-induction arguments, which makes the forward-induction reasoning less transparent than it could be. Moreover, in some games such concepts will be unable to filter out the "natural" forward-induction strategy for a player. Consider, for instance, the example "Painting Chris' house," which we discussed at the beginning of this chapter. In that game, the "natural" forward-induction strategy for you is to choose a price of 300. For Barbara, rejecting her colleague's offer can only be part of a rational strategy for her if she subsequently chooses a price of 400. Hence, in the spirit of forward induction, it makes sense for you to believe that Barbara will indeed choose a price of 400 if you observe that she has rejected her colleague's offer. But then, your only optimal choice is to choose a price of 300 yourself. However, a forward-induction refinement of sequential equilibrium will never be able to select this price of 300 for you! The reason is that there is only one sequential equilibrium in this game, in which you believe – after observing Barbara reject her colleague's offer – that Barbara will choose a price of 200. Hence, your only optimal choice under sequential equilibrium is to choose a price of 200 as well. That is, every forward-induction refinement of sequential equilibrium will uniquely select the choice 200 for you, which is not the natural forward induction choice for you in this game.

In contrast, *common strong belief in rationality* is a "pure" forward-induction concept, which is not encapsulated within any backward-induction concept such as common belief in future rationality. If we were to build a forward-induction concept that takes the backward-induction concept of common belief in future rationality as a starting point, then within the example "Painting Chris' house" we would never be able to select your natural forward-induction choice 300, as under common belief in future rationality there is only one strategy you can rationally choose, namely 200.

Explicable equilibrium

The forward-induction concept of *explicable equilibrium*, as proposed by Reny (1992b), is different from the concepts discussed above, as it does not take sequential equilibrium as a starting point, but rather the weaker concept of *weak sequential rationality*. Remember from the literature section in Chapter 8 that *weak sequential rationality* intuitively means that players *initially* believe in their opponents' future rationality, but do not necessarily believe so at later stages of the game. This is weaker than the *sequential rationality* condition imposed in sequential equilibrium, which states that players *always* believe in their opponents' future rationality.

Explicable equilibrium is very similar – in spirit – to the concept of *extensive form rationalizability*, and equivalently to *common strong belief in rationality*. It also defines a rationality ordering on the players' sets of strategies, and imposes that a player, at each of his information sets, looks for the "most rational" strategies for the opponents that could have led to this information set, and assigns positive probability only to such strategies. However, the rationality orderings are defined differently than under common strong belief in rationality, as explicable equilibrium takes weakly sequentially rational assessments as a starting point for the analysis.

Reny (1992b) showed, in his Proposition 3, that for *generic* games with perfect information, explicable equilibrium always leads to the unique *backward-induction outcome*, but not necessarily to the unique backward-induction strategies for the players. This result is similar to Theorem 4 in Battigalli (1997), which showed that the same holds for the concept of extensive form rationalizability – and hence also for common strong belief in rationality. Like Battigalli, Reny also proved this result by making use of the properties of fully stable sets as defined by Kohlberg and Mertens (1986).

Burning-money games

The example "Watching TV with Barbara," which we discussed in this chapter, can be viewed as an instance of a *burning-money game* – a type of game first studied by van Damme (1989) and Ben-Porath and Dekel (1992). A *burning-money game* is a game with two players in which player 1, at the beginning of the game, can publicly and voluntarily burn a certain amount of money – and thereby reduce his utility – before making a simultaneous move with player 2. In the example "Watching TV with Barbara" you have the option of starting a fight with Barbara at the beginning of the game, which would reduce your utility by 2. This can thus be seen as "publicly burning 2 utility units" before starting the game with Barbara.

In his paper, van Damme (1989) studied a burning-money game in which player 1 can only burn a given amount of money, whereas Ben-Porath and Dekel (1992) analyzed games in which player 1 can freely choose the amount of money he would like to burn. Both papers use the algorithm of *iterated elimination of weakly dominated strategies* – as studied in Chapter 7 of this book – to analyze the burning-money game. More precisely, they take the full decision problem at the beginning of the game, and iteratively remove all weakly dominated strategies from that decision problem.

Ben-Porath and Dekel (1992) found the following striking result: Suppose that in the simultaneous-move game between players 1 and 2, there is a combination of choices (c_1^*, c_2^*), which: (1) is strictly better for player 1 than any other choice-combination in that game, and (2) is strictly better for player 2 than any other choice-combination (c_1^*, c_2) that involves c_1^* for player 1. Then, iterated elimination of weakly dominated strategies leads to a unique outcome, where player 1 does not burn any money and player 1's most preferred choice-combination (c_1^*, c_2^*) is obtained in the game that follows. This is similar to our result in the example "Watching TV with Barbara," where under common strong belief in rationality you will not start a fight with Barbara, and you expect to watch your favorite program with Barbara.

Later, Shimoji (2002) showed that the result by Ben-Porath and Dekel (1992) also holds if we use *extensive form rationalizability* – or equivalently *common strong belief in rationality* – instead of iterated elimination of weakly dominated strategies. That is, in the burning-money games studied in Ben-Porath and Dekel (1992), common strong belief in rationality always leads to a unique outcome, in which you do not burn any money, yet receive your most preferred outcome in the game that follows.

In particular, in the burning-money games studied by Ben-Porath and Dekel (1992), the algorithm of iterated elimination of weakly dominated strategies always yields the same result as common strong belief in rationality. This is actually true for other dynamic games of interest as well, which is the reason that people have often used the algorithm of iterated elimination of weakly dominated strategies as a *forward-induction concept* for such dynamic games.

Bibliography

Anscombe, F.J. and Aumann, R.J. (1963). A definition of subjective probability. *Annals of Mathematical Statistics*, **34**, 199–205.

Armbruster, W. and Böge, W. (1979). Bayesian game theory. In *Game Theory and Related Topics*, ed. O. Moeschlin and D. Pallaschke. Amsterdam: North-Holland.

Asheim, G.B. (2001). Proper rationalizability in lexicographic beliefs. *International Journal of Game Theory*, **30**, 453–78.

Asheim, G.B. (2002). On the epistemic foundation for backward induction. *Mathematical Social Sciences*, **44**, 121–44.

Asheim, G.B. (2006). *The Consistent Preferences Approach to Deductive Reasoning in Games*. Theory and Decision Library. Dordrecht, The Netherlands: Springer.

Asheim, G.B. and Dufwenberg, M. (2003). Admissibility and common belief. *Games and Economic Behavior*, **42**, 208–34.

Asheim, G.B. and Perea, A. (2005). Sequential and quasi-perfect rationalizability in extensive games. *Games and Economic Behavior*, **53**, 15–42.

Asheim, G.B. and Perea, A. (2009). Algorithms for cautious reasoning in games. Working paper.

Aumann, R.J. (1974). Subjectivity and correlation in randomized strategies. *Journal of Mathematical Economics*, **1**, 67–96.

Aumann, R.J. (1976). Agreeing to disagree. *Annals of Statistics*, **4**, 1236–9.

Aumann, R.J. (1987). Correlated equilibrium as an expression of Bayesian rationality. *Econometrica*, **55**, 1–18.

Aumann, R.J. (1995). Backward induction and common knowledge of rationality. *Games and Economic Behavior*, **8**, 6–19.

Aumann, R.J. (1996). Reply to Binmore. *Games and Economic Behavior*, **17**, 138–46.

Aumann, R.J. (1998). On the centipede game. *Games and Economic Behavior*, **23**, 97–105.

Aumann, R.J. and Brandenburger, A. (1995). Epistemic conditions for Nash equilibrium. *Econometrica*, **63**, 1161–80.

Bach, C.W. and Heilmann, C. (2011). Agent connectedness and backward induction. *International Game Theory Review*, **13**, 195–208.

Balkenborg, D. and Winter, E. (1997). A necessary and sufficient epistemic condition for playing backward induction. *Journal of Mathematical Economics*, **27**, 325–45.

Baltag, A., Smets, S. and Zvesper, J.A. (2009) Keep "hoping" for rationality: A solution to the backward induction paradox. *Synthese*, **169**, 301–33 (*Knowledge, Rationality and Action*, 705–37).

Basu, K. (1990). On the non-existence of a rationality definition for extensive games. *International Journal of Game Theory*, **19**, 33–44.

Basu, K. (1994). The traveler's dilemma: Paradoxes of rationality in game theory. *American Economic Review*, **36**, 391–6.

Basu, K. and Weibull, J. (1991). Strategy subsets closed under rational behavior. *Economics Letters*, **36**, 141–6.

Battigalli, P. (1996). Strategic rationality orderings and the best rationalization principle. *Games and Economic Behavior*, **13**, 178–200.

Battigalli, P. (1997). On rationalizability in extensive games. *Journal of Economic Theory*, **74**, 40–61.

Battigalli, P. and Bonanno, G. (1999). Recent results on belief, knowledge and the epistemic foundations of game theory. *Research in Economics*, **53**, 149–225.

Battigalli, P. and Siniscalchi, M. (1999). Hierarchies of conditional beliefs and interactive epistemology in dynamic games. *Journal of Economic Theory*, **88**, 188–230.

Battigalli, P. and Siniscalchi, M. (2002). Strong belief and forward induction reasoning. *Journal of Economic Theory*, **106**, 356–91.

Ben-Porath, E. (1997). Rationality, Nash equilibrium and backwards induction in perfect-information games. *Review of Economic Studies*, **64**, 23–46.

Ben-Porath, E. and Dekel, E. (1992). Signaling future actions and the potential for sacrifice. *Journal of Economic Theory*, **57**, 36–51.

Bernheim, B.D. (1984). Rationalizable strategic behavior. *Econometrica*, **52**, 1007–28.

Bertrand, J. (1883). Théorie mathématique de la richesse sociale. *Journal des Savants*, **67**, 499–508.

Bicchieri, C. (1989). Self-refuting theories of strategic interaction: A paradox of common knowledge. *Erkenntnis*, **30**, 69–85.

Binmore, K. (1987). Modeling rational players, Part I. *Economics and Philosophy*, **3**, 179–214.

Binmore, K. (1996). A note on backward induction. *Games and Economic Behavior*, **17**, 135–7.

Blume, L.E., Brandenburger, A. and Dekel, E. (1991a). Lexicographic probabilities and choice under uncertainty. *Econometrica*, **59**, 61–79.

Blume, L.E., Brandenburger, A. and Dekel, E. (1991b). Lexicographic probabilities and equilibrium refinements. *Econometrica*, **59**, 81–98.

Board, O. (2002). Knowledge, beliefs and game-theoretic solution concepts. *Oxford Review of Economic Policy*, **18**, 433–45.

Böge, W. and Eisele, T.H. (1979). On solutions of Bayesian games. *International Journal of Game Theory*, **8**, 193–215.

Bonanno, G. (2010). AGM-consistency and perfect Bayesian equilibrium. Part I: Definition and properties. Working paper.

Borel, É. (1921). La théorie du jeu et les équations intégrales à noyau symétrique. *Comptes Rendus Hebdomadaire des Séances de l'Académie des Sciences* (Paris), **173**, 1304–8. (Translated by Leonard J. Savage (1953) as The theory of play and integral equations with skew symmetric kernels, *Econometrica*, **21**, 97–100.)

Borel, É. (1924). *Eléments de la Théorie des Probabilités*, 3rd edn. Paris: Hermann. (Pages 204–21 translated by Leonard J. Savage (1953) as On games that involve chance and the skill of players, *Econometrica*, 21 101–15.)

Borel, É. (1927). Sur les systèmes de formes linéaires à déterminant symétrique gauche et la théorie générale du jeu. *Comptes Rendus Hebdomadaire des Séances de l'Académie des Sciences* (Paris), **184**, 52–4. (Translated by Leonard J. Savage (1953) as On systems of linear forms of skew symmetric determinant and the general theory of play, *Econometrica*, **21**, 116–17.)

Börgers, T. (1994). Weak dominance and approximate common knowledge. *Journal of Economic Theory*, **64**, 265–76.

Börgers, T. and Samuelson, L. (1992). "Cautious" utility maximization and iterated weak dominance. *International Journal of Game Theory*, **21**, 13–25.

Brandenburger, A. (1992a). Knowledge and equilibrium in games. *Journal of Economic Perspectives*, **6**, 83–101.

Brandenburger, A. (1992b). Lexicographic probabilities and iterated admissibility. In *Economic Analysis of Markets and Games*, ed. P. Dasgupta, *et al.* Cambridge, MA: MIT Press, pp. 282–90.

Brandenburger, A. (2003). On the existence of a "complete" possibility structure. In *Cognitive Processes and Economic Behavior*, ed. N. Dimitri, M. Basili and I. Gilboa. London: Routledge.

Brandenburger, A. (2007). The power of paradox: Some recent developments in interactive epistemology. *International Journal of Game Theory*, **35**, 465–92.

Brandenburger, A. and Dekel, E. (1987). Rationalizability and correlated equilibria. *Econometrica*, **55**, 1391–402.

Brandenburger, A. and Dekel, E. (1989). The role of common knowledge assumptions in game theory. In *The Economics of Missing Markets, Information and Games*, ed. F. Hahn. Oxford: Oxford University Press, pp. 46–61.

Brandenburger, A. and Dekel, E. (1993). Hierarchies of beliefs and common knowledge. *Journal of Economic Theory*, **59**, 189–98.

Brandenburger, A. and Friedenberg, A. (2008). Intrinsic correlation in games. *Journal of Economic Theory*, **141**, 28–67.

Brandenburger, A. and Friedenberg, A. (2010). Self-admissible sets. *Journal of Economic Theory*, **145**, 785–811.

Brandenburger, A., Friedenberg, A. and Keisler, J. (2008). Admissibility in games. *Econometrica*, **76**, 307–52.

Brandenburger, A. and Keisler, J. (2006). An impossibility theorem on beliefs in games. *Studia Logica*, **84**, 211–40.

Chen, J. and Micali, S. (2011). The robustness of extensive-form rationalizability. Working paper.

Cho, I.-K. (1987). A refinement of sequential equilibrium. *Econometrica*, **55**, 1367–89.

Cho, I.-K. and Kreps, D.M. (1987). Signaling games and stable equilibria. *Quarterly Journal of Economics*, **102**, 179–221.

Clausing, T. (2003). Doxastic conditions for backward induction. *Theory and Decision*, **54**, 315–36.

Clausing, T. (2004). Belief revision in games of perfect information. *Economics and Philosophy*, **20**, 89–115.

Cournot, A.A. (1838). *Recherches sur les Principes Mathématiques de la Théorie des Richesses*. Paris: Hachette. (English translation: (1897) *Researches into the Mathematical Principles of the Theory of Wealth*, New York: Macmillan.)

Damme, E. van (1989). Stable equilibria and forward induction. *Journal of Economic Theory*, **48**, 476–96.

Dekel, E. and Fudenberg, D. (1990). Rational behavior with payoff uncertainty. *Journal of Economic Theory*, **52**, 243–67.

Dekel, E., Fudenberg, D. and Levine, D.K. (1999). Payoff information and self-confirming equilibrium. *Journal of Economic Theory*, **89**, 165–85.

Dekel, E., Fudenberg, D. and Levine, D.K. (2002). Subjective uncertainty over behavior strategies: A correction. *Journal of Economic Theory*, **104**, 473–8.

Dekel, E. and Gul, F. (1997). Rationality and knowledge in game theory. In *Advances in Economics and Econometrics: Theory and Applications* (Seventh World Congress, Vol. 1), ed. D.M. Kreps and K.F. Wallis, pp. 87–172.

Ellingsen, T. and Miettinen, T. (2008). Commitment and conflict in bilateral bargaining. *American Economic Review*, **98**, 1629–35.

Epstein, L.G. and Wang, T. (1996). "Beliefs about beliefs" without probabilities. *Econometrica*, **64**, 1343–73.

Farquharson, R. (1969). *Theory of Voting*. New Haven, CT: Yale University Press.

Feinberg, Y. (2005). Subjective reasoning: Dynamic games. *Games and Economic Behavior*, **52**, 54–93.

Finetti, B. de (1936). Les probabilités nulles. *Bulletin des Sciences Mathématiques*, **60**, 275–88.

Friedenberg, A. (2010). When do type structures contain all hierarchies of beliefs? *Games and Economic Behavior*, **68**, 108–29.

Fudenberg, D. and Tirole, J. (1991a). *Game Theory*. Cambridge, MA: MIT Press.

Fudenberg, D. and Tirole, J. (1991b). Perfect Bayesian equilibrium and sequential equilibrium. *Journal of Economic Theory*, **53**, 236–60.

Geanakoplos, J. (1992). Common knowledge. *Journal of Economic Perspectives*, **6**, 52–83.

Gradwohl, R. and Heifetz, A. (2011). Rationality and equilibrium in perfect-information games. Working paper.

Gul, F. (1997). Rationality and coherent theories of strategic behavior. *Journal of Economic Theory*, **70**, 1–31.

Halpern, J.Y. (2010). Lexicographic probability, conditional probability, and nonstandard probability. *Games and Economic Behavior*, **68**, 155–79.

Hammond, P.J. (1994). Elementary non-Archimedean representations of probability for decision theory and games. In *Scientific Philosopher*, Vol. 1, ed. P. Humpreys and P. Suppes. Dordrecht: Kluwer, pp. 25–49.

Harsanyi, J.C. (1967–1968). Games with incomplete information played by "Bayesian" players, I–III. *Management Science*, **14**, 159–82, 320–34, 486–502.

Heifetz, A. (1993). The Bayesian formulation of incomplete information: The non-compact case. *International Journal of Game Theory*, **21**, 329–38.

Heifetz, A. and Samet, D. (1998a). Topology-free typology of beliefs. *Journal of Economic Theory*, **82**, 324–41.

Heifetz, A. and Samet, D. (1998b). Knowledge spaces with arbitrarily high rank. *Games and Economic Behavior*, **22**, 260–73.

Hendon, E., Jacobsen, H.J. and Sloth, B. (1996). The one-shot-deviation principle for sequential rationality. *Games and Economic Behavior*, **12**, 274–82.

Herings, J.J. and Vannetelbosch, V.J. (1999). Refinements of rationalizability for normal-form games. *International Journal of Game Theory*, **28**, 53–68.

Herings, J.J. and Vannetelbosch, V.J. (2000). The equivalence of the Dekel–Fudenberg iterative procedure and weakly perfect rationalizability. *Economic Theory*, **15**, 677–87.

Hillas, J. (1994). Sequential equilibria and stable sets of beliefs. *Journal of Economic Theory*, **64**, 78–102.

Hotelling, H. (1929). Stability in competition. *Economic Journal*, **39**, 41–57.

Kakutani, S. (1941). A generalization of Brouwer's fixed point theorem. *Duke Mathematical Journal*, **8**, 457–9.

Keisler, J. and Lee, B.S. (2011). Common assumption of rationality. Working paper.

Kets, W. (2010). Bounded reasoning and higher-order uncertainty. Working paper.

Kohlberg, E. and Mertens, J.-F. (1986). On the strategic stability of equilibria. *Econometrica*, **54**, 1003–37.

Kreps, D.M., Milgrom, P., Roberts, J. and Wilson, R. (1982). Rational cooperation in the finitely repeated prisoner's dilemma. *Journal of Economic Theory*, **27**, 245–52.

Kreps, D.M. and Wilson, R. (1982). Sequential equilibria. *Econometrica*, **50**, 863–94.

Kripke, S. (1963). A semantical analysis of modal logic I: Normal modal propositional calculi. *Zeitschrift für Mathematische Logik und Grundlagen der Mathematik*, **9**, 67–96.

Kuhn, H.W. (1950). Extensive games. *Proceedings of the National Academy of Sciences of the United States of America*, **36**, 570–6.

Kuhn, H.W. (1953). Extensive games and the problem of information. In *Contributions to the Theory of Games*, Vol. II (*Annals of Mathematics Studies*, 28), ed. H.W. Kuhn and A.W. Tucker. Princeton, NJ: Princeton University Press, pp. 193–216.

Lewis, D.K. (1969). *Convention*. Cambridge, MA: Harvard University Press.

Luce, R.D. and Raiffa, H. (1957). *Games and Decisions*. New York: John Wiley and Sons.

McLennan, A. (1985). Justifiable beliefs in sequential equilibria. *Econometrica*, **53**, 889–904.

Meier, M. (2005). On the nonexistence of universal information structures. *Journal of Economic Theory*, **122**, 132–9.

Mertens, J.-F. and Zamir, S. (1985). Formulation of Bayesian analysis for games with incomplete information. *International Journal of Game Theory*, **14**, 1–29.

Miettinen, T. and Perea, A. (2010). Commitment in alternating offers bargaining. Working paper.

Monderer, D. and Samet, D. (1989). Approximating common knowledge with common beliefs. *Games and Economic Behavior*, **1**, 170–90.

Morgenstern, O. (1935). Vollkommene Voraussicht und wirtschaftliches Gleichgewicht. *Zeitschrift für Nationalökonomie*, **6**, 337–57. (Reprinted as (1976) Perfect foresight and economic equilibrium in *Selected Economic Writings of Oskar Morgenstern*, ed. A. Schotter, New York University Press, pp. 169–83.)

Moulin, H. (1979). Dominance solvable voting schemes. *Econometrica*, **47**, 1337–51.

Myerson, R.B. (1978). Refinements of the Nash equilibrium concept. *International Journal of Game Theory*, **7**, 73–80.

Nagel, R. (1995). Unravelling in guessing games: An experimental study. *American Economic Review*, **85**, 1313–26.

Nash, J.F. (1950). Equilibrium points in N-person games. *Proceedings of the National Academy of Sciences of the United States of America*, **36**, 48–9.

Nash, J.F. (1951). Non-cooperative games. *Annals of Mathematics*, **54**, 286–95.

Navarro, N. and Perea, A. (2011). A simple bargaining procedure for the Myerson value. Working paper.

Neumann, J. von (1928). Zur Theorie der Gesellschaftsspiele. *Mathematische Annalen*, **100**, 295–320. (Translated by S. Bargmann as (1959) On the theory of games of strategy. In *Contributions to the Theory of Games*, Vol. IV (*Annals of Mathematics Studies*, 40), ed. A.W. Tucker and R.D. Luce, Princeton, NJ: Princeton University Press, pp. 13–43.)

Neumann, J. von and Morgenstern, O. (1944, 1953). *Theory of Games and Economic Behavior*. Princeton, NJ: Princeton University Press.

Pearce, D. (1984). Rationalizable strategic behavior and the problem of perfection. *Econometrica*, **52**, 1029–50.

Penta, A. (2009). Robust dynamic mechanism design. Working paper.

Perea, A. (2001). *Rationality in Extensive Form Games*. Theory and Decision Library, Series C. Boston, Dordrecht, London: Kluwer Academic Publishers.

Perea, A. (2002). A note on the one-deviation property in extensive games. *Games and Economic Behavior*, **40**, 322–38.

Perea, A. (2006). Proper belief revision and rationalizability in dynamic games. *International Journal of Game Theory*, **34**, 529–59.

Perea, A. (2007a). A one-person doxastic characterization of Nash strategies. *Synthese*, **158**, 251–71 (*Knowledge, Rationality and Action*, 341–61).

Perea, A. (2007b). Proper belief revision and equilibrium in dynamic games. *Journal of Economic Theory*, **136**, 572–86.

Perea, A. (2007c). Epistemic foundations for backward induction: An overview. In *Interactive Logic Proceedings of the 7th Augustus de Morgan Workshop, London*, ed. J. van Benthem, D. Gabbay and B. Löwe. Texts in Logic and Games 1. Amsterdam University Press, pp. 159–93.

Perea, A. (2008). Minimal belief revision leads to backward induction. *Mathematical Social Sciences*, **56**, 1–26.

Perea, A. (2010). Backward induction versus forward induction reasoning. *Games*, **1**, 168–88.

Perea, A. (2011a). An algorithm for proper rationalizability. *Games and Economic Behavior*, **72**, 510–25.

Perea, A. (2011b). Belief in the opponents' future rationality. Working paper.

Polak, B. (1999). Epistemic conditions for Nash equilibrium, and common knowledge of rationality. *Econometrica*, **67**, 673–6.

Popper, K.R. (1934). *Logik der Forschung*. Vienna: Julius Springer Verlag.

Popper, K.R. (1968). *The Logic of Scientific Discovery*, 2nd edn. London: Hutchinson. (The first version of this book appeared as *Logik der Forschung*.)

Quesada, A. (2002). Belief system foundations of backward induction. *Theory and Decision*, **53**, 393–403.

Quesada, A. (2003). From common knowledge of rationality to backward induction. *International Game Theory Review*, **2**, 127–37.

Reny, P.J. (1992a). Rationality in extensive-form games. *Journal of Economic Perspectives*, **6**, 103–18.

Reny, P.J. (1992b). Backward induction, normal form perfection and explicable equilibria. *Econometrica*, **60**, 627–49.

Reny, P.J. (1993). Common belief and the theory of games with perfect information. *Journal of Economic Theory*, **59**, 257–74.

Rényi, A. (1955). On a new axiomatic theory of probability. *Acta Mathematica Academiae Scientiarum Hungaricae*, **6**, 285–335.

Robinson, A. (1973). Function theory on some nonarchimedean fields. *Papers on the Foundations of Mathematics. American Mathematical Monthly*, **80**, S87–S109.

Robles, J. (2006). Order independence of conditional dominance. Working paper.

Rosenthal, R.W. (1981). Games of perfect information: Predatory pricing and the chain-store paradox. *Journal of Economic Theory*, **25**, 92–100.

Rubinstein, A. (1982). Perfect equilibrium in a bargaining model. *Econometrica*, **50**, 97–110.

Rubinstein, A. (1991). Comments on the interpretation of game theory. *Econometrica*, **59**, 909–24.

Samet, D. (1996) Hypothetical knowledge and games with perfect information. *Games and Economic Behavior*, **17**, 230–51.

Samuelson, L. (1992). Dominated strategies and common knowledge. *Games and Economic Behavior*, **4**, 284–313.

Savage, L.J. (1954). *The Foundation of Statistics*. New York: Wiley.

Schuhmacher, F. (1999). Proper rationalizability and backward induction. *International Journal of Game Theory*, **28**, 599–615.

Schulte, O. (2003). Iterated backward inference: An algorithm for proper rationalizability. In *Proceedings of TARK IX (Theoretical Aspects of Reasoning about Knowledge)*.

Schwalbe, M. and Walker, P. (2001). Zermelo and the early history of game theory. *Games and Economic Behavior*, **34**, 123–137.

Selten, R. (1965). Spieltheoretische Behandlung eines Oligopolmodells mit Nachfragezeit. *Zeitschrift für die Gesamte Staatswissenschaft*, **121**, 301–24, 667–89.

Selten, R. (1975). Reexamination of the perfectness concept for equilibrium points in extensive games. *International Journal of Game Theory*, **4**, 25–55.

Selten, R. (1978). The chain-store paradox. *Theory and Decision*, **9**, 127–59.

Shimoji, M. (2002). On forward induction in money-burning games. *Economic Theory*, **19**, 637–48.

Shimoji, M. and Watson, J. (1998). Conditional dominance, rationalizability, and game forms. *Journal of Economic Theory*, **83**, 161–95.

Siniscalchi, M. (2008) Epistemic game theory: Beliefs and types. In *The New Palgrave Dictionary of Economics*, 2nd edn., ed. S.N. Durlauf and L.E. Blume. Palgrave Macmillan.

Stackelberg, H. von (1934). *Marktform und Gleichgewicht*. Vienna: Julius Springer.

Stahl, D.O. (1995). Lexicographic rationalizability and iterated admissibility. *Economics Letters*, **47**, 155–9.
Ståhl, I. (1972). *Bargaining Theory*. Stockholm, Sweden: Stockholm School of Economics.
Stalnaker, R. (1998). Belief revision in games: Forward and backward induction. *Mathematical Social Sciences*, **36**, 31–56.
Stinchcombe, M.B. (1988). Approximate common knowledge. Working paper.
Tan, T. and Werlang, S.R.C. (1988). The Bayesian foundations of solution concepts of games. *Journal of Economic Theory*, **45**, 370–91.
Tan, T. and Werlang, S.R.C. (1992). On Aumann's notion of common knowledge: An alternative approach. *Revista Brasileira de Economia*, **64**, 151–66.
Tsakas, E. (2011). Epistemic equivalence of lexicographic belief representations. Working paper.
Zermelo, E. (1913). Über eine Anwendung der Mengenlehre auf die Theorie des Schachspiels. *Proceedings Fifth International Congress of Mathematicians*, **2**, 501–4.

Index

Lightning Source UK Ltd.
Milton Keynes UK
UKOW05f2128240915

259126UK00010B/154/P

9 781107 401396